WITHDRAWN
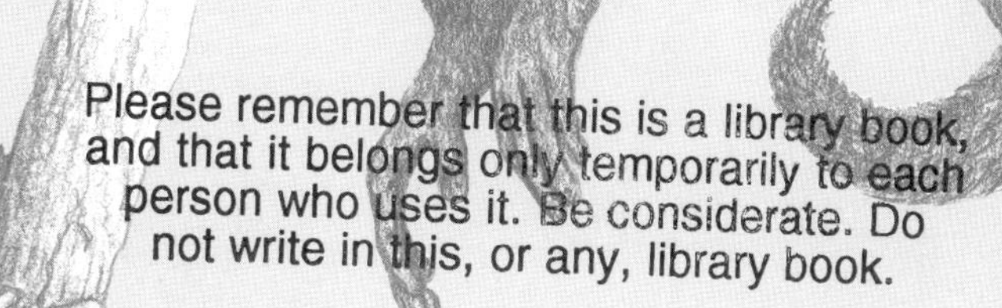

WITHDRAWN

THE VARIETY OF LIFE

WITHDRAWN

BY THE SAME AUTHOR

The Famine Business

Home Farm

(with Michael Allaby)

Future Cook

The Food Connection

Food Crops for the Future

Global Ecology

Last Animals at the Zoo

The Engineer in the Garden

The Day Before Yesterday

Neanderthals, Bandits, and Farmers

163182
VC Lib

THE VARIETY OF LIFE

A Survey and a Celebration of all the Creatures that Have Ever Lived

COLIN TUDGE

OXFORD
UNIVERSITY PRESS
Great Clarendon Street, Oxford OX2 6DP
Oxford University Press is a department of the University of Oxford
and furthers the University's aim of excellence in research, scholarship,
and education by publishing worldwide in
Oxford New York
Athens Auckland Bangkok Bogotá Buenos Aires Calcutta
Cape Town Chennai Dar es Salaam Delhi Florence Hong Kong Istanbul
Karachi Kuala Lumpur Madrid Melbourne Mexico City Mumbai
Nairobi Paris São Paulo Singapore Taipei Tokyo Toronto Warsaw
with associated companies in Berlin Ibadan
Oxford is a registered trade mark of Oxford University Press
in the UK and in certain other countries
Published in the United States
by Oxford University Press Inc., New York
© Oxford University Press, 2000
The moral rights of the author have been asserted
Database right Oxford University Press (maker)
First published 2000
All rights reserved. No part of this publication may be reproduced,
stored in a retrieval system, or transmitted, in any form or by any means,
without the prior permission in writing of Oxford University Press,
or as expressly permitted by law, or under terms agreed with the appropriate
reprographics rights organization. Enquiries concerning reproduction
outside those terms and in other countries should be sent to the Rights Department,
Oxford University Press, at the address above.
You must not circulate this book in any other binding or cover
and you must impose this same condition on any acquirer
A catalogue record for this book is available from the British Library
Library of Congress Cataloging in Publication Data
(Data applied for)
ISBN 0 19 850311 3
Typeset by Footnote Graphics, Warminster, Wilts
Printed by Butler & Tanner Ltd, Frome, Somerset

CONTENTS

ACKNOWLEDGEMENTS

FROM MY very first memories I am aware of my debt to teachers and friends who have shaped my interest in the natural world. At Dulwich I was fortunate to be taught biology by Douglas Hillyer, Brian Jones, Colin Stoneman, and Sammy Cole; and then, at Cambridge, learnt a great deal from my zoology supervisor, Dr Richard Skaer. From later times I am indebted in particular to Dr Donald Gould, who gave me my first job and remains a good friend despite that; Dr Michael O'Donnell, who gave me time to write my first book in the 1970s; Graham Chedd, who published my first article in *New Scientist*; Dr Roger Lewin and Dr Bernard Dixon, who have always been helpful and encouraging; and Dr Helena Cronin and my friends at the Centre for Philosophy at the London School of Economics, which over the past several years has been a source of endless stimulation.

As this book got underway I discussed general plans with Dr Neil Chalmers, Director of London's Natural History Museum, Sir Robert May at the Department of Zoology at Oxford, Professor Stephen Blackmore of the Natural History Museum, London, who was also Chairman of the UK Systematics Forum, and Dr Gren Lucas of the Royal Botanic Gardens, Kew; and am grateful to all of them for their encouragement and suggestions.

In the preparation of the book I owe a great deal to Dr Michael Rodgers, exemplary editor at Oxford University Press, with whom I have been discussing plans and changes of mind through the best part of a decade; Debbie Sutcliffe, who brilliantly organized the illustrations—and employed her fine talent to provide some of them herself; Pete Russell, for his excellent design; Jean Macqueen for her comprehensive index; and Sarah Bunney, whose efforts as a copy editor and supplier of ideas in her own field of palaeoanthropology lead me to conclude that no book of fact should be surrendered to the printers un-Bunney'd.

I am especially grateful to Gordon Howes, who drew the fish and reptiles, and as an erstwhile fish taxonomist at London's Natural History Museum has helped me to pick my way through the ideas and supplied many references and leads; and to Gina Douglas, of the Linnean Society of London, for her help in tracking down crucial references.

Finally, thanks to my daughter Amy (now teaching biology in London), who accompanied me and generally organized our grand tour of the United States in the autumn of 1995, when we visited as many as possible of the centres of serious systematic endeavour. Our drive from San Diego to New York via Yosemite, Badlands, Yellowstone, Niagara, and Cape Cod, plus all the greatest cities en route, cost us more in speeding fines than in gas but is among life's outstanding memories. Without Amy, I believe I would still be stuck in some motel in Montana or Ohio, trying to master the mysteries of America's free-enterprise telephone system.

Specifically this book has relied heavily—principally, in fact—on conversations and correspondence with the scientists listed below. I am endlessly grateful to them for their hospitality, for discussing their ideas so freely and often at great length, for supplying key references, and, in many cases, for introductions to others (including rivals!). Many have also read particular chapters and passages and sometimes demanded re-writes—each part of the book has been read by at least one world authority—but, of course, the mistakes are mine.

PART I, the introduction to classification and cladistics, has owed a great deal to my discussions with Dr David Swofford at the Smithsonian Institution, Washington DC, who continues to create some of the most valuable software without which modern cladistic analyses would be virtually impossible; Dr Christine Janis, at Brown University, Providence, Rhode Island, who has helped me over several years with many parts of the book; and Dr Peter Forey, of the Natural History Museum, London, for vital insights into the intricacies of cladistics.

For the opening three sections of PART II, I am particularly indebted to Professor Norman Pace, now at the University of California, Berkeley, who introduced me and my daughter to the pleasures of microbial chauvinism around the campus of Indiana University at Bloomington; Dr Matt Kane, who discussed his own research on the microbes that live symbiotically with termites and also introduced us to his colleagues at the Smithsonian Institution in Washington DC; Dr Bob Lindstrom of Yellowstone National Park, Wyoming, for the excellent tour *in situ* of thermophilic prokaryotes in the hot springs; Professor James Lake, of the University of California, Los Angeles, for much interesting discussion on broad fronts; Dr Bernard Dixon, for a down-to-earth view of bacteriology; but especially to Professor Carl Woese at the University of Illinois, Urbana, whose extraordinary research since the 1970s has transformed my—and the whole world's—perception of life's true diversity; and to Dr Mitchell Sogin, at the Marine Biological Laboratory, Wood's Hole, Massachusetts, who has shown that Linnaeus's two eukaryotic kingdoms must now be extended to a dozen or so more. I also greatly enjoyed discussions with Dr Suzanne Fredericq of the Smithsonian Institution, Washington DC, on the mysteries of red seaweeds—truly an alternative life form.

For *Section 4*, I am especially grateful to Dr Thomas Bruns at the University of California, Berkeley, for help both in preparing the chapter and in reading the text; to Professor Tom White and Dr Barbara Bowman of Roche Molecular Systems, Alameda, California, for deep and pleasant discussions on the molecular phylogeny of fungi; and to Dr Paula DePriest at the Smithsonian Institution, Washington DC, for fine discussions on lichens.

For the many sections on various aspects of invertebrate animals—*Sections 5–12*—I am indebted to the following: Professor Claus Nielsen, of the Zoological Museum, Copenhagen, for invaluable correspondence; Dr Simon Conway Morris, of the University of Cambridge, with whom I have enjoyed fine conversations over many years; Dr Judith Winston, of the Virginia Museum of Natural History, Martinsville, for discussion of bryozoans; Dr Sandra Romano, formerly at the Smithsonian Institution, Washington DC, now at the University of Guam, for her superb summary of the cnidarians; Dr David Reid, at the Natural History Museum,

London, who guided me through the many pitfalls of the molluscs; Professor Michael Akam, now at the University Museum, Cambridge, who taught me about Hox genes and introduced me to the problems of arthropod phylogeny; and his erstwhile colleague, Dr Michalis Averof, formerly at the Wellcome/CRC Institute, Cambridge, for helpful elaborations; Professor William Shear, of Hampden-Sydney College, Virginia, who gave me endless help with arthropods in general and chelicerates and insects in particular; Richard Hoffmann, of the Virginia Museum of Natural History, Martinsville, for providing valuable insights into myriapods; and Dr Robert Hessler and Dr William Newman at the Scripps Institution of Oceanography, La Jolla, California, who helped me to get to grips with the crustaceans (and Dr Hessler's dissections of tiny crustaceans are the most beautiful I have ever seen). I am also very grateful to Dr Richard Brusca at the University of Charleston, South Carolina, one of the co-authors of the invaluable 'Brusca and Brusca', for his cogent criticisms. Chelicerates were well served not only by Bill Shear but also by Dr Paul Selden at the University of Manchester; and by Dr Jeffrey Shultz at the University of Maryland at College Park. Richard Brusca also helped me with the echinoderms, and I am particularly grateful to Dr Greg Wray, at the State University of New York, Stony Brook, for his excellent insights into echinoderm development; his is a model of modern biological research.

For help with the various groups of vertebrates—*Sections 13–22*—I am particularly aware of my debt to Dr William Bemis, who has prepared brilliant course notes for his lucky students in the Department of Organismic and Evolutionary Biology at the University of Massachusetts, and very kindly gave me a copy when I visited him at Amherst in 1995. The trees and most of the 'Guides' in Sections 14 and 15 are based on these notes. For further insights into fish in general, I am also grateful to Dr John Maisey, at the American Museum of Natural History; and to Gordon Howes. Dr Axel Meyer, at State University of New York, Stony Brook, explained the many modern attempts to discover the relationships of lobefins and tetrapods—and my account also owes a great deal to my discussions with Christine Janis. I have enjoyed fine conversations over many years with Professor Michael Benton, of the University of Bristol, who has helped enormously with the vertebrate sections; and I am also grateful to Dr Nick Fraser, of the Virginia Museum of Natural History, Martinsville, for excellent discussions on Triassic reptiles.

For insights into the phylogeny of mammals (*Section 18*), I thank Christine Janis at Brown; Dr Michael Novacek and Dr Nancy Simmons at the American Museum of Natural History; Dr David Krause and Dr John Hunter at the State University of New York, Stony Brook; Dr David Archibald at San Diego State University, California, for discussions of Cretaceous mammals; Dr André Wyss at the University of California, Santa Barbara; Dr Don Prothero, of the Occidental College, Los Angeles, California, for discussions on perissodactyls; and Dr Alan Turner at the University of Liverpool, with whom I have enjoyed pleasant discussions on carnivore phylogeny.

My ideas on primates in general (*Section 19*) and hominids in particular (*Section*

20) have been shaped over years by conversations with Dr Chris Stringer and Robert Kruszynsky at the Natural History Museum, London; Dr Clive Gamble of the University of Southampton; Professor Bernard Wood, now at George Washington University, Washington DC; Dr Rob Foley of the University of Cambridge; and, in the particular context of this book, conversation with Dr Ian Tattersall at the American Museum of Natural History, New York.

My ideas on the classification of birds, living and extinct—*Sections 21 and 22*—have relied virtually entirely on my conversations with Dr Joel Cracraft and Dr Luis Chiappe at the American Museum of Natural History, and also with Dr Mike Hounsome at the Manchester Museum, Manchester; and Dr Michael Braun of the Smithsonian Institution, Washington DC.

When I began this book I had not looked seriously at plant classification since school days and have been privileged to be brought up to speed during my writing of *Sections 23–25* by Dr Peter Raven, Director of the Missouri Botanical Garden, St Louis; Dr Peter Crane, then Director of the Field Museum of Natural History, Chicago but now director of the Royal Botanic Gardens, Kew; Dr Liz Zimmer, of the Smithsonian Institution, Washington DC; and Dr Simon Owens, Dr Mark Chase, and Dr Nicholas Hind, at Kew.

Finally, I have had many discussions over the years on matters of conservation, the subject of the Epilogue (PART III), but am especially aware of my debt to Dr Georgina Mace, of the Institute of Zoology, London.

To all of the above: many thanks.

London C. T.
November 1999

ILLUSTRATION CREDITS

Illustrations for the Survey of all living creatures were provided by the following artists:

2 Bacteria and Achaea; **3** Eucarya

GRAHAME CHAMBERS

4 Fungi; **5** Animalia (some); **18** Mammalia; **21** Aves; **22** Neornithes; **23** Plantae **24** Angiospermae; **25** Compositae / Asteraceae

HALLI VERINDER

5 Animalia;

BIRGITTE RUBAEK AND BETH BEYERHOLM

(reproduced from C. Nielsen (1995): *Animal Evolution: Interrelationships of the Living Phyla*, Oxford University Press, 1995)

6 Cnidaria;

DEBBIE SUTCLIFFE

7 Mollusca; **13** Chordata; **14** Chondrichthyes; **15** Actinopterygii; **16** Sarcopterygii; **17** Reptilia

GORDON HOWES

8 Arthropoda; **9** Crustacea; **10** Insecta; **11** Chelicerata & Pycnogonida ; **12** Echnodermata;

JEREMY DIX

19 Primates; **20** Hominidae;

MAURICIO ANTON

The illustration of smut on page 176 is based on a photograph kindly supplied by IACR-Rothamsted

To my children

Mandy, Amy, and Robin

PART I

THE CRAFT AND SCIENCE OF CLASSIFICATION

CHAPTER ONE

'SO MANY GOODLY CREATURES'

WHEN I WAS at school and university in the 1950s and early 60s teachers and pupils took it to be self-evident that biology was about living creatures. It was also, of course, about processes—physiology, ecology, and above all evolution—but at the core of all our enquiries were the organisms themselves. We never stopped asking, 'What are they? What's out there?'.

So we worked steadily through all the known organisms group by group: the annelids (earthworms, leeches, and their like); the arthropods (crustaceans, insects, spiders, trilobites, and so on); echinoderms (starfish, sea urchins, and the rest); vertebrates (backboned creatures, such as fish, dinosaurs, and ourselves); the confusing host of creatures from amoebae to diatoms that in those days were lumped together as 'protists'; seaweeds; fungi; slime moulds; plants, which then, as now, were assumed to include some seaweeds but not others, plus mosses, ferns, conifers, and flowering plants; and the creatures loosely called bacteria, which in those days nobody seemed to link satisfactorily to the rest and were sometimes (grotesquely) thrust in among the plants. The general art, craft, and science of classification was and is called taxonomy; and modern taxonomy based on evolutionary principles is commonly and properly called systematics.

I loved this natural-historical foray among our fellow creatures. This to me was what biology was for; to admire and get to know what was out there. True systematics, too, required that creatures should be classified not by arbitrary criteria but according to their actual phylogenetic relationships, in so far as these could be judged—where phylogeny implies 'evolutionary history'. Classification, in short, was based on the phylogenetic tree, which, at least in general form and shape, resembles a genealogical tree of a human dynasty. Annelids thus emerged as metaphorical cousins of the arthropods, and echinoderms as metaphorical cousins of the vertebrates, although the annelid/arthropod lineage was only very distantly related to the echinoderm/vertebrate lineage. The phylogenetic tree was thus a summary, in graphic form, of evolutionary history. Once you grasped this principle you could see, in your mind's eye, how an earthworm and a bee—two unsurpassingly homely creatures—must have shared a common ancestor who lived, undoubtedly in the sea, at least 600 million years ago. In short, an eye for classification is a constant reminder that nothing on this planet is as homely as it seems—there are several thousand million years of evolutionary drama behind everything that moves and breathes.

Taxonomy, though, used to have a down side. In some courses (although not, I am pleased to say, in the ones that I was lucky enough to take), students who had been drawn to biology by the romance of it, the desire to associate with their fellow creatures—or even, perhaps, to find a cure for cancer or to feed the world—found themselves counting the bristles on the legs of shrimps, to see if shrimp A was more like shrimp B or shrimp C. Taxonomy could sometimes seem less than riveting. Meanwhile, as the 1960s, 70s, and 80s unfolded, wonderful things were happening in other branches of biology: in ecology, animal behaviour, and evolutionary theory. Most spectacular of all was the rise of molecular biology, which—together with classical genetics and the rapidly unfolding ideas of evolution—has now provided biology with a core of ideas that is beginning to feel as robust and satisfying as the classical and quantum theory that lies at the heart of modern physics.

There was also another kind of threat to the traditional craft of classification. Molecular biology is concerned primarily with the workings of DNA—the stuff of which genes are made: and genes shape the bodies of all living things. Although there are large and significant differences in the operation of bacterial DNA on the one hand, and that of 'eukaryotes' on the other (creatures like us, and fungi, and protozoans and oak trees, whose body cells contain distinct nuclei), DNA on the whole operates in remarkably consistent and similar ways across the whole spectrum of living things. In other words, molecular biology has increasingly emphasized the underlying unity of life. Many molecular biologists, therefore, simply do not care what kinds of cells they work on; any one, they have tended to feel, can serve as a 'model' for all the rest. Indeed a breed of molecular biologists has grown up who actually cannot tell the difference between a frog and a toad—or, indeed, when you boil it down, between a toad and a toadstool—because, quite simply, the difference does not seem to matter to them. DNA is DNA is DNA.

So taxonomy—systematics—has in many circles been marginalized these past few decades. My two daughters both studied biology at school and university and neither was taken on the thorough conducted tour of living things that I enjoyed (although my younger daughter, Amy, did receive excellent instruction on the classification of birds at her university course in Manchester). To many biologists classification seemed to suffer from a series of cumulatively fatal flaws. First, it could be dull. Second, the various major branches of biology were all acquiring an increasingly firm core of theory whereas the traditional craft of classification began to seem almost amateurish: a series of manoeuvres without convincing, rational foundation. It was natural history of the most nerdish kind, an exercise in train-spotting—or so it seemed to those who were not versed in it. Formal courses in biology are of limited duration and other subjects (ecology, behaviour, evolution, molecular biology) just seemed to be so much more interesting and 'relevant'. Besides, if it was truly the case, as molecular biology seemed to imply, that the perceived variety of living creatures is merely superficial—that we are all just variations on a theme of DNA—then why bother to distinguish one creature from another? Why differentiate between a frog and a fungus if, at the most fundamental level, they were both the same?

That is how many biologists came to feel about taxonomy during the last decades of the twentieth century. So why write a book on such a lame-duck subject?

THE UGLY DUCKLING

In truth, the art and science of taxonomy should never have allowed itself to slump into dullness. As I hope will become apparent in these introductory chapters, all who have sought to classify living creatures from Aristotle onwards have been immersed perforce in some of the most profound issues of biology—and indeed of philosophy and also, in earlier times, theology. But taxonomy is also a practical craft: practitioners of tropical medicine, for example, really do need to distinguish one mosquito from another; and minute variations in leaf sheen and stem bristle really can mark the difference between a plant that can save lives and one that is merely pretty. It is sad, therefore, that in some university courses the practicalities have dominated, and the intellectual currents have been lost to view.

Yet it was during the decades when taxonomy seemed to lose its pole position in biology that the science of systematics truly came of age. Biologists must, of course, classify creatures according to their observable characteristics (as discussed in more detail in Chapter 2); and to infer relationships they look both at living species and at the fossils of extinct kinds. The nature and the amount of data available to them have leapt ahead in the past few decades. In particular, palaeontologists seem constantly to unearth the most wondrous caches of fossils—Aladdin's caves of ancient creatures whose existence we had no right to anticipate. Even as recently as the 1960s, biologists still doubted whether they would ever find significant fossils from the Precambrian period—from the geological time, more than 545 million years ago, when no creature had evolved hard skeletons, so that fossilization seemed impossible. Now rich beds of Precambrian fossils are known on several continents. From the later Cambrian period—notably from the Burgess Shale in Canada, around 530 million years old, and studied in particular by Simon Conway Morris of the University of Cambridge—have come series of arthropod-like animals quite unlike those of today.

It is not unusual for creatures to diversify in many different directions when they first evolve, with most of the variants later dying out: we see the same pattern when fish first evolved, or mammals or birds, or what you will. But to have glimpsed the first wild fling of the arthropods is luxury indeed. Western Australia is yielding fossil fishes, 400 million years old, of astonishing detail. Palaeontologists now have a wonderful series of fossil birds—fragile creatures that do not fossilize easily—greatly enriching the picture of bird evolution that used to be provided only by *Archaeopteryx*. In 1998 came reports of dinosaurs with feathers—dinosaurs that are clearly related to birds, but certainly would not traditionally be classed as birds. Human fossils, always so elusive and confusing in the past, now show a satisfying, if more diversified sequence back to the plains of Africa around 4.5 million years ago; only one of several different trails led to *Homo sapiens*. In short, the fossil discoveries of recent years have been wonderful: and all these new, extraordinary creatures are grist to the taxonomists' mill. If we do not attempt to classify them, we simply will have no idea what they are.

Vast new caches of information, too, have come from a quite different source: study of the body's chemistry and, in particular, of DNA. For the stuff that reveals the underlying unity of Earthly life also reveals the true depths of diversity, and is providing new insights into systematic relationships. Traditional studies of anatomical features can deceive for all kinds of reasons, not least because different creatures that are unrelated often adapt to similar circumstances in similar ways, and so they come to resemble each other. Molecular studies can uncover such deceptions and, indeed, can be revelatory. They have confirmed, for example, how recently human beings shared a common ancestor with chimpanzees (probably around 5 million years ago), and how closely whales are related to cows. Yet all is not so simple, as the following chapters reveal: if and when molecular studies conflict with anatomy, it is often far from obvious which set of data should be believed.

Yet the recent influx of fossil and molecular data is only half the story. The greatest advance in modern taxonomy has been made in the underlying theory and the method: in particular by the method of cladistics developed from the 1950s onwards by a German entomologist, Willi Hennig. Without cladistics to keep taxonomists on the rails, the mass of new information could simply be an embarrassment. With cladistics it seems possible at last to unravel the true history of living creatures and hence to reveal their evolutionary relationships—and so to reflect those relationships in classifications.

One last input—a bonus, but a necessary one—is the computer. When cladistical methods are applied to masses of data and, in particular, to the potentially infinite mass of molecular data, the resulting calculations are of horrendous complexity. Without computers to help us out—and, more to the point, the kind of software developed specifically for the purpose by, for example, David Swofford at the Smithsonian Institution in Washington DC—the library of data would largely be beyond analysis, at least in ways that were liable to yield worthwhile answers.

This, then, is the first reason for taking systematics seriously. It is quite simply the core of modern biology. Its ambition is vaunting—to reveal and to present in summarized form the evolutionary history of all creatures on Earth. Its methods and its philosophy tax the deepest thinkers and deploy the most subtle of techniques. Those who mocked the dull feathers of taxonomy while other, flashier game was taking flight, misconstrued the nature of the beast: the ugly duckling has grown up. Yet even if this were not so, the fact is that all of humanity needs taxonomy. 'O wonder!', cries Miranda in *The Tempest*, 'How many goodly creatures are there here!' Indeed, there are—far more than Miranda could have dreamed of. For our own sake and theirs, we need to keep track of them.

KEEPING TRACK

The inventory of known living species stands at around 1.7 million—but there is no central inventory, so nobody knows for sure. There is a curious insouciance in this: we keep better tabs on the stars, which, for all they do for us, are simply points of light. It seems certain, though, that this figure of less than 2 million falls short of reality by at least an order of magnitude (at least tenfold, that is) and probably by several orders of magnitude (hundreds or thousands of times). Thus in the 1970s Terry Erwin of the Smithsonian Institution anaesthetized and then counted all the species of beetle in just one tree in Panama, perceived that the number of unknown species far outweighed the ones that had previously been identified, and through a sequence of reasoning that may seem a trifle tortuous but is widely agreed to be reasonable, calculated that the true number of all species on Earth is probably nearer to 30 million. If the proportion of the different kinds of creatures among the unknowns was about the same as in the ones we do know about, then most of that 30 million would be animals; most of those animals would be insects; and most of those insects would be beetles. God, as the great British biologist J. B. S. Haldane (1892–1964) once commented, seems 'inordinately fond of beetles'.

Some biologists feel that Terry Erwin got a little carried away, and that the true number of living species is probably nearer 8 million. But others feel he may not have been bold enough; that the true number is perhaps nearer 100 million. Many, too, feel that God is perhaps not quite so enamoured of beetles as Haldane supposed; it's just that they are attractive and much studied and so we know a lot about them—compared with what we know about the rest, that is. It has been suggested, after all, that every species of creature on Earth of reasonable size has at least one specialist nematode worm attendant upon it (nematodes are commonly called 'roundworms' and sometimes 'eelworms') living either as a parasite or at least a commensal (just a resident). Mites, too, diminutive relatives of spiders, are ubiquitous and much understudied, except for the few that do obvious harm—for example, to horticultural crops. There must be many more thousands of them as well. So the real total of biodiversity, on this planet at this moment, seems to be somewhere between 8 and 100 million, with 30 million standing as a reasonable guess.

Even 30 million, however, now seems a far too modest estimate. In the seventeenth century, a Dutch linen-draper and pioneer microscopist called Anton van Leeuwenhoek (1632–1723) showed that the world contains creatures too small to see with the unaided eye, which he called 'little animalcules'. Such creatures are nowadays collectively called 'microbes'—a useful term—and they are known to include creatures from three quite different categories: bacteria; the newly discovered, bacteria-like archaes; and creatures of the kind colloquially lumped together as 'protozoans' or, more broadly, 'protists'. In France, in the nineteenth century, Louis Pasteur (1822–95) showed how very important these microbes are, in brewing and pickling, and in causing disease. Huge industries—brewing, baking, and pharmaceuticals—have been based on the cultivation of microbes. Today these industries in modern guise are subsumed within 'biotechnology', and are spreading beyond their traditional bounds into all of industrial chemistry. Microbial diseases continue to dominate world health, because antibiotics and vaccines have removed only the top layers of infection, and only in some countries. Because they are so obviously important, microbes have been well studied, and the inventory of known bacterial and archaeal species now stands at around 40 000.

The figure of 40 000, however, is now known to be a desperately feeble estimate. Traditionally bacteria could be identified and indeed discovered only by cultivating them: taking a fragment of soil, say, putting it into culture, and seeing what grows. The present catalogue, then, contains only those bacteria that can be cultured. But modern biologists such as Norman Pace at the University of California at Berkeley are now able to pick out bacteria in the soil, or in any other medium, just by looking for their DNA; and so they are finding new types without culturing them first, and indeed without ever setting eyes on the intact organism. The DNA alone is the shibboleth; by their DNA shall they be known, at least for the time being. By such means microbiologists are now suggesting that the real number of bacteria and archaes in the world may exceed the number that are known—the compliant types that can be grown—by 10 000 to one. So the real number of bacterial and archaeal species may not be 40 000 but 400 000 000. Add this to Terry Erwin's estimate of macroscopic creatures and we begin to see that our grasp of 'biodiversity', and what it really implies, is tenuous indeed.

Is this the limit? Not quite. In fact, not by a long way; not at least if we take account of time. Thus it has often casually been suggested that the species now on Earth represent only about 1 per cent of all the organisms that have ever lived on our planet. It is easy to see how this might be so—but easy to see, too, that this again must be a serious underestimate. Thus the world now contains only two species of elephant—sole representatives of the mammalian order Proboscidea. But about 150 proboscidean species are known from the past 50 million years, including a huge array of 'true' elephants (members of the family Elephantidae) plus mastodonts, gomphotheres, deinotheres, and others. There are only five living species of rhinoceros, three in Asia and two in Africa, but the fossil inventory now runs to 200. The rhinoceros superfamily probably arose in Eurasia, which has yielded a huge array of ancient species. Many more lived in North America, which they could in principle have reached across the landbridges that at times in the past linked Siberia and Alaska, or Scandinavia / Greenland / Newfoundland. All the American and European rhinos are long gone and the modern African types, the black and the white, are johnnies-come-lately. Similarly, there are only four living species of hyaena—brown, striped, spotted, and the peculiar termite-eating aardwolf—but about 70 species are known from the time that hyaenas first appeared around 20 million years ago.

Thus, even from such a rapid recce of big, conspicuous, creatures, we can easily see how the number of extinct types could exceed the living species by a hundredfold. But then we might consider that all these conspicuous types are modern by the standards of the world. The specific lineages of elephants and rhinos extend back only 50 million years or so, and the hyaenas much less than that. But life is known to have appeared on Earth at least 3500 million years ago, and perhaps nearer 4000 million; 'only' a few hundred million years after the Earth itself was formed, about 4500 million years ago. Thus there has been life of some kind on this planet for at least 70 times longer than there has been any creature resembling an elephant. Elephants are slow-breeding and their generation times average around 30 years but, even so, there have been 70 times or so as many species in the past 50 million years as there are now. So how many more species of all kinds of creature might there have been over the past 3500 million years, given that most are small and some have a generation time measurable in hours? It would be surprising if the total number of species in the past did not exceed the present inventory by at least 10 000 times.

In short, the number of species that have lived on Earth since life first began could easily be about 400 million times 10 000, which is 4 million million, or 4000 (American) billions—roughly a thousand species for every year that life has existed on Earth.[1] Of course, these estimates may be out by an order of magnitude, or even by several orders of magnitude. But even if they were exaggerated a millionfold the total would still be vast; and far too great for any human mind to grasp.

What is more interesting in this world than our fellow human beings and other living creatures? Why do we know so few of what must be out there? What kind of philistines are we? Yet we can make a more practical point than this. We need to interact with other species whether we want to or not. They are our food and our environment: homes, scenery, soil, even the oxygen in the air is provided by courtesy of plants and photosynthesizing bacteria. We need actively to exploit our fellow creatures to survive. This is not an option: we have to exploit them unless we prefer to die. Therefore purely for selfish reasons (as well as for reasons that we may hope are less selfish) we also need to conserve them. Besides, even if we learnt to do without our fellow creatures—perhaps found some inexhaustible supply of food on some distant planet—they would not necessarily ignore us. We are flesh, too, for all our conceit, and many are more than happy to feed upon us. To contain, exploit, or conserve our fellow creatures we need to keep tabs on them.

But how can we, when there are so many? If we simply made a list of the known species now on Earth, we would need 1.7 million names. Because each living species is conventionally given a two-part name, according to the binomial system devised by the great Swedish biologist Carl von Linné (or Carolus Linnaeus in the Latinized form) in the eighteenth century, we would need 3.4 million words. An average-sized novel contains about 100 000 words; a fat, one-volume encyclopaedia has around 500 000. So we would need around seven fat volumes just to make a list of present-day known species. Clearly, if we seriously set out to find and list *all* the creatures

[1] This list does not include viruses. Admittedly, viruses can be considered 'living' up to a point; or, at least, they have many of the qualities of living things, and they are 'obligate parasites', which occur only in the company of other living organisms. But nobody knows what they really are. They are obviously not 'primitive', as is sometimes supposed, because they incorporate features of advanced life forms. But they are minimalist; reduced to just a few genes, which are able to take over the genetic apparatus of other organisms and turn it to their own purposes. Indeed the best guess seems to be that viruses are derived from the genomes of other living organisms, so that a virus that causes disease in mammals, say, very possibly arose originally as a group of mammalian genes. Of course, viruses can be and are classified. Clearly, though, it is very hard to devise a system of classification that relates them to other living things. They seem to have multiple origins, and effectively occupy a parallel environment. Interesting as they are, then, they do not belong in a book like this and will not feature again.

now on Earth, the seven volumes would probably extend to 70 or more; and if we could find out all the creatures that have lived in the past we would need a substantial library—just to make the barest list, without footnotes and explanation. How can we possibly cope?

Classify, is the answer. Put the different creatures into groups; and then nest those groups within larger groups; and so on and so on. Once classify—at least if you do it well—and any list, no matter how prodigious it is or is liable to become, becomes tractable. It is miraculous when you think about it, but it is the case. So here is a second reason for taking taxonomy seriously. We need to do it because we need to share this planet with a vast if as yet sadly unquantified host of other creatures, and we need at least to keep them under surveillance.

Yet there is one more reason to consider systematics, and I think it the best. The prime motive of science is not to control the Universe but to appreciate it more fully. It is a huge privilege to live on Earth and to share it with so many goodly and fantastical creatures—albeit a privilege of which we are grotesquely careless. In truth, if we did not need to exploit other species we might simply and unimprovably spend our lives in admiration of them; they are so extraordinary. But in order to get close to them we do have to give them names, and keep them in some kind of order in our minds. Those who have been brought up to believe that naming is an unpoetical pursuit—'Today we have naming of parts'—might care to consider that many of the greatest nature poets have also been keen naturalists, with a detailed knowledge of what is out there: Shakespeare, Wordsworth, John Clare, D. H. Lawrence—the list goes on and on. Indeed it seems we can never appreciate anything fully unless we first put a name to it and have some feel for what it is and where it comes from. A parable will make the point.

A VIRUS OF FINE ART

Imagine that in some pleasantly out-of-the-way art gallery, let's say the incomparable Dulwich in South London, there arises an information virus, analogous to the kinds that infect computers. This hypothetical virus does no harm to the paintings. It merely obliterates the signatures. But then it creeps out of the frames and into the catalogues and reference books, and into the minds of the visitors and critics and everyone who has any vestige of knowledge. It spreads from the Dulwich to the Tate and then to the National, and so to the smaller collections, the Courtauld's and the Wallace, then out to Britain's other cities and grand houses, and across the channel to the Rijksmuseum and the Louvre, the Prado and the Uffizi, and east to the Hermitage and into Asia, and across the Atlantic to the Metropolitan Museum of New York, the Guggenheim and the Frick, and all the great galleries of the United States, and out beyond there into all the world, all the time working its way through every work of reference, and all the memories in the minds of every art enthusiast. The paintings themselves remain unscathed. But soon, nobody in the world knows who painted what, or where, or when.

What would we lose through such a virus? According to one school of criticism, very little; for some have argued that works of art should be self-contained and need no extraneous information to be appreciated: no biography, no history, no referents of any kind. But most would be appalled by the loss. The paintings would still be there, right enough, as stunning as ever. But without any knowledge of who, when, and where—let alone of why—the paintings would have no meaning. We would lose all sense of development—of ideas, subject, style, and technique. There would still be clues to history and provenance: a preponderance of paintings in Italy of religious and mythological motif, bold in colour and with a distinct tendency to glow, would surely help future scholars to perceive that Italian paintings have a particular 'feel', and to infer that comparable pictures even in the furthest corners of the world probably came from Italy in the first place. Soon, with luck and fingers crossed, they would be grouping Dutch landscapes with Dutch landscapes, English portraits with English portraits, and so on and so on; although Dutchmen who chose to paint like Italians, like Cuyp, would give them pause.

With time, though, they would be able to ascribe particular groups of paintings to particular artists: Rubens, Rembrandt, Vermeer, Constable, Turner, Poussin, El Greco, Tiepolo would surely be among those who might reasonably be distinguished from the rest. But these authors would remain hypothetical, and would have to be given arbitrary names because the originals would be lost. Around such relative certainties, others might be grouped as 'school of'. But it would be a bold scholar indeed (and undoubtedly forever controversial) who suggested that all the known works of Picasso or Cezanne could each be ascribed to one hand; in fact no one might ever realize that that was the case.

As more decades passed, scholars would begin to see connections between groups that they had previously been at pains to separate: for example, that the human figures of Cezanne have much in common, in the way they are perceived and constructed, with those of Rubens. But without knowledge of dates the scholars would still be obliged to ask, 'Did Rubens influence Cezanne, or did Cezanne influence Rubens? Or were they perhaps contemporaries, drawing their ideas from a common source?'. These questions seem ludicrous from our present vantage point but in the post-virus age the matter would have to be addressed afresh; and, without historical knowledge, the answer is far from obvious. Is it more reasonable to infer that Cezanne's sketchier style is an allusion to Rubens, or that Rubens represents a more finished version of Cezanne's prototypes?

In short, such a virus would take away much of our insight; and without that insight, we would also lose much of the pleasure that painting holds. The critics who have argued that everything we need to know is contained within the frame would surely realize the error of their ways. The loss of historical knowledge would be as catastrophic as, say, the loss of an entire gallery. No one would doubt, either—would they?—that the first task of scholars in a post-virus age must be to recapture the lost biographies and histories. 'Who?' and 'when?' are the essential questions, and to a lesser extent 'where?'. When those questions are addressed, and not until, we can ask begin to ask 'why?' with some hope of a satisfying answer.

But animals and plants, fungi and bacteria, do not come with signatures and historical asides. They just *are*. Where they came from, and why they take their present forms, we have to work out for ourselves. As naturalists we begin like the art-scholars in a post-virus age. Whatever the practicalities may be, we need for aesthetic reasons to find out what is what. The reasons for studying systematics are hence threefold: intellectual, practical, and aesthetic. In this volume, I want to explore all three, but particularly the first and last. Why, though, write this particular book, in this particular way? Various critics in a spirit of helpfulness have suggested several other ways of going about it. Why haven't I always taken their advice?

WHY THIS BOOK IS LIKE IT IS, AND HOW TO MAKE BEST USE OF IT

As recorded in the Acknowledgements and in the Sources and further reading, a huge number of scientists and friends worldwide have helped me with this book. All have been extraordinarily kind but some have raised a few doubts. What's this book *for*?, they ask. More to the point, *who* is it for? Does it not fall between two stools—too detailed for the 'layperson' and too general for the professional? The phylogenetic trees are shown without detailed notes to explain precisely why they are arranged as they are—so are they really of use to undergraduates, or university teachers, seeking to set courses in the taxonomy of crustaceans, or echinoderms, or whatever? Then, again, systematics is developing rapidly, as established classifications are subjected to sharper and sharper cladistic and molecular analyses, and more and more fossils come on line, and bacteria and protists turn out to be ever more weird. New phylogenetic trees and classifications are published daily on the internet. What use, in such a field as this, is a book published between hard covers, with all the passage of time that this entails?

Cogent criticism—but I never actually doubted, not for a second, that this book should be done. For one thing, I have been writing about science for more than three decades and have never been convinced that the gap between the professional scientist and the non-scientist is anything like so vast or unbridgeable as it is commonly held to be. At least: a zoologist may know a thousand times more about animals than a non-zoologist (or about the particular animals they happen to study), but I have often been shocked to find how little zoologists tend to know about plants, or how little botanists commonly know about animals. In fact, although a few scientists are remarkably broad and put most of us to shame, most are extreme specialists, with chasteningly little knowledge of anything outside their own specialty. So a general account of plants that would satisfy a weekend naturalist—and indeed can be appreciated by a bright nine-year-old—should also be of use to a professional biologist who does not happen to be a plant taxonomist. There is no shame in this. It is hard to stay ahead in any specialty, and life is short. Botanists have very little time to think deeply about animals and generally have little reason to; and most zoologists have little time or reason to think seriously about plants.

But botanists do sometimes need to think about animals—or may simply be curious, like any naturalist; and zoologists do tend to like plants, even if they don't have much time for them. So there is a place for books on plants for zoologists, and books on animals for botanists; and a book that fills such a role will be the same as one intended for amateurs, who look at living things for pleasure. Is this really so? Many argue, after all, that although professional biologists must specialize, they still share many ideas and have a common vocabulary. They all know what DNA is, for example, and have a rough idea how it works. They don't need to be told the basic things that a non-biologist might not be aware of, and so the gap between the professional and the amateur remains.

There are several points here. First, most of the technical ideas and vocabulary in this book apply specifically to the modern science of systematics; and, I have found, most biologists who are not themselves systematists do not know these specialist terms. If you know any practising biologists (who might include doctors, for example) ask them what they understand by 'symplesiomorphy'. It is a pleasant word when you grow used to it—roll it around the tongue—and fundamental to the modern method of cladistics. But I would bet that at least nine out of 10 biologists have no idea what it means.

Even so, some will argue that nobody who is not a professional taxonomist need understand a word like symplesiomorphy. It is, the detractors will suggest, merely a piece of 'jargon' that should be confined to the specialist journals, and perhaps expurgated even from them. But again I will contend the precise opposite. First, we should distinguish between 'jargon' on the one hand and 'technical terms' on the other. The word 'jargon' derives from the French for the twittering of birds and implies incomprehensibility—and, of course, all insider groups from generals to cat-burglars develop their own semiotic pidgins that sound like twittering birds to the outsiders and indeed are largely intended to keep outsiders at bay. Technical terms, on the other hand, refer to things and phenomena that exist only in recondite contexts and therefore need to be described in their own specialist language. Symplesiomorphy can be explained in non-technical terms—it refers to characteristics that different creatures hold in common, but which are simply 'primitive' features that are also found in other creatures. All will become clear as this book unfolds, including the special meaning of the word 'primitive'. The concept of symplesiomorphy is crucial; and the word, despite its abundance of syllables, alludes to that crucial concept precisely and unambiguously.

Thus if you feel that systematics is worthwhile at all, then 'symplesiomorphy' is one of a short list of technical terms that is worth getting to grips with. This is meant to be an introductory text—an introduction to the true implications of biodiversity and of the means by which biologists try to come to terms with it; and the whole point of such a book is, I suggest, to guide readers from a position of essentially zero knowledge to the point where they can, if they choose, begin to read the specialist literature for themselves. Needless to say, the specialist literature is technical, so an introduction will fail miserably in its task if it does not explain the technicalities. So the technicalities are explained and used in this book; and the explanations, as I see it, are of equal value to the graduate biologist as they are to the weekend naturalist—unless the biologist happens already to be a specialist systematist. In fact, introductory books that avoid technical terms, as many seem to do as a matter of policy, sell the reader short. They do not actually introduce the subject at all: they merely provide a pastiche, and a filleted one at that.

What of the critics who suggest that this book is lacking in detail, so that it would be hard to base a university course on it? Well, this is not intended to be a textbook, but textbooks are not the only kinds of books that are useful. I hope that teachers might feel they can base general courses on biodiversity on this book; but if they want to offer a specialist course on a particular group, crustaceans or ferns or whatever, then if they are serious teachers they should, of course, go back to the specialist literature. On the other hand, I hope that both they and the students will still find this book useful for background and context.

I have deliberately not filled the phylogenetic trees with specialist notes because I am assuming that if anybody does want to base a course on any particular group, then they will indeed go to the specialist literature. But because I have not appended such notes, I am referring to the phylogenetic trees throughout this book simply as 'trees', rather than as 'cladograms'. A cladogram, as we will see, is a diagrammatic representation of a specific hypothesis, about the way in which particular creatures are thought to be related; and it must carry notes, to show the reasoning behind the hypothesis. A tree is a more general statement that may or may not be based on a cladogram, or on a series of cladograms. The trees in this book are all based on cladograms, but the reasoning behind the cladograms is not spelled out in detail.

So I have written this book at the level that I feel is right (and it is, in the end, a matter of 'feel'). I did not want to write a textbook to squeeze particular groups of students through particular exams. That's for other people to do. In the end, the only sensible course for authors is to write the kinds of books that they would like to read themselves—and hope that there are other people out there who have the same kinds of tastes. Unless the author comes from Mars, that ought to be true at least some of the time.

But still, some will ask, what use is a book that may take a decade to write (as this one has) and a year to publish, on a subject that changes so fast? New schemes of classification are now published on the internet, and the details change by the week. Well, the details may indeed seem labile but the deep ideas are far more constant. Carl Woese's notion from the 1970s that living things should be classified in three 'domains'—bacteria, archaes, and eukaryotes—will surely be with us for many decades to come, and perhaps effectively forever. Mitch Sogin's suggestion that the eukaryotes include at least a dozen lineages so distinct that they each deserve the rank of 'kingdom' seems to be founded just as firmly. Even if the details change we can be sure that biologists will never again accept the peremptory, eighteenth-century division of all living things into 'animals' and 'plants'. Cladistics seems to have an inexorable logic, and surely will stay with us; and molecular biology will continue to unfold for the next 1000 years. These are the kinds of ideas that the browser on the internet needs to understand if the deluge of new ideas is to make any sense at all.

A book like this has two functions, then: first, to provide the background to all the new data; and, second, to provide a database with which to compare the new findings. After all, if you don't know what people thought previously, you will not appreciate the innovations. Besides—and very importantly—not every newly published cladogram is actually true. Because different studies begin with different data, and the data in any one study can be interpreted in many different ways, that is obviously the case. The information in this book cannot be called the orthodoxy, because there is no generally agreed, all-embracing canon. Neither can its ideas be guaranteed to be correct; all ideas in science are hypotheses, awaiting improvement. But the ideas represented here are the best opinions of current world authorities both as published and as told to me. That is not a perfect representation of the whole truth: but in the real world, in real time, I don't know how to get closer to it.

Overall, then, I have had two broad aims in writing this book, which some may feel are hugely pretentious (though I am prepared to live with that charge). My specific aim is to help to restore the art, craft, and, indeed, the modern science of taxonomy, or systematics, back to the centre of biological teaching and thinking, which is where it ought to be. Systematics is the discipline that introduces us to the creatures themselves. Biology without quantified ecology and animal behaviour, and without the unifying grandeur of evolutionary theory and molecular biology, is just natural history. But biological theory without actual, living, breathing creatures, is just philosophy. I like natural history and I like philosophy, but it is only when we put the two together that we have biology, the subject that truly merits a lifetime's absorption.

My second broad aim is simply to point out that nature is wonderful, and that much of the wonder lies in its variety. Without classification, the variety is simply bewildering, and bewilderment gets in the way of thought. The act of classifying focuses our thoughts and the more we think, the greater the wonder becomes; for this, as Hamlet said in a somewhat different context, is the appetite that grows from what it feeds on. Classification, in short, is not a dull pursuit for obsessives. It is the essential aid to understanding, the means by which to come to grips with life's variety. It puts us in touch. This book has become a hobby. It is good to see it finished, but in many ways I am sorry it is over.

CHAPTER TWO

CLASSIFICATION AND THE SEARCH FOR ORDER

EVERYTHING THAT LIVES needs to acknowledge the existence of other creatures—to feed on them, to drive them away, to mate with them. All creatures also need to distinguish one kind from another, at least roughly: noxious from innocuous, edible from toxic, potential mate from impending predator. Worms can do this. Even protists, even fungi, even plants, can tell like from unlike, so that a flower will accept compatible pollen, and reject what is alien. Such differentiation implies the ability to classify: each creature places some of its fellows in one category and some in another, and responds accordingly.

Clearly, living creatures do not need to be conscious in order to impose such elementary classification. Plants are not conscious; fungi do not reflect. Still less do they need to attach names to the different categories of creatures that they acknowledge; only human beings, it seems, are skilled in the arts of naming things (**nomenclature**) even though some monkeys, such as vervets, warn their neighbours when enemies are about and make different sounds to connote different threats—for instance, a 'snake' noise or a 'leopard' noise. But even without human language, and blessed only with what seems modest intelligence, many creatures can clearly impose classifications of impressive subtlety. Thus songbirds know that kestrels and owls, and anything vaguely like them, belong in the general category of 'bird of prey', but that this category does not include pigeons: so they mob the former and ignore the latter. Pigeons in their turn have shown in laboratory trials that they can distinguish photographs of trees, or parts of trees, from pictures of things that resemble trees but are really something else, like telegraph poles or sticks of celery. The ability of pigeons to discriminate and to categorize is extraordinary.

Human beings, however, are able to classify to a more refined degree; and not only to categorize, but also to give names to all the categories they distinguish. They have done this in a remarkable variety of ways, according to a host of different criteria, depending on their needs and predilections.

THE MANY WAYS TO CLASSIFY

How people choose to classify their fellow creatures—what categories they put them in—reflects and in large part determines their attitude towards them. New Guinea cannibals had various special terms for their victims, along the lines of 'long pig'. Many languages worldwide contain the same kind of idea—that foreigners are not merely from a different place, but of a different species. In part, such terms are self-protective: it is easier to kill, enslave, or eat a person whom you have decreed is not a person at all. 'What's in a name?', asks Juliet. 'A rose by any other name would smell as sweet.' Well, up to a point—for once call it 'weed' and its sweetness cloys, and you may root it out with a clear conscience. The point is not merely emotional. Science, allegedly the most rational of pursuits, is also swayed by its own nomenclature. In the words of James Lake of the University of California at Los Angeles 'biological thought can be profoundly affected by classifications'.

The many and various classifications of living things that different groups of people have imposed for their own purposes often correspond in a rough-and-ready way to the classifications of biologists, but by no means always. Commercial foresters class timber trees as 'softwood' or 'hardwood'. Softwoods are usually conifers (which botanists call gymnosperms) and hardwoods generally derive from flowering trees (angiosperms). But this is not invariable and, besides, some softwoods are physically far tougher than some hardwoods. Cooks and biologists in general agree that fish are fish, although, as we will see, the term 'fish', even in biological circles, embraces a remarkable miscellany of vertebrate lineages. But cooks also invoke the category of 'shellfish': as motley a gathering of molluscs and crustaceans as ever were plucked from the seven seas. To American chefs, any fish prised from the Atlantic or Pacific that is vaguely gadiform (cod-like) is labelled 'scrod'. Among the many euphemistic neologisms of British fishmongers is the 'rock salmon', more properly known as the common dogfish, which is a kind of shark and, as such, is more distantly related to the salmon than the salmon is to the horse. Gardeners differentiate arcanely between 'wild flowers' and 'weeds' (a 'weed', in general, being any plant that grows where the owner of the ground does not want it to). Biologists, too, can be ad hoc in their classifications. A 'microbe' is any unicellular (single-celled) creature that is too small to see with the naked eye and thus includes creatures from all the three domains of life.

Through all classifications, whether devised by songbirds, professors of botany, or fishmongers, two separate sets of considerations run in parallel. The first is operational: how do you actually go about classifying? What criteria do you adopt to decide whether item A belongs in category X or category Y? A songbird weighing up a bird of prey perhaps looks for curvature of beak or length of claw—or, more probably, relies on overall impression or gestalt; and by such means it distinguishes a falcon from a duck. For chefs, succulence on the inside and a crunchy carapace on the outside are the marks of shellfish. The many criteria of classification used by scientists will become apparent as this book unfolds.

If the criteria are made explicit, and can be repeated by other people simply by following the instructions—if, for example, they depend on qualities that can be measured and are not judged simply according to personal taste—then those criteria can be called 'objective'. But objective criteria may also be arbitrary. I might, for example, decree that all insects with legs over 2 centimetres long should be placed in a new grouping called 'mega-insects'. The criterion would be perfectly objective, in the sense that it is explicit and repeatable, which would not be the case if I simply decreed that there should be a special category marked 'beautiful'. But although the grouping of mega-insects would be objectively defined it would also be arbitrary. Nothing very special distinguishes insects that just happen to be large, except their largeness.

Behind all classifications, too, there is inevitably a philosophy. All classifications impose some view of the world: they all make a statement. Thus a chef regards the Universe as a market: the things within it are divided up according to edibility and wholesomeness. A songbird is concerned with food, too, but also with enemies and potential mates. A tailor, organizing his spare buttons, probably aims simply for tidiness. Buttons are fairly neutral things, holding no particular threat or promise; but it is probably convenient to separate those for riding breeches from those for clerical evening wear. As we will see, the many classifications devised by scientists, historically, have pursued various philosophies.

Whatever criteria we adopt, however, and whatever we are classifying and for whatever purpose, all classifications tend to follow a common pattern. The items in question are first divided into big categories, and then each big category is divided into smaller categories, and so on. The result is a series of nesting groups called a **hierarchy**: little ones grouped within bigger ones within bigger ones still. All classifications, in short, are hierarchical.

Ad hoc ways of categorizing living things are useful. Long may they persist. Practical people doing practical things need to carve up the world in their own ways, and it is not for outsiders to cavil. The flesh of abalones (ear shells) and of lobsters are comparably tender even if those creatures are less closely related to each other than eagles are to sea squirts; so why not call them both 'shellfish'? A 'weed' is a useful category, as any farmer will attest. Each system of classification casts its own light upon the world; lay different classifications side by side and you illuminate from different angles, and truly begin to see in three dimensions.

But it is one thing simply to manipulate aspects of nature for our own convenience, and another to seek understanding. Of course, deeper understanding can lead to more subtle exploitation—and thus it is, for example, that increasing knowledge of microbes on the one hand and of the operation of DNA on the other, is currently bringing us the mixed but in general momentous benefits of biotechnology. But for true biologists, the real pleasure comes simply from knowing, and from the sense of coming closer to other living things. I am defining 'biologists' broadly: to include naturalists, in the old-fashioned sense, who like simply to observe nature, and scientists, who seek to explain its workings by proposing hypotheses and then testing them out. Some naturalists shy away from explanation, and some scientists who work on particular aspects of life seem to have little taste for the outdoors. But some biologists qualify both as scientists and as naturalists—like Charles Darwin, who, even if he had not formulated the idea of evolution by means of natural selection and so begun the modern age of biology, would still be remembered as one of the greatest of all field observers. My point, however, is that true biologists seek something more in nature than mere utility. They feel in their bones, and all their senses proclaim, that there is an *order* among living things; and it is this 'natural' order that they seek to reflect in their classifications. They feel, too, that there should be criteria for classification that are not simply arbitrary, but reflect some real and important affiliations.

Thus, what is often called 'natural classification' is based on what biologists construe as the underlying order of nature; or that is the intention. Biologists (broadly defined) have been seeking to devise such a classification at least since the time of Aristotle; and, in practice, their attempts have fallen into four main phases.

THE FOUR PHASES OF BIOLOGICAL CLASSIFICATION

The pivotal event in the history of biology was the publication, in 1859, of *The Origin of Species by Means of Natural Selection*, by Charles Darwin (1809–82). Indeed, biology as a whole can now be divided into two ages: the time before *Origin*, when ideas about evolution were generally absent or regarded with hostility or, at best, with a few promising exceptions, were ill-formed; and the time after, in which evolution by means of natural selection was and is firmly on the agenda. Darwin is to biology what Beethoven is to music; no one who came afterwards can ignore the inroads he made, even if some of them choose to take issue with some of his ideas.

Phases I and II of biological classification unfolded in the pre-Darwinian era; and III and IV are post-Darwinian. For convenience (different biologists tend to use the terms in different ways, so I had better make clear what I mean by them) I will apply the word **taxonomy** to all four phases of biological classification (from the Greek *taxis*, meaning distribution); and will reserve the term **systematics** for the two post-Darwinian phases. I will call the four phases 'ancient', 'classical', 'immediate post-Darwinian', and 'cladistic'. The cladistic age—phase IV—is the present.

PHASE I: ANCIENT TAXONOMY

The ancient phase is best represented by Aristotle (384–322 BC). Aristotle was a keen naturalist and a seminal logician, and he exercised both of these talents in his taxonomy. Thus he spelled out the key idea—that in order to classify in a formal, reproducible manner, you have to rely on something more than the gestalt that enables songbirds to differentiate between pigeons and hawks. You have in fact to define the criteria of your classification. In practice, you have carefully to describe the things in question, and then create groupings according to the features that they have in common.

But Aristotle also showed that in the case of living creatures this process is far from straightforward. You can produce quite different classifications depending on which features—or **characters**, as modern biologists prefer to say—you choose to focus on. But he also showed (another crucial insight) that some sets of characters seem to give more satisfactory results than others. Thus he quickly saw that a classification based on numbers of legs produces obvious anomalies; for example, human beings become lumped with birds. His alternative proposal—to divide animals into oviparous (egg-layers) and viviparous (by which he meant exclusively mammals)—has a more satisfactory feel even though, taken alone, it does lead to further problems (and it helps that he knew nothing of the duck-billed platypus, one of the few surviving mammals that lays eggs).

All in all, then, Aristotle did not arrive at anything resembling a modern classification, but he did confront crucial issues: the need to state clearly the criteria for classification; the fact that different criteria produce different results; and the inescapable need, therefore, to choose between different criteria. The classical taxonomists of phase II carried the whole process significantly forward—with many more examples, a more rigorous search for criteria by which to classify, and more attempts to discern the underlying order that could justify one classification rather than another.

PHASE II: CLASSICAL TAXONOMY

The 'classical' age of taxonomy occupied the three centuries or so before Darwin published *The Origin of Species*. 'Modern' science is commonly assumed to have begun in the late seventeenth century, with Isaac Newton and his contemporaries in physics and, for example, John Ray, William Harvey, Anton van Leeuwenhoek, Marcello Malpighi, and others in biology. But earlier centuries built up a huge head of steam—for as Newton said, 'I have stood on the shoulders of giants'—and the science of earlier centuries, including the natural history, should not be taken as lightly as is usually the case. We can reasonably if arbitrarily suggest that the age of classical taxonomy truly began in the sixteenth century.

The guiding philosophy throughout phase II was largely (though by no means exclusively) utilitarian. The taxonomists of the sixteenth and seventeenth centuries were commonly herbalists and horticulturalists; and in the eighteenth and early nineteenth centuries, the great ages of exploration, biologists sought exotic creatures that might be pressed into those arts. The greatest botanical gardens in the world at Kew, west of London, were established with a strongly economic motive—a concern that still persists, and is reflected, for example, in research on plants that might furnish drugs, and on tropical timber trees such as the dipterocarps of Southeast Asia and the araucarias of South America. Even the Zoological Society of London, founded in 1826, was originally intended to find useful animals that might prove suitable for introduction to Britain (although nothing is more meddlesome than the transport of species from country to country, as present-day Australians will attest, knee-deep as they sometimes literally are in rabbits, cats, foxes, and cane-toads).

Because the biology of phase II had such a strong economic and political thrust, the taxonomies of that period were, to some extent at least, directed to commercial ends. Thus a prime concern in the late eighteenth century was to provide **keys**, which would enable non-experts to identify plants (plants more often than animals, particularly useful ones) that had previously been discovered and described by specialists. The construction of keys is a very subtle business: the great French biologist, Jean-Baptiste Lamarck (1744–1829), who produced an erroneous but nonetheless ingenious hypothesis of evolution 50 years before Darwin, was a skilled key-maker. Key-making aids the business of taxonomy in some ways, because it depends on close observation and accurate description. It is less helpful in others—the diagnostic criteria whereby a plant may be identified are not necessarily those of greatest biological significance. Hence keys for identification are generally called 'artificial' keys.

But phase II was also a golden age of 'real' science, in which natural philosophers sought to find order in the Universe. Some scientists felt that the orderliness that they discerned replaced what had previously been seen as divine order, imposed by God; and this seems to be how modern people conventionally see the historical role of science. Yet many—perhaps most—scientists during the phase II years were extremely devout and held precisely the opposite belief. Isaac Newton himself typifies the attitude of many seventeenth-century scientists. He believed that the order of the Universe directly reflected the tidy mind of God. For him, scientific research was an act of reverence; for he perceived human intelligence as a gift from God, and felt there could be no better way to use that gift than to explore the mind of his maker. This kind of attitude persisted and often prevailed well into the nineteenth century. Richard Owen (1804–92), the most famous biologist in Britain before Darwin came along and the principal founder of London's Natural History Museum, was as devout as Newton, and would have agreed absolutely with his attitude to science.

Be that as it may, the seventeenth, eighteenth, and nineteenth centuries produced marvellous evidence of nature's order. Newton's own laws of mechanics vindicate the notions that nature does indeed operate according to laws, and that these laws are there to be discovered. Order emerged just as emphatically in chemistry. In 1803, the British chemist John Dalton formulated his idea that matter is made of different kinds of atom, each kind being characteristic of each element. Suddenly the huge corpus of chemical knowledge that had been accumulating for several centuries began to fall into place; although the culmination of Dalton's ideas, the arrangement of elements according to atomic weight in the periodic table, was not achieved until 1869, by the Russian Dmitri Mendeleyev, which is slightly after the end of our phase II. Never mind: from the sixteenth to the nineteenth centuries the idea was consolidated and triumphantly demonstrated on all sides: *nature is orderly*.

So long as taxonomists seek simply to classify organisms for practical purposes—for example, for designing keys—then they are functioning primarily as technicians. But the taxonomists of phase II also aspired to be scientists and so they wanted to devise a classification that truly reflected and encapsulated the presumed order of nature. But what order is there among organisms? Can living things be arranged, like atomic elements, in a logical, finite, unbroken series? Are their shapes and sizes controlled by laws, comparable with those that Newton revealed are directing the Universe as a whole?

The classical taxonomists of phase II searched hard for the order underlying living things, and came up with various hypotheses. Thus one of the outstanding taxonomists of the classical phase, the French botanist Antoine-Laurent de Jussieu (1748–1836), suggested initially that nature in general and the forms of plants in particular are 'continuous': that is, he supposed that living plants between them form an uninterrupted series, and that each species in the series grades by minute degrees into the next one. If there were apparent gaps, then this was because the intermediates had not yet been found. This is obviously not the case—at least if we consider only the living types; but the notion of natural continuity is a perfectly reasonable hypothesis, a starting point (and the important thing about hypotheses in science is not that they should be right, but that they should be testable; and that, through being tested, they lead to greater insight).

All in all, although they did not find the underlying order that they sought among living things—laws to match Newton's all-embracing overview of the Universe—the classical taxonomists did lay foundations of modern practices that persisted well into the twentieth century. Indeed to a significant degree their practices, if not their philosophies, are still with us. Their general approach was to organize living things according to their macroscopic or 'gross' structure (**morphology**) and as perceived under the microscope (for which the modern term is **ultrastructure**). They also took evidence from development: **embryology**. In general (how else?) the creatures that were perceived to be most similar were placed in the same group, and those that seemed less similar were placed in different groups. Importantly, the classical taxonomists did not assess the degree of similarity simply on general impression, but focused on particular characters. Thus they again encountered, in many different contexts, the problem faced by Aristotle: that organism A may have one set of characters in common with organism B, and another set in common with organism C. It was necessary, then, to decide which characters gave the more robust, or 'true', or consistent classification. More often than not, it seems, the classical taxonomists made the right decisions. Thus Jussieu perceived that the number of 'seed-leaves' or cotyledons in the seeds of flowering plants was of more significance than the mere form of their flowers; and, indeed, botanists still distinguish between 'monocots' and 'dicots', although, as described in Section 24, the nature of the distinction has been modified within the past few years. Jussieu also saw, however, that the forms of the flowers in general are more satisfactory characters on which to base classification than are the forms of the leaves—another insight that persists.

But the greatest and most influential of the classical taxonomists was Carolus Linnaeus (1707–78), also known as Carl von Linné. He deserves a brief section to himself.

THE APOTHEOSIS OF PHASE II—LINNAEAN CLASSIFICATION

Linnaeus was a Swede: one of a relatively small but highly significant coterie of Scandinavians who have made pivotal contributions to science. As a young man reading medicine at the University of Lund he became interested in the new idea that plants are sexual beings (nothing can be taken for granted!) and in 1730 he began a system of plant classification based on the sexual organs (the male stamens and the female pistils) of flowers. He was an explorer, too; in travels through Lapland in 1732 he discovered 100 new species of plants.

Linnaeus then moved to Holland where, in 1735, he published *Systema Naturae*, which he continued to expand and revise until the 1750s: a classification of animals, plants, and minerals. Linnaeus lived almost exactly 100 years before Darwin, and apparently knew nothing of evolution, but his observations and instincts led him to classify creatures in ways that to post-Darwinians often seem eminently acceptable. For example, Linnaeus named the orang-utan *Homo troglodytes*—implying affinity to human beings, which he called *Homo sapiens*. In fact, the orang-utan's affinity to us is implicit in its common name, which in Indonesian means 'man of the woods'.

Most significant, however, is that Linnaeus provided formal rules of grouping and ranking, which, with only minor modifications, should endure forever. First he proposed the **binomial** method of naming living creatures; positing the idea in 1749 and presenting it formally in 1753 in his *Species Plantarum*. In this system, each living creature is given two formal names when it is first described. The first of these names is generic, as in *Homo*; and the second is specific, as in *sapiens*. Creatures with the same generic name are said to belong to the same **genus**; and the specific name pinpoints the particular **species**. Hence among the close relatives of *Homo sapiens* are *Homo neanderthalensis* (Neanderthal man),[1] *Homo erectus* (literally 'upright man', who was the first human species to venture beyond Africa), and *Homo habilis* ('handy man': the first bona fide human being of all).

Such names are generally called 'latin', although in practice their roots tend to be both Latin and Greek, and often they are simply contrived, or constructed after people and places (as in *neanderthalensis*, named after the Neander Valley in Germany). The 'latin' or 'scientific' names of species are always written in italics; the generic name always begins with an upper-case letter and the specific name is always lower case, even when it is derived from a proper name. When it is obvious which genus we are talking about it is permissible to present the generic name simply as an initial, and so we can say '*Homo sapiens* and *H. neanderthalensis* and *H. erectus*'. But we should not simply drop in '*H. sapiens*' unless it is absolutely clear that the '*H.*' stands for *Homo*. After all, many different genera of all kinds have names that begin with '*H*' and many share the same specific qualifier: *vulgaris*, *africanus*, *rubra*, etcetera.

Linnaeus originally devised the binomial system as a kind of shorthand. He

[1] As I mention later in Section 20, some specialists in human evolution have preferred to think of Neanderthal man simply as a subdivision or 'subspecies' of *Homo sapiens* and therefore refer to Neanderthals as *Homo sapiens neanderthalensis*; the third name being a 'subspecific'. Others point out, however, that although Neanderthals and modern people lived side by side in Europe and the Middle East for at least 5000 years (and probably for more than 10 000 years), there was probably little flow of genes between them, so the two effectively remained separate. Thus they felt it was reasonable (as well as neater) to treat the two as separate species, and to give them different specific names. Matthias Krings of the University of Munich and his colleagues (who include the pioneer in this area, Svante Pääbo) have now shown that DNA extracted from Neanderthal bones differs significantly from that of modern humans, so that the two really did remain distinct, and definitely should be seen as separate species. So, *H. neanderthalensis* can now be considered 'correct'; and *H. sapiens neanderthalensis* is just wishful thinking.

inherited the already immense scholarship of the sixteenth, seventeenth, and early eighteenth centuries, and throughout that time biologists tended to append brief descriptive essays, written in Latin, to the new plants, animals, and other creatures that they had found. Linnaeus sought to draw up a coherent list of everything that had been discovered up to his own time, and to add more of his own. So he précised the essays of his peers and predecessors by recording a couple of words from each one. Hence, each species acquired two names; and, miraculously, two is all that is required to make a tidy system that should serve for all time, even if the list of known species reaches 100 million or more. Such problems as do arise have to do with nature's frequent refusal to conform with the biologists' conception of what a species actually is—as discussed later in this chapter.

Linnaeus also proposed a formal hierarchical system of classification. The binomial system already implies this: the small category of the 'species' nests within the broader category 'genus', just as the genus *Homo* embraces *H. sapiens*, *H. neanderthalensis*, *H. erectus*, and the rest. Linnaeus also placed the genera (plural of genus) within larger groupings, and then larger still. Originally he proposed just a few sizes—or **ranks**—of grouping: genera nesting within orders, orders within classes, classes within kingdoms.

Since Linnaeus's day taxonomists have added more ranks, so that now there are eight: species, genus, family, order, class, phylum, kingdom, and—the latest to be added, and the greatest—domain. In practice, eight ranks is not enough to reflect the full complexity of nature, and taxonomists have long felt the need to insert intermediate groupings, as in 'suborder' and 'superfamily'. This works up to a point—I will use this system in this book—but it can get out of hand, and some taxonomists have invented yet more rankings such as 'tribe' (sometimes adopted as a subdivision of a family), 'infraorder' (a subdivision of 'suborder'), 'parvorder' (a subdivision of infraorder), and so on. Sometimes, too, taxonomists have perhaps spent too much time discussing the proper ranking: whether a particular group should be regarded as an order or a subclass, for example. For these and other reasons, some modern taxonomists would like to abandon the Linnaean rankings all together, and simply show the relationships between creatures on a diagram. But many others (including me) feel that this would be going too far. I suggest that we should continue the Linnaean system, but that we should not be quite as literal as seemed appropriate in the eighteenth century; in other words, that we should adopt an approach that I call 'Neolinnaean Impressionism'. I expand this idea in Chapter 5.

As can be seen from a cursory glance at Part II, the names of the various ranks tend to have their own characteristic endings. In particular, families of plants nearly always end in '-aceae'—although there are some variations, as in Gramineae[2] (the grass family), Compositae (daisies and their relatives), and Labiatae (mint, thyme, and so on); and families of animals always end in '-idae'. This makes it very easy to

[2] Names like Gramineae are now considered old-fashioned. Modern botanists name families after the first-named genus in the group and add '-aceae'. So Gramineae is now Poaceae, and Compositae is now Asteraceae.

refer to members of a family informally. Thus zoologists refer to all the many living and extinct members of the horse family, Equidae, as 'equids'; while the particular members of the modern horse genus, *Equus*, are 'equines'. Similarly, members of the human family, Hominidae, are 'hominids'; and those of the particular genus *Homo* are 'hominines'.

Even in his own time, Linnaeus was acknowledged to be outstanding. Today his name lives on both in everyday biology and specifically at the Linnean Society of London, established in his memory in 1788. Yet Linnaeus was not a lone flower in a desert. The whole age of biology from the sixteenth century to the first half of the nineteenth was as vibrant and full of genius as the present. Many of Linnaeus's contemporaries and immediate successors are remembered and quoted still. While Jussieu did much to establish the major groupings of plants, his compatriot Georges Léopold Cuvier (1769–1832) outlined the phyla (*embranchements*) of the animals. In England, Sir Joseph Banks (1743–1820) became one of the first and most famous directors of The Royal Botanic Gardens at Kew (whose beginnings date to 1759); and the studies of vertebrate fossils by Richard Owen are still pertinent. Owen, incidentally, gave us the word 'dinosaur'.

Yet those fine biologists failed to find the underlying natural order that they knew they needed to give their classifications a solid foundation. For that order, did they but know it, lies in genealogy: in the fact that all living creatures are literally related, one to another, just as each of us is related to our sisters and our cousins and our aunts. That notion, which may seem obvious now, could not be realized fully until after 1859, when Darwin established the idea of evolution in *The Origin of Species*. This took the craft and science of biological classification into phase III. After Darwin, taxonomy truly became systematics.

PHASE III: DARWIN AND AFTER

Darwin was not the first person to think of evolution: of course not. But he it was who established that evolution is a fact—the natural way of things. He succeeded in this partly by providing the first plausible mechanism of evolution—'by means of natural selection'; and this made the idea generally credible. He also established the idea of evolution through the sheer mass of evidence that he brought to bear, and the inexorable logic of his arguments. It is not true (as it has been fashionable to argue) that the idea of evolution was simply 'in the air' when Darwin published the *Origin*, and that he merely dotted the i's and crossed the t's. Many people had toyed with the notion of evolution and some had thought seriously about it at least since the eighteenth century, but the prevalent thesis in the 1850s was that of divine Creation: the idea, as described in *Genesis*, that God had created all the creatures now on Earth, and all those extinct types that were then turning up in the fossil record, in the forms in which they are found, each one independently.

Neither is it true that everyone after Darwin immediately said, 'Of course! Living things must have evolved just as Darwin says!'. Many continued to defend the idea that each creature was created individually, and 'Creationism' is still a significant political force in parts of the United States and Australia. This remains the case even though the established church, at least in England, has long been happy to accept that the literal view of the origin of species as offered by Darwin can happily coexist with the narrative offered in *Genesis*, which many nowadays are content to accept as metaphorical (or as a kind of proto-scientific hypothesis). Many good Darwinian biologists, too, happily use the word 'creature' (as I do) without bothering to consider that it means, literally, 'a thing created'. The metaphor of creation remains pleasing. Neither is it true that all biologists rushed to embrace the idea of evolution once Darwin pointed it out, and that all theologians rejected it (despite the famous public spat between Thomas Henry Huxley, the friend and champion of Darwin, and Bishop Samuel Wilberforce, alias Soapy Sam, at Oxford in 1860). Some excellent scientists, including Darwin's own erstwhile mentor Adam Sedgwick and Richard Owen, resisted the idea of evolution, and some who accepted evolution doubted whether the specific mechanism of natural selection could achieve what Darwin claimed for it. On the other hand, many churchmen did accept the idea of evolution. (Darwin is buried in Westminster Abbey.)

It is the case, however, that after Darwin the idea of evolution was permanently and ineradicably on the agenda. All serious discussion of theoretical biology since Darwin has centred around the idea of evolution, and you are at least as likely to hear the name of Darwin in a discourse on molecular biology as anywhere. With Darwin's idea of natural selection the modern age of biology truly began: a little fitfully, a little untidily, often protestingly, but inexorably. Evolution by means of natural selection is a powerful idea and the more that natural philosophers examine it, the more it seems to be true. In the words of the great Russian/American biologist Theodosius Dobzhansky (1900–75), 'Nothing in biology makes sense except in the light of evolution'.

In a discussion of taxonomy, however, we do not need to worry about the particular role of natural selection in driving evolution. We merely need to acknowledge that evolution is the way of things. That fact—and particularly the way in which Darwin presented it—has profound implications for classification.

A NEW VIEW OF SPECIES AND THE ONE GREAT TREE OF LIFE

The main title of Darwin's seminal book has two clauses. The second is 'by means of natural selection'. But the first is 'The origin of species'. Importantly, Darwin's idea of what is a **species**, and what it is capable of, was radically different from that of most of his contemporaries and predecessors. Many of his scientific peers objected to and sometimes rejected his ideas not because he spoke of evolution, but because he challenged existing beliefs about species.

So what, first of all, is a species? Well, the conventional modern view, which has been summarized by the great German/American biologist Ernst Mayr, is that a species is a group of creatures that can breed together sexually to produce 'fully viable' offspring. Two creatures are said to belong to different species if their offspring (if any) are not 'fully viable'. 'Viable' implies the ability to live and reproduce; and 'fully viable' implies the ability to live and reproduce at least as well as your parents did.

In practice, there are various ways of falling below the 'fully viable' criterion, and, moreover, the divisions between different species are not always absolute. For example, horses and cows clearly belong to different species because, if they did ever mate, no offspring at all would result. For a whole sequence of reasons, the sperm of a stallion would simply fail to fertilize the ovum of a cow; or, if it did, the resulting embryo would soon die. On the other hand, a mare will produce an offspring known as a mule, if impregnated by a he-ass; and a she-ass will provide a hinnie if tupped by a stallion. Hinnies are feeble creatures, with a horse-like head on a donkey's body, but mules have donkey-like heads and a horse-like body and are extremely powerful beasts. Yet for all their strength and stamina mules are not sexually fertile, so they are not 'fully viable', so horses and asses clearly belong to different species. Carrion crows and hooded crows can and do mate in the wild to produce hybrid offspring that are fertile; yet those hybrids are clearly not fully viable because they do not compete ecologically with either parent, and finish up occupying a 'hybrid zone' between the territories of the two parent species. So carrion and hooded crows remain separate. In practice, then, Mayr's definition needs to be modified slightly if it is to be applied to all the phenomena of nature. Sometimes it really is difficult to say where one species ends and another begins. In general, though, and certainly as a starting point, his idea obtains: species are groups of creatures that breed sexually with each other, but not with creatures from other groups.

But although the biologists and naturalists of the nineteenth century would have been happy up to a point with such a definition, they also felt that 'species' had metaphysical connotations. For in essence—and sometimes overtly—they were followers of Plato. Plato maintained that all objects on Earth are merely rough facsimiles of some ideal archetype that exists only in heaven. Pre-Darwinian biologists felt, similarly, that earthly horses were merely imperfect replicas of some ideal horse, and earthly lions were rough and ready copies of the original, heavenly lion, and so on.

Thus the modern view of species is down-to-earth and practical; if two creatures will breed successfully together, then, in general, we say they are of the same species; and if they do not, then we do not. But the Platonic view has transcendental overtones. The species is conceived not merely as a population of creatures that behaves in a particular way, but as the manifestation of a heavenly ideal. In practice, too, the philosophy of Plato had long been absorbed within that of Christianity; so each of Plato's vaguely pantheistic 'ideals' was perceived in practice to be an idea of God.

The metaphysics matters enormously, as Darwin well knew. So long as we acknowledge that 'species' is merely a biological concept, then we are happy to accept any biological observation pertaining to the species that seems to make sense. For example, we would not be surprised to find that one species could change—evolve—into another; or indeed that any one species could give rise to several different lineages, each of which could evolve along separate lines, to produce several or many different lineages. Why not? If that is what the fossil record or any other evidence suggests, where's the problem? But if we believe, as the pre-Darwinians did, that each species represents a divine ideal, then the notion that a species might change into a different species becomes not only strange but blasphemous. We may now be bemused that this should have been a problem; but as Ernst Mayr commented in *Toward a New Philosophy of Biology* (1988), many of Darwin's contemporaries were less offended by his apparent rejection of God's Creation, than by his perceived abrogation of Plato.

Victorian biologists knew full well that different lineages within any one species can alter over time. The great agricultural revolution of the eighteenth century produced striking changes in pigs and cattle within a few decades. Darwin himself consorted with breeders of pigeons, and kept many breeds himself. The changes were brought about partly by cross-breeding but mainly by selecting the individuals of each generation that showed desirable traits—the process of 'artificial selection'; and this is what gave Darwin the idea of 'natural selection'. But Darwin's contemporaries assumed that artificial selection could produce only relatively minor variations. To be sure, the variations among domestic pigeons seem far from minor: tumblers, fantails, jacobins, and the rest are very different creatures in appearance and habits. But Darwin's fellow breeders assumed that the very first pigeon breeders must have started out with several different species of pigeon—so that tumblers, jacobins, and so on all had different ancestors.

Darwin, however, threw all this aside. He pointed out, for a start, that although there are many species of wild pigeon there is nothing in the wild that resembles the tumbler or the jacobin. He also observed that when the different pigeon strains are crossed, many of the offspring begin to show the characters of the common wild pigeon known as the rock dove. For this and other reasons he concluded that all domestic pigeons, no matter how various, had descended from rock doves. In short, he first observed that artificial selection could produce very large changes indeed.

But he went further. He went on to suggest that species barriers were not, in fact, inviolable; and that if selection proceeded for long enough then eventually the descendants would be so different from their ancestors that they could no longer breed with them (even if the ancestors were still around to be bred with) and that they would, in fact, form a new species. Indeed he suggested that any one species could become split into several distinct breeding populations and that each of these, over time, could evolve into new species that were clearly distinct not only from their common ancestor, but from each other. In other words, any one species, given time, could **radiate** to produce many species; and each of those daughter species could then radiate again.

Extrapolating, Darwin argued that all creatures on Earth may have originated from just one common ancestor so that each species is a member of an all-embracing tree: exactly like the family tree by which any of us might represent our fathers and grandfathers, our siblings and our fourth cousins twice removed. Trees of human families, showing the relationships between individuals, are an exercise in genealogy. But whereas our personal family trees merely show relationships between individuals, evolutionary trees show the relationships between entire species—or, indeed, between orders or classes or whatever. Genealogy carried out on such a grand scale is called **phylogeny** (from the Greek *phylos*, meaning 'tribe' or 'race'). Thus Darwin's evolutionary tree is a **phylogenetic tree**; genealogy writ large.[3]

Trees are by their nature hierarchical. Twigs spring from branches, which spring from boughs. The best and most economical kinds of classifications are also hierarchical. The phylogenetic tree is real: at least, it is intended to represent the unique history of each and every creature. If we can but find out that history, if we can but trace phylogeny, then we have precisely the 'order' of nature that the classical taxonomists sought, but could not find. The boughs on the great, unique evolutionary tree of life on Earth can be seen to correspond to the highest Linnaean rankings (in their modernized form)—the domains, kingdoms, phyla, classes, and so on; the smaller branches are the middle rankings—orders, families, genera; and the myriad twigs, at the very tips of the tree, are the species.

Perfect. Here we have tidiness and objectivity all in one: for what could be more 'objective' than the one, unique history of Earthly life? Darwin saw this, and indeed made the point explicitly in Chapter 13 of *The Origin of Species*:

From the first dawn of life, all organic beings are found to resemble each other in descending degrees, so that they can be classed in groups under groups. This classification is evidently not arbitrary like the groupings of stars in constellations.

And:

I believe that the *arrangement* of the groups within each class, in due subordination and relation to the other groups, must be strictly genealogical in order to be natural.

In short, the classification of living things should be based on phylogeny. That remains the dream.

So, we might reasonably conclude, taxonomists after Darwin must simply have heaved a sigh of relief and said to themselves: 'At last we can see the principles that should underpin our classifications! All we have to do is seek out the phylogeny!' But in practice, this was not quite the case.

[3] Some theoretical biologists, like my good friend Richard Webb at the London School of Economics, suggest that the genealogical trees of human families and prize cattle bear only a superficial resemblance to the phylogenetic trees that show the descent and relationships of different species and larger groups of organisms. This, of course, has some truth in it. Species, when you boil them down, are really just groups of individuals who happen to be able to breed together (according to the Ernst Mayr definition) and a phylogenetic tree shows these groups splitting, changing, and splitting again in a tree-like manner. By contrast, genealogical trees—pedigrees—show specific individuals coming together for the purposes of sexual reproduction and producing offspring like themselves, which later find mates of their own, and so on. Thus the underlying processes that drive a phylogenetic and a genealogical tree are simply not the same. Yet there are similarities: notably, both kinds of tree show relationships and lines of descent. It seems reasonable, then, to regard a genealogical, 'family' tree at least as a metaphor for a phylogenetic tree, which Darwin seems to have done (as shown by the quotes in this chapter). So I will stick with my notion that phylogeny is 'genealogy writ large' and leave the caveats as a footnote.

Some post-Darwinian taxonomists have largely ignored phylogeny, simply because they are practical people first and theoreticians second. They may or may not have been interested in evolutionary theory; but even if they were, they did not always feel that a classification based on phylogeny was necessarily of the greatest use to them or their employers. Classification, after all, is used in large part for the mundane but vital business of identification, not least of the pests and pathogens that threaten lives and economies; and sometimes, as the eighteenth-century devisers of artificial keys knew well enough, identification is easiest when corners are cut, and the creatures are categorized simply by appearance, and not necessarily by their true relationship. But classification based on true, phylogenetic relationships brings huge practical advantages of its own. For example, plants that are closely related phylogenetically often have similar chemical properties, such that the family Labiatae provides many an aromatic herb, from mint to marjoram, while members of the Solanaceae, which range from deadly nightshade to tomato and capsicum, provide many a drug and toxin. Thus, if you begin by classifying a creature according to its true relationships, then you can often guess its pharmacological and commercial qualities before you even look.

So many taxonomists who are motivated primarily by practical goals, as well as those of theoretical bent who truly care about evolutionary history, have sought to devise a classification based on phylogeny. But this is much more easily said than done. There are practical problems—simply to provide enough data, and to handle that data (see Chapter 4); and there are theoretical problems, which I discuss in the next chapter. As we will see, the theoretical problems remained unsolved for 100 years after Darwin, and it is this period of 100 years that I regard as phase III. The final solution, or at least it looks like the final solution, was provided only in the 1960s, by a German entomologist called Willi Hennig. Hennig seems to be little known outside taxonomic circles; but within this circumscribed field (though not outside it) he deserves to be placed alongside the giants, Linnaeus and Darwin.

CHAPTER THREE

THE NATURAL ORDER

DARWIN'S DREAM AND HENNIG'S SOLUTION

Charles Darwin showed that the order in nature sought by the classical taxonomists lies in phylogeny; in the literal, genealogical relationships between different species. Creatures could and should be placed in the same grouping if, and only if, they all shared a common ancestor—at least if we are seeking a taxonomy that truly reflects the order of nature. All creatures, Darwin suggested—bacteria, oak trees, worms, and human beings—belong to the same genealogical tree: ultimately, they all could trace their ancestry to a single, common source. The great weight of evidence since Darwin, including the most modern molecular studies, suggests that he was right. So for taxonomists who seek a classification based on the true order of nature —in other words, the taxonomists who can truly call themselves systematists—the task is simply to work out who is related to whom.

But 'simply' is not the right word. The road to phylogenetic insight is riddled with pitfalls and although some were recognized early on—in principle, even before Darwin—others were not fully appreciated until after the 1950s when German entomologist Willi Hennig began to develop the ideas and methods known as **cladistics** (named from the Greek *clados*, meaning 'branch'). Hennig first began to publish his thoughts on classification in 1950 but the works that are commonly perceived as seminal date from the mid-1960s, especially his book *Phylogenetic Systematics* (1966).

Even now, some traditionalist taxonomists regard cladistics with suspicion, while some accept its central tenets while rejecting less-essential features. But the core philosophy of cladistics has become the orthodoxy. Cladistics is certainly the guiding light in this book and all the trees in Part II are based on cladistic principles. In truth, I do reject some of the more eccentric criteria by which Hennig suggested relationships could be judged (he put too much store by biogeography, for instance) and question his more purist ideas on the naming of groups—or at least suggest some relaxation in the interests of easy communications. Overall, however, Hennig's basic ideas and methods seem to provide just the stroke of genius that was needed. Cladistics, in short, is taxonomy phase IV.

But before we look specifically at the insights and methods of Hennig we should ask why his input was necessary: why, in fact, it is so difficult to divine the phylogeny of living creatures. Surely, after all, it's just a matter of observation—describing and measuring the different characters, and seeing who is most similar to whom. Well, that is precisely what is required, and what is done: but out of this endeavour there arises a hierarchy of practical and theoretical problems. We can start to get some insight into the difficulties by means of a metaphor.

CONVERGENCE, DIVERGENCE, AND RADIATION—THE ALLEGORICAL TALE OF SMITH, SMITH, HARRIS, AND ROBINSON

Imagine you are whisked back in time to an English village inn in the early nineteenth century and fall in with a jolly group of young chaps, two farmers and two lawyers. The farmers are very obviously farmers, with broad shoulders and accents, red faces and moleskin trousers; and the lawyers are archetypal lawyers, somewhat dandified with canes, tail coats, and the authentic accents of an Oxford college. You pick up from the somewhat jumbled introduction in that somewhat raucous pub that two of the party are called Smith, and are brothers; and the other two are called Robinson and Harris.

It would be reasonable to assume, in the absence of better knowledge, that the two Smith brothers would look and behave similarly; so either the two farmers are Smiths, or the two lawyers. Later you are somewhat surprised to learn that the Smith brothers actually comprise one of the farmers and one of the lawyers. The other farmer is Robinson, and the other lawyer is Harris. In other words, the two who are literally related look different: and the two pairs who look similar are not related at all.

This amiable quartet between them illustrate almost all the main principles of systematics; and your initial misunderstanding demonstrates almost all the problems that beset those who seek to discern relationship from appearance. Let me emphasize, though, that this is simply a metaphor. Biologists in their role as systematists seek to establish relationships between species, or higher ranks of groupings; and the difference between species or groups of species are brought about by genetic changes of the kind that are called evolutionary. The apparent physical differences between lawyers and farmers—clothes, complexion, accent—do not involve genetic change, and are simply behavioural (social) rather than evolutionary. But the principles I want to discuss are comparable nonetheless.

First of all, what historical events explain the apparent differences between the men who are closely related, and the similarities between those who are not?

Well, the father of the two Smith brothers was himself a farmer, who had enjoyed some success. Like many a man of his period and social class he wanted his sons to 'get on', and he perceived that the path to social advancement lay through education. So he sent the boys to be tutored by the curate of a nearby village, an Oxford graduate who was making a reputation as a teacher. Smith the lawyer had taken well to book learning and eventually won a scholarship to Balliol College, Oxford; and because he had resided with his tutor (living-in was usual in those days) and spent some years in Oxford, his previously rustic accent was transformed. But his brother could not get on with books at all and had returned as soon as was decently possible to the family farm, which he eventually took over from his father.

The Smith brothers illustrate the phenomenon that evolutionary biologists call **divergence**. Two creatures with a common ancestor may rapidly adapt to different circumstances, and as they do so may change both their behaviour and their appearance. Any one lineage of creatures may diverge in many different ways to adopt many different forms and ways of life so that one lineage becomes many; and the sum of all the divergences is called a **radiation**—a graphic term. Part of Darwin's achievement was to see (after a little help from his friends) that the many various finches on the Galápagos Islands, with their different diet, habits, body size, and beaks, had all radiated from a single kind of finch that had probably been blown to the volcanic islands from mainland Ecuador, many years before. But often, when two creatures have diverged, as the brothers Smith diverged, it can be hard to see that they are in fact related.

Smith the lawyer and Harris the lawyer illustrate another of the problems that can cause much confusion: that of **convergence**. For Harris comes from a long line of merchant seamen. His male ancestors were colourful fellows, given in earlier times to tarred pigtails and a fine line in profanity; but Harris's own father had graduated to gold braid and, like Smith the lawyer's father, wanted a softer life for his son. So Harris junior, like Smith junior, had eventually won a scholarship to Balliol and had gone on to read law. At least in dress and speech, Harris and his friend Smith are now alike as two ship's biscuits; but their fathers, one in cocked hat and white stockings, the other in broad hat and gaiters, could hardly have looked more different. By adapting to a similar environment, the two younger men have converged.

Parallelism, or **parallel evolution**, may be seen as a special case of convergence. Suppose, looking ahead, that Smith the lawyer gives rise to three more generations of lawyers like himself; and that Harris does the same. Successive generations differ in appearance from their fathers as the fashions change—silver-topped canes give way to umbrellas, for example—but in each generation the Smith lawyers and the Harris lawyers resemble each other, as each continues to adapt to the demands of the same profession. The 'evolution' of the two lineages (using the term 'evolution' loosely) runs in parallel. But suppose, now, that at the end of the nineteenth century one of the Harrises discovers a taste for the military, and signs up for the Boer War. Now his dress and behaviour change. We have divergence again. But when World War I breaks out the boys of the Smith family also become soldiers—and we are back to convergence. Lineages of wild creatures can also converge, run in parallel, diverge, and re-converge again as the millennia pass. Whatever phase they are in you will face possible confusion if and when you seek to assess their relationships solely on the basis of features that may be highly conspicuous but in the end may be merely superficial.

FROM ALLEGORY TO REALITY—THE MAMMALS' 'DANCE THROUGH TIME'

Between them, the mammals illustrate just how stunning (and potentially confusing) the phenomena of divergence and convergence—and less often of parallelism—can be. For example, the two main groups of modern mammals are the marsupials and the eutherians, both of whom apparently descended from a common ancestor who lived about 120 million years ago, during the early Cretaceous period. Each of them has radiated wonderfully as the different lineages within them have diverged: the marsupials ranging from the rhinoceros-like diprotodonts, who died out some time after the Aborigines arrived in Australia, through kangaroos and koalas, and the eutherians including horses, cats, bats, whales, sloths, and human beings.

Each of the great mammalian groupings obviously includes body types and ways of life (sometimes called 'ecomorphs') that are not matched by the other. Thus, although there are hopping eutherians, none is quite like a kangaroo; and although there are swimming marsupials, none is like a whale, and the flying phalangers do not closely resemble bats. At least equally striking, however, are the convergences and parallels between the two great groups of mammal—so much so that in colonial days the various marsupials were often named after their perceived eutherian 'equivalents': marsupial 'mice', marsupial 'moles', the Thylacine 'wolf', and so on (although nowadays these creatures are re-acquiring their earlier, Aboriginal names).

Go back in time—and look also to South America, where marsupials once flourished at least as much as in present-day Australia—and you see even more convergences, both between the marsupials and the eutherians and within each of those groups. The sabre-tooth cats, which are among the most famous of extinct eutherians, are a good example. The North American *Smilodon* (sometimes erroneously called 'sabre-toothed tiger') is the best known but in fact there were many species of sabre-tooth. More relevant here is that sabre-tooths are now known to have evolved at least three times independently within the Felidae, the cat family—a three-tined parallel evolution. There are no sabre- tooths today, as things have turned out, but the modern clouded leopard of Asia (*Neofelis nebulosa*) also has large canine teeth and might be said to be reinventing the sabre-toothed mode for the fourth time.

But the parallels and convergences among the eutherians go even further. Until the Miocene epoch (which started some 24 million years ago) there existed another family of carnivores that looked almost exactly like cats but are now placed in a quite different family, the Nimravidae. The nimravids and felids as a whole provide a grand case of convergence. More striking, though, is that the nimravids also produced sabre-tooths. So there have been five separate groups of sabre-tooths (including the clouded leopards), from two eutherian families. But then, two different families of marsupials also produced sabre-tooths. *Thylacosmilus* of the South American marsupial family Thylacosmilidae was very like the felid sabre-tooths, with great curved canine teeth. *Thylacoleo*—the Australian marsupial 'lion' of the Thylacoleonidae—also had sabres that were formed, bizarrely, not from the canine teeth but from the incisors. So there have been at least seven separate inventions of the sabre-tooth among four different mammalian families, yet the most recent common ancestor of those four families lived deep in the time of the dinosaurs and presumably was a small, superficially shrew-like creature, as mammals typically were in those distant times.

Look at the evolutionary history of mammals as a whole and you see an extraordinary interplay of divergences (leading to radiations), convergences, and parallelisms that have produced what Elisabeth Vrba of Yale University has called 'the dance through time'. It is a burlesque dance, with each lineage trying on the clothes of many others, and each way of life essayed by many different lineages. Thus the 200 extinct species of rhinoceros mentioned in Chapter 1 did not all resemble modern rhinoceroses. Only a minority had horns on their noses, for example, and by no means all had the modern tank-like form. Some were small and agile, like ponies. One had a hippo-like shape—and, perhaps, was more hippo-like in habits than a hippo. The biggest mammal of all, sometimes called *Baluchitherium* or *Indricotherium* but nowadays with the preferred name of *Paraceratherium*, was a rhinoceros shaped like a giant giraffe that stood more than 5 metres tall at the shoulder, and could then stretch its mighty head and neck another metre or so skywards. Giraffes, indeed, are only one of several mammalian lineages to essay the giraffe-like form. Besides *Paraceratherium*, there have also been some very long-necked camels.

Among the carnivores the dance through time is positively giddy. Nowadays dogs look and behave like dogs—medium-sized, light-bodied, cursorial (running) creatures that commonly hunt in packs; bears are heavy and tend to be versatile and more or less solitary; and hyaenas are bone-crushers, except for the extraordinary feeble-toothed aardwolf of southern and eastern Africa, which lives on termites. But over the past 30 million years there have been dog-like bears, some small and possibly hunting in packs, while the formidable *Arctodus* was half as big again as a modern grizzly and had a short face with a gripping jaw and long running legs—a bear that functioned like a giant Rottweiler. There have been dogs that crushed bones like hyaenas, whose presence in North America may have prevented the immigration of true hyaenas when Beringia—the landbridge across the present Bering Strait—emerged during the ice ages of the past million years and so allowed the influx of Eurasian cats and ungulates (hoofed animals). But before hyaenas evolved as bone-crushers they evidently looked and behaved more like dogs.

But why do we interpret the fossil evidence in this way? Why should a biologist suppose that ancient hyaenas or bears that hunted like dogs actually belonged to the hyaena or bear lineages, and were not simply dogs themselves? Why suppose that an ancient bone-crushing dog was indeed a dog, and not a hyaena? The answer lies in the crucial concepts of homology and homoplasy. To illustrate these notions I return briefly to our four early-nineteenth-century drinking companions.

THE KEY ISSUES OF HOMOLOGY AND HOMOPLASY

Although in your first evening with Smith, Smith, Harris, and Robinson you guess that the two Smiths must either both be farmers or both be lawyers, various of their features—or, as a biologist would say, their **characters**—do not quite seem to square with that assumption. Thus, although Smith the farmer and Smith the lawyer (as they turn out to be) look so different, they tend to sit in much the same way, resting an elbow on the table with their head in their hand. When they remove their hats you see that they both have black curly hair, and each of them, too, has a gap between their front two teeth. The other farmer, by contrast, has a mop of straight, fair, Anglo-Saxon hair with very straight teeth whereas the hair of the other lawyer, although also blond, is definitely receding and his front teeth are slightly overlapping.

So when you finally learn the true relationships of the four, you are not quite so surprised as you otherwise might have been. Common sense and experience tells you that the similarities between lawyer Smith and lawyer Harris are the kind that are easily affected: a man may choose his clothes, and to some extent his accent, and a life indoors will blanch his complexion. But the similarities that link Smith the lawyer and Smith the farmer are not so malleable. For example, in the days before orthodontics it certainly was not easy to alter the disposition of your teeth. Indeed, long before Gregor Mendel's genetic experiments with peas in the mid-nineteenth century, everyone knew that characters such as the shape of the teeth were likely to be inherited. The teeth alone (plus the hair, the way of sitting, and so on) subconsciously alerted you to the idea that Smith the farmer and Smith the lawyer were related, and had inherited those particular characters from a common ancestor.

Characters that are inherited from a common ancestor are said to be **homologous**. In fact, although in the case of Smith the lawyer and Smith the farmer the characters they have in common are similar in appearance—for example, the gap teeth—this does not have to be the case. For example, the arm of a human being is homologous with the wing of a bird: both evolved from the forelimb of a common ancestor, who lived at least 300 million years ago (when the diapsid reptilian ancestors of birds diverged from the synapsid reptilian ancestor of mammals). In short, common inheritance is what counts, not similar appearance. Characters that two different creatures have in common are said to be **shared**.

On the other hand, the characters that Smith the lawyer shares with Harris are mere superficialities; the top hat, the silver-topped cane, and so on. They have acquired similar features simply because they have each, independently, adapted to the same way of life, and wear the uniform that goes with it. Such similarities—**adaptations** acquired independently to a common way of life—are said to be **homoplasies**, from which the adjective is homoplasious (or, as some say, 'homoplastic'). Among living creatures, we might say that the antlers of a deer and the horns of an antelope are homoplasious. They look roughly similar (except that antlers are generally branched) and they serve the same function, but they are formed from different tissues within the two animals.

In short, homologies reflect true relationship whereas homoplasies do not. In seeking true relationships, therefore, systematists must seek to differentiate the characters that are truly homologous from those that are merely homoplasious. This is obviously difficult because characters in different creatures that are truly homologous may nonetheless look different (hence we can say that individual characters diverge, as well as entire creatures) while homoplasious features tend to look the same, effectively by definition. Wild creatures, unlike men in pubs, are unable simply to tell you their life histories: systematists have to work it out for themselves. How they do this is a long and complicated story, and one that has often given rise to much controversy.

BACK TO REAL LIFE—THE SEARCH FOR HOMOLOGY

Biologists have traditionally used essentially commonsense approaches to help them distinguish homology from homoplasy. These methods are still indispensible, although they cannot be relied on absolutely. No matter how hard taxonomists may search for 'objective' rules, they can never entirely escape the need for human judgement. The quest to establish phylogeny is an attempt to reconstruct history, and in this there can be no royal road to truth. But by framing rules and principles we can certainly improve on random guesswork.

The fundamental idea in the search for true homology is that some features of living creatures are liable to remain unchanged from generation to generation whereas others seem liable to be swayed this way and that by the rigours of the environment. Characters that persist through the generations are said to be **conserved**. Conserved characters (like the gap-teeth of the Smith brothers) should provide clues to real relationships; whereas the characters that respond more readily to circumstance (like a man's working clothes) are much less likely to. This is just common sense.

But the notion that conserved features should signify homology more reliably than those that are less conserved seems to make little sense unless we assume that all living creatures have evolved, and indeed that they all evolved from common ancestors, as Darwin supposed. Yet the essential distinction between features that are highly conserved and those more labile was made well before Darwin's time. In the eighteenth century, as we saw in Chapter 2, Antoine-Laurent de Jussieu suggested that the number of seed-leaves in flowering plants indicate their relationships more reliably than the form of the flowers, and that flowers are more reliable than leaves. But Jussieu had no conception of evolution, or of common ancestry. The term 'homology' was coined in a biological context by Richard Owen, some decades before Darwin's *Origin of Species*: and until his later years, Owen, though a great biologist, was anti-evolutionist. So although the idea of 'homology' seems to make no real sense except in the concept of Darwinian evolution, the greatest biologists had a sense of what it implied even before Darwin. They knew there was an order in nature somewhere, and felt that the concept of homology would lead them to it. They just could not bring themselves to believe that the order lay in genealogy.

Natural history museums commonly illustrate the concepts of homology and homoplasy by reference to human beings, fish (such as a goldfish), and dolphins. How do we work out their relationships? Well, in overall form the dolphin and the goldfish are much the same: both have cigar-shaped bodies and propel themselves by fins. Indeed, dolphins are pragmatically classed as 'fish' in many societies. But closer inspection reveals that the fin of the dolphin is supported by an array of bones laid out very like the bones of the human wrist, hand, and fingers; it is like a human hand in an oven glove. By contrast, the fin of the goldfish is supported by a fan of thin rods like rays, which do not spring from a wrist. Of course, these facts alone do not prove that the dolphin and the human being are closer to each other than either is to the fish. The two cigar-shaped animals might have diverged from the fork-like human in very early times and then, independently, developed different details of form: scales versus rubbery skin, ray-fins versus hand-like fins, and so on. Human beings and dolphins might have developed hand-like forelimbs independently. But common sense alone suggests that this is unlikely. After all, the dolphin fin and the human hand are very different in function, so why on earth should they have a similar structure unless they both had been inherited from a common ancestor?

There is an important twist to this story. For, in truth, homology is a relative concept: there are degrees of it. If you trace back the ancestry of goldfish, dolphins, and human beings far enough, you find that they all evolved from a common, fish-like ancestor who lived somewhat more than 400 million years ago. This ancestor had two pectoral and two pelvic fins that went on to evolve into the ray-fins of goldfish and their allies on the one hand, and the limbs of mammals and other such creatures on the other. If we want to sort out the relationships between a goldfish, a dolphin, and a human being, then, we would indeed say that the forelimbs of the latter two are homologous, whereas the dolphin fin and the goldfish fin are merely homoplasious. But if instead we were working out the relationships between a goldfish, a dolphin, and an ant, we should conclude that the goldfish and the dolphin fins are homologous in the sense that they both evolved from the same ancient, pectoral structures, whereas the legs of ants are merely homoplasious. Context is all.

There are further complications. It has transpired in recent years that the general body form of all animals is coded by a set of homeotic genes called the Hox gene complex; and although there are considerable differences of detail, the Hox genes of a human being, say, are remarkably similar overall to those of a housefly, brine shrimp, or what you will. The job of Hox genes is not to provide the specific codes for individual organs—for example, they do not tell the animal how to make a liver or an eye. But they do tell the developing embryo where to place the liver, or the eye, or whatever. The eyes of flies and the eyes of human beings are about as different as two organs could be; no one would suggest that they are in any way homologous. Yet, it transpires, the individual Hox genes that tell the fly or the human to place two eyes in the front of the head are the same in the two animals (and, indeed, as far as is known, in all animals). This discovery amazed everybody: it is one of the most stunning insights of late twentieth-century biology. We will meet Hox genes again in Chapter

4 and again in Part II, Section 5. Note, meanwhile, that the eyes of flies and humans are not homologous, but the genes that tell those two animals where to place their eyes, are. The concept of homology is crucial in taxonomic studies, but it is not quite as straightforward as it may seem.

The example of the goldfish, the dolphin, and the human being I have given is a little too obvious. It is easy to see that dolphin fins are more like human hands than they are like goldfish fins. Other characters and other creatures are far less easy to explain. So systematists have devised a series of guidelines—rules and principles—to help them decide in general whether any particular character, shared by two or more different creatures, is homologous or homoplasious.

For example, complex characters that are shared by different creatures, like feathers, are far more likely to indicate homology than simple characters, like fur. Again, the roots of this idea are commonsensical. Common sense says that any particular complex organ, requiring many different genes for its construction, is likely to evolve only once; so that all creatures with such an organ must indeed have inherited it from the same ancestor. Feathers are immensely variable (compare a down feather of an ostrich with a flight feather from a sparrow): and yet all show the same complicated, basic structure. Look at young birds, in or out of the egg, and you find they all grow their feathers by the same route. If we explored the genetics of feathers we would surely find that in ostriches and sparrows and everything between essentially the same genes produce their feathers. Every source of evidence supports the notion of universal feather homology but common sense suggests that this is the case even in the absence of detailed evidence: it just seems inconceivable that such an intricate structure could have evolved more than once, in more than one group of ancestral reptiles.

Contrast fur. To be sure, fur is seen as a unifying feature of modern mammals. But some extinct reptiles, quite unrelated to mammals, probably had fur of a kind, including the flying pterodactyls. Bumble bees and moths have 'fur'. In fact, bumble-bee fur is made of chitin and mammal fur of keratin, which are quite different, so the two are obviously not homologous; and, of course, many other differences proclaim that the two kinds of creature are unrelated. So bumble-bee fur, mammalian fur, and pterodactyl fur are merely homoplasious. But note that 'fur', defined broadly, has a relatively simple structure: it's just a rod-like extension of the skin. We need not be at all surprised if more than one lineage of creature produced such structures.

Then again, many creatures have lost particular organs: for example, many animals that live in caves have lost their eyes, wholly or in part, and many island birds are flightless and sometimes almost wingless. Should a systematist leap to conclude that two eyeless species of fish are necessarily related, or two kinds of flightless bird, like a kiwi and dodo? No is the short answer. But why not? Well, just as it is easy in principle to evolve some simple structure in one form or another, so it is extremely easy to lose entire structures, whether simple or complex. For genes operate in hierarchies. A huge consortium of genes may be required to produce the eye of a vertebrate or the wing of a bird (and all the nerves and muscles that go with it) but each such consortium will be controlled by just a few 'master' genes, like a sergeant ordering a platoon. And just one mutation in such a master gene could, in theory, suppress the expression of an entire and perhaps huge genetic complex that has evolved over millions of years and guides the formation of an eye, an ear, a limb, a head, or what you will.

There are many reasons, too, why natural selection may sometimes favour the loss of an organ. Bird flight is wonderful—a tremendous asset. As we will see in Part II (Section 21), birds are phylogenetic dinosaurs—the only dinosaurs to survive beyond the Cretaceous period. Flight, surely, is what brought them through the crisis that caused the demise of the dinosaurs (although, admittedly, the other flying reptiles, the pterodactyls, also fell by the wayside). But flight requires enormous input: the flight feathers must be constantly groomed and re-grown; the giant flight muscles of the breast account for much of the mass of the bird. Flight can be dangerous, too: a bird that flies near the coast may be blown out to sea. Indeed, in some situations—notably on islands, which are too small to support big mammalian predators and where cross-winds are a constant hazard—flight brings very little net advantage and the costs of it become an embarrassment. Accordingly, many island birds independently have abandoned flight: there are or have been flightless geese and ducks, cormorants, pigeons (the dodo and solitaire), rails, auks (like the great auk), parrots (the kakapo and various extinct types), and so on. Similarly, eyes are not only useless in a coal-black cave but are positively hazardous—too easily wounded and so infected. Thus natural selection positively militates against eyes in cave animals. The loss, again, is genetically easy. Animals from many different lineages that live in burrows have much reduced limbs or have lost their limbs entirely. Some sedentary animals like mussels and barnacles have lost their heads. Many committed parasites have lost much of their entire bodies, and consist of little more than a gut and gonads.

Modern systematists have become aware, too, that small and apparently incidental details may indicate homology more reliably than the conspicuous features—precisely because the trivial characters are (or seem) irrelevant to the creature's survival, and therefore escape the day-to-day pressures of natural selection. Thus palaeoanthropologists—specialists in the study of human evolution—seeking to establish the relationships of ancient humans place much store by the number of cusps on fossil teeth, because the number of cusps does not greatly affect the function of the teeth, so any similarities seem more likely to reflect common ancestry than convergent adaptation. But the overall size of the teeth, or the thickness of the enamel, has less phylogenetic significance because such characters adapt rapidly to diet and often vary greatly between closely related lineages. Enamel thickness is not a good guide to relationships simply because it is so important to the animal's day-to-day survival, and thus is strongly influenced over short periods by natural selection.

Finally, as already hinted, systematists put far more store by some sources of information than others. In particular, they tend to suppose that similarities between embryos are liable to reflect homology—because embryos are not in general so subject to the selective pressures of the world at large; and they also tend to take molecular similarities very seriously. But I discuss these issues in more detail in Chapter 4.

Yet for all the rules and guidelines, however sensible and well tried, there is no royal road to truth. Feathers are as clear a character as anyone could wish for, but no one can say with absolute certainty that feathers have not evolved more than once. Taxonomists are obliged simply to surmise that this is the case; the need for human judgement is inescapable. But most anatomical characters that most taxonomists deal with most of the time are much more difficult to assess than feathers are; consider, for instance, all the weird bristles and flanges on the myriads of different arthropods known to science. The necessary judgements are made by experts. Who else is there to make them? But herein lie various kinds of snag. For taxonomy is relatively a small field, and any one group—particularly an obscure group—is liable to be studied by only one or a very few people. Thus those experts may have no peers, or very few: no one of comparable expertise to challenge their ideas. They are likely to develop their own methods of working, and their own criteria. Increasingly they tend to work on intuition—on 'feel'.

There is, of course, no substitute for expertise, and the loss of many specialist museum taxonomists through redundancy in recent decades is tragic. Nonetheless, science should not rely exclusively on individual experts, working in their own ways, however honest and brilliant they may be. The strength of science lies in mutual criticism. Every idea must avail itself to challenge. But if individual taxonomists work with only a few colleagues, or if their only colleagues are their own students and if they develop their own ways of thinking, who can ever question what they say? Experts who are not challenged become authorities not simply in the academic but virtually in the ecclesiastical sense. Their statements become dogma; and dogma, no matter what brilliance lies behind it, is in principle the antithesis of science.

Before we progress further with the theory of modern classification it is worth taking a brief diversion: a morality tale to show that it is one thing to differentiate clearly between homologies and homoplasies, but it can be quite another to steer a sensible course between them. However clear the principles, the practicalities of taxonomy are tricky.

A MORALITY TALE—SEALS, SEALIONS, WALRUSES, WEASELS, AND BEARS

Seals, sealions, fur seals, and walruses are carnivorous mammals that have long since taken to the sea. Traditionally, systematists have arranged them into three families: sealions and fur seals in the Otariidae; walruses in the Odobenidae; and true seals in the Phocidae. This is the arrangement favoured by G. B. Corbet and J. E. Hill in the third edition of *A World List of Mammalian Species* (1991), which comes as near as any volume does to being the standard reference for mammals. Traditionally, also, all these marine mammals were seen to belong to a single group, the Pinnipedia. In their turn, the pinnipeds were ranked either as a suborder of the Carnivora—the mammalian order that includes dogs, bears, weasels, raccoons, mongooses, civets, cats, and hyaenas—or as a discrete order, distinct from dogs and those other terrestrial carnivores (the latter alternative is favoured by Corbet and Hill). In these traditional schemes, all three pinniped families were assumed to have arisen from a common ancestor.

But some zoologists smelled convergence. Perhaps, they said, the similarities between phocids, otariids, and odabenids are merely superficial. For the seals' method of swimming is quite different from that of the walruses and sealions. Seals propel themselves through the water rather like a fish, with sinuous movements of the body culminating in the tail. Their front flippers are merely for steering. Walruses, sealions, and fur seals use their front flippers as underwater wings, essentially flying through the water, and allow their hinder regions to trail. Because the three groups move so differently, why assume that they emanated from the same ancestor? Wasn't more likely that walruses and sealions evolved from one ancestor, and seals from another?

So then the notion developed that the otariids and the odabenids arose from one ancestor, and the phocids from another. Otariids have external ears—reflected in the 'ot-' in their name—and so do odabenids; and this seemed to suggest that these two families are closely related. So the idea grew up that otariids and odabenids had evolved from bear-like ancestors, from the family Ursidae; while the seals were derived from the Mustelidae—the family of weasels, badgers, and otters. It all seemed very neat, and zoologists felt pleased with themselves—that they had seen through the surface features of pinnipeds, and spotted the deceptive convergence.

But André Wyss of the University of California at Santa Barbara argued in 1988 in quite the opposite fashion. He noted that although sealions and walruses use their flippers differently from seals, the flippers in all three groups have virtually the same bone structure. In all three, for example, the fifth digit is powerful while the third is weak. But if sealions and walruses had evolved quite separately from seals, is it not odd that they should have such similar flippers—especially as they use them in quite different ways? After all, mammals have returned to the water many times. In addition to pinnipeds and otters there are whales and dolphins (cetaceans), seacows and manatees (sirenians), the extinct pony-sized desmostylians, marsupials such as the otter-like yapok from South America, beavers and many other rodents, insectivores like the water shrew and the giant otter-shrew of Central Africa, and a long list of semi-aquatic types like the polar bear and hippopotamus. They have all adapted to the water in different ways, even though there are some general similarities, including streamlining and (in some cases) ways of holding the breath.

In short, said Andy Wyss, it would be a very odd coincidence if two separate groups of carnivores had adapted independently to the water by modifying their limbs in exactly the same way—and, moreover, had done so even though they use their modified limbs in quite different ways. It seemed to him much more likely that all the pinnipeds evolved from a common ancestor, and later evolved along separate lines to produce two different swimming styles. He also observed that the presence of external ears in walruses and sealions does not denote common ancestry. It is merely a primitive feature, a symplesiomorphy, that has been lost in the true seals—and such losses are common in evolution. In fact, close examination of similarities suggests that walruses and seals are probably closer to each other than either is to sealions. Thus, the integrity of the pinnipeds as a group seems reconfirmed; but the families may need rethinking.

Willi Hennig himself would surely have approved Andy Wyss's analysis, for he suggested himself as a rule of thumb that in the absence of strong evidence to the contrary it is better to assume that two similar characters are homologous than to assume convergence. Otherwise systematicists could draw up almost any phylogenetic tree they choose by suggesting that any similarities that seem inconvenient are simply due to convergence (leading to homoplasy), and can be ignored.

So, after all, we should be looking for one common ancestor for all pinnipeds. Present evidence suggests that the common ancestor was probably one of the Ursidae, the bear family. But bears are clearly related to raccoons (Procyonidae), weasels and otters (Mustelidae), and dogs (Canidae), and it still is not easy to say for certain that pinnipeds did indeed derive from bears rather than one of those other families. Further molecular studies should increase the degree of certainty. The lesson from this tale is that although expertise is absolutely indispensible if we truly seek to classify creatures according to their phylogeny, even the greatest experts can lead themselves astray.

In addition to expertise we need methods of working that are objective (which, in the end, means dependent on measurements that anyone can make) and explicit (so that everyone can follow the rules and see how everyone else reaches their conclusions). No method can guarantee to lead inexorably to the truth, for the truth we are seeking here is of an evolutionary kind, which means it is historical. We cannot re-create the past: we can only infer as best we can what might have happened. But good methods, based on sound logic, easily followed and universally applicable, should help us to avoid the main traps. The past few decades have seen two main attempts to impose objective, explicit rules and methods that would help taxonomists to infer phylogeny accurately, and so to base their classifications on it. One is cladistics, which is the guiding principle of this book, and will be discussed at length. The other is known as **phenetics**, or **numerical taxonomy**. Numerical taxonomy enjoyed a vogue from the 1960s onwards—after Hennig had first published his ideas on cladistics, but before they became widely known—but now has little following. Since phenetics is precladistic, conceptually if not chronologically, it might be seen as the last fling of phase III taxonomy. Because its ideas are still around it is worth a brief diversion.

A DIVERSION—PHENETICS OR NUMERICAL TAXONOMY

The basic idea of phenetics was to blitz the problems with as much data as possible. Taxonomists were exhorted simply to list all similarities between creatures, and the differences too; the more the merrier. Everything measurable was measured and compared, and the results analysed statistically; hence the alternative term 'numerical taxonomy'. This strictly numerical approach certainly seemed to have its attractions. Darwin himself comments in *The Origin of Species* (Chapter 13):

> We may err . . . in regard to single points of structure, but when several characters, let them be ever so trifling, occur together throughout a large group of beings . . . we may feel almost sure, on the theory of descent, that these characters have been inherited from a common ancestor.

Even though phenetics was an attempt to minimize reliance on authority, it is always good to find support in an authority like Darwin. Where he said 'several characters', we may reasonably read 'as many as possible'.

Common sense suggests that creatures are more likely to be closely related if they have many features in common, and less closely related if they have fewer features in common. But we have seen some of the problems: that creature A may have some features in common with creature B, and different features in common with creatures C and D; and that homology may be confused with homoplasy. Yet it seemed to the numerical taxonomists that if only they listed enough characters, then statistical analysis would solve such problems in passing. Thus, for example, human beings are more or less naked, as are dolphins and some pigs, while chimpanzees are hairy, so a simplistic classification based just on hairiness would link humans to dolphins and pigs and leave chimps on their own. But if we listed scores or hundreds of features of human beings, pigs, dolphins, and chimps, then we would surely see that humans have a huge number of attributes in common with chimps, compared with which our single resemblance to dolphins and pigs is clearly trivial. The statistics alone should lead us to the truth.

Yet phenetics is clearly cavalier. Its most exuberant practitioners did not bother even to differentiate between homologous characters, and homoplasies. If they included enough data, they felt, then such niceties would come out in the wash. The statistics alone would be the arbiter; no subjective authority was needed. But as Mark Ridley pointed out in *Evolution and Classification* (1986), the lists of similarities and differences can be analysed by various statistical techniques, and different techniques can produce different groupings. So the numerical taxonomists first had to choose which statistical approach to apply—and thus were obliged to reintroduce the subjectivity that they were so keen to avoid.

Let us put numerical taxonomy aside, then, and return to the main thread of our story: the path to cladistics. Hennig, like a good post-Darwinian, wanted to devise a taxonomy based on phylogeny. Also, like most post-Darwinians and some pre-Darwinians, he recognized the crucial distinction between homologies and homoplasies. But he also saw, as few others had done, that homology taken alone is not sufficient guide to genealogy. One other kind of distinction needs to be taken into account; whether particular features are unique to the creatures under consideration, or whether they are also shared by other creatures as well. We can illustrate the importance of this distinction by one last, fleeting visit to our friends Smith, Smith, Harris, and Robinson.

A LAST VISIT TO THE INN—SYNAPOMORPHY AND SYMPLESIOMORPHY

Let us cast back to our first visit to the hypothetical inn, when you assumed that either the two farmers or the two lawyers must be the brothers Smith; and focus for the moment on the two farmers. Why did you think they might be related? Because they looked and talked much the same. But now look around the pub. Apart from the occasional oddball—the lawyers Smith and Harris, and the publican himself, decked out in a braided waistcoat—virtually everybody looks like the farmers Smith and Robinson, simply because virtually everybody in rural England of the early nineteenth century was either a farmer or a farm worker. They all wear flat felt hats and big rough jackets, have bucolic complexions, and talk with rich, rolling accents, rolling 'r's and softened 't's. The fact that your new farmer friends share these characteristics absolutely does not indicate that they are more closely related to each other than to anyone else in the bar.

In the context of early ninteenth-century England, and in the language of biology, the garb and accents of a farmer are said to be primitive features or, in scientific Greek, are **plesiomorphies**. 'Primitive' in this context emphatically is not a pejorative term. It does not mean 'stupid' or 'inferior', or have any other of those modern connotations. It simply means 'close to the general state', or, more specifically, to the 'ancestral' state. The term plesiomorphy derives from the Greek root '*plesio*' meaning 'near'—near, that is, to the ancestral state. Primitive features that are shared by more than one creature are called **symplesiomorphies**, where '*sym*' implies 'common'. The felt hats and gaiters of Smith the farmer and Robinson the farmer are indeed shared homologies: they both inherited their garbs from some distant, prototype farmer. But that fact does not by itself indicate any specific family relationship between the two because those shared features are simply symplesiomorphies; features shared by all farmers.

In order to establish true family relationships between different individuals or creatures we need to find features that they have in common but are not shared by others. Such features are said to be **derived**—or, in Greek, are **apomorphies**, from the root '*apo-*' meaning 'far from', as in 'far from the ancestral state'. Shared derived features are called **synapomorphies**. If we look around the pub, we find that no one, apart from the two Smith brothers, has a gap between their front teeth. At least within the context of the pub, then, gap-teeth emerges as a synapomorphy. It does indicate true relationship.

Finally, we may note that the publican is the only individual with a braided waistcoat. The waistcoat is his particular badge of office, which marks him out from the rest. In scientific Greek, this is an **autapomorphy**.

So now we can abandon our friends in the pub for they have done their job: illustrated the basic theory that lies behind modern, cladistic classification. In order to establish the relationship between any group of creatures we have to examine their features or characters—morphological, molecular, behavioural, what you will—and see what they have in common. As taxonomists recognized well before Darwin, however, some characters are more informative than others: and in particular, as Richard Owen first formally described, we must distinguish between characters that are homologous and those that are merely homoplasious. But it was left to Hennig to point out that homology alone is not enough: homology is a necessary but not sufficient criterion. To establish true phylogenetic relationship we must distinguish between homologies that are apomorphic—'derived'—and those that are plesiomorphic—'primitive'. Shared derived features (synapomorphies) define true groupings whereas shared primitive features (symplesiomorphies) merely show that the creatures in question are all members of some much bigger, more general group. This latter distinction is the key feature of cladistics.

Again, however, a cautionary note is necessary. Like 'homology', the notion of symplesiomorphy and synapomorphy depend on context and vantage point. Thus birds are distinguished from all other vertebrates by their possession of feathers.[1] So for birds as a whole feathers are a synapomorphy and we can say beyond all doubt that a creature with feathers is, indeed, a bird. But if we want to discover the relationships between ostriches, parrots, and crows then the mere presence of feathers does not help us at all. They all have feathers just because they are all birds. At this level feathers are a symplesiomorphy. To differentiate between ostriches, crows, or parrots we need to seek out apomorphies that distinguish each from the others. In the case of parrots the hooked beak, the high-placed nostrils, and the feet with two toes pointing forward and two backwards are a synapomorphy that they share with all other parrots, but are not shared (at least in that combination!) by other birds. A macaw might care to reflect that the character that places it within the order Psittaciformes alongside cockatoos, parakeets, conures, budgerigars, and all the other parrots is the synapomorphy of the hooked beak. But for macaws, cockatoos, and budgies feathers are simply a symplesiomorphy, shared also by non-parrots as diverse as cassowaries and willow-warblers.

In truth, Willi Hennig was not the first or the only taxonomist to perceive the need to differentiate between primitive and derived features. But he was the first to set out a formal set of rules to help taxonomists make this distinction—and went on to suggest an entire method and philosophy of classification and nomenclature. This, of course, is the method of cladistics. Taxonomists who practise cladistics are called **cladists**. All the classifications in this book are based on the rules and tenets of

[1] This rather glib comment is now abnegated by research by Ji Qiang and his colleagues. In excavations in northeastern China, these scientists found two theropod dinosaurs with feathers. Birds are generally thought to have evolved from theropod dinosaurs (see Section 21), but the finding forces us either to extend the definition of birds or to admit that feathers are not the grand, unifying avian synapomorphy that has always been supposed. For present purposes, however, the traditional feathers-equals-birds hypothesis will do.

cladistics because, at least to this author at the time of writing, the logic of it seems inexorable: it does indeed seem to offer the best path through the bear-traps that await all who seek to classify by phylogeny.

Before we look more closely at the methods of cladistics another brief morality tale will illustrate the need for them.

ORANG-UTANS, CHIMPANZEES, AND HUMAN BEINGS

Cladists devise classifications on the basis of synapomorphies—shared derived homologies—and only on the basis of synapomorphies. Other features, no matter how conspicuous, are scrupulously ignored. Non-cladists have sometimes accused the cladists of arbitrary dogma. But studies by Jeffrey Schwartz of Pittsburgh University in the 1980s shows why purism is necessary.

Consider, said Schwartz, human beings, chimpanzees, and orang-utans. What is the relationship between them? They are all very similar and biologists assumed at the time that they shared a common ancestor 10–5 million years ago. They might be equally closely related one to another. That is, their shared common ancestor might have divided into three groups at one particular time to give rise to the three modern lineages. But as Willi Hennig pointed out (more of this later) ancestral groups seem much more likely just to split into two. So we can assume one of three scenarios. Either orangs split from the common ancestor first, leaving chimps and humans to divide later; or the chimps divided first, leaving the orangs and humans together; or the humans split off first, leaving the orangs and chimps together. If the first of these possibilities is the case, then modern chimps and humans are most closely related, and orangs are the odd ones out; if the second was the case, then orangs and humans are closest; and if the third, then orangs and chimps are closest.

In fact, most modern primatologists now believe that the first scenario is the case: orangs split earliest, so chimps and humans are most closely related. Earlier classifications dating from the 1960s and before commonly showed the human lineage splitting very early (even as far back as 25 million years or so), leaving orangs and chimps as close relatives. As has often been commented, it's as if palaeoanthropologists accepted Darwin's idea that humans and apes share a common ancestor, but wanted to put as much distance as possible between us and them. This latter idea is reflected in traditional classification: chimps and orangs were —and often still are—commonly placed together in the family Pongidae, while humans had their own family, the Hominidae.

By the mid-1980s, however, Jeffrey Schwartz was pointing out that both the main scenarios at the time—that humans are closest to orangs, and that chimps are closest to orangs—rested in part on ideas that simply were not valid; at least if we take cladistic theory seriously. It is at least plausible that humans and orangs are closest and that chimps diverged earliest. But this scenario has rarely (if ever) been seriously considered.

Why, after all, have biologists of recent years generally preferred one or other of the alternative scenarios? Well, the first seems intuitively plausible because orangs are so different in general appearance from chimps and humans. For one thing, they have huge long arms and the males have big flanged faces. On the other hand, orangs and chimps seem to belong together because they are both, well, so ape-like. Humans—smart, dome-skulled, upright and naked—are obviously different. But according to cladistic thinking, as Jeffrey Schwartz pointed out, the differences between orangs on the one hand and chimps and humans on the other, should simply be ignored. Synapomorphies are what matter. Observed differences should play no part in the classification. This is not just a piece of arbitrary dogma; nothing in cladistics is arbitrary dogma. The point is that long arms are an adaptation to extreme arboreal life and are known to be the kinds of characters that creatures can rapidly evolve, under heavy selective pressure. Humans and orangs might in principle have divided from chimps only a few thousand years ago—let alone a few million—and even in that short time the orangs could have evolved their long arms for life in the trees, while humans evolved stilt-like legs for life on the ground. We should not, in short, be deceived by divergences—not if we truly seek a classification based on evolutionary history. When you ignore such extravagences, said Schwartz, you find that humans and orangs have a great deal in common, including thick enamel on the teeth, whereas chimps have thin enamel.

Neither is it permissible to suggest that chimps and orangs must belong together just because they are both conspicuously ape-like. The ape-like features they have in common—arms longer than legs, hairiness, prominent faces, big canine teeth, brains that are big but smaller than a human's—are simply primitive features (symplesiomorphies): features that derive more or less unchanged from the presumed common ancestor of orangs, chimps, and humans. Again, we should not be seduced by such general features. In seeking to reconstruct the history, such characters should simply be ignored. No argument.

In fact, very few if any primatologists now accept Jeffrey Schwartz's polemical suggestion that humans might be closer to orangs than either is to chimps. Thickness of tooth enamel, for example, turns out to be a red herring: it varies enormously from species to species, depending on diet (just as arm length might vary according to the mode of locomotion). Molecular evidence now seems to confirm the propinquity of humans and chimps. All accept, however, that Schwartz's contribution to the debate was well worth while. It really was careless to link chimps with orangs or chimps with humans simply on the basis of general appearance. Note, too, how even in this apparently straightforward case, general appearance pointed to two totally opposite conclusions: that chimps are closer to orangs or that they are closer to humans. Consider how easily we might be led astray—and, beyond doubt, often have been lead astray—when seeking to classify creatures that are much less familiar than the great apes and ourselves: clams, or groups of fossil plants, or whatever. If we rely on general impression, if we are swayed simply by extravagant divergences, if we do not distinguish primitive features from derived, then we can be hopelessly deceived. Willi Hennig's cladistic rules can seem harsh and sergeant-majorish, often at variance with common sense. But as Jeffrey Schwartz's mind-game showed very clearly, the hair-shirt approach is necessary.

Now we should look in more detail at cladistics in action, starting with the essential concepts of the clade and cladogram.

CLADES AND CLADOGRAMS

In all classifications, however they are devised and whatever they are for, small groups of things that are considered to be very similar are nested within larger groups that have more general things in common, and so on and so on. In cladistic schemes, too, little groups nest within bigger groups. In cladistics, however, the things being classified are living creatures; the overriding philosophy is to base the classification on phylogeny; and the approach is to define the groups strictly on the bases of their synapomorphies—the shared characters that have been derived from a common ancestor and are exclusive to the group in question.

We can see in practice how this nesting works by looking at, say, the way that cladists would classify a human being. Human beings belong to the domain Eucarya. The key synapomorphy here is that in eukaryotes the genetic material, held on the chromosomes, is contained within a distinct nucleus: but in Bacteria and Archaea, the other two domains, there is no discrete nucleus. The domain Eucarya includes several kingdoms, and the one that humans belong to is the Animalia. In truth it is not easy to define the key synapomorphies of animals: but one that may well do the trick is the presence of the Hox genes, which guide the body plans of animals. Kingdom Animalia contains many phyla; and humans belong to the phylum Chordata, defined by the presence of a stiffening rod along the back, the chord, which in our particular subphylum, the Vertebrata, is formed into a bony vertebral column. Subphylum Vertebrata contains a range of classes and ours is the Mammalia. The key synapomorphies here include hair (of a particular kind; nothing like bumble-bee fur), warm-bloodedness (again of a particular kind), suckling of young, the way the bones grow, and so on and so on. Mammalia is divided into subclasses, one of which is the Eutheria: mammals that produce a definite placenta. We have a placenta, and so we are eutherians. The eutherian subclass includes many orders and ours is that of the Primates: again difficult to define precisely because few of its features are unique to mammals. Nonetheless, primates are defined by a suite of synapomorphies that include two pectoral breasts, flat nails instead of claws, complete bony eye sockets, and so on. Within the Primates we qualify as Hominoidea (which can be ranked as a superfamily)—the great apes, in whom the internalized tail, reduced to a coccyx, is a synapomorphy, clearly distinguishing us from allegedly 'tail-less' monkeys such as the macaques. The genus *Homo* is again difficult to define precisely, but a brain size above 500 millilitres in volume has commonly been taken, somewhat unsatisfactorily, as a synapomorphy. The genus *Homo* is held to include at least half-a-dozen species; and the synapomorphies that define our own, *Homo sapiens*, include the gracile (relatively lightweight) bones, chin, flat face, high forehead, and big rounded cranium to house a brain of around 1450 millilitres in volume. Brain size alone cannot truly be called a synapomorphy of *Homo sapiens* because Neanderthals also have big brains, but their foreheads are lower and their crania more flattened.

So the little groups nest within the larger groups, and so on and so on and the whole array can be presented graphically as a tree. This is true of all classifications. But the particular trees that cladists produce, in which the nesting groups are defined on the basis of their synapomorphies, are called **cladograms**. All the trees presented in Part II of this book are cladograms, or are modified slightly from cladograms, that have been put together by specialists. But I refer to them as trees rather than cladograms for the technical reason I gave in Chapter 1.

The cladogram, like any tree, is a branched structure (*clados* is Greek for branch). Little branches spring from bigger branches, which spring from bigger branches, and so on. In cladistic parlance, a point where one branch takes off from another is called a **node**; and the lines between nodes are **internodes**. Because the cladogram is meant to represent phylogeny, then we can say that each node represents the **common ancestor** of the branches that spring from it.

But now we come to an essential and sometimes tricky point of nomenclature. Human beings find it difficult to talk to each other in pictures: they need words to convey information. So they attach names to the different branches of the tree, which indeed are the names of the formal taxa. The big branches from which all the rest spring are the domains—such as Bacteria, Archaea, and Eucarya; the smaller branches that spring from them are the kingdoms—Animalia, Plantae, Fungi, and so on; and the tiny twigs at the tips of the tree are the individual species.

Cladists are more purist in their approach to naming than previous taxonomists have been. Essential to the cladist philosophy is the concept of the **clade**. A clade consists of the common ancestor of a group plus all its descendants.[2] Graphically, this means that a clade includes the creature represented by the node, and all the branches that spring from that node. Thus the species *Homo sapiens* is a clade. In principle it includes all the people who have ever lived, right back to the very first *Homo sapiens*. The genus *Homo* is a clade, of higher rank: it contains *H. sapiens* and all the other extinct species of *Homo*—*H. erectus, H. neanderthalensis*, and so on, right back to the

[2] Some further clarification may be necessary. Because all the creatures on Earth seem to belong to the same genealogical tree, and all seem to have descended from some ancient, long-gone ancestor (the first to operate with the trinity of DNA, RNA, and protein), then any two Earthly creatures can trace their descent backwards to a common ancestor. You and I share a common ancestor with rhinoceroses (an ancestor who may have lived around 80 million years ago) or with oak trees (who probably lived somewhat more than a thousand million years ago), and so on. But when cladists talk about 'common ancestors' they generally mean the 'most recent common ancestor'. Thus chimpanzees and humans share a common ancestor who probably lived around 5 million years ago. But all chimp and human species also include the ancestor who lived about a thousand million years ago, who also gave rise to oak trees (and many other millions of creatures as well). But the only common ancestor who really interests us is the most recent one. That one gave rise to the clade that includes all chimpanzees, all bonobos (which used to be called pygmy chimpanzees), and all humans, living and extinct. To say 'common ancestor', when we really mean 'most recent common ancestor', is just shorthand. I suggest that the term **clade founder** would be useful instead of 'most recent common ancestor', and adopt it from time to time throughout this book.

first common ancestor on the plains of Africa some 5 million years ago. The order Primates is a larger clade, containing *Homo* and all the apes, monkeys, lemurs, and tarsiers—right back to an ancestor who may have lived in the Cretaceous period, more than 65 million years ago, when the terrestrial world was still dominated by dinosaurs. Mammalia is a clade—all the primates, horses, and all the plethora of extinct types, right back to the first true mammal in the latest Triassic period, some 210 million years ago. Note that the different ranks of taxa—species, genus, order, and so on—are all clades. Little clades nest within bigger clades. A clade is a clade provided it contains all the creatures in the group, right back to and including the common ancestor.

But although this method of nomenclature is perfectly logical, it clashes somewhat with traditional methods of naming. In truth, no modern taxonomist, cladist or non-cladist, argues with the idea of the clade but some—including many cladists—do suggest that we should relax the method of naming somewhat. The point can be made by reference to the creatures universally known as reptiles. Everyone knows what a reptile is: it is a tetrapod (four-legged vertebrate) land animal (though some have lost their legs and many have returned to the water in one guise or another) with a thick scaly skin; it may lay eggs or may produce live young, but in either case protects each embryo with a series of membranes known as the allantois, the chorion, and the amnion. The protection of the embryo distinguishes reptiles from amphibians, which do not provide such membranes. Living reptiles include tortoises and turtles, snakes and lizards, crocodilians, and the tuatara. Extinct reptiles include dinosaurs and a host of other creatures besides. The group is enormously diverse, but we can easily see what all its members all have in common. Where is the problem?

Trouble begins if we want to suggest that the creatures we recognize colloquially as 'reptiles' should be placed in a formal taxon with the name 'Reptilia'. To be sure, most traditional textbooks refer to the class Reptilia. But Reptilia as traditionally defined is not a clade. After all, birds descended from reptiles—almost certainly from a dinosaur. Mammals descended from reptiles of a quite different kind, called synapsids. Thus if Reptilia is to be acknowledged as a clade then it must include all mammals and all birds. In cladistic parlance, if we want to use the word Reptilia at all, then you, and I, and horses and ducks and canaries, are Reptilia.

Many cladists stick to this rule of nomenclature and in scientific papers refer to the creatures that most of us simply call 'dinosaurs', as 'non-avian dinosaurs'—meaning that they are not including the birds, which descended from the dinosaurs. By the same token if they wanted to refer to all the creatures that most of us call reptiles they would have to say 'non-avian, non-mammalian reptiles'. The issue is important —naming matters—and is still causing some acrimony. The cladists' refusal to recognize any group as a formal taxon unless it is a true clade strikes many as pedantry and (unfortunately) tends to put them off cladistics all together. On the other hand, the cladists tend to argue that it is merely sloppy, not to say perverse, to define groups carefully on the basis of synapomorphies, and then to attach names to the groups that include some members of a clade but not others. In this book, although it is emphatically cladist in principle, I suggest a compromise between the two schools of nomenclature—one I hope will bridge the gap between the traditional and the new. This approach to nomenclature I call 'Neolinnaean Impressionism'; I discuss it at greater length in Chapter 5.

Meanwhile, three more essential technical terms need explaining. A grouping that is a true clade—containing the common ancestor and all descendants—is said to be **monophyletic**. A grouping that contains some members of a clade, but with some hived off into other groups, is said to be **paraphyletic**. The traditional group Reptilia is a perfect example of the latter category: it contains the ancestral reptile and all its descendants except the ones that have evolved into birds or into mammals. A grouping that contains creatures from various clades, all lumped together simply for convenience, is said to be **polyphyletic**. Just to emphasize the point, cladists do not recognize any grouping as a formal taxon unless it is monophyletic—meaning it must fulfil all the criteria of a true clade. But traditionalists will admit some paraphyletic groupings as formal taxa; and I will argue, with some modifications and caveats, that this should be allowed. Traditionalists have often admitted polyphyletic groups as well, but this is unacceptable. Polyphyletic groupings can be convenient—for example, when dealing with large, complicated assemblages of fossils, which seem loosely to fit together but are not thoroughly analysed. But they need not be treated as formal taxa, with Graeco-Latin titles, and spelt with a capital letter; they can simply be referred to informally, in lower case. Thus the group 'Pisces', which used to appear as a formal taxon in many traditional textbooks, is an extremely polyphyletic group—sharks, cod, and lungfish belong to very different lineages. But the informal term 'fish' is perfectly acceptable and useful, and will do instead of the spurious 'Pisces'.

Now, briefly, I will discuss a few technicalities, to show how cladists construct their cladograms in the first place.

OUTGROUPS, ROOTS, AND RESOLUTION

Suppose you were given the task of classifying a lizard, a squirrel, a baboon, and a horse. You have to say which is most closely related to which and then to say which groups nest within which. Such a task would be very easy but it illustrates the principles.

You begin by listing the characters of each. Actually even this apparently mechanical process is not quite so mechanical and straightforward as it sounds, however, because, as taxonomists point out, 'It depends what you mean by a character'. A character can be anything you care to take note of and you could, in theory, make an infinity of notes. You could say that the presence of claws is one character, and that the ratio of the lengths of the second to the third claws of each foot is another, and so on ad infinitum. This may simply seem ridiculous but there could well be times when the ratio of claw length indicates something important. It is impossible, then, even at this most basic level, to eliminate the element of judgement, and judgement inevitably involves subjectivity, or at least common sense. Nonetheless, with creatures such as these it is not too difficult to make a shortlist of interesting features.

Then you can begin to group the creatures according to the characters they have in common. Immediately you face the problem that Aristotle noted more than 2000 years ago: that creature A might have some features in common with creature B, but another set of characters in common with creature C. Thus the squirrel seems to have a lot in common with the baboon and the horse—fur, warm-bloodedness, and so on. The lizard is clearly the odd one out: scaly and cold-blooded. And, when you look closely at the internal anatomy, the embryology, and so on, you find that the squirrel, baboon, and horse also have a great many other features in common that are not shared by the lizard. But there is one striking anomaly. Whereas the squirrel and the baboon have five toes on each foot—they are **pentadactyl**—the horse has only one toe per foot. But the lizard does have five toes per foot. So why not group the lizard with the baboon and squirrel on the strength of their pentadactyly, and make the horse the odd one out?

This example is so simple that no one would have trouble with it. Traditional taxonomists of every stripe would simply point out that an overwhelming range of characters links squirrel, baboon, and horse, compared with which the number of toes is clearly trivial. A numerical taxonomist would count all the characters, analyse the groupings statistically, and show graphically that this is indeed the case. Yet both miss the point. The point is not the number of characters, but the nature of the characters. Fur, warm-bloodedness, and so on are synapomorphies of mammals that should be taken into account, and link the squirrel, baboon, and horse; clearly distinguishing those three from the lizard. Pentadactyly is merely a primitive feature of all modern tetrapods and tells us nothing about the particular relationships of particular tetrapods, one to another. Many tetrapods have abandoned strict pentadactyly: for example, frogs tend to have only four fingers per hand, cows have two, horses have one, while snakes and many other reptiles and amphibians have lost their legs all together and all the toes with them. All the same, pentadactyly is the primitive state of tetrapods.

This statement, however, is made with the benefit of hindsight. Biologists do not know a priori which shared features are merely primitive and which are unique to the group in question. No living creature comes with labels attached. So cladists need a way of working out from first principles which are the primitive features and which are derived. In practice, they do this by bringing in another creature, not part of the group under immediate scrutiny, called an **outgroup**.

The ideal outgroup is a creature that is related to the group under scrutiny, but is not part of that group; in our present example, a salamander would be a good choice. It is a tetrapod, like the animals being studied; but it is not a close relative of lizards or of any of the mammals. This may seem to be cheating, for how can the taxonomist decide what creature is appropriate as an outgroup unless he has some previous knowledge of the creatures under question, and of their relatives? In fact, some previous knowledge is always useful, but it is not vital. If the cladist chooses an inappropriate outgroup, then its inappropriateness will become apparent as the study progresses, and he can choose another one. If in the present case he chose an ant as an outgroup, he would soon find that it had so little in common with any of the creatures under study that it told him nothing at all. If, on the other hand, he chose a dog as an outgroup, he would soon find that in fact it grouped very neatly with the squirrel, baboon, and horse to the exclusion of the lizard, and so again was inappropriate. If the creatures in hand were truly unknown, he would just have to proceed by trial and error. Trial and error is always necessary when seeking to apply theoretical rules to the real world.

Anyway, let us suppose that the cladist adopts a salamander as the outgroup (either guessing straight away that the salamander is a good choice, or working his way towards it). Because the salamander is pentadactyl, like most of the creatures in the study group, this suggests that pentadactyly is indeed a primitive feature of tetrapods as a whole, and cannot help to distinguish the relationships of particular tetrapods (at least in the present context). He also finds that the salamander is cold-blooded and naked, like the lizard, suggesting that the warm-bloodedness and hairiness of the squirrel, baboon, and horse is a derived feature, which indeed may be a synapomorphy of the three. And so on.

Of course, with just one outgroup and such small numbers, these kinds of observations do not make the case. Perhaps all other tetrapods are warm-blooded, and the lizard and the salamander just happen to be exceptions. But if the cladist repeats the exercise with more outgroups—other amphibians, fish, and indeed ants (which, after all, belong to the kingdom Animalia)—then he soon establishes that cold-bloodedness and nakedness are the primitive state of tetrapods as a whole, just as they are of animals as a whole, and that warm-bloodedness and furriness are indeed derived features.

You might argue, of course, that it is foolish and pedantic to apply such elaborate procedures to the group of creatures in our present example. We all know that most animals are cold-blooded, and that modern tetrapods generally have five toes per foot unless they have become especially adapted to some particular way of life that demands fewer, and so on. But herein lies the point: taxonomists most of the time deal with creatures that are much less known, from deep-sea crustaceans to tropical fungi; and, as Jeffrey Schwartz showed, it is possible to go hopelessly astray even when dealing with the most familiar, if we do not sedulously apply the rules that can guide us past the pitfalls.

Note, though, that the cladist has two ambitions. The first is to place the various creatures in true, 'natural' groups—meaning that all the creatures in any one group should be more closely related to each other than to any creature in any other group; and the second is to show the phylogenetic relationship between the groups—the order in which the groups diverged from their common ancestors. These two ambitions are complementary, but are not the same. Thus a cladist may produce a tree that shows the relationship of different groups but does not show the order in which they evolved: such a tree is said to be **unrooted**. Conversely one that does show the order of play is said to be **rooted**. In practice, cladists progress from unrooted trees to rooted trees by applying whatever extraneous empirical knowledge they can bring to bear. Thus a knowledge of fossils would show a biologist of any kind that salamanders are, in fact, more similar to the common ancestor of all the tetrapods, than a squirrel or a horse is. The skeletons of modern salamanders are more like ancient tetrapod bones than squirrel skeletons are.

But for many groups there are no good fossils, and the taxonomist must infer relationships only from living creatures; in many other cases there are only fossils, and no direct living descendants. Even when there are fossils and living animals, it is good to have two independent sources of information: the fossil record and the logical analysis of living creatures. So how can evolutionary history be inferred simply from living creatures? Again, outgroups come to the rescue. If you look at enough tetrapods, and at enough other animals to act as outgroups of the tetrapods (including ants), then you begin to see that, for example, cold-bloodedness is the primitive state and that warm-blooded clades must nest within larger clades of cold-blooded creatures. This remains an inference, to be sure. No rules can be guaranteed to lead us to unequivocal truth. But it is a sensible inference. Guesses based on the rules are more likely to be right than those based on general impression.

Finally, you can see that in most of the trees in Part II, most of the time, the nodes give rise to only two branches, which then branch into two again, and so on. Only occasionally (or at least as little as possible) does a node give rise to three branches or more. A two-tined branch is a **dichotomy**; and a many-tined fork is a **polychotomy**.

Why so? Well, the reason derives originally from Willi Hennig's commonsensical natural history. The division of the branches at a node represents, in graphic form, an ancestral species dividing into two species. Why two rather than many? Because Hennig supposed that, in nature, this was the more likely case. At a time of speciation, he thought, an ancestral group is more likely to give rise to two descendant groups than to many. Of course, huge groups of diverse creatures may descend from a common ancestor: all mammals are assumed to have come from one single proto-mammal, all birds from one single proto-bird, and so on. This is how clades are defined. But the huge diversity is assumed to have come about by a series of dichotomies: one parent group giving rise to two daughter groups, which then split again into two, and so on and so on.

A tree showing only dichotomies is said to be **resolved**. One still beset by polytomies is felt to be in an imperfect, unfinished state and is said to be **unresolved**. In truth, resolution can be very difficult. Thus ornithologists in general have inferred that the most modern birds—cuckoos, eagles, storks, sparrows, and so on—all descended from a common ancestor, but it is very difficult to work out the order in which they emerged. If two different types look very different (like storks and sparrows, say) this may be because they diverged a long time ago, but it could be because they have each evolved very rapidly. Then again, the fossil record suggests that many different bird groups arose within a very short time of each other, around 60 million years ago; and when events occurred close together (perhaps within a few decades of each other) a long time ago, it is hard to discern the order of play. Many cladists, such as Joel Cracraft of the American Museum of Natural History, New York, who helped with the bird section in this book, point out that in our anxiety to resolve cladograms we should not jump the gun. If we think we know that 10 different groups of birds descended from one particular ancestor, but we do not know in what order the descendant groups appeared, then it is presumptuous to present a perfectly resolved cladogram. It is far more honest simply to present a 10-tined polychotomy, and hope and expect that future studies will allow further resolution. In general, polychotomies shown in this book represent areas of work in progress (some progressing more quickly than others).

These, then, are the broad ideas and concepts behind cladistic techniques. Now we should look briefly at a few implications; and in particular at the concept of 'sister group' and at the ticklish, crucial issue of ancestors.

SISTERS AND ANCESTORS

Two groups that arise as twin branches in a simple dichotomy from a single node are called **sister groups**. Note that sister groups need not be of the same size, or of equivalent rank, and often the disparity is spectacular. Thus the tree in Section 11 shows the pycnogonids as the sister group of the chelicerates. The pycnogonids include a few peculiar marine creatures that resemble spiders made out of pipe-cleaners while the chelicerates include the vast array of eurypterids, scorpions, spiders, mites, and so on; one of the most speciose of all animal phyla. No matter. The proposal is that pycnogonids and the chelicerates arose from a common ancestor; and that one sister remained marine and somewhat obscure while the other went on to conquer the land (and, for a time, the sea) in a manner surpassed only by the insects and tetrapods. Evolutionary history, like human history, is full of such prince-and-pauper discrepancies.

Now for an intriguing point of philosophy. Willi Hennig stated explicitly that he wanted to devise a taxonomy based on phylogeny. Ideally the trees—cladograms—should be summaries of evolutionary history, and the nodes should represent common ancestors of the descendant branches. But Hennig was a purist. He recognized full well that the evolutionary trees must always be inferences, inferred from an analysis of characters. Characters, after all, provide all the data there are: fossils do not have labels attached to say who they are related to, and living creatures will not tell us their family histories over a quiet drink. To make inferences we have to make certain assumptions: for example, that complex mechanisms are unlikely to evolve twice in the same form. But we should be as austere as possible. We should not infer more than is justified, or assume more than is necessary. William of Ockham, one of a distinguished coterie of medieval philosophers of science, made the same point in the fourteenth century: it is the principle of Ockham's (or Occam's) razor.

So now apply such thinking to the traditional phylogenetic trees. These usually show ancestors. Typically, the taxonomist first decides what the common ancestor of a particular group of creatures might have looked like, then reviews the known fossils, and then puts the fossil that most closely resembles the presumed ancestor into the allotted slot. This seems reasonable enough at first sight. But is it?

In this way, after all, *Archaeopteryx* has traditionally been presented as the common ancestor of all birds: all the living types, and all the extinct types as well. Certainly, *Archaeopteryx* is a good candidate. To some critics, indeed, it has seemed too good to be true. It is the right kind of age—late Jurassic, around 150 million years ago. It has the right combination of characters—some reptilian, like the long tail and the teeth, and some characteristically avian, including (notably) the feathers. It lacked the protruding breastbone, the keel, which in modern birds anchors the flight muscles; but its aerodynamic feathers and its wings suggest that it did indeed fly, and that the keel evolved later.

But now consider the facts of the case. *Archaeopteryx* is known from just six fossil skeletons (plus a feather), the first of which was discovered in the mid-nineteenth century in the Solnhofen limestone of Bavaria. During the late Jurassic, Solnhofen was a lagoon. Fossilization is an extremely rare event so for every dead *Archaetopteryx* that was turned into a fossil there were probably many thousands, perhaps millions, that did not. Besides *Archaeopteryx*, too, there might have been scores or even hundreds of other Jurassic genera of birds that have left no trace at all. Is it probable—even likely—that the particular Jurassic bird that happened almost by a miracle to come to light in Bavaria should be the ancestor of all subsequent birds? Do we really want to suggest that this creature's eggs gave rise to offspring whose descendants became ostriches and penguins, vultures and swans? When the matter is put like that, such an assumption seems absurd.

But if we cannot realistically show *Archaeopteryx* as the ancestor of modern birds, how should we show it? As the sister group, is the answer. It is reasonable to assume that *Archaeopteryx* shared a common ancestor with modern birds, and that no creature that is not a bird shared that particular ancestor. To show *Archaeopteryx* and all the rest of the birds as sister groups is an economical and unambiguous way of showing what we can reasonably assume to be the case, while making no unjustifiable claims. Of course it may be (by the remotest chance) that the particular *Archaeopteryx* fossils that we know about were the ancestors of all later birds. Showing them as the sister group in Section 21 does not preclude this possibility; it merely suggests that we have no right to assume this.

So all this explains why the trees in Part II, based on cladograms, look the way that they do: a series of dichotomies (where the data allow this) and hence a series of sister groups. Ancestors are presumed to be at the nodes: but the ancestors themselves must always remain hypothetical because we cannot safely assume that any particular creature we happen to know about was in reality the ancestor we seek. We can show likely candidates as putative ancestors by presenting them as sister groups.

I have mentioned William of Ockham. He should be invoked once more, in a slightly different context.

THE NOTION OF PARSIMONY

Cladistics is extremely thorough. It recommends that we look at as many creatures as possible—including as many outgroups as possible—and at as many discrete characters as can be discerned. The snag is the sheer quantity of data that is thus generated and, even worse, the number of possible ways in which any one collection of creatures could in principle be grouped—especially if, as is desirable, the groups are presented in fully resolved, dichotomous trees. Indeed the number of theoretically possible trees that could be derived from half a dozen creatures runs into many millions.

But the rules of cladistics and the commonsensical assumptions that lie behind those rules help us to discover which of the many possible trees that could be drawn is the most likely to be true. Suppose, for example, we were trying to classify birds and we decided, just to be thorough, that the colour black might be a useful, enlightening feature. We certainly should not assume a priori that it is not a useful feature. If we included black as a feature we would generate a whole range of trees in which black swans were linked with condors and cormorants, crows and black cockatoos. But virtually none of the other characters we took into account would group black swans with blackbirds: they tend to group black swans with white swans, and swans in general with geese and ducks. So we would infer that blackness is simply the kind of character that might arise independently, time and time again, in group after group. On the other hand, we find that although petrels and albatrosses appear in different groups if we classify birds on the basis of colour, they are linked if we classify according to the shape of the bones—for both petrels and albatrosses have peculiar, tubular nostrils. That is precisely the kind of feature that is unlikely to evolve more than once.

By such examination we find that of the many millions of theoretically possible ways of grouping birds, some are far more likely than others to reflect phylogenetic history. If we tried to sort through all the possible cladograms just by thinking about them ('by hand') we would find there was not enough time left in the life in the Universe to do so. Fortunately, however, enormously intricate computer programs have been devised that generate all possible trees very rapidly and quickly show which are the more and less likely. The most likely are, for example, the ones that do not require complex characters to evolve in the same form more than once. The ones that contain fewest such requirements are said to be the most **parsimonious**. Parsimony does not guarantee correctness: in the end we cannot second-guess nature, and if nature wants to do unlikely things, then it will. But we can assert that a more parsimonious classification is more likely to reflect the true phylogeny than one that requires many devious evolutionary diversions and ad hoc re-inventions, and likelihood in a subject like this is all we can hope for.

In general, the cladogram should be seen not as the one true evolutionary history of a group, like the genealogy of some Royal household, but as a hypothesis, which attempts to discover that genealogy by explicit methods and which—because it is explicit—is testable. Testable hypotheses, as the philosopher Sir Karl Popper pointed out, are the essence of all science. The mathematical search for parsimony is part of the testing process.

I end this chapter with one last point of historical interest: the controversial career of transformed cladistics.

CLADISTICS TRANSFORMED

As we have seen, Willi Hennig and the school he gave rise to developed the rules of cladistics largely as an antidote to some of the more free-wheeling approaches to classification that prevailed before: taxonomy based on the assertions of experts who sometimes worked in their own peculiar ways and could not be questioned, like medieval priests. No one denies the need for scholarship but scholarship in science (as opposed to medieval theology) has to be amenable to criticism, and, without clear rules and explicit philosophy, criticism is impossible. There is a general need for an explicit method; and the particular method of Hennig seems to be the best.

But in the 1970s (before many non-taxonomists had even heard of cladistics) some cladists felt that Hennig's purist rules were not quite purist enough. Certainly, they said, it is good to base taxonomy on phylogeny—evolutionary history—for this is truly the unifying fact that underlies the order of nature. Certainly cladistic methods aid in this. But as Norman Platnick of the American Museum of Natural History put the matter in 1979, 'Hennig presented his methods by reference to one particular model of the evolutionary process, whereas contemporary cladists recognize that neither the value nor the success of the methods is limited by the value or success of Hennig's particular evolutionary model'. In other words, Hennig intended his methods to apply to creatures that had got to be the way they are because they had evolved—but his method works perfectly well whether they have evolved or not; and indeed it can be applied in principle even to non-living entities, such as chairs and wardrobes. So long as there is some order underlying the similarities, cladistic technique will provide a classification that reflects that order. If we simply group living things on the basis of what we can see with no assumptions about their evolutionary relationships, then we ought to get the same results as we would if we did assume evolution. If we do not get the same results then that is itself salutary: it would suggest that our evolutionary assumptions needed rethinking. In any case (applying Occam's razor again) it is good to begin any exercise in science with the fewest possible assumptions.

So various systematists in the United States and at London's Natural History Museum set out to devise a version of cladistics that was even more purist than Hennig's, which they called **transformed cladistics**. Practitioners of transformed cladistics specifically did not assume evolution when framing their classifications. The less they assumed, they felt, the better.

In truth, the transformed cladists were not denying that evolution is the way of the world. They merely said that for the purposes of classification it was safer, at least in the first instance, not to assume the fact of evolution; and indeed to use the classification as a means of checking evolutionary assumptions. But, of course, such a position can readily be misunderstood, and sometimes the transformed cladists made their point too stridently and so invited misunderstanding. Be that as it may, some observers concluded through the usual passage of Chinese whispers that some leading systematists, notably in the American Museum of Natural History and the Natural History Museum, were 'denying' the need for evolutionary theory and hence (by small extrapolation) were denying evolution itself. This was sensation indeed, because those two institutions were surely strongholds of evolutionary thinking. In addition, transformed cladism emerged at a time when American Creationists were being particularly bullish. In the early 1980s, in Britain's *New Scientist* magazine, one professor of biology accused the transformed cladists of Marxist subversion. That particular controversy ran and ran.

Today, transformed cladism is still with us. Many still argue that it is always best to assume as little as possible and that there is no need to assume evolutionary relationships in advance, when framing a cladogram. No serious player denies evolution. But the evolutionary relationships should emerge from the cladogram; they should not shape the cladogram. Others see less need for such purism. In any case, the heat of the early 1980s has now subsided. Cladistics, whether some choose to 'transform' it or not, is, rightly, the prevailing orthodoxy; nesting groups defined exclusively by shared, derived, homologous characters.

I want to round off these introductory chapters with the attempt to reconcile Linnaean nomenclature with modern cladistic principles—'Neolinnaean Impressionism'. But first it is worth looking briefly at the kind of data that modern taxonomists take into account—fossils, molecules, and so on. It is truly wonderful; revelatory, indeed.

CHAPTER FOUR

DATA

EVER SINCE ARISTOTLE, taxonomically inclined biologists have faced three kinds of question. First, what is the underlying order within nature that a 'natural' classification should reflect? Second, how can we handle the data in ways that will lead us to a natural classification—so that (in modern parlance) we can distinguish between homologies and homoplasies, apomorphies and plesiomorphies? And finally —the most basic and mundane question of all—what kind of information is, in fact, available? Chapters 1–3 deal with the first two questions (and the brief answers are: 'phylogeny', as first described by Darwin; and 'cladistics', as first outlined by Willi Hennig). This chapter looks briefly at the data on which the great flights of taxonomic speculation are founded—speculation that we hope is leading us to the truth.

In general, taxonomists since ancient times have simply looked more and more closely at living creatures as their knowledge and confidence increased and, crucially, as instrumentation has advanced. Modern, serious taxonomy owes as much to the light and the electron microscope, to chemistry and molecular biology, and to the methods of nuclear physics that enable fossils to be dated accurately, as astronomy has owed to the astrolabe, the telescope, and the ever-growing battalions of satellites.

In brief, then, the data regarded as 'characters' have been based (roughly in order of their historical appearance) on gross structure, or **morphology**; on detailed structure, or **ultrastructure;** on study of development, or **embryology**; on fossil evidence, or **palaeontology**; on behavioural studies, which in general are **ethology**; on body chemistry, or **biochemistry**; and, finally, on the study of the DNA from which the genes themselves are constructed, **molecular biology**. Sometimes different sources of data seem to lead to different conclusions: for example, fossil data have sometimes seemed to conflict with molecular data. But in general, each source of data has its own strengths and weaknesses, and it is a mistake to suggest that any one source of information is always and unequivocally superior or more reliable than any other. The task now is twofold: on the one hand, to make sense of the sheer mass of data that is coming on line and, on the other, to reconcile different sources of data into one grand overview. The first of these problems is now being tackled by computer programs that exploit cladistic principles, as described in Chapter 3. The second is more difficult; and where there is conflict, biologists just have to argue it out.

For the rest of this chapter, however, I want to look briefly at the different sources of data, to see what they have contributed in the past and might contribute in the future. The sequence of ideas discussed is more or less historical.

MORPHOLOGY

Morphology is what you can see; but seeing is a lot harder than it seems. All tyro biologists spend half their lives trying to draw whatever is before them, and know how hard it is to represent what is there, and not what the textbook or some folk memory tells them ought to be there. There is deep psychology in here, as well as technical ineptitude; Claude Monet, the most impressionist of the French Impressionists, spoke of the need to develop 'the innocent eye', but Darwin himself pointed out how difficult it is to observe anything at all unless you have some idea of what it is you are supposed to be looking at. Thus objectivity and preconception are constantly at war in a way that is analogous to the simmering conflict between authority and methodology, discussed in Chapter 3. When the creature is unfamiliar, and perhaps long dead and damaged, the difficulties are magnified.

There can be conflict, too, between the morphological evidence that is before your eyes and, in particular, the molecular evidence. Fifteen years or so ago, some molecular systematists acted as if they had found the royal road to truth. Some of them, however, belonged to a modern school of biology that deals only with creatures that are already dismembered or indeed dissolved and for all intents and purposes, they had never seen a living animal. One very early molecular study suggested, for example, that horses are more closely related to monkeys than monkeys are to apes. In such cases no harm is done; we all know that monkeys are related to apes, and that horses are very different, because we can see the difference with our own eyes; and can simply dismiss such peculiar molecular results as anomalous. We know nowadays that it is a mistake to put too much store by any one piece of molecular evidence. But sometimes the issues are a lot more tricky. For example, modern molecular studies by Dan Graur at Tel Aviv University and his colleagues in Israel suggest that guinea-pigs are not related to other rodents. Other zoologists who know guinea-pigs well feel that this is simply nonsense. Although I feel in my bones that guinea-pigs are rodents I do not feel entitled to a serious opinion and merely suggest that morphological data cannot lightly be thrown aside, if only because we would surely be in trouble if we lost faith in our own eyes and common sense. On the other hand, the molecular studies by Mitch Sogin and others of what are commonly called 'protozoans' (as described in Part II, Section 3), have truly made a nonsense of traditional ideas based on the creatures' observable structures. By the same token, we cannot lightly dismiss the molecular insights into guinea-pigs and rodents.[1] We will just have to wait and see.

This last observation, however, raises another difficulty of morphology; that, in the end, the structure of a creature simply does not yield enough clearly distinct characters. To be sure, in the case of a vertebrate, you can measure all the bones, and then you can calculate all the different ratios between the different measurements,

[1] Some more recent molecular studies support Dan Graur's conclusions. But, then again, other molecular studies contradict Graur, and some suggest that Graur has obtained a deceptive result through looking at too few molecules. It is not my point here, however, to argue the wrongs and rights of a particular case; merely to point out that different techniques can lead to different conclusions.

and in fact can go on doing this ad infinitum; and I know of taxonomists in far-flung corners of the world who literally spend their entire lives making such measurements on just a few specimens. But, as we have seen, many of the features that the taxonomist might choose to measure tell us nothing at all of phylogenetic interest. Indeed the features that are most striking can be the least informative, because closely related creatures may diverge rapidly as each may adapt to a quite different way of life, whereas others that are only distantly related may converge, structurally, as each adapts to a similar way of life. Plesiomorphies are unhelpful, too. Often, too, what seems like a series of different characters turns out merely to be different aspects of a single character. For example, arthropods (such as insects and spiders) have an outer protective covering (exoskeleton) made of tough protein stiffened with the polysaccharide chitin, and they also lack cilia—the hair-like projections of cells with which most animals shift mucus around. The chitinous exoskeleton and the loss of cilia look like two perfectly good and discrete characters. But some zoologists have suggested that arthropods lack cilia *because* they have chitinous exoskeletons: cells covered in chitin cannot also possess cilia. So they could in principle be two aspects of the same character. (I do not say that I actually believe this; simply that the point illustrates a principle.)

All the same, as we will see, molecular studies really do provide a huge number of different characters to study—at least from living creatures. This is why, in the end, the guinea-pig/rodent issue seems more likely to be resolved by molecular studies rather than morphological ones. There is little more to find out about the gross structure of guinea-pigs; but there is a wealth of potential molecular data still to be uncovered.

The main difficulty in morphology is, as discussed in Chapter 3, to differentiate between homologies and homoplasies, and then to go one step further, and distinguish the synapomorphic homologies from the merely plesiomorphic ones.

ULTRASTRUCTURE

Ultrastructure is morphology writ small. The potential to look at detailed structure began in the seventeenth century with pioneer microscopists such as the Italian Marcello Malphighi, who described many fine structures within various body organs, and the Dutchman Anton van Leeuwenhoek, who discovered what he called 'little animalcules', which presumably were a mixture of bacteria and protozoans. (Nowadays, bacteria and protozoans together are colloquially called 'microbes'.) Microscopy came on apace in the eighteenth and nineteenth centuries and by the 1850s was well advanced; chromosomes, the structures in which genes are packaged, were discovered in the late nineteenth century (and nowadays the study of chromosome structure and behaviour, or **karyology**, provides another source of taxonomic data). From the 1930s onwards studies with the light microscope have been increasingly abetted by the electron microscope, which shows the fine structures of the cell in remarkable detail.

Both forms of microscopy are also supported in their turn by biochemical techniques by which different cell components can be separated and individually analysed. One of the most important of these is the ultracentrifuge, which spins at enormous speed. Mashed up tissue is placed in tubes, spun, and the different cell components separate according to their density. So the heaviest components finish up at the bottom of the tubes (furthest from the centre as the centrifuge spins), with the lightest at the top. The mass of the different components can then be expressed as a **sedimentation coefficient**, or S—a measure of the rate at which the components gravitate from the centre when spun around. Thus, as we will see later, the nucleic acid **RNA** divides when spun into various fragments that are distinguished by different values of S (16S and 18S being small and light and 23S and comparably sized fragments being larger and heavier).

Ultrastructural studies are in principle subject to the same kind of problems that beset those of gross morphology; that in principle structures may look similar because they have converged, or different simply because they have diverged rapidly. But such objections seem to have less force at this level. For example, one of the most conspicuous components of most cells are the **mitochondria** (singular: mitochondrion), which contain the enzymes that supply the cell with energy. Thus they are sometimes called 'the power-houses of the cell' although that particular metaphor seems all together too mechanical. 'Patisseries of the cell' has a pleasanter ring to it. Different organisms have differently shaped mitochondria. But there is no obvious reason why the shape of the mitochondrion should in any way be influenced by the external environment, in the way that, say, the long arms of an orang-utan are an obvious ad hoc adaptation to life in the trees. So if two creatures do have similar mitochondria, it is at least a reasonable bet that this reflects true relationship.

Systematics has benefited enormously from ultrastructural studies, but two areas perhaps deserve special mention. First, it has been known since the nineteenth century that whereas the cells of plants, animals, fungi, and the host of comparable creatures hold the bulk of their genes inside discrete 'nuclei', the cells of bacteria do not. This difference is now believed to be profound; so much so that the creatures with nuclei are placed in their own domain, the Eucarya, whereas the creatures without nuclei are informally called prokaryotes, although are in fact now grouped in two separate domains, the Bacteria and the Archaea. This is all discussed at length in Part II, Section 1. The point here is that these huge divisions—the deepest divisions recognized in modern taxonomy—are based, at least in the first instance, on features of ultrastructure.

Second, embryology has played a huge part in systematics; but particularly when embryos are small it is easy to be deceived. The light microscope enables embryologists to observe the development and 'fate' of individual cells; and when you can see that a given organ in two different lineages does or does not arise from the same set of cells, then this certainly helps you to decide between homology and homoplasy.

EMBRYOLOGY

In the nineteenth and early twentieth centuries taxonomists tended to feel much the same way about embryology as many of their late twentieth-century successors have felt about molecular data: that it is the royal road to truth. In both cases there is good reason to be enthusiastic—the insights from both sources have given systematics a tremendous shove, and solved problems that otherwise could not have been approached. But in both cases, there are caveats.

The joy of embryology is that any similarities really do seem likely, at least at first sight, to result from homology; if two animals both develop a particular organ from the same particular cells this surely is because they share a common ancestor, and not because natural selection particularly favours one route of development rather than another. The main snag here is that when embryos are very young, consisting of only a few cells, then the number of ways in which it is theoretically possible for those cells to develop seems somewhat limited, so they might all finish up doing much the same things because they have little alternative. Yet there usually are more options than is at first apparent. For example, the young embryos of the great division of animals known as the protostomes, which include arthropods, molluscs, and segmented annelid worms such as earthworms, generally begin dividing in a spiral fashion—and, indeed, this feature has done much to establish the relationship between molluscs and annelids. Embryos of the deuterostomes, which include vertebrates and echinoderms (such as starfish), begin dividing in a radial fashion, with each new cell stacked squarely on the one before. Again, this is one of several embryonic features that first helped to establish the otherwise unlikely looking relationship between vertebrates and echinoderms, which in more obvious ways could hardly be more different.

Even so, we should not get carried away. The acknowledged father of modern embryology was the German/Estonian biologist Karl Ernst von Baer, who was born in the classical age of taxonomy (1792) and died in 1876, some time after *The Origin of Species* was published. He observed that embryos from related animals resemble each other more as embryos than as adults—and the younger they are, the closer the resemblance. He called this the 'biogenic law'. In essence, at least in groups such as the chordates (the phylum that includes the vertebrates), his 'law' is largely true: thus young fish embryos look like little commas, with the beginnings of a backbone, and so do very young birds, dogs, and humans. The reason seems intuitively obvious: fish and humans both have heads and backbones; and the backbones are liable to be put in place before the limbs; and it is not until the limbs start to take shape that we begin to see sharp differences between the two.

Ernst Heinrich Haeckel (1834–1919), von Baer's intellectual successor, took the idea too far with his notion that 'ontology recapitulates phylogeny', which means that in the course of its development (**ontology**) an animal effectively re-enacts its own evolutionary ancestry. Up to a point, it may seem to—sometimes; thus human embryos do roughly resemble fish. But the correspondence is somewhat loose, and where it is true it tends to be so only for the reasons that von Baer pointed out: young embryos are likely to resemble each other more closely than the adults that they grow into.

But even von Baer's biogenic law is really only an observation—not a 'law' at all —and it holds only up to a point. For example, whereas the young forms of mammals or birds do not have to do anything except sit in the womb or the egg and grow and develop, the young forms of many other animals are free-living **larvae**, even though they may in some cases effectively be no more than embryos. Such larvae are exposed to the rigours of the environment, and are therefore under strong selective pressure to adapt; and so we find that the free-living embryos/larvae of closely related creatures may, in fact, take hugely different forms. Echinoderm larvae, for example, are immensely variable.

Overall, then, embryological information is extremely important, not to say vital. But, like all other information, it has to be treated with caution and balanced against other sources of data.

BEHAVIOUR

Behaviour by itself can never be taken as a reliable guide to relationship. There is, after all, only a limited number of ways to survive on this planet; and creatures with vastly different phylogenetic backgrounds may finish up doing the same things—living down holes, at the bottom of the sea, in trees, and so on. Contrariwise, closely related creatures may finish up doing vastly different things—such that otters are semi-aquatic whereas weasels, also from the family Mustelidae, are emphatically terrestrial. In short, behaviour is perhaps even more flexible, and hence subject to convergence and divergence, than any other set of characters.

Even so, various quirks of behaviour offer clues—and I mean quirks; little habits that do not by themselves seem to be of huge adaptive significance, but nonetheless persist. For example, both storks and New World vultures, such as condors, excrete over their own legs; a behaviour that has been associated with cooling, though this seems only to be a guess. Many ornithologists have suggested on other grounds that New World vultures are related to storks rather than to Old World birds of prey; and this peculiar habit seems to add to the case. But some modern ornithologists (such as Joel Cracraft of the American Museum of Natural History in New York, whose ideas I am following in this book) do not agree that condors and storks belong together and excretory behaviour alone certainly does not carry the day. On the whole, then, behavioural clues remain as gilt on the gingerbread.

FOSSILS

Most organisms when they die are chewed up and digested by scavengers, or rot, or are simply smashed to pieces. Those that lack a scaffold of bone or an armour of chalk or chitin-stiffened protein—including most of the vast array of creatures known as 'protists'—rarely form fossils at all. But soft-bodied creatures can leave representations of themselves. Impressions of them in soft mud can fossilize, so that the fossil is not the creature itself but a kind of mould of one (a mould that may in turn become filled with what is then a replica of the original). Such fossils can be remarkably informative and, indeed, tell us most of what we know about life in the Precambrian, more than 570 million years ago. Animals and plants that expire in tropical forests fall on to warm moist soil that heaves with 'detritivorous' invertebrates, and bacteria and fungi, and may disappear in days; a particular pity, because tropical forests harbour the most species. Only a relatively few groups in peculiarly favoured spots fossilize reasonably reliably; so we have huge consignments of fossil clams, which even in life are interred in sediment, but only a few fossil winkles, which, when dead, are ground to sand on the turbulent beaches. Groups that we know are ancient, and ought to fossilize reasonably easily (because they have hard parts and live in a wide variety of places), may nonetheless disappear from the fossil record, or simply not appear at all, over hundreds of millions of years. Hence, as described in Section 11, there are two groups of spiders today: the mesotheles and the opisthotheles. The mesotheles, of which there are only two living genera, are obviously the more primitive; for example, and notably, their 'abdomens' are segmented, like any ancient arthropod, and not fused into one great sac as in the familiar spiders of house and garden. Even so, the first known fossil of an unequivocal mesothele has only just been found—by Paul Selden of the University of Manchester. In fact, the new fossil hails from the late Carboniferous of France—as Selden says, 'thus extending the record of this group from 0 to 290 million years'. Spiders all in all are common and successful. If major groups of them can go missing since before the time of the dinosaurs, what other creatures, animal, plant, and otherwise, have escaped us altogether?

There are many other snags. Complete fossils, except of small creatures, are extremely rare: vital bits are often missing. What there is often badly damaged (vertebrate palaeontologists are apt to compare even some of their most prized fossils to 'road kills'). Most mammalian fossils are teeth. Teeth, as has often been commented, are the first to rot before death, but are the most resilient post mortem. It is easy to make mistakes with interpretations of fossils, but often the opportunity does not arise because the vital bits are missing. Thus it seems that our own genus *Homo* descended from small African predecessors called *Australopithecus* ('southern ape'). The first *Australopithecus* skull was described in 1925 by Raymond Dart (1893–1988); and although Dart suggested that the creature probably walked upright because the base of its skull suggested that it perched on the top of a vertical backbone, nobody knew that this was really so until Donald Johanson found the first australopithecine skeleton ('Lucy') in the 1970s.

Taken over all, then, the fossil record is inveterately 'spotty'. We know there are huge gaps. We can be certain that major species, and indeed some major groups, are missing all together. In many ways, and for many reasons, we have to be wary.

Yet the fossil record is invaluable. Without it we simply would not know that any creatures had ever existed on Earth apart from the ones that are living now. Some bright theoretical biologists—including Darwin, probably—might well have inferred that present-day living creatures must have evolved from earlier types that are long-since gone. But without the fossil evidence it would be difficult to convince the sceptics, and evolution would remain a minority hypothesis (and probably a heresy). Even if people did come to expect that other creatures had existed in the past, we could never imagine what they were really like. The scientists' predilection for 'parsimony' would ensure that our guesses were conservative—and no conservative speculation could lead us to imagine the eurypterids; or the trilobites, who perhaps include the ancestors of all modern arthropods; or the ammonites, once ubiquitous cephalopods; or the cycad-like bennettitaleans; or indeed the dinosaurs, which included some of the mightiest animals of all time and held sway over the world for 120 million years. Yet the reality—or marvellously intriguing glimpses of it—is revealed by fossils; and the reality far exceeds our imaginings.

For all its obvious imperfections, too, the fossil record does tell us much of what we want to know in pleasing detail, and sometimes—in some areas during some periods and for some lineages—it tells us far more than we have any right to hope for. The Burgess Shale fossils of early to mid-Cambrian Canada seem to show us an entire fauna, some 530 million years old, including several arthropod lineages unlike anything living today. The fossil fishes of Western Australia are stunning. Fossils show us in good detail how vertebrates first came on land; how mammals arose from among reptilian ranks—including the development of a middle-ear bone from the ancient reptilian lower jaw. It shows how birds probably arose from among the ranks of slim, fleet-footed dinosaurs (though, in truth, some palaeo-ornithologists still deny this). It shows how groups that now are rare—such as the Ginkgoales, now reduced to one genus, the *Ginkgo* or maidenhair tree—once formed great forests.

The fossil record shows us the true meaning of biogeography; for example, how the horse and camel families arose in North America and only later spread into South America and Eurasia—and went extinct in North America; and how the present-day large fauna of North America, including the bison, the moose, and the wapiti, are immigrants from Eurasia that entered the Americas across the intermittent Siberia–Alaskan landbridge of Beringia. For that matter, too, it reveals alder trees in Indonesia, in colder times, and hippos in the north of England, in balmier times. A soft-bodied creature called *Kimberella*, which left impressions in Precambrian mud, used to be described as a cubozoan jellyfish. But Mikhail Fedonkin of the Russian Academy of Sciences and Benjamin Waggoner of the University of California at Berkeley reported in 1997 that *Kimberella* was really an early, soft-shelled mollusc. If Fedonkin and Waggoner are right, this would mean that all the great related phyla—molluscs, annelid worms, and arthropods—must have arisen deep in the Precambrian, more than 570 million years ago. It is very difficult to measure 'appreciation'; but I am inclined to suggest that without fossils, our appreciation of life's variety and versatility would be reduced by about 90 per cent. In fact, if we accept Theodosius Dobzhansky's comment that 'Nothing in biology makes sense except in the light of evolution', then we must admit that the fossil record is pivotal. Without it, after all, it would be difficult to convince people—probably including most biologists—that evolution has been the way of the living world.

Yet the fossil record must be read with care. The presence of a fossil in a given stratum shows that the particular creature existed at that time; but its absence does not show that it did not exist at that time. As the adage is: absence of evidence is not evidence of absence. Thus, arachnologists never doubted that mesothelic spiders existed as far back as the Carboniferous period—it was just that they had never been found. Similarly, it seems that mesotheles are not the direct ancestors of opisthotheles, but are their sister group. This means at the very least that the common ancestor of the mesotheles and the opisthotheles must be older than the Carboniferous (not that this was in much doubt, either).

From a knowledge of evolutionary principles and the time it takes for species to evolve and disappear, it is possible to calculate the number of species from a given lineage that must have lived in the past, once we know the number that are alive now, and the age of the oldest known types. Thus there are 200 or so species of living primate, and the oldest known fossils are about 55 million years old; the principles tell us that there must, therefore, have been about 6500 primate species since the time of the oldest known type. In fact, only about 250 fossil primate species are known: so we know less than 4 per cent of the species that probably existed.

Because we know only 4 per cent of all the ancient primates, the chances of finding the very first primate are extremely remote—because the older the fossils are the less chance there is of finding them, and the first types would naturally have been rare. So we can guess that the first primate must have appeared quite a long time before the oldest known type; and, in practice, the fewer that we actually know, the further back the first one is liable to have lived. Thus palaeoprimatologists calculate that the first primate-like mammals possibly appeared during the late Cretaceous period, about 80 million years ago, when the dinosaurs still had 15 million years or so to run, even though no primate fossils are known from dinosaur times. In short, it is possible to infer a rich picture of the past even when the fossil record is known to be even more sparse than usual; and the primate fossil record is notoriously meagre.

Finally, knowledge of evolutionary history has been enormously enriched by methods of dating that have been steadily developed over the past 200 years. When fossils are buried in sediments the depth of their burial reflects their age—because the rate of sedimentation can be gauged. Such methods are now abetted by modern insights from radioactive isotopes, which 'decay' at known rates. Carbon-14 decay helps palaeontologists and archaeologists to date material from the last few thousand or hundred years; and potassium–argon dating offers insights into more ancient times. Once a few strata have been dated, then the age of many other areas can often be inferred. For example, radioactive and other methods enable geologists to date particular volcanic eruptions in the past; and chemical methods enable them to see which particles of ancient dust came from which eruption. Thus the presence of such particles in and around particular fossils effectively dates the fossils.

As I describe a little later in this chapter, modern molecular methods not only enable biologists to infer relationships but also to judge when any two lineages last shared a common ancestor. The fossil record, once dated, provides an independent check on the ideas developed from molecular studies. So, for example, molecular studies in the late 1960s suggested that human beings and chimpanzees shared a common ancestor as recently as 3 million years ago—although the fossil record, at least as read by many palaeontologists, once seemed to suggest that chimps and humans may have been going their separate ways for more than 25 million years ago. Newly discovered fossils—and a closer look at the ones that were already known in the 1960s—suggests that the molecular biologists were more or less correct, and that the common ancestor probably lived around 5 million years ago. Molecules and fossils working together, in short, are a powerful duo.

Finally, note in general how various lines of evidence suggest to palaeontologists that they ought to be looking for particular fossils in particular rocks even though no such fossils have been found in such places in the past. Thus the fact that mesothelic and opisthothelic spiders are sister groups, and mesotheles are known from the late Carboniferous, suggests that palaeo-arachnologists should look for spiders in rocks older than 290 million years. For primates, too, a short chain of reasoning suggests that the first primates of all are probably about 15 million years older than the oldest ones we know about. Molecular evidence in general can tell us —or at least provide some insight into—the age of the common ancestor of any two creatures; so we know what age of rock ought to contain that ancestor. Knowledge of the times at which particular lineages diverged in turn helps us to refine the phylogenetic tree. After all, when we say that creatures A and B are more closely related to each other than either is to C, we mean that A and B shared a common ancestor more recently than they did with C. If the rocks yield fossils that suggest that this is simply not the case, then we have to look at our proposed phylogeny again.

MOLECULES

Under 'molecules' I am including both 'biochemistry'—the long-established study of body chemistry in general, but in particular of proteins (including enzymes)—and the more recently developed 'molecular biology', which explores DNA in particular, the stuff of which the genes themselves are made. The two sets of approaches run into each other because they are allied conceptually, and studies may be carried out in the same kinds of laboratory, sometimes side by side. To appreciate these new techniques, it is necessary to understand a little basic molecular biology. A brief primer follows. Readers who have read this kind of stuff a million times before are advised to skip on to the next subheading, 'Molecular techniques in action'.

A BRIEF PRIMER OF MOLECULAR BIOLOGY

Francis Crick has summarized molecular biology in what he calls the 'central dogma': 'DNA makes RNA makes protein'. In slightly more detail: the idea of the gene was initiated by Gregor Mendel (1822–84) in the 1860s, working in a monastery in Brno—now in the Czech Republic. Indeed, Mendel did more than initiate the idea; all by himself and unaided (he was, in fact, inhibited by well-meaning but unimaginative advisers) he outlined most of the principal ideas on which the modern science of genetics was founded. But his work was simply ignored until rediscovered at the end of the nineteenth century; so the science of genetics—and, indeed, the words 'genetics' and 'gene'—did not finally get under way until the twentieth century.

For the first four decades of the twentieth century, however, nobody knew what genes actually are. At least, it had become clear that they reside within the chromosomes—intriguing cipher-like structures that appear and divide with great precision during cell division. But what the genes are made of, nobody knew. It was known that chromosomes contain a peculiar organic acid that had been discovered in the 1870s, and became known as deoxyribonucleic acid, or DNA; and that they also contain proteins. Those biologists who thought about the matter at all assumed that the protein fraction of the chromosomes contained the genes, and that the acid was, well, just there. But in general the 'classical' geneticists of the first half of the twentieth century thought about genes as if they were just beads on a string—the string being the chromosome. As chemistry this is crude; but the string-of-beads idea remains a very useful 'model', and most of the practical genetics that, for example, underpins livestock breeding, conservation studies, and medical genetic counselling, is still 'classical' in essence, even though it may also use modern chemical techniques.

In 1944, however, the Canadian-born American bacteriologist Oswald Avery and his colleagues showed beyond any reasonable doubt that of the two components of chromosomes, it is the DNA and not the protein that forms the genes. The proteins in chromosomes (it later became clear) are there partly to hold the whole thing together, and partly to help regulate gene function. Many biologists effectively ignored Avery's work but a few, including Linus Pauling in the United States, Maurice Wilkins in London (soon to be joined by Rosalind Franklin), and Francis Crick and James Watson in Cambridge, realized that DNA, this obscure also-ran organic acid, was, in fact, the most interesting molecule in all of nature. They set out to discover its structure. Remarkably quickly, Crick and Watson produced a model of the DNA molecule that they announced in *Nature* in 1953. Those two, plus Maurice Wilkins, received Nobel prizes in 1962; Rosalind Franklin had died, only 37 years old, in 1958.

The Crick and Watson model shows that each molecule—or rather 'macromolecule'[2]—of DNA consists of two enormously long chains, which are coiled around each other: the famous 'double helix'. Each chain consists of a series of subunits called nucleotides. Each nucleotide has three components: a molecule of the sugar, deoxyribose; a phosphate, which joins one nucleotide to the next one in the chain; and a 'base'. The deoxyribose and the phosphate are exactly the same in all the nucleotides, but the bases are of four types: adenine, thymine, guanine, and cytosine. These are generally known by their initial letters: A, T, G, C. Each nucleotide contains just one of the four.

The rough, general chemistry of DNA was known by the 1950s; Crick and Watson's great contribution was to show how all the components are put together in

[2] Molecules such as DNA and protein consisting of strings of smaller molecules joined together are commonly called 'macromolecules'. The word 'molecule' is then freed, and can be applied to each of the component parts.

a three-dimensional structure. It is extraordinary, though, that the only source of variation in the DNA molecule comes from the bases; indeed, it was because of this that early twentieth-century biologists tended to assume that genes could not be made of DNA, but had to be made of protein. Genes, after all, are in principle infinitely variable, and so are proteins. DNA, with its four plodding bases, seemed far too uniform to do the job.

As ever, however, nature proved to be way ahead of us; and it quickly transpired that, by a few little tricks, the four bases of DNA can do all that is required of a gene. They are in fact analogous to the letters of the alphabet. The alphabet has only 26 letters; but those 26, arranged in a potential infinity of ways, can produce an infinity of words, of which the extraordinary vocabulary of Shakespeare is but a minuscule fraction.

How, though, does the structure of DNA relate to its function? Well, it was known from early in the twentieth century that the job of genes is to make proteins; effectively, the rule is, 'One gene, one protein'. Proteins are the business end of life. They account for much of the structure of the body: contractile proteins form the muscles; some (small) proteins act as hormones; and, most importantly, most enzymes are proteins—enzymes being the catalysts that drive the metabolism of the whole body. It has also been known for some decades that proteins (like DNA) consist of long chains of amino acids; and although proteins are in principle infinitely variable, almost all of them in nature are compounded from only about 20 different amino acids. Again the same principle: the small(ish) range of amino acids are the alphabet from which the infinite vocabulary of proteins can be formed.

One problem was to explain how the four bases in DNA could provide a code that would correspond to the 20 amino acids in proteins. This was solved by Francis Crick and Sydney Brenner in the early 1960s. They showed that although the bases in the DNA occur in a continuous, unbroken chain, in practice they operate in groups of three (each group being known as a codon). In fact, with four bases, 64 different groups of three are conceptually possible: $4 \times 4 \times 4$. Because there are only 20 or so amino acids, there is, in fact, far more coding capacity in these groups of three than is needed. It transpires that some amino acids may be coded for by more than one different codon; and that some of the codons do not code for amino acids, but instead act as punctuation marks—effectively saying, 'This gene begins (or ends) at this point'.

Problem two was to show how genes could duplicate themselves so exactly during cell division. In fact, Crick and Watson provided the answer to this in 1953. They pointed out that the two chains of the DNA molecule are loosely linked together by chemical bonds between their bases. But the linkage followed strict rules. Adenine on one chain would bind only with thymine on the other; and cytosine would bind only with guanine. Hence all the way down the middle of the double chain we find 'base pairs': A–T and C–G. When DNA replicates, which means when genes replicate, the two chains break apart, and then each chain forms a new partner. Because the linkage is so strictly determined, each new partner must be a perfect complement to the other. Each chain, in fact, acts as a template to its new partner. Hence the duplication is exact—except for the odd mistake; and these mistakes are the main source of genetic mutations, which are the ultimate source of genetic variation in nature. The initial breaking apart of the two DNA chains, and the forming of new partners, are all supervised by enzymes.

So much for basic theory. How in practice does DNA go about its serious work of making proteins? With the aid of a second, somewhat simpler nucleic acid is the answer; one known as ribonucleic acid, or RNA. In essence, RNA is very similar to DNA except that its macromolecules tend to be much shorter; it has only one chain instead of two; the sugar in the nucleotides is ribose rather than deoxyribose; and instead of the base thymine, RNA contains a similar molecule called uracil. But the other three bases are the same as in DNA. RNA acts as the intermediary between DNA and protein; hence Crick's 'dogma': 'DNA makes RNA makes protein'. In practice, during protein formation, the two DNA chains separate along part of their length; then one of the chains forms a complement to itself, not of new DNA but of RNA. RNA, whose structure reflects that of the DNA, then encodes all the information needed to ensure that amino acids are joined together in the appropriate order to form the required protein.

A diversion is needed. Living things have one of two different kinds of cell. In the cells of bacteria and archaes (which used to be lumped in with bacteria but are now known to be very different) the DNA lies within the cell body in small bodies called **plasmids**, and also within a single chromosome that also just resides in the cell body. But in plants, animals, fungi, seaweeds, and the vast array of creatures loosely known as 'protists', most of the DNA is contained within several or many chromosomes, and all the chromosomes are held within a special body in the cell called the **nucleus**. Bacteria and archaes are called **prokaryotes** (from the Greek *karyon*, meaning 'kernel', or indeed 'nucleus'); and organisms whose cells have nuclei are called **eukaryotes** (where *eu-* is Greek for 'good' or 'true').

In eukaryotes (and I'll confine this discussion to eukaryotes) DNA within the nucleus produces molecules of RNA, which then travel out to bodies in the surrounding 'cytoplasm' called ribosomes; and there, in the ribosomes, proteins are put together. In practice, there are three main kinds of RNA (indeed, there are a couple more kinds as well, but we don't need to worry about the others). These three are messenger RNA, or **mRNA**; transfer RNA, or **tRNA**; and ribosomal RNA, or **rRNA**. Messenger RNA carries the code from the DNA that corresponds to the protein that is to be produced. Transfer RNA molecules are small, and their job is to ferry individual amino acids to the ribosomes where they can be assembled into proteins, in the sequence laid down by the mRNA. Ribosomal RNA resides permanently within the ribosomes, where it oversees and generally takes part in the assembly of all the proteins.

Although RNA operates within the cytoplasm, all of it is first produced within the nucleus, for it is produced by DNA. It follows that DNA has several functions. The part of the DNA that produces mRNA is, in fact, the bit that codes for protein;

for the mRNA acts as the intermediary. But some of the DNA has instead to produce tRNA; and some has to produce rRNA. Thus there are protein-producing genes (which in reality are mRNA-producing genes); and there are also genes for tRNA and rRNA.

That is the very basic story, as it became evident in the 1950s and 60s. But since then there have been several more huge and unexpected insights; and from these have come the techniques that are now of such value in establishing phylogeny.

THE MOLECULAR TECHNIQUES IN ACTION

First, it has become clear that most of the DNA within the chromosomes of eukaryotes does not code for proteins (mRNA) or for tRNA or rRNA. In fact, the function of most of it is unknown—although the peremptory term 'junk DNA', which was at one time customary, has now been dropped because molecular biologists feel it is presumptuous to suggest that it does nothing at all, simply because its role is unclear. But the DNA that is of no apparent function is not subject to natural selection (because it does not matter whether it changes or not) and so can be highly variable.

Second, it has become obvious these past few decades that structures within the cytoplasm, such as the mitochondria that occur in (almost) all eukaryotic cells, and the chloroplasts that are found within the cells of plants, seaweeds, and some protists, contain DNA of their own. In practice, mitochondria contain only a few per cent of the total cell's DNA; but these 'mitochondrial genes' nonetheless have a significant influence on the cell's behaviour. Most of the time mitochondrial genes are inherited only through the female line, because mitochondria reside in the cytoplasm, and almost all of the cytoplasm in the young embryo comes from the egg rather than the sperm. In addition, as described more fully in Part II (Section 3), mitochondria and chloroplasts are thought to be descended from bacteria that first lived as parasites or commensals (residents) in the first eukaryotic cells. Accordingly—albeit surprisingly—mitochondrial DNA, or **mtDNA**, resembles prokaryotic DNA, and is different in significant details from the DNA of the nucleus.

Third, it has become evident from the 1960s onwards (the American chemist Linus Pauling was a pioneer in this, too) that DNA changes steadily over time, largely because when the strands break apart and form new partners as they do every time a cell divides, mistakes creep in, and the wrong nucleotide is inserted. These small changes are **mutations**; and although it has always been assumed that most mutations are harmful, while a few prove to be beneficial, the surprise has been that most mutations simply do not matter. They do not seem to affect the survival of the organism one bit. This observation has given rise to the idea of **neutral evolution**: the organism is changing—evolving—over time not simply through Darwinian selection but simply because, in reality, its DNA is less stable than has been supposed. The point is, however, that these changes tend to affect all the organisms in any one lineage in much the same way, because sexual congress leads to swapping of genetic material. But over time, the DNA within two different lineages that once shared a common ancestor will become more and more different, as their DNA continues to change in its own random way. Because the changes happen at a fairly steady pace (by no means rock-steady; but as good as a bad wristwatch) it becomes possible to judge how much time has passed since the two lineages shared a common ancestor. At least, it becomes possible to do this by cross-checking with the fossil record, once the fossils have been dated. Hence these constant changes, caused by mistakes in copying, provide modern biologists with a **molecular clock**.

Fourth, it has become abundantly clear (and the general principle has long been accepted) that the genome—the total collection of genes in the cell of an organism—is hierarchical in nature. The function of some high-and-mighty genes is to control the activities of others. In particular, zoologists have rightly become hugely excited by a family of genes that are found within all animals—the **Hox genes**, which were introduced in Chapter 3. Their role is to determine what each section of the animal actually becomes, which they do by turning on (or switching off) the genes that form the various parts of each section. It should soon become possible effectively to define the different phyla of animals in terms of their Hox genes: how many Hox genes they have, and how they are laid out and operate. The overall point is, however, that all the different components of the genome each can contribute information that can help to establish phylogenetic relationships; and in some cases this molecular information is virtually all there is.

It is intuitively obvious that the closeness of relationship—the **phylogenetic distance**—between any two organisms can be gauged by the difference between their DNA. But there are two bonuses. Because the changes in DNA occur at a fairly predictable rate, the degree of difference also gives direct insight into the time at which any two lineages last shared a common ancestor; and when this estimate is matched against the fossil evidence, the insight gained is considerable.

It also transpires that different regions of the DNA change over time at very different rates. Some regions change significantly within a few generations so that the perceived differences can and do reveal relationships even within family groups. These regions are used in forensic science for **genetic fingerprinting**, and by conservation biologists to judge who is related to whom within single populations. Such studies have shown, for example, that the chicks of any one female hedge sparrow will commonly have been sired by several different males (and that the males commonly sire chicks by several different females).

Mitochondrial DNA, or mtDNA, changes rapidly, too, but not as rapidly as the hypervariable regions in nuclear DNA. Analysis and comparison of mtDNA enables biologists to trace the relationships of races within species; and, for example, has been used to establish the relationships and possible peregrinations of the different populations of human beings. The kind of DNA that actually makes proteins (or, rather, makes the mRNA that makes proteins) tends to vary more slowly than mtDNA. Comparisons here enable biologists to gauge the phylogenetic distance of different groups within classes: for example, to establish the relationship between the different orders of birds (within the class Aves).

At the very slow end of the scale, it seems that ribosomal or rRNA—or the DNA that codes for rRNA—is very highly conserved indeed. Significant differences between the rRNA of any two organisms would indicate that they parted company a thousand million years or so previously. Hence rRNA is commonly used to establish the relationships between kingdoms and domains. In fact, rRNA is formed as one big molecule, which is then broken down into several pieces of characteristic size. Eukaryotic ribosomes invariably contain a small piece of rRNA called 18S rRNA. In prokaryotes, the equivalent piece of rRNA is smaller, a mere 16S. In phylogenetic terms, this apparently trivial molecular difference is huge; and this helps to confirm that prokaryotes and eukaryotes should be placed in different domains. Carl Woese's suggestion that bacteria and archaes should be placed in different domains, even though they are both prokaryotes, also derives primarily from differences in their rRNA; and so to does Mitch Sogin's suggestion that eukaryotes should be divided into at least a dozen kingdoms (see Part II, Section 1).

In general, molecular methods for judging phylogeny are of two conceptual types. Some methods merely compare two equivalent molecules from two different organisms without analysing either of them in detail. One such method is **DNA–DNA hybridization**. The technique is somewhat complex in detail but, in essence, single strands of DNA from two different organisms are mixed, and the extent to which they cleave together shows how closely those organisms are related. After all, if the two organisms were from the same species, then the bonding between their respective DNAs should be more or less absolute; but if the two species are very different—say human DNA and wasp DNA—strands of DNA from each would cling together only here and there. By such means in the 1980s, Charles Sibley and Jon Ahlquist explored the relationships of huge numbers of birds. In this book I am following the ideas of Joel Cracraft rather than of Sibley and Ahlquist, primarily because he uses cladistic methods and Sibley and Ahlquist on the whole did not; but Sibley's and Ahlquist's studies are immensely valuable and no bird classification can be considered truly robust that does not take their findings into account.

In the same general way, too (straight comparison rather than detailed analysis), biologists for several decades have been judging the phylogenetic distance of organisms by immunological means. They take a protein from the creature in question—say a fish—and inject it into a rabbit, which produces antibodies against it. Then they mix the same protein from the same fish with the antibodies, and also (but separately) mix an equivalent protein from another, related fish with the antibodies. The antibodies should cling eagerly to the proteins of the first fish, against which they were originally formed. How eagerly they seize on the protein from the second fish depends on how similar the two proteins are—which depends in turn on how closely the fish are related.

It is possible to write an entire book on molecular phylogeny, as John Avise has admirably demonstrated in *Molecular Markers, Natural History, and Evolution* (1994). My point here is simply to show what a variety of methods there are, and that different methods have different strengths and weaknesses. The general advantage of molecular techniques is that they are universal. Fish are fish and birds are birds but both have DNA. Traditional fish biologists group their subjects according to fishy criteria—scales, fins, and so on—while bird taxonomists look at such features as feathers and syrinxes (vocal organs). So by traditional criteria it is very difficult for anyone to judge whether a group of fish that the fish specialist calls a 'family' does or does not embrace the same range of diversity as there is in a bird 'family'; for how do you compare variations in an aspect of fish anatomy with variations in bird anatomy? But if the relationships are judged by molecular means (in addition to the traditional morphological studies, not instead of them!) then, in principle, at least, the degree of variation within the two groups can be measured on the same yardstick: the range of variation in the DNA.

The second great advantage of molecular studies is the sheer volume of information that they can yield. In principle, the position of every base in the DNA chain can be informative—effectively each one is a 'character'—and the 100 000 or so genes of the human being, say, contain about 3000 million bases. Of course, such huge quantities of data easily turn into an embarrassment of riches, so that molecular biologists must co-operate with very sharp mathematicians and computer aficionados to make sense of it all; and then the sense that they do make depends in large measure on which method of analysis they use, for different approaches can lead to different conclusions. But I will say more about this in Chapter 5.

Needless to say, there are some problems with molecular studies. An outstanding disadvantage springs from the fact that there are only four nucleotides. Thus any one base can be altered only in one of three ways—from A, say, to C, T, or G. Repeated mutation at the same site is likely to cause any one base to revert back to what it was originally. Thus there is a limit to the number of mutations that any one stretch of DNA can reveal, and when this limit is reached the stretch is said to be 'saturated'. It is difficult, too, when comparing stretches of DNA from different creatures to line them up properly—that is, to be certain that the piece of DNA from one creature is, in fact, homologous with the piece from the other. If not, then the comparison is nonsense. But although there are such difficulties, the amount of information that can be derived from molecular studies seems virtually boundless. These are exciting times, and there is a great deal more to come.

In general, phylogeny is difficult; it is in essence an attempt to reconstruct the evolutionary history of all the creatures on Earth, over a period of nearly 4000 million years, with extremely limited data, and with a huge number of obvious and not-so-obvious pitfalls along the way. Systematists attempting this task need all the help they can get. Present techniques, with modern molecular and palaeontological studies building on the traditional and still-vital approaches of classical times, and with the crucial assistance of cladistic theory on the one hand and modern maths and computers on the other, seem to be getting us there.

CHAPTER FIVE

CLADES, GRADES, AND THE NAMING OF PARTS

A PLEA FOR NEOLINNAEAN IMPRESSIONISM

WILLI HENNIG and his fellow cladists have sought to reform the ways in which taxonomists identify the different groups of creatures. They have also pursued a subsidiary agenda: to reform the conventions of **nomenclature**—the way the groups are named. First they insist that only clades, or monophyletic groups, should be acknowledged as formal taxa. Second, they propose that each new branching point (node) in the phylogenetic tree should be treated as a new Linnaean rank. As with all cladism the thinking is eminently logical, and perhaps in a generation or so cladistic nomenclature will prevail. Logical or no, however, these new rules raise problems and at present there is conflict (albeit mostly amicable, though not always) between those who insist on the cladistic approach to naming and those who feel that the traditional system, adhering more closely to the ideas of Linnaeus, has advantages that we should hang on to.

In this chapter I will offer a compromise, which I like to call (perhaps pretentiously, but why not?) 'Neolinnaean Impressionism'. In general I feel that we should welcome the logic and the explicitness of the cladists but that, when it comes to naming, we should temper logic with pragmatism. Indeed, we should distinguish clearly between the two phases of modern systematics. The first is to discover the phylogeny of creatures, and to summarize that phylogeny in the form of a phylogenetic tree; and the second is to break up that tree into convenient pieces, and name them accordingly, which is nomenclature. The first phase is an exercise in truth. We really want to know the histories of creatures, by what route they came to be as they are, and who they are related to. Nothing must compromise the search for truth; this is what science is about. This remains the case even though we know that 'truth' is forever elusive—especially in a subject like phylogeny, where we are trying to reconstruct events that sometimes happened many millions of years ago. But the second part—the naming—is an exercise in information retrieval, alias librarianship. We should not name things in ways that compromise the truth but we should acknowledge, nonetheless, as librarians do, that information may sometimes be conveyed most economically and tidily if we sometimes allow pragmatism to prevail over strict logic. So long as the phylogenetic tree itself is unaffected by our attempts to attach names to it, then no harm is done if our nomenclature cuts a few corners.

Perhaps, too, we need two levels of language in which to express classification: one to serve the specialists, who need to be precise and detailed, and another, more user-friendly, for non-specialists. There are precedents. Computers operate several levels of language simultaneously, some that enable the computer to talk to itself or to other computers, and others that enable it to communicate with human beings. Several taxonomists have already devised systems of classification for specialist purposes. These do tend to be difficult for the outsider to follow (and, in fact, they have not generally caught on among specialists either). But if we are to provide a user-friendly way of describing the groups and conveying their relative positions then we need to relax a little.

These, then, are the thoughts that lie behind 'Neolinnaean Impressionism'. But what, first, are the problems raised by strictly cladistic nomenclature? And why is it incompatible—at least in part—with the traditional method of Linnaeus?

GRADES AND CLADES, MONOPHYLY AND PARAPHYLY

It is easy to see why cladists want to base formal classification strictly on clades, and to throw out all taxa that are not monophyletic. Such a system is eminently logical. The phylogenetic tree, painstakingly extrapolated from cladograms, has an innately hierarchical structure—and hierarchy is the essence of all classification. Little clades nest within big clades. Why not regard the big clades as high-ranking taxa (phyla, classes, and so on) and the smaller clades as low-ranking taxa (families, genera, species)? Why compromise such neatness? Or as E. O. Wiley and colleagues put the matter in *The Compleat Cladist* (1991): 'Going to all the trouble of finding the groups and then throwing them away does not make sense to us'.

All naturalists since classical times, however, and indeed all human beings who are aware of other living things, also intuitively acknowledge the concept of the **grade**. 'Grade', to be sure, is a much vaguer concept than 'clade'. It simply refers to creatures that have 'reached a certain stage of evolution'—although that in itself is a highly contentious kind of phrase. But all of us know in our bones that reptiles (spelt informally, and colloquially, with a lower-case 'r') all seem to belong together. In a general way, they all seem to be the same kind of creature. They are all tetrapods; and if we look closely we find that they are all amniotes, too, whose embryos are surrounded by three membranes, one of which is the amnion. Although some reptiles have returned to water, at least in part—turtles, terrapins, and sea snakes are examples—reptiles are basically landlubbers that for the most part reproduce on land (the exceptions being creatures such as the long-extinct ichthyosaurs, which clearly gave birth to live young in the water, as do modern dolphins).

Creatures that we feel to be of the same grade do not have to be closely related. Thus we could say that 'worms' are a grade although the creatures we call 'worms' belong to many different lineages on distant parts of the phylogenetic tree. But all the creatures we informally call reptiles did apparently descend from the same ancestor, which was itself a reptile. Thus the reptiles would form a clade—except that two different reptilian lineages each went on to develop new sets of apomorphies, which, as every traditional biologist agrees, puts each of them into a new grade. Thus one reptilian lineage developed the fur and other features that define mammals, while another lineage, quite separately, went on to evolve the feathers and other features that define birds.

Mammals are a clade, and therefore the cladists are happy to acknowledge the traditional taxon Mammalia: and birds, too, are a clade, universally ascribed to the formal taxon Aves. Mammalia and Aves are, in fact, subclades within the grand clade of the Amniota. But the traditional class Reptilia is not a clade. It is just a section of the clade Amniota: the section that is left after the Mammalia and the Aves have been hived off. It cannot be defined by synapomorphies, as is the proper way. It is instead defined by a combination of the features it has and the features it lacks: reptiles are the amniotes that lack fur or feathers. At best, the cladists suggest, we could say that the traditional Reptilia are 'non-avian, non-mammalian amniotes'. The traditional Reptilia is, in fact, paraphyletic: a clade from which subclades have been removed.

There is more than mere nomenclature at stake. Cladists suggest that if we retain the traditional class Reptilia we severely undermine the attempt to base a taxonomy on phylogeny. For the members of a taxon that is truly defined by phylogeny should be more closely related to each other than they are to the members of any other taxon. Lions are more closely related to tigers than either is to dogs, and so we place lions and tigers in the family Felidae and dogs alongside foxes and jackals in the Canidae. Indeed, you do not have to be a cladist to accept that, as a general principle, a taxon is a nonsense unless the creatures within it are more closely related to its fellows within the taxon, than to other creatures outside the taxon.

But consider birds and dinosaurs. From the time of Thomas Henry Huxley (1825–95), many biologists have believed that birds evolved from dinosaurs.[1] Among living reptiles, crocodiles are the closest relatives of dinosaurs. So crocodiles are the closest living relatives of birds—and, indeed, they are much closer to birds than they are to, say, tortoises or even to lizards. Yet traditional taxonomy places turtles, lizards, and crocodiles in the same class, Reptilia, while putting birds in a separate class, Aves. If our classification really is to be based on phylogeny then this seems a nonsense. Wiley and colleagues' point, quoted above, seems to be perfectly justified.

But the cladists' insistence that all formal taxa must be monophyletic, although logical, raises infelicities of its own. Many vertebrate palaeontologists believe that all dinosaurs descended from a common ancestor that was itself a dinosaur, and so we can recognize a clade called Dinosauria. But if birds truly descended from dinosaurs then, according to cladistic nomenclature, they must be included within the formal taxon Dinosauria because they form part of the dinosaur clade. So many modern biologists find themselves obliged to comment that birds are dinosaurs, which they are, phylogenetically speaking. But how, then, should we collectively refer to the

[1] Although some biologists, such as Alan Feduccia of the University of North Carolina, continue to argue that this is not the case.

dinosaurs that we all recognize as dinosaurs, such as *Tyrannosaurus* and *Diplodocus*, when we do not want to refer also to modern birds? In the scientific literature these days the expression 'non-avian dinosaurs' is common. This, many would say, is a contortion: we were better off in the days when dinosaurs were dinosaurs and birds were birds.

For the fact remains that although 'reptile' is a somewhat vague term, hard to define in formal terms, we nonetheless all know what it is. Grades do not express phylogeny, but they express something real and useful nonetheless. Even cladists, when they are not engaged in formal problems of taxonomy—when, in fact, they are just being naturalists—talk about 'reptiles'. I have never heard anyone in conversation refer to 'non-avian, non-mammalian, amniotes'.

So: should we be good cladists, and throw out the formal taxon Reptilia? Or should we go along with tradition, acknowledging that Reptilia is not a clade but following the instinct that, nonetheless, it is a useful grouping? Well, it would certainly be confusing to jumble monophyletic groups and paraphyletic groups without a by your leave. We might get away with it in this present instance, because we all feel we know what reptiles are, and are confident that mammals and birds arose from among the reptilian ranks. But if we were dealing with more recondite creatures such a mixing of categories would be hopeless.

Yet there is a simple compromise, which many other authors have already adopted. That is to retain traditional paraphyletic groups when there is a strong case for doing so, but to signal the fact that they are paraphyletic by some typographic adornment. Paraphyly is commonly indicated by adding an asterisk. So the traditional Reptilia becomes Reptilia*. By the same token, the traditional class Amphibia becomes Amphibia*, because some ancient amphibian or other gave rise to all the amniotes; and the phylum Crustacea becomes Crustacea*, because it may have given rise to the insects and myriapods (centipedes and millipedes). If we believe, as some (but not all) zoologists do, that myriapods gave rise to insects, then they should be called Myriapoda*.

Such a system raises problems of its own, as cladists point out. In particular, if we recognize a formal taxon Reptilia*, then it becomes difficult to decide whether fossils seemingly on the evolutionary path that led from reptiles to mammals should be classed as Reptilia* or as Mammalia. I would simply answer that Reptilia* is a term designed for user-friendliness. It is intended to provide easy access to the information (a piece of librarianship) but it is not intended to unravel the deep problems of specialists. As suggested earlier, specialists may have to evolve their own languages for their own purposes. In the fullness of time, someone might demonstrate that we can abandon paraphyletic groupings without making it more difficult to convey information, and then the paraphyletic groups really can go the way of all flesh. But in this book, at the present time, it seems appropriate to retain a few well-honed paraphyletic groups and to asterisk them appropriately.

Note, finally in passing, that by this convention Reptilia without an asterisk is synonymous with Amniota, and includes birds and mammals, whereas Reptilia* means non-avian, non-mammalian amniotes. Let me offer one last example where it seems sensible to retain paraphyletic grouping. The mighty invertebrates called eurypterids crop up throughout this volume. They were magnificent creatures, up to 2 metres long, that loosely resembled scorpions—indeed most biologists assume that they were closely related to scorpions although, as you will find in Section 11 in Part II, some authorities feel that the similarities are due simply to symplesiomorphies, and that there is no special relationship. The fossil record of eurypterids stretches from almost 500 million years ago (deep in the Ordovician period) until about 250 million years ago (which is well into the Permian). Whether or not the eurypterids are closely related to scorpions, palaeontologists generally accept that something over 400 million years ago some eurypterid or other gave rise to the arachnids—the group that includes scorpions, spiders, mites, and many others. In other words, arachnids sprang from the ranks of the eurypterids just as birds emerged from among the dinosaurs.

Although the fossil record cannot be relied on absolutely, it offers a very fair bet that eurypterids existed long before the first arachnids appeared—possibly for about 100 million years before—and that they persisted for at least another 150 million years after arachnids had appeared. The eurypterids were a huge and varied group. There must have been many thousands of species of which we almost certainly know only a small fraction. But biologists generally agree that the taxon Arachnida (conventionally recognized as a class or subclass) is monophyletic, meaning that all of its members arose from just one ancestral species of eurypterid. All the other eurypterids went on being eurypterids until they went extinct.

Traditional biologists have no problems in naming the eurypterids. They simply place the creatures that they feel have eurypterid features in the formal class Eurypterida, and creatures that have the synapomorphies of arachnids are placed formally in the class Arachnida. But if we want to maintain that the term Eurypterida refers to the lineage of creatures that lasted from about 500 to 250 million years ago, and budded off the Arachnida along the way, then, clearly, we have to acknowledge that Eurypterida is a paraphyletic grouping. If we want to define a clade that includes the Eurypterida, then that clade would have to include the Arachnida. So, in referring to eurypterids as opposed to arachnids, we would have to say 'non-arachnid eurypterids'. We are back to the infelicity of 'non-avian dinosaurs'. The cladistic solution to this is to break up the Eurypterida into its subgroups, each of which would be monophyletic. This can be done—although the term Eurypterida, in its traditional formal sense, would have to go.

This brings us back to the difference between specialists and non-specialists, the purpose of language, and the desirability of communication. It is good that zoologists who are not invertebrate specialists recognize the existence of eurypterids at all; and excellent that non-zoologists should do so. Eurypterids were supremely significant creatures but life is too short for most of us to learn eurypterid phylogeny in detail, with all the names that would have to be appended if we slavishly followed cladistic nomenclature. For normal discourse with non-specialists it seems reasonable to call the Eurypterida the Eurypterida*: maintaining the traditional meaning, but signalling the paraphyletic status of the group. What could be simpler?

POLYPHYLY AND INFORMALITY

Even non-cladists would generally prefer to banish polyphyletic groups from formal taxonomy. Such groups represent ignorance of true phylogeny, or at best serve as a pending file that needs to be sorted out when somebody has the time—and suitable techniques as well. Groups such as 'Pachydermata', which were supposed to include all thick-skinned mammals—elephants, hippos, and rhinos—are clearly a nonsense, albeit a pleasant reminder of more amateurish and leisurely days. But the issue of polyphyly does hold some interesting twists. In particular, the margins between polyphyly and paraphyly can become blurred.

For instance, the traditional grouping 'Cyclostomata' ('circle mouths') includes the living hagfishes and lampreys. Hagfishes and lampreys are both naked, superficially eel-like creatures that lack jaws; indeed they are the only living jawless vertebrates. Vertebrates that have jaws, like sharks, codfish, dinosaurs, and human beings, descended from vertebrates without jaws. All the jawed vertebrates together form a clade called the Gnathostomata ('jawed mouths'). If we consider only the living vertebrates, then hagfishes and lampreys taken together form a paraphyletic group: sole representatives of the group from which the Gnathostomata evolved. In the system adopted in this book, then, we could represent them together as the Cyclostomata*.

But if we consider fossils as well as living groups, then the picture changes. For the extinct jawless vertebrates include a whole range of armoured 'fishes' that clearly belonged to several different lineages but were collectively called 'ostracoderms' ('bony skin'). Ostracoderms, together with hagfishes and lampreys, form a group traditionally called the Agnatha ('without jaws') but in the system adopted here should be called Agnatha* (because the gnathostomes descended from them). Hagfishes, however, seem to be the most primitive of all the known agnathans, whereas lampreys seem to be closely related to gnathostomes. So, on the phylogenetic tree, several groups of ostracoderms seem to separate hagfishes from lampreys. If, therefore, we take the extinct agnathans (ostracoderms) into account, cyclostomes are polyphyletic. Hence, if we include fossil forms, we can see that hagfishes and lampreys have been plucked from separate lineages, and bundled together simply because they share both homoplasies (including eel-like bodies) and symplesiomorphies (such as jawlessness).

Biologists often want to refer to groups of creatures that are not closely related but have qualities of various kinds in common. Then they can simply apply informal names, which are spelt with a lowercase initial letter. Thus zoologists now recognize that the single-celled creatures that were once dumped into the alleged phylum Protozoa in fact belong to many different groups—and represent many different kingdoms. But it is still convenient to refer to single-celled creatures with animal-like qualities informally as 'protozoans'. 'Algae', too, were once acknowledged as a division (a zoologist would say 'phylum') of plants but although some organisms of algal grade are plants, some (such as the brown and red seaweeds) emphatically are not. Nonetheless, 'algae' remains a useful informal term. Protozoans and single-celled algae together can informally be called 'protists'. It is reasonable to speak of 'worms' although worms hail from many taxa. 'Fish' is a useful colloquial term that embraces creatures from several discrete lineages. Coelacanths are 'fish', but they are phylogenetically closer to human beings than they are to skate, which are also 'fish'. 'Microbes' is a perfectly good informal term meaning any creature too small to see without a microscope. Even 'cyclostome' is a good informal term. You never know when you might want to refer to hagfishes and lampreys together. In fact, any group of creatures can be grouped informally in any way you like. That is the advantage of informality. Of course, any formal taxon that is normally spelt with an initial capital, as in Mammalia or Insecta, can also be referred to informally, if de-Latinized and spelt with a lowercase initial, as in 'mammals' or 'insects'.

Finally, although polyphyletic groupings should have no place in formal classification, there is sometimes a case for painting them on the phylogenetic tree informally. The ancient jawed fishes called palaeonisciforms are a case in point. Clearly they are a complex group, with many lineages, and in their day they were ecologically extremely important. They also seem to include the ancestors of the modern sturgeons and paddlefish. The palaeonisciforms demand to be studied, but the details are for specialists. The rest of us can simply acknowledge that the palaeonisciforms existed, and in their time were various and ecologically important, and have given rise to living descendants. It seems reasonable (and is accepted practice by cladists and non-cladists alike) just to place them and similar groups into phylogenetic trees and classifications informally, spelt with a lowercase initial. If we regard classification as an exercise in information retrieval, then we can think of such groups as pending files.

The cladists' insistence that all taxa should be monophyletic is not their only clash with tradition. Following Willi Hennig they also suggest that each branching point (node) in the phylogenetic tree should be taken to signal a new rank. This raises more problems.

HOW MANY RANKS?

The notion that each new node should be taken to represent a new rank is perfectly logical, and it looks at first sight as if this cladistic ruling should chime very neatly with the traditional Linnaean approach. Both systems are innately hierarchical. The twigs of the phylogenetic tree represent species, which nest within branchlets, which represent genera, which in turn nest in larger branches, which are the families, and so on all the way up to the mighty boughs—the domains. In practice, however, cladistic and Linnaean nomenclature do not fit together anything like so tidily as we might at first expect.

The first difficulty is that of numbers. As we have seen, Linnaeus originally proposed only a few rankings, including species, genus, order, class, and kingdom. Later taxonomists raised this to seven: species, genus, family, order, class, phylum, and kingdom. In the 1970s, Carl Woese gave us the eighth, right at the top: the domain.

Because the eight-rank system clearly extends Linnaeus's original plan it seems reasonable to call it 'Neolinnaean'. Eight ranks seems a comfortable number, which gives rise to a comfortable suite of names: class Mammalia, order Perissodactyla, family Equidae, and so on. Yet the number of significant nodes between the trunk of the tree of life and the species at the twigs can easily run to 30 or more—and in reality it could run into hundreds. It is very difficult to accommodate 30 (or more) nesting clades into eight recognized Neolinnaean rankings. Attempts to fit hogsheads into pint pots are innately procrustean.

In practice, Neolinnaeans long since invented the prefixes 'sub' and 'super' to extend the number of rankings, as in superorder and suborder, superfamily and subfamily. This in theory increases the number of standard ranks to 24, although fewer than this in practice because 'superdomain' would be a nonsense, and some of us feel that 'superkingdom' and 'superphylum' are at least infelicitous (because 'kingdom' and 'phylum' are meant to have a superlative quality); and 'subgenus' does not sit too happily, although it turns up from time to time.

By standard and generally inoffensive Neolinnaean means we can get up to about 20 ranks, but when taxonomists feel the need to go beyond 20 they have often felt obliged to invent intercalary rankings. Thus the large mammalian family Bovidae (cattle, sheep, antelopes, and so on) is commonly divided first into subfamilies and then into tribes before finally fragmenting into genera. Charles Sibley and Jon Ahlquist, in their classic book *Phylogeny and Classification of Birds* (1990), divide various bird orders into suborders, infraorders, and then parvorders before they reach superfamily, which then, of course, must be split several times more. 'Microclass' has even been coined. But once we move beyond about 20 subdivisions (and, preferably, far fewer) the classification becomes bewildering; and we may reasonably ask whether a classification that bewilders is really a classification at all.

The number of ranks is only part of the problem. More serious is that different parts of the tree seem to lend themselves to different degrees of subdivision; and that different taxonomists, with different ambitions and interests, can easily finish up producing classifications that seem seriously at odds. They may all recognize the same groupings—and happily admit, for example, that mammals are warm-blooded creatures with fur that suckle their young. But, because of their perceived need to give each nesting clade its own name and rank, they may finish up ascribing vastly different ranks to the same grouping.

Thus all the mammal specialists that I know, and everyone else who takes any kind of interest in natural history, thinks of the mammals as a class. If Mammalia is not a class, then what is? Consider the excellent book by John Long from the Western Australian Museum at Perth called *The Rise of Fishes* (1995).[2] The author sets out a classification for the entire clade of 'fishes'—including not only those creatures that we colloquially call 'fish', but also the four-legged creatures, the tetrapods, which descended from fish and thus must be included in the fishy clade. He begins the

[2] Although I am criticizing the classification offered in *The Rise of Fishes*, this is an excellent book, which I recommend unreservedly. Except for this aspect of the classification.

whole series with the jawless fishes, the Agnatha, which he ranks as a 'superphylum'. This seems to give plenty of leeway but is itself a bit suspect because most zoologists like to credit the group Chordata with the rank of phylum; yet Chordata is clearly a much higher ranking than Agnatha. We might argue, too, as above, that 'superphylum' is not felicitous. But superphylum is what Long suggests.

Traditional zoologists took 'Agnatha' to include only the jawless fishes. After all, the term 'Agnatha' means 'without jaws'. But agnathans are the ancestors of the jawed vertebrates, the Gnathostomata. In *The Rise of Fishes*, Long gives his superphylum Agnatha monophyletic status, in true cladistic style, and it is therefore taken to include all the jawed vertebrates. So if we take the meanings of the terms at face value, then you and I and Rottweilers are jawless. Be that as it may, in Long's classification the superphylum Agnatha is further broken down into the infraphylum Vertebrata and then into the subphylum Gnathostomata.

Then the subphylum Gnathostomata is broken into four great classes: Chondrichthyes, the cartilaginous fish, including sharks and rays; the extinct Placodermi and Acanthodii—two more great fish groups; and, finally, the Osteichthyes, or bony fish. The bony fish in turn include the modern ray-finned fish, the Actinopterygii; the lungfish, or Dipnoi; and the lobefins, or Crossopterygii. Because the Osteichthyes are considered a 'Class', these three huge subgroups must be given the rank of 'subclass'.

The four-legged vertebrates—tetrapods—descended from the lobefins. So we can identify a clade Tetrapoda that nests within the larger clade, Crossopterygii. But what rank should we ascribe to the Tetrapoda? John Long ranks the Crossopterygii as a subclass. We have now run out of standard Neolinnaean ranks. So Tetrapoda must be given some arbitrary ranking, like 'division'.[3] If the classification retains its cladistic purity, then the Tetrapoda become synonymous with the Amphibia, because cladists would not admit Amphibia as a valid taxon unless it is taken to include reptiles, birds, and mammals. Within the Tetrapoda (or Amphibia) we would now need to recognize the subgroup of Amniota—which in cladistic parlance becomes synonymous with the Reptilia, which in cladistic convention must be taken to include the birds and mammals. Amniota (or Reptilia) would then have to be ranked as an 'infradivision', or some such. Reptilia would then have to be broken into subgroups (parvodivisions?), one of which would be the subgroup known as the Synapsida, from which the mammals descended. The Mammalia would then be a subgroup of the Synapsida, presumably with the rank of subparvodivision, or perhaps parvodivisionette. One cannot help but feel that we were all better off in the innocent days when the Mammalia was simply a class.

In short, cladistics and Linnaean classification (or Neolinnaean, as we might call it in its eight-ranked form) have not proved quite so compatible as might first have seemed the case. Some modern systematists feel that Neolinnaean classification should be abandoned all together. The only representation worth having, these radicals suggest, is the phylogenetic tree itself. Particular creatures would be classed

[3] Botanists traditionally use 'division' as a formal ranking, more or less equivalent to the zoologist's 'phylum'. But zoologists use 'division' for ad hoc purposes.

simply by identifying their positions on the tree. Classification would thus become an exercise in coordinate geometry.

But this cannot be right. Specialists might find such a system useful but if they adopted it they would thereby abandon any possibility of communicating with non-specialists. That would be deeply regrettable, not to say cultural suicide, and possibly economic suicide in these days when everyone, even specialist scientists, are supposed to be 'accountable'. It would also run into a nonsense that has been identified by Peter Raven, the director of the Missouri Botanical Garden at St Louis: that the more expert taxonomists become in any particular group of creatures, the less they actually need a classification. People have most need of classifications when they are seeking to get to grips with creatures they do not know well.

'Neolinnaean Impressionism' is my own proposed solution to all these problems. The system is 'Neolinnaean' and not simply 'Linnaean', because it extends Linnaeus's ideas in two main ways. First, it increases the number of formal ranks to eight. Second, it distinguishes between monophyletic and paraphyletic groups, and signals the latter with an asterisk. It also admits polyphyletic groups in some circumstances (for example, the palaeonisciform fishes), although these are not acknowledged as formal ranks. A few of the more felicitous intercalary rankings may be included, such as 'subclass', or 'tribe', or even 'cohort', but these remain informal.

The system is 'impressionistic' precisely because it is pragmatic, rather than strictly logical. In particular, the eight Neolinnaean ranks are taken to be sacrosanct, to provide fixed points in the classification. Classifications must be stable if they are truly to be useful, and the eight formal ranks provide the stability. But intercalary rankings can be introduced whenever it seems desirable to do so. Sometimes it might seem desirable to give these intermediate rankings some semi-formal status, such as subclass or superorder, cohort, or tribe. But, I suggest, we need not feel obliged to nominate the status of these intermediate groupings at all. Thus within the phylum Arthropoda it seems reasonable to link the Crustacea, Insecta, and Myriapoda within a clade called Mandibulata. The 'mandib' root implies jaws, which are common to all three groups but not to chelicerates or trilobites. Similarly, all the four-legged vertebrates form a clade that is commonly called Tetrapoda. But I feel no great urgency to ascribe any formal ranking to the Mandibulata or the Tetrapoda. We can see that each of them fits somewhere between the formal ranks of phylum and class.

In fact, many taxonomists with varying philosophies have also suggested that intercalary groupings can be left without formal ranking, at least sometimes. In similar vein the late Colin Patterson of London's Natural History Museum and D. E. Rosen point out that any new fossil that is ascribed to a new clade must—according to cladistic convention—be given a new ranking; and that each new group may thereby affect the ranking of all the others. Confusion would quickly reign, created partly by the ever-increasing complexity and partly by the loss of stability. It is better, said Patterson, simply to leave these new groupings without formal ranking, and just call them all 'plesions'. Patterson may not agree but the pragmatism thus implied seems to me to be in the spirit of Neolinnaean Impressionism.

PART II

A SURVEY OF ALL LIVING CREATURES

HOW TO USE THIS BOOK

Part II is the *raison d'être* of the book: a survey of all living things. As you see, it is divided into 25 sections, each of which treats one group (or 'taxon') of creatures. Between them they cover virtually all the main groups of creatures that have ever lived on Earth: the omissions include some protist groups that have not yet been satisfactorily classified. But because it is not possible nor in the end desirable to give equal weight to each group, I have treated the different groups in greater or lesser detail.

Overall, then, Part II has the general plan of a school atlas. Every school atlas has a map of Asia, but only the ones intended for Asian students contain detailed maps of, say, Hong Kong or Tamil Nadu. Those for use in the USA would present just one map of Asia, but might perhaps home in on Wisconsin or even on Manhatten; while a British school atlas would typically offer just one map of North America but focus on, say, Buckinghamshire. In the same way, I treat all of the animals in one section and then focus on a few subgroups.

For example, phylum Mollusca is given a whole section to itself (7) while the Chaetognatha—quaint creatures though they doubtless are—are simply given a paragraph in the animals section (5). Phylum Chordata, by contrast, occupies nine sections in all (13–22), which some might feel is over-egging the pudding. Again, although I do not treat the extinct class Placodermi in great detail, I do discuss all the living chordate classes at length; and within the class Mammalia (18) I home in first on the order of the Primates (19) and then on our own family, the Hominidae (20). Similarly the kingdom of the Plantae has its own section (23) and I then home in on the Angiospermae or flowering plants (24) and end Part II with a look at the largest of the angiosperm families, the Compositae (25).

Each of the 25 sections is dominated by a 'tree'. These trees in general concept resemble trees of human families: the presumed ancestors are to the left and the descendant groups are to the right. Living groups, or extinct groups that have simply expired without issue (like, say, the extinct plants known as the bennettitaleans), are called 'crown' groups: and most crown groups are accompanied by an illustration in the tree, and illustrated again in the text within or near the appropriate paragraph.

The text in each of the 25 sections is separated into two parts by a strong heading that says A GUIDE TO (the group in question). The first part is a general introduction to the particular group; and the second describes the different members of the group in more detail. These descriptions are linked to the trees: mostly they follow the appropriate tree from top to bottom.

In Section 1, titled 'From two kingdoms to three domains', the tree (page 102) serves two functions. First, it shows the entire tree of living (and extinct) creatures, following Charles Darwin's suggestion that all creatures on Earth did indeed descend from a common ancestor, so that they all share the same tree. Second, as in all good atlases, this first tree provides a key to all the other sections in the book. Thus, by glancing at this first tree you can see exactly where, say, the phylum Chelicerata or the family Compositae fit in the grand scheme of things.

You will also perceive some inconsistency in the way the different groups are labelled, both in the text and in the trees. Thus some groups are named in formal scientific 'Latin', as in Ascomycota, which is one of the principle groupings of the Fungi; whereas others are presented in vernacular English, as in 'true yeasts', which are a subgroup of the Ascomycota. There are other adornments on some of the names, notably daggers (†) and asterisks (*). The reasons for at least some of these peculiarities are somewhat technical, as I explain below.

CONVENTIONS FOR NAMING TAXA

In accord with the principles of nomenclature outlined in Part I, groups in this book are named to the following convention:

- *A taxon name beginning with a capital letter and with no typological adornments,* as in Mammalia, refers to a monophyletic group, or clade. Reptilia without qualifying marks must also include mammals and birds; Crustacea without qualifiying marks must include insects and myriapods as well. But in this book I do not, in fact, use the terms Reptilia or Crustacea in this sense. Reptilia is synonymous with Amniota; and if we accept that crustaceans gave rise to insects and myriapods, then Crustacea would be synonymous with Mandibulata.

- *A taxon name beginning with a capital letter and qualified by an asterisk,* as in Reptilia*, implies a paraphyletic grouping, or 'stem' group. Reptilia* is what most people mean by 'reptiles': Reptilia without Mammalia and Aves, alias 'non-mammalian, non-avian amniotes'.

- *A dagger sign appended to a taxon name,* as in Placodermi†, shows that the group is extinct.

- *Informal names such as protist or worm are spelt with a lowercase initial letter* (except at the beginning of a sentence) and with no attempt at latinization.

- *Polyphyletic groups are regarded as informal groups, and are spelt with a lowercase initial letter,* and in the vernacular.

- *Formal taxa can also be referred to informally by spelling their names with a lowercase initial letter* and removing the Latin endings, as in mammals or actinopterygians.

I

FROM TWO KINGDOMS TO THREE DOMAINS

BIOLOGISTS AND NON-BIOLOGISTS alike have often implied that classification has no intellectual content: that it is simply a glorified exercise in filing. How shocking! In truth, when we look at the development of taxonomy over the past 400 years, since the first serious probings in the sixteenth century, we see a shift in overall perception of living things and of our place among them that is as profound as the sea change in cosmology that began with Nicolaus Copernicus. Before Copernicus, cosmologists assumed that Earth was the centre of the Universe—for this idea came from the classical astronomers, and conformed with the Christian view of the relationship between Man and God. But in the early 1500s Copernicus suggested that the Sun, not Earth, is 'fixed': that the planet is subordinate to the star. Galileo in the early seventeenth century extended the idea and by the late twentieth century cosmologists conceived of Earth as an ordinary planet orbiting a mediocre star near the edge of just one among billions of galaxies. As the geocentric Universe faded away so, too, at least in its primordial form, did the anthropocentric Universe.

Conceptually, biology generally trails behind physics. To be sure, Darwin's idea of evolution has overtaken the eighteenth-century, anti-Enlightenment conceit of special creation; but well into the twentieth century many biologists instinctively felt that human beings represent the culmination of evolution, and that our species is not simply the centre of evolutionary unfolding but is in effect the point of it: indeed, to a large extent, biologists merely substituted the word 'evolution' for 'creation' in an otherwise traditional account of how life came into being. But in the late twentieth century we can see that our contribution to the genealogical tree of life is as peripheral and minuscule as that of Earth to the Universe. The tree as we see it now is truly vast. Since life first began on Earth, it has probably produced hundreds or even thousands of billions of twigs, where each twig is a species; and *Homo sapiens* is just one of them. Furthermore, and more significantly, the tree has at least three great boughs, and each bough has many branches, and our twig is part of just one of them. In short, our species has been as comprehensively peripheralized by biology as it has by cosmology; and the biological discipline that has brought about this shift is that of systematics, which I take to be classification based on phylogeny. By counting what is out there and asking who is related to whom, modern taxonomists, inspired by evolutionary theory, have drawn a tree of life on Earth that is as vast and intricate as the modern cosmologists' view of the entire Universe, and should be as humbling. If that is not an intellectual contribution it is hard to see what is.

As outlined in the introductory chapters, among the many conceptual and practical advances in taxonomy over the past four centuries, three have been outstanding. The first great leap into theoretical modernity was made by Linnaeus, with his all-embracing nesting hierarchy of taxa and his binomial system for naming species. The second was by Darwin, who proposed that all living things could be seen as members of the same vast genealogical (or phylogenetic) tree, springing from a single trunk. And the third has been by Willi Hennig, who introduced the rigour of cladistics, enabling us to handle the swelling data logically, in the way that is most likely to provide the most accurate phylogenetic tree.

But running in parallel with these theoretical advances—truly in parallel, for the two exercises have been quite different—there has been an empirical, a factual advance. Thus the lists of known species have grown from a few thousand in the eighteenth century to the 1.7 million or so known today—with the realization that there could be at least another 30 million to follow and probably many more, once we get to grips with the microbes. But also, and in this context more importantly, biologists have slowly come to appreciate that the sheer range of creatures is far greater than had ever been envisaged in the past. As a result, biologists have come to realize that the features—characters—that strike us most immediately, and seem so obviously to define the various categories of living things, are often of little fundamental significance. Plants, animals, and fungi, for example—what could be more different? And yet, so modern insights suggest, the differences between these three are trivial compared with the differences between them and other living things; and especially when compared with the differences to be found between those other living things.

All of biology over the past 400 years has contributed to this increasing appreciation of life's variety. But in retrospect at least we can distinguish about half a dozen sets of insights, of quite different kinds and made in quite different contexts, that have contributed especially to our modern view of biodiversity, and of how living things should be classified. The history is, frankly, a mess: showing once more that ideas in science do not come on stream in a logical order as is sometimes supposed, edging towards some inevitable conclusion, but pop up here and there and are sometimes overlooked for decades and sometimes for centuries, sometimes suffering abuse along the way, before their threads are finally pulled together. So if the following historical account seems untidy—well, that's just the way history is.

INSIGHTS INTO LIFE'S VARIETY

Once more, we can reasonably start our overview of taxonomic history with Linnaeus. He divided all living things into two great kingdoms: animals and plants. Such a division is intuitively obvious: animals are conspicuous and mobile, and plants are conspicuous, green, and immobile; and Linnaeus, it seems, was happy to conform to common observation. He included fungi among the plants and we might reasonably ask why he did not apparently consider them different enough to be given their own kingdom because they are nothing like plants, except in their immobility. But he did not. His two-way split was rough and ready, but it was a start; and it gave us the concept of the **kingdom**, which has retained its significance.

Another hugely significant step had been taken in the century before Linnaeus, by Anton van Leeuwenhoek, the Dutchman who appeared in Part I as one of the first great microscopists. He worked exclusively with simple microscopes (single lenses held between metal plates) but nonetheless submitted 165 letters on microscopical matters to Britain's Royal Society between 1672 and 1723. He had no Latin but the Royal Society thought so much of his work that they were content to translate his observations from the Dutch. Leeuwenhoek at first looked at the details of largish organisms: bees, moulds, and lice. But then in 1674, in standing rainwater, he discovered what he called 'little animalcules': probably protozoans. Later, he also evidently discovered bacteria; not least within his own stools, although, he said, he found 'animalcules' in them only when 'the stuff was a bit looser than ordinary'.

Thus, did he but know it, Leeuwenhoek discovered two vast categories of creature, which, we can say with twenty-first-century hindsight, clearly do not fit into Linnaeus's two categories. Perhaps Linnaeus's own attitude towards them would have been different if he himself had been a microscopist; but—perversely for his time—he preferred the naked eye. Whatever the reason, in Linnaeus's time until well into the twentieth century Leeuwenhoek's animalcules, in all their extraordinary variety, have generally been rammed into the Linnaean kingdoms of Animalia and Plantae. Yet they thrust themselves into people's consciousness. Even in Leeuwenhoek's own time Peter the Great, no less, travelled to Delft to see the animalcules, and at least since the time of Louis Pasteur we have come to appreciate more and more their roles in the making of wine and pickles, in causing disease, and indeed in the world's ecology in general. Yet only in the past few years has their true phylogenetic status been appreciated. Leeuwenhoek's animalcules have now proved to be much more various than plants and animals, and in many cases vastly more different from plants and animals than plants and animals are from each other.

A third group of insights that have proved to be of crucial significance in taxonomy was made in the early decades of the nineteenth century, again by microscopists. First, the German botanist Matthias Schleiden (1804–81) reported in 1838 that the tissues of plants are constructed from individual cells, as a house is constructed from bricks; and that each cell has its own nucleus. In the following year another German, Theodor Schwann (1810–82), announced that the same is true of animal tissues. Then, in 1841, the Polish–German embryologist Robert Remak (1815–65) described cell division. Now we know that the nucleus contains (most of) the genetic material of the cell—that is, the DNA; and that this DNA becomes visible during cell division as it bunches together to form chromosomes. It is now clear, however, that although animals and plants have bodies constructed from a multitude of nucleated cells, many other organisms are constructed quite differently. In modern taxonomy, differences in cell structure, and to a lesser extent in cell number, are matters of pivotal significance.

But the importance or indeed the relevance of Schleiden and Schwann's work to taxonomy, or of Leeuwenhoek's for that matter, was not fully appreciated until the past few decades. As Lynn Margulis and Karlene Schwartz describe in *Five Kingdoms* (1988), most biologists until deep into the twentieth century were content simply to follow Linnaeus, and to divide living things into 'Animalia' and 'Plantae'. To be sure, since the mid-nineteenth century some systematists have realized that some organisms (including Leeuwenhoek's animalcules) are very different from plants and animals, and have sometimes proposed third or even fourth kingdoms to accommodate them. But, commented Margulis and Schwartz, 'most biologists either ignored these proposals or considered them unimportant curiosities, the special pleading of eccentrics'.

Yet these 'eccentrics' included some of the world's great biologists, not least the German Darwinist Ernst Haeckel (1834–1919), who proposed a third kingdom to include what he called the 'Protista'. He shifted the boundaries of the 'Protista' from time to time, but, in general, his new kingdom largely coincided with Leeuwenhoek's conception of 'animalcule' and with the modern concept of 'microbe': single-celled organisms of all types. Thus Haeckel's 'Protista' included protozoans (like *Amoeba*) plus a group that he called the Monera. His Monera included the bacteria and what in those days were called 'blue-green algae'—although they are really photosynthetic bacteria and should properly be called Cyanobacteria (where *cyano* means blue). Oddly (or at least it seems odd to me) Haeckel acknowledged that his Monera were profoundly different from protozoans, in that the latter have a nucleus in their cells and the former do not. Modern biologists consider the presence or absence of a nucleus to be one of the most significant divisions within all living things; but Haeckel quite consciously put those without a nucleus, and those with, in the same taxon. Thus his kingdom Protista is both polyphyletic and paraphyletic, because we must assume (as Haeckel did) that the nucleated organisms arose from some of the non-nucleated types.

That the distinction between nucleated and non-nucleated is of prime significance does not seem to have dawned fully until after the 1930s: at least, in 1937 a French marine biologist called Edouard Chatton wrote a somewhat obscure paper in which he suggested that organisms whose cells lack nuclei should be called *procariotique*, and organisms whose cells are nucleated should be called *eucaryotique* (where *pro* means 'before', *eu* means 'good' or 'true', and *karyon* means kernel, or indeed nucleus). Nowadays biologists distinguish very clearly between prokaryotes[1] and eukaryotes.

[1] Americans speak of 'procaryotes' and 'eucaryotes', and formally of 'Procaryota' and 'Eucarya'. The British prefer a 'k', as in prokaryote and eukaryote, perhaps to acknowledge that the original root is Greek and not Latin. I suspect that the 'c' will prevail in the long term but I was brought up with 'k' and continue to use it in this book—except in the context of Eucarya, which is the formal name proposed by Carl Woese for the domain that includes all eukaryotes. Formal names should not be subject to local variation.

Further advance in formal taxonomy came with the American biologist Herbert F. Copeland (1902–68) of Sacramento City College, California, who split Haeckel's Monera into two. Copeland still retained the term Monera, but used it now only to refer to the prokaryotes—that is, bacteria in the traditional sense—while he placed the eukaryotic microbes, plus various algae like the seaweeds, in a new kingdom, the Protoctista. Then, concluding this phase of taxonomic development, in the 1950s the American Robert H. Whittaker of Cornell University proposed his five-kingdom system: Animalia, Plantae, Fungi, Protoctista, and Monera. This (with some modifications) is the system that Margulis and Schwartz adopted in their *Five Kingdoms*, and is indeed still widely promulgated.

Beyond doubt, Whittaker's five-kingdom system has taken us several conceptual stages beyond Linnaeus's two kingdoms. First, at a general level, it breaks free of the idea that there is something sacrosanct about the concept of 'kingdom'; and often in science the greatest problem of all is to break away from some deep-rooted preconception, as Darwin discovered when he suggested that species boundaries are not infrangible. Second, it acknowledges once and for all that fungi are not to be rammed in with the plants: that they are profoundly different organisms. Third, by the same token, it rescues protozoans from their former, uneasy categorization among animals, and some of the seaweeds and other 'algae' from their forced association with plants. Finally, and very significantly, it acknowledges the specialness of prokaryotes (even though Whittaker afforded them only one kingdom, against four eukaryote kingdoms). All in all, it seems to provide a satisfying overview.

Yet Whittaker's five-kingdom system should now be seen only as the culmination of phase III taxonomy (as outlined in Part I, Chapter 2). It does not incorporate the data now provided by molecular biology for the simple reason that it was first proposed at least 15 years before such data came seriously on to the scene. Neither does it acknowledge the power of cladistics, or the need for cladistics. Thus we can now see that Whittaker's Monera contained two very different groups of organisms, and that his Protoctista is a very mixed bag indeed and is paraphyletic to an unacceptable degree. Whittaker's system is still taught but it really should be regarded now only as a step, albeit an important step, on the road to modernity. The modern picture is very different.

THE MODERN OVERVIEW

On the methodological front, the most profound advance since Whittaker's five kingdoms is Willi Hennig's cladistics. On the technical front, the advent of molecular methods has been revelatory. On the reality front—how in practice we should classify living things in the light of the new approaches and data—the greatest advance must be associated with the name of Carl Woese, of the University of Illinois. For Woese observed in the 1970s that the molecular differences between organisms that hitherto had all been called 'bacteria' (which Whittaker and others called 'Monera') were profound; that they should be divided into quite distinct groups that Woese at that time called the Archaebacteria and the Eubacteria (*archae* meaning ancient).

Indeed, the Archaebacteria and the Eubacteria were more different from each other than either of them were from the eukaryotes. Woese might simply have added another kingdom to the list of five, but he saw that he needed to be much more radical than this; he perceived that the distinction between the Archaebacteria, the Eubacteria, and the eukaryotes as a whole was far more significant than the differences between any two eukaryote kingdoms. To do justice to phylogenetic reality, said Woese, we need to devise a new ranking, above the level of the Linnaean kingdom. Felicitously, he proposed the rank of **domain**. Thus, he suggested, we should divide all living things into three domains: the Archaebacteria, the Eubacteria, and the Eucarya, which included all the eukaryotic creatures—protists, fungi, plants, and animals. Later he tidied up the names, and the three domains are now widely known as the Archaea (colloquially 'archaes'), the Bacteria, and the Eucarya. Note that Bacteria in the new sense is different from 'Bacteria' in the old sense, for the old 'Bacteria' included the archaes.

So under Linnaeus we had two kingdoms; and no one asked about the state of their cells because the concept of the cell emerged only in the early nineteenth century. Then with Robert Whittaker we had five kingdoms, of which one was prokaryotic and four were eukaryotic. Now, with Woese, we have three domains, of which two are prokaryotic and only one is eukaryotic. The picture has grown deeper, and the balance has clearly shifted.

In the light of emerging molecular evidence, the classification of the two prokaryote groups is being completely rethought. These are early days, and the findings so far are discussed in Section 2. The eukaryotes, too, are being re-thought, not least by Mitch Sogin and his colleagues at the Center for Molecular Biology at the Marine Biological Laboratory, Woods Hole, Massachusetts. Again, growing molecular evidence is showing profound differences between eukaryotes that are not evident from their gross structure, or even from their ultrastructure; and also that organisms that on the face of things seem to be obviously different are actually largely similar, and clearly shared a common ancestor only (relatively) recently. Because eukaryotes occupy most of the rest of this book there seems little point in dwelling on details, but it does seem worthwhile to point out a few salients.

First, the term Protista, which Haeckel proposed to include all single-celled organisms, no longer has status as a formal taxon. But it does seem to be a useful term that can be used informally or indeed colloquially to describe single-celled eukaryotes (or those with just a few cells). Thus the adjective 'protist', with a lower-case initial letter, refers nowadays to a grade of eukaryotes. The term 'microbe' is also worth keeping, again as a purely colloquial term, to refer to any creature that is too small to see with the naked eye—and thus includes the protists and virtually all of the prokaryotes. The term 'Protoctista' seems to serve no purpose. It clearly does not describe a formal taxon that can be defended in the face of cladistic analysis and molecular or ultrastructural data, and seems to have no obvious niche as an informal adjective because it includes both protists and seaweeds, which in general are not protistan. Clearly, it is not a particularly helpful grouping.

Eukaryotes as a whole seem to divide not into two, or three, or four kingdoms, but perhaps into as many as 20; all at least as different from each other as the plants are from the animals, and in some cases a great deal more so. Five of these kingdoms contain a predominance of organisms that are multicellular and large: animals, plants, fungi, brown seaweeds (including kelps), and red seaweeds. Informally, I suggest that these might be alluded to as **mega-eukaryotes**, although the term is not exactly felicitous. The green seaweeds are now accommodated within the plants (or, at least, many leading plant taxonomists now regard green seaweeds as plants; this seems a strong idea to me and so is the one followed in this book). Then there are four kingdoms of slime moulds and slime nets, of which three (the Acrasiomycota, Myxomycota, and labyrinthulids) are outstanding. These are peculiar organisms that are single-celled (protistan) for only part of their lives, but at other times form colonies that are effectively multicellular and macroscopic. The remaining dozen or so eukaryote kingdoms all contain creatures that in grade are predominantly protist.

This rough-and-ready division of eukaryotes into 'mega-eukaryotes', 'slime moulds', and 'protists', has no phylogenetic significance but I do feel it is a useful *aide-mémoire* in an area that otherwise can start to feel too complicated. (This *aide-mémoire* is discussed in more detail in Section 3 on eukaryotes.)

Note, however, that the Plantae, Animalia, and Fungi as now conceived all contain some protists (which in this classification merely means 'single-celled eukaryotes'). Thus Animalia as now conceived includes the Choanoflagellata, which seem to be closely allied to the common ancestor of all the other animals; Plantae includes various groups of single-celled green algae, which again seem close to the plants' common ancestor; and Fungi includes yeasts, which are not primitively single-celled, but have adopted the protistan mode secondarily.

Finally, note in passing that the plants, animals, and fungi, which so long have been considered to represent the great divisions of life, are, in fact, quite closely related to each other. They all shared a common ancestor fairly recently—not much more than a thousand million years ago. The common ancestors that link other groups of eukaryotes, and link plants, animals, and fungi to those other groups, lived much further back than that. Note, too, as a final irony, that fungi now seem to be closer to animals than either group is to plants. Old-fashioned taxonomies that shove fungi in with the plants hence make a double mistake.

The new perception of the three domains has profoundly changed our conception of ourselves. Thus until well into the twentieth century most biologists saw our own class—the Mammalia—as the self-evidently 'leading' group among the Vertebrata, which they saw as the principal phylum among the Animalia; indeed, it was common until recent decades to class all non-vertebrate animals together not simply as 'invertebrates' (which is a fair-enough informal term) but in the putatively formal group, the 'Invertebrata'. The Animalia, in turn, was seen to embrace a half of all living things. Now biologists see that mammals are just one class among a dozen or so of vertebrates, and that vertebrates are just one phylum among 36 or so of animals, and that animals are just one kingdom among a dozen or so of Eucarya, and

1 · *KEY TO PART II*

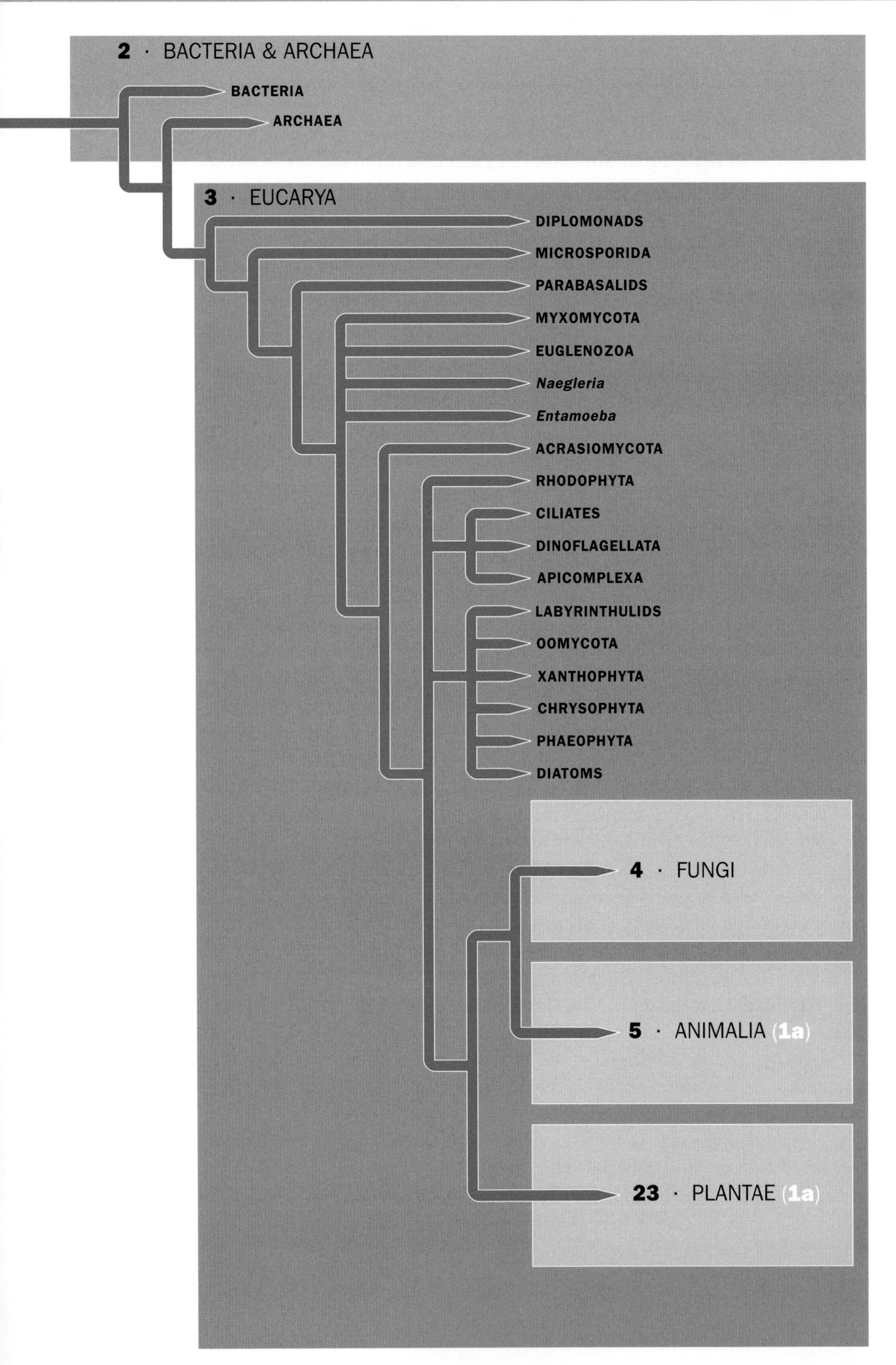

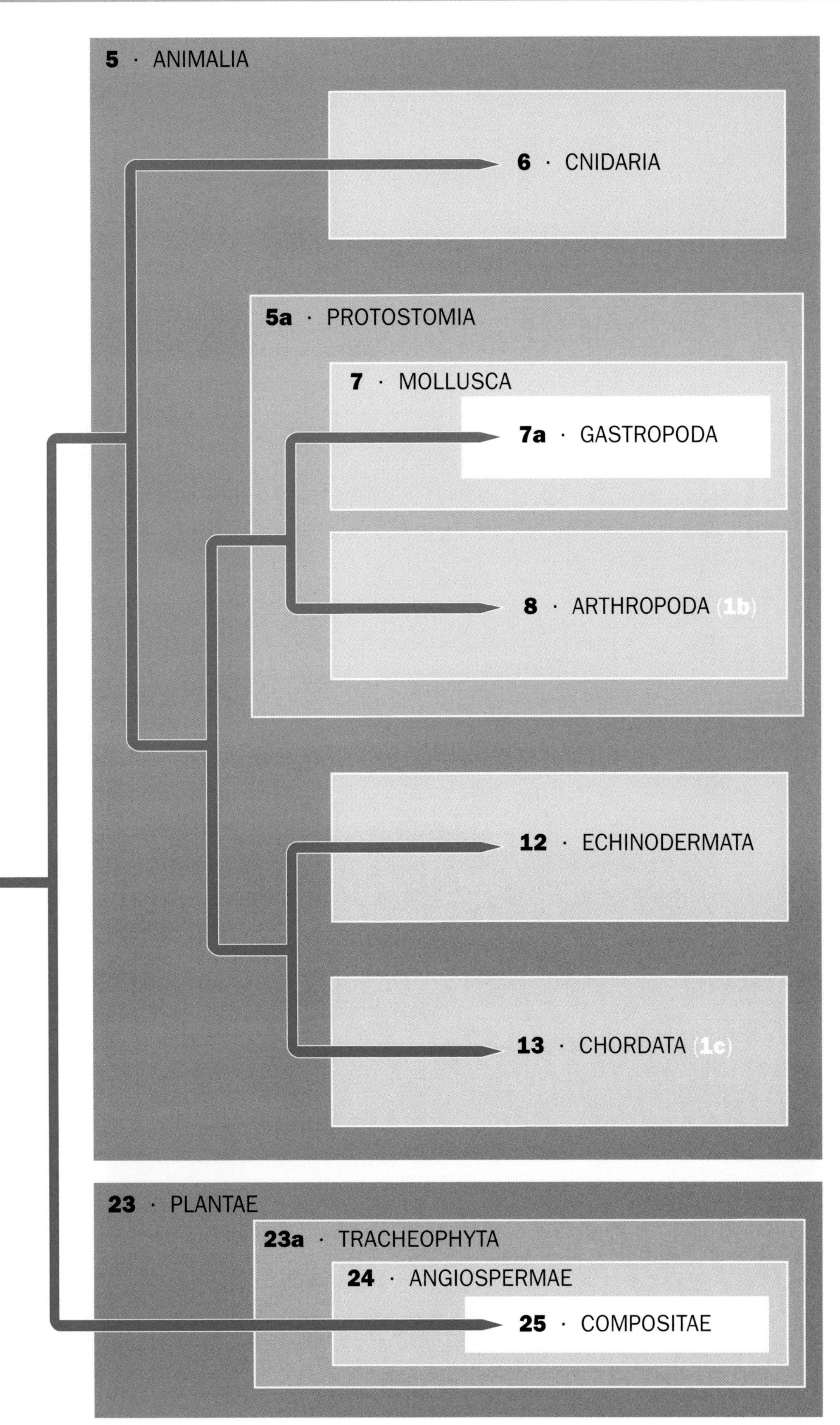
5 · ANIMALIA
6 · CNIDARIA
5a · PROTOSTOMIA
7 · MOLLUSCA
7a · GASTROPODA
8 · ARTHROPODA (1b)
12 · ECHINODERMATA
13 · CHORDATA (1c)
23 · PLANTAE
23a · TRACHEOPHYTA
24 · ANGIOSPERMAE
25 · COMPOSITAE

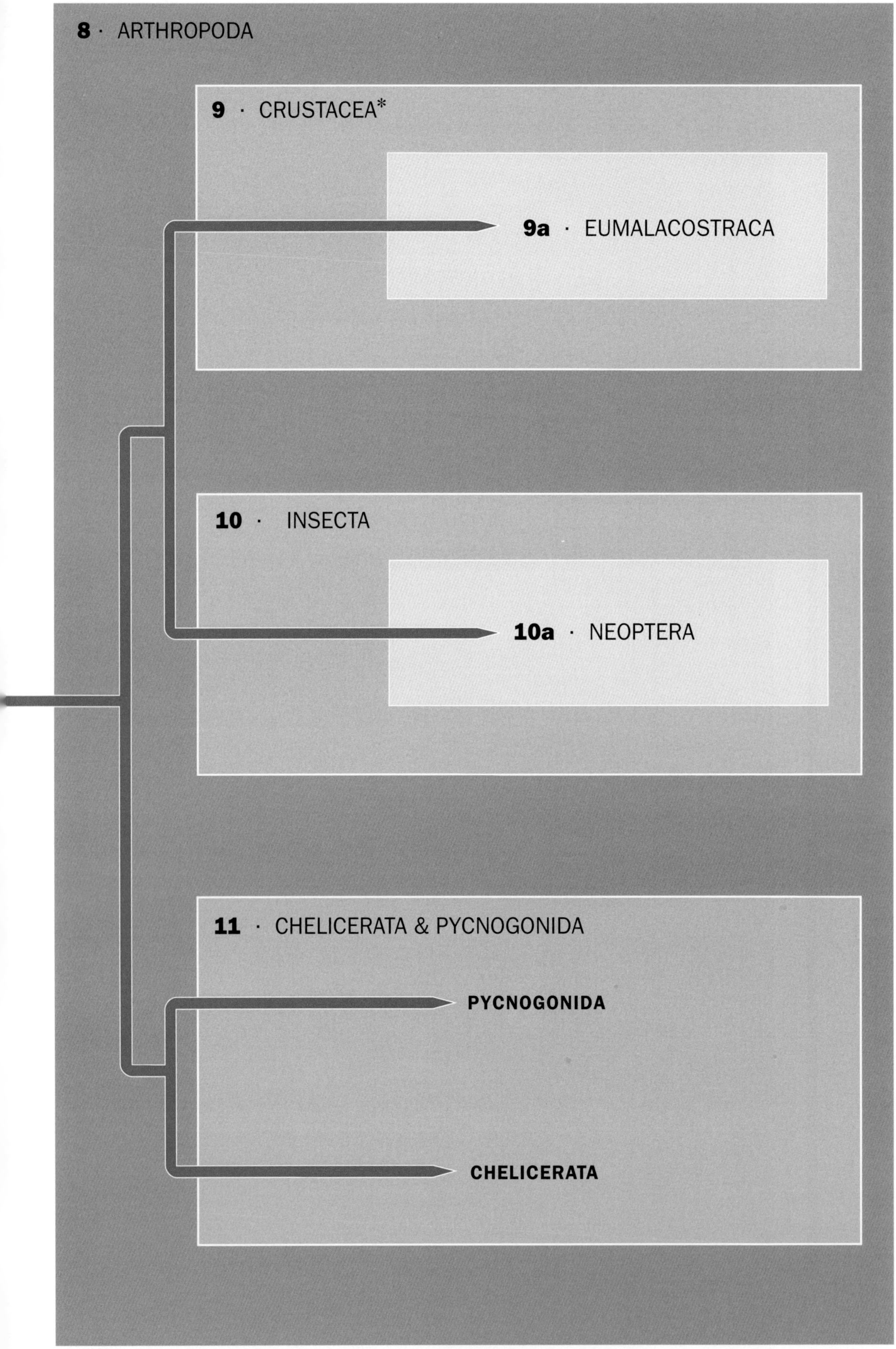

8 · ARTHROPODA

9 · CRUSTACEA*

9a · EUMALACOSTRACA

10 · INSECTA

10a · NEOPTERA

11 · CHELICERATA & PYCNOGONIDA

PYCNOGONIDA

CHELICERATA

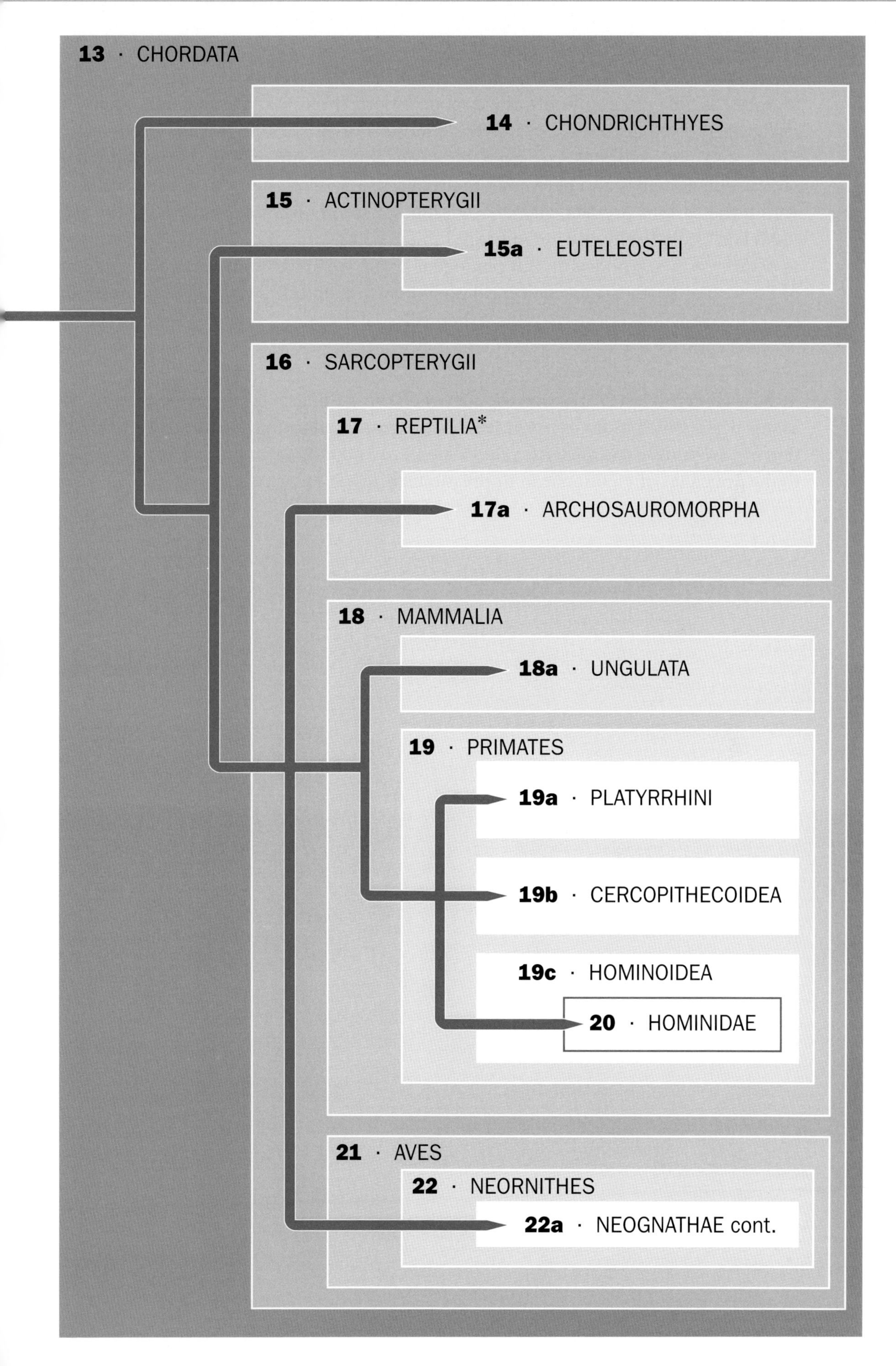

13 · CHORDATA
14 · CHONDRICHTHYES
15 · ACTINOPTERYGII
15a · EUTELEOSTEI
16 · SARCOPTERYGII
17 · REPTILIA*
17a · ARCHOSAUROMORPHA
18 · MAMMALIA
18a · UNGULATA
19 · PRIMATES
19a · PLATYRRHINI
19b · CERCOPITHECOIDEA
19c · HOMINOIDEA
20 · HOMINIDAE
21 · AVES
22 · NEORNITHES
22a · NEOGNATHAE cont.

that the Eucarya represent only one domain among three; and, furthermore, that the really deep divisions between different kinds of organisms lie not between vertebrates and 'invertebrates', or even between animals and plants, but within the prokaryote domains, and between the prokaryote domains and the eukaryotes, and within the various protists, and even within apparently humble groups like the red seaweeds. Ecologically and individually, human beings are, of course, wonderful, just as planet Earth is wonderful. But phylogenetically we are an outpost, a tiny figment of life, just as Earth is a cosmological nonentity that no other intelligent life-form in the Universe would bother to put on their celestial maps.

So now we can at last embark on the survey of living things: taking phylogenetic trees as our basis, which as far as possible are cladograms; adopting the three-domain system of taxonomy as proposed by Carl Woese; and naming groupings (taxa) according to the principles of Neolinnaean Impressionism.

2

THE DOMAINS OF THE PROKARYOTES

BACTERIA AND ARCHAEA

THE PROKARYOTES—Bacteria and Archaea—are too small individually to see with the naked eye but together they outweigh the macroscopic creatures at least tenfold. They are on our skin and in our guts, and sometimes, in disease, in our flesh. They are in the air, in all the world's water, and, living or dead, form much of the substance of compost and soil. If all macroscopic creatures were banished, and all minerals were dissolved away, the remaining prokaryotes would still provide a spectral outline of our planet, both land and oceans. Elephants require a continent to roam but prokaryotes can fit into any kind of space: a thousand typical bacteria might, in principle, stand line abreast across the head of a pin; a single gram of soil may contain 100 million individuals. Most of life on Earth—most of the mass, and most of the variety—is prokaryotic.

As already described, Anton van Leeuwenhoek first revealed prokaryotes in the seventeenth century with his home-made, single-lens microscopes—though he called them 'little animalcules', which must have included eukaryotic protists as well. His 'little animalcules' became known collectively as 'microbes' (any creature too small to see with the naked eye); and the modern discipline of 'microbiology' truly began in the second half of the nineteenth century when Louis Pasteur of France studied bacteria and yeasts, first in an industrial context (for example, the souring of milk) and then in relation to disease, from the sicknesses of silkworms to anthrax and rabies. The German physician Robert Koch (1843–1910), with his famous 'postulates' of 1890, then showed how to establish whether, and to what extent, particular bacteria actually *cause* particular diseases.

Classical microbiology has also revealed that prokaryotes are extremely various in their metabolism—the range of nutritional and respiratory strategies far exceeds that of plants, animals, and fungi combined and, indeed, the metabolic strategies that plants, animals, and fungi use generally evolved first in prokaryotes. Many resist conditions that we would regard as extreme. Thus some bacteria are able to grow thick walls and so to form spores and, in this form, some can resist boiling. Among these super-resistant types are *Clostridium botulinum*, source of the exceedingly dangerous toxin botulin; and *Bacillus cereus*, formerly acknowledged only as a 'commensal' (or harmless resident) of the gut, but now known also a source of food poisoning.[1] Other bacteria can withstand—or prefer—extreme acidity or salinity. Many breed remarkably quickly: the gut bacterium *Escherichia coli* (always called simply '*E. coli*') may replicate every 20 minutes in ideal conditions and if such conditions could be sustained could produce a mass greater than that of the Earth within 3 days.

Bacteria can swap genetic information without the elaborations of sex as manifest in, say, animals and plants: in particular, they may simply pass bundles of DNA from one to another, sometimes between lineages that are phylogenetically distant. Because they breed quickly and exchange genetic information promiscuously they can evolve rapidly, adapting continuously to novel circumstances from plastics to oil slicks. For various reasons, however (including their failure to embrace the more intricate elaborations of sex), their ability to evolve beyond a somewhat simple grade has proved limited—except in the lineage that produced the eukaryotes.

Because they are so small, various and versatile, prokaryotes are prevalent in all habitats and biological contexts. Viable spores have been found, by NASA balloons, more than 30 kilometres above Earth; several times higher than jumbo jets are wont to fly. They have also been found in the deepest ocean rifts and, more astonishingly, living in rock, feeding on rock, more than a kilometre below the ground. Nearer the surface, they are principal creators of soil fertility: as agents of decay they recycle the nutrients tied up in the corpses of animals and plants, while a range of bacteria called **nitrogen-fixers** are able to capture gaseous nitrogen from the atmosphere and turn it into soluble nitrogen-containing compounds, such as the ammonium radical, which, for plants, are food. Leguminous plants such as beans and acacias carry nitrogen-fixing *Rhizobium* bacteria in nodules in their roots. So do some other plants: alder, for example, has captive colonies of nitrogen-fixing *Frankia* that eke out the provender of dank river banks. *Azolla* is a tiny aquatic fern that floats like duckweed in paddy fields and carries nitrogen-fixing cyanobacteria called *Anabaena* within its tiny leaves, and the excess nitrogen compounds thus produced leak out and help, significantly, to fertilize the rice.

Bacterial decay can mean putrescence, but also implies the sweet fermentations of cheese and pickles. In the guts of animals, bacteria may be essential aids to digestion; indeed, the entire nutritional strategy of ruminants such as cows and sheep, and other specialist herbivores such as rabbits and elephants, is built around the cultivation of gut microbes. In less-benign mode bacteria invade the flesh itself and are supreme agents of disease, either through their direct depredations or, as with diphtheria, through their production of toxins. Vaccines and antibiotics have done much to contain them, so that syphilis has lost its sting and people no longer expect to die of septicaemia ('blood poisoning'). But some bacterial diseases such as bacterial meningitis are as terrifying as ever. Tuberculosis has again become a global menace. Legionnaire's disease is an ecological oddity. The causative bacterium, *Legionella*

[1] Fortunately most of the more familiar sources of food poisoning, such as *E. coli*, *Salmonella*, and *Shigella*, are destroyed by boiling.

pneumophila, is ubiquitous in rivers and lakes, yet harmless; but it causes potentially lethal disease when inhaled as an aerosol—for example, from air-conditioning systems.

Prokaryotes, in short, are everywhere. They influence our lives in a thousand ways. They can live perfectly well without us and our kind, and did so for several thousand million years before we came on the scene. Sometimes they kill us. Yet we could not live without them.

All this was seen and described by classical microbiologists. Yet modern research, using techniques of molecular biology, is now showing that classical studies have uncovered only a fraction of what is there to be seen. Most obviously, the new techniques are revealing the enormous and unsuspected depth of genetic variation within prokaryotes, in addition to the more obvious metabolic variation; that, and the deep difference between prokaryotes in general, and eukaryotes. This is what led Carl Woese to propose that we needed Neolinnaean rankings at the level of the domain; and that the division between the two prokaryote domains, Bacteria and Archaea, was at least as great as between those two domains and the Eucarya.

In addition, molecular biologists have devised quite new ways of searching for micro-organisms in the wild; and these techniques are suggesting that the true inventory of wild prokaryotes may be many, many times longer than classical microbiology could have suspected. Traditional microbiologists had to seek out new kinds of bacteria by looking for them down the microscope. If they could not see them then they could not demonstrate that they were there, and certainly could not add them to their species lists. But in order to observe bacteria down the microscope the traditionalists first had to cultivate them in the laboratory. So they would take samples of, say, soil, and put them in various kinds of nutrient broth; and each broth, depending on its chemical composition, would encourage different samples of the bacteria within to grow. In short, says Woese, classical microbiology is the study of bacteria that can be cultured. Those that did not respond to laboratory cosseting were overlooked.

Nowadays, however, microbiologists such as Norman Pace, now at the University of California at Berkeley, apply chemical probes to the raw substrate (for example, soil), of the kind that are designed to latch on to particular sections of DNA or RNA that may lie within. The presence of these nucleic acids in turn denotes the presence of living organisms. These techniques are astonishingly sensitive; they can reveal individual, scattered prokaryotes in relatively vast and confused samples. Metaphorical needles leap from allegorical haystacks as big as a small city. The range of creatures now being revealed in the wild that had not been made apparent by the classic techniques is beginning to beggar belief. Indeed, Professor Pace suggests that classical microbiology probably described only about one in 10 000 or perhaps only one in 100 000 of the bacterial/archaeal variety that actually exists. Furthermore, because we have missed so much, we obviously have an extremely biased and partial view of what it is that prokaryotes really do in the wild, and what they contribute to the world's ecology as a whole.

Studies so far suggest that of the two prokaryote domains, the Bacteria contain about 10 times as many types as the Archaea. But this discrepancy, as Woese observes, reflects mainly the amount of time that the two groups have been studied, and the numbers of microbiologists studying them. Studies in the Antarctic by Edward DeLong and his colleagues from the University of California at Santa Barbara now suggest that the oceans are teeming with archaes; indeed, that oceanic archaes may be the most widespread organisms on Earth. Those marine archaes are metabolically active, and because their total biomass is enormous we must assume that their metabolic contribution to the world's ecology is commensurately huge. But what they do is anybody's guess. It is astonishing that they could have been missed: the oversight, says Gary Olsen of the University of Illinois, 'is equivalent to surveying one square kilometre of the African savannah and missing 300 elephants'. But it is all a question of instrumentation. Without eyes, we would miss elephants; without telescopes, we would miss most of the stars; and without RNA probes we have, hitherto, missed one of the most significant components of the ocean biomass. But biomass does not necessarily imply diversity. The oceans might contain a billion billion billion replicates of just a few archaeal species—analogous, on a quite different scale, to the great shoals of anchovies and krill.

Even more profoundly, classical microbiology has revealed within the past few decades that prokaryotes commonly live within hot springs, or within the similar hydrothermal vents that run along the beds of the world's oceans. These **thermophiles** do not merely tolerate hot water, like the kitchen bacteria whose dormant phases may survive boiling: they thrive in it. Many reproduce most quickly at or around the boiling temperature of water. Some, living at even greater oceanic depths where pressures are greater and boiling point is raised, breed most rapidly at several degrees above normal boiling point.

But the new search methods—seeking out microbes by nucleic acid probes—now show that these thermophiles are not mere oddities, anomalously adapted to conditions that we consider extreme. The thermophiles include bacteria and archaes from a wide range of prokaryote kingdoms, and often from several distinct groups within a single kingdom. In fact the ability to live at high temperatures is shared by so many prokaryotes, of such different ancestry, that it should probably be seen not as a special adaptation but as a primitive character. That is, life on Earth probably began under such 'extreme' conditions. Present-day thermophiles are not late adventurers but are conservatives, who have retained the ancient ways. We call our everyday temperatures 'normal' and regard ourselves and comparable creatures as 'normophiles'. But we are not 'normal' by general standards. We are weird, cold-adapted creatures. As unbiased historians of life we should call the bacteria of the hot springs 'normophiles' and ourselves 'cryophiles'—'lovers of the cold'. It is too late now to play with names; but the shift of emphasis is nonetheless profound.

One last revelation, essential to this theme, is that the observable forms of prokaryotes, and the metabolic differences between them, do not correspond easily with the phylogenetic differences as revealed by the structure of their DNA and RNA.

Traditional taxonomy of prokaryotes has been based on appearance and metabolism; so if we seek a classification based on phylogeny, we have to change the traditional taxonomy. This does not mean that traditional taxonomies should simply be abandoned, because they are based on easily observable features and therefore can be of great practical value. But biologists who seek to base taxonomy on history and ancestry must rethink.

To appreciate this point, we should first look in slightly more detail at the range of metabolism within the prokaryotes—which effectively means within living systems as a whole, because eukaryotes do little or nothing metabolically that bacteria and archaes did not venture first.

THE MANY WAYS OF BEING A PROKARYOTE

As the tree accompanying this section shows, prokaryotes are far more variable in appearance than is often imagined: some are spherical, some are rod-shaped, some have whip-like appendages called **flagella** (or flagella-like projections) that can confer remarkable turns of speed and bursts of energy, some (the mycoplasmas) lack cell walls altogether. Many form colonies that can be extremely beautiful and sometimes huge. Well back in the Precambrian era prokaryotes formed huge structures in the oceans like modern-day coral reefs—living cells in an ever-growing matrix of minerals —that find their present-day counterpart in the stromatolites that still stand like petrified puffballs on beaches of Western Australia, and in the beautiful calcareous terraces around the hot springs in Yellowstone in the USA, New Zealand, and Turkey. Yet the true scope of their variety is not to be seen in their bodily forms: it lies in their metabolism and their mode of operation.

Most obviously, prokaryotes feed in a remarkable variety of ways. All creatures need two kinds of input from their environment: raw materials, from which to fashion the fabric of their cells; and a source of energy. Different creatures take in their raw materials in different forms, and whatever form is appropriate to that creature is a nutrient. For plants, carbon dioxide is a nutrient; for us, it is not. Whatever the form, all organisms need to take in a basic shortlist of elements that includes carbon, hydrogen, oxygen, nitrogen, phosphorus, sulphur, and a few metals such as sodium and potassium. They also need various permutations from a much longer list including non-metals such as iodine and chlorine and the metals magnesium, iron, calcium, copper, molybdenum, and many more. Indeed, any one organism may in practice partake of a generous proportion of the periodic table.

In **heterotrophs** like ourselves and other animals the tasks of acquiring food and energy are conflated: the complex organic molecules within our bread and beef provide us both with nutrients (raw materials) and, as those molecules are broken down, with energy. But in **autotrophs** the source of energy and the source of nutrients are clearly distinct. Plants, for example, are autotrophs; for them, the primary source of energy is sunlight, which they harness though the process of photosynthesis and use to convert carbon dioxide and water into complex organic molecules.[2] Plants also acquire some of their energy by breaking down some of those organic molecules after they have made them—for example, emergent shoots of potato are nourished by starch within the tuber, just as an animal draws on reserves of glycogen. But the point is that the potato plant creates the starch for itself, whereas the animal synthesizes its glycogen from organic molecules that it has eaten, and which have already been made by plants or by other animals that had eaten plants.

There are heterotrophs and autotrophs among prokaryotes, too, but a much wider range of both than are seen among animals, plants, and fungi. Indeed, prokaryotes invented most of the basic ways of feeding (all except phagocytosis, the wholesale engulfing of one cell by another, which seems to be a eukaryote innovation; or, at least, it was confined to the ancestral prokaryote that gave rise to the eukaryotes). Prokaryotes, too, have developed some ways of gathering nutrients and energy that remain unique to them, because the eukaryotes never adopted them. In practice, ways of gathering nutrients and energy fall roughly into four broad categories, some of which are essayed by eukaryotes, and all of which are practised by prokaryotes, as follows:

- **Photoautotrophs** are organisms that practise photosynthesis. They use light as their source of energy and atmospheric carbon dioxide (CO_2) as their main source of carbon. Plants, seaweeds, and various protists are the most conspicuous practitioners of photosynthesis. But prokaryotes invented it, and species from five different prokaryote kingdoms practise it. Eukaryotes acquired the technique only by turning entire prokaryotes into chloroplasts.

- **Photoheterotrophs** are organisms that use light as their main source of energy, but acquire most of their carbon in organic form. We could argue that some plants aspire to photoheterotrophy up to a point, including insectivorous angiosperms (for example, pitcher plants, which get at least some of their carbon by digesting insects), and a range of parasites such as *Striga*, a tropical relative of the foxglove, which acquires its carbon (and water) by tapping in to the roots of maize, legumes and other plants and stealing organic metabolites. But these are special cases. Photoheterotrophy, in general, is the speciality of some purple and other bacteria.

- **Chemoautotrophs** are an extremely significant group, exclusively prokaryotic. Like the photosynthesizers, they use carbon dioxide as their principal carbon source. But they acquire their energy from chemical sources. Some, known as 'sulphur

[2] All living things on Earth are built from compounds of carbon, with other elements incorporated. It may be that of all elements, only carbon has the chemical versatility required to build molecules that can manifest life; so perhaps all life throughout the Universe is carbon-based. Be that as it may, chemists use the term 'organic' to mean 'a compound that contains carbon'. So 'organic nitrogen', for example, means nitrogen incorporated into a carbon-containing compound, as in an amino acid; whereas 'inorganic nitrogen' means nitrogen gas (N), nitrate (NO_3^-), ammonia (NH_3), or what you will. But most biologists tend to think that simple carbon compounds such as carbon dioxide are not exactly 'organic', even though they do contain carbon. So they tend to say that heterotrophs take in their carbon in 'organic' form (for example, in the form of fat, protein, or carbohydrate) as opposed to carbon dioxide, which is the form favoured by autotrophs. In this account, I am treating carbon dioxide and other such materials as 'inorganic' carbon, even though some chemists would say this is a nonsense.

bacteria', oxidize reduced forms of sulphur, such as hydrogen sulphide (H_2S), elemental sulphur (S), or sulphites (SO_3^{2-}), while others reduce sulphate (SO_4^{2-}). Some oxidize compounds of nitrogen, such as ammonium (NH_4^-) or nitrite (NO_2^-). Some oxidize other inorganic materials, such as hydrogen (H_2), carbon monoxide (CO), or ferrous iron (Fe^{2-}). Some archaes generate methane gas as they metabolize (methane being carbon in the highly reduced form of CH_4): these are called **methanogens**.

Wonderfully (at least by comparison with eukaryotes), chemoautotrophs can grow on a medium that is strictly mineral, without input of light; and some of them are hence called **chemolithotrophs** (from the Greek *lithos*, a rock). Chemolithotrophs have now been found within stone statues, slowly rotting them away; all known cases so far are in Germany, because German microbiologists did the research, but statues worldwide are doubtless afflicted. More exotically and naturally, chemolithotrophs have also been found growing in the pores of rocks a mile beneath the Earth's surface. If there is life on other planets, it seems most likely to be found in such environments, which should be highly protected, uniform, and enviably hospitable to those creatures that are adjusted to it. Growth need not be rapid; such creatures might in theory time their metabolism in thousands of years. The notion that life can continue only on the surface of the land or in the surface waters (in what is commonly accepted as 'the biosphere') is another piece of what Norm Pace calls 'eukaryote chauvinism'.

- **Chemoheterotrophs** derive their energy from chemicals and use organic compounds as the main source of carbon. Animals, protozoans (animal-like protists), fungi, and many prokaryotes are chemoheterotrophs. Although traditional textbooks tend to say that most bacteria are chemoheterotrophs, this judgement seems more than premature. In truth, traditional knowledge of prokaryotes is based on those that can be cultured in the laboratory; and chemoheterotrophs are the easiest to culture. As outlined, new techniques are revealing many times more prokaryotes in nature than have ever been grown and it is certainly not obvious that most of the hitherto unknowns are chemoheterotrophs.

Prokaryotes also vary enormously in their method of respiration. Some, **anaerobes,** hate oxygen; indeed, they are poisoned by it. Some, **microaerophiles,** require oxygen, but only in small amounts, and (in general) are poisoned by too much. Others are **aerobes**: using oxygen in much the same ways as animals do to 'burn' organic molecules and release the energy they contain. In general, aerobic respiration is more efficient than anaerobic—the organic fuels are processed more thoroughly—but oxygen is lively stuff and destroys organisms that are not equipped to cope with it. But there was virtually no free oxygen in the atmosphere until cyanobacteria, and then plants began to practise their particular form of photosynthesis, which generates oxygen.[3] This seems to have occurred about 2500 million years ago,

[3] Other organisms that practise photosynthesis include some purple bacteria, green sulphur bacteria, and green non-sulphur bacteria. But their method of photosynthesis does not generate oxygen. Oxygen-generating photosynthesis seems to have evolved after non-oxygen generating photosynthesis and, because photosynthesis probably evolved only once, the oxygen-generating form presumably evolved from one of the forms that does not generate oxygen.

some thousand million years after life first appeared, and before that all organisms were obliged to be anaerobic. Presumably, the anaerobic prokaryotes of today are either direct descendants of the earliest types who evolved in an oxygen-free atmosphere, or have lost the skills of aerobiosis in the way that many birds have lost the power of flight.

Finally, some archaes and bacteria are most at home in extremely saline conditions; for example, in soda lakes or within the Dead Sea—the kinds of conditions that, in everyday life, cooks create for the preservation of food precisely because they seem to be inimical to microbial life. Saline lovers are called **halophiles**. The most extreme halophiles are archaes.

These, then, are the background points. The tree shows the main kingdoms of Bacteria and Archaea as now worked out in particular by Carl Woese, and is based on his studies of their small subunit ribosomal RNA (16S rRNA): the molecule that is likely to reflect the huge phylogenetic distances most accurately. Microbiologists will immediately see that this classification differs markedly from traditional taxonomies that are based upon phenotype—that is, upon appearance and life style.

A GUIDE TO THE PROKARYOTES

In the prokaryotic tree I have had to bend the rules of this book a little. I wanted to provide portraits of all the creatures mentioned. But some of the prokaryotes that seem most significant phylogenetically are at present known only from their nucleic acids, and sometimes in part from traces of their metabolism. They have not all been seen as intact organisms, and those that have not, cannot be portrayed. In such cases just the names are provided.

Note first of all that phenotype (observable physical features) really is an extremely poor guide to phylogeny. Photosynthesis, for example, is practised by species from five different bacterial kingdoms; and the closest relatives of many photosynthetic species are often heterotrophs. Indeed in any one group you may find all the modes of nutrition that are apparently so radically different, while anaerobes often emerge as the sisters of aerobes, and thermophiles may be cheek by jowl with normophiles. It is clear, too, that there are at least three major groups of sulphur metabolizers.

Of the traditional, phenotype-based groupings, only a few are confirmed by analysis of their RNA. For example, the photosynthetic cyanobacteria do form a coherent group and so do the spirochaetes and their relatives, known for their corkscrew shape and the long flagella-like projections[4] that drive them along.

Finally, traditional microbiologists drew a sharp distinction among the bacteria between the so-called 'Gram-positives' and the 'Gram-negatives'. These categories spring from the work of the Danish bacteriologist Hans Christian Gram (1853–1938), who showed in 1884 that if bacteria are first dyed and then treated with a solvent such

[4] These outgrowths should not simply be called 'flagella' because, unlike 'true' flagella, they are enclosed within the outer sheath of the organism.

as alcohol or acetone, then some (the Gram-positives) retain the dye whereas others (the Gram-negatives) release it and so are bleached by the solvent. This difference is now known to reflect profound differences in cell-wall structure. But whereas the new molecular studies confirm that the Gram-positives are indeed a coherent group (albeit with two deeply divided branches—and now including some types that don't retain dye in their walls!), the Gram-negatives are a very mixed bag. It is as if zoologists had classified animals into 'Birds' and 'The Rest'. The birds would form a good group but non-ornithologists might well feel that the distinctions between gorillas and oysters had been somewhat glossed over. Nonetheless, 'Gram-positive' versus 'Gram-negative' is still a useful practical distinction in laboratory and clinic.

In the light of these background comments we can now explore the tree, as derived from Carl Woese's studies of prokaryote RNAs. The tree again shows all three domains of life. Note once more that the **Archaea** are the sister group of the **Eucarya**, while the **Bacteria** are the sister group of the Archaea + Eucarya. It is difficult to say how many kingdoms of Bacteria there are for at least three reasons: (1) the definition of 'kingdom' is somewhat arbitrary; (2) the biologists who are now re-aligning prokaryote classification often tend to reject Linnaean rankings altogether (as do many modern systematists) and do not take the question of ranking seriously; and (3) studies of new ecotypes—for example, the thermophiles—and molecular methods of study are opening up huge areas that have yet to be investigated in detail. Here, however, following Carl Woese, I have identified 10 major groupings of Bacteria, which we can reasonably call kingdoms. At present there seem to be three kingdoms of Archaea, but future studies seem liable to reveal more.

THE KINGDOMS OF BACTERIA

Proteobacteria (purple bacteria and mitochondria)

The kingdom Proteobacteria is a huge and varied group of Gram-negative bacteria commonly called the 'purple bacteria'. Traditionally the Proteobacteria have been divided into four groups, labelled **alpha**, **beta**, **gamma**, and **delta**; but RNA studies have now revealed a fifth, labelled **epsilon**, which appear to be a deep division within the deltas. The epsilons include the genus *Wolinella* and the agent of stomach ulcers, *Campylobacter pylorus*. Many proteobacteria practise photosynthesis that is distinct from the photosynthesis of other bacteria, and uses bacterial chlorophyll a. Note that photosynthetic genera turn up in three of the groups—alpha, beta, and gamma—and that all three of them also contain many types that are not photosynthetic; indeed, the closest relatives of the photosynthetic types are often non-photosynthesizers. Because photosynthesis is unlikely to have evolved more than once it is presumably the primitive condition of the group, but has been lost many times (just as many different kinds of bird have lost the ability to fly). More generally, throughout the proteobacteria we find that the closest relatives of heterotrophs may be chemolithotrophs. Sometimes, too, anaerobes and aerobes turn out to be closely related.

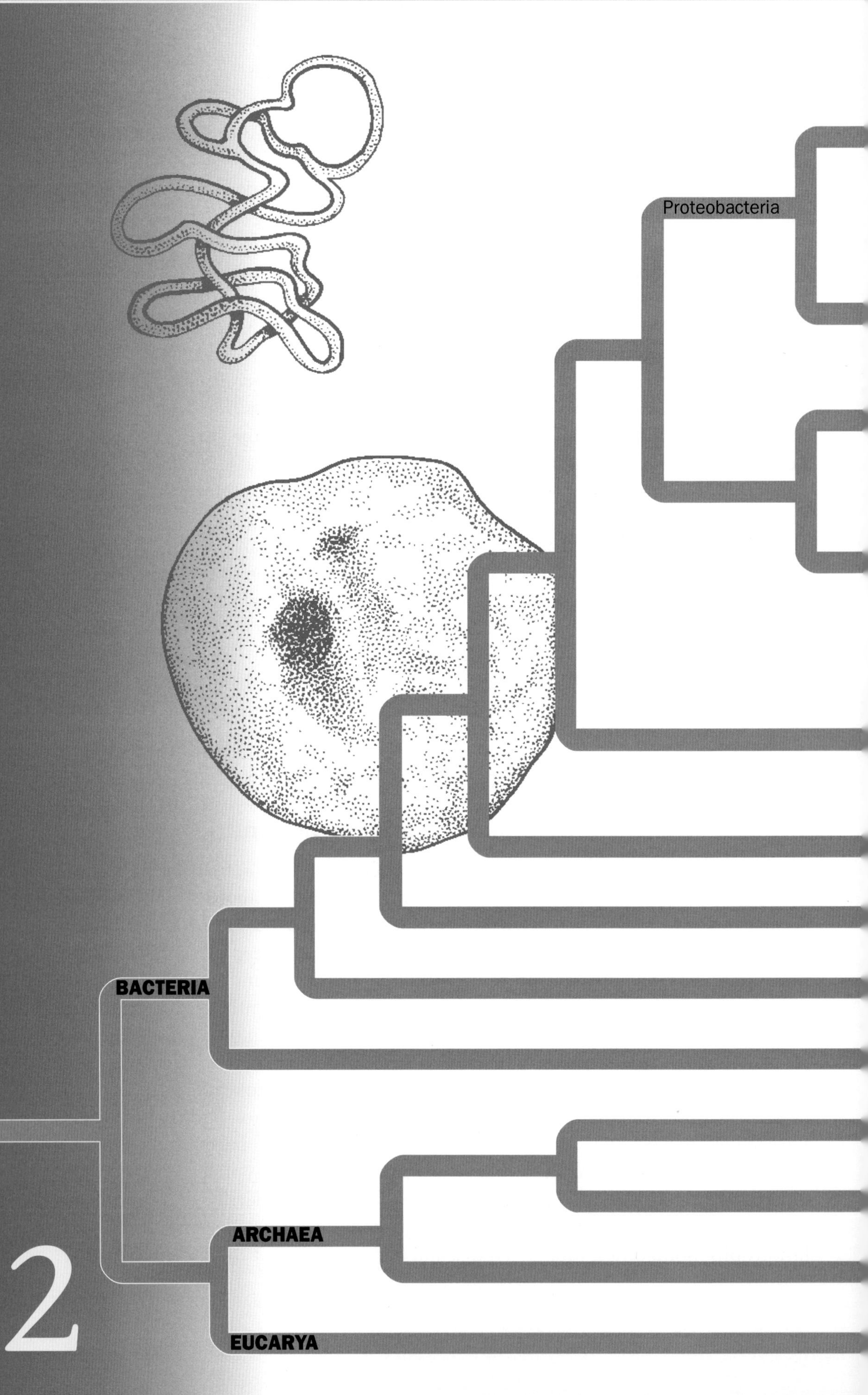
Proteobacteria
BACTERIA
ARCHAEA
EUCARYA
2

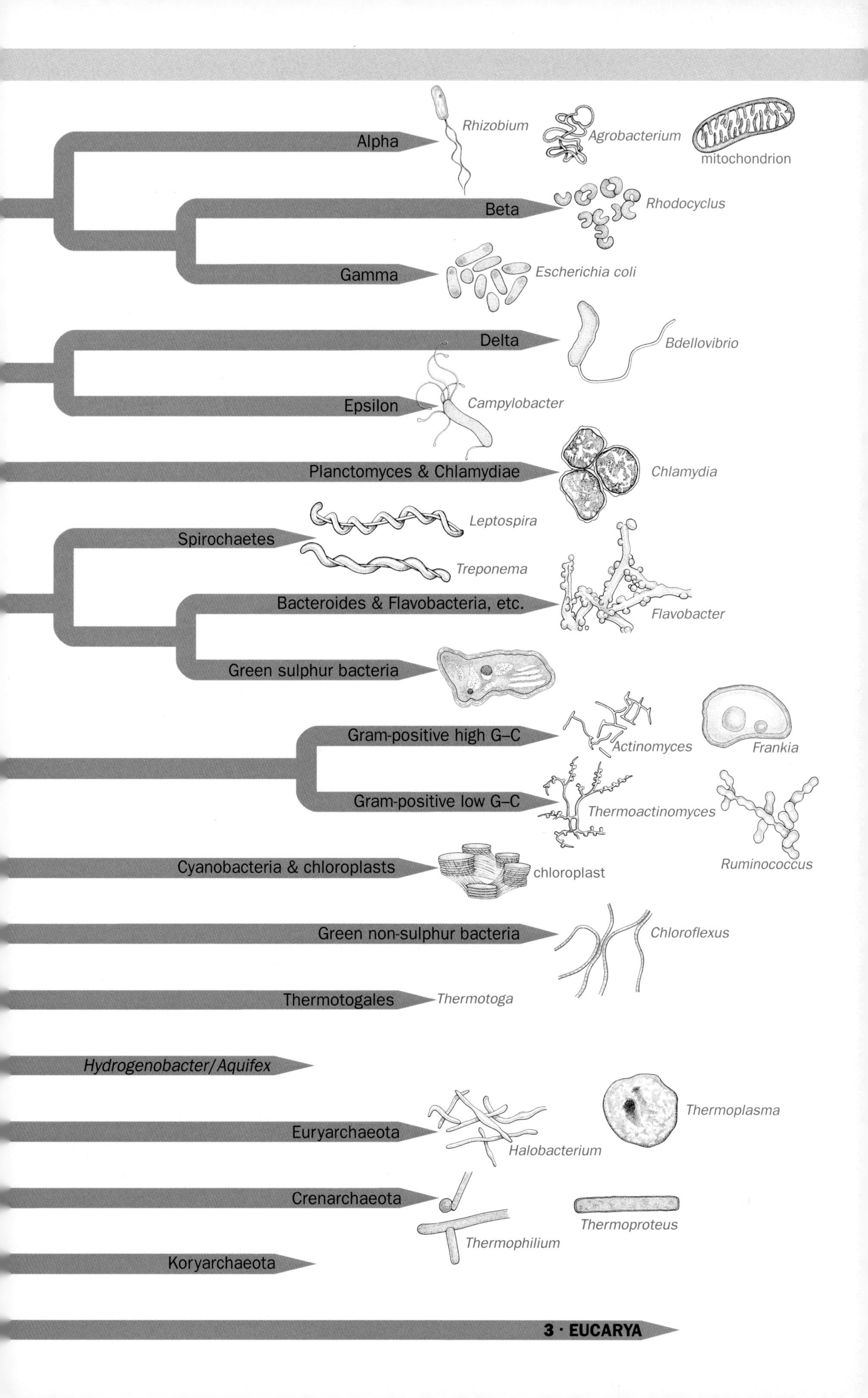

Alpha
Rhizobium
Agrobacterium
mitochondrion
Beta
Rhodocyclus
Gamma
Escherichia coli
Delta
Bdellovibrio
Epsilon
Campylobacter
Planctomyces & Chlamydiae
Chlamydia
Spirochaetes
Leptospira
Treponema
Bacteroides & Flavobacteria, etc.
Flavobacter
Green sulphur bacteria
Gram-positive high G–C
Actinomyces
Frankia
Gram-positive low G–C
Thermoactinomyces
Ruminococcus
Cyanobacteria & chloroplasts
chloroplast
Green non-sulphur bacteria
Chloroflexus
Thermotogales
Thermotoga
Hydrogenobacter/Aquifex
Euryarchaeota
Thermoplasma
Halobacterium
Crenarchaeota
Thermoproteus
Thermophilium
Koryarchaeota
3 · EUCARYA

Aerobic respiration (unlike photosynthesis) does seem to have arisen several times independently—indeed, has done so several times independently among the alpha proteobacteria alone.

Alpha proteobacteria

Various alpha proteobacteria form close associations with eukaryotes. Thus *Rhizobium* forms nodules within the roots of leguminous plants, within which they 'fix' atmospheric nitrogen: this (in the form of ammonia) becomes a nutrient for the plant, and the bacteria in turn receive organic carbon. The related *Agrobacterium* is a pathogen of plants, which forms tumours; and indeed is used in modern genetic

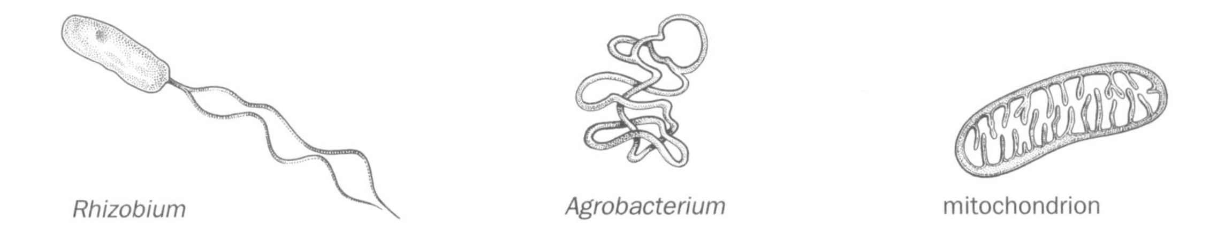

engineering to carry novel genes into new hosts. The rickettsias are intracellular pathogens of animals. These three groups are closely related. All in all, then, says Carl Woese, we need not be surprised that the mitochondria typical of eukaryotic cells probably arose also from alpha proteobacteria; they, too, after all, dwell intimately within the eukaryotic cell.

Beta proteobacteria

The beta proteobacteria include a mixture of well-established and newly defined genera—such as *Rhodocyclus*, which until recently was loosely classed as a 'purple non-sulphur' bacterium. The betas also include photosynthesizers and non-photosynthesizers; some that metabolize sulphur and some that do not; the soil bacterium *Nitrosomonas*, which oxidizes ammonium (NH_4^+) to nitrite (NO_2^-) within the soil.

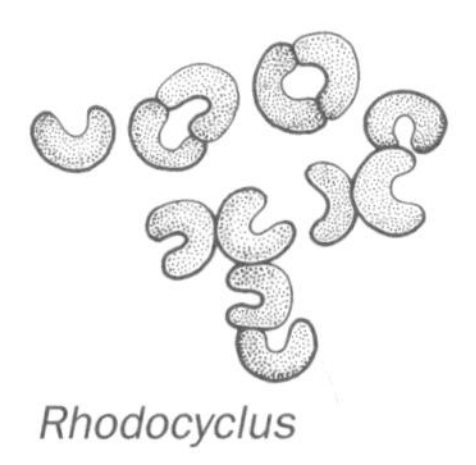

Gamma proteobacteria

The gamma proteobacteria include three distinct feeding types: photosynthesizers, many of which find their closest relatives among non-photosynthesizers; plus heterotrophs and chemolithotrophs, which again may be closely related phylogenetically despite their great difference in lifestyle. The photosynthetic gammas include those of the purple sulphur type, like *Chromatium*. Non-photosynthetic types include *Legionella*, the cause of legionnaires' disease; the enterics such as *Escherichia coli*,

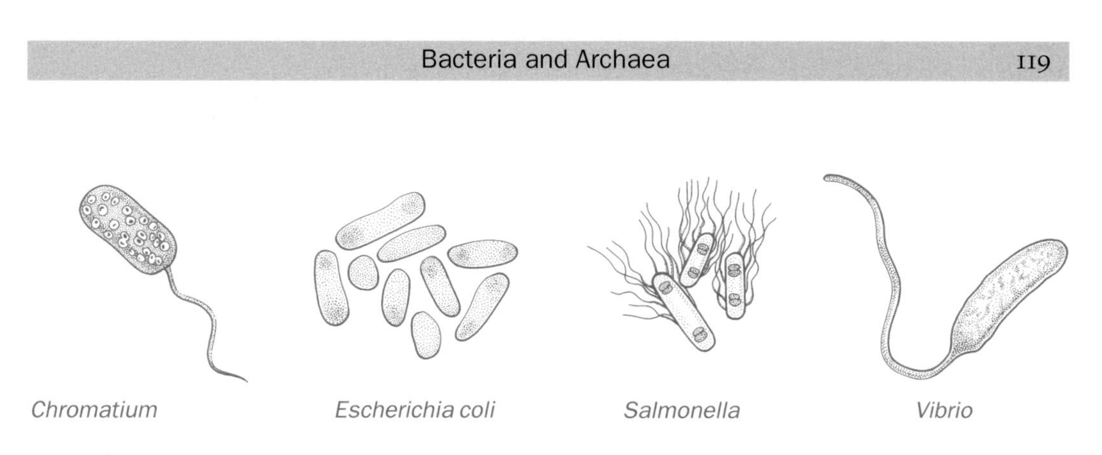
Chromatium Escherichia coli Salmonella Vibrio

Salmonella, and their kind; *Vibrio*, which includes *V. cholerae*, the agent of cholera; *Oceanospirilla*, fluorescent pseudomonads, and so on.

Deltas and epsilons

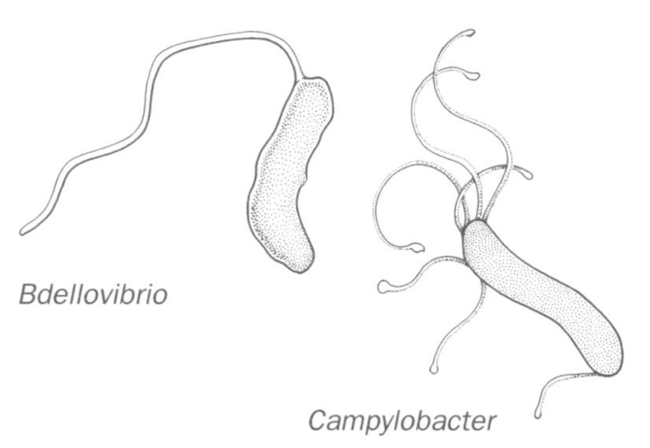
Bdellovibrio

Campylobacter

Among the delta proteobacteria we also find three disparate phenotypes. First there are reducers of sulphur and sulphate. Then there is the clade of the myxobacteria, which are remarkably reminiscent of the cellular slime moulds—which are eukaryotes, and totally unrelated. That is, at some stage of the myxobacterial life cycle the individual cells aggregate to form stalked structures bearing colourful fruiting bodies. The deltas include the bdellovibrios, which are among the most athletic prokaryotes. They are parasites of other Gram-negative bacteria, and attack their hosts like guided missiles, rushing into them at 100 cell-lengths per second—which is equivalent to a 30-centimetre-long rabbit running about 110 kilometres an hour, or a man at half the speed of sound. The attacking *Bdellovibrio* then attaches by its non-flagellar end and bores its way in by spinning at more than 100 revolutions a second. Phylogenetically speaking, the myxobacteria and bdellovibrios seem to be aerobic descendants of some anaerobic sulphur-metabolizing ancestor. Finally, the epsilon proteobacteria, such as *Campylobacter pylorus*, emerge as a deep division of the deltas.

Planctomyces and Chlamydiae

The kingdom **Planctomyces** contains such genera as *Pasteuria* and *Pirella*. They live in fresh and brackish waters and the sea, attaching themselves to the substrate with a holdfast (like kelps). They reproduce by budding. More to the point, from a taxonomic point of view, they are the bacteria that lack the polymer murein in their cell walls. The rRNA of these organisms is very different from other bacteria, but probably because of rapid evolution rather than early divergence. In the kingdom **Chlamydiae** are obligate intracellular parasites that affect birds and human beings and probably (they are not well studied) a lot of other species as well. At the time of writing, there is a serious, smouldering epidemic of *Chlamydia* among koalas in Australia.

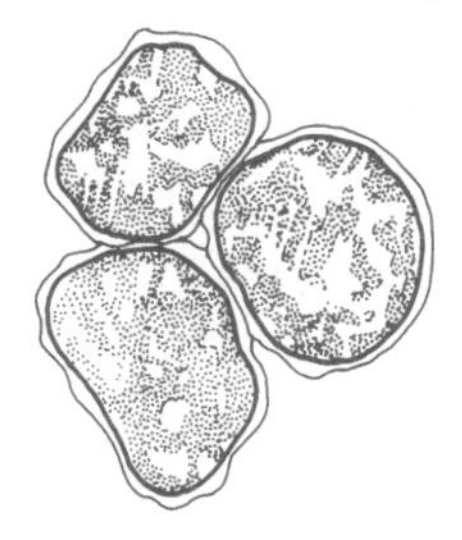
Chlamydia

Spirochaetes

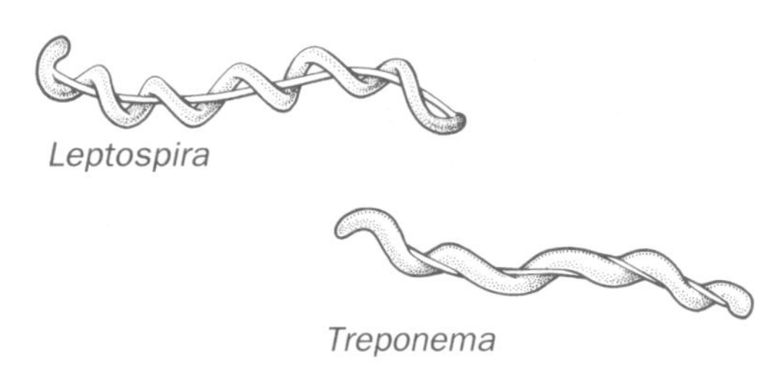
Leptospira

Treponema

The kingdom Spirochaetes is one of the few classical taxa that does bear scrutiny when its RNA is probed. Spirochaetes divide clearly into two groups. One group includes *Leptospira*, which is responsible for leptospirosis or Weil's disease; and the other includes *Treponema*, which is the agent of syphilis and yaws, *Borrelia*, the agent of relapsing fever, and *Spirochaeta*.

Bacteroides, Flavobacteria, and their relatives

Flavobacter

Again, this group is a tremendously mixed bag phenotypically. It links, for example, the anaerobic Bacterioides with various aerobes such as the gliding bacterium *Cytophaga*. *Cytophaga* is mostly non-pathogenic, but *C. columnaris* causes epidemics in fish hatcheries.

Green sulphur bacteria

Only four species of green sulphur bacteria are known, all within two genera—*Chlorobium* and *Chloroherpeton*. Their general phenotype is considered to be very primitive and little is known about them except that they seem very different from all other bacteria; they have emerged as sisters to the Bacteroides / Flavobacteria group.

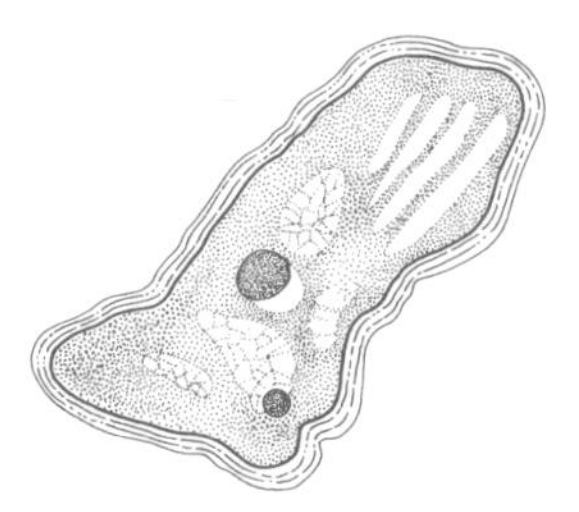
green sulphur bacterium

Gram-positive bacteria: high G–Cs and low G–Cs

The Gram-positive bacteria share the Gram-positive cell wall but, apart from that, they seem a phenotypically mixed bag: a group containing many familiar and economically important species and known both for their extreme nastiness and for their benignity. There are two major subdivisions: 'high G–Cs' and 'low G–Cs', plus two small groups that are clearly related but hard to place.

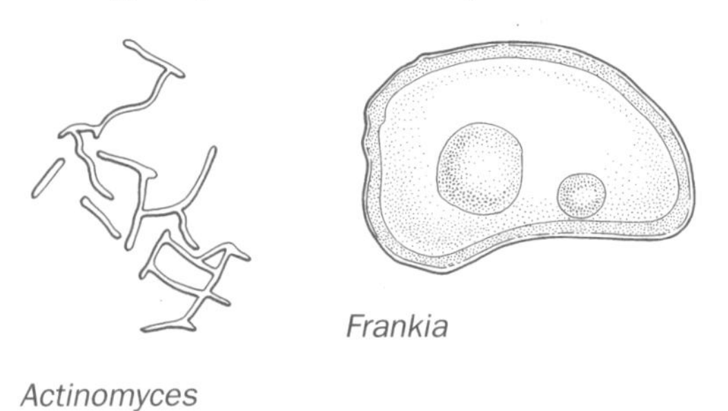
Frankia

Actinomyces

High G–Cs have DNA containing more than 55 per cent of the bases guanine and cytosine. They are mostly aerobic, and include such luminaries as *Actinomyces* and *Streptomyces*, noted as sources of antibiotics; *Thermomonospora*, a thermophile; and *Frankia*, which acts as a nitrogen-fixer within the roots of nonleguminous plants such as the alder. The genus *Mycobacterium* includes the agents of tuberculosis and leprosy—*M. tuberculosis* and *M. leprae*.

Low G–Cs have less than 50 per cent guanine and cytosine among their DNA bases. They include the genus *Bacillus*, which seems to be harmless most of the time but includes *B. cereus*, an occasional cause of food poisoning, and also—more seriously—*B. anthracis*, the agent of anthrax. *Staphylococcus* is the cause of boils while *S. aureus* is notorious as a highly dangerous, opportunist invader of surgical wounds. *Streptococcus* includes pathogenic species that cause pneumonia and bacterial meningitis. The extremely variable, anaerobic *Clostridium* has spores that resist boiling and, among other things, is the source of botulism, gas gangrene, and tetanus. *Mycoplasma* is one of the curious genera that have lost their cell walls altogether and seems to be a degenerate *Clostridium*. *Lactobacillus* is also a low G–C Gram-positive. It is one of the fermenting agents used to make yoghurt.

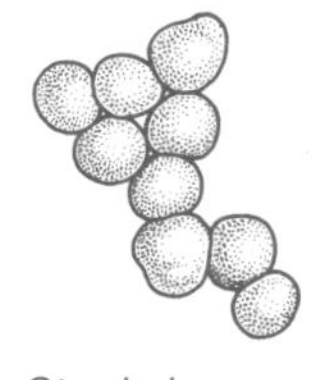
Staphylococcus

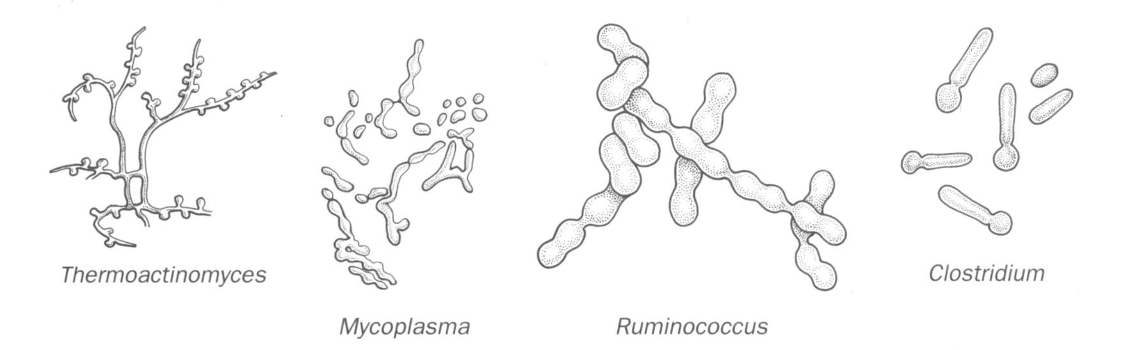
Thermoactinomyces *Mycoplasma* *Ruminococcus* *Clostridium*

Within the Gram-positives, the anaerobic types seem closest to the ancestral roots, and aerobiosis seems to have evolved more than once within the group. Indeed the evolution of the group seems to parallel the infusion of oxygen into the Earth's atmosphere. Thus *Lactobacillus*, *Streptococcus*, and mycoplasmas are basically anaerobic, but can tolerate and even sometimes use a little oxygen, while *Bacillus* is basically aerobic although some grow well anaerobically. Carl Woese speculates that the anaerobes and microaerophiles evolved while the atmosphere was still low in oxygen, and *Bacillus* evolved later.

Cyanobacteria—and chloroplasts

The Cyanobacteria were traditionally and erroneously called 'blue-green' algae. Ecologically they are supremely important in their own right, evident everywhere as a green scum, and photosynthesizing on a huge scale worldwide. Many, too, are nitrogen-fixers, including *Anabaena* as mentioned above. Beyond doubt, too, some long-gone cyanobacterium was the

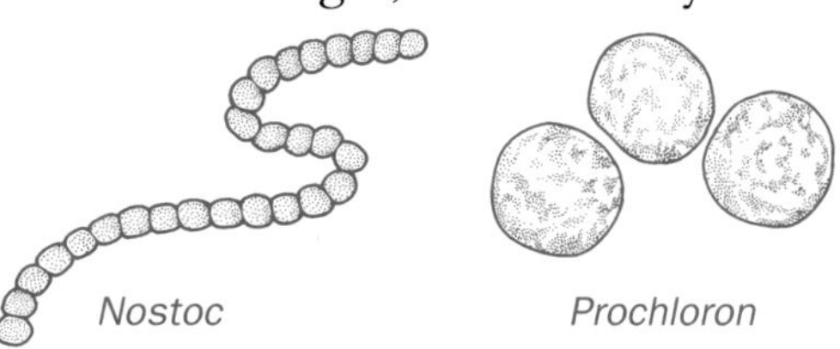
Nostoc *Prochloron*

ancestor of all chloroplasts, as now found in brown seaweeds, red seaweeds, and plants. Which of the living cyanobacteria is most closely related to the chloroplast ancestor is discussed in Section 3.

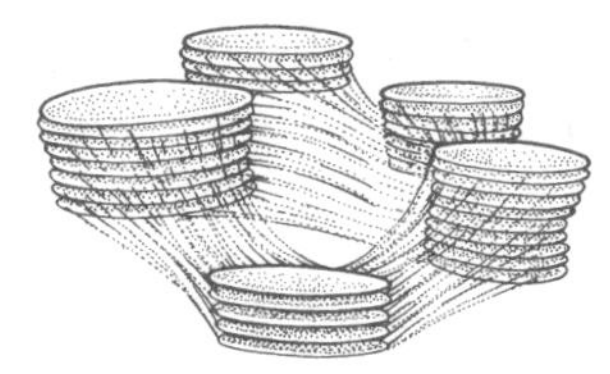
chloroplast

Green non-sulphur bacteria and their relatives

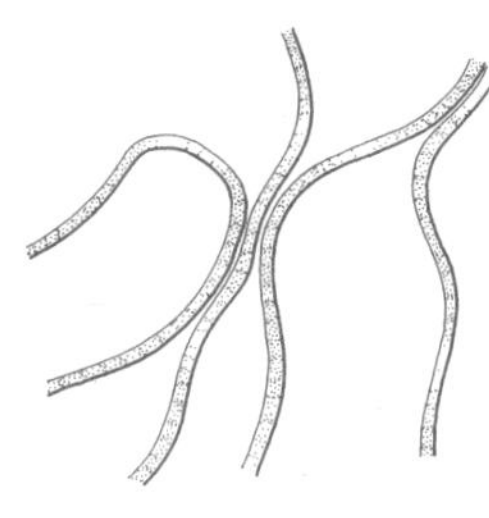
Chloroflexus

Only four species of green non-sulphur bacteria are known, from three genera: *Chloroflexus aurantiacus*, which is a thermophilic phototroph; *Hereptosiphon*, which includes two mesophilic (warm but not too hot) gliding species; and *Thermomicrobium roseum*, which is a thermophile. Sisters to these bacteria (it now transpires) is a clade that includes the radiation-resistant micrococci and related types, such as *Deinococcus* and its relatives; and *Thermus aquaticus*, which is ubiquitous in hot springs.

Thermotogales

New microbial studies, based not least on direct investigation of nucleic acids in the wild, suggest that there are many more high-ranking taxa of bacteria to be described. One recent find is *Thermotoga maritima*, which has unique lipids (fats) that have so far defied complete description, and whose rRNA is so different from the other bacteria that it seems to deserve an entire kingdom to itself, and is shown as the sister group of all the rest. According to Carl Woese, '*Thermotoga* would appear to represent a vast unexplored "other world" of thermophilic Bacteria'.

Aquifex and *Hydrogenobacter*

Branching even more deeply than *Thermotoga* in the bacterial tree is a group including *Aquifex* and *Hydrogenobacter*—microaerophiles that derive their energy by combining oxygen with hydrogen to form water. But because this 'other world' of bacteria is so unexplored, and must indeed be vast, further discussion of the deepest branches of the bacterial world will have to wait for future editions of this volume.

THE KINGDOMS OF ARCHAEA

The Archaea are clearly distinct from the Bacteria. Specifically, among other things, they have remarkable fats (lipids) that are like nothing else in nature (being branched and ether-linked). In general, as Carl Woese has noted, 'Archaea display their own characteristic version of every major macromolecular function'. Until they turned up in such vast numbers in the ocean, it seemed that almost all of them were adapted to conditions that we, from our animal-centric standpoint, would regard as extreme. (Admittedly Edward DeLong and his colleagues found them in the Antarctic; but that environment is pretty mild compared, for example, with hot springs and hydrothermal vents.)

Before the maritime types became evident, it seemed as if the Archaea divided reasonably neatly into four metabolic/ecological groups, each of which represented a different 'extreme'. One of the three are the halophiles, which prefer extreme salinity; and these incidentally are the only archaeal practitioners of photosynthesis, which they carry out with the aid of a unique photosynthetic pigment. The second metabolic group includes the extreme thermophiles, which have no known relatives that are not thermophiles. The archaean thermophiles metabolize sulphur. Third, there is a range of methanogens, which generate methane from carbon dioxide—and are the world's most important generators of methane (a potent greenhouse gas). Finally, a group represented by *Archaeoglobus* are thermophilic reducers of sulphate.

In practice, however, these four physiological phenotypes are divided between just two main kingdoms: the Euryarcheota and the Crenarchaeota. The kingdom **Euryarchaeota** includes all the methanogens and the extreme halophiles, which seem to have descended from one of the methanogens. The euryarchs also include some extreme thermophiles, such as *Thermoplasma*.

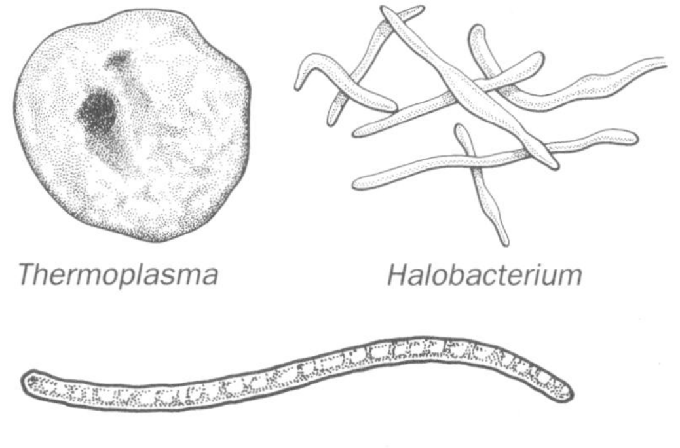

Most of the extreme thermophiles, however, belong in the kingdom **Crenarchaeota**; indeed they grow best at around the boiling point of water, and some grow most rapidly when the temperature is above boiling point. Overall, the crenarchs tend to be fairly uniform in phenotype. They all grow anaerobically although some can also grow aerobically, and most require sulphur for energy although some that use sulphur can in fact do without it. The newly discovered marine types seem to be divided fairly evenly between the euryarchs and the crenarchs.

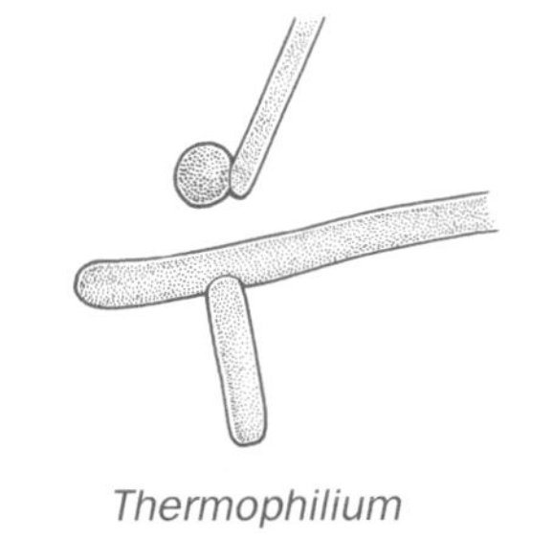

The third kingdom shown on the tree, the **Koryarchaeota**, is so far known only from DNA picked up from the wild, but, judging from the source of that DNA, they are probably thermophiles. Perhaps there are many more archaeal kingdoms yet to be discovered. Overall, there is clearly an awful lot to be found out about them. Phylogenetically, the Archaea represent one of the huge branches of nature and remarkably little is known about them. Their ecological impact must be commensurately huge but as yet can only be guessed.

HOW THE THREE DOMAINS ARE RELATED

Carl Woese and Norman Pace suggested long ago that it seems most appropriate to regard the Archaea as sister to the Eucarya, and the Bacteria as sister to the Archaea + Eucarya, as shown in the tree accompanying this section. Over the past decade, however, Woese and Pace's view of these grand relationships has been challenged by James Lake of the University of California at Los Angeles. He suggests that one particular group of Archaea should be seen as sisters to the Eucarya, and that this sister group should be renamed the Eocytes. The remaining archaeas should then be left within Carl Woese's original Archaea. Thus Lake effectively envisages the Eocytes as a fourth domain.

Nobody doubts that there are deep phylogenetic divisions within the Archaea, as Woese himself was the first to point out; and when you boil this discussion down to its simplest, the issue is whether the split that divides living archaes into 'eocytes' and the rest occurred after the split with the eukaryotes or before. If after, then the existing Archaea can properly be considered as a monophyletic group. If before, then the group 'Archaea', as now understood, would be paraphyletic (just as Reptilia* is paraphyletic because it is clearly ancestral to Mammalia and Aves). Some living archaeas would be specifically related to the eukaryotes, and the others would be the sister group of the eukaryotes plus those related archaeans. Woese and Pace favour the first scenario, meaning that all existing archaes form the monophyletic sister group of the eukaryotes; Lake favours the latter scenario.

The discussion continues but is immensely technical, and only a few people in the world are qualified to take part in it. It would be absurd for a non-specialist to comment. But the Woese–Pace scenario has precedence and the burden is on James Lake to show that they are wrong. Most specialists tend to favour the Woese–Pace view; and that is the picture presented here.

EVOLUTION OF THE PROKARYOTES

Within the Bacteria, note the relationships overall: *Aquifex*/*Hydrogenobacter* appear as the sister group of all the rest; then comes *Thermotoga* (in the kingdom Thermotogales); then the strange, green non-sulphur bacteria; then the cyanobacteria (and chloroplasts); then a huge grouping, with the two Gram-positive kingdoms on one side, and the vast array including the spirochaetes and the purples on the other. But note Woese's comment on the 'vast unexplored other world'. The list of species can include only those that we know about; and those that we know about are the ones grown in the laboratory. The work of Norman Pace and his colleagues suggests that the number of types still to be discovered is many times greater than the number so far known; and the continuing search in hot springs, carried out not least by biotechnology companies, is revealing that this is indeed the case. In short, the next 20 years or so is liable to show that our present view of bacterial diversity is very skewed indeed—although the picture presented here is far more realistic than that of the 1980s.

Why, though, does there seem to be so little correspondence, overall, between the phylogeny and the phenotypes? How is it that photosynthesis occurs in many different groups, or that aerobes so commonly rub shoulders with anaerobes, and thermophiles with normophiles, and so on? Well, as Carl Woese points out, this is very hard to explain if we assume—as traditional biology tended to assume—that the first organisms were heterotrophs, and that autotrophs evolved from those heterotrophs. If such were the case, we would have to produce one of several explanations, each of which seems unlikely. We could, for example, assume that all photosynthetic prokaryotes are closely related to each other—but this is what the RNA evidence expressly denies. We might suggest that photosynthesis evolved more than once. But photosynthesis is a tremendously intricate process that depends on the molecule chlorophyll in its various forms, and it seems unlikely that such a molecule could have originated repeatedly. (In the same way, palaeo-ornithologists dismiss the notion that feathers have evolved more than once.) Or we could surmise that different groups of prokaryotes have passed on the ability to photosynthesize from one to another; but this, too, is rendered unlikely by other evidence.

So, says Woese, it is much more economical to assume that the common ancestor of all living prokaryotes was an autotroph, from which the various forms of heterotroph have evolved independently. The transition from autotrophy to heterotrophy is easy to envisage; it merely involves the loss of a few enzymes and attendant systems—and such losses occur readily in evolution. In just such a way, many kinds of bird have lost the power of flight, and many parasites in many different groups of animals have lost the refinements of brains and senses. Loss—'degeneration'—is evolutionarily easy. In addition, the common ancestor was probably an anaerobe. It does seem that the ability to cope with and then to use oxygen has evolved several times independently, and this is not too difficult to envisage. At least, as the first cyanobacteria-like prokaryotes began to lace the atmosphere with oxygen, the selective pressure to deal with it would have been extremely high—so that organisms would have had to adapt to it (which they have done in various ways), stay away from it (in the way that many remain within airless marshes and sulphurous springs), or die. We can also reasonably envisage that the first organisms were thermophilic. In short, in the language of cladistics, autotrophy, anaerobiosis, and thermophilia are primitive features among the living prokaryotes; and it simply is unsurprising that these features turn up in group after group—or indeed, as is the case, that they typically occur among the most primitive members of their group.

Yet, in practice, it is hard to envisage the first organisms being autotrophs. Autotrophy requires considerable refinements, and must have taken several hundred million years to evolve. Probably, therefore, the first organisms of all were heterotrophs, and the first autotrophs emerged from them. Nonetheless, Woese's scenario holds, for it remains economical to assume that those primeval autotrophs were indeed the ancestors both of today's autotrophs and of all the heterotrophs that are alive today. Any of the primeval heterotrophs that survived the coming of the first autotrophs would surely have been outcompeted by the heterotrophs that evolved later. In short, today's heterotrophs represent a second wave of heterotrophs.

Finally, both Woese and Pace ask why there are various anomalies among the features of Archaea, Bacteria, and Eucarya. For example, the RNA shows that archaes are the closest relatives of eukaryotes; yet the structure of the outer cell membranes would suggest that bacteria are closer to eukaryotes. The reason, Woese and Pace suggest, is that in the beginning living systems were not divided into different organisms. There was just a living 'syncytium', which Woese has called the 'progenote': a more-or-less continuous living 'slime' that spread all over the globe wherever hot rocks met water, and in practice such a nexus was ubiquitous. Then we have merely to envisage that many of the major chemical systems that are now present in the three domains evolved before the primeval syncytium had divided into discrete organisms. The ultimate form of the different organisms, once they did arise, was then a matter of chance.

This, then, is a lightning introduction to two-thirds or, more probably, to at least 90 per cent, of the diversity of life on Earth; diversity that is encapsulated not in elephants or oak trees or us, but in creatures too small to see. Nonetheless, I devote the rest of this book to the minority. Eukaryotic chauvinism is difficult to shake off.

3

THE REALM OF THE NUCLEUS

DOMAIN EUCARYA

THE DOMAIN of the Eucarya includes most of the big and conspicuous creatures: seaweeds, animals, plants, and fungi and other fungus-like organisms. Prokaryotes can also be conspicuous *en masse*—especially in stromatolites—but, for the most part, large individual body size is a prerogative of eukaryotes.[1] Yet many eukaryotes are miniatures, too: in fact, it seems probable that most eukaryotes are protist in form, which means they are either single-celled or contain just a few co-operating cells. Eukaryotic cells in general are far bigger than prokaryotic cells—like a rhinoceros compared with a rabbit—but, even so, most of the single-celled eukaryotes are only just visible to the naked eye, at best. It is simply an ecological fact that most species of any kind, eukaryote or prokaryote, are small—because the world provides more potential niches for small creatures than for large. Elephants need an entire continent while protists thrive in specks of mud.

The eukaryotes of protist grade are extremely various. Some are busy, mobile heterotrophs. Some photosynthesize, and many of the photosynthesizers are mobile. Some live a fungus-like existence, though most of them are also mobile. In more innocent days, in fact until just a few decades ago, the most mobile protists were placed in the putative taxon 'Protozoa', which was given the rank of phylum or sub-kingdom, and then thrust into the kingdom Animalia; those that photosynthesize were bundled with the seaweeds into a group called 'Algae', which in turn was tucked in among the plants; while anything vaguely fungus-like was *ipso facto* a fungus (and in really old-fashioned classifications, the fungi, in turn, were deemed plants).

R. H. Whittaker's suggestion in the 1950s that all the protists, together with the seaweeds, should be placed in a new kingdom—Protoctista—was a useful interim measure, which at least nudged biologists out of a mind-set that was essentially eighteenth century. But now, the same kinds of molecular studies that have revealed the extraordinary variety of the prokaryotes are showing that Whittaker's 'protoctists',

[1] As noted on page 98, the British spell 'eukaryote' with a 'k', while Americans spell it with a 'c'. Such discrepancies are common enough, as all authors know. What's confusing in this case, however, is that the formal name of the domain, as first laid down by the American Carl Woese, is 'Eucarya'. So Brits like me are obliged to write 'Eucarya' formally, and 'eukaryote' informally.

too, are astonishingly diverse. For instance, every student of biology is introduced to *Paramecium*, the charming, ciliated 'slipper animalcule' that races around in drops of pondwater; and also to *Plasmodium*, the fearsome parasite of malaria, which probably first evolved in birds but spread to primates and still kills about a million people a year. In traditional classifications, both were placed in 'Protozoa'; and in Whittaker's scheme they both became 'Protoctista'. Yet RNA studies now reveal a greater phylogenetic distance between a paramecium and a plasmodium than there is between a human being and an oak tree. The terms 'protozoan', and the more general 'protist', can still be used as informal adjectives. But now it is clear that such terms refer only to general form, and have no more phylogenetic validity than 'tree' or 'toadstool'. It is no longer sensible to pretend that the protistan eukaryotes form a valid, monophyletic taxon that merits a formal label such as 'Protozoa', 'Protista', or 'Protoctista'. Indeed, if human beings and oak trees are placed in separate kingdoms then so, too, should paramecia and plasmodia.

So progressive biologists are placing paramecia and plasmodia in separate kingdoms. Indeed, whereas taxonomists once recognized just two kingdoms of eukaryotes (animals and plants) or three (animals, plants, and fungi), or four (animals, plants, fungi, and 'protoctists'), some modern biologists now acknowledge 20 kingdoms or more, and most of those are protists. Our own kingdom, the Animalia, has thus been dramatically demoted: from a conceptual 50 per cent of the whole to less than 5 per cent. Thus humans have been driven from the centre of the biological stage just as the astronomy of Copernicus and Galileo shifted planet Earth from the centre of the Universe. Overall, the classification of life has become messier and so more difficult to get to grips with. But the underlying diversity revealed by the new molecular studies is dazzling. It shows that life on Earth is even more extraordinary than has become obvious through the past few centuries of study and, for such a revelation, a little untidiness is a small price to pay.

THE SPECIALNESS OF EUKARYOTES

The large size of the eukaryotic cell relative to the prokaryotic cell—the metaphorical rhino against the allegorical rabbit—reflects far greater structural complexity. Taken as a group, prokaryotes are more biochemically versatile than eukaryotes. But taken individually, and considered as a piece of engineering, the eukaryotic cell is more intricate than the prokaryotic cell. Eukaryotes clearly evolved from prokaryotes. Complex systems always evolve from simpler ones; although complexity may sometimes be lost, secondarily, so that some simple organisms have, in fact, derived from more complex ones.

The most obvious single difference between prokaryotic and eukaryotic cells, and the one from which the two grades of creature derive their names, lies with the **nucleus**. When the cell is suitably stained the nucleus appears as a discrete entity against the background of cytoplasm, like a yolk in a fried egg (though this metaphor is purely visual). The nucleus is surrounded by membranes (whose structure is clearly visible under an electron microscope) and within it resides most of the genetic material of the eukaryotic cell. But prokaryotes lack nuclei. They package their genes in various forms within the cytoplasm; and although the genetic material of the prokaryote may occupy a distinct zone, this area is never enclosed within discrete membranes, as in a eukaryote. 'Eukaryote' is a typically chauvinist term. We regard nucleated cells as 'good' because that is the kind we have ourselves; even though, of course, prokaryotes were first on the scene by about a thousand million years.

DNA, whether in eukaryotes or prokaryotes, consists of long spiral-shaped molecules—the famous 'double helix' as first described by James Watson and Francis Crick in the early 1950s. But the two grades of organism package their DNA somewhat differently. In eukaryotes, the DNA is coiled around a core of proteins called **histones**, which among other things provide mechanical strength; and the whole protein–DNA complex is called a chromosome. The DNA of prokaryotes in general lacks this toughening core of histones—although, as we will see later, some archaes do have histones, which suggests that they are related to eukaryotes and gives a clue to eukaryote ancestry.

In fact, the nuclear membrane and the histones are both examples of what may be seen as the underlying, general difference between eukaryotic and prokaryotic cells. Eukaryotic cells have an intricate **cytoskeleton** (literally, 'cell skeleton'), which makes the whole structure tougher and more versatile; and prokaryotes at best have only the rudiments of a cytoskeleton. Thus in eukaryotes the outer cell membrane folds in on itself and forms a kind of three-dimensional latticework called the **endoplasmic reticulum**, which infiltrates and supports the entire cytoplasm. The nuclear membrane is itself formed from a part of the endoplasmic reticulum; and so, too, are the membranes that fold around (or in some cases form) other specialist structures (**organelles**) within the eukaryotic cell. Prokaryotic cells, lacking a cytoskeleton, are generally supported by an outer cell wall. Eukaryotic cells are naked in their primitive state, although some, such as those of plants and fungi, have re-evolved new kinds of cell walls.

The histones can be seen as a component of the cytoskeleton. Other protein components of the cytoskeleton are **contractile**, meaning that they can shorten forcefully, like muscles. One of these is **actin**, which is one of the main proteins of muscle (the other being myosin). The other contractile cell protein is **tubulin**, which apparently is derived from the endoplasmic reticulum. Both these contractile proteins are characteristic of eukaryotic cells but are found in only a few prokaryotes—and those that do contain these proteins probably have some special relationship with eukaryotes. The contractile proteins enable the eukaryotic cell to move in ways that most prokaryotes cannot. In particular, many eukaryotic cells can move like an amoeba, pushing out extensions, or **pseudopodia**, and sometimes using these to move around. Indeed, **amoeboid movement** may be seen to be primitive for eukaryotes. Many different protozoans are amoeboid in form—and although all 'amoebae' used to be placed together in the same 'class' of 'Protozoa' it is now clear that different amoebae may be more distant from each other phylogenetically than horses are from mushrooms, and the different amoeboid types should accordingly be placed in different kingdoms. But once we see that amoeboid movement is simply a primitive eukaryotic character, this becomes unsurprising; primitive characters have a way of popping up throughout a phylogenetic tree, just as the primitive tetrapod five-fingered 'hands' pop up throughout the vast and varied tree of amphibians, reptiles, and mammals.

Because they are endowed with actin and (primitively) they are not restricted by cell walls, eukaryotic cells have enormous scope for movement, which is denied to most prokaryotes. The outer membrane can move in on itself, bearing large molecules with it, in the process called **endocytosis**; and spit things bodily out again, which is **exocytosis**. Many eukaryotic cells can reshape their whole selves by changing shape in an 'amoeboid' manner, and by such amoeboid movements, or **phagocytosis**, they can engulf large particles for food. In such a way, protist eukaryotes often feed on prokaryotes. Prokaryotes evidently cannot practise endocytosis or the grander-scale phagocytosis: they must take their nourishment effectively in solution.

Finally, when eukaryotic cells divide they do so by extremely precise procedures called **mitosis** and **meiosis**. Mitosis is the form of cell division practised when somatic (body) cells divide into two. The somatic cells of many eukaryotes—for example, those of oak trees and human beings—contain two sets of chromosomes, and are then said to be **diploid**. In mitosis, this chromosome number is carefully preserved—that is, the chromosomes double and then are halved again—so that each daughter cell is diploid as well, and is in effect a facsimile of the parent cell. Meiosis is the form of cell division used during the formation of gametes—eggs and sperm. Gametes in general contain only one set of chromosomes—and are said to be **haploid**. So, during meiosis, the chromosome number is halved. Before and after mitosis and meiosis, the nuclear membrane first breaks down and then reappears.

Overall, mitosis and meiosis result in a wonderfully exact division of genetic spoils between the daughters; the division of eukaryotic cells is a precision process, almost military. The point is, however, that during mitosis and meiosis the chromosomes, toughened with histones, are hauled about athletically and even violently by tubulins. If they were not toughened by histones, they would surely break: and if the eukaryotic cells lacked tubulins, they could not manhandle the chromosomes into their required positions. Prokaryotes, generally lacking histones and tubulins, simply cannot and do not practise mitosis and meiosis. Typically, the cell division of prokaryotes is a more variable and apparently ad hoc affair.

In any eukaryote that practises sex (which is probably most of them) haploid cells (or successive generations thereof) alternate with diploid cells (or successive generations thereof). Thus in most animals, like human beings, the body cells are diploid but spermatozoa and unfertilized eggs are haploid; diploidy is restored when sperm and egg combine to form a one-celled embryo. Many other creatures (like many protozoans) are generally haploid, but may combine sexually to form a diploid stage that may last only a short time before dividing to produce haploid offspring. In green seaweeds, mosses, and ferns, there is distinct **alternation of generations**; a discrete haploid stage alternates with a diploid. As it happens, the leaves of mosses are haploid, those of ferns are diploid, while the thalli (fronds) of green seaweeds may be either.

The intricate cytoplasm of the eukaryotic cell, interlaced with endoplasmic reticulum, generally contains several kinds of specialized structures—organelles. Almost all eukaryotic cells have **mitochondria**, which contain the enzymes that break down carbohydrates to provide energy. I say 'almost all' because the peculiar protistan parasite *Giardia* lacks mitochondria, though whether it has lost them secondarily, or never had them in the first place, is unknown. The cells of many eukaryotes also include **plastids**, the commonest kind of which, **chloroplasts**, contain the green pigment chlorophyll, which is the key molecule in photosynthesis. Many eukaryotic cells also carry one or more and sometimes many whip-like projections on their surface. These are of two main kinds: long or short. The long ones are traditionally called **flagella** (singular, flagellum), while the short ones are **cilia** (singular, cilium). Eukaryotic flagella and cilia turn out to have the same basic structure, which in general differs from that of the 'flagella' of some prokaryotes; so Lynn Margulis of the University of Massachusetts has suggested the term **undulipodia** to cover both the kinds of projection found in eukaryotes. 'Undulipodia' is a useful collective term but the traditional terms seem to be worth keeping. 'Cilia' is unambiguous and I, for one, am happy to say 'eukaryotic flagella' or 'prokaryotic flagella' when there is a chance of confusion, and just to say 'flagella' when there is not.

Overall, then, the eukaryotic cell is very different from, and far bigger and more complex than, the typical prokaryotic cell. So how did the intricate eukaryotic cell evolve? There is an obvious conceptual problem. Eukaryotic cells must have evolved from prokaryotic cells because, in general, we have to assume that complicated structures evolved from simpler ones and, in any case, there is much specific evidence to suggest that this is so. But present-day prokaryotes in general seem to be very different from eukaryotes: no nucleus, no true mitosis or meiosis, no endoplasmic reticulum, no mitochondria, no plastids, and no definite evidence of the typically eukaryotic tubulin or actin. In fact, there seems to be a huge lack of continuity. How can the gap be bridged?

HOW THE EUKARYOTIC CELL EVOLVED

Intuitively, as a first hypothesis, we might guess that the eukaryotic cell arose from some single prokaryotic ancestor, which, over time, simply became more complicated. We can imagine, for example, the outer membrane folding inwards to form a rudimentary endoplasmic reticulum, which then became more and more elaborate, folding back upon itself to form organelles and other inclusions. In particular, we can see how the endoplasmic reticulum could have folded to enclose all the genetic material, and hence form the nucleus. We can see, too, how natural selection would have favoured formation of a nucleus, because the nuclear membranes create an environment within them that in general is more favourable to DNA than is the environment of the cytoplasm. Similarly, many prokaryotes bundle their DNA together into 'nucleoids', as if to protect them from the general cytoplasmic hurly-burly.

Thus, elaboration of some ancient, somewhat eukaryotic-like prokaryote could explain much of what we see in modern eukaryotic cells, including the nucleus. But, for various reasons, the organelles mentioned above—mitochondria, plastids, and undulipodia—do not seem to be so easily explained away. Indeed, in 1910 the Russian biologist C. Mereschkowsky proposed that these organelles could have evolved from bacteria that had invaded the original, proto-eukaryotic cell and stayed to become a part of it. Perhaps they came in as parasites, and perhaps the proto-eukaryote engulfed them for food; but whichever was the case, they became permanent lodgers. This is a pleasing notion but it seemed too strange, and apparently made little impact until recent decades when Lynn Margulis became its champion.

Now, thanks to Margulis, most biologists believe that this quaint and even bizarre idea is true. The bulk of the eukaryotic cell, including the nucleus, does indeed seem to derive from some large ancestral prokaryote acting as the 'host'; while mitochondria and plastids are both apparently derived from invading bacteria. Eukaryotic cilia and flagella might also in theory have evolved from bacterial invaders, perhaps specifically from spirochaetes, but there is no evidence to support this particular idea so the origins of undulipodia remain mysterious, and I will not discuss them further. In general, though, the eukaryotic cell seems to have originated not from one prokaryotic ancestor, but from a coalition of several. Eukaryotic cells with mitochondria have at least two prokaryotic ancestors—one (at least) providing the cytoplasm and nucleus and the other providing the mitochondria. And eukaryotic cells that also contain plastids have at least three prokaryotic ancestors, who came together and began working in concert.

Mereschkowsky and Margulis favoured the coalition idea largely on morphological grounds: mitochondria and plastids look like some prokaryotes. Molecular studies now seem to prove the case beyond doubt. For, odd though this may seem, both mitochondria and plastids contain genes; and the fine structure and organization of these genes, and the RNAs and the enzymes and other proteins that those genes produce, are more like those of bacteria than of eukaryotes. It seems inescapable, then, that the genomes of mitochondria and plastids derive from ancient bacterial invaders. It is also clear that most of the genes of the original invaders have either been lost, or have passed into the nucleus of their 'host' cell; thus mitochondria generally contain far less DNA than occurs in a typical prokaryote. Yet the DNA that does remain within mitochondria and plastids is highly influential; for example, mitochondrial DNA makes some of the proteins concerned with respiration, and produces some of the enzymes necessary for mitochondria to replicate. Mitochondrial genes also have some odd effects on the organism as a whole. In some plants, for example, some mitochondrial genes cause male sterility.

Crucially, however, the genes in mitochondria and plastids work closely with those in the nucleus, so that the activities of the organelles and the rest of the cell are tightly co-ordinated. Thus we see plastids and mitochondria replicating at the same time as the nucleus divides, and as the cell as a whole replicates. The degree of co-ordination is such that some vital enzymes are made by consortia of genes, some supplied by the organelle and some by the nucleus. For example, one of the key enzymes in respiration, ATP synthetase, is produced by nuclear and mitochondrial genes working together.

In summary, then, the evidence that mitochondria and plastids have evolved from prokaryotic invaders—and specifically from bacterial invaders—is overwhelming. But in the 2500 million years or more since this invasion first took place, many of the invaders' genes have passed to the nucleus (or been lost) so that the genes of the invaders and the 'host' work perfectly together. It is because the genomes of invaders and host are effectively mingled that the eukaryotic cell can truly be considered to be an entity, even though its origins are multifarious. So, for example, the relationship of plastid to nucleus is in absolute contrast to that of an alga and a fungus when they combine to make a lichen. In lichens, the genomes of the two protagonists (symbionts) remain emphatically separate, and although the relationship is stable it remains one of armed truce, with alga and fungus both striving, by chemical and mechanical means, to keep the other in check.

Such, then, is the evidence for coalition in the eukaryotic cell. The next question is 'who?': that is, which prokaryotes formed the original consortium?

THE COLLABORATORS

The economical assumption is that the eukaryotic cell arose only once, and that all eukaryotes arose from that first ancestor. If this is so then we need at least three prokaryotic ancestors. One acted as the 'host', and provided the cytoplasm, endoplasmic reticulum, and most of the nucleus. One evolved to become the mitochondrion. Yet another gave rise to plastids—or perhaps there was more than one plastid ancestor, given that any one eukaryote may contain several kinds of plastid (the chloroplast is only one example), and chloroplasts themselves vary considerably from group to group.

There is no living prokaryote that precisely resembles the host, the mitochondrion, or the plastid. It would be surprising if there were, because these three entities have been evolving in concert for a very long time, have clearly changed enormously since they first got together, and mitochondria and plastids have both lost their ability to live independently under ordinary conditions. The best we can hope to find are living prokaryotes that also evolved from the various ancestors of the eukaryotic cell—bearing in mind that these descendant prokaryotes have also been evolving in their own niches for several thousand million years since the original eukaryote coalition took place. In short, we should look for prokaryotes that show at least some intriguing similarities to eukaryotes: in their physical structure, in their chemistry, and in their DNA and RNA. There are some plausible candidates.

THE ANCESTRAL HOST: A HEAT-LOVING ARCHAE

It has long been clear that the ancestral host—the supplier of cytoplasm, endoplasmic reticulum, and nucleus—was an archae, not a bacterium. Carl Woese showed that eukaryotic rRNA is far more similar to archaeal than to bacterial rRNA (which is not a conclusive argument in itself because bacterial rRNA might in theory have changed more rapidly than archael rRNA over the past few thousand million years; but is at least suggestive). Archaes tend to live in conditions that we regard as 'extreme', including hot, acid springs. But many do not—and in any case we have seen that thermophiles often have 'normophilic' relatives. Besides, it is at least conceivable that eukaryotes may themselves have originated as thermophiles. The point is, though, that we need not be dismayed if the most eukaryote-like of the living prokaryotes turn out to be thermophiles. In fact, among modern archaes, the two that seem to resemble the putative ancestor of eukaryotes most closely are the acid-loving thermophile *Thermoplasma* and the sulphur-dependent thermophile *Sulfolobus*.

Thermoplasma seems the more promising of the two. Unlike most prokaryotes —or, at least, unlike most of those that have so far been studied—it lacks a cell wall, and its shape is indefinite. Instead, its cytoplasm seems to be held in place by what might be a cytoskeleton; and this cytoskeleton seems to include a contractile protein that might be actin-like. There are many 'seems' in this, but these are early days. *Thermoplasma*'s DNA and RNA are in some ways special to itself (for example, its rRNA is distinctive) but in other ways they are intriguingly eukaryote-like. Thus, the genes of eukaryotes contain odd sequences of DNA that have no obvious function, and appear simply as interruptions—introns. Some of the genes of *Thermoplasma* also contain introns; notably, the genes that code for tRNAs, rRNAs, and the enzyme DNA polymerase (which is essential for the replication of DNA). Perhaps most suggestive of all, the chromosomes of *Thermoplasma* contain histone-like proteins; and the only other known prokaryote with anything resembling histones is the archae *Methanococcus*.

Sulfolobus also has flat, irregular cells by which it sticks itself closely to particles of sulphur, from which it derives its energy. Several of its enzymes seem very eukaryote-like. Most intriguing, however, is that its membranes seem to contain steroids; or at least they contain steroid receptors. Steroids play a huge part in the life of eukaryotes, and, for example, provide the basis of our own sex hormones. But apart (perhaps) from *Sulfolobus*, steroids are uniquely eukaryotic.

The histone-like proteins in the chromosomes of *Thermoplasma* serve to strengthen the strands of DNA; the actin-like proteins (if they are really present) help it to flatten itself to its substrate, thus enhancing the contact; and the putative steroids in the membranes of *Sulfolobus* help to strengthen the membranes, which is also part of their function in eukaryotes. All this general toughening-up and versatility of shape presumably help *Thermoplasma* and *Sulfolobus* to survive in an environment that, chemically speaking, can be called violent. But such features have also proved essential to eukaryotes. The histone-reinforced DNA can be hauled about bodily in mitosis and meiosis, which in turn enables huge molecules of DNA, containing vast amounts of genetic information, to be divided accurately. The tough, steroid-reinforced membranes can be pulled here and there by actin, which in turn allows amoeboid movement and phagocytosis—the consumption of bulk food. Features that evolve in response to one environment (in this case, hot acid) and are then pressed into different service in a new environment (the cool everyday world that we inhabit) are called 'pre-adaptations'. We can suggest, in short, that some acidophilic, thermophilic archae was pre-adapted, in many ways, for later life as a eukaryote.

By contrast, mitochondria and plastids did not originate as archaes. Their nucleic acids and their general form proclaim that their ancestors were bacteria. So, which ones?

PROTEOBACTERIA AS ANCESTORS OF MITOCHONDRIA

The bacterium that is most commonly mooted as the 'sister' of the mitochondrial ancestor is *Paracoccus*, which is a proteobacterium or 'purple' bacterium. Many purples are photosynthesizers but a few—like *Paracoccus*—can also live as heterotrophs. Some purple bacteria, like *Rhodopseudomonas spheroides*, do respire in ways that resemble mitochondria when they are grown in the dark.

Mitochondria enable modern eukaryotes to respire aerobically—that is, to use oxygen to 'burn' carbohydrates to supply energy. But the purple bacteria that first hitched up with the 'host' archae to form the earliest eukaryotes probably did not have that function at all. Indeed the host at the time was probably growing anaerobically, in some hot acid spring. Anaerobic organisms in general do not merely eschew the use of oxygen: they are often poisoned by it, for oxygen is extremely lively stuff, chemically, and anything that is chemically lively is wont to be highly destructive. So the original 'role' of the invading purple bacterium in the first eukaryotic cells was, probably, to help to deal with the oxygen that was being pumped in ever greater amounts into the atmosphere by some of the world's earliest prokaryotic photosynthesizers. One of several ways to 'detoxify' this noxious oxygen was to attach it to organic molecules and hence produce relatively harmless carbon dioxide and water. This particular method of disposal, it turned out, also generated energy that could be harvested, and aerobic respiration was born. The prize is great for organisms that can deal with oxygen in this way, because the liveliness of oxygen is an asset for those that can cope with it. Aerobic respiration is fast and efficient.

CYANOBACTERIA AS ANCESTORS OF PLASTIDS

Plants (including green seaweeds), the two kingdoms of the red and the brown seaweeds, and about half a dozen kingdoms of protists, possess plastids. Chloroplasts—the ones containing chlorophyll—are not the only kind of plastids; some (called chromoplasts) carry other pigments such as carotenoids, which lend the reds and yellows to fruit, flowers, and some leaves, whereas others (leucoplasts) in the leaves of plants store starch. But chloroplasts are the ones that deserve our attention; they are the principal kinds, and we may reasonably suppose that the minor types have been derived from them.

Chloroplasts themselves differ among different groups of eukaryotes. Those of rhodophytes contain phycobiliproteins in addition to chlorophyll, which is what makes red seaweeds red. Chlorophyll itself comes in various forms, which organic chemists distinguish by different letters. For example, the chlorophyll in red algae is chlorophyll a; plants contain chlorophyll a and b; some protists contain chlorophyll c; and brown seaweeds (phaeophytes), chrysophytes, dinoflagellates, haptophytes[2] and some others contain chlorophylls a and c, but never b. We can be sure that chloroplasts arose from prokaryotes. But did all chloroplasts in all eukaryotes arise from the same prokaryotic ancestor, or did the different kinds of chloroplasts (with different chlorophyll) each arise from different prokaryotes?

Many biologists have favoured the latter scenario. Thus, for example, the cyanobacterium *Synechococcus* contains chlorophyll a and phycobiliproteins, just like red seaweeds; so it seems very likely that the red seaweeds acquired their chloroplasts from a *Synechococcus*-like ancestor. Another cyanobacterium, *Prochloron*, contains chlorophylls a and b, but no phycobiliprotein. Surely a *Prochloron*-like bacterium gave rise to plant chloroplasts?

In practice, however, molecular studies (based on ribosomal RNA) suggest that chloroplasts did all derive from the same ancestor, and that that ancestor was like *Synechococcus* rather than *Prochloron*. *Synechococcus* has short, rod-like cells. Many of its members live in the surface waters of oceans with other small drifting organisms (forming the plankton), where it is estimated to provide up to 10 per cent of all primary productivity—meaning 10 per cent of all the organic molecules at the bases of all the oceanic food chains. But other *Synechococcus* species are thermophilic, enjoying temperatures around 74 degrees Celsius, and such thermophiles, living around 2500 million years ago, could well have formed the fruitful liaison with the thermophilic archae *Thermoplasma*.

To be sure, *Synechococcus* does not contain chlorophyll b, as found in plants; but it does have phycobiliproteins, which are not found in plants. But Betsey Dexter Dyer and Robert Alan Obar point out that chlorophyll b differs very little from chlorophyll a, and that the chemical transition from one to the other is straightforward. Once the ancestor had chlorophyll a, it could easily make chlorophyll b. It need not bother us that plant chloroplasts lack phycobiliproteins. Losses of primitive features are commonplace in evolution. Intriguingly, too, the eukaryotes that retain the primitive phycobiliproteins—red seaweeds—are also the earliest eukaryotic photosynthesizers to occur in the fossil record.

On present evidence, then, it seems that different chloroplasts in all eukaryotes derive from the same ancestor even though, at first sight, this seems unlikely; and that the common chloroplast ancestor was a cyanobacterium resembling the modern *Synechococcus*. But we need not assume that this ancestor invaded only once; that is,

[2] Haptophytes are a group of unicellular algae that do not appear on the tree as their position remains uncertain.

we need not assume that one and one only photosynthetic eukaryote appeared 2000 million or so years ago and that all subsequent lineages descended from it. It is quite possible, in theory, that different eukaryotic lineages acquired their cargoes of *Synechococcus*-like symbionts independently, and at different times. Indeed, there is much evidence to suggest that this is the case. For instance, chloroplasts in different eukaryotic lineages are surrounded by different numbers of membranes. In plants (including green algae) and red seaweeds, the plastids have two membranes; in euglenoids and dinoflagellates they have three; and in cryptomonads, haptophytes, chrysophytes, xanthophytes, and some other types they have four. It seems reasonable to suggest that plants and red seaweeds acquired their chloroplasts in the form of free-living *Synechococcus*-like cyanobacteria, which they then engulfed in a two-layered membrane formed from the endoplasmic reticulum; but that those with multiple membranes acquired their chloroplasts ready-made from other eukaryotes that already possessed them, and had already surrounded them with membranes of their own.

Note that the distribution of photosynthesizers among the eukaryotes is spotty. They turn up here and there, in many different groups, and their closest relatives, commonly, are non-photosynthesizers—so that photosynethetic euglenoids clearly shared their early evolutionary history with trypanosomes, which are blood parasites; and plants are close to animals and fungi. There is a similarly spotty distribution of photosynthesis among prokaryotes, although for different reasons. The different photosynthetic eukaryotes presumably acquired photosynthesis independently of each other, by taking photosynthesizing prokaryotes on board at different times. But it seems most likely that photosynthesis initially evolved among prokaryotes only once, and that its spotty distribution in modern types is explicable in two ways. First, early prokaryotes presumably passed the genes from one to another that were needed to fashion the photosynthetic apparatus—so that different groups acquired the skills from each other by horizontal transmission. Second, we can regard photosynthesis as a primitive feature of living prokaryotes, and photosynthetic types may often be closely related to non-photosynthetic types simply because the latter have lost the ability. Such secondary loss is common in nature. Nonetheless, it seems a little unsatisfactory—somewhat ad hoc—to explain an apparently similar phenomenon (the spotty distribution of photosynthetic types in eukaryotes and prokaryotes) in two different ways. So what is the excuse for this?

The fact is that in all matters of life's history biologists must deal only in likelihood, for we cannot rerun the video and see what really happened. Although the two instances look the same, they are different, and we need to make different assumptions. Thus it is clear that photosynthesis first evolved in prokaryotes. But it is obvious, too, that photosynthesis is an immensely complicated business, and it seems unlikely that such complexity could have evolved, from scratch, more than once. So however spottily the photosynthesizers now appear on the prokaryote tree, it still seems probable that they all derived from the same primordial photosynthetic ancestor. But the first eukaryotes merely had to acquire cyanobacteria ready-made.

Among present-day creatures we see many such symbioses that are similar, at least in principle. Fungi link up with algae—and/or cyanobacteria—to form lichens; among animals, too, sponges, corals, flatworms, clams, sea slugs, and sea squirts have all taken various photosynthetic protists on board. To be sure, none of these relationships involves any mingling of genomes, as has happened with plastids and host nucleus, but the point is made: symbiosis in principle is easy. So although the distribution of photosynthesizers is comparably spotty in prokaryotes and eukaryotes, the two cases warrant different explanations.

One last note. Alfred, Lord Tennyson, in the decade before Charles Darwin published *Origin of Species*, coined the expression 'Nature red in tooth and claw'. Darwin, in *Origin*, argued that natural selection provides the mechanism of evolutionary change, and emphasized that competition is the spur to natural selection. Herbert Spencer, in the decade after *Origin*, coined the expression 'survival of the fittest', which emphasizes nature's competitiveness; and Darwin later adopted that expression. Thus the belief has been encouraged that all creatures are selfish and that because they are selfish they must constantly be at war; and the notion that our own, eukaryotic lineage was formed in the beginning by a coalescence of different organisms, a co-operation, seems counter-intuitive.

But Darwin is widely recognized as the greatest biologist who has ever lived and he knew perfectly well, and emphasized, that different species in nature often do co-operate. Certain moths, for example, feed only from particular orchids, and as they feed they spread the orchid pollen; and the orchids, in turn, produce copious nectar to reward the moths, in intricate receptacles that only the moths can reach. This is co-operation of the kind known as **mutualism**, a special form of symbiosis (which, in turn, implies 'joining of life'). It is easy to see intuitively how such co-operation comes about. The moth does not care about the orchid. But it benefits by being a specialist feeder, because such specialism increases efficiency. The orchid does not care about the moth. But by producing the nectar that attracts particular moths, and ensuring that no other insect can sneak in to steal the nectar, it guarantees the moth a good feast and ensures that the pollinating moths will focus their attentions on itself. In short, co-operatives spring up *because* the creatures involved are selfish. Sometimes self-interest is served most efficaciously by destroying, eating, or perhaps just ignoring other individuals (whether of the same or different species); and sometimes self-interest is best served by co-operating. Whatever works best—enhances personal survival and reproduction most efficiently—natural selection will favour.

So we need not be surprised that eukaryotes first arose as coalescences of different organisms. But we can take comfort from it; from the thought that the need for personal survival does not necessarily lead to strife, but can lead equally well to co-operation.

So much for the generalities. Now we can look briefly at the organisms themselves. My ideas on eukaryotes have been primarily inspired by Mitch Sogin's seminal molecular studies at the Center for Molecular Evolution, Marine Biological Laboratory, Woods Hole, Massachusetts; and the tree is based fairly slavishly on his ideas, although it also borrows from Betsey Dexter Dyer and Robert Alan Obar's book *Tracing the History of Eukaryotic Cells* (1994).

A GUIDE TO THE EUCARYA

Although each of the terminal branches of the tree represents a discrete kingdom, most are named informally: either with a term like 'ciliates' or 'diplomonads' or, in some cases, simply with the name of a genus—such as *Naegleria*. This is because the classification of eukaryotes as a whole is in a state of transition: the data are not yet in; the ideas have not yet settled down; the divisions shown here are not universally accepted (or even familiar to many biologists); and the formal processes that used to ensure that lofty rankings such as kingdoms were named with proper solemnity and—with luck—consistency has yet to be brought to bear.

So the names are a bit of a rag-bag. Where possible they borrow established groupings from traditional classifications, but the present groupings rarely correspond precisely with the traditional ones, and where they do the original names may be unsuitable for various technical reasons. For example, in one fairly standard traditional classification (there was never a universally agreed system) the 'euglenoids' are ranked simply as a suborder and called 'Euglenoidina'. The ending '-ina', which seems like a diminutive, does not seem to match the new kingdom status. Biologists need to agree (1) that the euglenoids should indeed be ranked as a kingdom; (2) exactly what organisms should be included within that kingdom; (3) whether the root 'euglen-' will prove most suitable once all the members are included; and (4) what the word ending should be ('-ata', '-ida'?).

Where a kingdom is marked just with the name of a genus, this is because few organisms have yet been examined by the modern, primarily molecular techniques, and exposed to cladistic rigour, and the ones that have been examined have proved to be very different from all the others. As more organisms are explored, the genera that now look so lonely at the ends of their branches will doubtless acquire a halo of relatives and then the grouping can be suitably named—perhaps, and perhaps not, by adapting some term from traditional systems. But some at least of the more progressive taxonomists who are devising the new classifications are not particularly keen on traditional Linnaean classification, or on the Graeco-Latin nomenclature that goes with it. As I explained in Part I (Chapter 5), I think this is a pity, and that some 'Neolinnaean' compromise is definitely called for. If it is, then with luck, in 20 years or so, we will have a classification that is both phylogenetically valid, and labelled with proper pomp.

Meanwhile, however, the research is still much in progress and the naming has hardly begun, and we must make do with what we have. I, of course, believe that the new classification as shown here is right in essence even though it is still developing, but it surely will not prove robust in every detail. In any case, however rough-and-

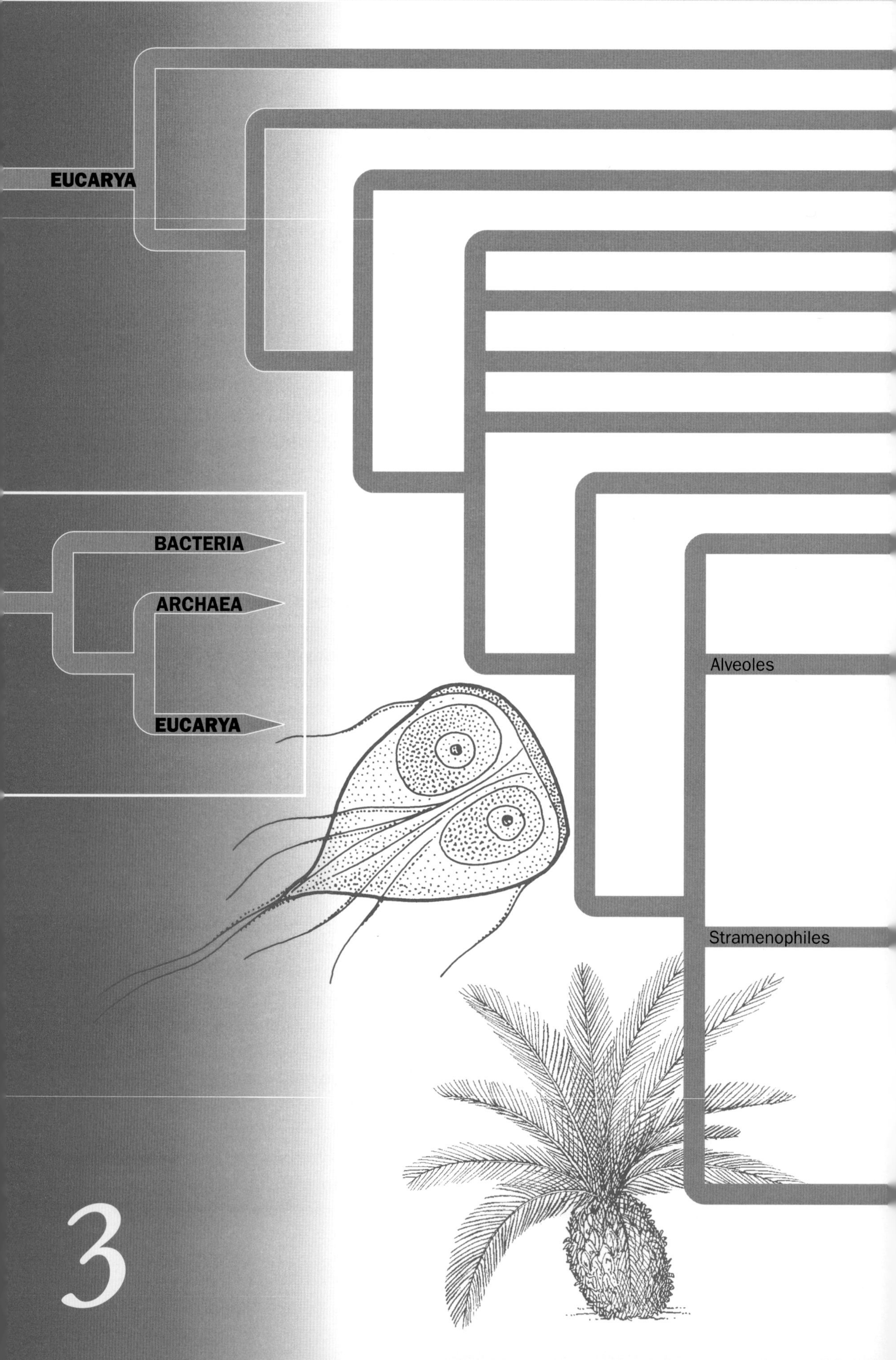
EUCARYA
BACTERIA
ARCHAEA
EUCARYA
Alveoles
Stramenophiles
3

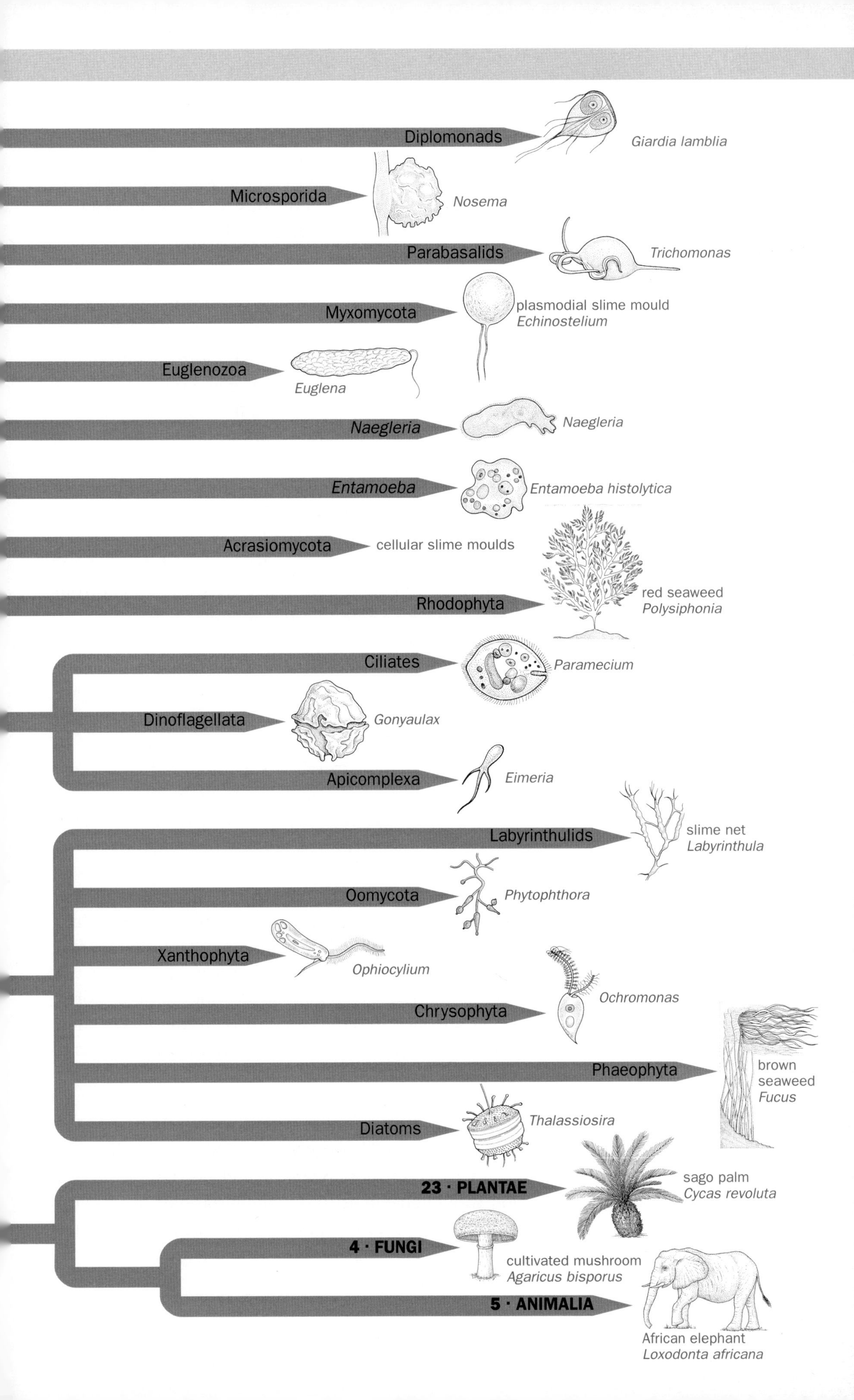

Diplomonads
Giardia lamblia
Microsporida
Nosema
Parabasalids
Trichomonas
Myxomycota
plasmodial slime mould
Echinostelium
Euglenozoa
Euglena
Naegleria
Naegleria
Entamoeba
Entamoeba histolytica
Acrasiomycota
cellular slime moulds
Rhodophyta
red seaweed
Polysiphonia
Ciliates
Paramecium
Dinoflagellata
Gonyaulax
Apicomplexa
Eimeria
Labyrinthulids
slime net
Labyrinthula
Oomycota
Phytophthora
Xanthophyta
Ophiocylium
Ochromonas
Chrysophyta
Phaeophyta
brown
seaweed
Fucus
Diatoms
Thalassiosira
23 · PLANTAE
sago palm
Cycas revoluta
4 · FUNGI
cultivated mushroom
Agaricus bisporus
5 · ANIMALIA
African elephant
Loxodonta africana

ready some biologists may feel this treatment to be, I feel it would be positively perverse to opt for anything more traditional. Whatever this tree may be losing in eighteenth-century solemnity, it surely is gaining in truth. So let us see what it has to offer.

You will surely be struck, but I hope intrigued rather than dismayed, by the complexity of this tree; and I hope suitably stunned by the thought that each of the terminal branches does indeed represent a kingdom, as profoundly different from all the rest as plants are from animals—and in many cases, much more so. Of course, some biologists will feel that to rank each branch so highly is going too far; that it is even a little flashy. Why can't the groupings shown here be ranked as phyla or classes, or even less, as in the past? But although it is generally hard to define any ranking precisely, certainly above the level of species, I do believe that the rankings ought to convey some serious fact about life; they should not simply be arbitrary. Here, the branches are distinguished by significant differences in ribosomal RNA—the kind of differences that denote very ancient, very 'deep', branching. Indeed, although *Naegleria* and *Entamoeba* seem to have much in common—they are both parasitic protozoans that behave as amoebae for at least part of their lives—they are more distant from each other, phylogenetically speaking, than plants are from animals.

For my part, I love the contrast between what is there to be seen—the creatures' phenotypes—and the underlying, evolutionary reality. The two ways of looking at living organisms, phenotype and clade, complement each other, like two angled spotlights that show a stage in three dimensions. Life is enhanced even more by the thought that our own wonderful kingdom, the Animalia, does not represent a half of all Earthly life as Linnaeus and all traditional naturalists supposed, but is just one branch of a vast, bushy tree; that we have thought ourselves to be more significant than we are only because we are animals ourselves and because we are so hugely influential ecologically—whereas some other creatures, though they might represent entire kingdoms, could disappear almost without any other creature noticing.

We cannot, however, take this tree as the last word: far from it, perhaps. Many, many more creatures remain to be exposed to molecular and cladistic rigour, including some that are significant ecologically, such as the planktonic Foraminifera and Radiolaria whose ancient skeletons form the chalk and flint rocks of some of the world's most dramatic landscapes; and many more molecules should be explored (ideally, complete genomes) before anyone can claim certainty. Thus the analysis that should emerge in the next few decades might be even more complicated than this—although there might also be more clustering, which in places would makes things easier.

But although this tree reflects reality more certainly than ever before and is therefore to be welcomed, it does not readily meet one of the requirements of classification. It is not easily remembered; most people could not take it in at a glance, as an ideal classification should enable us to do. Nine-year-olds, who relish complicated lists of dinosaurs and European football teams, will take the eukaryotes in their stride, but I sympathize with their teachers, who will find it much harder. So before we embark on a more formal look at the groups, let me first offer a highly simplified, highly subjective, phylogenetically invalid but nonetheless heuristic *aide-mémoire*.

AN INFORMAL OVERVIEW OF THE EUKARYOTES

I suggest that to be kind to our brains we can simply divide the 20-plus kingdoms that are shown in the eukaryotic tree—and, in all probability, any more that are likely to be defined in the future—into three big, informal groups: 'mega-eukaryotes', 'fungoids', and 'protists'. I spell them each with a lower-case initial letter to emphasize their informality; these names are mere colloquialisms, like 'tree', 'shellfish', and 'toadstool'. As will become apparent, each of these ad hoc groups includes a miscellany of kingdoms that are scattered throughout the eukaryote tree. But each group has a superficial, phenotypic coherence of the kind that human minds are equipped to remember.

MEGA-EUKARYOTES

The mega-eukaryotes include five kingdoms: plants, animals, fungi, red seaweeds, and brown seaweeds. The green seaweeds, which are quite different from the reds and browns, are now included with the plants. Mega-eukaryotic organisms are multicellular. In many other organisms—including many protists, and even including many prokaryotes—individual cells may co-operate to an extent to form colonies, and in some of those colonies the different cells become specialized to the point where they cannot survive without the others. But in true mega-eukaryotes this process of co-operation is taken to extremes. Many mega-eukaryotes contain billions and billions of cells, of many different types and specializations, all tightly co-ordinated with the others so that the whole is not a mere colony, like a stromatolite, but is an obvious, coherent entity, like a blue whale or an oak tree.

It is because they are multicellular that mega-eukaryotes are able to achieve large body size. The quality of largeness can be seen as an ecological niche, which the mega-eukaryotes have exploited. Bigness has many drawbacks—it is complicated, and big organisms commonly reproduce more slowly than small ones. But it has advantages, too: notably, big photosynthesizers, like plants and wracks, can find themselves a good place in the sun, and big heterotrophs can eat whatever is smaller. So natural selection has favoured large body size even though it brings complications. But nothing is ever for nothing in the natural world, and big creatures are, in their turn, parasitized unmercifully by the ones that have continued to exploit the possibilities of smallness. As Jonathan Swift observed, 'a flea hath smaller fleas that on him prey; and these have smaller fleas to bite 'em, and so proceed *ad infinitum*'. Despite such irritations, bigness has proved to be an eminently exploitable niche. Some biologists are impressed by the realization that there are fewer species of big creature than there are of small, as if this showed that large body size is somehow misguided. In truth, the world contains a greater variety of small creatures simply because there is more room for them.

In practice, the extent to which the different cells in a multicellular organism sacrifice their own autonomy differs from kingdom to kingdom, and even from group to group within kingdoms. Thus gardeners can reproduce many kinds of plants (though by no means all) by taking cuttings, because the tissues in the cutting retain the quality of 'totipotency': meaning that they are capable of providing all the tissues needed to make an entire new plant, including roots if roots are missing or shoots if shoots are missing; and the shoots in turn can produce flowers and so start a new generation. But although you could reproduce some animals in such a way—such as flatworms and starfish—you cannot take a cutting from a cow or a fish.[3] All the body cells in such organisms have lost their autonomy: the muscle cells cannot live except in the body as a whole, and are committed to remain as muscle cells; and the liver cells, brain cells, and so on are equally constrained. Once a vertebrate has passed the early embryo stage, only the gametes retain totipotency, and this can be realized only when two gametes are joined together.

In addition, the nature of the connections between the cells—the degree of physical contact, and the kind of information that passes between them—differs from kingdom to kingdom, and in different groups within kingdoms. For example, and obviously, plant cells are surrounded by thick cell walls made largely of cellulose, and are linked by threads of cytoplasm running through the walls, whereas animal cells are naked so that the cytoplasm has much broader contact with its neighbours, and there is a greater variety of channels between them. In fungi, typically, the main body of the organism is a **syncytium**; that is, there is no clear separation into individual cells, and many different nuclei share a continuous thread of cytoplasm. In red seaweeds, the different cells seem to engage in a certain degree of internecine strife, as if, after the best part of a billion years, they had not yet settled the rules of co-operation. In short, different mega-eukaryotes have solved the problems of cellular co-operation in different ways and come to different solutions—but in all cases the end result is a big, multicellular organism.

But the individual mega-eukaryote kingdoms should not (in the opinion of many modern biologists, including me) be defined by the fact that most of their members are multicellular and big. A kingdom, like any other taxon, should be formally defined according to clade; and all the members of a true clade can trace their origins to a common ancestor that itself is a member of the group. When the mega-eukaryote kingdoms are defined in this proper, formal way then some of them at least should be seen to include some single-celled members—of the kind that phenotypically speaking are protists. Thus the kingdom Plantae, as now defined, should include various groups of single-celled green algae; and the kingdom Animalia should, I believe, be taken to include the protozoans known as choanoflagellates.

[3] In 1995, Keith Campbell and Ian Wilmut of Roslin Institute, Edinburgh cloned sheep from cultured embryo cells, and in the following year produced 'Dolly' from cells cultured from an adult ewe. Thus they showed that totipotency can be restored to at least some differentiated animal cells; and Dolly was, in effect, produced from a 'cutting'. But such asexual reproduction does not occur naturally in mammals, whereas something comparable does occur naturally in plants.

These facts emphasize that 'mega-eukaryote' is only an *aide-mémoire*: it should not itself be mistaken for a formal classification.

Plants, fungi, and animals between them occupy all the subsequent chapters in this book. The brown and red seaweeds do not have their own section (*mea culpa*), but I describe them briefly later in this section (pages 150–1 and 154–5).

FUNGOIDS

The fungoids include the true Fungi plus various groups of other organisms that have fungus-like characteristics but are not true Fungi. The true Fungi form a distinct clade and should be defined as described in Section 4.

Besides the Fungi themselves, four of the kingdoms shown on the eukaryote tree are fungoid. Between them they represent three distinct phenotypes: Myxomycota and Acrasiomycota, which are slime moulds; labyrinthulids, which are the slime nets; and the mould-like Oomycota. There are other fungoid groups, such as the Plasmodiophoromycota and Dictyosteliomycota (more slime moulds) and the Hyphochytridiomycota (which are superficially similar to the chytrids, which are true fungi); but it is not clear at the time of writing how these more esoteric groups relate to the types listed here and in any case, ecologically speaking, they are minor. Non-specialists can be content simply to acknowledge that they exist.

The slime moulds are extraordinary creatures. They demonstrate that the difference between multicelled organisms and single-celled organisms is not so absolute as we might suppose. Thus, for part of their life cycle, they exist as single-celled amoebae; and indeed, as discussed later, in traditional classifications as late as the 1950s both the acrasiomycotes and the myxomycotes were commonly placed in the same taxon as other amoebae, such as the laboratory favourite *Amoeba* and the parasitic *Entamoeba*. But, to reproduce, slime-mould cells form a multicellular fruiting body that is quite differentiated. Indeed, it is like a miniature toadstool, with a basal disc, a stalk, and a sporangium at the top containing spores.

As you can see from the tree, however, the two major slime-mould kingdoms are phylogenetically very distant from each other, and this difference is reflected not only in their RNA but also in their structure and the details of their way of life. I describe slime moulds and the other fungoid groups briefly on pages 148–9, 150, and 153–4.

PROTISTS

The 'protists'[4] include all those kingdoms whose members are mostly single-celled, although some form colonies and others may contain more than one nucleus. The list presented here is not quite exhaustive: omissions include the haptophytes and the

[4] Lynn Margulis and Karlene Schwartz, in their 1988 edition of *Five Kingdoms*, follow Robert Whittaker and place all the protists within a single kingdom, the Protoctista, which they break down into 27 phyla. Their classification is not based on cladistic analyses and makes little use of molecular data and so cannot be taken as a reliable guide to phylogeny, but it does provide an accessible list of protists. It would complicate this book to no great purpose to try to cross-refer to their listing, but I will refer to their admirable descriptions from time to time.

ecologically important Foraminifera and Radiolaria. But we just have to wait for the necessary studies to be done, so that these and others can be fitted in.

Many of the protists superficially resemble animals because they may be highly motile, and they feed heterotrophically; these types are traditionally called protozoans, which is still a perfectly good informal adjective. Others photosynthesize, and these can be referred to informally as 'single-celled algae'. As previously noted, some protozoans (the choanoflagellates) have now been accommodated within the Animalia, and some single-celled algae (like the chlorophytes) have been restored to the Plantae. Also, traditional classifications treated Protozoa and Algae as formal taxa, but this was never satisfactory because some protozoans simply seem to be algae that have lost their chloroplasts (although we could say that the algae are protozoans that have acquired them). Indeed, some groups such as the dinoflagellates and euglenoids include some members that are photosynthetic and some that are heterotrophic, and some individual euglenoids may be photosynthetic when raised in sunlight or live as heterotrophs in the dark.

Although traditional classifications of protists tended to have little respect for phylogeny—neither the theory nor the data were available to make this possible—they did have one great advantage. They were designed to be user-friendly. Indeed, they were intended primarily to help practical biologists, in medicine, marine biology, and various branches of industry to identify the staggering miscellany of creatures that they found, and put them into some kind of order. So traditional classifications can still be useful—as a complete reference that has been specifically designed as an *aide-mémoire*. In any case, we cannot relish the shock of the new, molecular-cladistic, phylogenetic classification, until we know what we should be contrasting it with. For both these reasons, then, I will take a few paragraphs to outline the way that the protists were described in the zoology textbook that I used as an undergraduate in the early 1960s, the excellent and classic Borradaile, Eastham, Potts, and Saunders, always known as 'BEPS' (first published in 1932; my edition, the third, was published in 1959).

BEPS, as was then customary, placed all the protists in a subkingdom of the Animalia, which they called Protozoa. The book then divided the Protozoa into four classes: Mastigophora, otherwise known as Flagellata; Rhizopoda, also called Sarcodina; Sporozoa; and Ciliophora.

BEPS's Mastigophora, as its alternative name denotes, included all the protozoans that have one or more flagella, at least at some time in their lives. This huge grouping thus included creatures that are now distributed among no fewer than seven of the kingdoms shown here, including *Giardia* (diplomonads); *Trichomonas* (parabasalids); *Euglena* and its relatives (euglenoids); *Trypanosoma* (kinetoplastids); the dinoflagellates as a whole; and the chrysophytes. The traditional Mastigophora also included creatures such as *Chlamydomonas* and *Volvox*, which, as described in Section 23, are now placed in Plantae; and choanoflagellates, now classed as Animalia (Section 5).

BEPS's Rhizopoda contained all the protozoans that move primarily as amoebae, and hence included organisms that are now spread among four of the kingdoms shown in the tree here: the two groups of slime moulds—myxomycotes and acrasiomycotes; *Naegleria*; and *Entamoeba* + *Amoeba*. The old Rhizopoda also included groups that are not yet included in the modern cladogram, such as the plasmodiomycotes (more slime moulds), Foraminifera, and Radiolaria.

BEPS's Sporozoa included parasites of the kind that reproduce by generating large numbers of 'spores', and thus it embraces two of the kingdoms shown here: Microsporida and Apicomplexa.

BEPS's Ciliophora embraced the clade that here is labelled 'ciliates'. Ciliophorans (or ciliates) propel themselves by cilia, at least when young, and never move in an amoeboid fashion. *Paramecium* and *Stentor* are famous laboratory examples.

But that's enough reminiscing. I should now like to introduce the modern, formal classification—and see how the literal, phylogenetic relationships contrast with the superficial, phenotypic resemblances. As always, related creatures may essay many different forms, while any one body form may be adopted by many different lineages.

THE EUKARYOTE KINGDOMS

Giardia: the kingdom of the diplomonads

The genus *Giardia* includes *G. lamblia*, an unpleasant parasite about twice the size of a red blood cell that lives in the human gut. Note its extreme position at the basal end of the tree. RNA studies show *Giardia* to be really weird—occupying the same position relative to all other eukaryotes as sponges do from all other animals; the sister of all the rest. But it is phenotypically peculiar, too. Notably, *Giardia* is emphatically bilaterally symmetrical, with two nuclei staring from the microscope slide like eyes, and eight flagella by which it swims, swaying, more or less in a straight line. It has a depression on one side that serves as a sucker, so it sticks itself to the wall of its host's duodenum. As a parasite, its role seems equivocal; many people who carry it apparently remain healthy, but it seems to multiply quickly and cause problems in the wake of any other debilities. It is a disease of poor hygiene; passing out of one host in the faeces and ingested by the next. It is found worldwide, and has been described as 'the bane of campers'. As for the term 'diplomonad' (which clearly reflects its two-sidedness): my 1959 BEPS applies it just to a single order of Mastigophora, which included *Giardia*. Whether the term will persist (however modified) as the name of the kingdom that includes *Giardia*, remains to be seen.

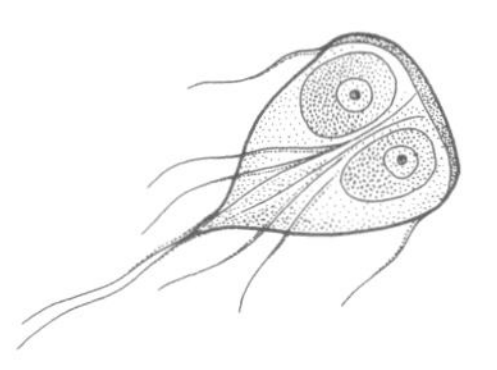

diplomonad
Giardia lamblia

Nosema: kingdom Microsporida

Microsporidans all live as parasites inside the cells of their hosts, and like other 'sporozoans', they reproduce by sudden bursts of division to generate an instant swarm. An economically important example is *Nosema*, which produces the disease 'pébrine' in silkworm larvae, as first described by Louis Pasteur. BEPS treated Microsporida (aka Microsporidia) as a suborder within its proposed taxon Sporozoa;[5] a grouping the authors also included *Plasmodium*. But *Plasmodium* is now placed within the Apicomplexa, which is enormously distant from the Microsporida.

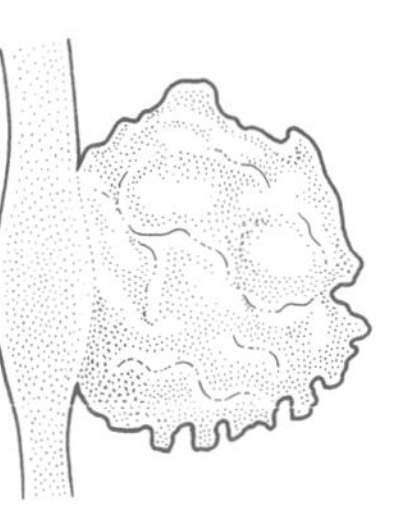
microsporidan *Nosema*

Trichomonas: the kingdom of the parabasalids

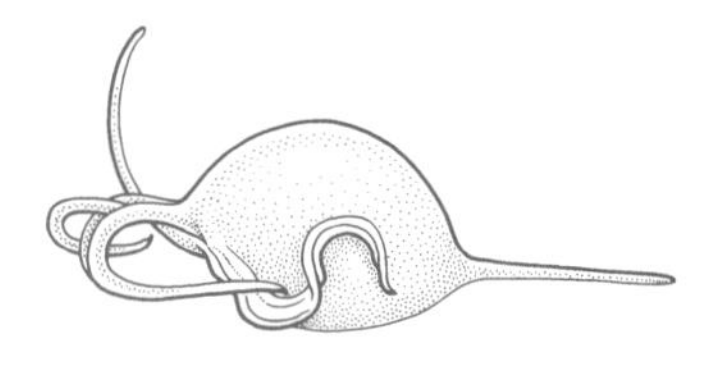
parabasalid, *Trichomonas*

The best-known *Trichomonas* is *T. vaginalis*, which lives in the vagina of women (particularly when the vagina is inflamed) and occasionally in the genito-urinary tract of men. *Trichomonas vaginalis* cannot live outside the human body and how it passes from host to host is not entirely obvious. *Trichomonas foetus* causes abortion in cattle, but there is no evidence that *T. vaginalis* has comparable effects. For BEPS, *Trichomomas* was a typical mastigophoran—a pear-shaped organism with one backward-pointing flagellum joined to the body by a membrane, as in trypanosomes, and four others pointing forwards. But RNA studies reveal its position: far from other 'flagellates', and way over towards the ancestral end of the eukaryotes. BEPS does not recognize the term 'parabasalid'.[6]

Plasmodial slime moulds: kingdom Myxomycota

'Myxomycota' comes from the Greek *myxa* meaning mucus, and *mykes*, fungus (a hangover from days when slime moulds were sometimes ranked as Fungi). Informally the myxomycotes are sometimes called 'true' slime moulds and often 'plasmodial slime moulds'. Like the acrasiomycotes, they take three distinct forms in the course of their life. Most conspicuously they form a patch of wet slime on fallen logs. The patch does not move bodily like a slug, as acrasiomycotes do at the comparable stage, but it does shift position by differential growth. Under the microscope, however, all is activity: you can see the protoplasm streaming throughout the mass, apparently propelled by proteins that resemble the actin and myosin of animal muscles, a motion that presumably serves to distribute nutrients throughout the organism. It is evident, too, that although the slimy structure has many nuclei, they are not separated by cell membranes: the nuclei all share a continuous sheet of cytoplasm.

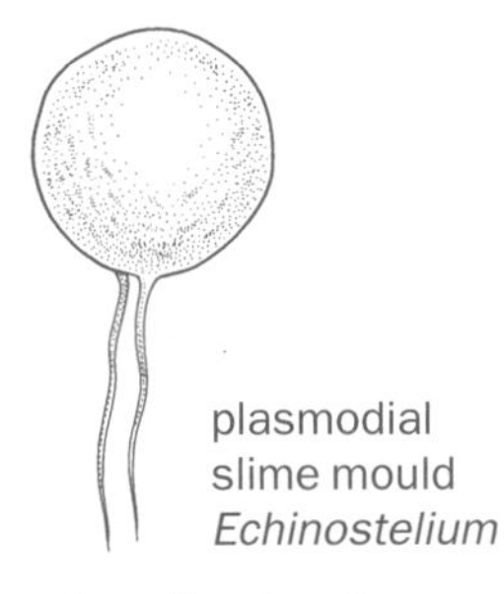
plasmodial slime mould *Echinostelium*

[5] Margulis and Schwartz treat Microsporida as a phylum, but prefer the term 'Cnidosporidia'.

[6] But Margulis and Schwartz rank the parabasalids as a class within their proposed phylum 'Zoomastigina'; a polyphyletic grouping that to a large extent overlaps BEPS's Mastigophora.

The term 'plasmodial' effectively means 'syncytial'; and this, of course, is very different, phenotypically, from the cellular slug of the acrasiomycotes. When conditions dry the creeping protoplasm may gather to form a mound from which grow the stalked sporangia. From them—as in acrasiomycotes—offspring emerge that are single-celled; although these cells may be either amoeboid, or have flagella. The single cells do not coalesce to form a slug, as in acrasiomycotes, but instead join in pairs and fuse—a true sexual act—to form a zygote, which multiplies to form the plasmodial phase once more.

Note again how far the Myxomycota is to the ancestral end of the eukaryotic tree—and how distant it is phylogenetically from the superficially similar Acrasiomycota.

Euglenoids and kinetoplastids: kingdom Euglenozoa

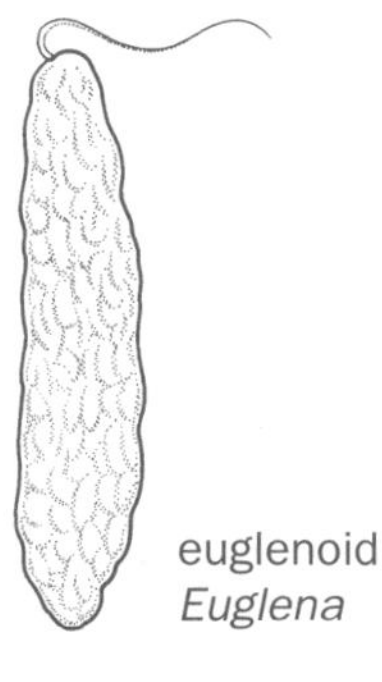
euglenoid *Euglena*

Euglenoids are protozoans like quaint little cigars with a gullet at the front and a forward-pointing flagellum. Some, like *Euglena viridis*, have chloroplasts, and photosynthesize. Others lack plastids, like *Peranema*, which lives in the faeces of frogs and toads having first been swallowed and passed through the gut.

In contrast, the kinetoplastids include the genus *Trypanosoma*, which are blood parasites. *Trypanosoma brucei* is carried by tsetse flies and causes sleeping sickness in Africa; and *T. cruzi* is carried by various species of bugs (not to be confused with beetles) and causes the equally debilitating Chagas' disease in Latin America.

BEPS regarded both *Euglena* and *Trypanosoma* as mastigophorans; 'flagellates'. As it turns out, RNA studies show that the groups to which these two genera belong are indeed related.

The kingdom of *Naegleria*

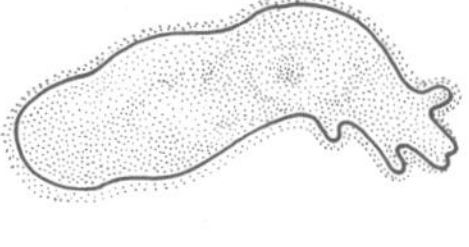
Naegleria

For much of the time *Naegleria* lives as an amoeba in dirty water, feeding on bacteria; but if the water is clean, and lacking organic foodstuffs, it develops a pair of flagella and swims off in search of nutritious effluvia. BEPS placed *Naegleria* in their class Rhizopoda, along with all other amoebae. RNA studies show it to be remarkably different, at least from the amoebae listed here.

The kingdom of *Entamoeba*

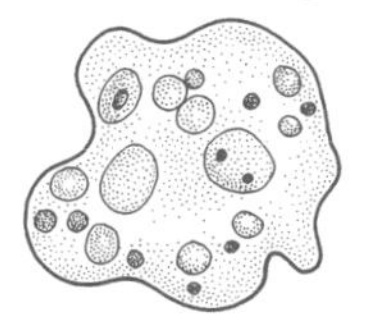
Entamoeba histolytica

Entamoeba histolytica is one of the commonest of all the parasitic amoebae. Phenotypically it is an amoeba that frequently lives inside the human colon, feeding on gut contents and causing no harm; but sometimes it invades the wall of the colon itself, causing the ulcers that are typical of amoebic dysentery. To find a new host it forms cysts, which are passed out with the faeces and are reingested by someone else; so entamoebiasis is yet another infection of poor hygiene. In traditional classifications, as in BEPS, *Entamoeba* was just another rhizopod, but RNA studies again confirm its uniqueness.

Cellular slime moulds: kingdom Acrasiomycota

'Acrasiomycota' come from the Greek *akrasia*, meaning 'bad mixture', and *mykes*, meaning fungus. Acrasiomycotes are informally called 'the cellular slime moulds' because their multicellular fruiting bodies are compounded from discrete cells; that is, each nucleus commands its own discrete territory of cytoplasm, surrounded by a cell membrane. Like fungi, acrasiomycotes may live heterotrophically in fresh water, on damp soil, or on rotting vegetation, especially fallen logs. In the amoeboid stage, the individual cells move and feed as independent amoebae, ingesting whole food—such as bacteria—by phagocytosis. Amazingly, however (and this is one of the most bizarre tricks in nature), when conditions are right, the amoebae congregate to form a multicellular and mobile organism that for all the world resembles a miniature slug. This slug is the migratory stage, which crawls around until it finds a suitable place to settle down, when it forms itself into a bag-like structure containing spores, and then finally grows into the miniature 'toadstool' phase from which the spores are dispersed.

Most multicellular organisms are formed by division of the single-celled embryo (zygote); the daughter cells simply remain in contact. It is rare indeed for multicellular organisms to form themselves, albeit temporarily, from cells that previously have been living and foraging independently.

Red seaweeds: kingdom Rhodophyta

Red seaweeds are formally known as Rhodophyta (which is Greek for 'red plants') but, phylogenetically, they are very far removed from plants. They are weird, wonderful, and extremely ancient; the first definitely photosynthetic multicellular organisms to appear in the fossil record are rhodophytes. Their antiquity is reflected in their variety: 4000 species are known, which is not a huge number, but RNA studies show that the phylogenetic distance between the different species is far, far greater than exists among plants. Traditionally red seaweeds have been divided into two classes, the Bangiophycidae and the Florideophycidae, but modern studies by Wilson Freshwater, Suzanne Fredericq, and others suggest that the former class is really several classes.

red seaweed *Polysiphonia*

Almost all the rhodophytes are marine and they are found all over the world, but especially in the tropics. All of them reproduce sexually; in most, the haploid stage dominates, but, in some, haploid and diploid generations alternate. All of them have reddish plastids containing chlorophyll a (but not b or c) and phycobilins.

In form, red seaweeds are marvellously various: some are filaments, some flattened discs, some are cushions, some are elaborate and erect, some feathery like filmy ferns. Many acquire a coating of calcium carbonate (which is one reason why their fossil record is so good), and some of these form circular crusts while others grow into calcified trees like corals. Some genera include some species that are calcified and others that are not.

In short, red seaweeds were among the first big organisms on the world scene, they are very different from anything else, and they are extraordinarily various. Compared with red seaweeds the remaining eukaryotes are either egregiously small and simple or else are phylogenetically circumscribed johnnies-come-lately. Look upon red seaweeds and admire.

The next three kingdoms—ciliates, dinoflagellates, and apicomplexans—are clearly related, and can be grouped formally as the **Alveoles**. Indeed we might in principle say that the Alveoles are a kingdom and that these three taxa are subkingdoms. But the issue of ranking does not seem to have been resolved—the problem being, as already hinted, that the people who know most about the relationships between the eukaryote groupings are not particularly interested in the niceties of Linnaean ranking. For the time being I will persist in calling these groups 'kingdoms' and leave 'Alveoles' without a formal rank. This is in the spirit of Neolinnaean classification. We may hope, however, that loose ends will be tied up in the coming decades.

The kingdom of the ciliates

The ciliates as understood correspond to the group that are commonly called Ciliophora. They all have cilia—sometimes a great many, all over the surface; and most have more than one nucleus, including some small micronuclei, plus one or more macronuclei. Several thousand species are known, living both in fresh water and the sea, and feeding mainly on bacteria. *Paramecium*, called the 'slipper animalcule' because it is indeed slipper-shaped, is the well-known laboratory example.

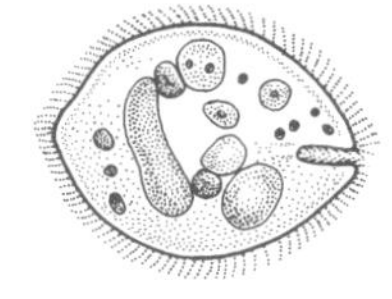
ciliate
Paramecium

Note that the RNA studies show ciliates, dinoflagellates, and apicomplexans grouping together—even though the three are so different phenotypically and in lifestyle.

Dinoflagellates: kingdom Dinoflagellata

Dinoflagellates (Greek *dinos*, 'whirling' and *flagellum*, whip) are nearly all marine and planktonic, especially in warm seas, although some live in fresh water. They look like

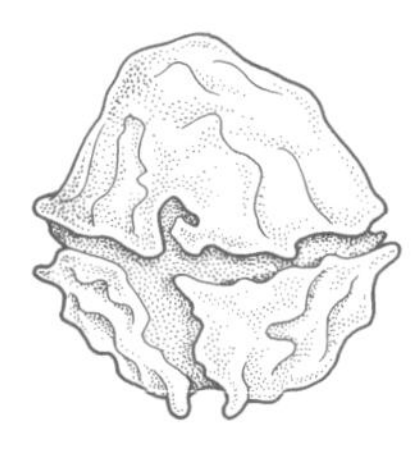
dinoflagellate
Gonyaulax

tiny, cracked, but mobile meringues, encased in a hard test and propelled by a single, long flagellum. Most remain as single cells although some form colonies. Some photosynthesize—they generally have chlorophyll a and variants of c and other pigments, including orange carotenes and xanthins. Many photosynthetic types form symbiotic relationships with corals, anemones, and clams. But some dinoflagellates are heterotrophs. Many produce powerful toxins and when, in certain conditions, they multiply in large numbers or 'bloom', causing 'red tides', they can be poisonous to marine animals (and ultimately also to humans if affected shellfish are eaten). Altogether, dinoflagellates are highly influential members of the plankton and of the marine biota as a whole.

Plasmodium: kingdom Apicomplexa

The apicomplexans—like the microsporidans—are all parasites of animals that can reproduce in bursts by producing a host of spores. Many cause much misery and are economically significant, like the blood parasite *Babesia*, which is carried by ticks and causes red-water fever in cattle; *Coccidia*, a gut parasite, which causes the often fatal cocciodiosis in poultry; *Eimeria*, another gut parasite widespread among both vertebrates and invertebrates; *Toxoplasma*, which in humans may among other things be passed from mother to fetus and then have severe effects on the infant brain; and *Plasmodium*, carried by mosquitoes and the cause of malaria. Malaria is one of the principal infections of human beings and also of other primates; but the disease may have arisen in birds, among whom it is widespread.

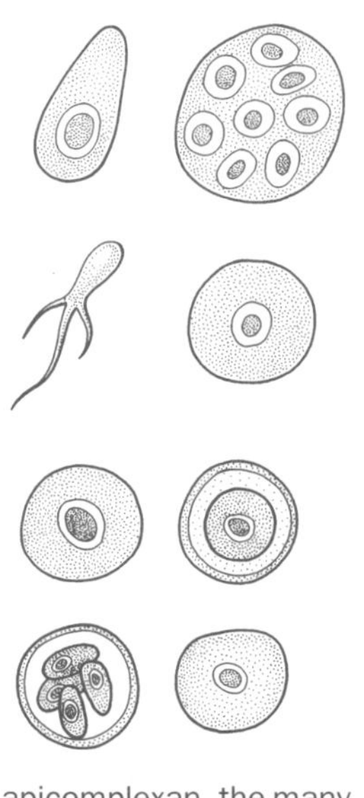
apicomplexan, the many forms of *Eimeria*

BEPS placed apicomplexans together with the microsporidans within the putative taxon Sporozoa. But the apparent similarity of the two groups is presumably due to convergence, one of the principal banes of all taxonomists: two entirely different organisms solving life's problems in similar ways. As the tree here depicts, apicomplexans are phylogenetically close to the dinoflagellates and the ciliates—despite appearances.

The next six kingdoms—labyrinthulids, Oomycota, Chrysophyta, Xanthophyta, Phaeophyta, and diatoms—all seem to be related in a grouping that Mitch Sogin calls the **Stramenophiles**

Slime nets: the kingdom of the labyrinthulids

The name labyrinthulid derives from the Greek *labyrinthulum*, meaning 'little labyrinth'.[7] The entire kingdom has only two known genera: *Labyrinthula* and *Labyrinthorhiza*. Both are mostly marine—they form transparent colonies on eel grass, *Zostera*, and the green seaweed, *Ulva*—but some types live in fresh water and soil. Their lifestyle is quite extraordinary. They form colonies of cells that shuttle, as Lynn Margulis says, 'like little cars in tracks' along narrow trails, which they lay down in front of themselves as they proceed, like army pathfinders. These trails are apparently formed from mucopolysaccharides (the stuff of mucus) and of actin-like proteins. The cells move apparently randomly until they happen upon food—a mass of yeasts, say—when they home in *en masse*. As a protection against desiccation older cells can mass together and form a tough membrane around themselves, waiting for better, wetter times. Then they rupture to produce a new colony of rampaging cells. Sometimes 'blooms' of *Labyrinthula* break out on diseased *Zostera*, though whether they cause the disease is unknown; at such times shellfish suffer, too. Slime nets are not widely known, but they demonstrate yet again how many strange ways there are to solve the problems of life on Earth.

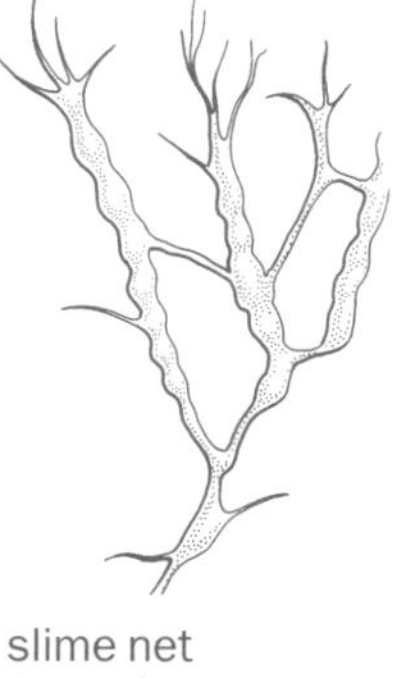
slime net
Labyrinthula

Potato blight and its relatives: kingdom Oomycota

The oomycotes (Greek *oion*, egg) are very fungus-like; indeed they have common names such as 'water moulds', 'white rusts', and 'downy mildews'. Like fungi, they produce threads called hyphae, and digest their food by secreting enzymes and absorbing the nutrients thus released. Like fungi, they live either as saprobes (digesters of dead organic material) or as parasites. Despite their similarities to fungi, however, they are clearly distinct—as was clear long before the fact was confirmed by analysing their RNA. For example, oomycote cell walls are made of cellulose, whereas those of fungi are of chitin; a very different biochemistry. Taken all in all, though, the phenotypic similarity between oomycotes and true Fungi—which presumably reflects convergence—is remarkable.

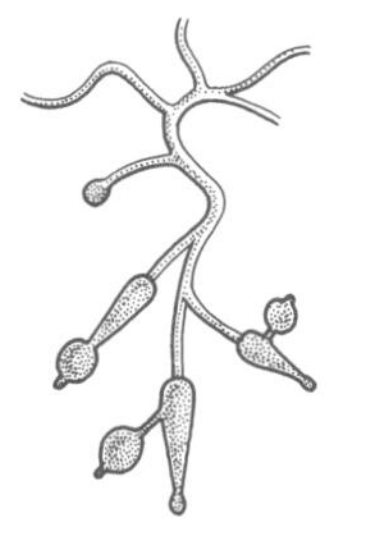
oomycote
Phytophthora

Like labyrinthulids, oomycotes can be economically important. *Phytophthora* is an oomycote: the agent of potato blight, which caused the great famines of Ireland and Western Scotland in the 1840s, and is still a pest to be reckoned with. *Pythium* causes 'damping off' in seedlings, a bane of horticulturalists.

[7] In the 1988 edition of *Five Kingdoms*, Lynn Margulis and Karlene Schwartz label slime nets 'Labyrinthulomycota'. As they point out, however, '- mycota' is not apt, because the slime nets are not fungi and, besides, their name is too long for comfort. 'Labyrinthulids' is too informal but will do for the time being; the more formal 'Labyrinthulata' would be pleasant.

Xanthophytes: kingdom Xanthophyta

xanthophyte
Ophiocylium

Xanthophytes (Greek *xanthos*, yellow) tend to form colonies, sometimes with discrete cells and sometimes syncytial. They are yellowy-green in colour, the yellow resulting from various xanthins and the green from chlorophylls a, c, and e. About a hundred different species are known, which live, very successfully, in fresh water.

Chrysophytes: kingdom Chrysophyta

Chrysophytes are the 'golden algae': *chrysos* is Greek for 'golden'. There are many different kinds and they live virtually everywhere in fresh, cool lakes and ponds, though some are marine. Many live as single-celled organisms but many others form colonies that may be elaborately shaped. The marine types fashion intricate skeletons of silicon by which the genera can be identified, and which provide a good fossil record that extends back to the Cambrian. Chrysophytes may be sisters to the phaeophytes, the brown seaweeds, which include the kelps and wracks.

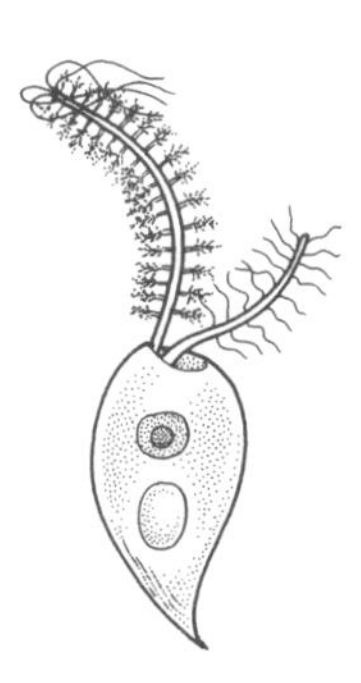

chrysophyte
Ochromonas

Brown seaweeds: kingdom Phaeophyta

Phaeophyta is Greek for 'dusky (brown) plants'—but they are not plants, although the colloquial term 'seaweed' seems to me of great descriptive value, and well worth invoking. About 1500 species are known. Most are marine; none has invaded the land and there is no good reason to suppose that they might ever have done so, though many withstand prodigious desiccation on the shore. If they had come on land and the plants had not, then perhaps the world would now be dominated by forests and plains of phaeophytes, just as it is now festooned in conifers and angiosperms. As things have turned out, the giant kelps—some up to 100 metres long—do form great forests in shallow oceans, such as off the coast of California where sea otters feed and play; and the *Sargassum* weed forms a huge, rich, floating habitat as big as a county in the Sargasso Sea, to the north of the West Indies. The wracks that dominate rocky coasts worldwide largely belong to the genus *Fucus*. They are wonderfully melancholic: Henry V's soldiers on the night

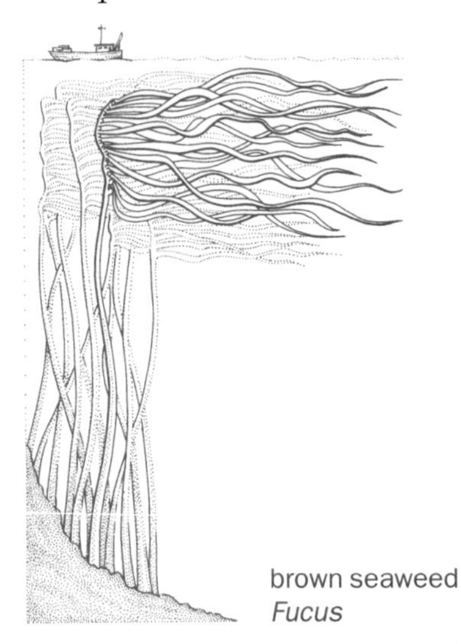

brown seaweed
Fucus

before Agincourt felt themselves 'even as men wrack'd upon a strand that seek to be washed off at the next tide'.

Many brown seaweeds are differentiated into **thallus** (a blade; plural 'thalli') and **holdfast** (which looks like a root but is just for gripping); and the thalli of some brown seaweeds are further differentiated, and may even contain sieve tubes to conduct water and the products of photosynthesis. Phaeophytes photosynthesize with the aid of chlorophylls a and c (not a and b as in plants) and they derive their brown or olivey colour from the pigment xanthophyll. They generally reproduce sexually with the aid of eggs and motile spermatozoa, each with two flagella; and also asexually by spores.

The brown seaweeds seem to be particularly closely related to the chrysophytes and perhaps the two should be united in a single kingdom, just as the protist chlorophytes and others have been united with the plants. As things stand, however, the phaeophytes deserve to be acknowledged as a magnificent kingdom in their own right.

The kingdom of the diatoms

Diatoms—all 10 000 or so known species of them—are among the most important members of the marine plankton. They are extraordinary creatures, with **tests** (shells) made in two halves like an old-fashioned but ill-fitting pillbox, the cells either solitary or forming simple colonies. Most are photosynthesizers and are brown in colour, with chlorophylls a and c plus carotenes and a range of xanthins; but a few are saprobes. Their accumulated skeletons form 'diatomaceous earth', with commercial uses from abrasives to filters.

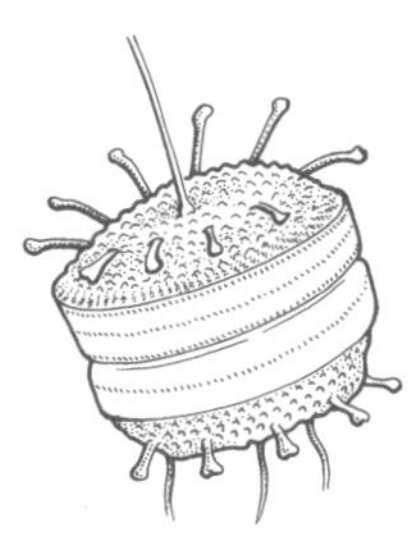

diatom
Thalassiosira

The remaining three kingdoms—**Plantae**, **Fungi**, and **Animalia**—occupy the rest of this book, and need no description here. But a few reflections seem in order on the eukaryotes as a whole.

Plantae
(sago palm)

Fungi
(cultivated mushroom)

Animalia
(African elephant)

THE VARIETY OF EUKARYOTES

The contrast between traditional classifications based on phenotype (overt physical characters) and modern classification based on RNA and hence (we hope) on true phylogeny, beautifully illustrates a range of important evolutionary principles. Thus we see that although some clades diverge markedly and radiate into many different forms, others converge, or simply retain primitive features. All of these proclivities complicated the endeavours of traditional systematists, who did not have access to molecular data, and were forced to base their classifications on phenotype.

So it seems that the amoeboid tendencies that once seemed to unite *Naegleria*, *Entamoeba*, and the various slime moulds are simply a primitive eukaryotic character; the kind of feature that systematists were wont to consider significant until Willi Hennig pointed out the error of their ways. On the other hand, the special similarity of Myxomycota and Acrasiomycota presumably results from convergence; that is, these two (and some other minor groups not shown on the tree) independently evolved the slime-mould way of life. The ostensible similarity of the microsporidans and the apicomplexans also presumably represents convergence; but note once more how huge is the phylogenetic distance between them. The various single-celled algae —euglenoids, dinoflagellates, chrysophytes, xanthophytes, diatoms, plus the chlorophytes and others that have now been put back with the plants—seem linked both by plesiomorphy (primitiveness) and convergence (although when you look closely at these groups, the resemblances between them are slight).

In contrast, creatures that are closely related may have totally different ways of life; examples of divergence, or radiation. Thus the euglenoids include many photosynthetic, free-living, pond-dwelling algae whereas the affiliated kinetoplastids include the parasitic agents of sleeping sickness and Chagas' disease. For that matter, animals, plants, and fungi are closely related—'closely', when set against the huge spectrum of difference within the entire eukaryotic domain—but the contrasts in their ways of life could hardly be more striking.

It follows that creatures of ostensibly similar habit may be widely scattered over the tree, while others that seem widely different may be the closest of relatives. Note, in this regard, the huge phylogenetic distance between the myxomycotes and the acrasiomycotes, and that *Naegleria* and *Entamoeba* belong in separate kingdoms although earlier biologists often supposed that they belonged together. By contrast, euglenoids and kinetoplastids group together although they may live such different lives.

Note, too, how broad is the spectrum of eukaryotes, and how oddly some groups are positioned on it. *Giardia* is remarkable: way out on its own. Myxomycotes, though they are convincingly multicellular at some stage of their lives, emerge from among a cluster of protist clades; showing once more that the route to multicellularity is not a straight, metalled road, and that multicellularity is not the last word in phenotypic development that renders the protistan lifestyle redundant. Some groups may try it while others, though evolving later, may not. And note how far to the primitive end of the tree lie the red seaweeds; a whole raft of protist kingdoms stands between them and the brown seaweeds, and plants, animals, and fungi.

Finally, note how closely the plants, animals, and fungi are linked together, and what a narrow portion of the spectrum they occupy. Ever since science began to move into its modern phase, at least four centuries ago, biologists have assumed that these three groups between them spanned all of life. Now they are seen as just three clades among at least 20 (and probably a great many more) in a domain that is just one of three (or more). The demotion is wonderful; our own kingdom has been ousted from the centre of life's stage just as Copernicus and Galileo banished our planet from the centre of the Universe. In addition, plants, animals, and fungi are so close, phylogenetically speaking, that it would make more sense to link them all within one kingdom than it would to put *Naegleria* and *Entamoeba* back together. Furthermore, it is not yet certain whether fungi are more closely related to plants or to animals; but they seem to be nearer to animals. Certainly, the old-fashioned notion that fungi should be thrust in among the plants, which is still promulgated in some extant texts, can now be seen to be grotesque. All in all, animals, plants, and fungi provide the ultimate warning: that lifestyle is a very poor guide to relationship.

This, then, is a brief sketch of the domain to which we ourselves belong. Each of the kingdoms deserves its own section in this book but, alas, only three of them—fungi, animals, and plants—are to be given their proper deserts. First, the fungi.

4

MUSHROOMS, MOULDS, AND LICHENS; RUSTS, SMUT, AND ROT

KINGDOM FUNGI

ONE OF THE LARGEST and oldest inhabitants of the state of Michigan, USA, if not of the whole world, is a honey fungus, of the species *Armillaria gallica*.[1] It covers 600 hectares, is estimated to weigh more than 100 tonnes, and is probably around 1500 years old. The world's heaviest recorded fruiting body of any kind was a polypore bracket fungus from England's New Forest that weighed 45.4 kilograms; and the biggest that is growing at the time of writing is a bracket fungus at Kew Gardens in London that is more than 1.5 metres long and 1.1 cubic metres in volume.

[1] A diversion on the common names of fungi. So many people have been involved with fungi in so many contexts, and they are so diverse and in other ways problematical, that their common names cause even more confusion that usual. In fact, even their 'Latin' names can be confusing, although they are intended to end all ambiguity. Thus the organism that the British call 'honey fungus' is more commonly known in the US as 'honey mushroom'; in either case it is *Armillaria gallica*, although it was previously known as *A. bullosa*.

More generally, Americans seem to refer to any mushroom-shaped fungal fruiting body as a 'mushroom', but the British tend to reserve that term for the several edible species in the genus *Agaricus* (a genus that was previously called *Psalliota*). The British tend to refer to other mushroom-shaped fungi as 'toadstools', which is the term I favour here. Some people feel that 'toadstool' should allude exclusively to toxic species, but I see nothing wrong with the concept of 'edible toadstool'. Thus I regard mushrooms (defined in the narrow British way) as edible toadstools. More generally still, people tend to use the word 'fungus' in different ways. Thus mycologists traditionally refer to any fungus-like organism as a 'fungus', with a lowercase 'f'; and distinguish members of the kingdom Fungi by giving them a capital F. As discussed in Section 3, however, about eight different kingdoms are fungus-like (although the number eight cannot be taken as fixed, for the higher classification is still being worked out). Being phylogenetically inclined, I prefer to reserve the word 'fungus', whether with upper- or lowercase 'f', exclusively for members of the kingdom Fungi, and to refer to the rest as 'fungus-like', although this is obviously inelegant. 'Fungoid' would not be a bad term for the fungus-like kingdoms but is not in common use in this context.

I cannot hope to resolve such inconsistencies of language and am not sure that it is desirable to try: inconsistency is inevitable when any vernacular is alive and active, and the alternative would be to impose something equivalent to the Académie Française. The best any of us can hope to do—as I am trying to do here—is to state clearly how we are using the different terms ourselves.

The biggest recorded edible puff-ball lived in Canada and was 2.6 metres in girth; because average puff-balls produce around 7 billion billion (10^{18}) spores, this one could, in principle, have populated several galaxies. Fungus specialists—mycologists—recognize about 60 000 species but the true figure may be around 1.5 million, with the unknown majority inhabiting the tropical forests that as yet are hardly explored. We tend to underestimate fungi, just as we tend to overlook prokaryotes; but, like prokaryotes, they are huge players in the world's ecology and in human economy.

Most fungi are saprobes, breaking down whatever is organic but dead (the old term 'saprophyte' is inappropriate because '-phyte' means plant and fungi emphatically are not plants). So, together with a host of animals, protozoans, and bacteria they are the world's great recyclers, converting organic material into chemical forms that plants can re-absorb through their roots—and also releasing an estimated 8500 million tonnes of carbon dioxide into the Earth's atmosphere every year. Many feed on wood—mighty trees and the fabric of houses; some digest only the cellulose of which wood is mainly composed; but most digest both the cellulose and the lignin that binds the cellulose fibres together and so strengthens the timber. The group called the Basidiomycota, which includes the mushrooms and most of the **fruiting bodies** that the British call 'toadstools', are particularly efficient rotters of wood.

Many of the fruiting bodies that fungi produce for the purposes of reproduction and dispersal are delectable and nutritious. Some, like truffles, morels, and shiitake, are among the most treasured of all foods. But others are toxic, some deadly; and some—many—bring hallucinations and dreams that at times have dominated the philosophies and shaped the social hierarchies of entire cultures. Moulds, in the phylum Ascomycota, are principal players in the creation of great cheeses: *Penicillium roquefortii* turns soured sheep's milk into roquefort, *P. camembertii* helps to turn ordinary cottage cheese into camembert, and so on. A close relative, *P. chrysogenum* (also called *P. notatum*) was the source of the world's first and still most widely used antibiotic, penicillin. 'Antibiotic' means 'anti-life'; and many fungi produce antibiotics of many different kinds that help them to compete with other organisms, including bacteria and other fungi. Baker's and brewer's yeast is also an ascomycote, of the species *Saccharomyces cerevisiae*. It turns the sugar in cereals to carbon dioxide (plus alcohol), and so leavens bread—and also turns the sugar in cereals to alcohol (plus carbon dioxide) and so makes beer. In short, many small fungi have long been the source of human delight and cottage industry; and now they are leading players in the new age of biotechnology.

But fungi are often parasites, too, and significant agents of disease. Various yeasts and other fungi, some not too distantly related to *Saccharomyces*, cause diseases in human beings that are sometimes merely irritating, like ringworm, but in some cases are deadly, commonly taking hold in the lungs and sometimes spreading into the body. Such infections are becoming more and more serious as AIDS continues to spread and enfeebles the human immune response. Other fungi in the form of rusts, smuts, rots, and mildews are the most important pathogens of plants—in the wild, of course, but also among cultivated plants from garden roses to the world's greatest crops. Farmers in the Sahel expect to lose up to half their sorghum crop each year through mildew alone. It is estimated that half the crops in storage in the Third World are spoiled by fungal attack. Mycotoxins, including the aflatoxins produced in groundnuts and cereals by *Aspergillus flavus* (another ascomycote), kill many thousands of people every year. (As I mentioned in Section 3, however, the agent of potato blight, which caused the potato famine in Ireland and western Scotland in the 1840s, is *Phytophthora infestans*; not a true fungus at all but an oomycote, which is a quite separate kingdom, although it is fungus-like. Another oomycote, *Pythium*, causes the 'damping off' of seedlings that so irritates gardeners.)

There is another, quite different side to fungi: their many and crucial, 'mutualistic' relationships with plants—**mutualism** being a special form of symbiosis in which both partners benefit. Fungus–plant symbioses take two principal forms. First, about 13 000 species of fungi—roughly a fifth of all known species, and including 40 per cent of the ascomycotes—form close associations with green algae (and to a lesser extent with cyanobacteria) to form **lichens**. Sometimes as many as four or five different organisms may partake in the formation of just one lichen. But only about 40 different species of alga or cyanobacterium are known to form lichenous relationships with fungi—so any one photosynthetic species may join up with a wide array of fungi. The key point is that 'lichen' is not the name of a clade of organisms, but simply of a way of life. The lichenous relationship has obviously evolved at least three times among existing fungi and possibly many more, and is comparable ecologically with the relationships that corals and giant clams form in the sea with various photosynthetic residents.

Second, many fungi—including many of the ones that form mushroom-like 'toadstools' in the autumn woods—form close associations with the roots of plants that are called **mycorrhizae**. The mycorrhizal relationship is again mutualistic: the fungus supplies nutrients of all kinds by absorbing minerals and digesting organic material from the surroundings, while the plant provides additional products of photosynthesis. Altogether about 85 per cent of plant species are said to form mycorrhizal relationships. The great trees of both temperate and tropical forests grow largely by virtue of their mycorrhizae. Many orchid seeds cannot germinate without a suitable infestation of mycorrhizal fungi, and some orchids retain the relationship through life and have lost the power of photosynthesis.

When we consider lichens and mycorrhizae together we see that the mutualistic relationship between plants and fungi extends right through the plant kingdom, from the lowliest algae to the most derived of coniferous and flowering plants. It seems, too, that some of the oldest plant fossils have fungal associates that are essentially mycorrhizal. For such reasons, many biologists now suggest that plants could not have invaded the land at all without fungal assistance; and it is most unlikely that fungi could have invaded the land without plants, because terrestrial fungi ultimately derive their food from plants just as animals do. In short, lichens and mycorrhizae may be taken as modern exemplars of the kinds of relationship that historically were crucial to the development of all terrestrial life. Note, too, in passing, that all the great kingdoms, including fungi, originated in water, even though some of them (notably fungi) now seem primarily terrestrial.

All the many, diverse manifestations of fungi make sense once we consider their mode of nutrition, and the way in which their anatomy—what zoologists would call their *Bauplan* (a German word meaning 'building plan')—is adapted to serve their way of feeding. For fungi are eukaryotic heterotrophs that do not practise phagocytosis. 'Heterotroph' means that they need to take in their carbon in organic form—already compounded into complex molecules by some obliging autotroph. 'Phagocytosis' is the ability of a cell to wrap itself around its food, as an amoeba does. All animals have cells that can practise phagocytosis; it is a key feature. But fungi have adopted a quite different strategy. Typically, the main body of a fungus is composed of a mass of threads called **hyphae**, which collectively form the **mycelium**. The hyphae penetrate the substrate, so that the fungus lives inside its food, in constant, intimate contact. Because hyphae are thread-like in form their surface area, and hence the area of contact with the surroundings, is huge. Yet hyphae are not mere absorbers, as in general is true of plant roots. They actively secrete enzymes, which digest whatever is digestible in their surroundings, and then absorb the resulting soup. These enzymes are collectively called 'exoenzymes' (because they operate outside the body that produces them) and include digesters of cellulose, called cellulases. Fungal hyphae may also excrete many other compounds as well, including many agents we call antibiotics.

Overall, then, the heterotrophic, non-phagocytic nutritional strategy of fungi is very efficient indeed. This is how (with assistance from beetles and bacteria) fungi can convert entire trees to dust and mush within a few years. A mycelium of a single individual may cover an entire field, or form a mycorrhizal relationship with an entire copse of mighty trees. Such mycelia can throw up impressive crops of mushrooms or toadstools within a few days. It can seem like hours, though that is an illusion; one day there are none, and the next the fields and the woods seem full of them. The speed with which they fructify—produce fruiting bodies—enhances the aura of magic conferred by their ability to poison and bring dreams. Yet it is not so surprising when you consider the sheer mass of active fungal material beneath the surface.

Many fungi, also, at least at some stage in their lives, form single-celled or at least very small entities that are protistan in form, and typically reproduce by budding. Such a form is called a 'yeast'. Note, however, that although 'yeast' is the name of a form, and many different fungi from several different groups may adopt the yeast-like manner, there is also a particular clade of Ascomycota that largely specialize in this mode. These are often called the 'true yeasts', which is how I refer to them here. *Saccharomyces* is one of them.

Overall, the lifestyle of fungi simply involves digesting whatever organic material surrounds their hyphae (or their isolated cells, in the case of yeasts). They also absorb the necessary minerals, phosphates, nitrogen and the rest, often in inorganic form, and some also need arcane supplements such as a vitamins and/or growth promoters. But whatever the details, their trick is to feed on their immediate surroundings. If they are surrounded by and are digesting dead organic material, like a log of a tree or a dead horse, we call them **saprobes**: agents of decay. If whatever is around them and they are feeding upon is alive, then the fungus is said to be a 'parasite'.

Yet a fungus does not 'care' whether its surroundings are alive or dead; in principle, anything organic is fair game. Living creatures live in perpetual danger of digestion by fungi, which they avoid only by mounting an active defence. Plants do this in all kinds of ways; for example, by producing resins or specific fungicides. Animals also produce a range of chemical and mechanical repellents but some of them, notably vertebrates, have a dedicated immune system as well, expressly designed to attack specific parasites. If the immunological guard is dropped then the creature is dead meat—literally, from the dispassionate viewpoint of the fungus. I have seen hibernating bats taken over by thick white mould that sprouted hideously from their entire body surface. Hibernating bats tend to stir themselves at least once in the course of the winter, and the signal that awakens them, apparently, is a full bladder. But their single flight uses energy enough to maintain them in hibernation for three weeks. Many fail to survive the winter simply because they run out of energy. I find it hard to believe that bats would compromise their chances of survival so dramatically simply to go for a wee; natural selection would surely have fostered greater economy. The true reason for their mid-winter exertions, I suggest, is to arouse the immune system. If the immune system takes time off, then a hibernating bat, slung head down in its cold cave, is meat on the hook. So are we all. The alacrity and efficiency with which generally innocuous 'yeasts' take over the lungs of patients who are immunocompromised shows how potent and ever-present is the fungal threat.

Fungi are also masters of reproduction. Typically they spread themselves via spores, which are of two kinds. The mitotic spores or **mitospores** are diploid, which means they contain two sets of chromosomes. In the simplest cases they germinate to produce a new individual, which is a form of asexual reproduction. The meiotic spores or **meiospores** are haploid, meaning they contain only one set of chromosomes. Meiospores sometimes simply germinate to form a new organism, which is asexual reproduction again; but sometimes two meiospores function as gametes, and fuse together to form a diploid cell that then goes on to form a new individual. This is sexual reproduction.

Two of the four widely acknowledged phyla of Fungi, the Ascomycota and the Basidiomycota, include species that produce big fruiting bodies. These big bodies tend either to be **epigeous**, which means they grow above ground and typically take the form that I call 'toadstool'; or are **hypogeous**, growing underground like truffles. The spores of hypogeous types are scattered by mammals, which first dig down to find them—just as pigs scatter truffle spores. Christopher Johnson of the James Cook University in Queensland has observed that hypogeous types tend to grow in places where the surface is hard or otherwise inhospitable—for example, in deserts or under permafrost. Then, it is more cost-effective for the fungus to produce an aroma to attract dispersive mammals (pigs in France; various small marsupials in arid Australia) than it is to burst through the surface. Of course, the best known of all hypogeous types—truffles—do grow in cosy woodland; but the generalization seems sound nonetheless.

Although the two forms look so different, it is easy to see how one form could grade into another; the hypogeous types, morphologically, are like epigeous types that have not erupted above the ground (and there is an intermediate form called **subhypogeous**). This suggests that the difference between the two could be brought about by just a few genes, or, indeed, by a single gene: one that arrested the process of erupting through the surface. In practice, the Ascomycota and the Basidiomycota each produce both epigeous and hypogeous forms. Thus it is that two fruiting bodies that look similar may in fact belong to quite different phyla; for example, the true truffle, *Tuber*, is an ascomycote, whereas the fungus commonly called 'false truffle', *Rhizopogon*, is a basidiomycote. But we also find—again unsurprisingly, once we see what is going on—that very different-looking fruiting bodies may be produced by closely related fungi. Thus Tom Bruns of the University of California at Berkeley and his colleagues, using mitochondrial DNA techniques, have found that *Rhizopogon* is very closely related to *Suillus*, which is a toadstool (Tom Bruns would call it a 'mushroom') in the basidiomycote family of the Boletaceae—the cep family. Mutation of just a few genes that control overall development could quickly bring about this radical change of form. Although the fruiting bodies of ascomycotes and basidiomycotes may adopt a similar range of forms, they differ in their detailed structure—which is how the two great groups are distinguished, as described in the guide below.

Many fungi produce mitotic and meiotic spores on different kinds of fruiting body—the former being exclusively asexual, the latter commonly sexual. The two kinds of fruiting body may look very different from each other so that the same fungus in different phases of its life cycle can look quite different; and because one form cannot necessarily (or commonly) be induced to turn into the other, it is not always obvious that asexual and sexual forms belong to the same species. Furthermore, some fungi seem to have abandoned sexual reproduction altogether, and are known only from their asexual fruiting bodies, bearing mitotic spores.

All this has had bizarre consequences. Different phases of the same fungus, bearing different kinds of fruiting body, have often been treated as different species, and sometimes have been assigned to quite different genera or even to different phyla; but if and when it becomes clear that this has happened, then the name of the species that is known to reproduce sexually (via meiotic spores) takes priority. For practical purposes, however, mycologists may continue to refer to the purely asexual form by its previous name.

But because fungi are traditionally classified according to the microanatomy of their sexual bodies, mycologists have been obliged to compile a kind of pending file to include all those types that are known only in their asexual state, for their relationships with the fungi that do reproduce sexually may be far from obvious. This pending file now includes about 25 000 species—almost half the known fungal species—and is commonly labelled 'Fungi Imperfecti' or 'Deuteromycota'. Of the two terms, the cod-Latin Fungi Imperfecti seems preferable because it is obviously ad hoc and is intended simply to refer to a mixed bag of organisms that have been bundled together purely for convenience. The Greek Deuteromycota sounds as if it is intended to be a clade, which is entirely spurious.

Membership of the Fungi Imperfecti fluctuates. New discoveries constantly add to the ranks. Others are removed, sometimes because they prove to be sexual after all, and can be properly assigned to formal groups; but also because fungi are of such obvious economic and ecological importance that they are now attracting high-class and well-supported molecular studies (of their DNA and RNA), which are revealing the true relationships between fungi imperfecti and the rest. For example, Barbara Bowman and her colleagues from Roche Molecular Systems at Alameda in California have shown that various pathogens of humans that had been placed among the Fungi Imperfecti, including *Histoplasma*, *Blastomyces*, and *Coccidioides immitis*, all fit neatly among the ascomycotes, which previously had been suspected but was far from certain. *Coccidioides immitis*, in particular, is so reduced in form that mycologists have sometimes doubted whether it was a fungus at all. Bowman and her colleagues have also shown that the closest relatives of these pathogens are often non-pathogens (just as is often true of bacterial pathogens). Knowledge of true relationships is of huge practical importance because related organisms are liable to have similar biochemistry even if their way of life is greatly different, and the non-pathogenic relatives of the pathogens, once identified, can be used as models by which to investigate the pathogens themselves with a view to controlling them.

In summary, the classification of the kingdom Fungi has not proved easy. More than 56 000 species have been described and more come to light all the time; the true number could easily exceed a million. Different lineages often adopt the same general form; any one species may take several forms that may look quite different (yeast-like or mycelial; asexual and sexual fruiting bodies); and, as we have seen, the sexual bodies that provide the main diagnostic characters are often absent or at least unknown. But molecular studies seem to be sorting out many of the problems that seemed insoluble by traditional means—and, encouragingly, the molecular findings are in general supporting the traditional conclusions, which means that the new research generally dovetails neatly into the existing ideas.

Thus the molecular studies confirm that the Fungi as defined in this book is indeed a clade—though it is quite distinct from various other kingdoms of eukaryotes that merely resemble fungi, but have at times been included among them. Within that clade, however, there is still much sorting to be done. Notably, the system has grown up whereby the kingdom of Fungi is divided into four great groups, which we can reasonably call phyla: Chytridiomycota (chytrids), Zygomycota, Ascomycota, and Basidiomycota.[2] Recent molecular evidence, however, indicates that the Zygomycota and Chytridiomycota have been mixed up together, and that the Zygomycota

[2] Most people will be more familiar with the terms 'Zygomycetes', 'Ascomycetes', and 'Basidiomycetes', rather than 'Zygomycota', 'Ascomycota', or 'Basidiomycota'. In general, the '- mycota' endings refer to the names of the phyla, while the '-mycetes' endings may apply to the names of classes within the phyla.

as presently defined are polyphyletic: that is, they include several distinct lineages, some of which are more closely related to known chytrids than to other zygomycotes. Renaming will be required. By the rules of Neolinnaean classification I should write the polyphyletic Zygomycota in the vernacular and with a lowercase initial letter—'zygomycotes'; but as the issue has yet to be settled I will continue to refer conservatively to Zygomycota. At least the two crown phyla, Basidiomycota and Ascomycota, seem secure enough: a sister pair of clades.

Each of the phyla themselves are traditionally divided into a great many orders, and although some of these seem to be valid clades others are breaking up under molecular scrutiny. How the different orders relate to each other seems even less settled; and I have abandoned formal groupings between the ranks of phyla and order (superorders, classes, and so on) altogether for the time being because there is such a state of flux, and so many opinions. But much has been achieved; and the new molecular studies are supporting traditional taxonomy much of the time, so we can expect more clarity within a few years.

The tree presented here is compounded from papers by Tom White, Barbara Bowman, John Taylor, Eric Swann, and Tom Bruns from the University of California at Berkeley and from Roche Molecular Systems at Alameda (near Berkeley); and Andrea Gargas and Paula DePriest at the Smithsonian Institution, Washington DC. These mycologists, in turn, are building on (though often much modifying) traditional classifications as presented, for example, in the eighth edition of *Ainsworth and Bisby's Dictionary of the Fungi* (1995), which I have also made use of. The traditional names of the phyla are presented formally, although we should acknowledge the polyphyletic status of Zygomycota. But the subdivisions within the Ascomycota and Basidiomycota are in some cases labelled informally, as in 'true yeasts'. These groups do seem to be emerging as true clades; but although they generally correspond very closely with traditional formal taxa they do not always correspond exactly, and it seems easiest simply to say what is in each clade (for example, 'true yeasts') than to adjust some pre-existing formal name. Perhaps when the boundaries of the clades are established robustly, formal names can be re-applied. My apologies to the purists, then, who like mellifluous Graeco-Latin. I prefer it myself. But we will just have to wait until the information is solid enough to justify the customary aggrandisement.

A GUIDE TO THE FUNGI

Not every eukaryote that looks like a fungus should be placed in the kingdom Fungi. Indeed the grouping labelled 'Fungi' in this book is one of only eight kingdoms that have fungus-like qualities. The seven non-fungal but fungus-like kingdoms[3] are discussed in the previous section; as a reminder, they include the Acrasiomycota and Dictyosteliomycota, which are both cellular slime moulds; the Myxomycota and Plasmodiophoromycota, which are plasmodial slime moulds; the labyrinthulids, the

[3] See footnote on page 159 on nomenclature.

'slime nets'; and the Hyphochytridiomycota and Oomycota, which have fungus-like hyphae.[4] Many mycologists study those non-fungal but fungus-like organisms; but the fact that mycologists opt to study them does not make them Fungi, as some have occasionally argued. Taxa must be defined phylogenetically, and not according to the groups that specialists choose to look at.

But if these other groups are in so many ways fungus-like, then why are they not categorized as fungi? Well, different biologists have tried to draw the line around the kingdom of the Fungi in different places. In *Five Kingdoms*, Lynn Margulis and Karlene Schwartz admit organisms to the Fungi only if, at all stages of their life cycle, they lack flagella (or, as they prefer to call them, eukaryotic flagella or 'undulipodia'). Such an arrangement would exclude the Chytridiomycota, whose gametes do have flagella. Flagella are, however, a primitive feature of eukaryotes (or at least of the eukaryotic clade that includes fungi, plants, and animals) and the lack of them, in Zygomycota, Ascomycota, and Basidiomycota, tells us effectively nothing about the phylogenetic relationship between those three groups and the Chytridiomycota. Cladistic rules are valuable not as Teutonic edicts, but because they provide the most reliable guide to true phylogeny. Merely to draw a line around the zygomycotes, ascomycotes, and basidiomycotes is arbitrary. On the other hand, some mycologists would like simply to declare that any organism with hyphae is a fungus, and so would include the oomycotes and the hyphochytridiomycotes. But, again, there is no good reason to suppose that oomycotes, hyphochytridiomycotes, and fungi are related; and we sacrifice a great deal of insight if we do not at least try to build our classification around true phylogeny.

What binds the four traditionally accepted phyla together in Fungi—Chytridiomycota, Zygomycota, Ascomycota, and Basidiomycota—are two, albeit seemingly recondite features of biochemistry. Thus, in all four groups the thin walls that surround the cells contain chitin, the nitrogen-stiffened polysaccharide that is also shared by some animals, such as arthropods and nematodes; whereas the groups that are merely fungus-like typically have cellulose cell walls, more like those of plants. Second, the true fungi use a distinctive set of enzymes to metabolize the amino acid lysine, quite different from the groups that are merely fungus-like. Such characters may seem mere niceties but they are precisely the kinds of features that tend to indicate true relationships; enzymes, after all, are proteins, and are the direct products of genes, and it is hard to see how all the acknowledged fungi would share such a distinctive pattern of enzymes unless they had similar genes; and hard to see how those genes would be so similar unless they derived from common ancestry. Molecular studies (based upon small subunit—18S—ribosomal RNA) now confirm that the four phyla are indeed linked in a single clade.

Molecular studies also suggest that fungi are phylogenetically close both to

[4] The eukaryote tree (pages 140–1) shows only the two principal slime-mould kingdoms—Acrasiomycota and Myxomycota—and also omits Hyphochytridiomycota and Plasmodiophoromycota. The relationship of these minor kingdoms to the rest does not seem to be clearly established at the time of writing.

Chytridiomycota
Zygomycota
FUNGI
Ascomycota
Basidiomycota
4

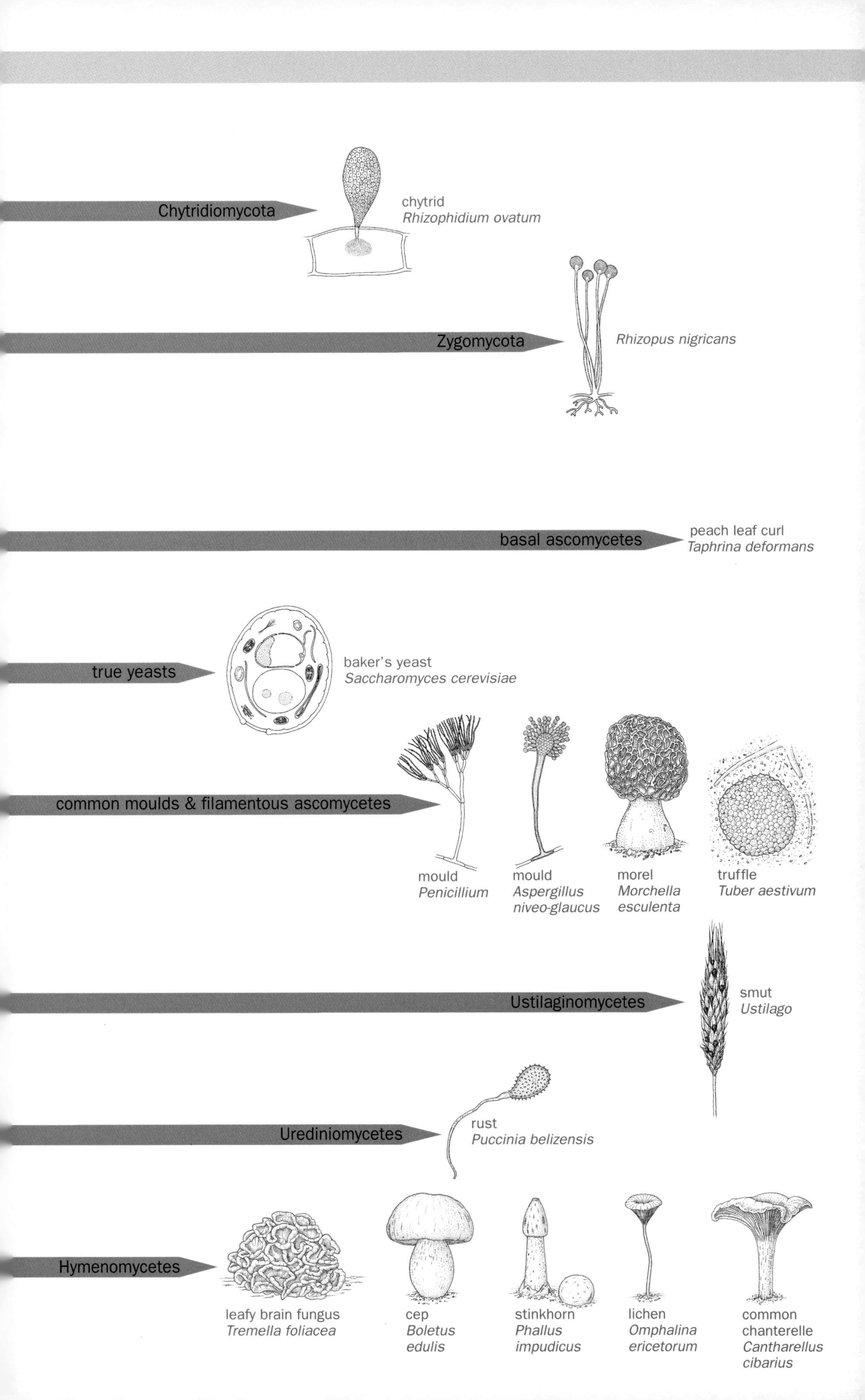
Chytridiomycota
chytrid
Rhizophidium ovatum
Zygomycota
Rhizopus nigricans
basal ascomycetes
peach leaf curl
Taphrina deformans
true yeasts
baker's yeast
Saccharomyces cerevisiae
common moulds & filamentous ascomycetes
mould
Penicillium
mould
Aspergillus
niveo-glaucus
morel
Morchella
esculenta
truffle
Tuber aestivum
Ustilaginomycetes
smut
Ustilago
Urediniomycetes
rust
Puccinia belizensis
Hymenomycetes
leafy brain fungus
Tremella foliacea
cep
Boletus
edulis
stinkhorn
Phallus
impudicus
lichen
Omphalina
ericetorum
common
chanterelle
Cantharellus
cibarius

animals and to plants, so that the three together form a clade. It also seems that the three kingdoms separated more or less simultaneously from each other and from other eukaryotes, about 1000 million years ago. The differences in DNA between fungi and animals are roughly the same as between fungi and plants, so the three are probably best presented as a trichotomy—but, if anything, the fungi seem marginally closer to animals than to plants. The presence of chitin in fungal cell walls has long suggested this animal affiliation; even so, it is counterintuitive. Certainly the traditional inclusion of fungi within the plants is way off beam.

Whatever the relationship between the chytridiomycotes (or 'chytrids') and zygomycotes may finally prove to be, they are both generally simpler in form than the ascomycotes and basidiomycotes; for example, their hyphae lack cross-walls (**septa**), so they are said to be 'aseptate'. Indeed, they can literally be considered more primitive and are sometimes referred to as 'lower fungi'. The Ascomycota and Basidiomycota emerge as sister groups and can reasonably be thought of as 'higher fungi'. Between them the ascomycotes and basidiomycotes include most of the organisms that most people think of as 'fungi'—that is, all the mushrooms, 'toadstools', puffballs, brackets, and agents of devastation such as dry rot, plus the yeasts and lichens, and the smuts and rusts, the great enemies of farmers. The hyphae of ascomycotes and basidiomycotes are commonly divided by septa (that is, are 'septate') and their sexual fruiting bodies are commonly far more elaborate than those of zygomycotes or chytrids. But their sexual bodies differ in micro-anatomy. Ascomycotes produce their meiotic spores within tiny capsules, called **asci**, like peas in a pod; whereas basidiomycotes produce theirs, commonly in groups of four, on the tips of small truncheon-like structures called **basidia**. The big fruiting bodies, such as truffles or mushrooms, contain many thousands of asci or basidia arranged on intricate supporting structures, seen, for example, in mushrooms and puffballs.

A few general features of the tree are particularly striking. We have already seen how fungi of similar general habits may occur in several different phyla and often—though this cannot be shown in detail—in several or many different orders within the same phylum. Thus many basidiomycotes have a 'toadstool' form; but so do some ascomycotes like the morel, *Morchella*. Truffles, too, are ascomycotes: but basidiomycetes of the genus *Rhizopogon* produce remarkably truffle-like underground fruiting bodies called 'false truffles'. Similarly, although we may acknowledge 'true yeasts' as a discrete clade among the ascomycotes, others in other phyla may also be yeast-like.

Lichens occur both among the ascomycotes and the basidiomycotes. Furthermore, different groups of lichens have clearly evolved independently at least twice in each of those phyla, and probably even more often. Lichens may occur in the same family as fungi that are not lichenized: thus the basidiomycote family Tricholomataceae includes the beautiful mushroom shiitake, and also some lichens. Although some biologists have suggested that lichens represent some kind of 'ideal' co-operative state to which all organisms in some sense 'aspire', Paula DePriest of the Smithsonian Institution in Washington DC points out that no such happy state should be envisaged. The relationship between the fungus and its algal or cyanobacterial residents clearly takes different forms in different lichens, and, in some, looks like barely contained warfare.

Many species of zygomycote, ascomycote, and basidiomycote form mycorrhizae, which differ in the way they interact with the root. Thus in some, **ectomycorrhizae**, the hyphae penetrate the root but remain between the cells; whereas others, **endomycorrhizae**, penetrate the root cells themselves. The most elaborate kinds of mycorrhizae form intricate, branching structures within the root cells and are called **arbuscular**;[5] these are formed by members of the zygomycote order Glomales, although other zygomycotes (of the genus *Endogone*) form ectomycorrhizae. All in all, zygomycote mycorrhizae are the most widespread botanically and geographically.

Note, too, how fungi that cause human diseases—*Pneumocystis*, *Candida*, *Histoplasma*, and *Aspergillus*—are scattered among the Ascomycota, as Barbara Bowman and her colleagues point out: the closest relatives of those pathogens are often non-pathogens. Clearly, pathogenicity has evolved independently many times among the ascomycotes.

And so on. Always the same pattern: any one lineage may essay many different forms; any one form may be essayed by many different lineages. But (just to make the crucial point yet again) unless we have a classification that truly reflects phylogeny, then we miss this glorious interplay of radiation and convergence.

'Chytrids': phylum Chytridiomycota

There are about 100 genera of Chytridiomycota, alias chytrids (pronounced 'kit-rids'), and traditionally they are shown as the sisters of all the other fungi, which means they are presumed to be closest to the ancestral fungal form. Indeed, in various molecular studies, the particular genus *Blastocladiella* has emerged as the sister of all the other fungi under scrutiny. This makes perfect sense, for a number of reasons. First, chytrids are primarily aquatic, and we can assume that the kingdom Fungi did indeed originate in water, just like the Animalia and Plantae. Second, although chytrids feed and grow via hyphae, either as saprobes or as parasites, and although they produce fruiting bodies on the surface just as a basidiomycote may send up toadstools, they also produce motile gametes—that is, gametes with flagella, like spermatozoa. This is in contrast to other fungi, but in the manner of most other eukaryote kingdoms; which suggests once more that motile gametes must be a primitive feature. Once the gametes have fused, the resulting zygote becomes a resting structure, useful for dispersal and for tiding over unfavourable times: a strategy also seen in zygomycotes such as *Mucor*, or 'pin-mould'. Hence in their biochemistry,

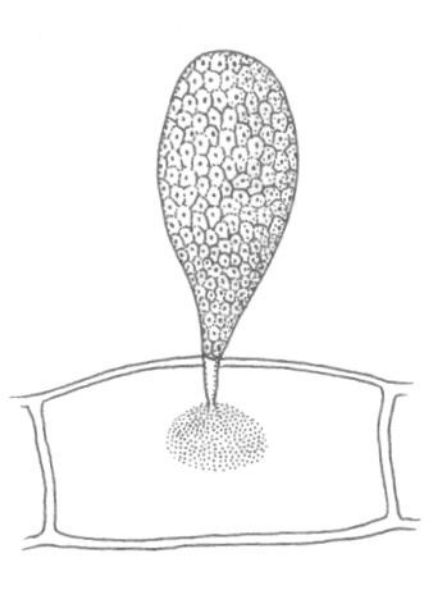

chytrid
Rhizophidium ovatum

[5] The traditional term is 'vesicular arbuscular mycorrhiza' or VAM, because in many types the arborescences are accompanied by swellings of the hyphae called vesicles. But in others there are no vesicles and the general term 'arbuscular mycorrhizae' is increasingly preferred.

general body plan, and at least some details of behaviour the chytrids emerge as bona fide fungi; while in their ecology and sexual behaviour they precisely fit the role of 'missing link' between the fungi and other kingdoms.

As you can see, however, I have not presented the Chytridiomycota as sisters to all the rest; but instead have shown the Chytriodiomycota + Zygomycota as the more primitive sisters of the Ascomycota + Basidiomycota. The reason is that both the Chytridiomycota and the Zygomycota may need to be re-thought in the light of recent studies by Tom Bruns at Berkeley, California. For when Bruns and his colleagues looked at 18S rRNA sequences in a variety of fungi they found that in some analyses *Mucor*, universally accepted as a zygomycote, came out closest to *Blastocladiella*, which tends to emerge as the most primitive of chytrids; whereas *Endogone*, another apparent zygomycote, seemed to group with various other chytrids. This suggests that the two traditional groups, Chytridiomycota and Zygomycota, may in fact be mixed up together. Bruns points out in his original paper that 'the conflict with traditional classification is not strongly supported'. Even so, fungus enthusiasts should watch this space; and, for the time being, it seems sensible to present the phylogeny as shown.

Chytrids live mainly in fresh water but a few are estuarine, some live in soil, and some live as anaerobes in the guts of herbivorous animals; some are saprobes and some are parasites—on nematodes, insects, plants, and other fungi (including other chytrids). Mary Berbee of the University of British Columbia at Vancouver, and John Taylor at Berkeley conclude from molecular studies that the chytrids as a whole are an ancient group; they seem to have diverged from terrestrial fungi around 550 million years ago, near the start of the Cambrian, and perhaps even before. But they also point out that chytrids that live in the guts of mammals must have undergone more recent radiations. Thus, chytrids such as *Caecomyces* and *Piromyces* occur in the hind guts of a wide variety of species, whereas others, like *Neocallimastix*, occur only in ruminant stomachs. They conclude that the ruminant stomach specialists must have split away from the hind-gut dwellers after the first therian mammals appeared, around 150 million years ago, and before the first ruminants came on the scene around 40 million years ago.

Pin-moulds and others: the Zygomycota

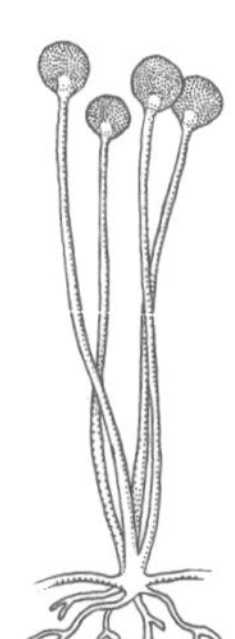
zygomycote *Rhizopus nigricans*

As we have seen, the Zygomycota may have to be re-classified: the group seems to be polyphyletic, and may well be mixed up with the chytrids. For the time being, however, Zygomycota is still widely recognized as a valid taxon, which we can call a phylum.

Be that as it may, more than 1000 species of zygomycote are known from all over the world. As with chytrids, their hyphae are not divided by septa. To reproduce asexually they produce spores inside **sporangia**, which are commonly borne aloft on special hyphae called **sporangiophores**—and it is these that form the characteristic mould. They also have a quaint method of sexual reproduction: two hyphae, apparently of opposite mating type, grow towards each other; they finally touch; the touching ends swell up, and the swollen ends coalesce to form a single sphere that becomes a thick-walled resting body, or **zygosphere**; nuclei from both sides intermingle and fuse in pairs and then undergo meiosis to form haploid spores.

Ainsworth and Bisby's Dictionary of Fungi recognizes two classes of zygomycote. The first, the **Trichomycetes**, includes fewer than 200 species with a relatively simple structure that mostly live as parasites or commensals in the guts of arthropods. The second, the **Zygomycetes**, includes nearly 900 known species, which *Ainsworth and Bisby* divides into seven orders. Of these, the **Mucorales** include the saprobic moulds such as *Mucor* and *Rhizopus*, and the athletic *Pilobolus*, alias 'the hat thrower', which grows on dung and shoots its sporangia about 2 metres into the air. The **Zoopagales** are parasites of small animals such as nematodes and of protists such as amoebae, and include *Cochlonema*, *Endocochlus*, and *Stylopage*. The **Entomophthorales** are also parasites, mostly of insects. Many zygomycotes are mycorrhizal. Members of the genus *Endogone* form ectomycorrhizae while those of the order **Glomales** form the sophisticated arbuscular form of mycorrhizae, in which branches of the hyphae form intricate structures within the root cells of the host. Botanically and geographically, zygomycote mycorrhizae are the most widespread of all—and so, in the form of the Glomales, the zygomycotes emerge as the most important symbionts of plants.

Spores of what seem to be zygomycotes—and indeed are probably relatives of *Glomus* in the Glomales—have been found in the company of rhizomes (underground stems) from some of the world's oldest fossils of vascular plants: of the genus *Rhynia* (one of the rhynophytes; see Section 23) dating from the early Devonian, about 400 million years ago. Given that the Glomales are such great makers of mycorrhizae, this adds fuel to the notion that plants and fungi have been working together (however uneasily) ever since they both ventured on to land. But we should not assume that these Glomales are the oldest fungi on land, or anything like, for among the same group of plant fossils are hyphae that apparently come from modern-looking fungi, such as ascomycotes or basidiomycotes. Thus the split between the zygomycotes (as represented by Glomales) and ascomycotes + basidiomycotes must have taken place before the early Devonian period.

True yeasts, moulds, morels, and truffles: phylum Ascomycota

The Ascomycota are the largest class of fungi with more than 32 000 species that range from single-celled yeasts to great fructifying mycorrhizal morels and truffles. But they are united and can be diagnosed by their possession of the ascus, the pod-like container of spores. Ascomycotes also reproduce asexually—the hyphae simply segment to form **conidiospores**; and some, so far as is known, only reproduce asexually.

Ascomycotes are both saprobes and parasites. Many are mycorrhizal and almost half of all the known species form lichens.

Ainsworth and Bisby apportions this plethora of species among 46 orders (plus about 29 families that cannot yet be confidently ascribed to any accepted order). The order **Sordariales** includes the saprobic *Neurospora*, the common bread mould, and a favourite with geneticists. *Elsinoë* of the **Dothideales** is an agent of scab in the tropics, not least of citrus and lima beans. The order **Saccharomycetales** includes *Saccharomyces cerevisiae*, which serves both as brewer's and baker's yeast, and also the versatile genus *Candida*, which includes both *C. albicans*, a pathogen of humans and animals, and *C. utilis*, the food yeast. **Pezizales** is a distinguished order that includes *Tuber*, the genus of the truffles and also *Morchella*, the morel; *Peziza*, which gives the order its name, has quaint cup-shaped fruiting bodies, like *P. badia*, the common brown elf cup. The order **Ophiostomatales** contains the agent of Dutch elm disease *Ophiostoma ulmi* (and the even deadlier *O. novo-ulmi*), which, in the 1970s, changed the face of the English countryside, where the elm was once known as 'the Wiltshire weed'. The spores of *Ophiostoma* are spread by beetles, which etch characteristic 'galleries' beneath the bark of the trees they are feeding on. Such galleries have been found in fossil wood dating from the Cretaceous period, which suggests that such beetles evolved in the early Cretaceous, perhaps 140 million years ago. *Ophiostoma* could have arisen at about the same time. Most ascomycotes disperse their ascospores forcibly, but Mary Berbee and John Taylor suggest that the ancestors of *Ophiostoma* came to depend instead on beetles to spread their spores, and lost the ability to scatter their spores by themselves.

Ainsworth and Bisby notes that different authors have grouped the many ascomycotan orders in various supraordinal groupings, but that none of the groupings seems phylogenetically robust. Various molecular studies at Berkeley, Roche, and the Smithsonian suggest, however, that the ascomycotes can reasonably divided into three great groups as shown here:

- The **basal ascomycetes** are a mixed, informal grouping of ascomycotes that simply seem to be somewhat primitive. They include the yeast-like *Schizosaccharomyces*, the pathogenic invader of human lungs; *Pneumocystis*; and the agent of peach leaf curl, *Taphrina deformans*. Somewhere among the basal types must lie the sister group of the yeasts + the moulds and the other hyphal ascomycotes—but as Mary Berbee and John Taylor comment, it is not clear which of the living types most closely resembles the first ancestral ascomycote.

- The **true yeasts** can be seen to be ascomycotes even though many of them remain permanently in a single-celled, essentially protist state, without producing hyphae at all, because at least some of them produce asci. Yeasts include *Saccharomyces*, genus of brewer's yeast, and *Candida*. But although yeasts have an extremely simple structure (essentially 'protist'), this simplicity is derived, not primitive.

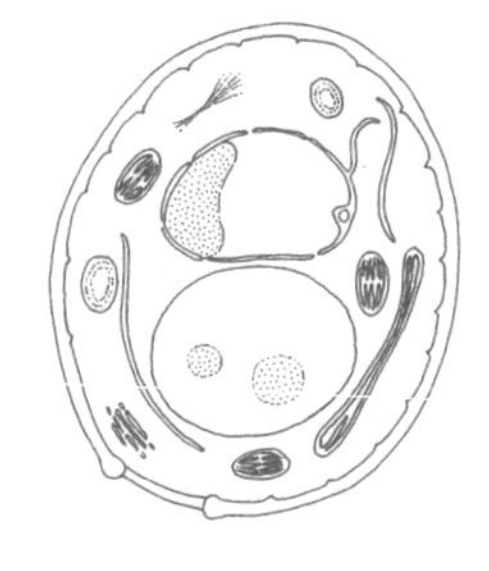

baker's yeast (resting cell)
Saccharomyces cerevisiae

Yeasts seem to be descended from species that had hyphae, and Berbee and Taylor suggest that the first ones appeared about 240 million years ago.

mould
Penicillium

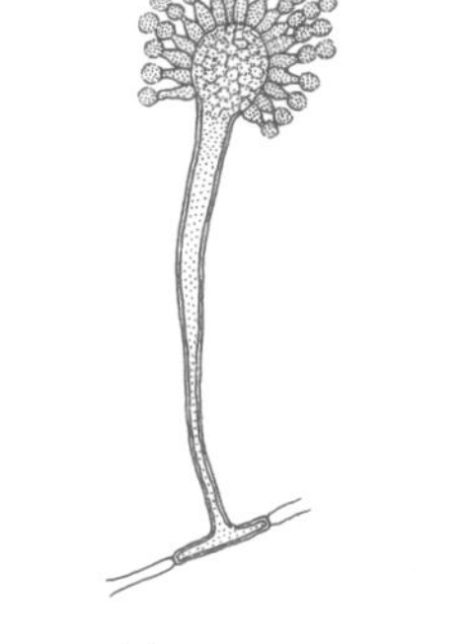
mould
Aspergillus niveo-glaucus

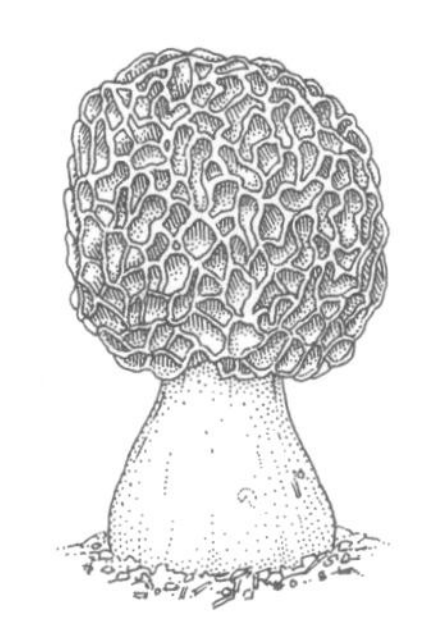
morel
Morchella esculenta

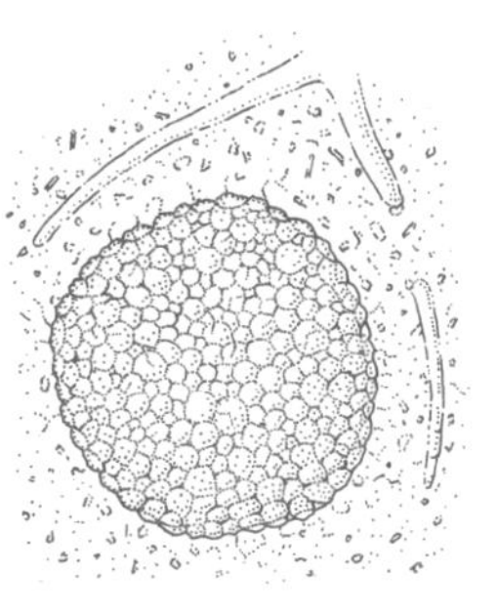
truffle
Tuber aestivum

• The **moulds**, such as *Neurospora*, and the macro-fungi such as *Tuber*, and *Morchella*, are sisters to the yeasts. In addition, about 13 orders among these macro-fungi include at least some types that are lichenized—and four orders are exclusively lichenized. Lichenization has evolved at least twice among modern ascomycotes, and probably more.

Although molecular studies suggest that the ascomycotes as a whole date only from the Carboniferous period, about 330–310 million years ago, there are some fossil spores that resemble ascospores, and yet date from the earlier Silurian period. But, then, the Ascomycota have long been considered sisters to the Basidiomycota; and, perhaps, say Berbee and Taylor, these ancient spores were produced by the fungi that were the common ancestors to both those modern groups.

Smuts, rusts, jellies, mushrooms, and brackets: phylum Basidiomycota

The Basidiomycota are another huge group containing more than 22 000 species. They are seen as sisters to the ascomycotes, the two forming a clade that can be seen as the sister to the more primitive zygomycotes + chytrids.

The characteristic reproductive body of basidiomycotes, and the character that defines the group, is the basidium—the truncheon-like bearer of spores described above. In the parasitic smuts and rusts the basidia are contained in simple structures that poke above the surface of the host; but in the more elaborate Hymenomycetes (the grouping that includes the big familiar fungi like the mushrooms), the basidia are held neatly on gills or in pores, typically beneath a protective umbrella or shelf-like bracket. But some basidiomycotes can assume a yeast-like state.

Smuts and rusts are major parasites of crop plants, and rusts, in particular, may have enormously complex life cycles that they vary according to climate and the state of the host.

Over the years different mycologists have recognized many different subgroupings (effectively classes) within the Basidiomycota. Here, I am following the account by Eric Swann and John Taylor at Berkeley, California, who base their classification on 18S rRNA. As shown on the accompanying tree, Swann and Taylor divide the Basidiomycota into three great groups, which we could call classes. The **Ustilaginomycetes** include most of the smuts; the **Urediniomycetes** include the rusts, plus others that include some traditionally classed as smuts; and the **Hymenomycetes** include all the big basidiomycotes—mushrooms, toadstools, brackets, and the 'jelly fungi'.

These groupings correspond closely with traditional classifications, except that the smuts are usually seen to be monophyletic whereas Swann and Taylor conclude that some of the traditional 'smuts' are really closer to the rusts. But exactly how the three groupings relate to each other has proved highly controversial. Some regard rusts (roughly the Urediniomycetes) as sisters to the Ustilaginomycetes + Hymenomycetes, whereas others have treated the rusts and hymenos as sisters, with the smuts (which roughly means the Ustilaginomycetes) as the outsiders. Swann and Taylor suggest that the jury is still out. In other words, it is safest to treat the three as a trichotomy, as here.

Ustilaginomycetes

Smuts, most of which seem to belong in Swann and Taylor's Ustilaginomycetes, include more than 1000 known species in 63 genera. In general, they are parasites of monocots, which include the grasses, which in turn include the cereals, the world's most important crop plants. Thus *Ustilago segetum* attacks barley, oats, and wheat, while *U. bullata* attacks brome grasses (*Bromus*). Smuts form mycelia when they are in their parasitic phase, but take a yeast-like form when cultured. *Sporobolomyces*, the shadow or mirror yeast, does this.

smut, *Ustilago*, in an ear of wheat

Urediniomycetes

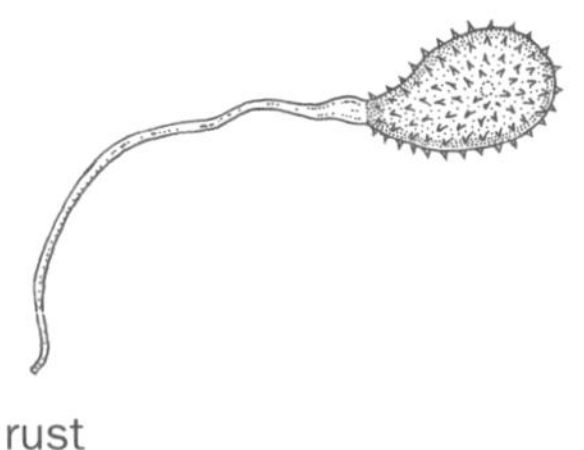

rust
Puccinia belizensis

Rusts, which make up much of the Urediniomycetes, are hugely important pathogens of plants. The genus *Puccinia* again includes a wide range of species that attack cereals: *P. graminis* is the black stem rust; *P. hordei* is the brown leaf rust of barley; *P. recondita* affects rye and wheat; *P. coronata* is the crown rust of oats; and *P. striiformis* is the yellow stripe of cereals. *Uromyces* is a common rust of leguminous plants, such as peas and beans, while *Phragmidium* attacks roses, blackberries, and their relatives.

Hymenomycetes

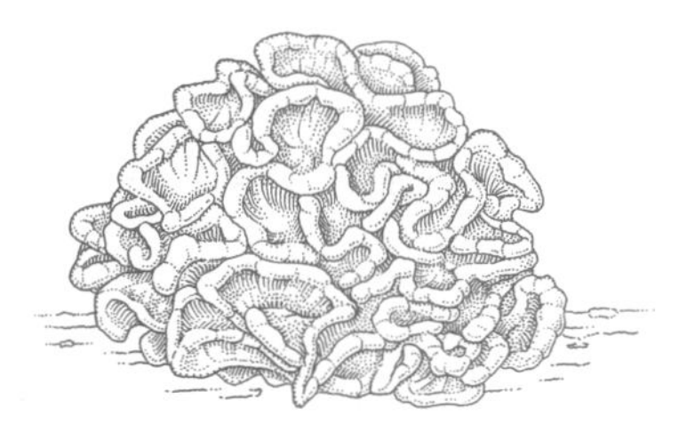
leafy brain fungus
Tremella foliacea

Traditional studies based on anatomy, and various molecular studies including that of Swann and Taylor, tend to agree in dividing the Hymenomycetes into two main groups. Sister to all the rest are the jelly fungi, in the order **Tremellales**. Some of the Tremellales, like *Tremella*, *T. gelatinosum*, the 'jelly hedgehog' are macro-fungi, though they have a distinctly jelly-like consistency; but others have a yeast-like form. The remaining hymenomycetes form a clade with nearly 14 000 known species. These have been grouped in various ways—*Ainsworth and Bisby*, for example, recognizes 31 orders—but these are up for discussion that is both fierce and esoteric, and it seems safest to stay out of it. Molecular studies are suggesting that many of the traditional groupings based on anatomy are simply wrong (at least in the sense that they do not reflect phylogeny) but there are far too few such studies to provide a comprehensive or robust alternative to the traditional systems, and those that there are lead to different conclusions depending on how they are analysed.

In general, the hymenomycetes are a wonderfully varied lot. Some take a mushroom-like or toadstool-like form; some are bracket fungi, which the Americans call shelf-fungi; many form mycorrhizae; and many are lichenous—with many lichenous types being in the same family or even in the same genus as non-lichenous types. The traditional orders (as laid out in *Ainsworth and Bisby*) will probably need to be revised in the light of molecular studies but it seems to make sense to use them for the time being, until there is a clear and comprehensive alternative.

Thus, outstanding among the traditional orders is the **Agaricales**, which contains the genus *Agaricus*, whose 200 or so species include *A. campestris*, the field mushroom, and *A. bisporus*, the cultivated mushroom; *Amanita*, another genus with about 200 species, many of which are poisonous and sometimes deadly, including the deathly white *A. virosa* the 'destroying angel' (a whited sepulchre indeed), and *A. muscaria*, the fly agaric, which with its white-spotted red cap is everyone's idea of a toadstool; *Coprinus*, the ink-caps, with another 100 or so species; *Armillaria*, whose 40-odd species include *A. mellea*, the exceedingly destructive honey fungus; and—both in the family Tricholomataceae—*Lentinula edulis*, the shiitake, and *Omphalina*, which is a lichen. Similarly, order **Cantheralles** includes *Cantharellus cibarius*, the gloriously esculent chanterelle, and *Multiclavula*, which is another lichen.

cultivated mushroom
Agaricus bisporus

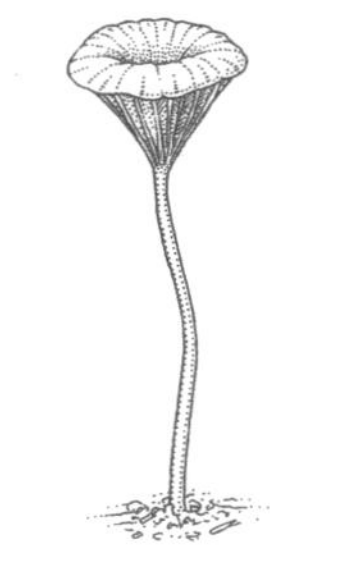
lichen
Omphalina ericetorum

The lovely toadstools *Cortinaria* and *Russula* are in the **Cortinariales** and **Russulales**, respectively. The **Boletales**

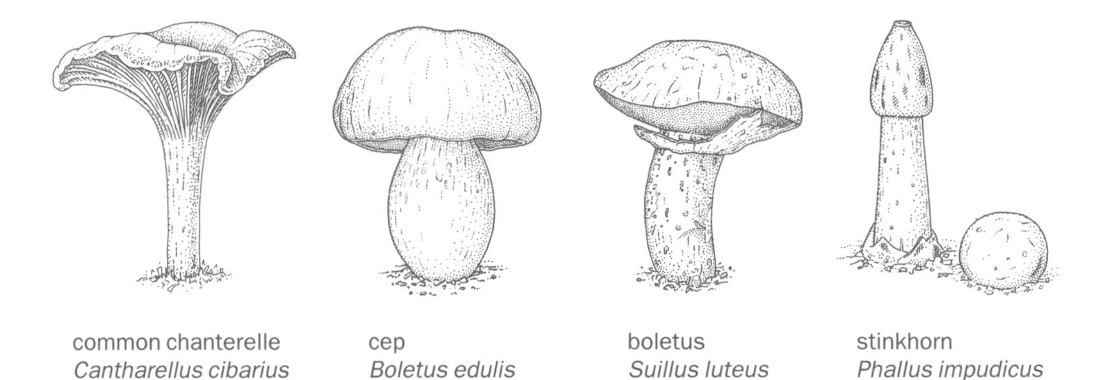

common chanterelle *Cantharellus cibarius*   cep *Boletus edulis*   boletus *Suillus luteus*   stinkhorn *Phallus impudicus*

includes *Boletus*, with *B. edulis*, the magnificent cep or 'penny bun' boletus, plus many less-succulent and even poisonous types such as *B. luridus* and *B. satanas*, the 'devil's boletus'. Among the Boletales, too, is *Suillus*, now shown to be a close relative of the false truffle, and *Serpula lacrymans*, the much-feared agent of dry rot. **Ganodermatales** includes *Ganoderma applanatum*, a common parasite of beeches; **Lycoperdales** includes the puff-balls and earth-balls such as *Lycoperdon*, the puff-ball, and *Langermannia gigantea*, the giant puff-ball; and **Phallales** is the order of *Phallus*, the stinkhorns.

One leading modern mycologist has described the traditional order **Poriales** as 'a hopeless polyphyletic garbage bag'. In the fullness of time, then, and preferably within the next few years, we should see 'Poriales' banished, and its members re-assigned to new groupings defined more phylogenetically. Meantime we may note that the members traditionally bundled into this garbage bag include a host of bracket fungi, as in the genus *Polyporus*—for example, *P. squamosus*, the Dryad's saddle; *P. sulphureus*, the sulphur-yellow polypore; and *P. betulinus*, the birch polypore or razor-strop fungus. *Pleurotus ostreatus* was also traditionally placed in the Poriales. This is the oyster fungus or oyster cap, which damages wood but is grown commercially for food.

polypore fungus
Polyporus brumalis

Of the genera mentioned above, *Amanita*, *Cortinaria*, *Russula*, and *Boletus*, are great producers of ectomycorrhizae. Many others form lichens, including *Dictyonema*, *Multiclavula*, and *Omphalina*; and some families, such as that of the shiitake, Tricholomataceae, contain both lichenized and non-lichenized species. Indeed some genera contain some species that are lichenous, and some that are not, including *Arthonia*, *Arthothelium*, *Bacidia*, *Mycomicrothelia*, *Omphalina*, and *Toninia*.

Mary Berbee and John Taylor suggest that the basidiomycotes diverged from the ascomycotes around 390 million years ago. Late Devonian hyphae dating from around 360 million years ago could be basidiomycotes, although the first unequivocal fossils of basidiomycotes date from about 290 million years ago—hyphae that were found among the tissues of a late Carboniferous fern. The three main lineages that evolved into the modern rusts, smuts, and hymenomycetes seems to have diverged about 340 million years ago, with genuine rusts first emerging around 310 million years ago, presumably as parasites of early vascular plants. The jelly fungi seem to have diverged from the other hymenomycetes around 220 million years ago. The oldest known bracket fungus is *Phellinites*, dating from about 165 million years ago, and molecular clock evidence suggests that the hymenomycetes radiated in the Cretaceous, around 130 million years ago, after the flowering plants (angiosperms) had become an important part of the flora. The earliest known mushroom is *Coprinites dominicana*, dating from only the Eocene epoch, about 40 million years ago, but mushrooms do not fossilize well and they are probably much older than that.

Now we can turn to the kingdom which, so at least some evidence suggests, is closest to the fungi: the animals.

5

THE ANIMALS
KINGDOM ANIMALIA

WE ARE ANIMALS. So are sponges, sea anemones, worms of all kinds, snails, octopuses, barnacles, flies, spiders, starfish, sharks, dinosaurs, sparrows, and dogs. Truly ours is a wondrous kingdom. Yet it is not easy to define what animals are. The general features they have in common are readily identifiable: they are active heterotrophs (they do not photosynthesize like plants and seaweeds) and they tend to be large—most are at least big enough to see with the naked eye, whereas most other active heterotrophs (like most of the heterotrophic protists and prokaryotes) are 'microbes'.

But in this modern, cladistic age such loose inventories will not do. Clades cannot be defined by negatives: we cannot simply say that animals are large creatures that do not photosynthesize. Neither can they be defined by general features such as heterotrophy, for there are many other kinds of heterotroph. Clades cannot generally be defined by size—and besides, as we shall see, at least one group of protists should be included among the animals. If the kingdom Animalia is to emerge as a true clade we need to find synapomorphies—features that all animals have (or may be presumed to have had in some ancestral state), which they all derived from a common ancestor, but are not simply common to other creatures as well.

Because it has been hard to find such features, many zoologists have questioned whether the alleged kingdom Animalia is really a clade at all. No one can doubt that animals arose from among the ranks of the protists; but did they all arise from the same protist, or from more than one? The sponges have been particularly bothersome: some naturalists in the past have suggested that they are not animals at all, and some modern biologists (such as Lynn Margulis and Karlene Schwartz in *Five Kingdoms*) still suggest that they did not descend from the same protist as other animals—which would make the kingdom Animalia polyphyletic.

Traditional biologists have identified a few convincing characters. It seems, for example, that all animals from sea anemones to aardvarks produce the protein collagen, a tough connective material found not least in skin. Other creatures do not apparently make collagen. For such reasons it now seems that we can reasonably declare the contest closed. The Animalia, including sponges, does seem to be monophyletic—and we have an idea who the common protist ancestor was: it was probably similar to the still-extant choanoflagellates. The only question now is whether the modern choanoflagellates should themselves be included within the Animalia, or should be regarded as a sister group (presumably a sister kingdom) to the Animalia. In this section I am largely following Claus Nielsen, who, in *Animal Evolution* (1995), treats the choanoflagellates as the animals' sister group. As you can see, however, I prefer to place the choanoflagellates within the Animalia. It seems to me that the two groups together do fulfil the requirements of a clade: they are both presumed to share the same 'clade founder', who would be some choanoflagellate-like protist. The main reason for separating the two, as I see it, is that choanoflagellates are protists (sometimes unicellular and sometimes colonial) whereas the rest of the animals are multicellular, or metazoan. But this is a difference only of grade and unless there are some very special reasons for acting otherwise, taxa should be defined only by phylogeny, and not by grade. It seems to me perfectly acceptable to include protist-grade creatures within the Animalia, just as it is to include protist-grade green algae within the Plantae. Nielsen himself included the choanoflagellates within the Animalia in an earlier publication, and B. S. C. Leadbeater and Irene Manton said in 1974 that choanoflagellates can 'be treated as animals, related to sponges'.

The living choanoflagellates—Choanoflagellata—include about 140 species, all marine, some solitary and some colonial. Each choanoflagellate cell carries a long cilium at one end, which is surrounded by a collar consisting of a circle of projections called **microvilli**—like a stockade. As it beats, the cilium draws water through the gaps between the circle of microvilli; nutritious particles in the current of water are thus caught on the outside of the collar; and these can then be ingested by the main body of the cell.

Virtually identical cells called **choanocytes** form the digestive tissue of sponges, and indeed the task of most of the cells in the body of the sponge is to create a platform from which the choanocytes can operate. The similarity between choanocytes and choanoflagellates indicates as clearly as any biological feature ever can that Choanoflagellata and Porifera (sponges) share a common ancestor. In fact, the colonial choanoflagellates form balls of cells (superficially like the colonial alga *Volvox*—page 185), and in some of these the cilia of the cells face outwards while in others they face inwards. In the latter, the choanoflagellate colony closely resembles the feeding mechanism of sponges, which consists of a hollow chamber lined with choanocytes. No other animal has true choanocytes, but most other animal phyla have 'collar cells' that resemble choanocytes, which in some cases at least could well be homologous with them. This similarity is one of several that suggest continuity between choanoflagellates, sponges, and all other animals.

Yet remarkably little of phylogenetic relevance seems to be known about choanoflagellates; 'remarkably' given that they are, probably, the world's first and oldest animals, and therefore seem to be of supreme interest. Apparently, at the time of writing, no one knows whether choanoflagellates have collagen. We should know such things. If choanoflagellates really are the world's first animals then they are extremely important. It is as if we were ignoring the last remaining population of *Homo erectus*.

In the end, however, it may prove best to define animals not according to some overt anatomical feature, or by some quirk of biochemistry, but by direct reference to their genes. For all kinds of animals examined so far contain a particular group of extremely important genes called the Hox complex, which seems to be unique to them. Hox genes thus seem to define animals as a whole; and, as we shall see later, the differences between them are already helping zoologists to re-assess the relationships between the different animal phyla.

THE HOX GENES

All eukaryotes so far investigated—not just the animals!—have a family of genes of the kind known as homeobox, and these contain a particular sequence of bases, which (it turns out) produces a protein that is able to control the activity of other genes—either switching them on or switching them off. Moreover, we now know that animals have a category of genes that are called homoeotic; and the function of homoeotic genes is to determine the fate of each particular region of the body. For example, homoeotic genes in a fly determine that the first segment of the thorax produces functional wings and the head produces antennae, and so on; and very similar genes in all other metazoan animals have similar functions. Not all homeobox genes are homoeotic in function, but many are; and not all homoeotic genes are homeobox in structure, but many are. Animal genes that are both homeobox in structure and homoeotic in function are called Hox genes. In general, the visible complexity of animals is reflected in the number and complexity of their Hox genes. Thus whereas a fly or a cephalochordate has only one set of Hox genes, vertebrates have four complete sets, each slightly different from the others, which between them carry a great deal more information.

In 1994, Jonathan Slack, Peter Holland, and C. F. Graham, all three then at the University of Oxford, suggested that animals as a whole might reasonably be defined through their possession of Hox genes, and that the different phyla might—at least in part—be defined by the pattern of expression of those genes. Moreover, they suggested that the elusive concept of body plan or *Bauplan*, at least at the level of the phylum, should be linked to the pattern of Hox-gene expression. Slack, Holland, and Graham were, however, referring exclusively to metazoans. If we accept that choanoflagellates are animals too—or even if we choose only to admit that they are the sister group of animals—then it surely would be interesting to ask whether choanoflagellates also have homeobox genes; if so, what those homeobox genes actually do in choanoflagellates; and whether the things that homeobox genes do in choanoflagellates suggest that they have the potential to act as Hox genes in metazoans.

It seems apt to predict that choanoflagellates do have genes that could pass as precursors of Hox genes; and that it was partly because choanoflagellates had such genes that they were able to evolve into metazoans. The complete hypothesis would be, in short, that the outstanding synapomorphy of the great kingdom of the Animalia is the possession of Hox genes (homeobox genes with a homoeotic function), and that choanoflagellates will prove to merit their status as true animals because they will be found to have convincing homologues of Hox genes. If all this proves to be the case, then it will be satisfying indeed.

Traditionally the classification of animals, and their presumed phylogeny, has been based on a study of visible, anatomical features and with special reference to embryology. Since the mid-1990s, however, DNA studies have prompted some radical revisions and great areas of animal taxonomy are now in a state of flux. These studies have been based, first, on the DNA that makes 18S ribosomal RNA (that is, the DNA that codes for the RNA of ribosomes—see Part I, Chapter 4) and, most recently, on the Hox genes. At least, the broad divisions remain, effectively as first defined in the nineteenth century. The big groupings of the Eumetazoa, the Bilateria, and the division of the Bilateria into Protostomia and Deuterostomia are still as they have been for 100 years or more, and most of the individual phyla as listed here have been recognized for decades. As is soon to be described, however, the molecular research now shows that some of the phyla that were previously classed as deuterostomes are really protostomes; and that the relationships between the different protostome phyla are not at all as the traditional studies suggested.

The tree shown here is a synthesis, put together from several contrasting sources of information. The general shape is based on Claus Nielsen's taxonomy in *Animal Evolution*, which is broadly traditional—based on morphological and embryological data—although, as we will see, Nielsen introduces a few novelties of his own. But I have also included a newly discovered phylum—the intriguing Cycliophora, whose single known representative, *Symbion*, is believed to be related to the rotifers. I have also rearranged the protostomes (and transferred some phyla from the deuterostomes to the protostomes) as suggested by the new molecular studies.

Animal taxonomy is in such a state of upheaval that very few zoologists will agree with all the details of the tree shown here. On the other hand, they are most unlikely to agree absolutely with each other. I cannot claim that this or any other tree represents the absolute and unequivocal truth, and neither can anybody else. I do claim, however, that this tree represents a reasonable summary—the best I am able to put together—of data that at present are copious but far from complete, and are in significant ways conflicting. Future studies will doubtless bring further changes; but the present state of knowledge is far from trivial, and the summary presented here does provide a solid foundation for future studies.

A GUIDE TO THE ANIMALIA: CLADES AND GRADES

In the bad old days zoologists were content to divide all animals into two great groups: the 'Vertebrata', which include fish, frogs, dinosaurs, birds, and, of course, ourselves; and the 'Invertebrata', which included all the rest—worms, snails, insects, and so on. The Vertebrata is a perfectly good clade but the 'Invertebrata' is not: as the tree shows, the latter is now taken to include at least 30 phyla, some of which (notably the echinoderms, such as starfish and sea urchins) are much more closely related to the vertebrates than to most other invertebrates. But the informal term 'invertebrate' is still useful (like 'protist' or 'worm') even though 'Invertebrata' is archaic.

The tree shown here lists 32 phyla altogether (of which the Vertebrata is one—although it is generally ranked simply as a subphylum of the larger grouping, the Chordata). Most authorities today recognize roughly this number. The differences lie in the details. Thus many authors divide the segmented worms—grouped together here as the Annelida—into several different phyla. I have also somewhat cavalierly presented a phylum labelled 'Bryozoa', which in fact contains two distinct groups, the Ectoprocta and the Entoprocta. Some zoologists maintain that the similarities between these two are purely superficial, and that they are not closely related. Nielsen, however, argues the other way around: that the similarities denote true relationships while the differences are superficial, and he presents the ectoprocts and entoprocts as sister groups that together constitute the Bryozoa. So I have put them together in a single phylum—and, indeed, many traditional texts combine the entoprocts and ectoprocts within a single putative phylum, Bryozoa.

PROTIST TO METAZOAN

Of more significance is that the Animalia as defined here is split into a series of very deeply divided clades, most of which contain several or many different phyla; and some of those divisions also represent a major shift in grade.

The most striking shift in grade lies between the **Choanoflagellata**, which are protists (and are often excluded from the Animalia), and the multicellular animals, which constitute the great clade of the **Metazoa**. Within the Metazoa we see several more shifts of grade. First, there is a clear transition between the **Porifera** plus **Placozoa** on the one hand, and the **Eumetazoa** on the other. In the Porifera and Placozoa, the different body cells seem to have an almost plant-like relationship one to another whereas in the Metazoa the cells generally co-operate to form tightly co-ordinated organisms.

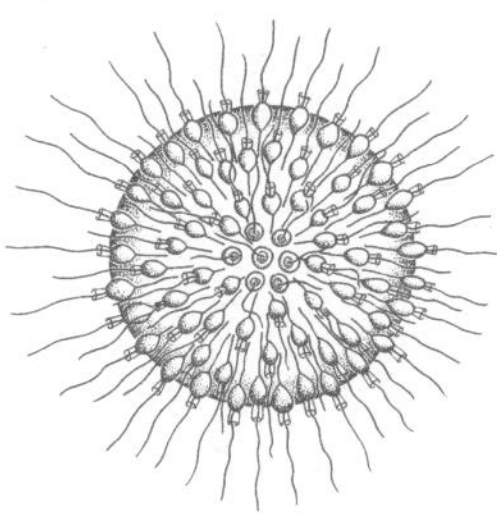

choanoflagellate
Sphaeroeca volvox

Sponges and placozoans: phyla Porifera and Placozoa

Members of the phylum **Porifera** are structured like no other creature. At its simplest, a sponge consists of a hollow cup of tissue with perforations in the sides called **ostia** (singular ostium) and an open top, the **osculum**; although in most living sponges this basic theme is greatly elaborated. Each cup is lined with choanocytes whose beating cilia produce a flow of water—in through the

sponge
Euplectella aspergillum

5 · KINGDOM ANIMALIA

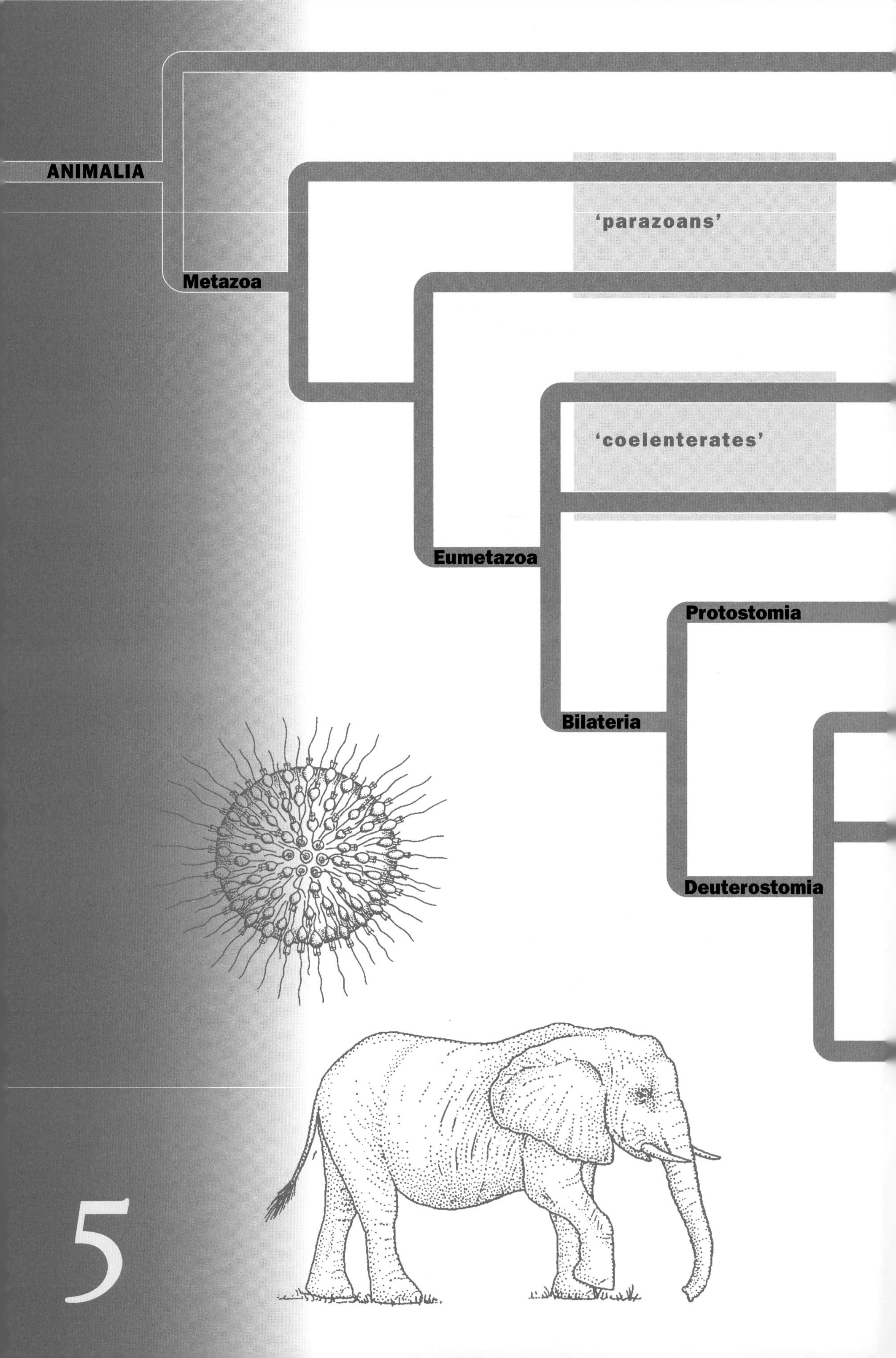

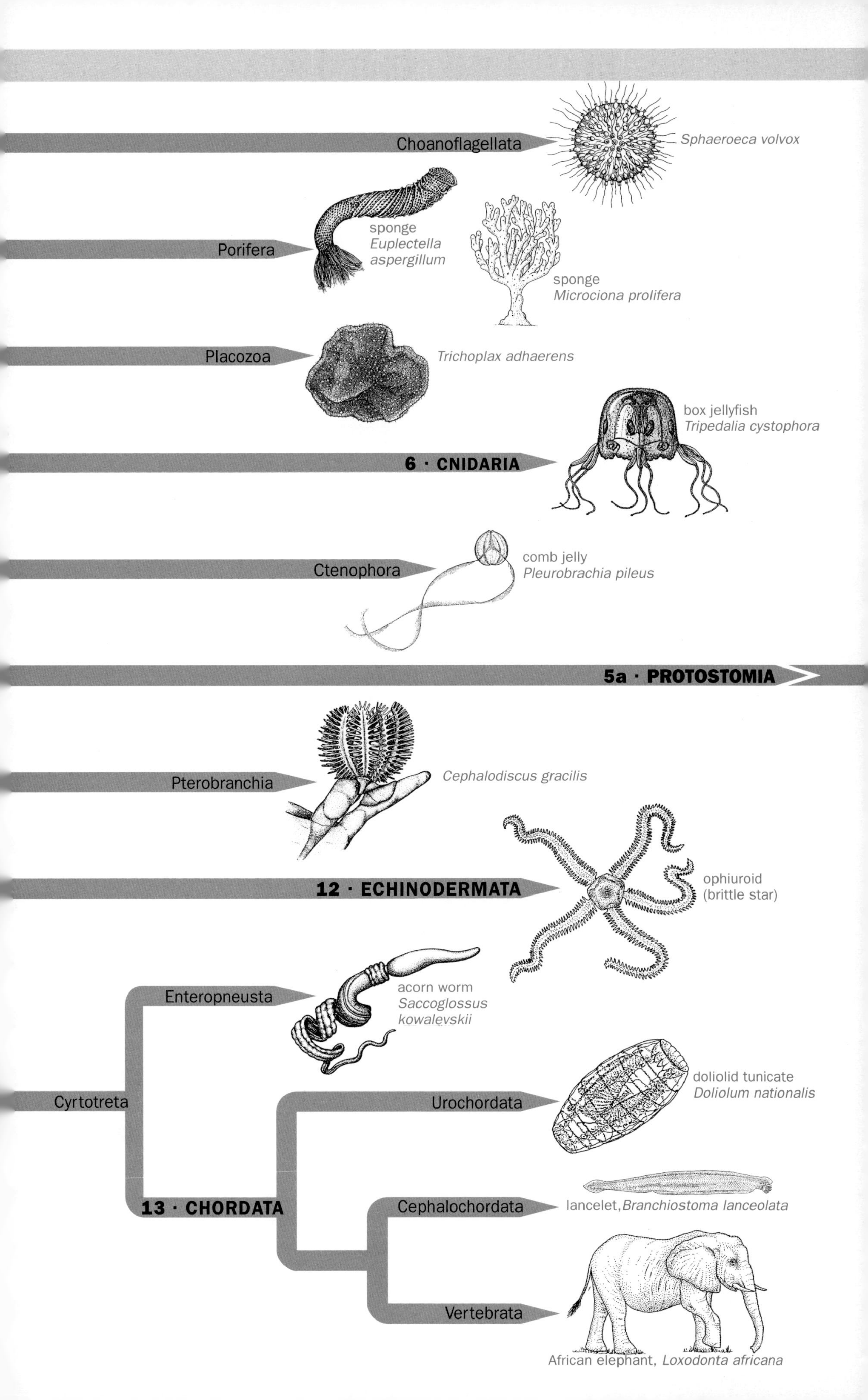
Choanoflagellata
Sphaeroeca volvox
sponge
Euplectella aspergillum
Porifera
sponge
Microciona prolifera
Placozoa
Trichoplax adhaerens
box jellyfish
Tripedalia cystophora
6 · CNIDARIA
Ctenophora
comb jelly
Pleurobrachia pileus
5a · PROTOSTOMIA
Pterobranchia
Cephalodiscus gracilis
12 · ECHINODERMATA
ophiuroid
(brittle star)
Enteropneusta
acorn worm
Saccoglossus kowalevskii
Cyrtotreta
doliolid tunicate
Doliolum nationalis
Urochordata
13 · CHORDATA
Cephalochordata
lancelet, Branchiostoma lanceolata
Vertebrata
African elephant, Loxodonta africana

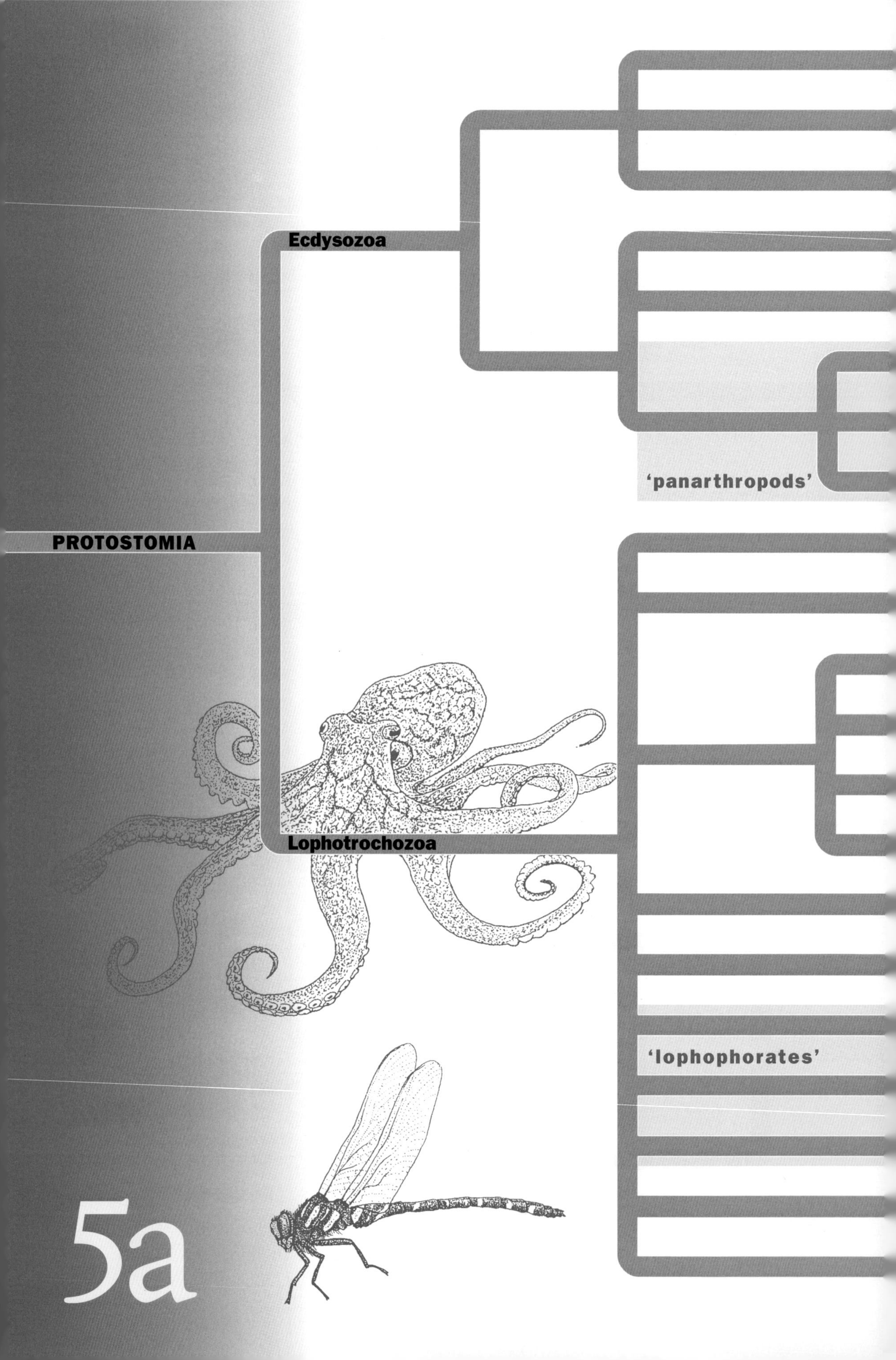
Ecdysozoa
'panarthropods'
PROTOSTOMIA
Lophotrochozoa
'lophophorates'
5a

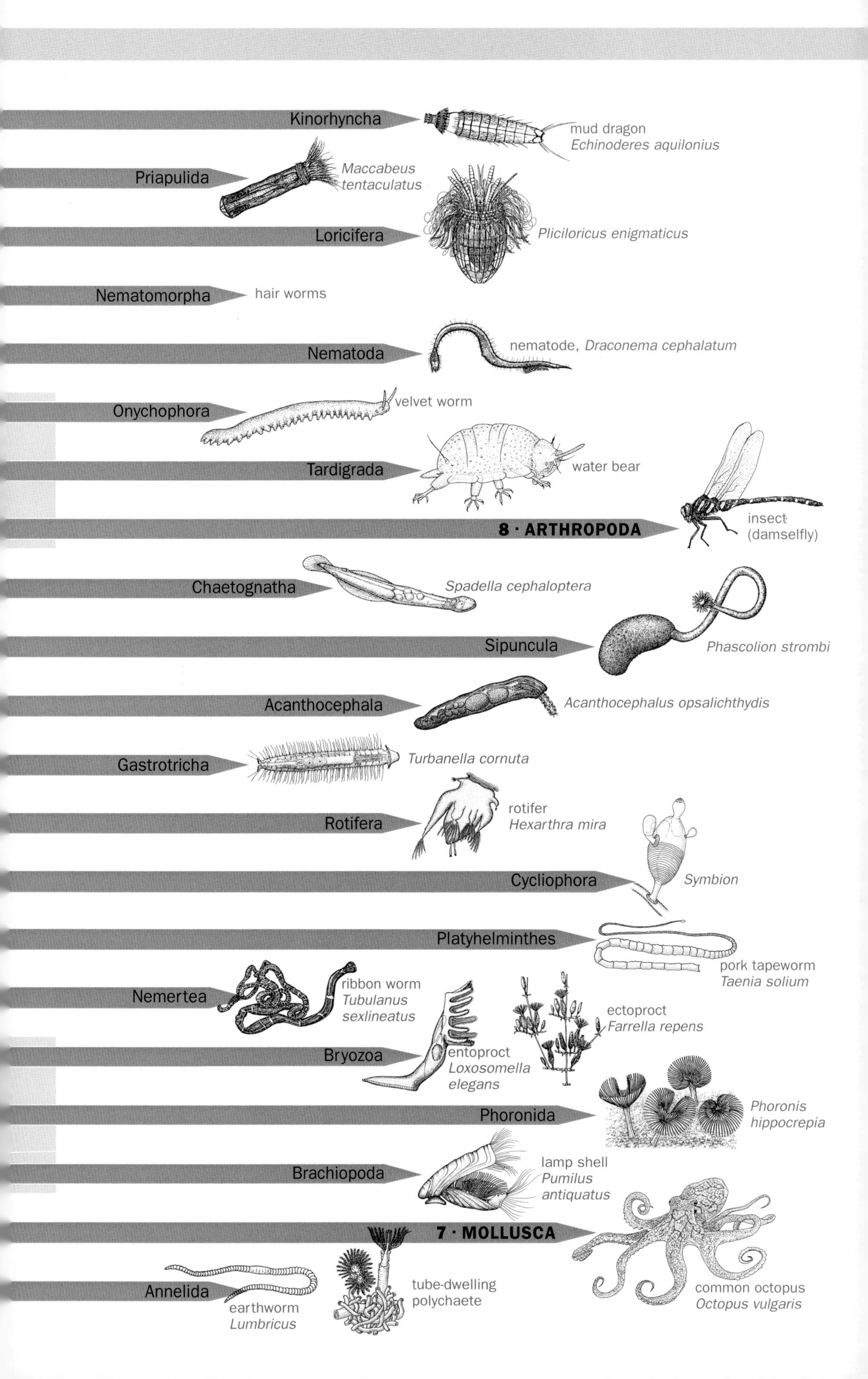

Kinorhyncha
mud dragon
Echinoderes aquilonius
Priapulida
Maccabeus
tentaculatus
Loricifera
Pliciloricus enigmaticus
Nematomorpha
hair worms
Nematoda
nematode, Draconema cephalatum
Onychophora
velvet worm
Tardigrada
water bear
8 · ARTHROPODA
insect
(damselfly)
Chaetognatha
Spadella cephaloptera
Sipuncula
Phascolion strombi
Acanthocephala
Acanthocephalus opsalichthydis
Gastrotricha
Turbanella cornuta
Rotifera
rotifer
Hexarthra mira
Cycliophora
Symbion
Platyhelminthes
pork tapeworm
Taenia solium
Nemertea
ribbon worm
Tubulanus
sexlineatus
ectoproct
Farrella repens
Bryozoa
entoproct
Loxosomella
elegans
Phoronida
Phoronis
hippocrepia
Brachiopoda
lamp shell
Pumilus
antiquatus
7 · MOLLUSCA
Annelida
earthworm
Lumbricus
tube-dwelling
polychaete
common octopus
Octopus vulgaris

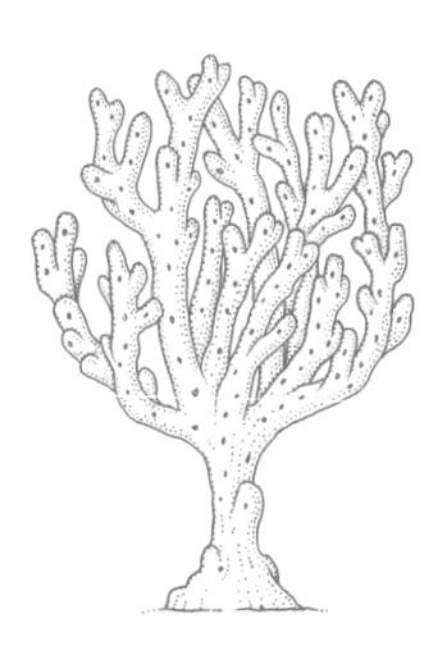

sponge
Microciona prolifera

ostia and out through the osculum. The choanocytes capture nutritious debris from the water, just as choanoflagellates do. In living sponges this basic theme is greatly elaborated. Almost all have a skeleton formed from the tough protein collagen, and many are reinforced by calcareous (calcium-based) or siliceous (silicon-based) spicules, which are of various shapes: some needle-like, and some with several spikes. These spicules have often persisted in the fossil record and the oldest known examples date back 600 million years—well back into the Precambrian—but the group could well be much older. Sponges remain highly successful. They are widespread with about 5000 living species.

The phylum **Placozoa** is now reduced to a single known species: *Trichoplax adhaerens*. *Trichoplax* is round and flat, just 1 or 2 millimetres across, and creeps over the surface of seaweeds in tropical seas, feeding mainly on protists but occasionally on larger fare. Placozoans are simple creatures without the clearly defined shape and co-ordination of eumetazoans, but various features of their cells and the connections between them resemble those of eumetazoans.

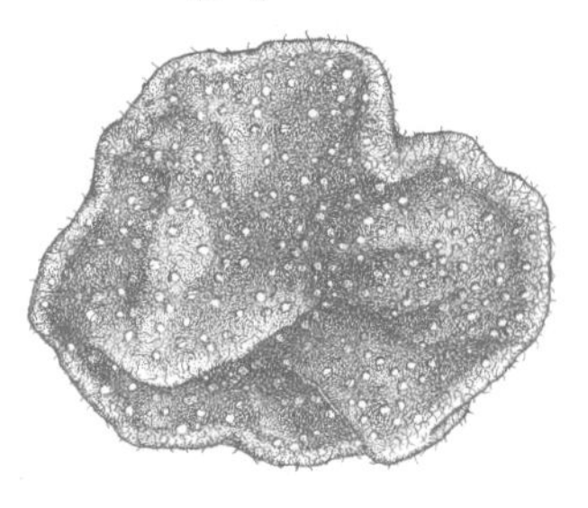

placozoan
Trichoplax adhaerens

All in all, sponges and placozoans do look and behave as multicelled organisms but in some ways they are more like colonies; their cells retain some quasi-independence. Thus if the cells of placozoans are separated, they can re-organize themselves into complete organisms, as slime moulds do. The sponge and placozoan level of organization is often called **parazoan**. Some zoologists treat Parazoa as a formal subkingdom but such a grouping would be paraphyletic and has no place in a cladistic taxonomy. But the informal adjective 'parazoan' is useful, like 'protist'.

TRULY MULTICELLED ANIMALS: THE EUMETAZOA

Now comes a second great shift of grade: between the parazoan Porifera and Placozoa on the one hand, and the Eumetazoa on the other. All the Eumetazoa are presumed to have descended from a common ancestor that was itself a eumetazoan, so Eumetazoa is a clade that can legitimately be treated as a taxon spelt formally with a capital 'E'. Eumetazoa might reasonably be ranked as a subkingdom.

The cells of eumetazoans truly co-operate to form unmistakable tissues and organs. Nutrients and 'information' (for example, in the form of hormones) flow efficiently between the cells so that only a relatively small proportion of them need be concerned directly with the acquisition of food, and these specialists can feed the rest. Freed from the basic chore of acquiring food the cells of eumetazoans can become highly specialized; and with such division of labour comes greater overall efficiency. In particular, some of the cells become specialized as nerve cells, which, together with the hierarchy of hormones, co-ordinate the action of the whole organism. Thus, no matter how many cells a eumetazoan animal acquires or how large it becomes, it can operate as one integrated unit. No animal is better co-ordinated than an ocean whale and none is faster to react than a squid, though either may be huge. The nervous system and the specialist muscle cells complement each other; it is their co-operation that gives animals their dash, which, in the eyes of naturalists since classical times, is what distinguishes animals from plants. The brain and hands of human beings are an advanced example of the nerve–muscle co-operation that begins with the lowliest eumetazoans.

The tree shows the Eumetazoa divided into three great branches: the **Cnidaria**, the **Ctenophora**, and the **Bilateria**. The Cnidaria (the 'c' is silent) include the sea anemones, jellyfish, and corals; and the Ctenophora (silent 'c' again) include 80-odd known species of sea gooseberries and comb jellies—aptly named creatures that commonly are transparent and gooseberry-like, floating in the plankton with long trailing tentacles. The Bilateria include all the other animals—worms, snails, starfish, sharks, human beings, and so on. There is yet another shift of grade between the Cnidaria and Ctenophora on the one hand and the Bilateria on the other.

RADIALLY SYMMETRICAL ANIMALS WITH TWO LAYERS OF CELLS: PHYLA CNIDARIA AND CTENOPHORA

The cnidarians and ctenophores have three outstanding features. First, their bodies are **radially symmetrical**. As we will see in Section 6, this is not always literally true; but when it is not, this is because the basic radial symmetry has been modified. A sea anemone, at least looked at superficially, is shaped like the flower of the same name.

Second, even though these animals may be huge—some individual jellyfish are more than a metre across and some have tentacles more than 10 metres long—the bodies of cnidarians and ctenophores are built up from just two layers of cells: **ectoderm** on the outside and **endoderm** on the inside, lining the cavity that forms the stomach. Cnidarians and ctenophores are said, in fact, to be **diploblastic**.

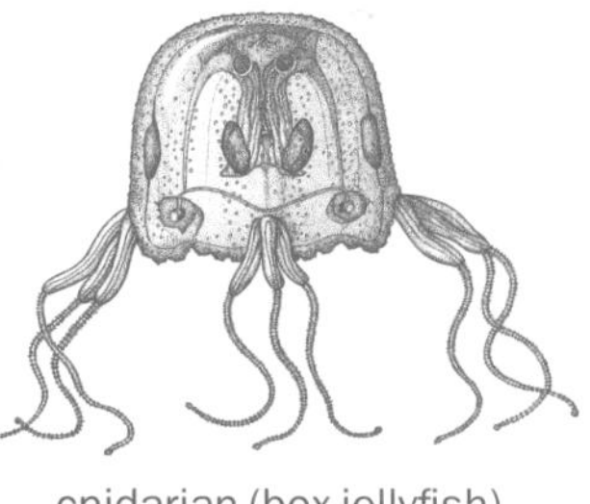

cnidarian (box jellyfish)
Tripedalia cystophora

Third, in both cnidarians and ctenophores the gut is 'blind'; it has only one opening. Indeed the overall shape of cnidarians and ctenophores is like a sack, with the opening serving as mouth and anus.

A brief diversion is called for. First (although there are many variations on this basic theme) metazoans in general pass through an early embryonic stage called the **blastula**, which consists simply of a hollow ball of cells. The cell layer around the outside, like the skin of a balloon, is only one cell thick. But then a part of the hollow ball starts to turn inwards, as if it has been poked. The inturning is called **invagination**; and the point where the ball is indented is called the **blastopore**. The invagination continues until the single-celled ball of cells has become a cup of cells—but the walls of the cup now obviously have two layers. This cup, with its two cell layers, is called a **gastrula**; and the inner cell layer becomes the endoderm and the outer layer the ectoderm.

The immediate point is that in creatures as diverse as anemones, corals, and jellyfish, and in sea gooseberries and comb jellies, the basic body plan is like that of a gastrula. The great mass of material in a jellyfish or a sea gooseberry is made up not of cells but of **mesogloea**, which is just a jelly-like material between the two layers of cells, and is produced in the main by the ectoderm. In essence, then, cnidarians and ctenophores, huge though some of them may sometimes be, are basically like gastrulae padded out with jelly, or mesogloea.

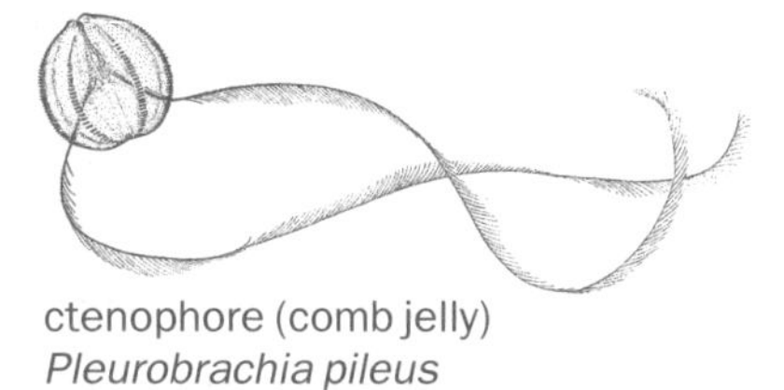

ctenophore (comb jelly)
Pleurobrachia pileus

Because of their general similarities, zoologists have commonly assumed that the Cnidaria and Ctenophora are closely related. The two are often grouped together formally as the 'Radiata', or the 'Diploblastica', and they have been thrust together to form the putative phylum 'Coelenterata'. Even when they are not placed in the same group, they are often shown as sister groups. But the features they share—radial symmetry, diploblasty, and a blind gut—are merely primitive, and there is no good reason to assume that the two groups are closely related. As discussed later, Claus Nielsen suggests that they may be very different—and he presents the ctenophores as the sister group of the deuterostomes. The trichotomy shown here is meant to reflect the present state of uncertainty: the relationship between cnidarians, ctenophores, and bilaterians is simply unclear.

BILATERALLY SYMMETRICAL ANIMALS WITH THREE LAYERS OF CELLS: BILATERIA OR TRIPLOBLASTICA

All animals 'above' the level of cnidarians and ctenophores basically have bilateral symmetry: a head end (or at least a front end, because some have no distinct head), a tail end, and two sides that are mirror-images of each other. Hence, 'Bilateria'. Echinoderms, such as starfish, appear to be radially symmetrical but basically they are good bilaterians, with bilaterally symmetrical larvae and bilaterally symmetrical ancestors. The alternative term 'Triploblastica' implies that their bodies—unlike those of cnidarians and ctenophores—are composed not from two layers of cells, but from three layers. The third layer, the **mesoderm**, is interposed between the endoderm and the ectoderm and offers enormous new scope in body design. In ourselves, for example, the muscles—including those of the heart—and the blood vessels are mesodermal, and mesodermal tissue is integrated in one way or another in virtually all the body organs.

This is the appropriate point to discuss Claus Nielsen's proposal—that ctenophores and cnidarians are not close relatives; and that the ctenophores are, in fact, related to the deuterostomes.

On the face of it, to be sure, the resemblances between cnidarians and ctenophores seem obvious; they are like gastrulae padded out with mesogloea. Mesogloea, however, does not consist solely of jelly; it often contains cells. In cnidarians, these mesogloeal cells are clearly derived from the ectoderm or endoderm; but Nielsen argues that in ctenophores they originate in a way which suggests that they are a true mesoderm. In other words, Nielsen effectively regards ctenophores not as diploblasts, but as rudimentary or conceivably 'degenerate' triploblasts. The basic gastrula-like form of ctenophores, Nielsen suggests, is just a primitive feature: in cladistic parlance, all that unites the ctenophores with the cnidarians is a symplesiomorphy, and symplesiomorphies do not denote special relationship. More specifically, Nielsen suggests that the ctenophores are sisters to the deuterostomes, the great group of triploblasts that include starfish and vertebrates; and he combines the Ctenophora and Deuterostomia in a great clade, Protornaezoa. The early embryos of ctenophores divide in a way that seems to support their affiliation with the deuterostomes.

I would like here to throw in some general zoological speculation. For it seems to me at least possible that neither the traditional nor Nielsen's hypothesis is correct—although both may be correct in part. After all, the initial divisions between the cnidarians, ctenophores, and triploblastic animals took place in the Precambrian, before about 600 million years ago, and we know very little of life from those times. The animals that lived at that time had soft bodies that did not easily fossilize, and the rocks that may have contained them are ancient, and largely worn away. We do know, however, that whenever palaeontologists stumble across some special cache of fossils from some previously unexplored period, they invariably find that life in the past was far richer than they imagined. By analogy it seems at least possible—though I would rather say probable—that in Precambrian times there were many different diploblastic lineages, each of which would qualify as a separate phylum. It also seems likely (at least to me) that many of those diploblastic lineages might have essayed some form of triploblasty, which, after all, does not seem in principle to be a particularly difficult trick to pull. It merely requires that one embryonic cell should multiply to form a new cell mass, or layer, between the endoderm and ectoderm, and should then be worked on by natural selection to form convincing structures. It may well be that present-day ctenophores show just one of the many early ventures into triploblasty: that the great clade of the Bilateria, or Triploblastica, represents another; and that most of the other putative triploblasts—and perhaps there were scores of them—have simply gone extinct.

If such a scenario is correct, then the two grand interpretations—the traditional view and Nielsen's—might both be partly right, but also both wrong. Perhaps we should envisage that the eumetazoan tree was bushy at its base, with several distinct diploblastic clades, several more that were clearly triploblastic, and several more that were somewhere in between. It just happens to be the case that only one clade from each eumetazoan body plan has survived: the one true, remaining diploblastic group is the Cnidaria; the one remaining clade of true triploblasts is the Bilateria; and the Ctenophora are the only remaining 'intermediates'. If this is so, then the three groups—Cnidaria, Ctenophora, and Bilateria—should be presented as a trichotomy. I present them here as a trichotomy not because I feel that my own speculation must be correct but as a compromise between the orthodox view (which links ctenophores with cnidarians) and Nielsen's (which links them with deuterostomes). A trichotomy is a statement of uncertainty. But it is a clear statement: it shows where the uncertainties lie.

Be that as it may: the main point is that the third great shift of grade as observed among living creatures and known fossils takes us from creatures that are radially symmetrical and diploblastic, to those that are bilaterally symmetrical and triploblastic: in other words, to the Bilateria. The bilaterians in turn branch into two great clades, the **Protostomia** and the **Deuterostomia**. These are clades, of equal grade. They are well-established divisions based largely on embryological studies, and they have stood the test of time. But as already intimated, there is a lot of discussion at present on which phylum belongs on which branch.

Perhaps most radically, some zoologists now suggest that the flatworms (Platyhelminthes) and others belong neither in the protostomes nor the deuterostones. In 1999, Mark Martindale and Matthew Kourakis published a tree showing the flatworms as the sister group of the protostomes + deuterostomes. This implies that the flatworms emerged before the protostomes and the deuterostomes divided one from another. This idea makes good sense (although it clashes with many other, traditional ideas—including the one that flatworms and molluscs are particularly closely related), but it is still only a proposal. Tradition has it that *all* bilaterians are either protostomes or deuterostomes, and that is how they are shown here.

CHARACTERISTICS OF BILATERIANS

The Protostomia includes a miscellany of unprepossessing unsegmented worms but finds its ultimate expression in the annelids (segmented worms such as the earthworms and lugworms), the molluscs (from cockles and snails to squids and octopuses), and the magnificent arthropods (crabs, lobsters, insects, spiders, scorpions, and so on). The other great branch, the Deuterostomia, also has its share of lowly creatures but culminates in the echinoderms—such as starfish and sea urchins—and, perhaps the most astonishing creatures of all, the vertebrates.

The two great clades of Protostomia and Deuterostomia are primarily defined not by the obvious differences between their most spectacular representatives—lobsters and elephants, say—but by more fundamental features of embryonic development. Such characters, not forced to adapt to the everyday exigencies of the environment, are thought to reflect phylogenetic relationships most reliably. They are:

- Animals in both groups begin life as single-celled fertilized eggs called zygotes, which then undergo **cleavage**—meaning that the cell divides and re-divides to form a multicelled embryo. But the pattern of cleavage is different in the two great groups. In deuterostomes cleavage is 'radial', so that the cells in the early embryo are stacked neatly on top of each other like tea chests in a pantechnicon. But at least in some protostomes the pattern of division is 'spiral', so the stack looks twisted (as if the pantechnicon had taken the corners too quickly). Again there are many variations, brought about not least by the presence of yolk. But the generalization seems to hold, at least roughly.

• As development proceeds, the gut arises in different ways in the two groups. We have seen that all animal embryos at some stage form a hollow blastula, and that this ball becomes indented to form a gastrula. The point of indentation is called the blastopore, and in both groups the blastopore usually becomes involved in the formation of the gut. In protostomes the blastopore pinches inwards to become a slit; and then the slit closes in the middle; and one end of the slit that is still open becomes the mouth while the other end becomes the anus. But in deuterostomes the blastopore simply becomes the anus and the mouth develops as an entirely separate opening. This is the difference that gave the two great groups their names: 'stome' means mouth; 'proto' means 'preliminary'; and 'deutero' means 'second'. In protostomes the blastopore is the preliminary mouth; but in deuterostomes the mouth requires a second opening. In practice there are many variations on this simple pattern and the only generalization that seems truly to hold is that in deuterostomes the blastopore never becomes the mouth, whereas in protostomes it sometimes does.

• Some protostomes and deuterostomes (but by no means all) still go through a planktonic larval stage. Superficially, these larvae look much the same in both groups—like little hot-air balloons driven along by bands of cilia. But when you look more closely you see differences, notably, but not exclusively, in the way the bands of cilia are laid out. The larvae of protostomes are called **trochophores**, while those of deuterostomes are **tornarias**. The differences may look small but they suggest nonetheless that the two great groups have followed quite separate evolutionary paths.

• The nervous systems of protostomes and deuterostomes are arranged differently—as is very clearly seen in the contrast between highly derived representatives of their clades, such as a lobster (protostome) and a fish (deuterostome). In the lobster the main nerve cord is a paired structure that runs along the belly of the animal, whereas in the fish the nerve cord is a single structure, running along the back. There are huge variations within the great array of protostome and deuterostome animals, and not all have such a clearly defined layout as lobsters and fish; but the general pattern is clear enough. Intriguingly, in 1822, the great French biologist Étienne Geoffroy Saint-Hilaire (1772–1844) suggested that the basic body form of a lobster (protostome) was really like that of a vertebrate (deuterostome) lying on its back; implying (albeit nearly 40 years before Darwin) that the two had a common ancestor and that the two body forms had each evolved as an inversion of the other. Most subsequent zoologists have felt that this idea was merely quaint but a recent study of genes that control development, by E. M. Robertis and Yoshiki Sasai of the University of California at Los Angeles, suggests that, in essence, Geoffroy might well have been right.

GRADE SHIFTS WITHIN THE TRIPLOBLASTICA

We can also see shifts of grade within each of the two great triploblastic clades. Three developments are of particular significance: of heads (**cephalization**); of a space (or spaces) between the outer body wall and the gut, called a **coelom** (pronounced 'see-loam'); and the division of the body into repeated modules—**segmentation**.

HEADS: CEPHALIZATION

A head is a specialized region at the front of the body, which, in particular, contains a definite aggregation of nerves, or a brain; and also is commonly equipped with organs of special sense, such as eyes. Some bilaterians that in other ways seem lowly, such as turbellarian flatworms, may nonetheless have definite heads. Many other creatures, apparently more derived (which simply means they deviated more radically from the ancestral state) have lost their heads, as earthworms have done (although earthworms do still have a brain). Some creatures, particularly those that live as parasites or are simply sessile (immobile) may become headless in the course of development, as barnacles do. Nothing is sacred to natural selection: wings, eyes, brains, heads, if they get in the way, they will be selected against. Animals with heads are said to be 'cephalized'.

SPACE IN THE BODY: THE COELOM

As described above, one of the most significant shifts in the design of animals was the development of a third cell layer, the mesoderm, which enabled a variety of new organs, or parts thereof, to be formed between the ectoderm and the endoderm, without interfering with those outer layers. Some animals, such as the flatworms (Platyhelminthes), have stuck with this simple arrangement, so the animal in cross-section consists simply of a 'skin' (ectoderm), internal organs (largely mesodermal), and gut (endoderm). But in the creatures we tend to think of as 'higher' animals, part of the mesoderm forms a sheet of tissue that wraps around to form a kind of bag, or a series of bags, between the gut and the outer body wall. Such a 'body space' is called a coelom. Deuterostomes characteristically have a series of three separate coelomic cavities (protocoel, mesocoel, and metacoel).

The acquisition of a coelom may not seem particularly significant; but it means among other things that the movement of the outside of the animal becomes quite independent of the activities of the gut. Thus in a fish, for example, the outside becomes a muscular tube (mesodermal muscles and dermis with ectodermal epidermis) that may be capable of violent movement—for catching prey or for escape—while the gut can go on happily digesting the last meal, and shuffling it from end to end, unaffected by the contortions of the outer body wall. At the same time the complex organs are neatly and safely held in the coelomic cavity (or cavities) between the gut and the body wall. In practice, then, the coelom is a hugely significant innovation: it allows quite new freedom of movement and increase in complexity and body size.

All the big and flashy creatures are **coelomate**: annelid worms, molluscs, arthropods, echinoderms, and chordates; and so, too, are some 'lesser' groups.

Did the protostomes and deuterostomes each descend from different, non-coelomate (**acoelomate**) ancestors, and acquire their coeloms independently? Or did they descend from a common ancestor that already had a coelom? If the latter, then the present-day acoelomate animals, like the flatworms, must have lost their ancestral coeloms—unless, of course, we accept the idea that flatworms evolved before the protostomes and deuterostomes split from each other. Nielsen proposes that protostomes and deuterostomes evolved independently from non-coelomate ancestors, and developed their coeloms independently.

The final step to elaboration is to take whatever structures you have and multiply them up *en bloc*. In other words, the simple body evolves into a more or less repetitive structure that is said to be segmented. Once the segments are in place the possibilities are endless.

DESIGN BY MODULE: SEGMENTS

The protostomes and deuterostomes may or may not have evolved their coeloms independently; but they are at least traditionally presumed to have evolved their segmented body plans independently. Embryologically and genetically speaking, segmentation seems a simple trick to pull: basically it's a question of repeating the entire set of structures necessary for survival—tubes for excretion (commonly known as **nephridia**), muscles for movements, nerves for co-ordination, paddles or other appendages for locomotion, and so on. This whole elaboration can, in principle, be achieved simply by doubling an organizational gene (such as a Hox gene); the doubling of genes in general happens all the time. Hence, segmentation is a neat and economical way of turning a small body into a large one; and of increasing complexity and efficiency, since different segments or groups of segments can then become modified in different ways for different specialist functions.

In some segmented animals, such as earthworms, the segments are much the same all the way down; a pattern that is said to be **homonomous**. If the different segments vary along the way the pattern is said to be **heteronomous**. Often groups of adjacent segments co-operate to form distinct body regions with discrete functions—like the head, thorax, and abdomen of an insect; and such specialized regions are called **tagmata** (created by the process of **tagmatization**).

Many creatures are conspicuously segmented—for example, earthworms and insects. But in others the segmentation, once achieved, is much modified. Thus human beings descended from clearly segmented chordate ancestors (which perhaps roughly resembled the living lancelets—see Section 13), but although parts of our body clearly are segmented (notably the backbone) in other parts differential development of different regions makes the underlying segmentation apparent only to the anatomist. The human head, for example, does not look segmented; yet segmentation is evident in the arrangement of the cranial nerves that move and sensitize the face. Molluscs such as snails generally appear unsegmented but it still is uncertain whether they did or did not descend from segmented ancestors.

This, then, is a brief overview of the shifts in grade throughout the animal kingdom: protist to metazoan; metazoan to eumetazoan; diploblastic to triploblastic; radial symmetry to bilateral symmetry (albeit with some reversions, as in echinoderms); cephalization; acoelomate to coelomate; unsegmented to segmented (again with reversions, as perhaps in molluscs); and tagmatization. Now let us look in more detail at the Bilateria. I will first explain the general layout of the tree as shown here; and then discuss the ways in which the present layout differs from more traditional treatments (including that of Claus Nielsen). Then I will briefly survey the phyla themselves.

CLASSIFICATION OF THE BILATERIA: TRADITIONAL AND MODERN

Zoologists broadly agree that the Bilateria should be divided into the phyla that are shown in the tree; although many would divide the Annelida—segmented worms—into several different phyla, as described later. There is also universal agreement that those phyla should be divided into the two great monophyletic divisions—the Protostomia and Deuterostomia—although, as mentioned, some propose that the flatworms (Platyhelminthes) emerged before the protostomes and the deuterostomes split from each other, so that they are really the sister group of the protostomes and deuterostomes combined. Zoologists also agree for the most part on which phyla belong in which of these great divisions. In particular, no one disagrees that molluscs, annelids, arthropods, and nematodes are protostomes; while echinoderms and chordates (the group that includes the vertebrates) are deuterostomes.

But the new molecular studies in particular raise a whole spectrum of difficulties, of which some may be details but others are extremely significant. Three issues in particular are outstanding.

Note first of all that the Protostomia is split into two great subdivisions, the **Ecdysozoa** and the **Lophotrochozoa**. Each group can be defined by several synapomorphies. Most obviously, all the ecdysozoans practise **ecdysis**: that is, they shed their exoskeletons all in one go, and then grow a new one (and thus it is that garden sheds accumulate the rejected integuments of spiders). Ecdysozoans also lack cilia. More important, however, is that the new groupings have been defined by molecular studies —first of 18S rRNA and then of their Hox genes.

The tree shows which phyla are placed in the Ecdysozoa, and which in the Lophotrochozoa. Previous studies, based on morphology, generally arranged the protostome phyla quite differently. For example, Claus Nielsen divides the Protostomia into two subdivisions, which he calls the Spiralia and the Aschelminthes (although he acknowledges that the Aschelminthes is probably not an entirely monophyletic grouping). But Nielsen's subdivisions differ in many striking ways from the new arrangement shown here. Thus within the Spiralia Nielsen includes three subgroups. One contains the Sipuncula, Annelida, Mollusca, Onychophora, Tardigrada, and Arthropoda; the next contains the Bryozoa (meaning the Ectoprocta and the Entoprocta); and the third contains the Platyhelminthes and the Nemertini. Nielsen's Aschelminthes includes the Chaetognatha, Gastrotricha, Nematoda, Nematomorpha, Priapula, Kinorhyncha, Loricifera, Rotifera, and Acanthochephala—and we should now add the newly discovered Cycliophora.

All this might seem a little too esoteric but a few points are particularly significant. For one thing, Nielsen has no doubt—and neither had most zoologists over the past 100 years or so—that the Arthropoda (together with the Onychophora and Tardigrada) are closely linked both with the Annelida, and with the Mollusca. Indeed, in a draft version of this book, before the molecular data became available, I showed the Annelida, Mollusca, and Panarthropoda as the three tines of a trichotomy (the Panarthropoda is a group suggested by Nielsen, which includes the Arthropoda, the Onychophora, and the Tardigrada). In addition, Nielsen clearly separates the Annelida, Mollusca, and Arthropoda from the Nematoda: he classes the annelids, molluscs, and arthropods as spiralians, and the nematodes as aschelminths.

Now, as you can see, the molecular studies show the nematodes as relatives of the '**panarthropods**', which on anatomical grounds alone is surprising, for superficially at least they look nothing like each other. On the other hand, the annelids, which seem so closely to resemble the arthropods—both are so clearly segmented, primitively with appendages on each segment—now appear to be far apart: the arthropods as ecdysozoans, the annelids as lophotrochozoans. But annelids and molluscs still seem to be closely allied (which, again, is not obvious from their anatomies).

Of special interest, too, are the Bryozoa, the Phoronida, and the Brachiopoda. Members of these three phyla have a peculiar feeding structure called the **lophophore**, which roughly resembles a sugar scoop fringed with cilia, and many zoologists in the past have proposed that these **lophophorates** form a clade, often known formally as the Lophophorata. Some of these zoologists have linked the lophophorate phyla with the protostomes whereas others have associated them with the deuterostomes, and others still have placed them in a third category, neither one nor the other. But some authorities—including Nielsen, whom I have followed in many respects in this account—do not consider that the lophophorates form a true clade. Nielsen argues that lophophores have evolved more than once, and that they are simply an albeit remarkable example of convergence. Indeed, Nielsen divides the lophophorate phyla as sharply as seems imaginable. He places the Bryozoa among the protostomes, and the Phoronida and Brachiopoda among the deuterostomes.

As you can see, the new molecular studies revive the traditional idea that the lophophorate phyla are related—but now they emerge as protostomes, among the Lophotrochozoa, alongside the molluscs and annelids. The three together do not seem to form a true clade, however, so the formal term 'Lophophorata' is inappropriate. In particular, the Bryozoa seem to be clearly divided from the other two lophophorate phyla. I have simply shown all the lophophorate phyla as part of a large polychotomy of lophotrochozoans, again reflecting the present state of uncertainty.

Taken all in all, the new classification of the Protostomia (and the defection of the Phoronida and the Brachiopoda from the Deuterostomia) is somewhat shocking; not because it offends the traditional view but because it apparently runs so counter to the anatomical data. In particular, annelid worms (like earthworms) and arthropods (like woodlice) are obviously segmented, and broadly speaking have appendages of a kind on each segment. They truly seem to have the same body plan. Onychophorans (sometimes called velvet worms) have a structure that is somewhat annelid-like, but also resembles an arthropod; and when I was at school the three groups —Annelida, Onychophora, and Arthropoda—seemed to form a logical sequence. Now onychophorans retain their link with arthropods, but annelids are entirely separated. Their apparently similar, sharply segmented body plan either arose more than once or—which is also more than possible—it is very primitive, but has been lost in the various groups (some related to annelids, and some to arthropods) that are not segmented. On the other hand, nematodes look nothing like arthropods—either when regarded superficially or examined minutely. Nematodes are peculiar, unsegmented worms (as explained later) with as little resemblance to a lobster or a bee as it seems possible to conceive. Yet the two groups seem to be related.

So why has the traditional classification been so upturned? How can so much obvious and widely accepted morphological data be put aside? The answer again lies with modern molecular studies. Among the animals, indeed, we seen one of the most striking clashes between the traditional sources of information, and the new. This is not the place for great detail, but the modern ball was set rolling in the mid-1990s by biologists in James Lake's department in the University of California at Los Angeles. They reported two main pieces of work, both based on studies of 18S rRNA.

The first study, by Kenneth M. Halanych and his colleagues, suggests that the lophophorates as a whole should be classed with the protostomes. This study did *not*, however, suggest that the lophophorates form a coherent clade (the putative 'Lophophorata'). The phoronids and the brachiopods seemed to group with the annelids and molluscs, while the Bryozoa emerged as the sister group of all those creatures put together. It seems, then, either that lophophorates have evolved more than once—separately in the Bryozoa on the one hand, and the Phoronida and Brachiopoda on the other—or that lophophores are a primitive feature of all the Lophotrochozoa, but were later lost in annelids and molluscs. The second paper, by Anna Marie Aguinaldo and her colleagues, showed that nematodes, arthropods, and other such creatures (like the onychophorans and tardigrades) form a coherent group, which they called the Ecdysozoa. Put the two pieces of work together, and we have the classification shown here: the Protostomia split into the Ecdysozoa and the Lophotrochozoa.

Much later—in 1999—came supportive evidence from a quite different group of molecules. Renaud de Rosa of the Université Paris-Sud and his colleagues looked at the Hox genes in various members of the putative Lophotrochozoa (notably a brachiopod and a polychaete annelid); and in various creatures that have been placed in the Ecdysozoa—including a priapulid, a nematode, and a fruit fly (which is, of course, an arthropod). On the evidence of the Hox genes, the two putative lophotrochozoans, and the three putative ecdysozoans, did indeed group together.

At this point, the molecular evidence is not definitive; more species from more phyla need to be studied. But it certainly seems good enough to risk the classification shown here. Thus, although the broad shape of the animal tree follows Nielsen, I have re-arranged the phyla of the Bilateria according to the molecular data; hence the tree mainly (though with additions) follows the paper by Anna Marie Aguinaldo and her colleagues in their 1997 *Nature* paper.

Let us now look at the Bilateria phyla, starting from the top with the protostome group, the Ecdysozoa.

PHYLA OF THE PROTOSTOMIA

THE ECDYSOZOA

Phyla Kinorhyncha, Priapulida, and Loricifera

The **Kinorhyncha** are commonly called 'mud dragons': they include about 150 species of creatures that live on the sea bottom, all less than a millimetre long. The **Priapulida** are marine creatures that look like short sticks: only 17 species are known. The **Loricifera** is a newly discovered phylum, but more than 100 species have been described from marine sediments of all types. Again, the loriciferans are microscopic.

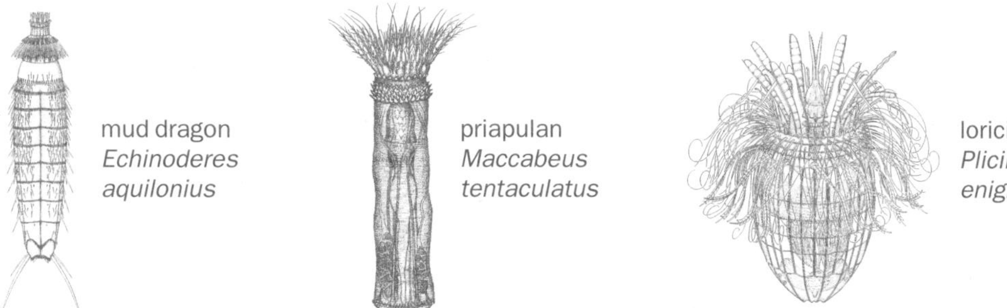

mud dragon *Echinoderes aquilonius*

priapulan *Maccabeus tentaculatus*

loriciferan *Pliciloricus enigmaticus*

Roundworms and hair worms: phyla Nematoda and Nematomorpha

Phylum **Nematomorpha** comprises the 'hair worms', so-called because they are remarkably thin, although they can be more than a metre long. They are parasites, mainly of arthropods. The nematomorphs are commonly shown as the sister group of the nematodes, although it is not clear that this is justified; and I am showing them as one tine of a trichotomy alongside the nematodes and panarthropods.

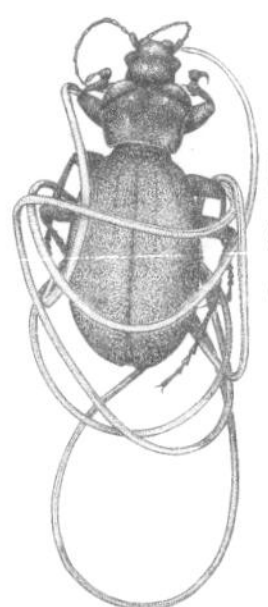

hair worm attacking a beetle *Gordius aquaticus*

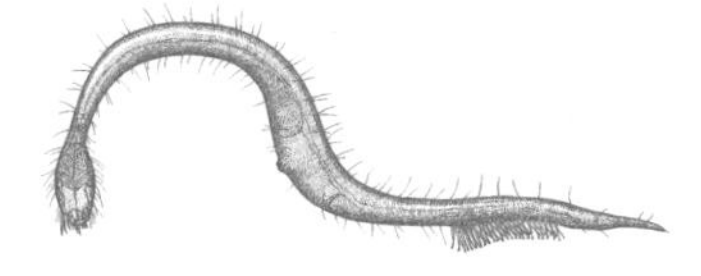

nematode
Draconema cephalatum

The phylum **Nematoda** has some of the most ecologically and economically significant of all animals. Nematodes in general are simply thin, round white worms with tapering ends (it can be hard to see which end is which): a simple body plan that varies remarkably little across the entire phylum. But this simplicity is also extremely successful. New marine research is revealing that nematodes are perhaps the commonest taxon of the deep-sea floor; but they are better known as ubiquitous and notorious parasites of animals, plants, and fungi that have also adapted to a host of other recondite habitats. One species is said to be peculiar to German beer-mats. Examples of direct importance include *Ascaris lumbricoides*, the pork roundworm (which also commonly infects people); and, far more significantly, the minute *Wuchereria bancrofti*, which is carried by mosquitoes and causes elephantiasis, and the hookworms, such as *Necator*. Overall about 20 000 species of nematode have been described but the true species list should probably run to many millions; indeed it has been suggested that every other kind of animal on Earth has at least one nematode species acting as its own personalized parasite. One remarkable feature of nematodes is that their bodies seem to contain a fixed number of cells; though why this should be so, and what advantages such precision brings, is unknown.

Now we come to three phyla that Nielsen groups together as the Panarthropoda: the **Onychophora**, or velvet worms; the **Tardigrada**, or water bears; and the **Arthropoda**. I discuss the onychophorans and tardigrades in Section 8 (pages 256–9) and the Arthropoda, which include the crustaceans, insects, chelicerates (arachnids and their relatives), and the extinct trilobites, in Sections 8–11. Together with the vertebrates and cephalopods, arthropods are simply the most extraordinary animals that have ever lived.

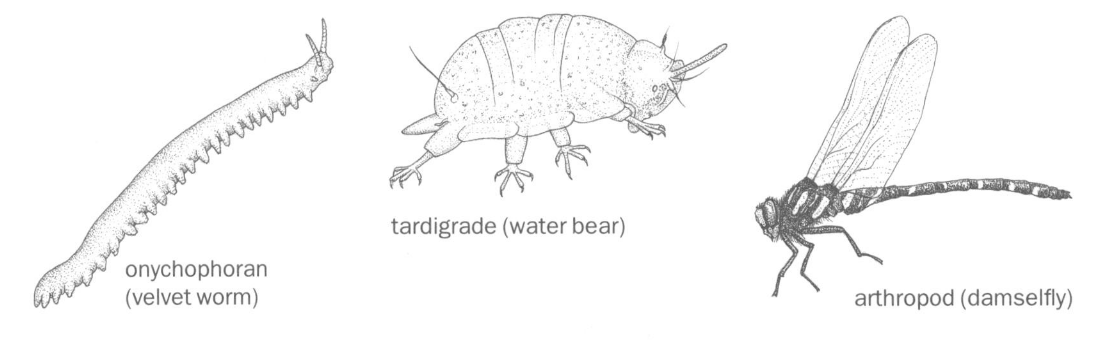

onychophoran (velvet worm)

tardigrade (water bear)

arthropod (damselfly)

THE LOPHOTROCHOZOA

Arrow worms and sipunculids: phyla Chaetognatha and Sipuncula

In the phylum **Chaetognatha** are rather charming, remarkably fish-like marine animals with one or two pairs of lateral fins and a large, horizontal tail fin. The

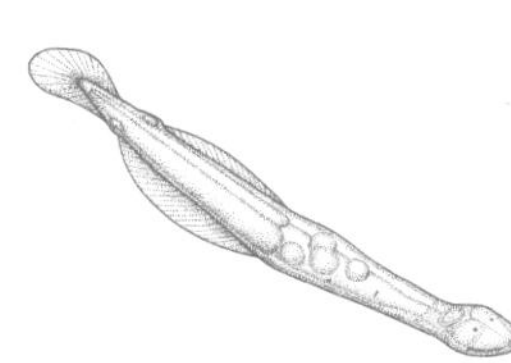

chaetognath
Spadella cephaloptera

mouth is tucked underneath and fitted with teeth made of chitin (a tough, horn-like substance) and grappling hooks with which the animals catch their small copepod (crustacean) prey. They poison their victims with a toxin produced by the bacteria they carry in their heads. This is not quite so bizarre as it sounds, for many creatures make comparable use of organs, cells, or materials produced by other creatures; for example, some molluscs borrow the stinging cells of jellyfish. About 200 species of chaetognaths are known but there are probably many more in the deep sea.

sipunculan
Phascolion strombi

The phylum **Sipuncula** includes about 320 species of astonishingly unprepossessing creatures, all marine, some roughly resembling sea cucumbers and others like sprouting potatoes, with their tentacled mouths at the end of the sprouts. But they have a coelom and a brain and various features of development. suggesting to many zoologists that they are related to the molluscs, annelids, and arthropods. Indeed, Claus Nielsen shows the sipunculids as the sister group of those three major phyla. The recent molecular studies show, however, that annelids, arthropods, and molluscs do not form a natural grouping, and so the position of the sipunculids remains uncertain.

Now comes a group of four phyla that are traditionally considered to be related (although the Cycliophora is a newcomer to the fold), and I have grouped them together in a polychotomy.

Phyla Acanthocephala, Gastrotricha, Rotifera, and Cycliophora

The **Acanthocephala** are worm-like creatures, which, though lacking a gut, vary in length from about 2 millimetres to almost 1 metre. As juveniles they are parasites in arthropods but as adults they live in the guts of vertebrates; and thus they have some economic significance. Phylum **Gastrotricha** includes about 430 species of generally worm-like, aquatic creatures, which are small to microscopic. The rotifers, phylum **Rotifera**, are aquatic creatures looking vaguely like John Wyndham's triffids and are notable largely for their smallness: the 1800 or so known species are less than a millimetre in length.

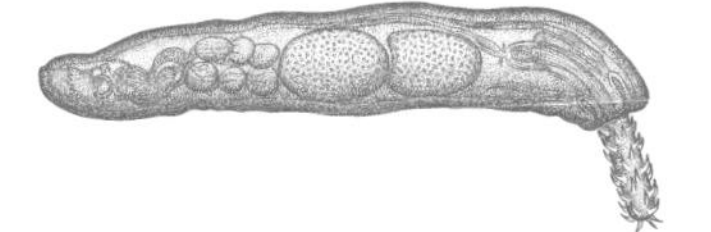

acanthocephalan
Acanthocephalus opsalichthydis

gastrotrich
Turbanella cornuta

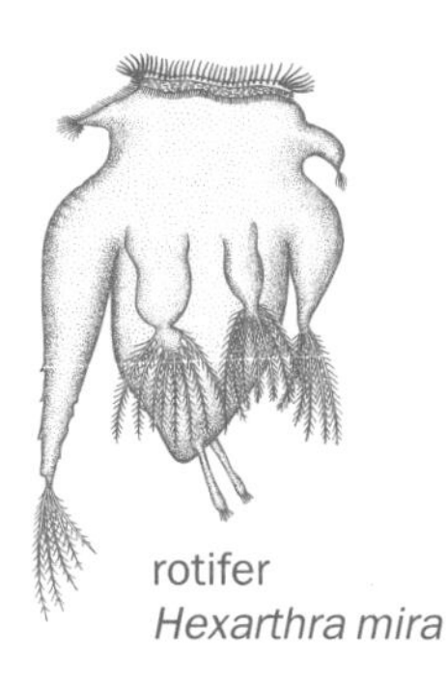

rotifer
Hexarthra mira

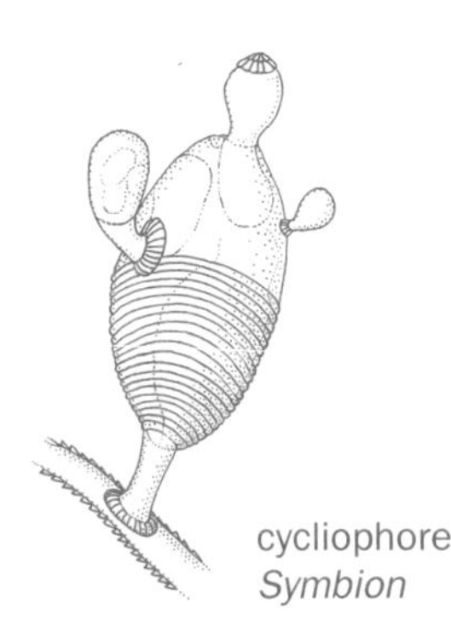
cycliophore
Symbion

The only known representative of the phylum **Cycliophora** is *Symbion pandora*, which was described as recently as 1995 by two Danish biologists, Peter Funch and Reinhardt Kristensen. *Symbion* is a tiny beast—a third of a millimetre—and sessile (fixed in one position): it was found clinging to the lip of a Norway lobster dragged from the North Sea. The female does the clinging and the dwarf males remain permanently attached to her. Recent molecular studies suggest that *Symbion* is related to the acanthocephalans and rotifers, and thus is it shown here. To be sure it is a beast for specialists but it reminds us all how little we really know: that an entire new phylum can appear over the side of a boat in an area that has been fished regularly for centuries.

Flatworms: phylum Platyhelminthes

The basic body plan of the Platyhelminthes, or flatworms, is relatively simple; notably, they lack a true coelom. About 20 000 species of flatworms are known, which traditionally are grouped in three classes. The first, the **Turbellaria**, are mostly free-living and aquatic, and include the laboratory favourite *Planaria*—a charming but much-abused creature, forever having its head sliced in two to demonstrate how each half can grow a new mirror-image of itself. The second class, the **Trematoda**, includes the flukes, such as liver fluke, and the third is the **Cestoda**, the tapeworms; both these groups are obviously of enormous economic and social importance. This classification is rather rough and ready; phylogenetic reality is far more complex and so, in an ideal world, should be the taxonomy.

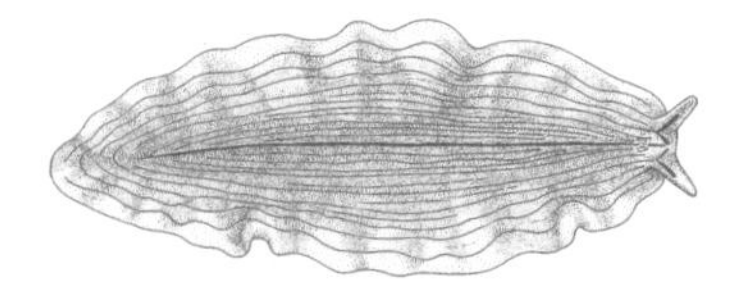
marine free-living turbellarian
Prostheceraeus vittatus

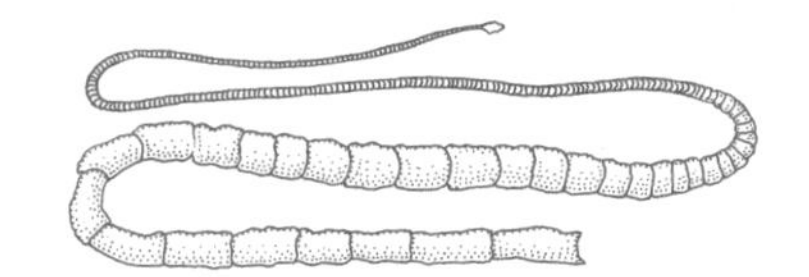
pork tapeworm
Taenia solium

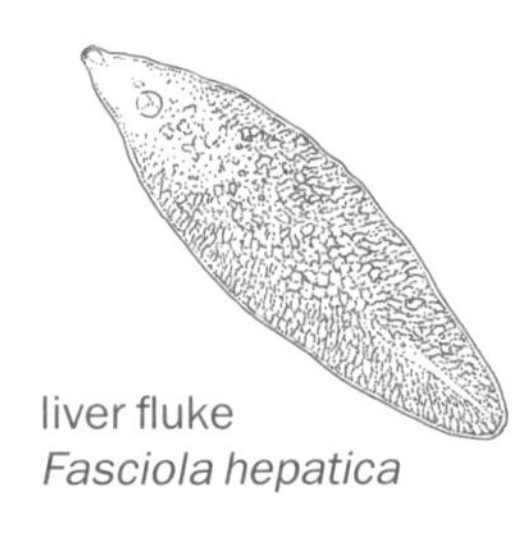
liver fluke
Fasciola hepatica

Ribbon worms: phylum Nemertea

Phylum Nemertea (which Nielsen calls the Nemertini and are also known as Rhynchocoela) includes the 900 or so known species of ribbon worms. Most nemerteans are marine (and mostly bottom-dwelling, or benthic), although some live in fresh water while a few live in moist places on land. They are mostly cylindrical but some (especially the swimming types) are flat—and one reported specimen was almost 60 metres long! Ribbon worms have an 'eversible proboscis': that is, a hollow proboscis that they can shoot out, propelled by hydrostatic pressure within. Nemerteans have often been presented as close relatives of the Platyhelminthes and even as the sister group, but others are not convinced. Thus some zoologists argue that the space within the proboscis that contains the fluid that allows it to evert is a true coelom, and that nemerteans cannot therefore be closely related to the platyhelminths but instead are closer to the molluscs and annelids. Nielsen argues that the space in the nemertine proboscis is not homologous with the coelom of annelids and molluscs. Again, molecular studies should clarify this issue: meanwhile, I am showing the nemerteans as part of the unresolved polychotomy of the Lophotrochozoa.

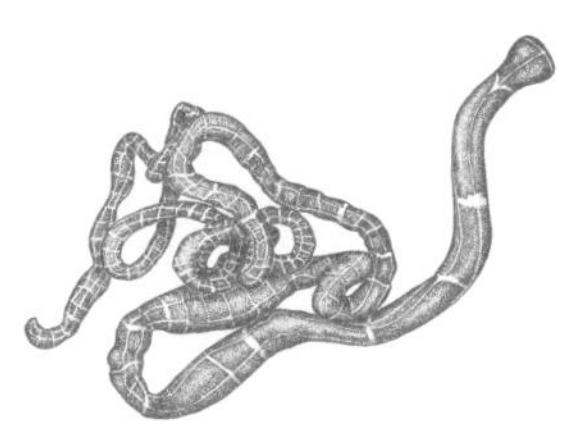
ribbon worm
Tubulanus sexlineatus

Now we come to a group of phyla that have always been controversial: they are known collectively as the lophophorates (and are sometimes given the formal name of Lophophorata)—the Bryozoa, Phoronida, and Brachiopoda.

'Moss animals': the Bryozoa (phyla Entoprocta and Endoprocta)

'Bryozoa' means 'moss animals'; as already intimated, the grouping really embraces two separate phyla, the Entoprocta and the Ectoprocta. Bryozoans may be solitary or colonial, and at first glance the individual animals (**zooids**) look like corals, although their body structure is quite different. Some of the colonies are branched, so that they resemble small seaweeds; and others are encrusting, forming mats on the surface of seaweed or stones. On close inspection the encrusting types commonly have a honeycomb-like structure, and naturalists know them as 'sea-mats'. Some look moss-like, and hence are called 'sea mosses'. In some ectoproct colonies the zooids are polymorphic: some specialize in defence, some in reproduction, some in cleaning, some serve as anchors, some feed and some do not. The Entoprocta has about 150 species, while the Ectoprocta has about 4000, plus an extensive fossil record. But the two groups of animals are quite different in detail, and Nielsen is being highly controversial in treating them as sisters.

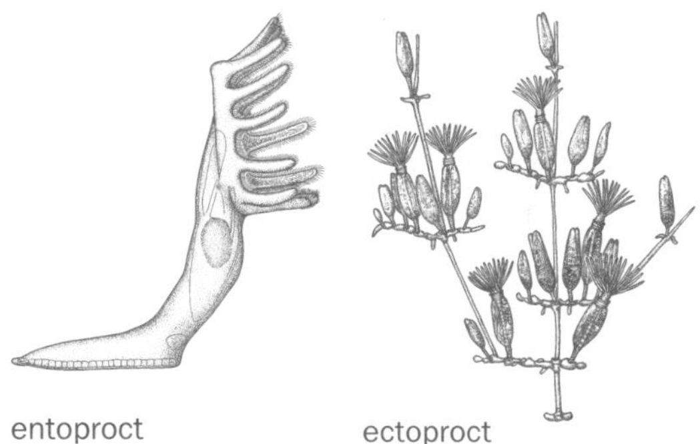
entoproct
Loxosomella elegans
ectoproct
Farrella repens

Phoronids: phylum Phoronida

Phylum Phoronida has only 12 recognized species in two genera (*Phoronis* and *Phoronopsis*). Phoronids are actually quite charming animals that live in chitinous

phoronid
Phoronis hippocrepia

tubes on the bottom of the sea, with their lophophores (mentioned above in connection with ectoproct bryozoans) poking out of the top like fans, sieving nutritious particles. Superficially they resemble some polychaete annelids, but their body plan is quite different. Among other things, their gut is U-shaped, with the anus opening near the mouth. Phoronids are generally solitary but some occur in masses or are even temporarily colonial. Their lophophores have enormous powers of regeneration: lose them, and they regrow them.

Lamp shells: phylum Brachiopoda

lamp shell
Pumilus antiquatus

The Brachiopoda superficially resemble bivalve molluscs, like cockles: in particular, they have cockle-like shells secreted by a mantle (although the shells are sometimes chitinous rather than calcareous, and the bottom one is stuck to the substrate by a stalk, while the upper one is free-moving). But brachiopods in basic structure are almost as different from bivalve molluscs as it is possible to be: here is yet another stunning example of convergence. In particular, brachiopods have a lophophore that they keep tucked inside the shells, though sometimes with the tips of the tentacles protruding. As I mentioned on page 197, many zoologists have regarded brachiopods and phoronids as sister groups, but Claus Nielsen rejects this. Still less does he feel that the brachiopods, phoronids, and ectoprocts form a natural grouping.

The brachiopods are still very much with us—there are about 300 living species—but they have had a far more glorious past. About 12 000 fossil species are known, going way back to the Precambrian era. Brachiopods are one of nature's early, successful animal designs. To be sure, they are immobile, or sessile, but it is only our prejudice that says that animals are supposed to run around. Particularly in the sea, many animals are quite content to stay put and let the food come to them. Motion is relative—hence the many variations on the theme of sessility.

The next phylum in the tree, the **Mollusca**, includes the snails and their relatives among the gastropods; the clams and relatives among the bivalves; and the wonderful dashing cephalopods, including the squids and octopuses, whose intelligence may match that of mammals. The molluscs are such an extraordinary and successful group that they have their own section (7). The phylum **Annelida** also deserves its own section but space forbids that (plus the fact that

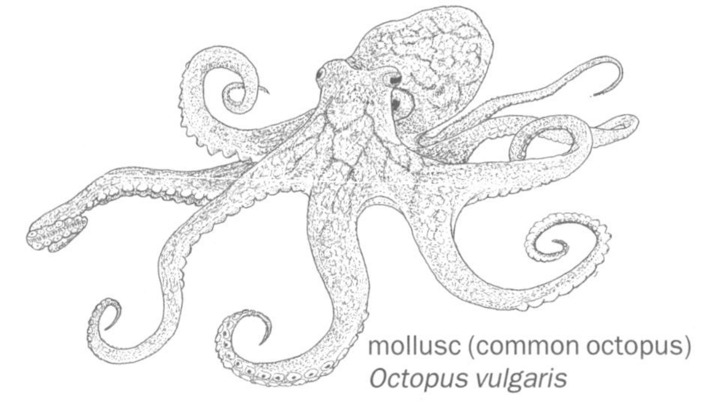
mollusc (common octopus)
Octopus vulgaris

often been presented as close relatives of the Platyhelminthes and even as the sister group, but others are not convinced. Thus some zoologists argue that the space within the proboscis that contains the fluid that allows it to evert is a true coelom, and that nemerteans cannot therefore be closely related to the platyhelminths but instead are closer to the molluscs and annelids. Nielsen argues that the space in the nemertine proboscis is not homologous with the coelom of annelids and molluscs. Again, molecular studies should clarify this issue: meanwhile, I am showing the nemerteans as part of the unresolved polychotomy of the Lophotrochozoa.

ribbon worm
Tubulanus sexlineatus

Now we come to a group of phyla that have always been controversial: they are known collectively as the lophophorates (and are sometimes given the formal name of Lophophorata)—the Bryozoa, Phoronida, and Brachiopoda.

'Moss animals': the Bryozoa (phyla Entoprocta and Endoprocta)

'Bryozoa' means 'moss animals'; as already intimated, the grouping really embraces two separate phyla, the Entoprocta and the Ectoprocta. Bryozoans may be solitary or colonial, and at first glance the individual animals (**zooids**) look like corals, although their body structure is quite different. Some of the colonies are branched, so that they resemble small seaweeds; and others are encrusting, forming mats on the surface of seaweed or stones. On close inspection the encrusting types commonly have a honeycomb-like structure, and naturalists know them as 'sea-mats'. Some look moss-like, and hence are called 'sea mosses'. In some ectoproct colonies the zooids are polymorphic: some specialize in defence, some in reproduction, some in cleaning, some serve as anchors, some feed and some do not. The Entoprocta has about 150 species, while the Ectoprocta has about 4000, plus an extensive fossil record. But the two groups of animals are quite different in detail, and Nielsen is being highly controversial in treating them as sisters.

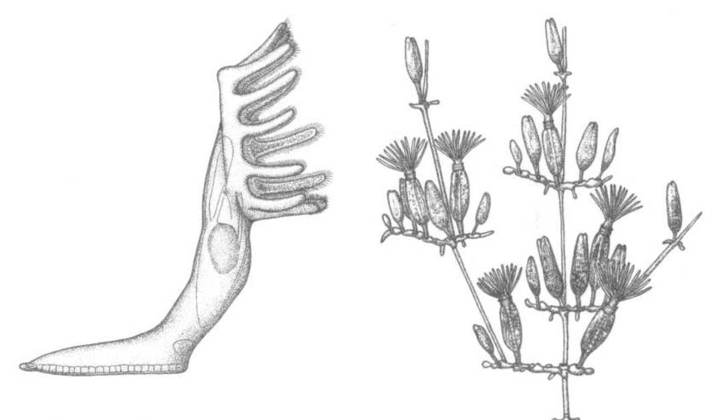

entoproct
Loxosomella elegans

ectoproct
Farrella repens

Phoronids: phylum Phoronida

Phylum Phoronida has only 12 recognized species in two genera (*Phoronis* and *Phoronopsis*). Phoronids are actually quite charming animals that live in chitinous

phoronid
Phoronis hippocrepia

tubes on the bottom of the sea, with their lophophores (mentioned above in connection with ectoproct bryozoans) poking out of the top like fans, sieving nutritious particles. Superficially they resemble some polychaete annelids, but their body plan is quite different. Among other things, their gut is U-shaped, with the anus opening near the mouth. Phoronids are generally solitary but some occur in masses or are even temporarily colonial. Their lophophores have enormous powers of regeneration: lose them, and they regrow them.

Lamp shells: phylum Brachiopoda

The Brachiopoda superficially resemble bivalve molluscs, like cockles: in particular, they have cockle-like shells secreted by a mantle (although the shells are sometimes chitinous rather than calcareous, and the bottom one is stuck to the substrate by a stalk, while the upper one is free-moving). But brachiopods in basic structure are almost as different from bivalve molluscs as it is possible to be: here is yet another stunning example of convergence. In particular, brachiopods have a lophophore that they keep tucked inside the shells, though sometimes with the tips of the tentacles protruding. As I mentioned on page 197, many zoologists have regarded brachiopods and phoronids as sister groups, but Claus Nielsen rejects this. Still less does he feel that the brachiopods, phoronids, and ectoprocts form a natural grouping.

lamp shell
Pumilus antiquatus

The brachiopods are still very much with us—there are about 300 living species—but they have had a far more glorious past. About 12 000 fossil species are known, going way back to the Precambrian era. Brachiopods are one of nature's early, successful animal designs. To be sure, they are immobile, or sessile, but it is only our prejudice that says that animals are supposed to run around. Particularly in the sea, many animals are quite content to stay put and let the food come to them. Motion is relative—hence the many variations on the theme of sessility.

———

The next phylum in the tree, the **Mollusca**, includes the snails and their relatives among the gastropods; the clams and relatives among the bivalves; and the wonderful dashing cephalopods, including the squids and octopuses, whose intelligence may match that of mammals. The molluscs are such an extraordinary and successful group that they have their own section (7). The phylum **Annelida** also deserves its own section but space forbids that (plus the fact that

mollusc (common octopus)
Octopus vulgaris

annelid classification still seems to be in a particular state of flux). The following is a very abbreviated description.

Segmented worms: phylum Annelida

The 15 000 or so annelid species were traditionally divided into three main groups—the Polychaeta ('many bristles') such as the lugworms, beloved by anglers for bait; the Oligochaeta ('few bristles'), which include the earthworms; and the Hirudinea, the leeches. But it now seems that the oligochaetes and the leeches form a clade together, while the rest are 'polychaetes'. The oligochaetes, however, clearly comprise two distinct lineages, one of which is closer to the leeches than it is to the other oligochaetes. So the old group 'Oligochaeta' is certainly paraphyletic and should no longer be recognized as a formal taxon, although the adjective 'oligochaete' is still useful. The remaining 'polychaetes' are a very mixed bag. Within them, indeed, Claus Nielsen includes several lineages that some other authorities prefer to treat either as separate phyla, or at least as distinct groupings that as yet are not assigned to any recognized phylum. These include the Gnathostomulida, the Pogonophora (in which Nielsen includes the Frenulata and the Vestimentifera), the Lobatocerebridae, the Myzostomida, and the Echiura. Somewhere among these polychaetes, though it is not clear where, is the sister to the clade of oligochaetes + leeches.

earthworm, *Lumbricus*

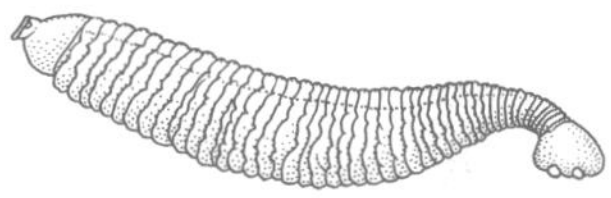

leech, *Pontobdella muricata*

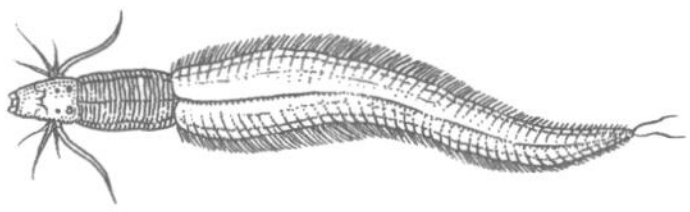

free-living polychaete *Nereis irrorata*

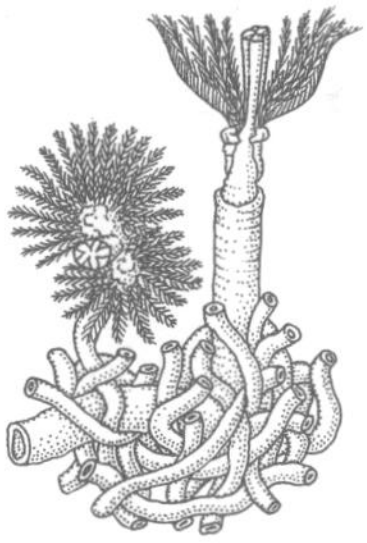

tube-dwelling polychaete

Annelids look intriguingly like onychophorans (which are sometimes known as 'velvet worms') and onychophorans look intriguingly like arthropods, and when I was at school in the 1950s annelids, onychophorans, and arthropods were represented as a neat trinity. The bodies of annelids and onychophorans are supported by internal pressure—a hydrostatic skeleton—whereas the shape of arthropods is conferred by their hard exoskeleton; but the overall segmented form is common to all three. Although cladistic parlance was not then used (Willi Hennig had published, but was still unknown in Britain), nowadays we would say that the onychophorans were seen as the sisters of the arthropods, and the annelids were sisters to the arthropods + onychophorans. In addition, classical zoologists had long recognized a strong relationship between annelids and molluscs, because although the adults look very different, their development and planktonic larvae (when they have them) can be remarkably similar. So in my schooldays molluscs appeared as the sisters of annelids + onychophorans + arthropods.

As already indicated, however, molecular evidence has now upset that cosy, commonsensical appraisal. The annelids and molluscs still seem to be closely associated, and the onychophorans and arthropods can reasonably be placed together with the tardigrades as shown here—in the grouping that Nielsen calls the Panarthropoda. But annelids and arthropods seem to be far apart, so the traditional annelid–onychophoran–arthropod trilogy is well and truly split asunder.

That completes our lightning tour of the protostomes. I end this summary look at the kingdom Animalia with the second great division of eumetazoans, the Deuterostomia.

PHYLA OF THE DEUTEROSTOMIA

To recap, deuterostomes are bilaterians whose embryos typically divide by radial cleavage, whose planktonic larvae (when they have them) are tornarias, whose bodies basically include three coelomic sacs, whose mouths always form separately from the blastopore (while the blastopore sometimes becomes the anus), and whose central nerve cord runs along the back.

As always, the Deuterostomia contains a few groups that have been hugely successful, and have had an enormous ecological impact: notably the phylum Echinodermata (the starfish, sea cucumbers, and so on); and the Vertebrata, one of the three groups that make up the phylum Chordata. These animals are treated separately later (Sections 12–22). Here we should look briefly at two of the ecological also-rans—the Pterobranchia and the Enteropneusta—primarily, again, to make the point that they exist and to show that the more-familiar groups represent just a few of the variants that evolution has essayed.

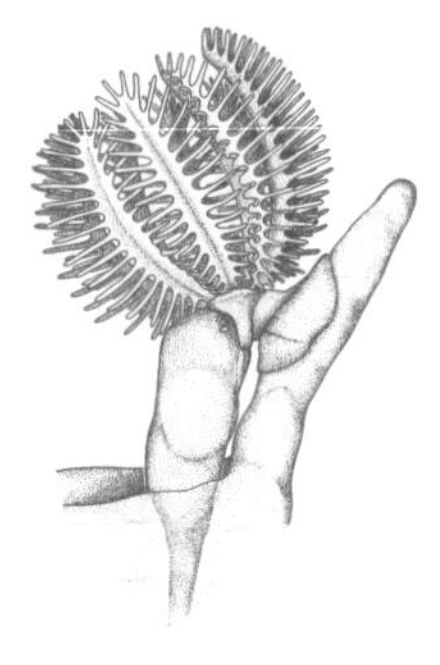

pterobranch *Cephalodiscus gracilis*

The **Pterobranchia** is a very small phylum with just two widely acknowledged genera, *Rhabdopleura* (with four species) and *Cephalodiscus* (with 15–20 species); there is perhaps a third genus, but it remains controversial. Again, the pterobranchs tend to form colonial systems of tubes from which they poke their feeding tentacles; but the individual zooids of *Cephalodiscus* can move around inside their tubes and can even leave and set up shop elsewhere if conditions at home prove unfavourable. Like brachiopods, pterobranchs are ancient creatures who have enjoyed a more propitious history. They seem to be close relatives of the graptolites, which are prominent in the fossil record from the Cambrian to the Carboniferous. Indeed, in 1993, P. N. Lilly reported a new species of *Cephalodiscus* from the southern Pacific, which he called *C. graptolitoides,* that he says is probably a graptolite. In other words, graptolites are not extinct: they are merely ancient pterobranchs that have failed to appear in the known fossil record since the Carboniferous.

This is not so unusual. Coelacanths and mesothele spiders are other living groups that have similarly gone missing for tens or even hundreds of millions of years, as we will see in the appropriate sections.

Pterobranchs and enteropneusts are often linked together in a grouping (sometimes called a phylum) called the Hemichordata, in which graptolites are also sometimes included. Nielsen regards the Hemichordata as a polyphyletic grouping and hence with no taxonomic validity (but 'hemichordates' still turn up in respectable books and this issue is again unsettled—again, perhaps, it could be resolved by molecular studies).

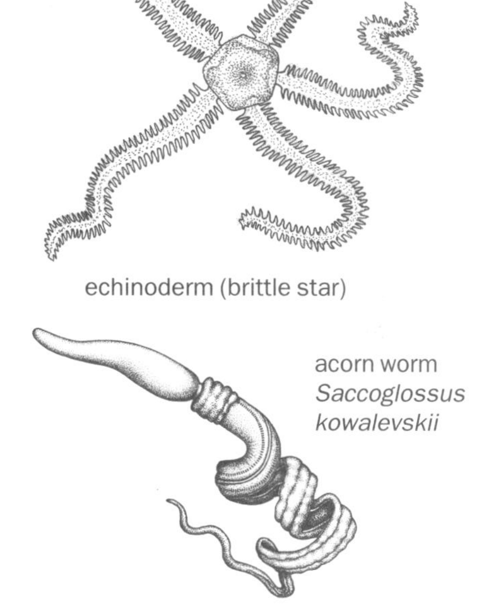

echinoderm (brittle star)

acorn worm *Saccoglossus kowalevskii*

The **Echinodermata** are dealt with in Section 12, and so we come on to the phylum **Enteropneusta**, commonly known as the acorn worms, of which the best known textbook example is *Balanoglossus*. There are about 70 species, which either creep or live in burrows. All are worm-like, but have the three-coelomed structure typical of deuterostomes.

Nielsen links the Enteropneusta and the three chordate phyla in a new grouping, the **Cyrtotreta**. What unites them all, he says, is the fact that the **pharynx**—the front end of the gut—is perforated by slits on either side that run right from the endodermal interior to the outside world. In other words, the cyrtotretes have **gill slits**. Primitively, these gill slits seem primarily to be a feeding apparatus: water flows in through the mouth and beating cilia, lining the slits, waft it through, and catch nutritious particles as it flows past. Perhaps in all cyrtotretes the gill slits also have a respiratory function—which becomes the main function in most of the vertebrates that retain them (that is, the fish). But only in the vertebrates and the salps, which is one of the groups of urochordates (tunicates or sea squirts), is the flow of water aided by muscular pumping.

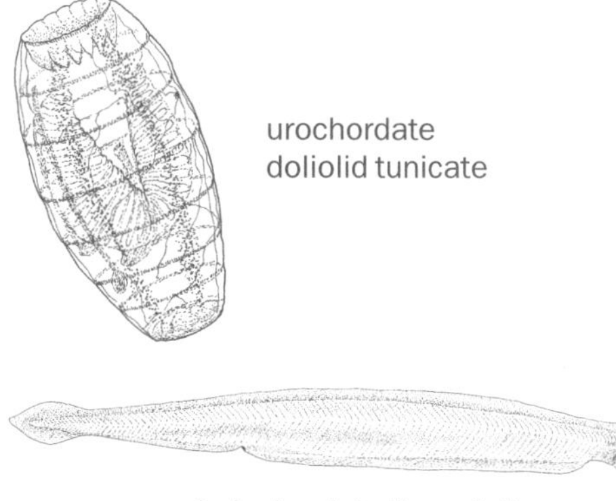

urochordate doliolid tunicate

cephalochordate (lancelet)

In practice, as you can see, Nielsen treats the enteropneusts as sisters to the three groups that he links within the **phylum Chordata**. The Chordata include two humble groups—the **Urochordata** and the **Cephalochordata**—plus the one that has produced the largest, cleverest, and generally most spectacular of all animals, the **Vertebrata**. The Chordata as a whole,

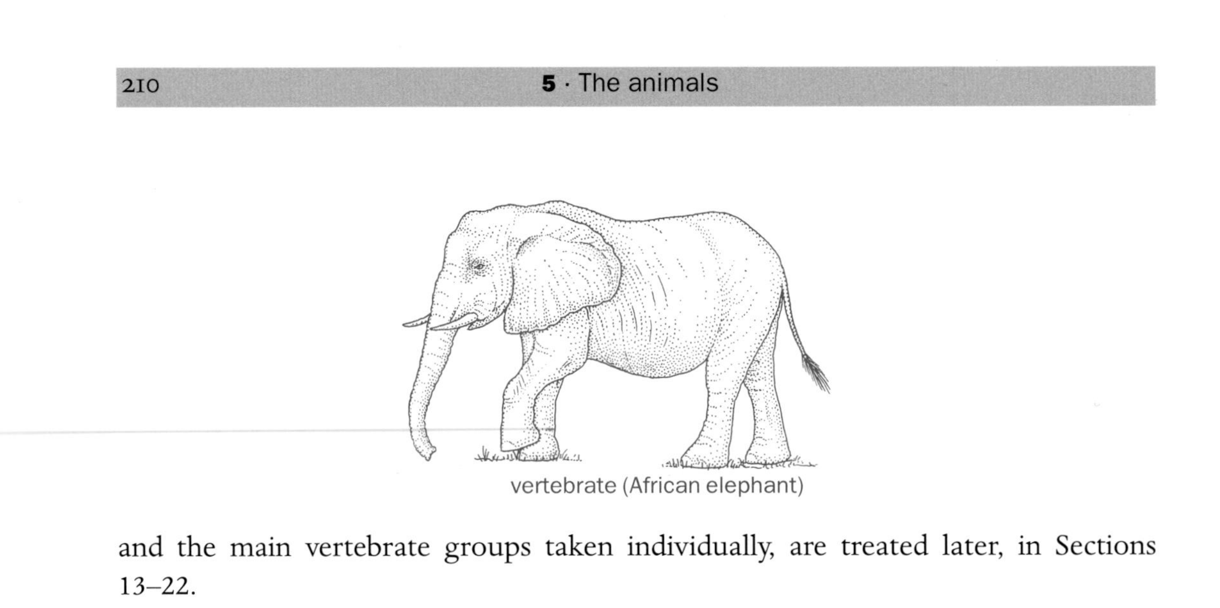
vertebrate (African elephant)

and the main vertebrate groups taken individually, are treated later, in Sections 13–22.

So now we can look at the main animal groups in more detail, beginning with a phylum of creatures that are primitive in structure but of huge significance ecologically: the Cnidaria.

6

ANEMONES, CORALS, JELLYFISH, AND SEA PENS

PHYLUM CNIDARIA

THE MOST ANCIENT of the living eumetazoan animals, and the simplest in body form, are nonetheless among the most conspicuous and ecologically significant throughout the world's oceans (and they even have a small presence in fresh water). For the earliest cnidarians (the 'c' is silent) appear among the Ediacarian fossils of southern Australia, which date from the Precambrian, about 600 million years ago. Their basic body form is a simple sac—just a stomach or **coelenteron** whose walls have only two layers of cells, with tentacles around the single opening that serves both as mouth and anus. Yet corals and jellyfish in particular are hugely influential predators, and are food to other creatures, while corals are the principal components and artificers of islands and reefs that provide the most species-rich habitats outside the tropical rainforest. In short, cnidarians demonstrate that an animal may be ancient and simple, and yet be enormously successful and influential.

The success and variety of cnidarians—there are 8000–9000 living species—springs in large part from their **dimorphism**: they have two body forms that fill two different, complementary ecological roles. The **polyp** is the form of the anemone: the mouth and tentacles typically face upwards[1] and what should be the top, or apex, of the animal is stuck to the substrate, or thrust into it if the substrate is soft. Jellyfish, on the other hand, have the form called the **medusa** (plural medusae): the animal floats, and the mouth and tentacles face downwards.[1] Polyps in general are sessile, which means they stay where they are, feeding from passing fish or plankton like insectivorous plants; while medusae either are fairly passively planktonic, or actively swim. Both polyps and medusae basically show radial symmetry, but some are obviously divided into quarters (that is, are **quadriradial** or **tetramerous**) whereas others, like the anthozoans with their slit-like mouths and other internal elaborations, are really bilaterally symmetrical.

[1] I say 'upwards' although, of course, anemones often cling to overhanging rocks and so in practice face downwards; and some jellyfish, such as *Cassiopeia*, swim upside down, with mouth and tentacles facing upwards. But the principle applies.

All polyps are able to reproduce asexually, sometimes generating more polyps and sometimes producing medusae, and in both cases doing so in various ways; while medusae reproduce sexually, generally shedding eggs and sperm into the water that fuse to produce embryos that develop into dispersive larvae called **planulae** (singular, planula). Often there are other juvenile and intermediate forms as well, bearing various names. Many cnidarians, including many cubozoans, scyphozoans, and hydrozoans have life cycles that include both polyps and medusae (as well as planula larvae and sometimes other juvenile forms) and they are said to practise **alternation of generations**.[2] But some have suppressed the polyp phase all together, including some hydrozoans and scyphozoans; while others, including the vast class of the anthozoans, have abandoned medusae, and their polyps reproduce both sexually and asexually.

Both polyps and medusae may be elaborate in structure: polyps often protect themselves in skeletons that may be chalky (calcareous) or horny (typically made of chitin-reinforced hardened protein, as in insect armour); in anthozoan and scyphozoan polyps the coelenteron is partially divided by vertical partitions called **mesenteries**, and so the inner, absorptive surface is expanded in the same way as in the stomachs of ruminants (as evident in the patterning of tripe); in some polyps the tentacles are branched; in jellyfish the edges of the mouth are typically extended into 'arms'; and so on. Yet the underlying form of both polyp and medusa is ultimately simple; indeed, as Claus Nielsen emphasizes in his book *Animal Evolution*, it is essentially an elaboration of the kind of body form that all metazoan animals pass through early in embryonic development, as outlined in Section 5. This form, the gastrula, is a hollow ball—like a tennis ball—whose walls consist of just two layers of cells (that is, is diploblastic); and one side of the ball has been pressed in, or 'invaginated', to form a cup. Put tentacles around the rim of the cup and you have the basic form either of a polyp or a medusa, depending only on which way up you hold it. In most living cnidarians, however, the two layers of cells in the body wall are separated by a layer of jelly-like material, the mesogloea, which in some kinds contains additional cells (though never forming a true mesoderm) and in general increases bulk and versatility. The jelly of jellyfish is mesogloea.

Their capacity for asexual reproduction enables polyps to form colonies, which many of them do; and this hugely extends their morphological and ecological range. Thus colonial corals may form enormous reefs—the individual polyps being typically connected by threads of tissue running through the calcareous walls that each polyp forms around itself. In sea pens, which are hydrozoans, one elongated polyp forms a central stem; this is anchored in the sediment and buds off rows or whorls or crescents of lateral polyps. Sometimes medusae also remain attached to their parent polyps, to

[2] As described in Section 23, plants also practise alternation of generations. But the two phenomena are not directly comparable. In plants, a diploid generation (in which the cells of the organisms contain two sets of chromosomes—one derived from the male parent and one from the female parent) alternates with a haploid generation (in which the cells contain only one set of chromosomes). In cnidarians, as is usual in animals (although there are variations), both generations are diploid, and only the gametes (eggs and sperm) are haploid.

contribute to the colony. In siphonophores, which are also hydrozoans, medusae and polyps may combine to form veritable floating cities that include up to a thousand individuals specialized for different functions—some feeding, some reproducing, some defending; that is, they are highly **polymorphic**. The man-o'-war jellyfish of the Atlantic are just such structures. One large individual, probably a modified polyp, forms a gas-filled balloon that buoys the man-o'-war, while a big, modified medusa forms the main swimming bell.

But the outstanding feature of cnidarians—the apomorphic character that unites them all and is distinct from anything else in the animal or any other kingdom—is the stinging cell. Called the **cnidocyte** (or the **nematoblast**) it contains an astonishing structure known as the **cnida** (or **nematocyst**).[3] This takes many different forms—about 36 variants have been described—but even at its simplest the cnida is among the most remarkable of structures to be fashioned by a single cell and at its most elaborate it is truly among the most wondrous inventions of nature.

In essence, the cnida, or nematocyst, consists of a long hollow tube, which in the resting state is pushed inwards (invaginated), as you might push in the finger of a rubber glove; but because the tube is long it is generally coiled round and round. Its tip is typically barbed (the barbs point inwards when the tube in invaginated, but outwards when it is extended) and typically has a hole at the end, through which toxin is extruded. When the cell is stimulated, generally through mechanical or chemical contact with a trigger, the tube comes shooting out, barbs and all, penetrates the skin of the unfortunate creature that has made contact and discharges its poison. The toxins can be fearsome. Some quite modest-looking jellyfish, such as the cubozoan 'sea wasps', have been known to kill human swimmers.

Equipped with cnidae, deployed on tentacles that in some big jellyfish trail more than 10 metres, cnidarians are formidable predators. All are carnivorous. Some vary the method of feeding; some corals, for example, trap organic debris on skeins of mucus-like fly-paper. Many, too, including the reef-building corals, benefit from photosynthetic protists that generally live within the body cells (intracellularly) but sometimes in the mesogloea. Some freshwater hydrozoans such as *Chlorohydra* harbour chlorophytes; these are unicellular green algae that botanists these days tend to classify as true plants (see Section 23); while marine types harbour cryptomonads and dinoflagellates, which are colloquially referred to as **zooxanthellae** and belong to quite different eukaryote kingdoms (see Section 3). In any case the protists are given a good place in the sun while the polyp commandeers at least some of their photosynthetic products: a fine example of mutualism. Many corals depend on their photosynthetic commensals and hence live only in the upper, photic zone of the sea.

This, then, is a lightning sketch of the cnidarians; ancient and simple but various, successful, and highly influential predators that are sometimes lethal even to some of the biggest and fiercest of other animals. They are remarkable creatures.

But the taxonomy of cnidarians continues to cause problems. Zoologists are

[3]Some authorities, including Claus Nielsen, treat the cnida and the nematocyst as synonyms, whereas others consider the nematocyst as just one class of cnida.

agreed that the phylum Cnidaria should be divided into four classes—Anthozoa, Scyphozoa, Hydrozoa, and Cubozoa; and they also seem to divide up each of those classes into roughly the same lists of orders. That is a very considerable start. But they have disagreed in just about every possible way on how those classes are related to each other, and on which is closest to the ancestral form and thus should be considered the sister of all the rest. And within each class, they have disagreed on relationships of the orders. For example, two excellent modern textbooks—Richard and Gary Brusca's *Invertebrates* (1990) and Claus Nielsen's *Animal Evolution* (1995)—offer classifications that in important respects are virtually opposite. Thus the Bruscas suggest that Hydrozoa are the primitive group, sisters to all the rest; and that, among the rest, the Anthozoa are sisters to Scyphozoa + Cubozoa. Nielsen shows Anthozoa as the primitive sister to all the rest: and Scyphozoa as sister to the Cubozoa + Hydozoa. Thus, among other things, Hydrozoa is seen as the most primitive group in one classification and as the most derived in the other. Both versions are based on interpretations of 'classical' morphological data. I am very grateful to Sandra Romano, then at the Smithsonian Institution, Washington DC, and now at the University of Guam, who has acted as guide and pointed me at the most recent papers describing both molecular and morphological data. But I am having to pick my way through the data (with Sandra Romano's help): for example, no one study that has been published recently applies the same molecular techniques to all the different cnidarian groups. I believe, however, that most cnidarian specialists would agree that the cladistic tree shown here is at least sensible, even though they will not all agree with all of its details.

A GUIDE TO THE CNIDARIA

Polyps ecologically are like insectivorous plants and renaissance scholars thought they were plants; only in the eighteenth century were they admitted as animals. Even so, Linnaeus grouped them together with sponges and a few other creatures as 'zoophytes', which literally means 'plant-animals'. In the nineteenth century Jean-Baptiste Lamarck, who is best remembered for his fallacious views on evolution but was a fine taxonomist, classed the medusoid cnidarians, the ctenophores, and the echinoderms together as 'Radiaires', or 'Radiata'. In the early nineteenth century, too, the excellent Norwegian naturalist Michael Sars (1802–69) showed that medusae and polyps were different forms of the same organisms; and that various odd creatures that hitherto had been assigned to different genera were larval and juvenile forms of cnidarians.

Thomas Henry Huxley, friend and champion of Charles Darwin, did much to elucidate the cnidarians during and after his exploratory trip aboard HMS *Rattlesnake* in the 1840s. But others seem to have beaten him to publication. Thus in 1847 the German zoologist Karl Leuckart (1822–98) coined the term 'Coelenterata', in which he included the Porifera, the Cnidaria, and the Ctenophora, which others later split into three separate phyla. Many zoologists since have recombined the Cnidaria and

Ctenophora into a phylum, the Coelenterata, but all now agree that the Cnidaria and Ctenophora are quite separate. Others tend to use the term 'Coelenterata' as a synonym of 'Cnidaria': and I confess that I still tend to refer to cnidarians as 'coelenterates'. On balance, because the term 'Coelenterata' has such a multifaceted past, whereas 'Cnidaria' has always meant one thing and one thing only, the latter is obviously to be preferred. A pity, though: 'coelenterate' (which tends to be pronounced 'sill-enterate') has a nice ring to it.

As intimated, zoologists broadly agree that Cnidaria breaks naturally into four classes, but are not agreed on the relationships between them. Which is the basal group—the one closest to the ancestral type? Is it the Anthozoa, as shown here, or the Hydrozoa, as has sometimes been suggested?

In this discussion, much hinges on whether polyps or medusae are the ancestral form: after all, Hydrozoa have both medusae and polyps, but Anthozoa (which include the corals and anemones) have only polyps. Nonetheless, we might hypothesize that neither polyps nor medusae are ancestral: that the first ever cnidarians had a structure that combined elements of both (if such a thing can be envisaged) or, indeed, was quite different from either; that both the extant forms are derived; and that the original form has simply been lost. But there is no evidence for such a view and for various reasons it seems more likely that one or other of the extant forms did evolve first.

The Bruscas argue that medusae are the probable ancestors because they are the sexual phase, the producers of sperm and eggs. If this were so, then Anthozoa would probably not be the most basal of all the cnidarians because they lack medusae. Besides, say the Bruscas, anthozoan polyps have the most elaborate structures. Others, however, including Nielsen, argue that polyps are ancestral and that medusae are a later invention, evolved as a specialist sexual-dispersive phase. This restores anthozoans as plausible basal types, and suggests that hydrozoans could well be derived. Indeed, says Nielsen, hydrozoans have several characters suggesting that they are highly derived, including details of their nerve-net—such as a synapse that resembles that of 'higher' animals. (If the Hydrozoa are derived, however, then they and the 'higher' animals must have evolved their sophisticated synapses independently—a remarkable example of convergence.) These are just a sample of the arguments on both sides and, on the face of things, they seem equally plausible. But emerging molecular studies and morphological evidence based on cladistics seem to suggest that, after all, anthozoans are the basal types—and hence the arrangement shown in the tree.

What of the relationship between the three more derived groups—Scyphozoa, Hydrozoa, and Cubozoa? If we really knew the answer, we should present two of these three as sisters, and the third as the sister to the other two combined. But modern molecular studies and cladistic morphological studies give equivocal results: it simply is not clear whether cubozoans are sisters of scyphozoans or of hydrozoans. So in the present state of knowledge, it is prudent to present the three as a trichotomy, as shown here.

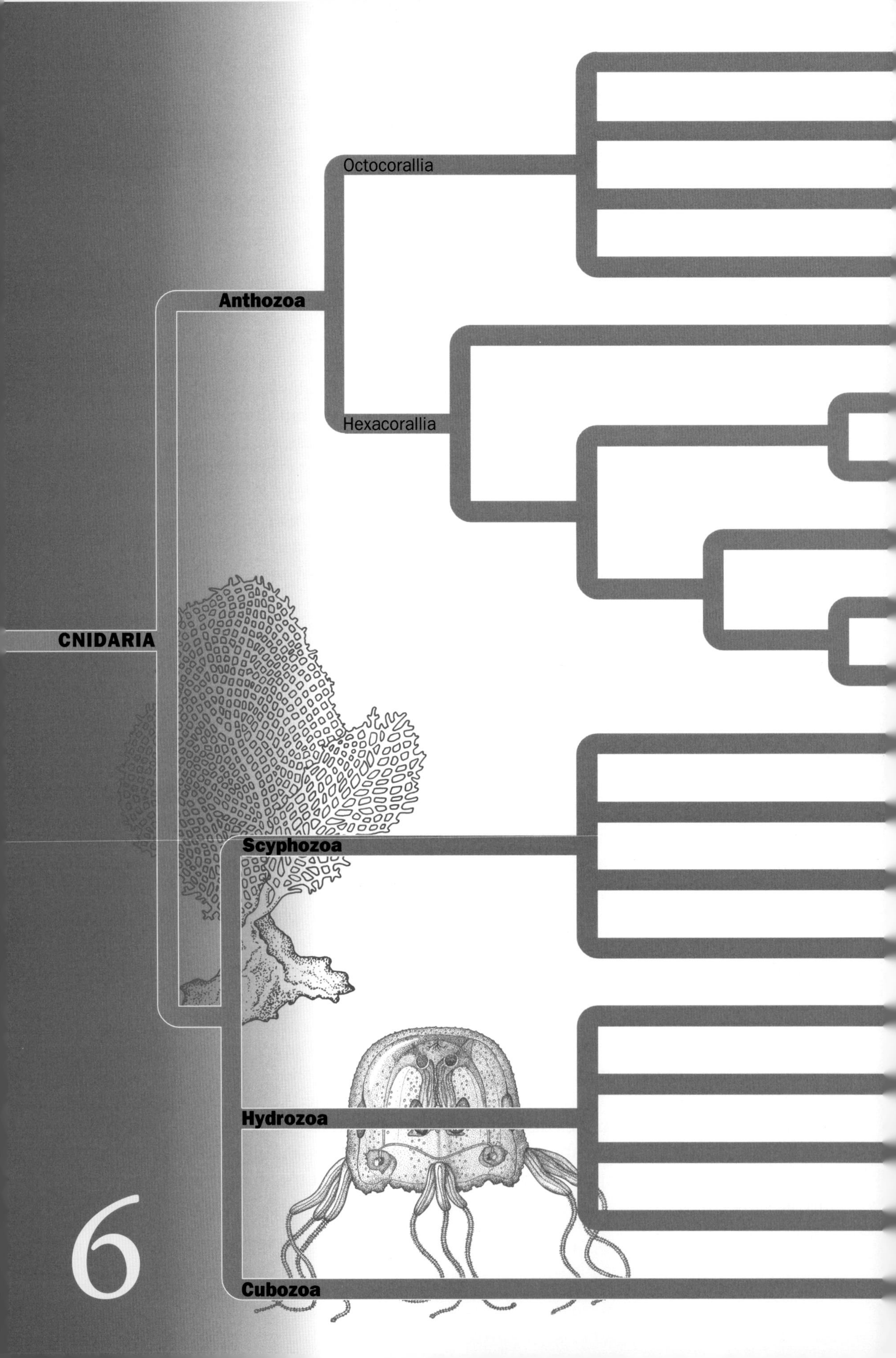
Octocorallia
Anthozoa
Hexacorallia
CNIDARIA
Scyphozoa
Hydrozoa
Cubozoa
6

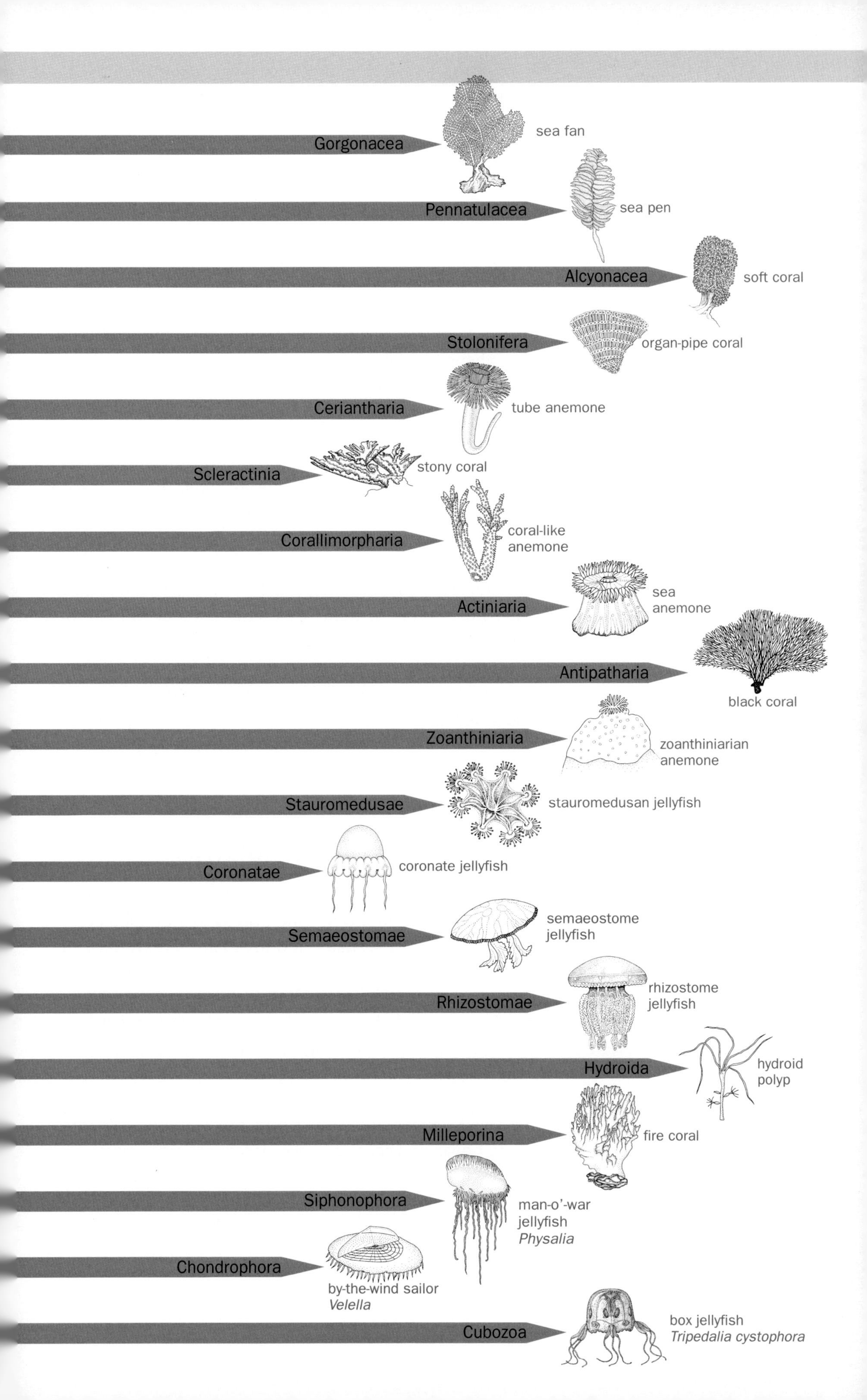

Gorgonacea
sea fan
Pennatulacea
sea pen
Alcyonacea
soft coral
Stolonifera
organ-pipe coral
Ceriantharia
tube anemone
Scleractinia
stony coral
Corallimorpharia
coral-like anemone
Actiniaria
sea anemone
Antipatharia
black coral
Zoanthiniaria
zoanthiniarian anemone
Stauromedusae
stauromedusan jellyfish
Coronatae
coronate jellyfish
Semaeostomae
semaeostome jellyfish
Rhizostomae
rhizostome jellyfish
Hydroida
hydroid polyp
Milleporina
fire coral
Siphonophora
man-o'-war jellyfish *Physalia*
Chondrophora
by-the-wind sailor *Velella*
Cubozoa
box jellyfish *Tripedalia cystophora*

The phylogenetic tree of the cnidarians is big and bushy, which reflects their extreme antiquity. The tree shows all the four classes—**Anthozoa**, **Scyphozoa**, **Hydrozoa**, and **Cubozoa**—with the largest, the Anthozoa, divided into two subclasses. It also shows 18 out of the 30 or so orders that between them illustrate the range of form and include all the types that a non-specialist is liable to encounter.

Note, however, a phenomenon that emerged so starkly with the prokaryotes: that any one clade may contain creatures of quite different phenotype; and that creatures of apparently similar phenotype may be found in several, and sometimes many, different clades. Thus non-specialists generally recognize just three basic forms of cnidarian: the solitary polyps, which are commonly and loosely called 'anemones'; the colonial polyps, which tend to get lumped together as 'corals'; and the medusae, which in general are referred to as 'jellyfish'. Clearly, at least two quite distinct classes—the Scyphozoa and the Cubozoa—are generally called 'jellyfish', while the Hydrozoa also contains a range of creatures that are popularly called 'jellyfish', including the man-o'war (*Physalia*) and the by-the-wind-sailor (*Velella*). 'Corals' manifest as 'true' or stony corals, plus soft corals, thorny corals, fire corals, milleporan corals, and so on—between them spread across several orders of Anthozoa and Hydrozoa. The tendency of evolution to produce similar phenotypes in many different taxa is seen in all of nature, even among mammals in which, for example, various groups besides the Felidae have been cat-like, while several only loosely related groups, including antelopes, deer, pronghorns, and the extinct South American litopterns, have re-invented the form of the rapid, hoofed, plains-runner. But in cnidarians (and prokaryotes) this tendency is particularly marked because the underlying body plan is so simple and seems to allow only limited variation.

Sea anemones, corals, and sea pens: class Anthozoa

Anthozoa is the largest grouping of cnidarians, with about 6000 species in three subclasses. Anthozoans have no medusae: all are marine polyps, either solitary or colonial, with the polyps reproducing both sexually and asexually. The mouth of each polyp is slit-like and leads into a pharynx, which extends into a coelenteron that is partitioned longitudinally by mesenteries. The mesogloea is thick.

The Anthozoa is divided into two subclasses. In the first, the **Octocorallia** or **Alcyonaria**, the tentacles occur in multiples of eight; and in the second, the **Hexacorallia** or **Zoantharia**, the tentacles come in multiples of six. Many 'classic' classifications showed a third subclass, the Ceriantipatharia, which included two orders, the Antipatharia (the black or thorny corals) and the Ceriantharia (the tube anemones). It is now clear, however, that these two orders are not closely related, do not form a clade, and certainly do not, therefore, form a valid subclass. In fact, as shown in the tree, the two orders seem to fit neatly within the Hexacorallia—with the tube anemones being basal, and the black corals apparently highly derived.

Sea fans, sea whips, sea pens, and sea pansies: subclass Octocorallia (or Alcyonaria)

The octocorals include many different orders, here represented by just four—the Gorgonacea, Pennatulacea, Alcyonacea, and Stolonifera. They are all colonial, with the polyps connected by sheets of tissue, or cords called **stolons**. Each polyp bears eight tentacles, which are usually pinnate (feathery).

- Order **Gorgonacea** includes the 18 or so families of sea fans and sea whips that typically form brightly coloured colonies up to several metres across, bound by a skeleton that is generally horny (from the hardened protein gorgonin) but is sometimes calcareous. In one family, the Isidae, calcareous and horny segments alternate along the branches, so that they are flexible.

- Order **Pennatulacea** comprises the sea pens and sea pansies; an elongated, central primary polyp forms a stalk up to about a metre long, and a swelling (**peduncle**) at the bottom to anchor it into its habitual soft sediment, while secondary polyps branch off the sides. Pennatulaceans are often luminescent.

- Order **Alcyonacea** comprises the soft corals—fleshy, flexible, often massive, with the ends (the distal portions) of the polyps retractable into the more compact bases.

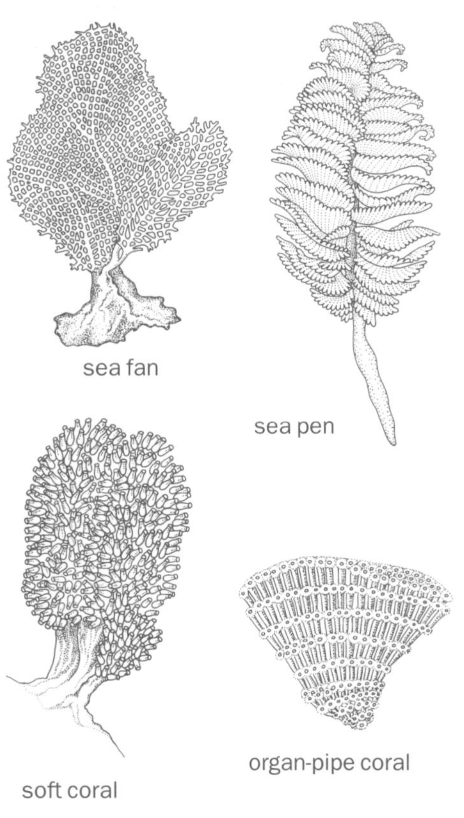

- Order **Stolonifera** comprises colonies of simple polyps that rise from a ribbon-like stolon trailed and wrapped over the substrate, the whole being covered with a horny skeleton. Stoloniferans are best known for the organ pipe 'corals', *Tubipora*.

Sea anemones, true corals, black corals, and tube anemones: subclass Hexacorallia (or Zoantharia)

Hexcorallians (or Zoantharians) are either solitary (as the anemones are) or colonial (like the reef-building corals) and are either naked or have calcareous or chitinous skeletons. The mesenteries that divide their coelenterons are paired, and usually in multiples of six; and hence the alternative name Hexacorallia. Zoantharians have one or several rows of tentacles, bear cnidae of many different kinds, and often carry enormous numbers of symbiotic zooxanthellae in their inner cell layer (endoderm). Because the hexacorallians include the 'true' or reef-building corals, in the order Scleractinia, they are highly influential ecologically—and geographically. Australia's Great Barrier Reef is as long as Great Britain.

Systematics have often shown just four orders of Hexacorallia: the Zoanthiniaria (which look like colonial sea anemones), the Corallimorpharia ('coral-like anemones'), the Actiniaria ('true' sea anemones), and the Scleractinia (the 'true' or stony corals). But in 1974 Hajo Schmidt in Heidelberg, Germany, classified the Anthozoa according to the form of their nematocysts, and suggested that two other groups also belong among the hexacorals: the Ceriantharia (tube anemones) and the Antipatharia (black or thorny corals). Molecular studies published in 1996 by Scott C. France of the University of New Hampshire and his colleagues, based on mitochondrial rRNA, now support Schmidt's idea. So the arrangement shown here is as proposed by Schmidt: Ceriantharia is shown as sister to all the rest; Corallimorpharia and Scleractinia are sisters; and Antipatharia and Actiniaria are sisters. As noted above, Ceriantharia and Antipatharia used to be considered sisters, and were placed together in their own subclass, the Ceriantipatharia. But the Ceriantipatharia was evidently a false grouping—based on appearance, rather than true relationship—and is now defunct.

• Order **Ceriantharia** is sister to all the other hexacorallians, which implies that it is closest to the ancestral type. The cerianthanians are the tube anemones: elongated, solitary beasts that live in soft sediments encased in tubes that are fashioned from specialized cnidae and mucus. The mouth is surrounded by short tentacles, with long, thin tentacles outside them.

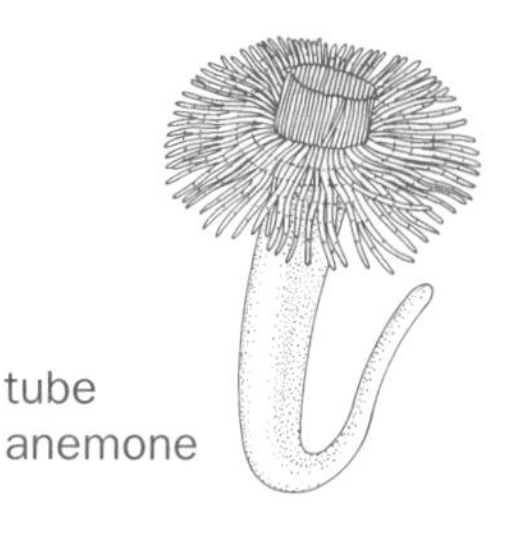
tube anemone

• Order **Scleractinia** includes the 'true' or stony corals. Ecologically the stony corals are the most significant of all the cnidarians and as creators of land mass are surely among the most significant of all animals. Scleractinians are either colonial or solitary. They wall themselves in calcareous (aragonite) exoskeletons that are sometimes delicate and sometimes massive. In the reef-building corals calcareous blades (called **septa**) extend into the mesenteries, giving the denuded skeletons a filigree exterior. The scleractinians include more than 2500 living species in 24 families. The single genus *Acropora* includes more than 150 species.

stony coral

• Order **Corallimorpharia** comprises the 'coral-like anemones'. Again, they are either solitary or colonial and lack a skeleton; examples include *Amplexidiscus* and *Corynactis*. Corallimorpharia emerges as sister to the Scleractinia and the two together are sometimes called the **Madreporia**.

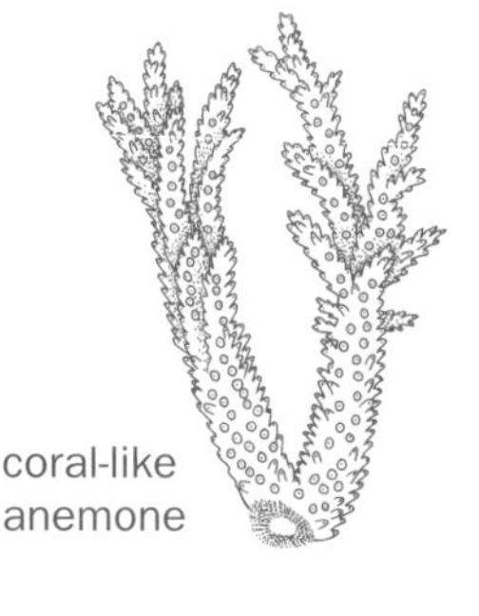
coral-like anemone

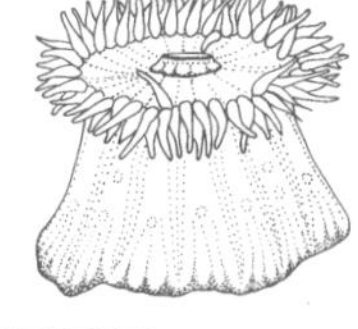
sea anemone

• Order **Actiniaria** is a huge and various group, containing about 41 families of 'true' sea anemones. They are either solitary or they form clones, but they are never colonial. They never have calcareous skeletons, but some secrete a chitinous skeleton, and most harbour zooxanthellae. Usually the column (the main body of the animal) is perforated, to extrude water as they contract; and the column is often adorned with warts, verrucas, pseudotentacles, or vesicles. The tentacles are hollow and either finger-like (digitiform) or branched. All the familiar anemones of seaside and aquarium are actiniarians, including *Adamsia*, the hitch-hiker, *Actinia*, the common beadlet anemone, and the attractive, feathery ('plumose') *Metridium*.

• Order **Antipatharia** comprises the black or thorny corals that are sometimes enormous gorgon-like structures up to 6 metres tall, with a hard skeleton that is usually brown or black from which poke small polyps, each bearing six (but up to 24) tentacles that are non-retractable. The skeleton is studded with thorns, hence the common name. Antipatharians live mainly in deep water in tropical seas.

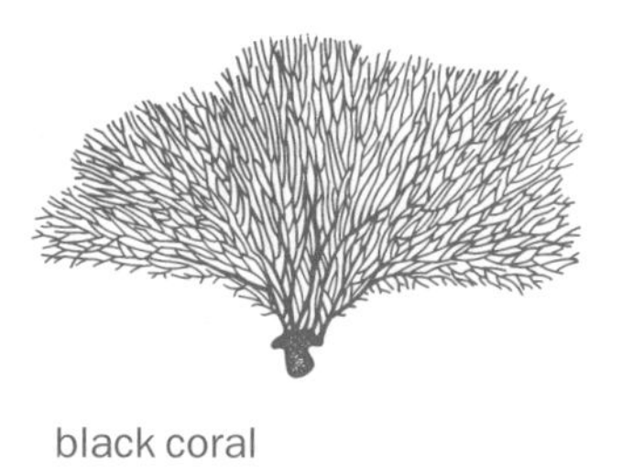
black coral

• Nematocyst structure suggests that the order **Zoanthiniaria** is sister to the Antipatharia. Zoanthiniarians resemble sea anemones but are usually colonial, with the polyps linked by a basal mat of tissue or a stolon. They lack a proper skeleton, but many kinds incorporate sand or spicules of sponges into their thick body walls. Usually they are copiously endowed with zooxanthellae, and many are epizoic, which means they hitch a lift on other animals—as, for example, *Adamsia* does on mollusc shells occupied by hermit crabs.

zoanthiniarian anemone

Jellyfish: class Scyphozoa

Scyphozoa contains most of the world's jellyfish: the medusoid stage dominates whereas the polyps are small, inconspicuous, and in some groups are even lacking. When they are present, however, the polyps have a wonderful way of generating new medusae: they split them off from the top by a series of transverse divisions as if by some industrial process in the manner known as **strobilation**. The coelenteron of the medusa is divided by four longitudinal mesenteries so that it is four-chambered, and, indeed, the whole animal generally has four-sided (tetramerous) symmetry. The medusa carries the gonads, which are born by the gastrodermis (the inner cell layer). All jellyfish are marine. Some are planktonic, floating near the surface; some—demersal forms—swim near the bottom; and some remain permanently attached to the bottom. There are about 200 species of jellyfish, in four orders.

• In the order **Stauromedusae** there are no polyps, but the medusa behaves like a polyp; that is, the surface away from the mouth (the 'exumbrella') attaches to the substrate by an adhesive disc in the centre. But—as is typical of medusae, but not of polyps—their reproduction is exclusively sexual. Stauromedusans live in shallow water at high latitudes.

stauromedusan jellyfish

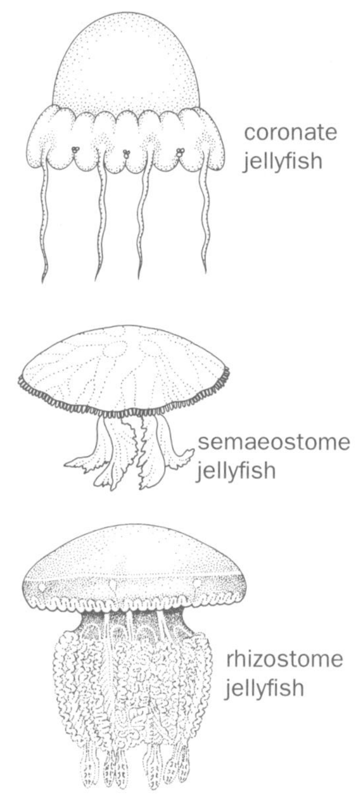

• Medusae in the order **Coronatae** are divided into upper and lower regions by an encircling groove, like a crown (hence the name), and the margin of the bell has an in-and-out, scalloped appearance. Coronates are active swimmers, generally at the sea bottom.

• Order **Semaeostomae** contains most of the jellyfish most of us are likely to see either in temperate or tropical seas, including the familiar *Aurelia* of the temperate seaside and the sometimes huge and strikingly orange *Cyanea*. The corners of the mouth in semaeostome medusae are drawn into four broad, frilly lobes.

• Medusae in the order **Rhizostomae** lack a central mouth. Instead, four oral lobes around the mouth are fused at their edges to produce several sucking 'mouths' or **ostioles** that open into a complicated system of canals within eight branching, arm-like appendages. Rhizostome medusae may be small or large and they swim vigorously through mainly tropical seas, powered by well-developed muscles beneath the umbrella. One example is *Cassiopeia*, which swims upside down.

Hydra, men-o'-war, and fire corals: class Hydrozoa

The hydrozoans are an immensely various group include both medusae and polyps, either of which may be the dominant form (and either of which, contrariwise, may be suppressed or lacking). The medusae are usually small and transparent; often they are retained on the polyps. The polyps are usually colonial and often are modified for different purposes—for feeding, breeding, or defending. Many have an exoskeleton that is usually chitinous but may be calcareous. On a technical point, the mesogloea is without cells. Hydrozoa includes about 2700 species in seven extant orders, of which four—Hydroida, Milleporina, Siphonophora, and Chondrophora—seem worth particular mention.

• Order **Hydroida** includes more than 55 families, divided among two suborders. In the first of these, the Anthomedusae, the polyps are the dominant phase and sometimes there are no medusae at all. The polyps are naked, without an exoskeleton. They include colonial seashore types such as *Hydractinia* and *Tubularia*, and also the solitary, medusa-less, freshwater species known to all zoology students, *Hydra*. In the suborder Leptomedusae the polyps are always colonial, with specialist reproductive and feeding individuals, and are encased in an exoskeleton, and most species lack medusae. *Obelia* is a well-known leptomedusoid.

hydroid polyp

• Order **Milleporina** comprises the 'fire corals', which all belong to one genus, *Millepora*. The colonies again contain more than one kind of polyp, the skeletons of which resemble those of corals and are massive or encrusting. Like the true corals, the milleporans rely heavily on the zooxanthellae that live within their cells. They do produce free medusae, but these are small and without a mouth or tentacles.

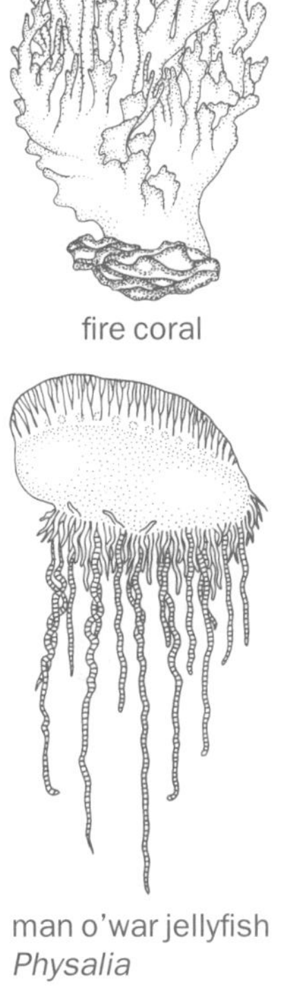

fire coral

man o'war jellyfish
Physalia

• Order **Siphonophora** includes some of the most extraordinary creations of nature—the man-o-war jellyfish, such as *Physalia*. Siphonophoran jellyfish are floating or swimming colonies including both polyps and medusae, which between them show the greatest degree of polymorphism of any cnidarian; only a few other creatures, including some ants, match them for variety of form and function. The balloon that buoys *Physalia* (which, as mentioned, is probably a modified polyp) is filled with gas that is oddly rich in carbon monoxide; the swimming bell is a medusa. In siphonophorans we see how groups of individuals, each relatively simple but closely co-operating and adapted for different purposes, can produce colonies that effectively function as if it were a single, complex organism.

by-the-wind sailor
Velella

• Some zoologists doubt whether the order **Chondrophora** belongs in the Hydrozoa at all; others have at times regarded them as highly modified siphonophorans. It is, after all, not easy to see whether they are colonies, composed from specialist feeding, reproductive, and defensive individuals, or represent a single but remarkably specialised polyp. Either way, they are highly derived creatures. They include *Velella*, the 'by-the-wind-sailor'. *Velella* floats like a jellyfish (although it is a polyp) and carries a triangular sail on the top, by means of which it indeed lives up to its name: the cnidarians never stop pulling surprises.

Sea wasps and box jellyfish: class Cubozoa

Because cubozoans are obviously jellyfish they have often been classed as a subgroup of the scyphozoans, or as sisters to the scyphozoans. Clearly, however, they are a class in their own right; but modern molecular studies fail to show whether they are more closely related to the Scyphozoa or to the Hydrozoa. Hence, for the time being, the three classes should be shown as a trichotomy, as in the tree here. Cubozoans have both a polyp and a medusa—the former metamorphosing into the latter. The term 'box jellyfish' is certainly apt because the medusa is square in cross-section with

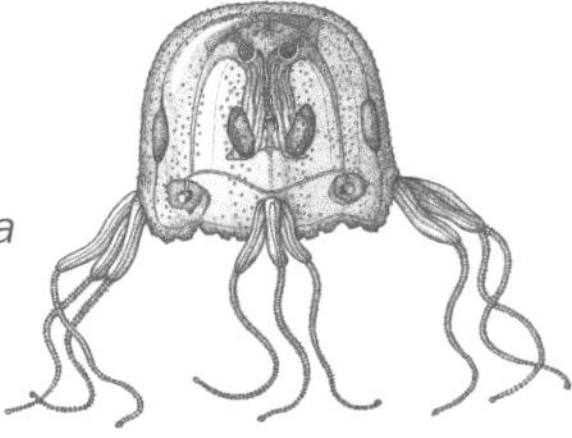

box jellyfish
Tripedalia cystophora

four flat sides. So is the soubriquet 'sea wasp' for their sting is extremely toxic and has even been known to kill human beings. Cubozoans live in all tropical seas.

These, then, are the cnidarians. It sometimes occurs to me that if creatures like this could think at all, then they would surely suppose that they represented some kind of evolutionary ultimate; after all, some are vast, some create entire islands, some sail the width of the world's oceans, some can kill creatures as big as a man, and some form organisms, or organism-like colonies, of enormous complexity. What more could evolution possibly achieve?

Well, one achievement has been the introduction of a third layer of cells between the epidermis and endo-, or gastrodermis, thus creating creatures that are triploblastic, rather than diploblastic. This middle cell layer—the **mesodermis**—provided the material for a whole new range of specialist organs, and of a far more efficient musculature. Triploblasty, in short, opened a whole new range of possibilities in physiological versatility, movement, behaviour, and, ultimately, in intelligence. There is no space in this book to dwell at length on the earliest and simplest of the triploblastic forms, which collectively and colloquially can be called 'worms'. Instead we will focus on some of the pinnacles of triploblastic development, beginning with the molluscs.

7

CLAMS AND COCKLES, SNAILS AND SLUGS, OCTOPUSES AND SQUIDS

PHYLUM MOLLUSCA

All taxa are unique, which is why they are acknowledged as taxa in the first place and given their own special names; but some, despite what the grammarians say, are more unique than others. There is nothing quite like a mollusc, to the extent that no one can say for certain who the molluscs' closest relatives are—whether segmented worms like annelids or simply non-segmented worms like platyhelminths. Furthermore, although the Mollusca do share a discernible body plan and few zoologists doubt their monophyly, they are extraordinarily various. They are also prone to convergence and the relationships between different groups are not yet certain. Indeed, within some groups, it is not always clear where one species ends and the next begins.

What is certain, though, is that molluscs are wonderful. At the lowly and/or degenerate and specialist end they include some of the most torpid of all 'higher' creatures—eyeless and literally brainless, stuck irredeemably to rock or thrust head-down in the mud to gather their nutrient as passively as plants—as well as some of the most dashing, powerful, and spectacular of all creatures on Earth, like some of the octopuses[1] and squids. The English zoologist J. Z. Young (1907–97), who for many years studied the behaviour of octopuses at the marine research station in Naples, opined that they are as intelligent as dogs. What seems to have held them back, indeed, so that they do not quite dominate the sea as fishes and whales do, is nothing more than a small physiological accident; that the pigment that carries oxygen in molluscan blood is not the iron-based haemoglobin favoured by vertebrates, but copper-based haemocyanin. Haemocyanin is not nearly so efficient. It gathers oxygen well enough, but does not readily surrender it again to the suffocating

[1] On a note of pedantry, the pus in 'octopus' is Greek, not Latin: and the anglicized plural is therefore '-puses' and not '-pi'.

tissues. It is better at storing oxygen than distributing it, like a librarian who hates people to borrow books. So cephalopods (squids and octopuses) are spectacularly lacking in stamina; given to brilliant, often jet-propelled rushes, but not to the steady haul or the repeated flashes of the vertebrate. A shame—or perhaps not. Perhaps if it were otherwise, this book might have been written by a cuttlefish.

Despite this design fault molluscs are extremely influential. Even modest ones like the shell-less and worm-like aplacophorans (solanogasters and caudofoveates) can be very numerous on deep-sea sediments while others—mussels, clams, whelks, snails, cephalopods—support vast portions of the oceanic food webs and, indeed, have been, and are, the staple provender of many a human population (including Maoris and Australian Aborigines). Oysters were a vital standby to many working people in Europe before pollution all but wiped them out, while cockles, whelks, and periwinkles (colloquially known as 'winkles') marinated in vinegar live on as vestigial delicacies. But oysters, abalones, and escargots, in particular, have also achieved the status of gourmet foods, closely followed by scallops, calamaries, and mussels. Molluscs are also key filter feeders and detritivores, hastening the recycling of nutrients in the sea, whereas others like squids and whelks are significant predators. Shells have often been currency, like cowries, while others are among the world's great natural icons, like the conch in William Golding's *Lord of the Flies* or the giant squid in Jules Verne's *Twenty Thousand Leagues Under the Sea*, a creature perceived to be as violent and malevolent as Herman Melville's *Moby Dick*, albeit just as libellously. Molluscs can also be significant pests: for example, shipworms, *Teredo*, burrow into the timbers of ships and modern piers and, far more importantly, snails carry the liver fluke, *Fasciola*, and the blood fluke, *Schistosoma*—the agent of bilharzia.

Unsurprisingly, given their ecological variety and success, molluscs are wonderfully speciose. About 100 000 species are known—but the 'about' is emphatic. For example, David Reid of London's Natural History Museum has shown that some species of periwinkle are extraordinarily polymorphic: they vary enormously not only in colour but in the size and shape of their shells, depending on where they live. On shores pounded by heavy waves natural selection favours small winkles with relatively big, clinging feet, whereas on more sheltered shores selection favours bigger, more solid individuals that are less resistant to waves but put up more of a show against predatory crabs. By looking at their whole anatomy Reid found that periwinkles with very different shells, which appear to represent many different species, in fact may all belong to the same one. The penis is particularly informative: it is a spectacular organ in gastropods, proportionate in size to, say, the human arm. On the other hand, traditional taxonomists have sometimes lumped what are now proving to be several species into one; and thus Reid and others have found that the four traditionally recognized winkles of Europe in fact represent seven species.

Add to all this the fact that many molluscs live only in sediment in deep ocean and other such remote places and so remain to be discovered—the first living monoplacophoran (*Neopilina*) was reported only in 1952—and we see how hazardous it is to guess the real number of living molluscan species. Among fossils, about 60 000 species are known. The record is rich in parts and poor in others. Clams spend their lives in sediment and so, like ancient Egyptian aristocrats, spend their lives preparing for fossilization, whereas winkles, once dead, are battered to pieces on the shore.

The word 'mollusc' was coined felicitously in the seventeenth century from the Latin *molluscus* meaning 'soft', and indeed is borrowed directly from the name of a thin-shelled nut called a 'mollusca'. Its inventor, however, also applied the term to the crustacean barnacles—emphasizing that although molluscs are apparently so distinctive, and the presently defined phylum Mollusca is a true clade, it has not always been easy to decide what is a mollusc and what is not. Anemones, jellyfish, sea squirts, sea cucumbers, polychaete worms, and barnacles are among the creatures that have at times been dragged into the molluscan fold, while brachiopods (lamp shells) were not finally prised loose from it until the end of the nineteenth century.

In the days before Willi Hennig and cladistics, however, systematists perhaps found it harder than they might to define the Mollusca accurately because they did not distinguish so formally between general features (symplesiomorphies), which do not reveal special relationships, and 'shared derived characters' (synapomorphies), which are special to the group in question and do reflect true relationships. The soft body, for example, which gives molluscs their name, is simply a symplesiomorphy: most invertebrates are soft-bodied. But molluscs can be defined by three outstanding synapomorphies: the **shell**, and the **mantle** that secretes it—the 'shell–mantle complex'; the peculiar file-like feeding apparatus, the **radula**; and the characteristic molluscan gills, the **ctenidia**. In the following description primitive and derived features are bundled together, for ease of reading; but these three are particularly important.

The soft bodies of molluscs are supported internally by water pressure: a **hydrostatic skeleton**, as in annelid worms and the cells of herbaceous plants. Hydrostatic pressure supports the typically molluscan **foot**, which is archetypically used for creeping as in chitons, monoplacophorans, and snails; and, by shifting the internal fluid around, they are able to change their body shape dramatically, so that snails, for example, can emerge from and retreat into their shells, and many bivalves extend and retreat their foot.

The fluid in the molluscan body is contained mainly in a space that is not a true coelom, but a **haemocoel**, which has a different embryological origin. The coelom itself is confined to small spaces around the nephridia ('kidneys'), heart, and gonads; but some zoologists have argued that these so-called 'coelomic' fragments are not homologous with the coelom of, say, annelids and should not be regarded as coelom at all; indeed, that molluscs are, strictly speaking, 'acoelomate'. (This is a minority view, but it shows once more that animal relationships in general are not sewn up.) The internal organs of molluscs are gathered together to form a **visceral mass**.

Most obviously, molluscs are known by their shells: calcium-based ('calcareous') carapaces secreted by the mantle's characteristic fold of skin that covers part of the body like a horse blanket. Yet there are two entire molluscan classes collectively called the aplacophorans that apparently never evolved a true shell but are studded instead with calcareous spicules. These are the Caudofoveata (caudofoveates) and the Solanogastres (solanogasters), both of which look more like worms than typical molluscs. The two aplacophoran classes are generally considered to be primitive, and their half-hearted armour seems to represent an early stage in shell evolution (though we should not, of course, assume that living caudofoveates or aplacophorans would ever evolve true shells, even if they were left to themselves for the next billion years; we can properly perceive mechanical progress in evolution, but not destiny). All the same, many different groups of molluscs, particularly among the gastropods and cephalopods, have independently lost their shells or reduced them to vestiges in the same way that many different birds have independently lost the power of flight.

The mantle, the source of the shell, has also been pressed to other purposes in different molluscan groups. In marine gastropods the space beneath, the 'mantle cavity', conceals the openings of the 'kidneys' (or their equivalent), the reproductive organs, the ctenidia, and sometimes various sense organs. In land gastropods, like snails, the gills are gone and the mantle is vascularized to form a lung. In many bivalves and in cephalopods the margins of the mantle are fused to form a tube, or **siphon**, which in some bivalves becomes a conspicuous snorkel and in cephalopods provides jet propulsion. In nudibranchs and other free-swimming gastropods, and in cephalopods, the mantle may be brightly coloured and many cephalopods communicate by changing colour. In giant clams the edge of the mantle harbours colonies of photosynthesizing dinoflagellates of the genus *Symbiodinium*: a fine symbiotic relationship in which the clam benefits from the products of photosynthesis while the protists find a good protected home in the sun, plus nutrients such as nitrogen.

In bivalves, like clams and mussels, the ctenidia are also pressed into a new service (although they continue to serve for respiration). Doubled back on themselves, greatly enlarged, and subdivided to form a kind of meshwork, they form the surface that extracts small particles food particles from the current of water passing between the shells: a highly efficient form of filter feeding.

Characteristic of the molluscs, though not possessed by all of them, is the peculiar feeding device known as the radula, which is like a heavy-toothed file. In some molluscs digestive enzymes are held in solid form as a **crystalline style**, from which the enzymes are rubbed off as if from the lead of a pencil. Finally the types that have larvae (which means most of those that live in the sea) show that the molluscs are true protostomes, belonging to the same great animal clade that includes annelids (segmented worms) and arthropods (insects, lobsters and their relatives).

The first tree shows all the eight living classes of Mollusca and the generally accepted subclasses, plus the outstanding orders. It is based primarily on papers from the *Origin and Evolutionary Radiation of the Mollusca* (1996) edited by John D. Taylor of London's Natural History Museum (NHM). The notes in the 'Guide' are based on Taylor's book and on works by Richard and Gary Brusca, Pat Willmer; Claus Nielsen; and discussions with David Reid and others at the NHM.

A GUIDE TO THE MOLLUSCA

Zoologists agree that the Mollusca, as now understood, is indeed monophyletic: a clade that includes all known descendants of the first molluscan common ancestor, and none that is not. This leaves three questions. Who were the ancestors of the molluscs? Which of the living phyla are most closely related to the molluscs—which, indeed, can be regarded as the molluscs' living sisters? Finally, how is the great molluscan clade subdivided? The picture has been highly contentious; and, just as everything seemed reasonably settled, new molecular studies from James Lake's department at the University of California, Los Angeles (UCLA), have thrown completely new light.

The first two issues can be treated together, for if we know the ancestry then we can probably also identify the living relatives. We do know—nobody seems seriously to doubt—that molluscs are protostomes; and we can reasonably infer that all protostomes descended originally from some kind of flatworm that would be classed as a platyhelminth. So the most interesting question has been how the molluscs relate to the other protostome phyla. There are many of these, of course, but the two that are outstanding in ecological impact and in numbers of species are the annelids and the arthropods. To simplify the discussion, then, we can focus on this single question: how do the three truly great protostome groups—the molluscs, annelids, and arthropods—relate to each other?

To begin with, most zoologists have taken it virtually as read that the three great protostome phyla are closely allied. Modern, adult molluscs generally look very different from modern, living annelids and arthropods but there are intriguing similarities. Notably the early, planktonic larvae that many marine molluscs still produce are very similar in appearance and in mode of development to those of annelids; and similarities of embryos, and particularly in details of development, are traditionally taken to reveal true relationship (although this is increasingly contested). To be sure, annelids and arthropods are emphatically segmented—the body of an earthworm, for example, is mainly a series of more or less identical units, like carriages in a train—while molluscs are not. But, say traditionalists, primitive molluscs do show signs of segmentation, as seen in the seven or eight shells of chitons and the serial 'kidneys' and gills of monoplacophorans. Claus Nielsen in *Animal Evolution* speaks of 'indications' that the ancestral mollusc specifically had eight segments.

Overall, then, traditional zoologists have been content with the notion that annelids, arthropods, and molluscs all share a common ancestor; that that common ancestor was segmented; and that it derived from some kind of platyhelminth.

So what is the relationship between molluscs, arthropods, and annelids? Evolutionary biologists generally agree—not because they know it for a fact, but because it seems most likely—that new branches generally arise as dichotomies, and not as polychotomies. In other words, we should not envisage that the common flatworm ancestor gave rise to the three major phyla simultaneously. Instead, we should expect first one branching, and then a second. If this was what happened, then two of the three living phyla would emerge as sisters and the third would be the outgroup—sister to the other two combined. So, the traditionalists have asked, is Mollusca the sister to the Annelida + Arthropoda? Is Arthropoda sister to Mollusca + Annelida? Or is Annelida sister to Mollusca + Arthropoda? If we consider only these three major

7 · PHYLUM MOLLUSCA

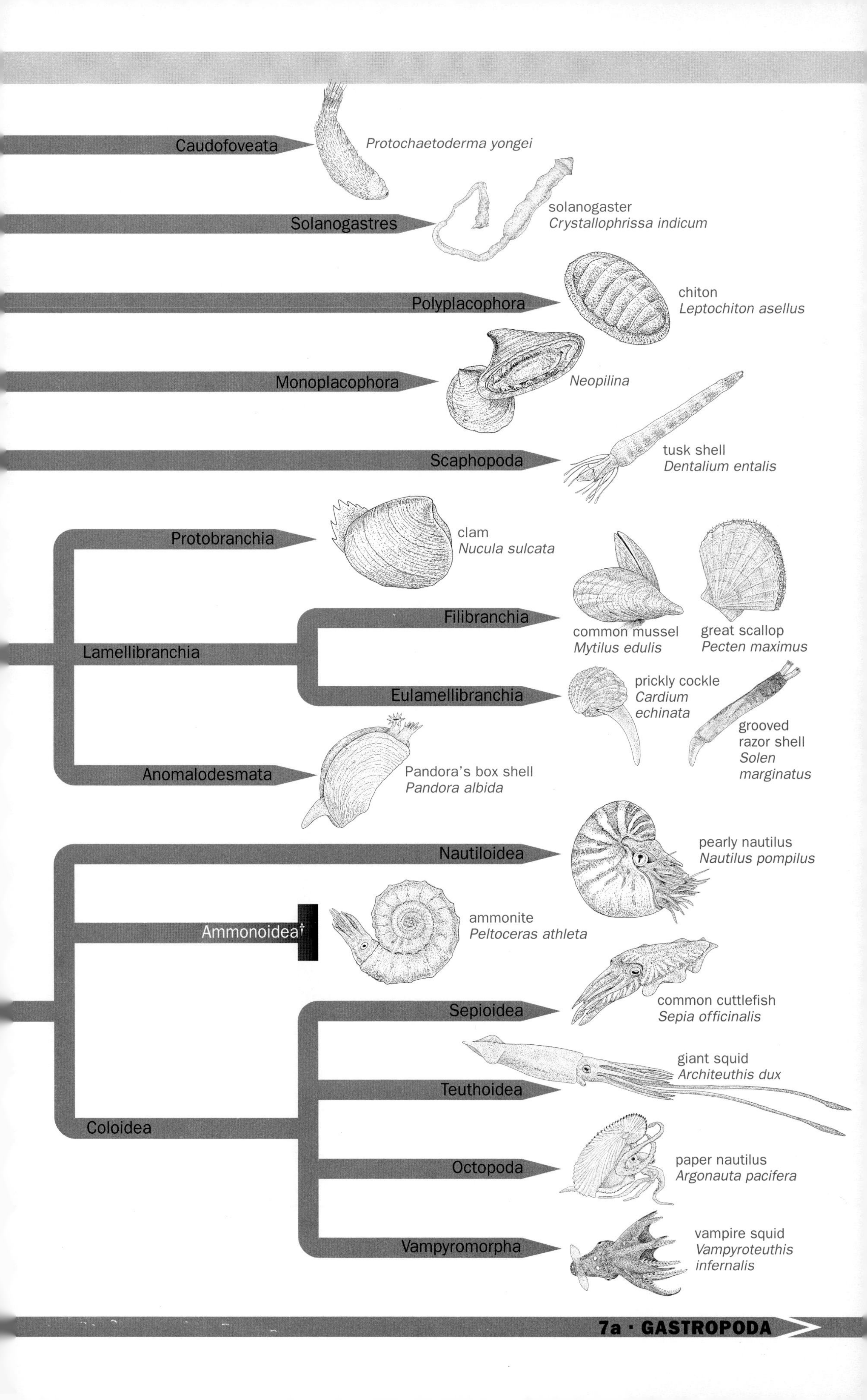
Caudofoveata
Protochaetoderma yongei
Solanogastres
solanogaster
Crystallophrissa indicum
Polyplacophora
chiton
Leptochiton asellus
Monoplacophora
Neopilina
Scaphopoda
tusk shell
Dentalium entalis
Protobranchia
clam
Nucula sulcata
Lamellibranchia
Filibranchia
common mussel
Mytilus edulis
great scallop
Pecten maximus
Eulamellibranchia
prickly cockle
Cardium
echinata
grooved
razor shell
Solen
marginatus
Anomalodesmata
Pandora's box shell
Pandora albida
Nautiloidea
pearly nautilus
Nautilus pompilus
Ammonoidea†
ammonite
Peltoceras athleta
Sepioidea
common cuttlefish
Sepia officinalis
giant squid
Architeuthis dux
Teuthoidea
Coloidea
Octopoda
paper nautilus
Argonauta pacifera
Vampyromorpha
vampire squid
Vampyroteuthis
infernalis
7a · GASTROPODA

GASTROPODA
Patellogastropoda
Vetigastropoda
Neritopsina
Caenogastropoda
Heterobranchia
7a

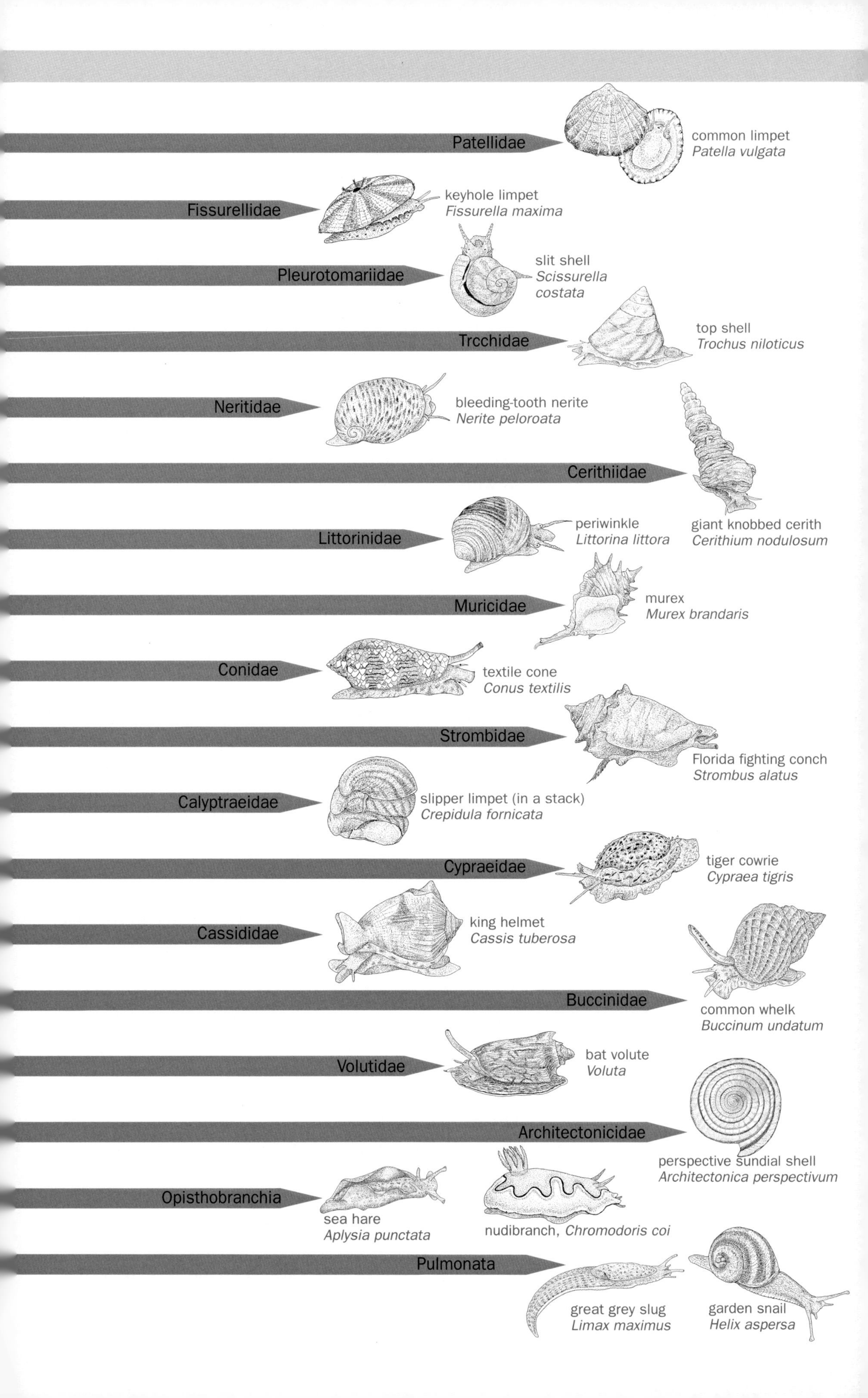
Patellidae
common limpet
Patella vulgata
Fissurellidae
keyhole limpet
Fissurella maxima
Pleurotomariidae
slit shell
Scissurella
costata
Trcchidae
top shell
Trochus niloticus
Neritidae
bleeding-tooth nerite
Nerite peloroata
Cerithiidae
Littorinidae
periwinkle
Littorina littora
giant knobbed cerith
Cerithium nodulosum
Muricidae
murex
Murex brandaris
Conidae
textile cone
Conus textilis
Strombidae
Florida fighting conch
Strombus alatus
Calyptraeidae
slipper limpet (in a stack)
Crepidula fornicata
Cypraeidae
tiger cowrie
Cypraea tigris
Cassididae
king helmet
Cassis tuberosa
Buccinidae
common whelk
Buccinum undatum
Volutidae
bat volute
Voluta
Architectonicidae
perspective sundial shell
Architectonica perspectivum
Opisthobranchia
sea hare
Aplysia punctata
nudibranch, Chromodoris coi
Pulmonata
great grey slug
Limax maximus
garden snail
Helix aspersa

groups, and ignore any minor phyla that may be (and probably are) tucked in between them, then these are the only three possibilities.

We need not bother with the last of these. No one, to my knowledge, has ever suggested that arthropods are closer to molluscs than either of those groups is to the annelids. This leaves us asking whether annelids are closer to the arthropods, or to the molluscs.

Common observation and a lot of detailed anatomical study have favoured the notion that annelids are close to arthropods, and that the molluscs are the outgroup. Annelids and arthropods are both emphatically segmented, and Nielsen combines them in a superphylum grouping, 'Euarticulata'. As described in Section 8, the onychophorans and the tardigrades traditionally seemed to provide a conceptual and perhaps even a literal link between the two phyla. In short, arthropods could easily be perceived as armoured annelids with legs. Tradition had it, then, that some flatworm-like creature (probably classifiable as a true platyhelminth) became the common ancestor of molluscs, annelids, and arthropods; that this lineage first divided, to give molluscs on the one hand and annelids + arthropods on the other; and then a second division split the arthropods from the annelids. QED.

Until recently, perhaps the most radical dissenting view has come from Pat Willmer. In *Invertebrate Relationships* (1990), she defies the traditional view that molluscs descended from segmented ancestors. Instead, she says, molluscs descended directly from a flatworm, but not from the same flatworm that gave rise to annelids and arthropods. Indeed, molluscs can be seen as elaborated flatworms with shells. To be sure, some organs, such as the nephridia, do occur in 'serial' repeats in some primitive molluscs such as chitons and solanogasters, although nephridia are commonly coalesced into a single kidney in derived molluscs such as snails. But, says Willmer, such repetition need not denote true segmentation. After all, flatworms also have repeated pairs of excretory organs—not because their bodies are segmented but because they lack an efficient blood system, so that each different part of the body needs its own waste-disposal unit. Primitive molluscs have circulatory systems, but have retained this flat-worm feature. All in all, Willmer's arguments show that traditional views, based on anatomy, can leave plenty of room for dispute.

But all traditional arguments now seem to be overturned by reports from Kenneth M. Halanych and his colleagues, and from Anna Marie Aguinaldo and her colleagues, at UCLA. As described on pages 200–1, they studied 18S RNA from a range of animals and showed that the protostomes split naturally into two great groups, the Ecdysozoa and the Lophotrochozoa. As shown in the tree in Section 8, this new grouping cuts right through the traditional grouping of the Mollusca, Annelida, and Arthropoda. The traditional association of molluscs with annelids remains—both are lophotrochozoans—but the arthropods seem to be quite different: they emerge as ecdysozoans. Thus, annelids and arthropods might *look* similar, but the resemblance seems to be yet another remarkable example of convergence.

At least zoologists generally agree that the phylum Mollusca divides naturally into the eight classes shown here, so even if the relationship between the classes is rethought in the future this should remain as a worthwhile catalogue. The following notes describe all the classes and subclasses as generally recognized, and a selection of the orders that ecologically or in other ways are the most significant.

THE CLASSES OF MOLLUSCS

The eight acknowledged classes of Mollusca are the **Caudofoveata**, **Solanogastres**, **Polyplacophora** (chitons), **Monoplacophora**, **Scaphopoda** (tusk shells), **Bivalvia**, **Gastropoda**, and the **Cephalopoda** (or Siphonopoda). Two of these classes—caudofoveates and solanogasters—are the worm-like 'aplacophorans'. Three—the polyplacophorans, monoplacophorans, and scaphopods—are ecologically minor, but do at least look like molluscs. The remaining three—the bivalves, gastropods, and cephalopods—are of huge importance, however 'importance' is measured.

As you can see, there is uncertainty at the base of the tree, as manifested in the trichotomy. The aplacophoran caudofoveates and solanogasters are assumed to be primitive (as opposed to derived and degenerate, although both groups are highly specialized for life in and on the sediment); and each of them is shown as a deep branch in its own right. The third deep branch contains all the remaining molluscs (the ones that actually look like molluscs).

Within the main molluscan group, the multishelled chitons (polyplacophorans) emerge as the sister of all the remainder. This 'remainder' includes five classes that again, in the present state of knowledge, must be shown as subdivisions of a trichotomy. The monoplacophorans are peculiar, and form a branch on their own. The scaphopods and bivalves together form the second branch; a key feature that unites them is the form of the mantle, which effectively wraps around the foot on each side, at least when the foot is unextended. The gastropods and cephalopods seem to belong together and form the third branch. Snails and squids may look very different (and have very different ways of life) but their body plan is intriguingly similar. In both, the mantle covers only a part of the rear of the body, where the visceral mass is confined, leaving the head and foot free for action.

Worm-like molluscs without shells: class Caudofoveata

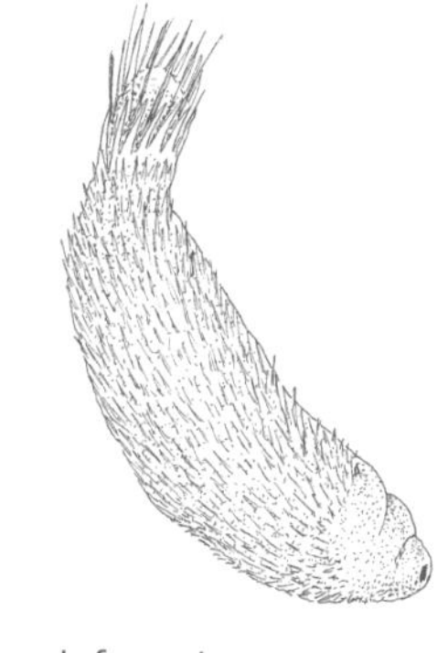

caudofoveate *Protochaetoderma yongei*

The caudofoveates are like cylindrical worms, their bodies enclosed in a cuticle presumed to be made of chitin (a tough horn-like substance), and covered in calcareous spicules like scales, rather than a shell (and hence their alternative name, 'Chaetodermomorpha'). They live in sediment (a lifestyle known as 'infaunal') under deep sea, upside down in burrows. Many molluscs, including many bivalves such as clams, have pursued the infaunal way of life, which does not seem hugely enviable. Caudofoveates have no eyes, tentacles, or statocysts (organs of balance), nor a crystalline style, nor a foot, although they do have a pair of ctenidia (gills). About 70 species are described, but little is known of their ecology.

More worm-like molluscs without shells: class Solanogastres

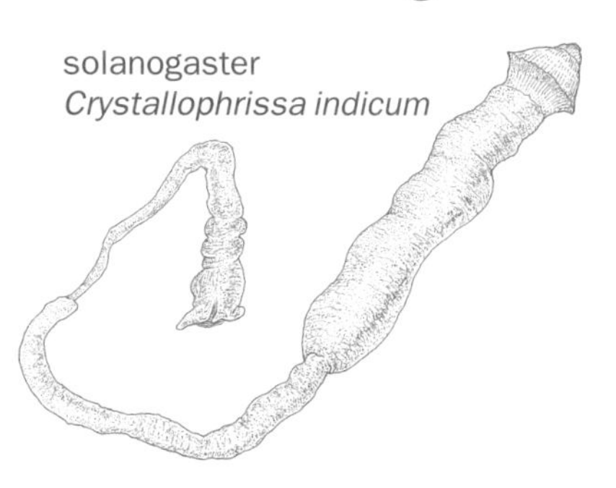

solanogaster
Crystallophrissa indicum

The solanogasters are also worm-like and spiculed rather than shelled. Although some have flattened bodies, others are cylindrical. Again they lack eyes, tentacles, statocysts, nephridia, or a crystalline style; some have a radula but others do not. They do not have a flattened foot, but they do have a ventral groove that secretes mucus, is hydrostatically operated, and is believed to be homologous with the foot of other molluscs.

Unprepossessing though they may be, however, in their own way solanogasters are successful. About 250 species have been described, mostly living within the sediment or on top of it (the lifestyle known as 'epifaunal'), and locally they can be abundant. Many feed on cnidarians, and some live within cnidarians.

Chitons: class Polyplacophora

chiton
Leptochiton asellus

Chitons resemble big, somewhat flattened slugs—but their backs are protected by eight (though sometimes seven) calcareous plates, like an armadillo. The mantle is thick and fleshy and extends beneath the plates all around to form a girdle; and in many groups the girdle extends back over the edge of the plates or even covers them entirely. Generally the girdle carries spines, scales, or bristles. Chitons are well endowed with ctenidia (with six to more than 80 pairs) but have only one pair of nephridia, and are without eyes, tentacles, or a crystalline style. But they do have a radula. Most chitons live as grazing herbivores on rocky shores, in the intertidal zone, but a few live in the deep sea. Altogether about 600 species are known, spread among three orders.

I have presented chitons here as a 'minor' class, which is how zoologists tend to think of them. Indeed there are not many species by the standards of, say, gastropods and they do not make a huge ecological impact. Taken individually, however, many are impressive, substantial beasts.

Neopilina: class Monoplacophora

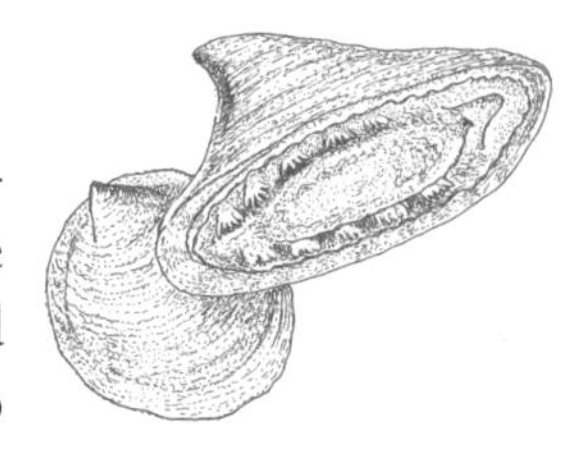

monoplacophoran
Neopilina

Until 1952 zoologists had no reason to suppose that monoplacophorans had survived beyond the early Palaeozoic, the time of the earliest known fossils. Then a Danish expedition found *Neopilina galatheae*; and another 11 species, including two more genera, have turned up since.

Monoplacophorans have a peculiar anatomy, which is

why they must be placed in their own class, whose relationship to the others is far from certain (as reflected in the trichotomy). They have a single cap-like shell, which gives them an unexceptional, limpet-like appearance. But then they have two pairs of gonads, six to seven pairs of kidneys (metanephridia), a strangely laid-out nervous system, no eyes, and tentacles only around the mouth; but they do have a crystalline style and a radula, though only a small head. None of the living species is more than 3 centimetres long, and they mostly live deep in the sea.

Tusk shells: class Scaphopoda

In scaphopods the shell is long, cylindrical, and tapering—like a tusk, in fact—and is open at both ends. The mantle is large, and extends over the entire undersurface. The head projects from the larger of the two shell openings; it is rudimentary and eyeless, but has a proboscis and clusters of club-like tentacles that catch and manipulate prey, and is served by both radula and crystalline style. Tusk shells also have a somewhat cylindrical foot; and they lack ctenidia. About 350 species are known, spread among eight families; and although many scaphopods live in deep waters, shells of *Dentalium entalis* do wash up on European shores.

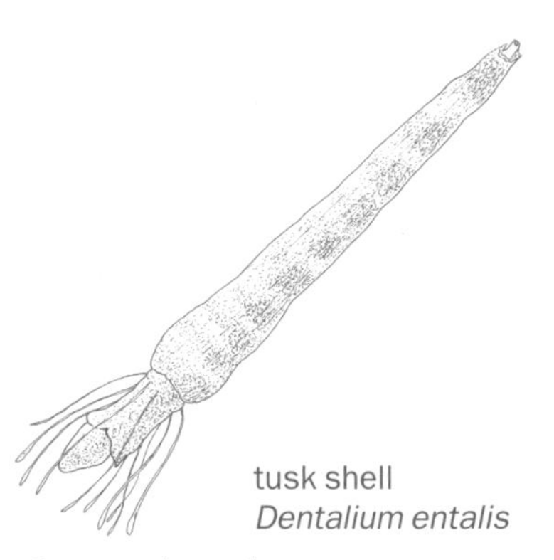

tusk shell
Dentalium entalis

Clams, mussels, and oysters: class Bivalvia

Bivalves (or Pelecypoda or Lamellibranchiata) form a large proportion of the creatures known to chefs as 'shellfish'. Typically the shell is in two pieces, the 'valves', which are hinged along the back by an elastic ligament and by shell 'teeth', and closed by powerful adductor muscles. The head is rudimentary, without eyes or radula, and the foot is typically flattened from side to side and has no sole. There is just one pair of ctenidia but these are big and folded back on themselves to form a highly efficient surface for filter feeding. The mantle is large, and often fused at the edges to form siphons that draw in a current of water and pass it out again (inhalent and exhalent), which is necessary both for respiration and feeding.

Bivalves are extremely numerous and various, with more than 8000 species, mostly in the sea but also in fresh water. But there is little agreement on their classification. I am following Richard and Gary Brusca, who divide them into three main subclasses, which at least is tidy. Two of those subclasses are relatively minor: subclass **Protobranchia** includes primitive bivalves with simple ctenidia that are not folded back on themselves; and subclass **Anomalodesmata** includes such types as

protobranch bivalve
Nucula sulcata

Pandora's box shell
Pandora albida

Pandora. The largest and most important subclass by far is the **Lamellibranchia**, which the Bruscas divide into two superorders according to the fine structure of their gills:

- Superorder **Filibranchia** (or Pteriomorpha) has more feathery gills, and is the more primitive. Filibranchs tend to be cemented to the substrate like oysters, or attached to it firmly by **byssal threads** like mussels or, in some cases, have secondarily freed themselves. Families include the Mytilidae, such as *Mytilus*, the mussels; Arcidae, which are clams of the kind called arc shells, such as *Arca*; Ostreidae, which are the true oysters, such as *Ostrea*; Pinnidae, the pen shells, such as *Pinna*; and Pectinidae, which are the scallops, such as *Pecten*.

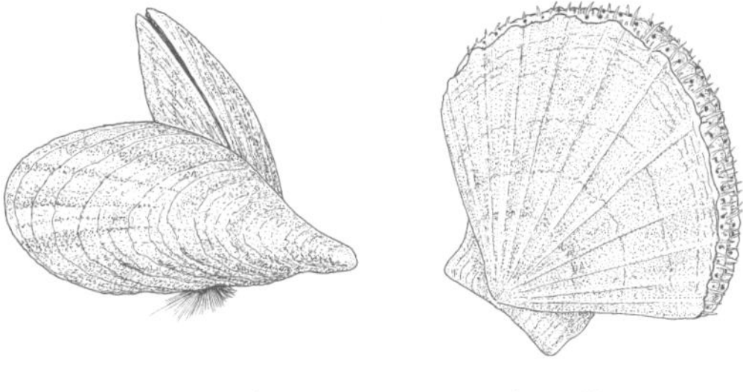
common mussel *Mytilus edulis*
great scallop *Pecten maximus*

- Superorder **Eulamellibranchia** (or Heterodonta) has more tightly structured gills. The mantle is more or less fused to form distinct openings to allow water in and out, and these openings are often drawn out to form siphons. The group is divided into various orders that reflect some interesting relationships. Thus the order Veneroida

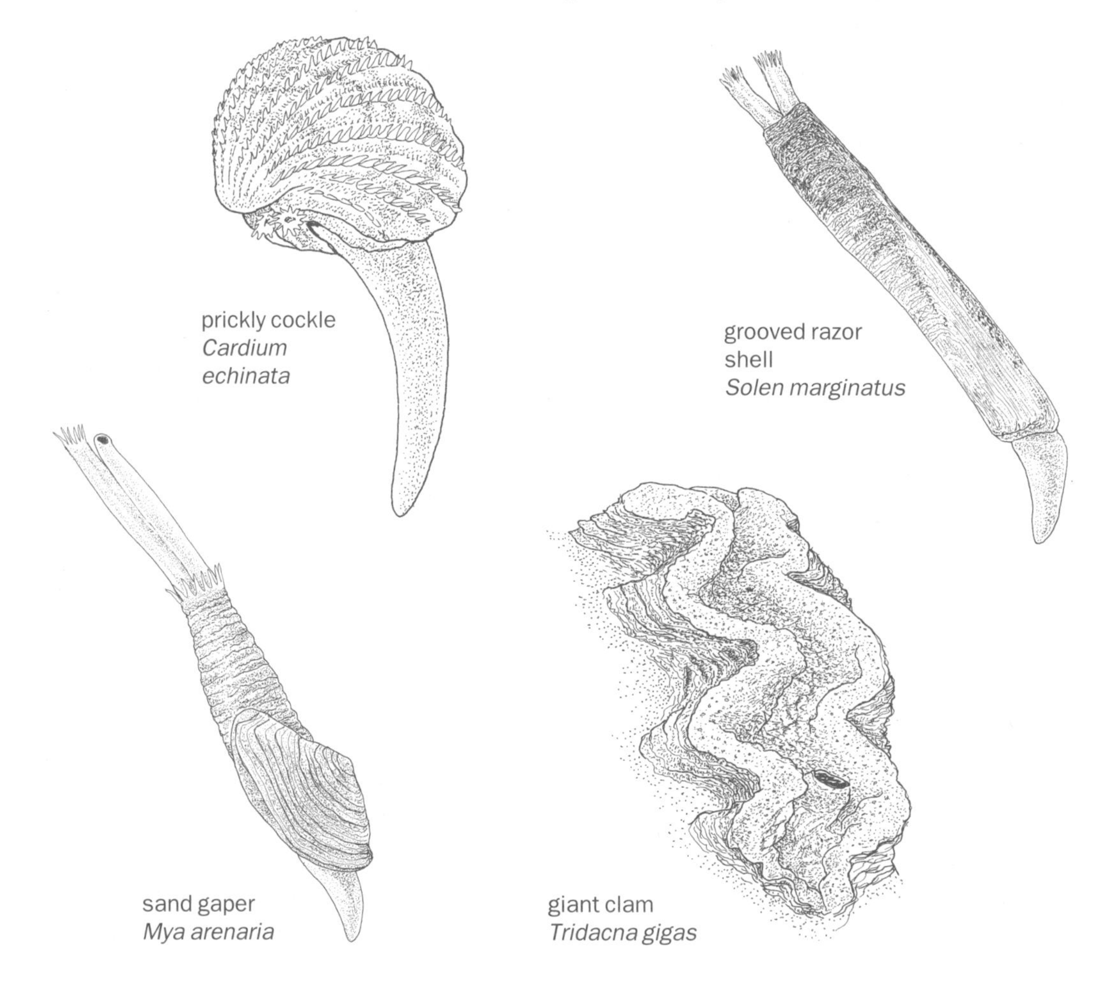
prickly cockle *Cardium echinata*
grooved razor shell *Solen marginatus*
sand gaper *Mya arenaria*
giant clam *Tridacna gigas*

includes the Cardiidae, the cockles, such as *Cardium*; Tridacnidae, the giant clams, such as *Tridacna*; Solenidae, which are the razor shells, such as *Solen*; and Veneridae, the venus clams, like *Chione*. Order Myoida includes thin-shelled burrowing forms with well-developed siphons, such as the Pholadidae, the piddocks, like *Pholas*; the Teredinidae, the destructive shipworms, like *Teredo*; and the Myidae, which are the gaper clams, such as *Mya*. Order Paleoheterodonta includes the family Unionoideae. which contains the freshwater swan mussel, *Anodonta*.

From snails to sea hares: class Gastropoda

The outstanding feature of the gastropods is **torsion**: at some point in development the part of the body containing the internal organs, the visceral mass, rotates through 90 to 180 degrees relative to the foot. As a result, the gut and nervous system are twisted, and the mantle and mantle cavity typically come to lie over the head. As so often seems to be the case, however, some groups have reversed what seems to be a principal evolutionary trend and have untwisted themselves (a process known as **detorsion**). In addition to torsion (but a quite different and unrelated phenomenon, a device to accommodate a mass of viscera in a basically tubular shell) gastropod shells are typically coiled; although many groups, independently, have greatly reduced their shells or lost them all together. Gastropods typically have a large, muscular, creeping foot which they can withdraw into their shells (if they have shells) and they then, typically but not invariably, close the opening to the shell with a lid, the **operculum**. The head is equipped with **statocysts** (organs of balance) and eyes, although the eyes are often reduced or lost. Gastropods do, however, have one or two pairs of tentacles, plus a radula and a crystalline style (which, again, is lost in many predatory groups). Land snails, in particular, have lost the ctenidia, and instead typically breathe via blood vessels in the mantle, which thus forms a lung.

In traditional classifications, as still found in most textbooks, the gastropods were divided into three subgroups (which for convenience we can regard as subclasses). The subclass Pulmonata contained the land-living snails and slugs; the Opisthobranchia included the nudibranchs and sea hares; and the Prosobranchia contained all the rest—including the obviously primitive limpets but also types that seem far more derived, such as periwinkles (Littorinidae) and whelks (Buccinidae).

But, as shown in the tree here, this traditional picture has been transformed in the past few years; as described by Winston Ponder and David Lindberg in *Origin and Evolutionary Radiation of the Mollusca* (1996). Most obviously, the three original subclasses have grown into five; and although two of the original subclasses retain their names (the Pulmonata and the Opisthobranchia) they do not retain their original Linnaean ranking. Most importantly, however, the members of the old 'Prosobranchia' have been completely rearranged. Now, the true limpets are placed in their own subclass, the Patellogastropoda, which is seen as the (primitive) sister to all the rest. The rest of the old 'prosobranchs' are distributed among the other four of the new subclasses: Vetigastropoda, Neritopsina, Caenogastropoda, and Heterobranchia.

The five gastropod subclasses

The subclasses Caenogastropoda and Heterobranchia are, as you see, the two most-derived sister groups. The Heterobranchia now embraces one group of refugees from the old Prosobranchia—here represented by the sundial shells, in the family Architectonicidae; and also includes the two remaining groups that were previously given subclass status—the Opisthobranchia and the Pulmonata. These two groups can now be seen as superorders, because each of them contains several widely acknowledged orders, embracing many families. But the two superorders are now seen to be much more closely related to each other than, say, winkles are to limpets, even though snails and sea slugs were once given separate subclass status whereas winkles and limpets were bundled into the same subclass.

• Subclass **Patellogastropoda** includes the true limpets, now seen as sisters to all the rest. Patellidae is a typical family, including the familiar European limpet *Patella*. Limpets have simple, cone-shaped shells, which, as far as can be seen, have always been simple cones. This is not an example of detorsion; they are not descended from coiled-shelled ancestors. This is one of the features that makes them clearly different from all the rest.

common limpet
Patella vulgata

• Subclass **Vetigastropoda** includes the common top shell *Trochus* in the family Trochidae—small examples of which are probably mistaken for periwinkles. It also includes several types with peculiar perforations in their shells: like *Fissurella*, the key-hole limpet, in the Fissurellidae; and *Entemnotrochus*, the slit shell, in the Pleurotomariidae.

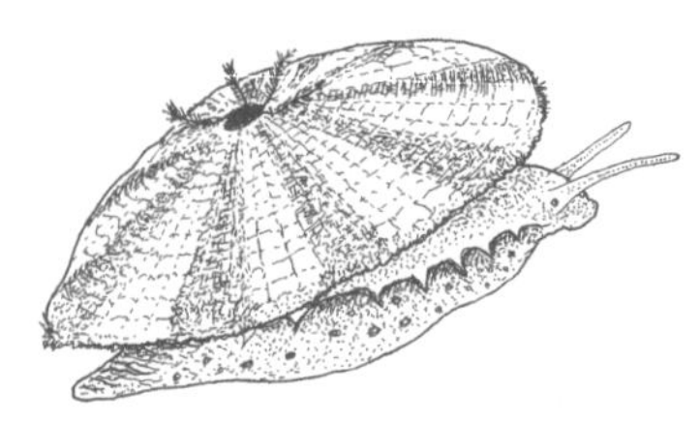

keyhole limpet
Fissurella maxima

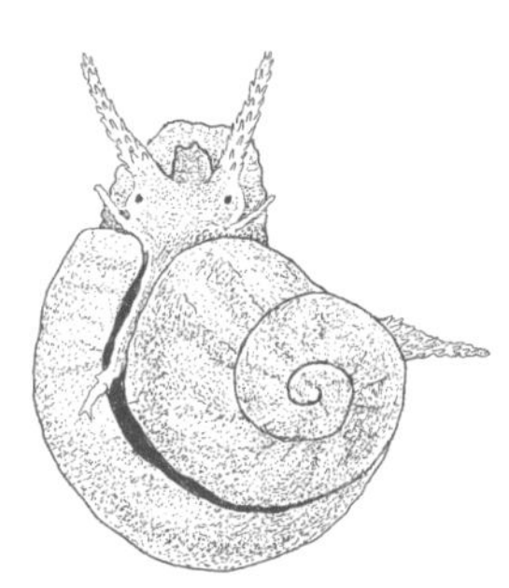

slit shell
Scissurella costata

top shell
Trochus niloticus

• Subclass **Neritopsina** is the sister to the Caenogastropoda + Heterobranchia. *Nerita*, commonly called a nerite in the family Neritidae, is a good example.

bleeding-tooth nerite
Nerita peloroata

• Subclass **Caenogastropoda** includes a huge array of mostly marine gastropods: the ceriths, such as *Cerithium*, in the family **Cerithiidae**; the periwinkles, such as *Littorina*, in the

giant knobbed cerith
Cerithium nodulosum

periwinkle, *Littorina littorea*

murex, *Murex brandaris*

textile cone, *Conus textilis*

Florida fighting conch
Strombus alatus

slipper limpet (in a stack)
Crepidula fornicata

tiger cowrie, *Cypraea tigris*

king helmet, *Cassis tuberosa*

common whelk
Buccinum undatum

netted dog whelk
Nassarius reticulata

bat volute, *Voluta*

Littorinidae; the murex, *Murex*, in **Muricidae**; the cone shells, such as *Conus*, in **Conidae**; the wonderful conches, *Strombus*, in **Strombidae**; the familiar slipper limpets, like *Crepidula*, in **Calyptraeidae**; the cowries, *Cypraea*, in **Cypraeidae**; the helmet shells, like *Cassis*, in **Cassididae**; the whelk, *Buccinum*, and *Nassarius*, the dog whelk or mud snail, in **Buccinidae**; and the volutes, *Voluta*, in **Volutidae**.

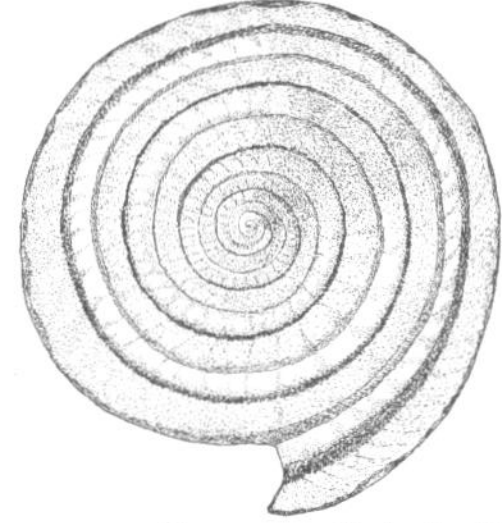
perspective sundial shell
Architectonica perspectivum

• Subclass **Heterobranchia** includes the sundial shells such as *Architectonica*, in the family **Architectonicidae**. But most of the heterobranchs belong either to the **Opisthobranchia** or the **Pulmonata**. I have shown these two groups as superorders because each of them contains a great many families.

Sea slugs and their kind: superorder Opisthobranchia

In opisthobranchs, the body is generally uncoiled (detorted) to some extent and the shell is either external or internal; but in many different lineages, independently, the shell has been reduced or lost all together. Many opisthobranchs are extremely beautiful and brightly coloured. Some do not crawl like other gastropods but glide through the water by undulations of their extended mantles, which have been compared to the skirts of a Spanish dancer. Some capture the stinging cells (nematoblasts) of cnidarians and incorporate them into their own surface tissues; a fine protection. Traditional classifications recognize 100 or so families of opisthobranchs, arranged among nine orders. These orders include the **Anaspidea**, the sea hares; **Nudibranchia**, the 'true' nudibranchs; **Thecosomata**, the shelled pteropods; and **Gymnosomata**, the naked pteropods.

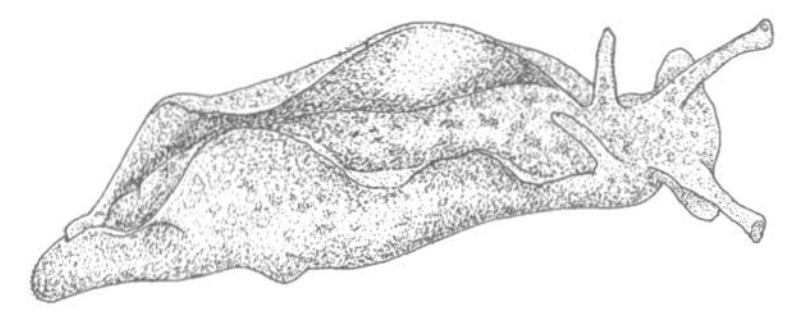

sea hare, *Aplysia punctata*

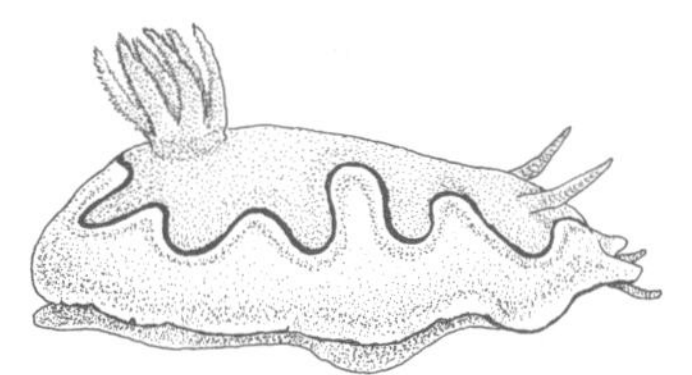

nudibranch, *Chromodoris coi*

Snails and slugs: superorder Pulmonata

Among the pulmonates are the familiar snails and slugs of gardens. They have lost their ctenidia (with one possible exception) and their mantle cavity forms a lung with a contractile opening. They are hermaphrodites, usually without larvae, that live mainly on the land or in fresh water although a few are marine. Most have shells, which are coiled; but various groups have lost their shells.

The Bruscas describe three pulmonate orders. The **Archaeopulmonata** are primitive pulmonates that are mainly littoral (living on the shore), like *Otina*. Archaeopulmonates have spirally coiled shells and generally no operculum. The order **Basommatophora** includes intertidal and freshwater forms, including the freshwater limpets. Finally, and by far the most importantly, the order **Stylommatophora** includes the 15 000 or so recognized species of snails and slugs, which live on land and have eyes on the tips of sensory and retractible stalks. Well-known genera are *Helix*, which includes the garden snail and the Roman snail (which is pressed into service as escargots); *Cepaea*, made famous for the genetic studies of its extraordinarily various shells, some plain and some stripy; *Limax*, which is the typical

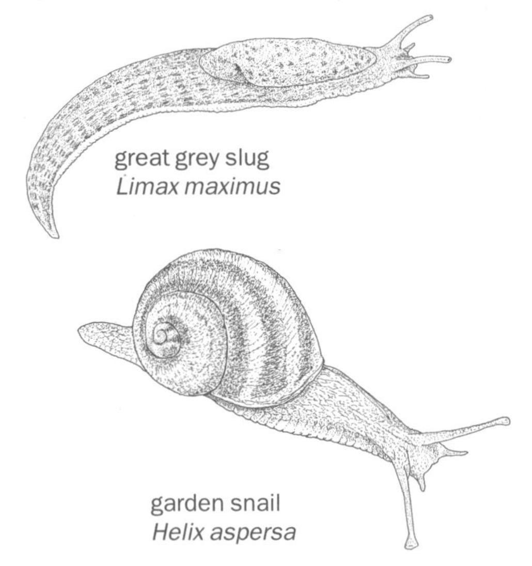

great grey slug
Limax maximus

garden snail
Helix aspersa

garden slug; and *Achatina*, a giant, edible land snail, made famous by its disastrous introduction to Tahiti and the neighbouring Society Islands, where it became a major pest of local crops—so numerous that growers commonly collected wheelbarrows full from small gardens. Cavalier government biologists then introduced a predatory snail *Euglandina*, to clear up the *Achatina*, but it went on instead to annihilate the local snails of the genus *Partula*. Introduction is a dangerous game.

Squids, octopuses, and other aristocrats: class Cephalopoda

It is no longer considered respectable to speak of 'pinnacles of evolution'; but, if it were, then we would have to say that the Cephalopoda, also sometimes called Siphonopoda, occupy a pinnacle among animals alongside the arthropods and vertebrates. Many of the 650 or so living species are big, swift, and intelligent: glorious creatures. Primitively, cephalopods do have shells whose structure is unique to themselves; it is a series of closed chambers, increasing in number with age, and the animal occupies the youngest chambers. In living cephalopods the shell is coiled, but some fossil types had straight, multichambered shells, like dunce's caps; and some of those ancient types were huge, with shells several metres long. In many living cephalopods, however, the shell is reduced or absent.

Cephalopods have large heads, with the mouth surrounded by muscular tentacles, sometimes with suckers and/or hooks and sometimes not: the name 'cephalopod', of course, means 'head-foot'. The mouth, tucked out of sight within the encircling tentacles, is equipped with a horny (chitinous) and sometimes calcium-reinforced (calcareous) beak, like a parrot's; and also with a radula. They have big camera-like eyes, superficially like our own (but differing markedly in basic design). Part of the mantle forms a muscular siphon, through which water is forced to provide jet propulsion: a rare form of locomotion in nature. Many protect themselves by extruding a cloud of ink from the **ink sac**, and many can change colour (sometimes as camouflage and sometimes by way of communication) by action of pigment-filled cells called **chromatophores**. All are marine, either swimming (pelagic) or living at the bottom (benthic). They are significant predators—for example, of crabs—and they, in their turn, are significant prey of sperm whales.

The three cephalopod subclasses

There are two living subclasses, **Nautiloidea** (*Nautilus*) and **Coloidea** (cuttlefish, squids, octopuses, and vampire squids). But, on the tree, I have also painted in the **Ammonoidea** as a third subclass, because although ammonites are long extinct (and their exact relationship to living types can never be ascertained by molecular means) they are extremely common in the fossil record and for a long time were highly significant creatures. Because this book acknowledges eurypterids, placoderms, dinosaurs, and 'seed-ferns', it must also recognize ammonites.

Nautilus: subclass Nautiloidea

pearly nautilus
Nautilus pompilus

Subclass Nautiloidea probably arose in the Ordovician, around 450 million years ago, and around 17 000 fossil species have been described; but now there is only one genus, *Nautilus*, with six known species in the Indo-Pacific. They are beautiful creatures with a many-chambered shell coiled like a crozier and 80–90 suckerless tentacles, which are protected by a fleshy hood; four tentacles are modified in males as agents of copulation. Their beaks are both chitinous and calcareous. They have two pairs of ctenidia, which accounts for the alternative name of the taxon, Tetrabranchiata ('branch' also meaning gill). Nautiloids lack an ink sac or chromatophores; their eye is like a pinhole camera, without cornea or lens; the statocyst is simple; and the nervous system is diffuse. Taken all in all, nautiloids put up a good show of being primitive.

Ammonites: subclass Ammonoidea†

The ammonites resembled nautiloids at least superficially, and were clearly to some degree related. Although ammonites are now extinct they appeared first after the nautiloids, and, after the Devonian (around 350 million years ago), they largely replaced them. They clearly became abundant; their coiled shells are among the few fossils known to everyone, turning up by the million on many a beach. Yet, unlike nautiloids, ammonites failed to survive beyond the Cretaceous–Tertiary boundary, 65 million years ago.

ammonite
Peltoceras athleta

Cuttlefish, squids, and octopuses: subclass Coloidea

Subclass Coloidea includes the cuttlefish, squids, octopuses, and vampire squids. Some have a shell but when they do it is generally reduced, and in many it is absent all together. The head and the foot are merged into a single structure. They have either 8 or 10 tentacles, 2 of which in males are modified for copulation. They have only one pair of gills, which accounts for their alternative name of Dibranchiata—in contrast to the Tetrabranchiata (the four-gilled nautiloids). Their beaks are chitinous. Their eyes and statocyst are complex and their nervous system is well developed and concentrated. They have both chromatophores and ink sac. They look, and are, of a higher grade than the nautiloids. There are four coloid orders: Sepioidea, Teuthoidea, Octopoda, and Vampyromorpha.

- In the order **Sepioidea**, the cuttlefish, the body is short and flattened and the animals swim using lateral fins. If there is a shell at all it is internal; commonly taking the form of the flat, calcareous cuttlefish 'bone' that serves the living animal as a regulator of buoyancy, like the swim-bladder of a teleost fish. Cuttlefish 'bone' also

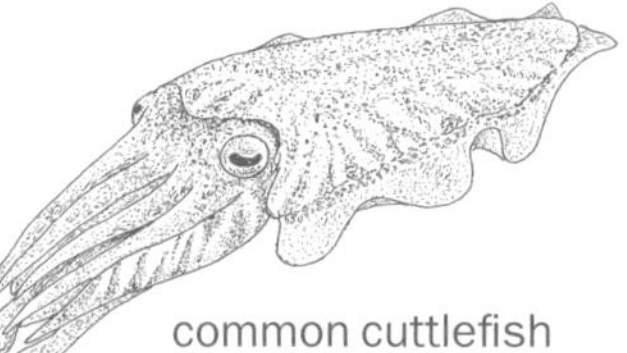
common cuttlefish
Sepia officinalis

turns up in abundance on beaches and is given to budgerigars to trim their beaks. Other cuttlefish have internal shells that are coiled and chambered, nautilus-like. Cuttlefish have eight coiled arms with two long tentacles, which bear suckers only on the spoon-like tips.

• Order **Teuthoidea** contains the squids. They have long, tubular bodies, again with lateral fins, and the internal shell is reduced to a flattened proteinaceous rod. Squids have eight arms and two extra non-retractable tentacles, making 10 appendages in all and accounting for the alternative ordinal name, Decapoda. The tentacle suckers often have hooks. Only the cetaceans and the biggest fish match the biggest squids for size and speed.

giant squid, *Architeuthis dux*

• The octopuses are in the order **Octopoda**; their short round bodies are usually without fins, and the shell is usually vestigial or absent; but female argonauts secrete a papery, coiled egg-case. Octopods have eight arms joined by a web of skin. Most of the 200 or so known species live on the sea bottom.

common octopus, *Octopus vulgaris*

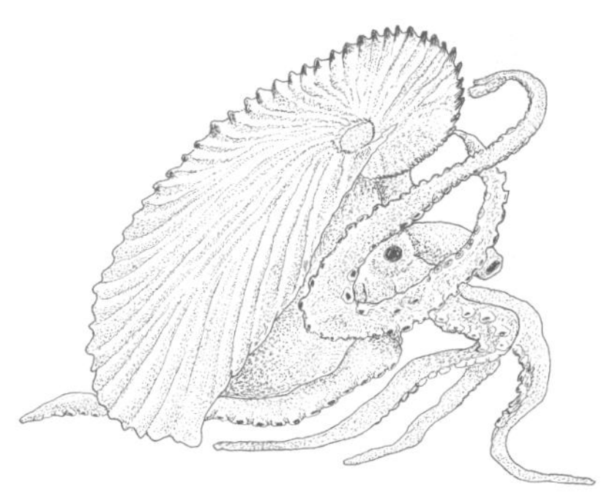

paper nautilus, *Argonauta pacifera*

• The sinister-sounding vampire squids are in the order **Vampyromorpha**. Their bodies are plump with one pair of fins, and the shell is reduced to a leaf-shaped, transparent vestige. They have eight arms, each with one row of suckers and all joined by an extensive web of skin; and they also have two tendril-like filaments,

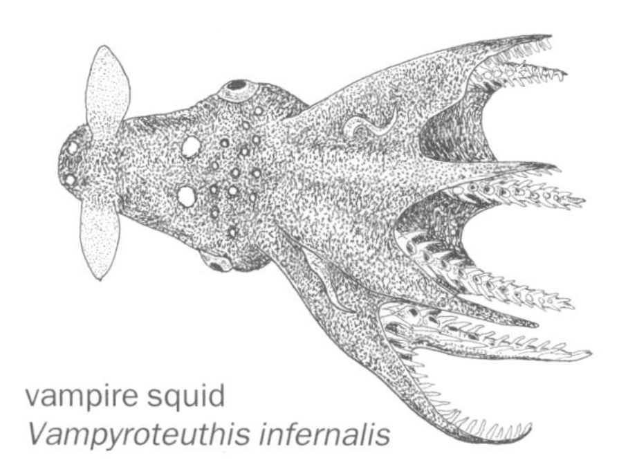

vampire squid
Vampyroteuthis infernalis

representing a fifth pair of tentacles, which they can retract. Vampire squids mostly live in deep water.

So much for the molluscs. The other truly outstanding phylum of protostomes is the arthropods, which occupy the next four sections.

8

ANIMALS WITH JOINTED LEGS

PHYLUM ARTHROPODA

MITES AND BARNACLES, robber crabs and fleas, horseshoe crabs and 'sea spiders', extinct trilobites and giant eurypterids: no other phylum of animals is so various as the arthropods, or so speciose, or compares in biomass, or has had a greater impact on the ecology of this planet over the past 600 million years since their first possible appearance in the Precambrian.

Trilobites teemed in the oceans from the start of the Cambrian, around 545 million years ago, until they petered out about 300 million years later; but then, and still, their maritime niches and many more besides were taken over by crustaceans. On land, insects and arachnids—best known for the spiders—are the most ubiquitous and often the most conspicuous of animals. Insects, notably beetles, include about a quarter of all known species of organism (although, as noted in Part I, Chapter 1, the number known must fall short of the real number by orders of magnitude, and the greatest diversity undoubtedly lies among the microbes). The scorpion-like eurypterids, which some zoologists see as extinct arachnids and others regard as sisters to the arachnids, were sometimes as big as a small rowing boat and for a time (until the jawed fishes got into their stroke) were the most formidable predators on Earth. The horseshoe crabs, the arachnids' closest living relatives, still swarm in their millions along the coasts of North America and Asia to remind us of arthropod life in its primeval days, long before the age of dinosaurs. Horseshoe crabs, eurypterids, and arachnids between them make up the great arthropod class of the chelicerates. Myriapods—centipedes and millipedes and their more recondite relatives—are less varied and of less ecological significance than chelicerates and insects; but centipedes are considerable predators nonetheless, and millipedes get through significant quantities of dead and decaying vegetation.

We often treat arthropods as pests; insects and mites, after all, are among the most potent destroyers of crops and bearers of disease. But if we removed all arthropods from the surface of the globe then all the most familiar ecosystems would collapse within weeks. Long chains of creatures would be robbed of their food supply; thousands of plants would be without agents of pollination; and—most importantly —the world would soon be littered with the festering remains of everything that dies, for the arthropod detritivores are among the most significant agents of organic recycling. The humblest creatures are in some ways the most significant (although the lines of influence within any ecosystem operate from the top down—from elephants down to insects—as well as from the bottom up).

In addition to the well-known groups, ancient sites dating back to the mid-Cambrian period, such as the Burgess Shale of Canada, have yielded a host of ancient arthropods or arthropod-like creatures whose relationships to each other and to established arthropods is far from clear. I will not discuss them here precisely because their position is so uncertain. Weird and wonderful though they were, however, there is nothing particularly mysterious about those Burgess Shale animals as has sometimes been claimed. When arthropods first appeared with their hard, protected bodies and their efficient and versatile limbs they were a revolution—a brand new body plan with immense potential to diversify in all kinds of directions; and in those early seas, with little serious competition, they did indeed radiate almost without restraint. In the end, however, just a few of the early lineages proved more efficient than the rest and they went on to radiate in their turn, eventually replacing all of the earlier, experimental types. But this same kind of pattern—an early radiation that produces a huge variety of forms, of which only a few come through—is seen in virtually all major groups: bony fish, reptiles, mammals, birds, hominids, flowering plants, what you will. There is nothing especially unusual about the Burgess Shale arthropods: we can expect extravagances at the base of new lineages.

Although they are so immensely varied, arthropods do seem to share a single, unifying body plan, and they have been recognized as a discrete phylum since the mid-nineteenth century. Most obvious is their body-armour—the tough outer **cuticle**, which is made from hardened protein ('scleroprotein') plus **chitin** (a polysaccharide toughened by links of nitrogen) and forms a complete exoskeleton that extends into the front and rear of the gut. The cuticle is hardened in places to form plates or **sclerites**, which are linked by flexible membranes that allow them to articulate. The typical arthropod body is clearly segmented or **metamerized**. When all the segments are the same, the body (or part thereof) is said to be **homonomous**; and when the segments vary, it is **heteronomous**. But even the most-homonomous-looking arthropods tend to be heteronomous to some extent; none is homonomous to the extent of, say, lugworms or earthworms. In most arthropods, too, the body is clearly **tagmatized**, meaning that different regions, or **tagmata** (singular, tagma), are specialized for different purposes—as in the head, thorax, and abdomen of insects.

Primitively, and typically, each arthropod segment carries a pair of jointed **appendages**. On the head these are modified to form **mouthparts** and sometimes **antennae** or **palps,** or what you will. In the trunk or thorax they form limbs and sometimes gills, and on the abdomens of crustaceans, insects, and chelicerates may be pressed to various other services (although they are often missing from some or all of the abdominal segments). In short, such rows of appendages offer the versatility of a Swiss army knife.

The body space between the gut and body wall, the coelom, which is typical of all animals above the level of the flatworm, is largely replaced in arthropods by a **haemocoel**; that is, there is still a space, and it is just as roomy as a coelom, but it arises in the embryo via a different developmental route. But whereas the coelom of an annelid, say, is divided at each segment to form a series of chambers, the haemocoel of arthropods forms just one (or a few) continuous blood-filled cavity—like a hull without bulkheads. In many arthropods, the blood is wafted around the haemocoel at low pressure by an open-ended heart; but others (typically those that are active and large, like horseshoe crabs, the house centipede *Scutigera*, and many decapod crustaceans) have well-organized circulations, with vessels. The arthropod nervous system is like that of annelids and molluscs, with a dorsal 'brain' and paired, ventral nerve cords (reflecting Geoffroy St Hilaire's notion that an arthropod is like a vertebrate lying on its back).

Five distinct groups of animals are generally recognized as the major arthropod taxa: the extinct Trilobitomorpha*; the Chelicerata, which include the horseshoe crabs (Xiphosura) and the Arachnida; the Crustacea*; the Myriapoda; and the Insecta. Myriapods are commonly supposed to be sisters to the insects and even to be ancestral to the insects; but these suggestions are now widely disputed, as I discuss later.

Two groups that have proved difficult to classify in the past are the peculiar Pycnogonida or 'sea spiders'; and the parasitic Pentastomida, whose name literally means 'five mouths' although they are commonly known as 'tongue worms'. But the pycnogonids now seem to be perfectly good chelicerates (but are shown in the tree here conservatively as sisters to the chelicerates), while the pentastomids seem to be highly specialized crustaceans (conservatively shown here as sisters to the crustaceans).

But there are two more living groups of arthropod-like creatures that are still difficult to place: the Onychophora or 'velvet worms', and the Tardigrada or 'water bears', which both have lobe-like feet and are commonly called 'lobopods' although they clearly are not closely related to each other. In the present state of uncertainty, it seems proper to treat the Arthropoda, Onychophora, and Tardigrada as an unresolved trichotomy: in other words, the three groups do seem to be related, but nobody knows who is more closely related to whom. Nielsen invokes the term 'panarthropod' to denote the Arthropoda, Onychophora, and Tardigrada together; and this is the course adopted in the present tree.

Some traditional treatments also show a larger taxonomy, linking the panarthropods, molluscs, and annelids. But research from the University of California at Los Angeles (UCLA), described in Sections 5 and 7 (pages 200 and 229), now shows that molluscs and annelids on the one hand and panarthropods on the other belong to different branches of the protostome tree (as shown in 5a, pages 188–9). It seems reasonable to presume that molluscs + annelids and panarthropods each arose from some flatworm.

Sections 9–11 deal separately with the Crustacea*, Insecta, and Chelicerata; and as this leaves the myriapods short-changed, I mention them at the end of this section (pages 265–7). My discussions in the USA and Britain have proved to me that it would be impossible to produce a classification of arthropods that pleased all authorities (although the disagreements do not seem to run quite so deep as they did a few years ago). So the present tree should be seen as a conservative compromise: it lays out the problems (as reflected in the trichotomies) but does not presume to stab at the answers.

A GUIDE TO THE ARTHROPODA

Three fundamental questions still hang over the phylogeny and hence over the taxonomy of the phylum Arthropoda. First, and most basically: Are the arthropods truly monophyletic—do they really form a clade? Second, how are the arthropods related to other phyla? And, finally—the most intricate issue—how are the various arthropod groups related to each other?

ARE THE ARTHROPODS REALLY A CLADE?

Did the hugely successful body plan of the arthropods arise only once, or several times? Is the Arthropoda really a coherent, monophyletic phylum? Should we write 'Arthropoda', as the name of a formal, genuine taxon, or simply treat the 'arthropods' as a grade, like 'fish' or 'microbes'? This discussion has raged for decades. The issues are far from simple.

For instance, it has been extremely difficult to decide whether apparently comparable features in different arthropods have been inherited from a common ancestor, or have become similar through convergence: different lineages independently adapting in similar ways to similar environmental pressures. Thus, myriapods and insects both breathe through tubes called **tracheae**, which they are traditionally assumed to have inherited from a common ancestor. But the two might well have acquired their breathing tubes independently, as adaptations to life on land; after all, some spiders and isopods (such as woodlice, known in the USA as 'sow bugs') also have tracheae. Such examples abound to an extent that has prompted some biologists to argue that the apparent unity of the Arthropoda is an illusion: that the grand similarities that seem to unite them are in general due to convergence. In truth, it has been argued, the group we call 'Arthropoda' is compounded from three separate phyla, which arose independently from different worm-like ancestors. Let me explain the points for and against.[1]

MONOPHYLY VERSUS POLYPHYLY

The chitinous exoskeleton, the jointed limbs, and the haemocoel are merely the most conspicuous of the features that arthropods have in common. There are many more. For example, where annelid worms have their mouths at the front, arthropods have moved the mouth underneath (to a ventral position) and a little rearwards, and have three 'pre-oral' segments in front of it. Presumably the first arthropods acquired ventral mouths so as to sweep up their food from the sea bottom—but why do all arthropods have *three* pre-oral segments if the different arthropod groups arose

[1] The traditional zoological case for both monophyly and polyphyly was excellently described by Pat Willmer of the University of St Andrews in her book *Invertebrate Relationships* (1990).

separately? Again: at least some members of the main arthropod groups have **compound eyes**, usually in addition to simple ones: crustaceans, insects, trilobites and—among the chelicerates—horseshoe crabs.

The similarities extend to details. Thus the cuticle generally has three layers, and each of the two innermost layers are themselves layered, with stacks of chitin rods on top of one another, each slightly angled to the one below and above. All arthropods grow by moulting their exoskeletons, and moulting is triggered by steroid hormones called **ecdysones**. Arthropods lack cilia, though these are a common feature of most other animal phyla. All arthropod muscles are striated—like, say, the human bicep; none is 'smooth', without striations, as in the human gut.

Long and impressive though it is, however, the catalogue of similarities among arthropods does not by itself demonstrate monophyly. As discussed in Part I (Chapter 3), the issue is not whether all the arthropods shared a common ancestor. Of course they did. At some point we ourselves shared an ancestor with the arthropods and the snails and indeed with oak trees. More importantly: Was that last comment ancestor an arthropod itself? If it was not: if it was simply a worm-like creature; and if that worm-like ancestor also gave rise to more 'worms' and other such beasties; and if each of the different arthropod groups arose from a different wormy descendant of the grand, common ancestral worm; then we would have to say that the various arthropod groups arose independently, from different wormy lineages. That would be polyphyly.

So, as always, the most significant question is not, 'What do the arthropods have in common?'. It is, 'What do they have in common that is unique to themselves?'. What, in fact, in the vocabulary of cladistics, are the synapomorphies, the unique derived characters?

WHERE ARE THE ARTHROPOD SYNAPOMORPHIES?

The list of uniquely arthropod characters is nothing like so impressive as the general list. The main arthropod feature—the chitinous cuticle—is shared by various other invertebrate groups, including annelids and nematodes. Furthermore, the cuticle of nematodes extends a short way into the foregut and hindgut, just as in arthropods. Nematodes also use ecdysone-like hormones to trigger their moults; so chitin, moulting, and ecdysones seem to go together. Even the particular pattern assumed by the chitin in the hardened sclerites may tell us little about ancestry. Chitin seems naturally to orientate itself in this way, just as salt naturally forms crystals of characteristic shape as a rock-pool evaporates to dryness. These are just quirks of chemistry; the way that materials of a particular kind behave when conditions are appropriate.

True, in arthropods the cuticle is thickened and hardened to form sclerites, which *are* unique to them. But is this such a tremendous deal? After all, only small genetic changes might in theory produce such thickening in the cuticle of a segmented worm; and as Pat Willmer points out in her book *Invertebrate Relationships*, there would have been tremendous selective pressure to acquire such reinforcement

8 · PHYLUM ARTHROPODA

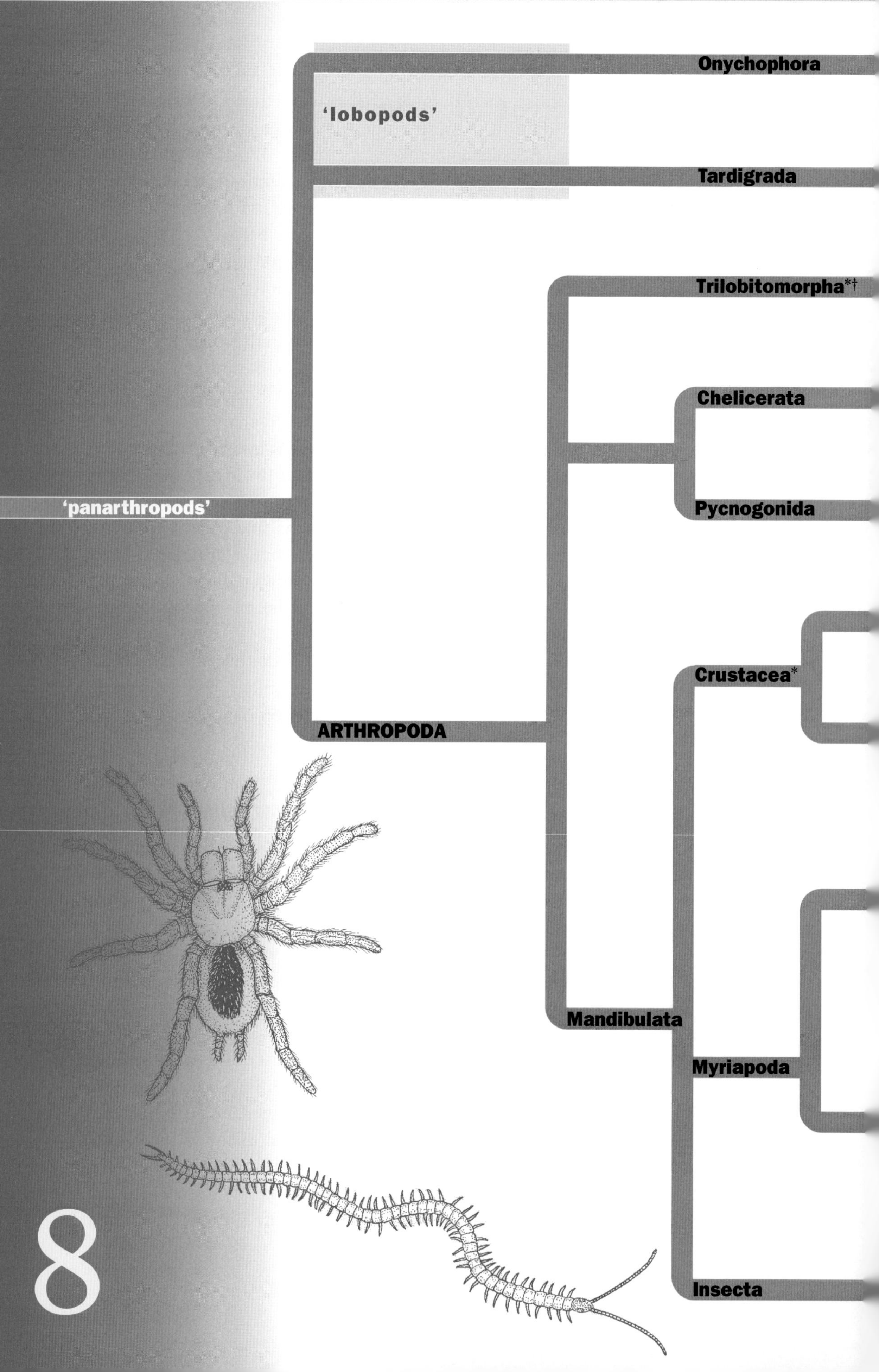

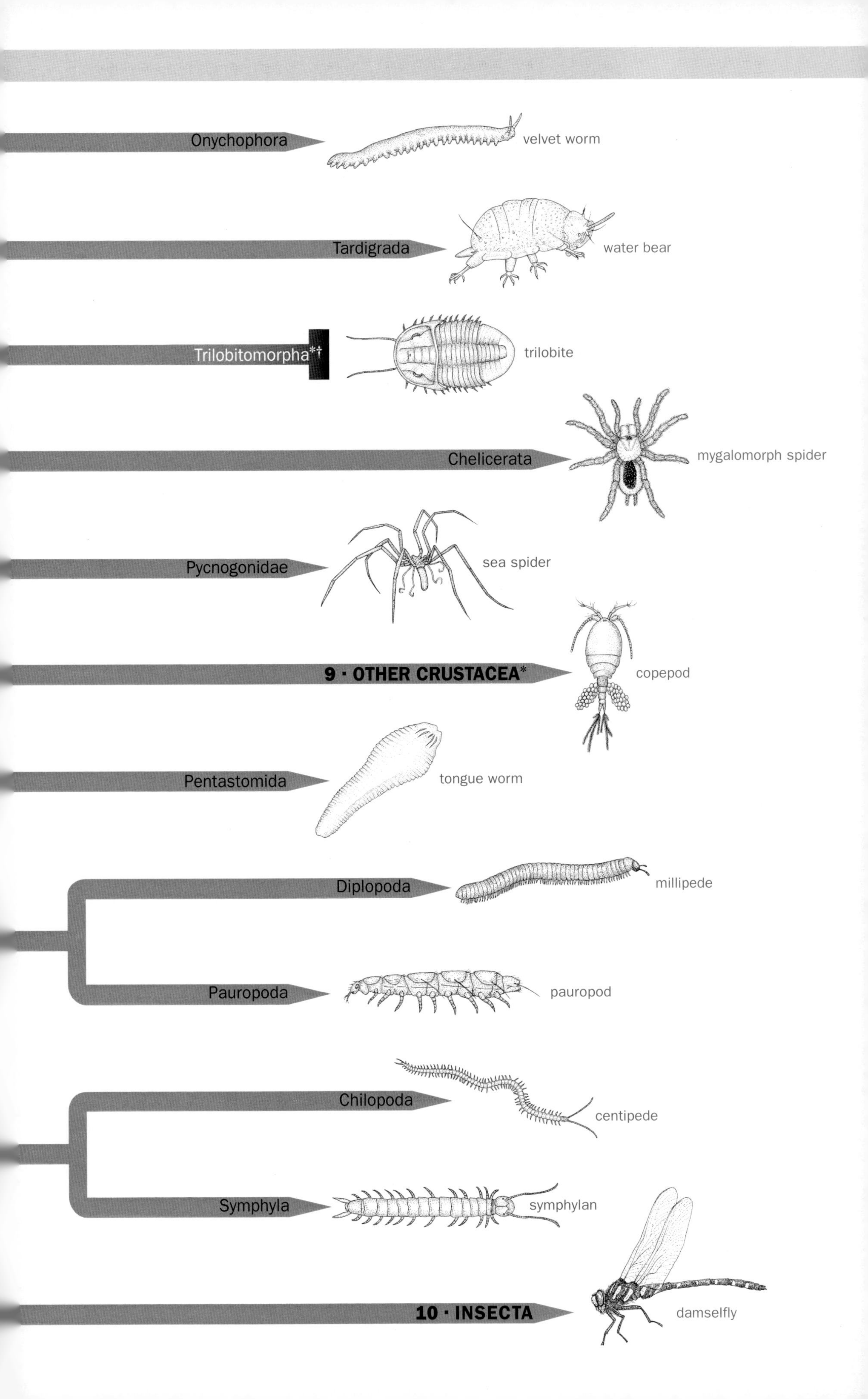
Onychophora
velvet worm
Tardigrada
water bear
Trilobitomorpha*†
trilobite
Chelicerata
mygalomorph spider
Pycnogonidae
sea spider
9 · OTHER CRUSTACEA*
copepod
Pentastomida
tongue worm
Diplopoda
millipede
Pauropoda
pauropod
Chilopoda
centipede
Symphyla
symphylan
10 · INSECTA
damselfly

at the turn of the Precambrian and Cambrian around 545 million years ago—for once one creature was armed and armoured, others would be obliged to follow suit. Entire battalions of segmented worms of different types may independently have toughened up—and hence become 'arthropodized'.

Then, again, many items on the list of shared features may all result from that one fundamental character: incarceration within the stiff, chitinous exoskeleton. Ecdysone hormone may simply be an automatic accompaniment; a necessary adjunct. When the surfaces are covered in cuticle, cilia cannot operate: hence, perhaps, their absence. (On the other hand, we might note that the loss of cilia is not a trivial thing, because in other animals they serve all kinds of functions, not least in reproduction, as they help to shuffle the gametes around. So the loss of cilia is sometimes taken as evidence *for* arthropod monophyly, because—the argument goes—this is not a trick that evolution would be likely to pull very often!)

The stiff exoskeleton might also explain why arthropod muscles tend to be striped; for striped muscle works best when extensibility is restricted (that is, when there is little room for manoeuvre). With a stiff exoskeleton, too, there is no need for the coelom to provide hydraulic support: so the internal compartments of the coelom can be lost—and it is better if they are, to equalize the pressures within. Hence the haemocoel. In short, given the chitinous armour, other features seem to follow. On the other hand, ultrafine exploration reveals intriguing contrasts between different arthropod groups. Thus, while insects harden their cuticles by 'tanning' with quinones, chelicerates including *Limulus*, the horseshoe crab, bind proteins together by disulphide links. Crustaceans adopt a third strategy: they use some tanning, but then add inorganic carbonate. The chemistry of ecdysone also varies from group to group. In short, as Pat Willmer comments, it is as if everything that can be different *is* different.

The greatest champion of arthropod polyphyly was the British biologist Sidnie Manton, who was most active in the 1970s. She focused on the features that seem most characteristically arthropod: the mouthparts and the limbs. She argued that despite appearances the appendages of different groups have a fundamentally different design; and, crucially, that it is impossible to derive the different designs from each other, or to derive all of them from a common, functional, ancestral form.

Take mouthparts. To be sure, crustaceans, insects, and myriapods all have jaws and in some traditional schemes are combined within the supposedly monophyletic 'Mandibulata' (as, indeed, I have combined them here). But, said Manton, the mouthparts of crustaceans on the one hand and insects and myriapods on the other have fundamentally different designs. Thus, she said, in crustaceans the mandible is constructed only from the base of the primitive mouth appendage to form a **gnathobase** (where 'gnatho-' means 'jaw'). By contrast, the mandibles of insects and myriapods are 'whole-limb' appendages: that is, they incorporate *all* the segments of the original, ancestral appendage. Manton also invoked the concept of 'plausibility'. As she said, it is extremely difficult to see how a gnathobase jaw could have evolved from a whole-limb jaw, or vice versa, if we assume that the intermediate types were themselves functional. It is also hard to envisage a common ancestral jaw that could have given rise to both kinds of jaw—assuming again that the early, intermediate types were functional. But, of course, the intermediate types must have been functional, or the lineage would have died out. Thus, she said, a close relationship between animals with gnathobase jaws and those with whole-limb jaws is simply implausible.

Indeed, Manton felt there could be no close relationship between crustacean appendages in general, and those of insects and myriapods. Thus, she said, crustacean limbs in their primitive state are **biramous**—meaning that each appendage has two branches, endites and exopodites. We can see this forked structure clearly and conveniently enlarged in the abdominal appendages of lobsters in whom one branch forms the walking leg and the other forms the gill, albeit tucked out of sight beneath the overhanging carapace. By contrast, the appendages of insects and myriapods are **uniramous**, with no sign of an exopodite. Again, Manton saw no plausible evolutionary route from one design to the other.

For these reasons, said Manton, the crustaceans on the one hand and the insects and myriapods on the other could not have descended from a common, arthropod ancestor. Indeed they were different phyla. The Crustacea formed one phylum, and the insects and myriapods she placed in a new phylum 'Uniramia', named after the form of their appendages. She also included the Onychophora within the Uniramia, for she argued that they, too, had 'whole-limb' jaws and uniramous appendages. This last refinement remained controversial, however, for even some of her supporters were reluctant to link velvet worms with insects and myriapods. Finally, Manton grouped the chelicerates and trilobites together within a third 'arthropod' phylum. Chelicerate limbs appear uniramous but ancestral types seem to have been biramous. *Limulus* has distinct outer rami on its hindermost limbs, and so did the trilobites. But this 'outer ramus' arises from a different limb segment from that of the crustaceans, so, she said, the limbs of crustaceans are clearly not homologous with those of chelicerates + trilobites.

In Britain, at least, Manton had great influence. For a time, my favourite museum presented displays of 'Uniramia', Chelicerata (with trilobites), and Crustacea as separate phyla. But many remained unconvinced. Over the past few years, too, various lines of evidence, though not necessarily forming a coherent or totally convincing story, have nibbled away at some of her premises and conclusions. So most zoologists now seem happy to accept that the main arthropod groups—trilobites, chelicerates, crustaceans, myriapods, and insects—do indeed form a true clade.

THE ROUTE BACK TO ARTHROPOD MONOPHYLY

Since Manton's day (although it was not so long ago) a great deal of new evidence has come to light. In particular, there are new fossils, while many fossils that were already known—and many living creatures—have been studied more closely. Molecular evidence in general is mostly post-Manton; and, although these are early days, it is now becoming possible to explore homologies by examining the individual genes that code for the particular characters. If two apparently similar features in two different animals prove to be coded by what is obviously the same gene, then it is a very good bet that they are truly homologous (although not necessarily synapomorphous).

These new studies throw doubt on Manton's proposal that 'uniramian' jaws are vastly different from crustacean jaws. Thus, new anatomical research suggests that insect jaws are not after all compounded from the whole ancestral appendage, but only from basal segments. Genetic studies suggest, too, that the gene encoding the tip of insect appendages is not expressed in the mandibles (jaws), suggesting that the tip itself has not been incorporated in the jaw structure. In short, insects, like crustaceans, seem to have gnathobase jaws.

It seems, too, that the grand distinction between uniramous and biramous limbs may not be so fundamental as Manton maintained. Thus Jarmila Kukalová-Peck of Carleton University in Ottawa has looked afresh at some of the beautifully preserved fossils that have been collected in Russia and studied over decades. She now argues that the legs of primitive—ancestral—insects were not uniramous at all. They carried extra appendages ('exites'), which have simply been lost. Such losses are a common feature of evolutionary history: after all, the horse's one-toed hoof is clearly derived from the five-toed ('pentadactyl') foot of the primitive tetrapod.

This argument does not by itself restore the notion of arthropod monophyly, but it does remove one of the principal reasons for doubting such monophyly. For if Kukalová-Peck is right then there seems to be no outstanding reason to separate the crustaceans from the insects + myriapods (+ onychophorans), which Manton grouped together as Uniramia. The jaws of crustaceans and of 'uniramians' both seem to be essentially the same, and the appendages of each and all of them could well have evolved from some ancestral limb that was polyramous. Kukalová-Peck's argument has yet another twist. She suggests that all true arthropod legs derived originally from an 11-segment ancestral limb with branches on several or all segments. This does indeed suggest arthropod monophyly. Yet she also argues that the limbs of onychophorans do not share this 11-segment pattern. So onychophorans presumably had a different ancestor, and are not part of the monophyletic phylum Arthropoda.

The arguments against Manton are not solid. But general evidence does seem to suggest that Arthropoda really is a clade, and the burden of proof was always with Manton to show that it is not. If her principal arguments are undermined, then the traditional impression of arthropod monophyly seems to be re-established, if only by default. In short, Arthropoda has re-emerged as a true clade; and the Onychophora are left to form a separate, possibly sister phylum.

With arthropod monophyly restored, we can move on to the next broad area of discussion. How are the arthropods related to other kinds of animal?

ARTHROPOD RELATIONSHIPS: THE TARDIGRADES AND ONYCHOPHORANS

As mentioned in the introductory paragraphs, the onychophorans and tardigrades remain difficult to place, but do seem allied to the arthropods. Because they will receive no mention anywhere else, they merit a few paragraphs here.

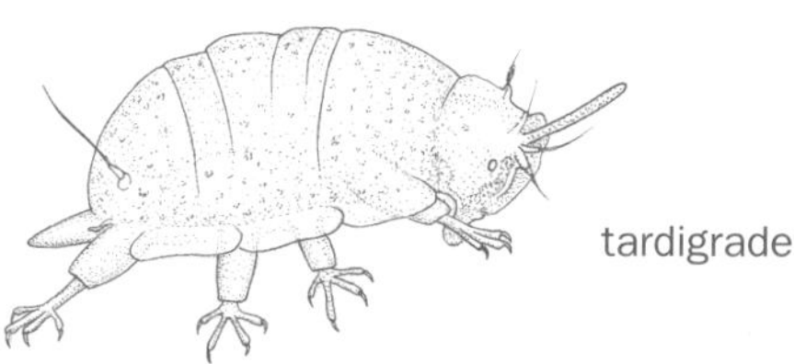
tardigrade

The **Tardigrada** are tiny—mostly under 0.5 millimetre, though one reaches 1.7 millimetres—rounded, eight-legged creatures known colloquially as 'water bears' because under the microscope they can indeed be seen to lumber along like aged grizzlies: *Tardus* is Latin for 'slow', and *gradus* is 'step'. But if the world is ever laid waste by ecological disaster or collision with yet another asteroid then the tardigrades are good candidates to carry the metazoan and more-or-less arthropod body plan through the crisis. In the wild they produce a thick-walled protective coating (cyst) that enables them to survive extreme desiccation—for decades, centuries, and perhaps even for millennia—and, experimentally, they have endured absolute alcohol, ionizing radiation, vacuum, and temperatures from –272 degrees to +149 degrees Celsius. Their natural habitats range from moss (many mosses are themselves great toleraters of dry conditions) to the deep ocean and even hot springs. Tardigrades were first described in the late eighteenth century (1773) and 400 species are now known in 8 families and 3 orders. They travel easily on the wind or on the feet of other animals, and some species are extremely widespread.

But what of their relationships? Like arthropods, their bodies are covered in a complex, multi-layered chitinous cuticle that is periodically moulted, and is thickened, mainly dorsally, to form plates—reminiscent of arthropod sclerites though not necessarily homologous with them. Like arthropods, too, their main body cavity is a haemocoel and their muscles are arranged in bands that are attached at points beneath the exoskeleton—unlike the continuous muscle sheets of annelids. But their legs are not jointed. They are hollow extensions of the body wall of the kind known as **lobopods**. In some, they are partially telescopic, and end in claws. Tardigrades may be 'degenerate' descendants of true arthropod ancestors or may simply represent a separate arthropod-like lineage. They are commonly given their own phylum, Tardigrada, and that is how they are treated here.

When the Reverend Lansdown Guilding first described an onychophoran in 1826 he supposed it was a slug—a mollusc; albeit a 'slug with legs'. The **Onychophora** are indeed like slugs or perhaps more like caterpillars, because they have between 14 and 43 pairs of 'legs'; although, with their textured, often bright-hued and iridescent surface they are also commonly known as 'velvet worms'. Eighty or so species are now known in two families, one of which lives throughout the tropics while the other is confined to the Southern Hemisphere. They are 1.5 to 15 or more centimetres long. Always they favour damp places.

But what exactly are they? No group has prompted more zoological debate. Some have said they are polychaete worms. Others suggest that they are 'degenerate' arthropods that have lost some arthropod features, notably the

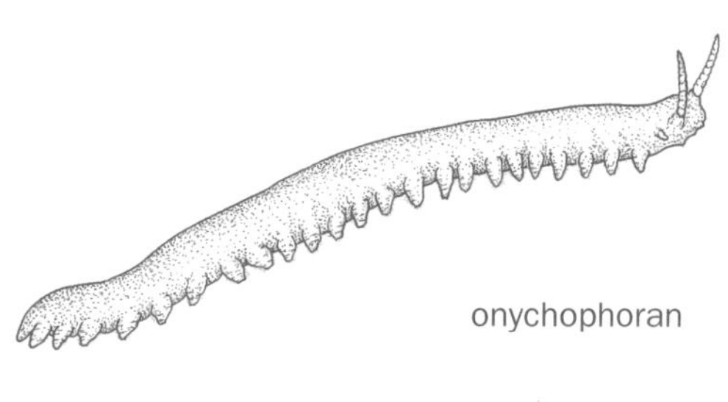
onychophoran

joint legs, as they adapted to a slug-like existence. As already mentioned, Sidnie Manton argued that they are close relatives of the myriapods and insects, and linked all three in a separate phylum, the 'Uniramia'. Often they are presented as 'intermediates' between annelids and arthropods, sometimes with an implication that they resemble arthropod ancestors.

Whatever their relationships, onychophorans do seem to combine annelid-*like* and arthropod-*like* features. Like annelids and quite unlike arthropods, their cuticle is only weakly sclerotized (thickened). Like many annelids they have unconvincing heads (that is, there is little cephalization) and their bodies are strongly homonomous. Their eyes also resemble annelids' and some of their organs have cilia. But, like arthropods, they moult their cuticle periodically as they grow. They also—like arthropods—have a large and well-developed brain (relatively speaking), they breathe via spiracles and tracheae, and they have a haemocoel around which the blood is swirled by a dorsal, open-ended heart. But the main interest focuses on the body appendages. Are the projections at the front homologous with arthropod antennae? More broadly, are the 'legs' homologous with those of arthropods? Some zoologists think so but others argue that they are simply extensions of the body wall: 'lobopods'. In any case the legs end in claws and hence the name: *onycho* is Greek for 'talon' and *phora* is 'bearer'.

The wrangling goes on. For now it seems safest to leave onychophorans in their own phylum—Onychophora—and to regard them as another experiment in arthropodization; perhaps an offshoot of the ancestors that gave rise to the 'true' Arthropoda, and perhaps an independent lineage. Molecular studies might resolve the issues, but preliminary results are confusing. For example, one such study places the onychophorans right in the middle of the arthropods; but most invertebrate zoologists simply feel this is a nonsense.

In practice, most authorities seem to accept that tardigrades and onychophorans are related to arthropods; but the nature of the relationship remains uncertain. Some suggest that onychophorans and tardigrades are themselves related, within a phylum that they call Lobopoda—'lobe feet'. But although both groups do have lobe feet they are very different in many other ways. Some feel that onychophorans are sisters to the arthropods, with tardigrades as sisters to the onychophorans + arthropods; but others feel that tardigrades have a special relationship with arthropods, and that the onychophorans are the outgroup. Because the issue is so unresolved I have simply presented a trichotomy on the tree. Such an arrangement does attest that either the onychophorans or the tardigrades could be sisters to the arthropods but does not say which, or propose any particular relationship between onochophorans and tardigrades.

Finally, when I was at school in the late 1950s we learnt with due solemnity and at great length that arthropods and annelids were close relatives, with onychophorans as obvious intermediates. It is indeed possible to draw a very plausible sequence between polychaete annelids, with their segments, appendages, and jaws, through the semi-armoured and semi-limbed onychophorans, to the arthropods. A general relationship between annelids and molluscs was also inferred (note, for example, the similarities of their larvae and their nervous systems). Overall, then, the onychophorans appeared as sisters to the arthropods; these two together were seen as sisters to the annelids; and this trio were sisters to the molluscs.

Here, though, we see yet another clash between traditional anatomical studies, and modern molecular research; and although the results of such contests are often equivocal (traditional scholarship is not instantly routed by molecular data), in this case it really seems that the molecules win. As described on page 200, and shown in tree 5a, the new studies from UCLA suggest that annelids belong among the lophophorates, whereas onychophorans are ecdyozoans. Traditional studies do give a clue to this; after all, onychophorans grow by moulting (ecdysis) and annelids do not. This detail, however, was traditionally thought to be less important than the overall similarity of body form. Now it seems that the 'detail' is all-important, while the overall similarity is just another, albeit remarkable, example of convergence.

So much for the relatives of the arthropods. What are the relationships within the Arthropoda?

HOW ARE THE ARTHROPODS RELATED ONE TO ANOTHER?

Zoologists over the years have generally arranged the five main arthropod groups, plus the onychophorans, in three main ways. (Tardigrades have not commonly featured in general discussions of arthropod taxonomy.)

First, as we have seen, Sidnie Manton proposed that the Arthropoda and Onychophora between them really represent three different phyla. The insects, myriapods, and onychophorans she placed in the Uniramia; Crustacea formed a second phylum; and trilobites with chelicerates formed the third. Most zoologists, however, both before Manton and after, believe that the arthropods are monophyletic, and that Arthropoda should indeed be recognized as a true phylum. This would mean that all the major groups within the Arthropoda—**Trilobitomorpha***, **Chelicerata**, **Pycnogonida**, **Crustacea***, **Myriapoda**, and **Insecta**—emerge as subphyla (although the pycnogonids should probably be absorbed within the chelicerates and the pentastomids within the crustaceans). So then we have to decide how the different subphyla are related.

Some have favoured the 'TCC tree', which divides the Arthropoda into two main branches. On one branch are the insects and myriapods; and on the other are the trilobites, chelicerates, and crustaceans (hence TCC). But others favour the idea that Crustacea*, Myriapoda, and Insecta are closely related, with the three forming a grouping (we can simply call it a 'division') called Mandibulata. There are common-sense reasons for supposing that the Crustacea* is ancestral to the other two. In both the mandibulate tree and the TCC tree the Arthropoda are presented as a true, monophyletic clade with the Onychophora as the outgroup.

I have explained that most zoologists now accept that arthropods are monophyletic after all; Manton's alternative suggestion was interesting but wrong. Now we have to choose between the TCC scenario, and the mandibulate idea. In either, we have somehow to accommodate the trilobites; so first let us look at them briefly.

The trilobites: Trilobitomorpha*†

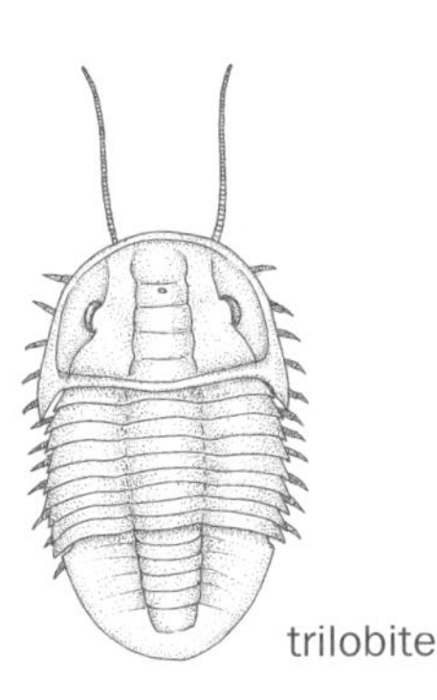
trilobite

The trilobites, like the dinosaurs, are icons of times long past. They flourished in the oceans of the Cambrian and Ordovician periods, beginning around 540 million years ago, with a few lasting until the Permian, 280 million years ago. They were extremely successful: abundant, widespread, and diversifying to produce almost 4000 known species. Most were bottom-dwellers only a few centimetres long, but some grew to 60 or 70 centimetres, and a few, less than 1 centimetre long and with spines that aided flotation, seem to have been planktonic. Most seem to have been scavengers or gobblers of detritus, but some at least may have been filter feeders and some perhaps were predators, lying in ambush in the sediment.

The trilobites were 'true', almost archetypal arthropods with a somewhat flattened segmented and armoured body (thicker on top than beneath) divided into three tagmata: the head or **cephalon**, usually with compound eyes; the **thorax**; and the **pygidium**, a series of apparently fused segments at the rear. Their name, which obviously means 'tri-lobed', derives from their apparent lengthwise division into three. The central 'lobe' is the body, and the two lateral lobes are outgrowths of the plates on the back, which form roofs over the appendages. These latter—typically 18 pairs or so along the thorax—are two-tined or 'biramous', though whether this biramousness is homologous with that of crustaceans or of primitive chelicerates has often been argued, without resolution.

For many millions of years, then, trilobites were by far the most important and ubiquitous arthropods around. They lived, too, throughout the crucial period just before the major modern arthropod groups are known to have appeared. They have many arthropod synapomorphies, such as compound eyes; but they also have the kind of primitiveness, including an impressive row of bifurcated appendages, that implies versatility—the potential to evolve into many other forms. As we have seen, Sidnie Manton linked trilobites with chelicerates; the TCC tree links them to chelicerates and crustaceans but not to myriapods and insects; and the 'mandibulate' tree (favoured here) links trilobites with everything else.

One problem has been that the existing subphyla really do seem to be very different from each other; so some zoologists find it easy enough to derive one or other of the moderns from the trilobites, but not all of them. But Robert Hessler at the Scripps Institution of Oceanography at La Jolla in California simply points out that the trilobites were extremely variable. Some of them were very chelicerate-like, and others were crustacean-like.

In fact, it seems sensible simply to regard the trilobites as a stem group. They initially radiated from some common clade founder, with different descendants evolving in many different directions: and some of those descendants survived to become the first chelicerates, and others survived to become the first crustaceans, and the rest went to the wall. If you insist on a strictly cladistic taxonomy, then you cannot admit the trilobites as a taxon at all, since—as the probable ancestors of chelicerates and crustaceans—they are not by themselves a clade. But if you take the more relaxed view, as I do in this book, and admit paraphyletic groups as taxa so long as we acknowledge that they are indeed paraphyletic, then we can reasonably refer to all the diverse trilobites as Trilobitomorpha*. The affix '- morpha' implies 'trilobite-like', which suitably reflects their diversity; and the asterisk denotes their paraphyly.

In short, trilobites seem to occupy the same position among the arthropods as the Amphibia* occupy among the tetrapods (except that some amphibians survive, whereas all of the trilobites are gone). In the tree here, the trilobite, chelicerate and crustacean + insect + myriapod lineages are presented as a trichotomy. But I assume that the stem creatures from which all these creatures arose—the handle of this three-tined fork—would themselves be identified as trilobites; or at least as 'trilobito-morphs'.

In principle, then, it seems easy enough to derive all living arthropods from trilobite-like ancestors. In other words, they could sit happily enough at the base of a TCC tree, or of a tree built around Mandibulata. They do not help us distinguish between the two. We must look more closely at the living groups.

RELATIONSHIPS OF LIVING ARTHROPODS

In truth, much traditional zoology and increasingly more modern evidence, largely molecular, suggests that **Mandibulata** is indeed a valid grouping—as shown in the tree. But, exactly how the three mandibulate groups are related to each other remains controversial, which is why, conservatively, I show them as a trichotomy.

Thus, common sense points to a broad relationship between crustaceans, insects, and myriapods. All, after all, have jaws and ostensibly comparable auxiliary mouthparts, and feel their way around by antennae. Chelicerates, which have chop-stick-like **chelicerae** instead of jaws and **palps** in place of antennae, seem obviously to be the odd creatures out. Now that Manton's argument has apparently been invalidated—the notion that crustacean jaws are not homologous with those of myriapods and insects—then there seems little reason to doubt that the jawed types do indeed form a clade, which we might as well call Mandibulata. That leaves two issues. First, what is the nature of the relationship between insects, myriapods, and

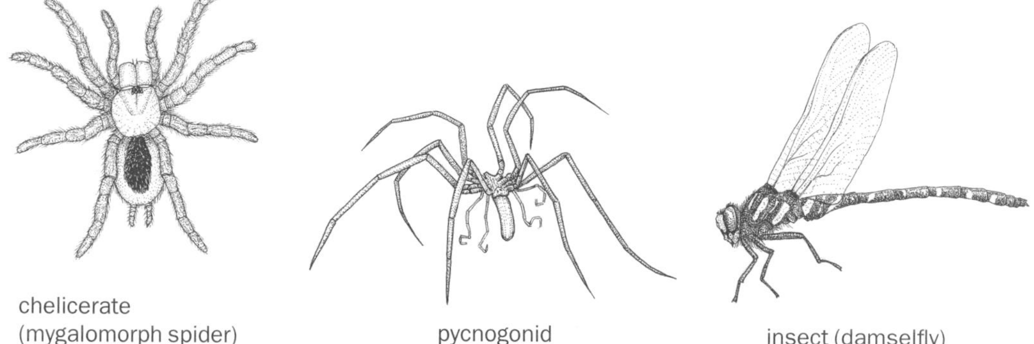

chelicerate (mygalomorph spider) pycnogonid insect (damselfly)

crustaceans? Second, do the insects and myriapods enjoy any special relationship?

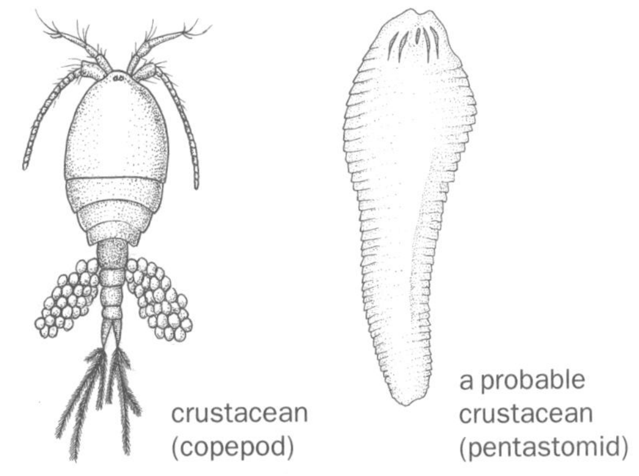

Both of these issues can become extremely complicated, once we start to follow them through. Yet we can at least keep the answer to the first of them fairly simple. If we accept that crustaceans, insects, and myriapods are indeed closely related, then it seems to follow perfectly reasonably that crustaceans—defined broadly—are the ancestors of the other two. After all, if we suggest that the three groups are related then we are also saying, by definition, that they all share a common ancestor. Then the question is, what did that common ancestor look like? The obvious answer is that it looked like a creature that any zoologist would say, without hesitation, was a crustacean.

There are various general reasons for supposing this. First, crustaceans appeared in the fossil record long before insects and myriapods did—and the time gap is about 100 million years, so it seems that they really did appear first even though the fossil record is notoriously unreliable. Second, crustaceans are fundamentally marine (none is truly terrestrial and only a minority live in fresh water) whereas insects and myriapods are fundamentally terrestrial. Insects and myriapods may not have been the first animals to venture on to land (chelicerates may well have beaten them to it); but the fossil evidence suggests that when they did come on to land they were already recognizable as insects and myriapods. Unless the fossil record is deceiving us, it follows that the ancestors of insects and myriapods must have been marine. When insects and myriapods came on to land marine arthropods were either crustaceans or trilobites—so crustaceans are certainly very plausible candidates. Then again, crustaceans are far more varied in body form than insects are, even though there are many more insect species. This is the kind of pattern you would expect if the insects were indeed a derived subset of the crustaceans.

Modern genetic studies are also suggesting a close relationship between crustaceans and insects, which supports the validity of the Mandibulata grouping. These include some delightful studies by Michael Akam and Michalis Averof at the Wellcome Laboratory in Cambridge, England. They studied the Hox genes of the brine shrimp, *Artemia*, which is, of course, a crustacean; and of various insects. As explained in Chapter 3 in Part I and again in Section 5 of Part II, Hox genes are the genes that are common to all animals and determine the fate of the different parts of the body during development. In arthropods, rows of Hox genes determine the fate of the body segments—the genes being arranged along the chromosome in the same order as the segments that they control. Akam and Averof found that the Hox genes of the shrimps and the insects are almost uncannily similar; and that each of the individual Hox genes did virtually the same job in the shrimp as it did in the insects.

In practice, it is not quite so easy as it seems to interpret Akam and Averof's initial results. For one thing, the body plans of insects are not so similar to those of the shrimp-like crustaceans as they might appear. In insects, for example, head and thorax are quite distinct, whereas in crustaceans they are commonly more or less fused into a cephalothorax; and the three segments of an insect thorax do not correspond in any simple way with the variable number of segments in a crustacean cephalothorax. I have also observed throughout this volume that similarity *per se* does not indicate special relationships. Perhaps, for example, the Hox genes of scorpions —which are chelicerates—may prove to be the same as shrimps' and insects', in which case the TCC hypothesis would look just as good as the mandibulate idea. We can at least say, however, that the genes that organize fairy shrimp bodies are astonishingly similar to those that organize insect bodies, which at least is compatible with the notion of special relationship.

In short, the economical idea—which fits the evidence—is that crustaceans are ancestral to insects (and to myriapods). If this is so then, by definition, the crustaceans are paraphyletic. If we insisted on a strictly cladistic taxonomy then we would be forced to say that the crustaceans cannot therefore be admitted as a taxon, with the formal name of Crustacea. But I suggest that this merely demonstrates yet again the folly of equating phylogeny with taxonomy. There is nothing wrong with paraphyletic groups provided we define them tidily, and provided we signal their paraphyly with an asterisk. Hence Crustacea* has the same noble status as Reptilia*: the great paraphyletic tetrapod group that went on to produce two new grades in the mammals and birds.

Indeed (this is speculative, but there is some evidence for it) insects and myriapods may well have evolved from crustaceans *after* crustaceans had divided into the various classes that we see today (as described in Section 9). In other words, insects and myriapods should probably be shown emerging from within the ranks of the Crustacea*, just as mammals and birds are shown emerging from among the ranks of the Reptilia* in Section 21. The only real difference is that we know which reptiles mammals are related to, and which particular dinosaurs are the sisters of birds, but we do not know which particular crustaceans are most closely related to insects and to myriapods. So at present, as in the tree, it seems safest simply to show the Crustacea*, Insecta, and Myriapoda as three tines of an unresolved polychotomy, and leave future editions of this book to tell a bolder story.

Finally, many zoologists have argued that insects and myriapods are closely related. Are they? And, if so, what is the nature of the relationship?

DO INSECTS AND MYRIAPODS HAVE A SPECIAL RELATIONSHIP?

Over the past few decades argument on insect and myriapod relationships has spanned the complete spectrum of logical possibility. The prevalent view at least since the 1930s has been that the myriapods are monophyletic, and that they enjoy a special

—sister-group—relationship with the insects. Indeed, zoologists commonly combined the two in a grouping to which they have given various names, of which the most satisfactory is 'Atelocerata', reflecting the fact that both kinds of animal have only one pair of antennae.[2] To be sure, the two groups look very different at first sight—myriapods with their elongated, homonomous bodies and multitude of legs (which is what 'myriapod' means) and insects resolutely six-legged, and with bodies clearly demarcated into head, thorax, and abdomen. But the two kinds of animal *feel* the same, with their horny, uncalcified armour. In more detail: both have unbranched (uniramous) limbs; both breathe via tracheae; and both have a kidney-like excretory system constructed from **Malpighian tubules**. Both, furthermore, have antennae—but lack any additional appendages that could correspond to the second antennae that are a characteristic feature of crustaceans.

It has often been argued, too, that the myriapods are not only closely related to the insects but are ancestral to them. One modern study suggests that the millipedes and pauropods are sisters; that the symphylans are sisters to the millipedes + pauropods; that the insects are sisters to the symphylans + millipedes + pauropods; and that the centipedes are sisters to the insects + symphylans + millipedes + pauropods. Thus, in this hypothesis, the insects emerge from right in the midst of the myriapods. Of course, if the myriapods are ancestral to the insects, then they again must be seen as a paraphyletic grouping, and called Myriapoda*. Furthermore, if this were true, then Atelocerata would again emerge, emphatically, as a valid clade.

But others declare that the Myriapoda is not a clade at all: that the four recognized classes within it—Chilopoda (centipedes), Diplopoda (millipedes), and the lesser-known Pauropoda and Symphyla—have nothing much to do with each other. Finally, some argue that whether or not the myriapods are monophyletic, they have nothing to do with the insects. Like the insects, they may have descended from crustacean ancestors. But even if they did, they did not descend from the same crustacean lineage as the insects did.

To discuss all these points to exhaustion would drive everybody mad, but here are a few of the points. I was brought up on the argument, dating from the 1930s, that myriapods are ancestral to insects; and, more particularly, that insects arose from myriapods by **neoteny**. Neoteny is the process by which juvenile forms of animals become sexually mature and hence produce a whole new lineage of creatures that resemble the larvae of the ancestors. Hence the tadpole-like axolotls developed from their salamander-ancestors by neoteny: they retain the gills of the strictly aquatic juvenile and yet become sexually mature. Some crustacean groups are thought to have evolved neotenously from others; and vertebrates might have begun as larval sea squirts. Thus the phenomenon is common in nature and has launched some dramatic lineages. In general, myriapods acquire more segments as they develop

[2] Other names include 'Antennata', which does not seem satisfactory because crustaceans and perhaps onychophorans also have antennae; 'Uniramia'—but Manton applied the same term to her hypothetical myriapods + insects + onychophorans; and 'Tracheata'—although various arachnids and some land-based crustaceans also have tracheae, so that does not seem to work either.

from embryos to adults. What more natural than to suppose that an insect is a myriapod with arrested development?

There was also a general reason for finding the neoteny theory plausible. The homonomous trunks of myriapods seem to be more simple, in overall design, than the emphatically tagmatized bodies of insects. Zoologists generally assume, usually reasonably, that complex structures evolve from simpler structures. So the idea that highly tagmatized insects had homonomous ancestors does have a certain common-sense appeal.

On the other hand, various studies seem to support the special relationship between insects and crustaceans, but suggest that the myriapods are outsiders. For example, it has long been observed that insects and crustaceans have very similar compound eyes, while the simple eyes of myriapods are different. More recent, ultra-structural studies now show that individual nerve cells in crustaceans and insects seem to follow very similar paths as they develop—and very quirky paths at that; while again, myriapod nerves seem to develop along different lines.

All in all, the issue of myriapod–insect–crustacean relationships is still very much open. Most evidence now seems to support a general relationship between the three—validating the grouping Mandibulata—but which of the three in the group is most closely related to which of the others, and who if anyone is ancestral to whom, is simply not known. Molecular studies should help to resolve the issues over the next few years, but biologists need to look at many more genes (for different genes suggest different relationships) and at many more outgroups. Until we have extensive studies of genes, including the Hox genes, not only from all the subphyla of arthropods but preferably from all the individual classes, and have comparably robust data from well-chosen outgroups (especially annelids and molluscs), then we simply cannot apply cladistic principles reliably. It will, of course, never be possible to apply molecular techniques to the extinct groups that could be so instructive—especially the trilobites.

So this is why I have presented the arthropod tree here in this mealy-mouthed fashion, showing onychophorans, tardigrades, and arthropods as a trichotomy, with trilobites, chelicerates (+ pycnogonids), and mandibulates as a second trichotomy, and crustaceans (+ pentastomids), myriapods, and insects as yet another trichotomy. The tree does not acknowledge the Atelocerata as a valid grouping: there may be no special relationship between insects and myriapods. But the tree does have two positive features. It does acknowledge the validity of the Mandibulata, and the monophyly of the Myriapoda. Too many people would raise serious objections if the tree was significantly more definite than this. No cladist likes to see a tree so 'unresolved'. But it is better to show the uncertainties clearly, than to feign certainty where none is justified.

The chelicerates, crustaceans, and insects all have sections of their own (9–11). Because the myriapods have not, let us look at them briefly before moving on.

Myriapods: class Myriapoda

The class Myriapoda includes four subclasses, two of which (the millipedes, **Diplopoda**, and the centipedes, **Chilopoda**) are familiar to everyone, while the other two (the

Pauropoda and the **Symphyla**) belong in that subculture of creatures that populates the leaf litter and other protected places and are known only to specialists.

The heads of myriapods are very insect-like in plan: the same number of segments with the same arrangement of appendages. The only striking difference is that myriapods retain the second maxillae as paired organs, and not as a fused labium. But the remaining segments are all much of a muchness; the body behind the head, the **trunk**, is largely (but by no means entirely) homonomous. As discussed on pages 263–5, zoologists since the 1930s have believed that the largely homonomous myriapod trunk is ancestral to the sharply tagmatized insect body; but some, now, are not so sure and suggest that tagmatized insects may have shared common ancestors with tagmatized crustaceans, and perhaps that myriapod homonomy is secondary—presumably an adaptation for life underground.

A few millipedes live in semi-arid environments but most prefer damp places and, overall, they lack the stunning ecological versatility of insects. Like most arachnids, but unlike most insects, myriapods transfer sperm by indirect means—basically by producing sperm in containers called **spermatophores**, which the female picks up. But, like many arachnids, myriapods may still practise elaborate courtship. Thus, most centipedes build a silken 'nuptial net' to hold the spermatophore and this becomes the centre of ritual lovemaking that culminates when the female finally inserts the spermatophore into her **gonopore**. Millipedes may copulate to transfer the spermatophore, embracing with legs and mandibles. Development in myriapods is direct—the emerging babies look like miniature adults, although the young of millipedes and of many centipedes have many fewer segments and legs than the adults. Only in the most primitive insects is development so direct.

Millipedes: subclass Diplopoda

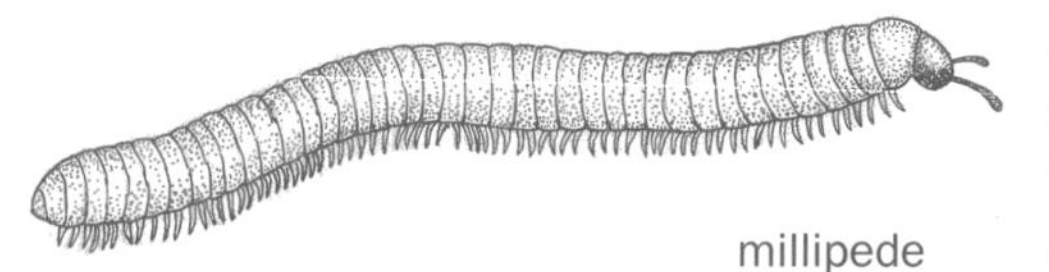
millipede

Millipedes move through their environment in various ways. Some, the 'flat-backed kind', wedge their way through the leaf litter. Others are dedicated tunnel borers—bulldozing their way through the soil in search of the dead vegetation that for most species (though not all!) is their preferred diet. As in crustaceans, millipede sclerites are often hardened with calcium, and in some species some sclerites form complete circles, for extra strength. As a burrowing millipede butts forward it ducks its head so the first trunk segment takes the brunt; this segment, the **collum**, is extremely thick and tough. Principles of engineering show that animals with many legs can generate more power—and millipedes may have hundreds: the record is held by a Southeast Asian species, with 325 pairs. Some millipedes even add extra segments, and legs, after they become sexually mature. But to avoid weakness due to excess length, most of the segments in a millipede are fused into pairs so that there appear to be two pairs of legs per segment; hence the name, Diplopoda. Millipedes are extremely successful: there are 10 000 known species worldwide.

Pauropods: subclass Pauropoda

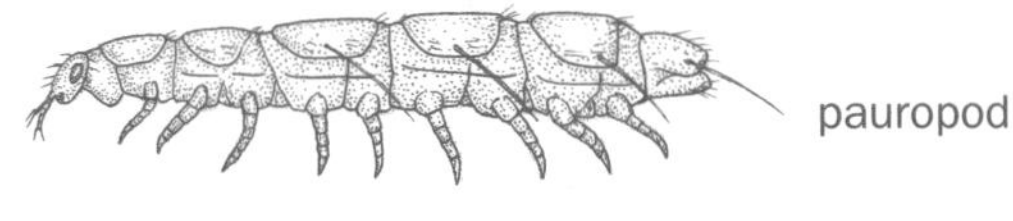
pauropod

Pauropods, like the wingless insects and several of the arachnid orders, have the *mien* of also-rans. They are tiny—just 0.5 to 1.5 millimetres long—with only 9 to 11 leg-bearing trunk segments; some of which, as in millipedes, are fused to form **diplo-segments**. They occur worldwide, but only in moist soils and leaf litter and (again like some arachnids) they usually lack tracheae or hearts, and oxygen simply diffuses through their soft (uncalcified) cuticles. They are eyeless, too; evidently resigned to lowliness. Yet 500 species are known, in five families.

Centipedes: subclass Chilopoda

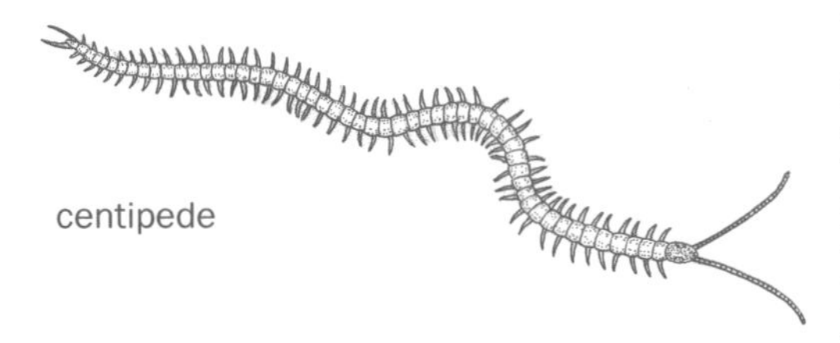
centipede

While millipedes tend towards vegetarianism, centipedes are carnivorous. Like millipedes, most move slowly through the soil or leaf litter, either making their own burrows, or exploiting existing crevices. But a few are surface-sprinting predators with long legs and poisoned raptorial fangs formed from the first trunk appendages. *Scutigera* dashes at 42 centimetres per second in pursuit of flies: a cheetah with the same ratio of speed to body length would break the sound barrier. The venom of *Scolopendra* will subdue small lizards. Some species rear to snatch insects on the wing. Yet there are curious variants—notably the geophilomorphs that burrow in the style of an earthworm, swelling the body and thrusting forward with peristaltic contractions of the trunk muscles. About 2500 species of centipede are known, in about 20 families. It is surprising there are not more.

Symphylans: subclass Symphyla

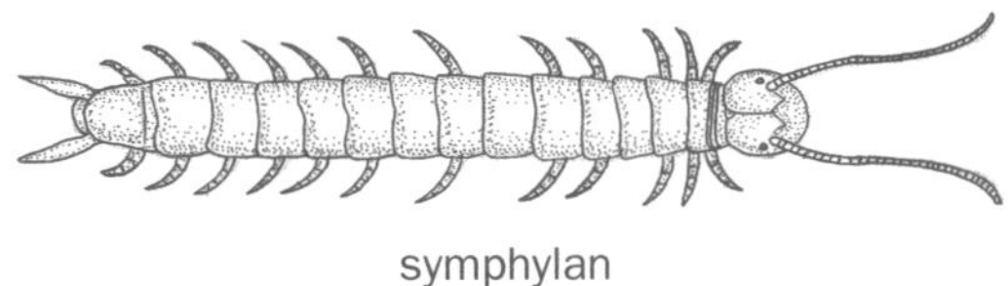
symphylan

Symphylans are the lowly sisters of centipedes, just as pauropods are the poor relations of millipedes. Some zoologists claim that they have a special relationship with insects—indeed that insect ancestors may have evolved from symphylan ancestors—but others emphatically deny this. Like pauropods, symphylans are eyeless and tiny: 0.5 to 8 millimetres. They have only 14 segments, of which the first 12 each have a pair of legs. They do have tracheae in the first three segments, but presumably derive much of their oxygen through their soft cuticles. They live in soil and rotting vegetation. The two known families contain 120 species.

In the next three sections I look at the main arthropod groups in more detail: crustaceans, insects, and chelicerates.

9

LOBSTERS, CRABS, SHRIMPS, BARNACLES, AND MANY MORE BESIDES

SUBPHYLUM CRUSTACEA*

You can sweep a net through an unpolluted pond and emerge with representatives of at least four crustacean orders from three different classes: *Daphnia*, a cladoceran 'water flea', *Gammarus*, a freshwater amphipod 'shrimp'; faintly comical copepods like space rockets with shopping baskets and a quizzical central eye; and ostracods, loosely enfolded in an overlarge carapace, busily gliding like clockwork meringues. On the seashore there are barnacles masquerading as molluscs, beachhoppers among the stranded kelp, crabs in the tidal pools and under rocks. Some of the glories of all nature are to be found at the fishmonger's: lobsters, crabs, and a host of creatures from several groups loosely categorized as 'shrimps' or 'prawns'. Shrimpishness is a primitive and common crustacean motif, recurring in many orders. Any garden harbours woodlice, which the Americans call sowbugs; they are the only crustaceans that have truly become terrestrial. But many a crab has come on to land as well (although they generally return to the sea to breed), including the wonderful robber crab—a hugely overgrown hermit crab that climbs palm trees to feed on the nuts. But there are many, even more recondite forms. The parasitic barnacle *Sacculina* has reinvented the form of the fungus, spreading through its host like a mycelium. Crustaceans range in size from a parasitic blob that can cling to the antenna of a copepod, to a mighty 7-kilo Maine lobster or a giant spider crab spanning 3 metres.

About 50 000 species of crustacean are known, but several times more might remain to be discovered. Even then they would be many times less speciose than insects—and yet, for three reasons, they are far more various in shape and way of life. First, as a group they are extremely ancient, dating well back into the Cambrian at least 500 million years ago, and so they have had plenty of time to evolve and radiate. Second, they are mostly aquatic and so escape the vicissitudes of gravity that set such limits on body form. They are spared desiccation, too, and can dangle their gills with abandon, as krill do. Then again, water provides a very different medium depending on the animal's size. On the small scale, viscosity matters; the water is like treacle. On the larger scale, inertia is what counts. Physicists speak of the contrast between a low and high Reynolds number. So in a small creature, a waggling appendage tipped with **setae** ('hairs') becomes an oar, an organ of propulsion. In a larger creature the same appendage merely sweeps the water while the animal stays still, and becomes a device for feeding by filtration. Thus large and small crustaceans live in different physical universes.

Third, crustaceans have an extremely successful body plan that lends itself to endless variety. The segments (up to 32 or so) are each fitted with a pair of appendages that are fundamentally biramous but, unrestrained by gravity, can take many different forms and serve for swimming, walking, offence, communication, reproduction, feeding, respiration, or, indeed, for several of these at once. The typical crustacean is a mobile Swiss army knife. Apart from some of the extreme parasites whose bodies have been remoulded, all crustaceans are clearly tagmatized into head and trunk and, in most, the trunk divides into thorax and abdomen—although the trunks of remipedes and cephalocarids look virtually homonomous. The number of segments varies in thorax and abdomen but the head always has five. The five pairs of head appendages form two pairs of antennae, one of mandibles, and two of maxillae. Most crustaceans protect their front ends with a **carapace** or a **cephalic shield**. Characteristic, too, is the planktonic **nauplius larva** that has only three pairs of appendages that later become antennae and mandibles. Other segments and appendages are added later. But many crustaceans bypass the nauplius stage, while others add several more larval stages between nauplius and adult.

The two trees I have included here are based in large part on the ideas of Richard and Gary Brusca, as presented in their text *Invertebrates* (1990); but are also guided by comments from Robert Hessler and William Newman at the Scripps Oceanographic Laboratory, La Jolla, California. The main tree shows all the principal classes and subclasses that are commonly recognized, but only a selection of the enormous number of orders. I have taken slight liberties with this tree for reasons shortly to be explained.

A GUIDE TO THE CRUSTACEA*

Because the Arthropoda has re-emerged as a perfectly good, monophyletic phylum, the Crustacea can be treated as a subphylum, like the Chelicerata and Insecta. By giving the Crustacea such a high Linnaean ranking, we are able to refer to the huge divisions—Maxillopoda, Malacostraca, and the rest—as 'classes', which seems very fitting.

The first of the two trees—showing the crustaceans as a whole—presents the subphylum as a clade: a true, monophyletic taxon. It certainly seems reasonable to suggest that all the creatures we acknowledge as 'crustaceans' evolved from the same crustacean ancestor, so that cladistic criterion at least is filled. But no one knows what this common ancestor actually was. Some, like Robert Hessler, suggest that the crustaceans arose from among the ranks of the miscellaneous 'trilobitomorphs', the long-gone trilobites and trilobite-like creatures I described in Section 8. But latest palaeontological evidence suggests that the crustaceans may have predated all other known arthropods, including the trilobites, and molecular evidence also indicates their extreme antiquity.

As noted in Section 8, there is now good evidence that the crustaceans, insects, and myriapods enjoy a special relationship, and, indeed, that between them they form a group that can be called the Mandibulata—'the jawed ones'.[1] It is at least possible, too (some would say probable), that the ancestor of insects, and/or of myriapods, was itself a crustacean. We do not know for certain that crustaceans include the ancestors of insects and/or myriapods but it seems likely that they did, not least on grounds of common sense. The fossil record suggests that crustaceans appeared about 150 million years before the insects; it seems extremely likely that the ancestors of insects lived in the sea; so it is at least a reasonable bet that the common ancestor of insects was either a crustacean, or some long-gone marine myriapod. Given the extreme antiquity of the crustaceans, this ancestor may even be ancestral to the chelicerates as well as to the myriapods and/or the insects—and, indeed, could be ancestral to the trilobites. In short, the relationships between the major arthropod lineages is at present disputatious, but crustaceans could turn out to be the basal group of them all. Molecular studies should throw more light on the matter. Even as things stand, however, it seems very likely that crustaceans have given rise to at least one other taxon, and so are paraphyletic—like the Reptilia*, which undoubtedly gave rise both to the mammals and to the birds. Thus, just as the name Reptilia* carries an asterisk, so, too, should Crustacea*.

The main tree shows the crustacean subphylum divided into six classes, whereas most accounts (including that of Richard and Gary Brusca) include only five. My sixth is the Pentastomida, the parasitic 'tongue worms', that the zoologists I have talked to seem to agree are degenerate crustaceans, which have special affinities with the Maxillopoda. In fact, latest molecular evidence is suggesting that pentastomids *are* maxillopods, and that they should simply be presented as a subclass or an order within the Maxillopoda. Presenting them as the sister group of the maxillopods seems a reasonable, middle-of-the-road position.

What of the relationship between the crustacean classes? As with insects and myriapods, much has centred on the issue of homonomy. Thus zoologists have tended to feel, mainly as a matter of common sense, that the body form of the ancestral type was more likely to be simple than complex. They have generally felt, too, that homonomy is simpler than heteronomy; and so, putting the two points together, they have tended to suppose that the common ancestor of the crustaceans ought to have a homonomous body. Yet as discussed in the sections on arthropods and insects (8 and 10), all is not necessarily so simple. As Michael Akam at the Wellcome

[1] The Linnaean ranking of the Mandibulata lies somewhere between a subphylum and a phylum. In the spirit of Neolinnaean Impressionism, it seems sensible to leave the grouping without formal rank.

9 · SUBPHYLUM CRUSTACEA*

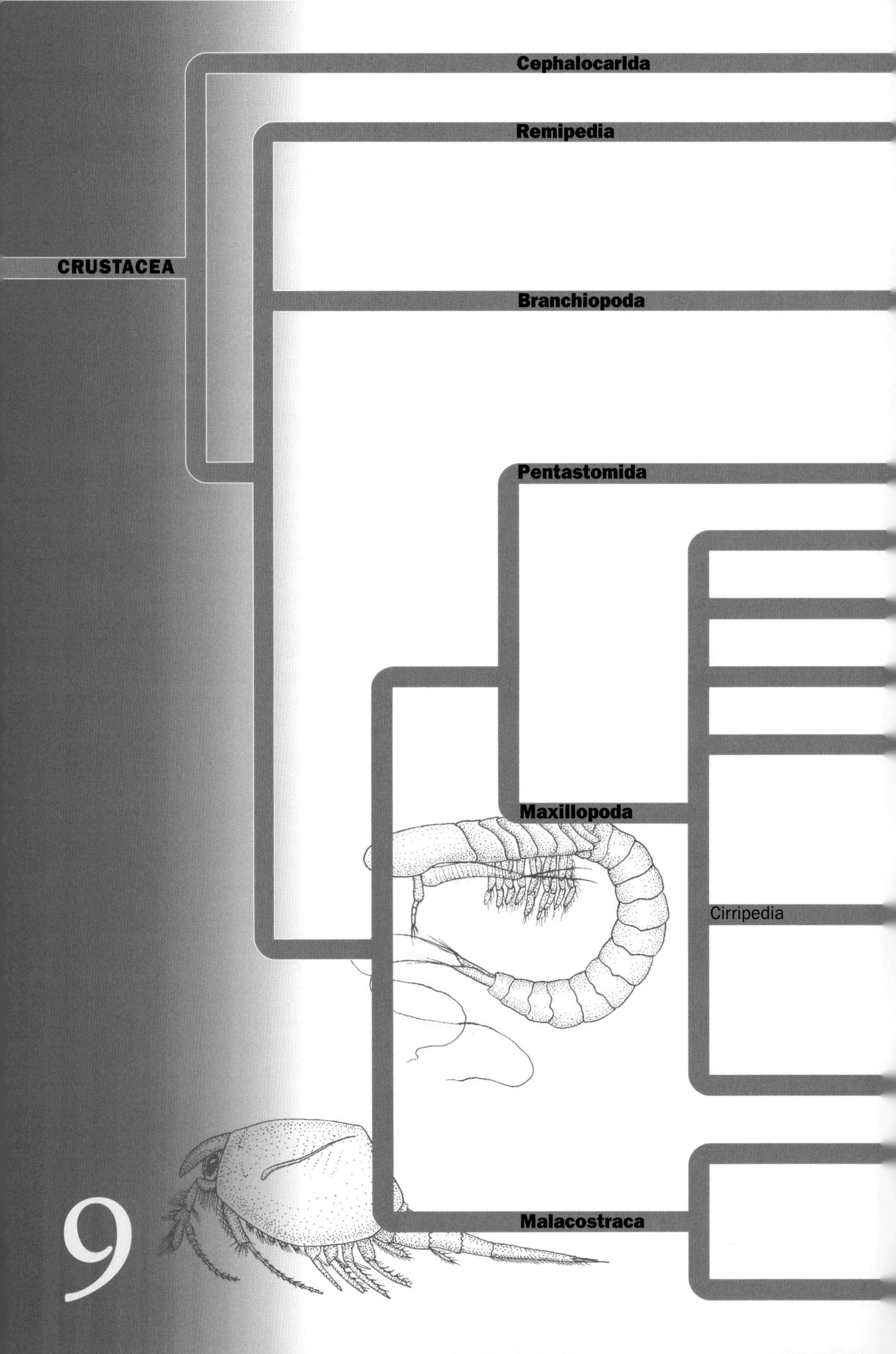

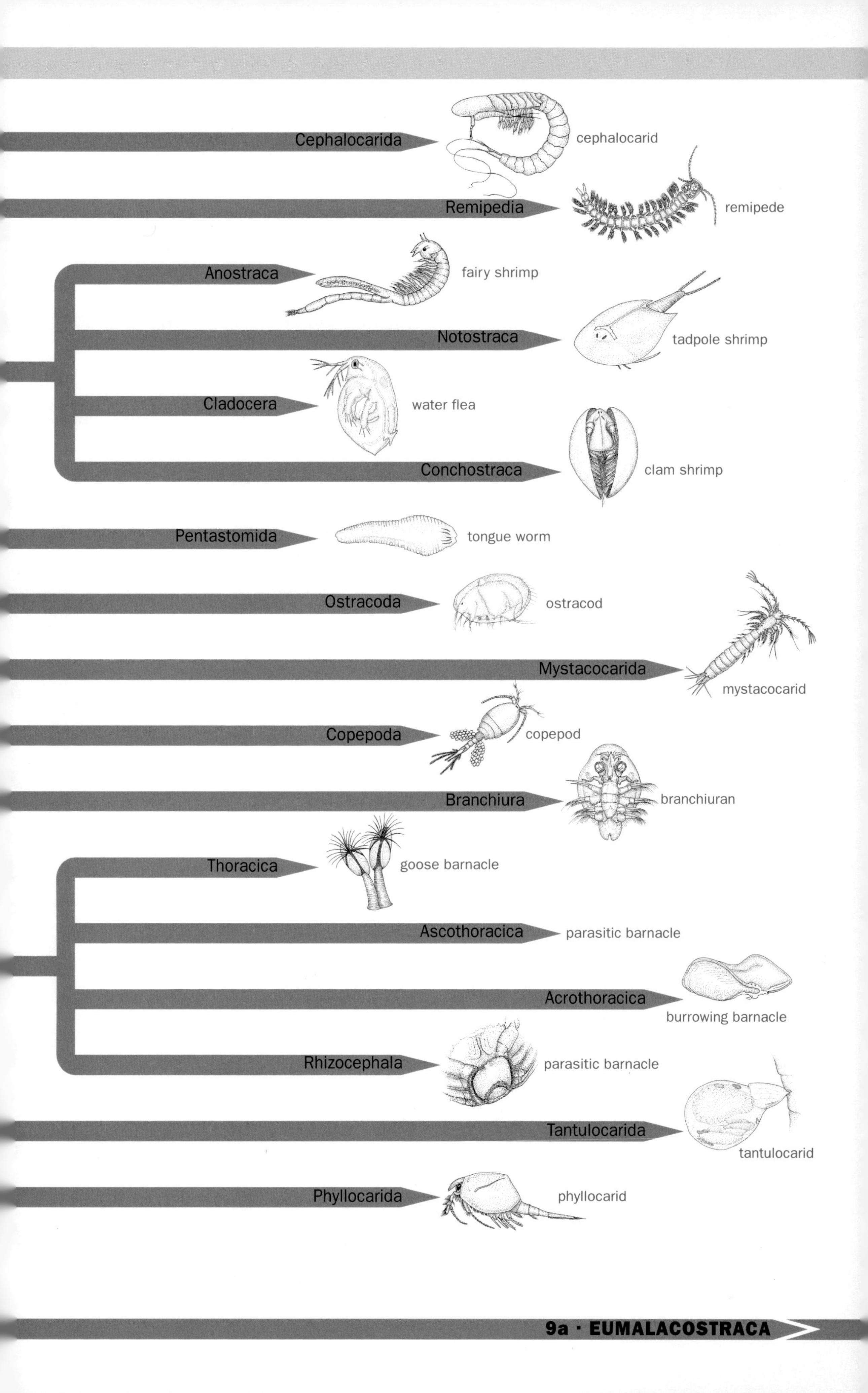
Cephalocarida
cephalocarid
Remipedia
remipede
Anostraca
fairy shrimp
Notostraca
tadpole shrimp
Cladocera
water flea
Conchostraca
clam shrimp
Pentastomida
tongue worm
Ostracoda
ostracod
Mystacocarida
mystacocarid
Copepoda
copepod
Branchiura
branchiuran
Thoracica
goose barnacle
Ascothoracica
parasitic barnacle
Acrothoracica
burrowing barnacle
Rhizocephala
parasitic barnacle
Tantulocarida
tantulocarid
Phyllocarida
phyllocarid
9a · EUMALACOSTRACA

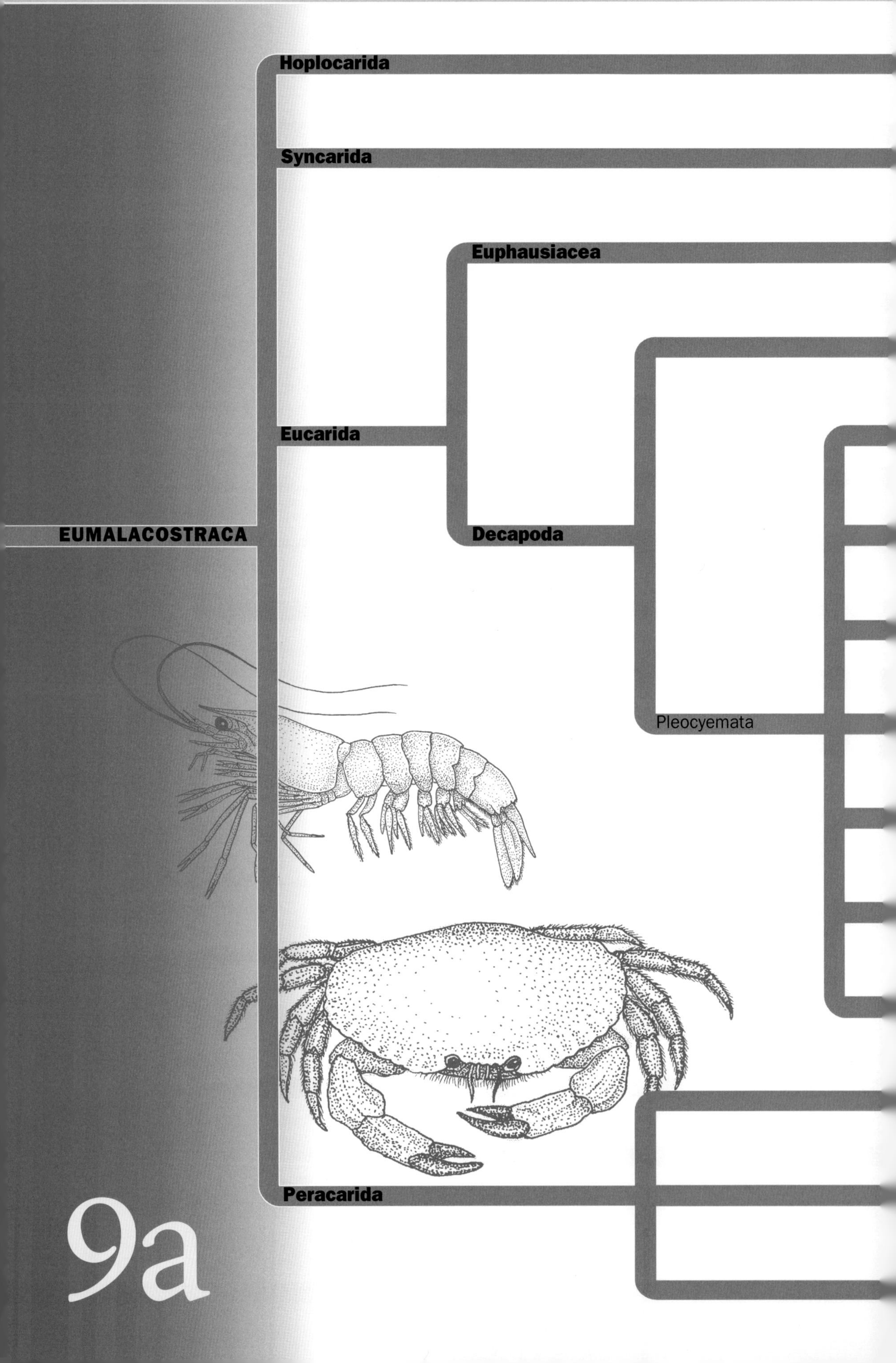
Hoplocarida
Syncarida
Euphausiacea
Eucarida
EUMALACOSTRACA
Decapoda
Pleocyemata
Peracarida
9a

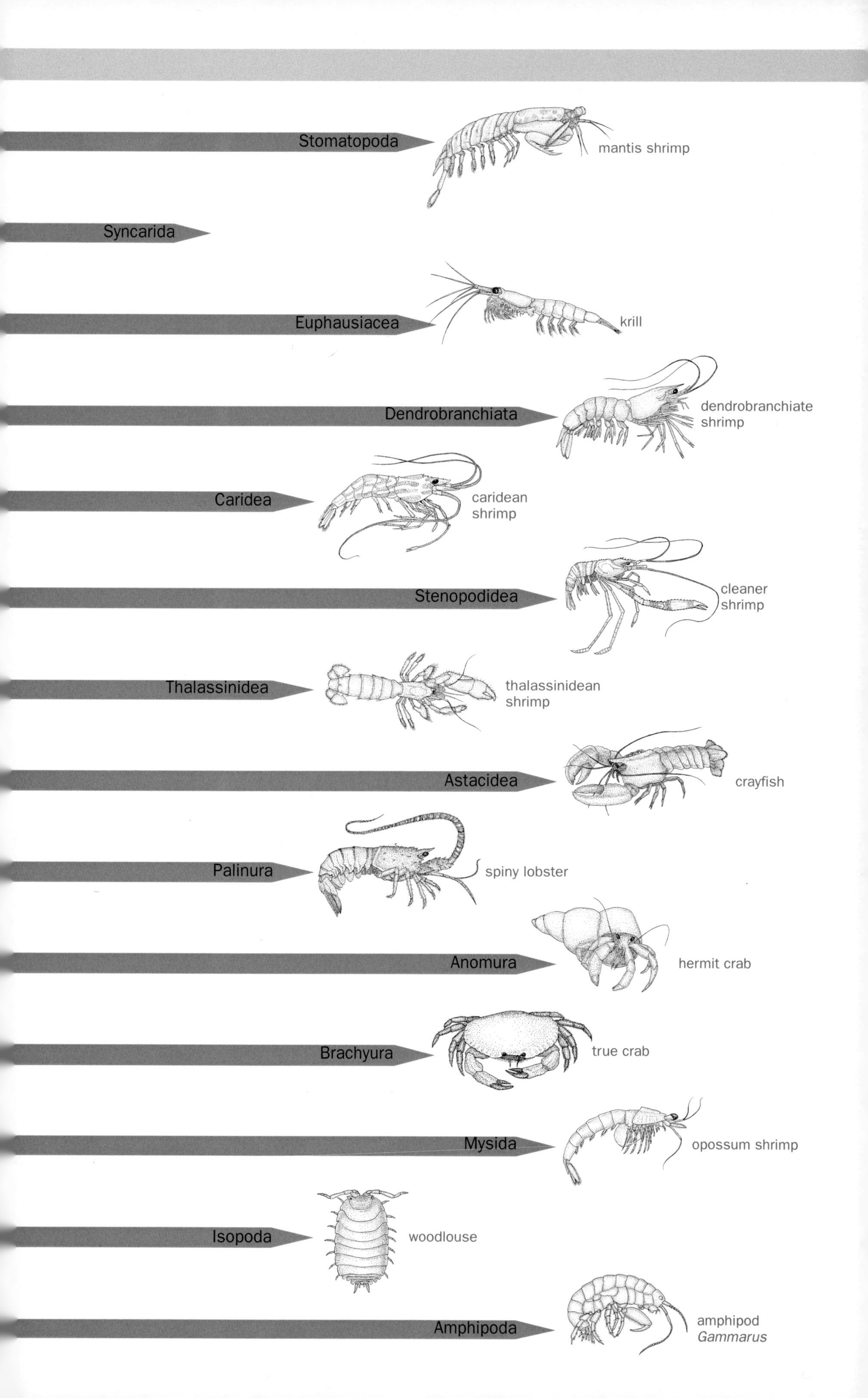
Stomatopoda
mantis shrimp
Syncarida
Euphausiacea
krill
Dendrobranchiata
dendrobranchiate shrimp
Caridea
caridean shrimp
Stenopodidea
cleaner shrimp
Thalassinidea
thalassinidean shrimp
Astacidea
crayfish
Palinura
spiny lobster
Anomura
hermit crab
Brachyura
true crab
Mysida
opossum shrimp
Isopoda
woodlouse
Amphipoda
amphipod
Gammarus

Laboratory in Cambridge, England, suggests, it could be easier for a homonomous body to evolve from a heteronomous one, than the other way around; so we might expect homonomy to arise on different occasions independently, as a secondary development, just as many different birds have independently simplified their wings and become flightless. The apparent homonomy of woodlice (order Isopoda), for example, is clearly secondary. So perhaps—just perhaps—the search for a homonomous crustacean common ancestor is a red herring.

Despite these caveats, most authorities have sought a homonomous crustacean ancestor, and there have been two outstanding candidates: the cephalocarids, and the recently discovered remipedes. Robert Hessler at the Scripps Oceanographic Laboratory has spent several decades looking intently at crustaceans and suggests that, of those two groups, the cephalocarids are by far the more promising candidates. Much about them is primitive. They could easily represent the basal crustaceans. Remipedes, on the other hand, have primitive-looking trunks but highly derived heads, and on these grounds alone Hessler suggests that they could not possibly be ancestral. Others dispute this; but, following Hessler, I am nonetheless presenting the cephalocarids as the sister group of all other crustaceans, so suggesting that of all living crustaceans they are most closely related to the common crustacean ancestor. I am presenting the remipedes simply as one tine in a deep trichotomy. At present, after all, nobody really knows where they fit, and polychotomies are intended to define areas of ignorance. Such a trichotomy should be seen as a holding operation: keeping the picture tidy until further evidence (which in large part is likely to be molecular) provides greater clarity.

The other two branches of this trichotomy should provoke less protest. Several recent cladistic studies (including that of Richard and Gary Brusca) suggest that the remaining crustaceans (that is, crustaceans minus cephalocarids and remipedes) fork into two main branches: the branchiopods on the one hand, and the maxillopods + malacostracans on the other. This is what is shown on the tree. The only deviation offered here is the inclusion of pentastomids alongside the maxillopods, as already discussed.

In general, most of the crustaceans that are best known—crabs, lobsters, shrimps, krill, woodlice, and barnacles—and all the biggest ones belong among the classes Maxillopoda and Malacostraca. These two are regarded as the most derived of all the crustaceans, and are commonly presented as sister groups. Take out the pentastomids, and the phylogeny shown here follows convention. In dividing the Maxillopoda into six subclasses, and the Malacostraca into two subclasses, I am following the Bruscas. All the subclasses are included—and, as you see, a tree of the supremely important and various divisions of the Eumalacostraca is given its own spread. The two trees together show only a selection of the maxillopod and malacostracan orders: the ones that contain the best-known crustaceans, are of greatest ecological significance, and between them show the greatest range of crustacean body types and ways of life. But the trees do show all the orders of the branchiopods, cirripedes, and eumalacostracans, as outlined by the Bruscas.

The Crustacea* illustrate beautifully the extraordinary unevenness of nature; that because of some small advantage which may or may not be easy to detect, some lineages radiate into thousands of species and scores of habitats, whereas others that seem similar remain as also-rans. Thus both the Remipedia and Cephalocarida—ranked as entire classes—are each represented by only a few species, whereas the Decapoda, a mere order, contains most of the most conspicuous crustaceans, from the commonest shrimps to lobsters and hermit crabs.

Cephalocarids: class Cephalocarida

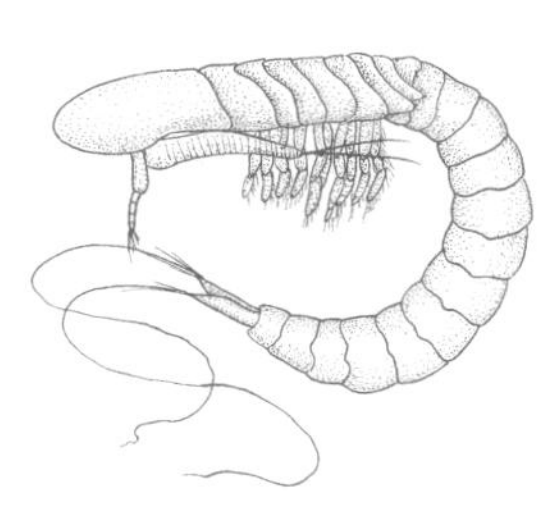

cephalocarid

Cephalocarids are slightly more shrimp-like than remipedes but still have fairly homonomous bodies and in many other respects seem more primitive. They are only 2.0–3.7 millimetres long and feed on detritus on the sea bottom. Only nine species in four genera are known but they are wide-ranging: from the intertidal zone to a depth of 1500 metres. Robert Hessler sees them as the sister group of all other crustaceans and that is how they are presented here.

Remipedes: class Remipedia

Remipedes first came to light only in recent decades, in a cavern in Grand Bahama Island, and because they represent an entire class of arthropod they illustrate how little we really know of the living world. They swim on their backs, beating their appendages. They are small—up to 3 centimetres—but are voracious predators, immobilizing their prey, like spiders, with poison injected through the second maxillae.

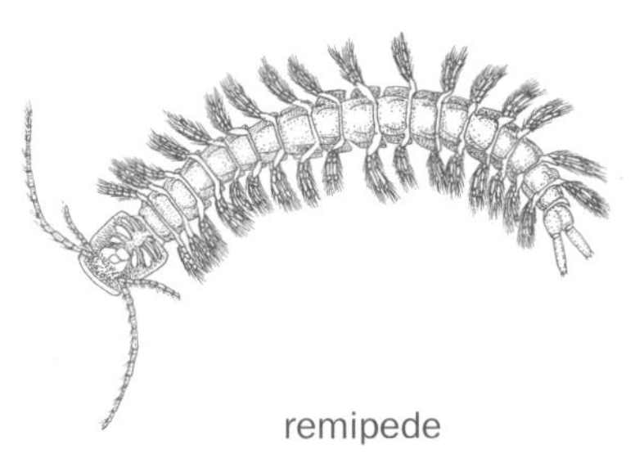

remipede

Most zoologists would suppose that these **maxillipeds** could not be homologous with a spider's chelicerae but if crustaceans are indeed as ancient as they now seem, and if they are ancestral to chelicerates, then such homology is possible. Most remarkable, however, are the millipede-like bodies of remipedes; but, as discussed, it is not clear whether this feature is truly primitive, or whether it indicates any particular affinity between the remipedes and ancestral crustaceans.

Shrimps and water fleas: class Branchiopoda

Overall, the branchiopods are so immensely variable that many have doubted their monophyletic status. But they are conventionally presented as a coherent group and, pending further evidence, this is how they are treated here. They are commonly divided into four orders: Anostraca, Notostraca, Cladocera, and Conchostraca.

• Order **Anostraca** includes the fairy shrimps, such as *Artemia*, the brine shrimp. They are wispy in appearance but are as tough as nails, living in ephemeral ponds where they must withstand desiccation and in salt lakes where they are important food for water birds.

• Order **Notostraca** comprises the remarkable tadpole shrimps, which have a broad carapace that grows only from the head but extends over the thorax so that they look like tadpoles. All nine living species live in temporary ponds inland, and lay eggs that withstand desiccation. Some feed on detritus, some scavenge, and some are predators. *Triops* is a crustacean pest: it burrows in rice paddies and dislodges the young plants.

• Order **Cladocera** contains the water fleas, like *Daphnia*, beloved by aquarists as fish food. The single carapace is folded along the midline so they look like miniature bivalve molluscs, and they swim by beating their antennae. About 400 species are known.

• Order **Conchostraca** includes about 200 species of clam shrimps, in which the bivalved carapace encloses both the head and all of the body, and even has concentric growth lines like a small clam.

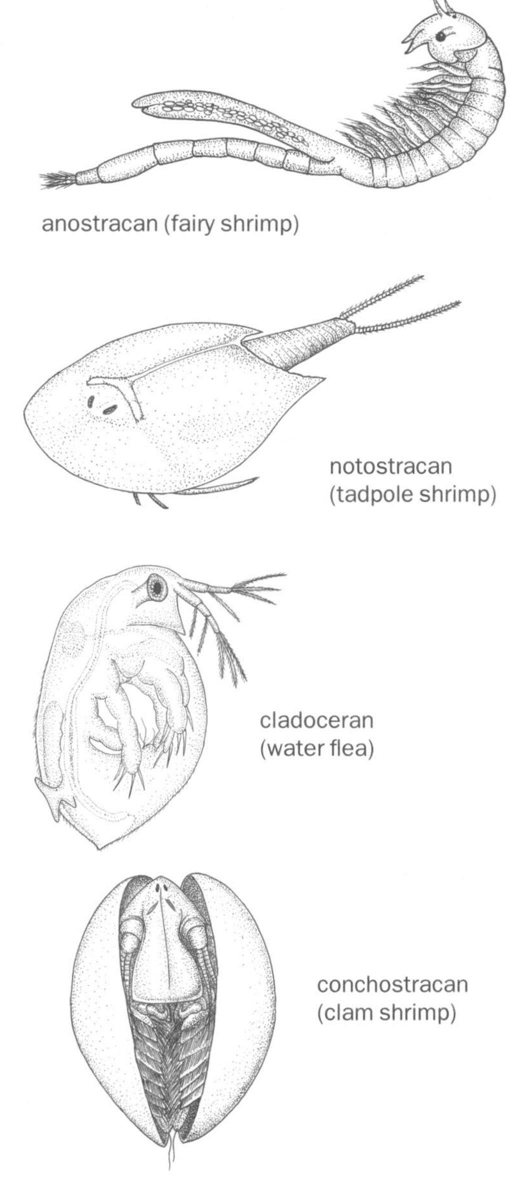
anostracan (fairy shrimp)

notostracan (tadpole shrimp)

cladoceran (water flea)

conchostracan (clam shrimp)

Tongue worms: class Pentastomida

Pentastomids, alias 'tongue worms', are weird little worm-like creatures, 2–13 centimetres long, which all live as blood-sucking parasites in the lungs or nasal passages of vertebrates—usually reptiles but sometimes mammals or birds, which serve as their 'definitive hosts'. Many also require an intermediate host: their larvae leave the definitive host by boring into the gut and passing out with the faeces, to be picked up by another vertebrate of more or less any type. The intermediate host is then generally eaten by another definitive host, which completes the cycle. Pentastomids are quite diverse. There are only 95 species, but they are various enough to be placed in seven families in two orders.

As in tardigrades and onychophorans, pentastomids have lobopod 'legs', which appear merely as extensions of the body wall; and again these legs end in claws. There are two pairs of legs—four in all; and these projections, together with the 'snout' on which many pentastomids bear the mouth, give a five-snouted appearance and hence the name 'pentastomid', which literally means 'five-mouthed'. Apart from that, and again in common with tardigrades and onychophorans, pentastomids show a mixture of features that seem both annelid-like and arthropod-like. Thus they have a porous, thin, and non-chitinous cuticle and a muscular body wall with annelid-like muscle sheets. But—like arthropods—they grow through a series of moults and their main body cavity is a haemocoel.

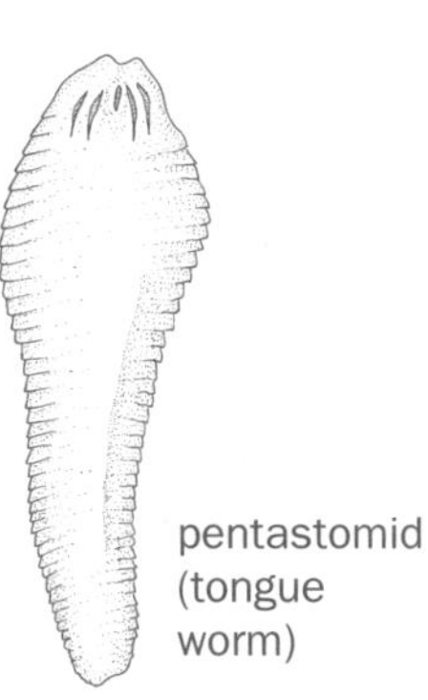

pentastomid (tongue worm)

Intriguingly, and surely revealingly, the first-stage larva of pentastomids resembles the nauplius larva of crustaceans. Accordingly, various zoologists in the 1970s and 80s suggested that they are crustaceans—probably allied with maxillopods—and that their soft-bodied, superficially annelid-like aspect simply reflects the degeneracy of parasitism. After all, many other crustaceans are parasitic, and some are far more altered than pentastomids. For all these reasons it seems reasonable to present them as the sister group of the maxillopods; and recent molecular data suggest that they are maxillopods.

Barnacles, ostracods, and copepods: class Maxillopoda

Sessile, mollusc-like barnacles, bizarre parasitic barnacles and other parasites, planktonic ostracods and copepods—the Maxillopoda are extremely various and are split into no fewer than six subclasses, further divided into superorders and orders. As we have seen, too, Maxillopoda should probably include the Pentastomida. But (pentastomids aside) all maxillopods do seem to have a basically similar body plan. All have highly shortened bodies, generally based on a body plan of 5–6–4: five head segments, six thoracic segments, and four abdominal segments. For this and other reasons the maxillopods are thought to be monophyletic.

Subclass Ostracoda

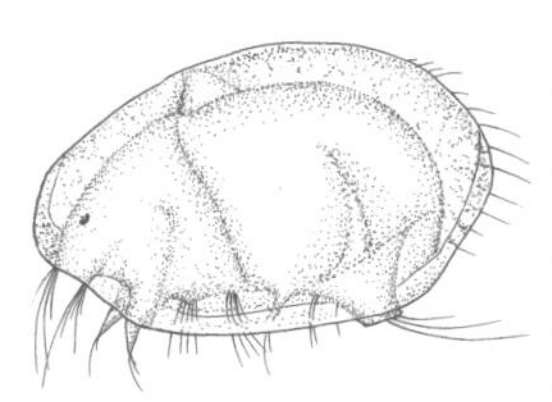

ostracod

Like conchostracans, ostracods have a carapace like a cockle shell that encloses body and head; a nice piece of convergent evolution. But ostracods deploy their appendages differently and lack the growth rings of the clam shrimps. Most of the 2000 or so species are benthic crawlers or burrowers (at the bottom of the sea), though many are planktonic, and they occur down to 7000 metres. A few live in moist habitats on land.

Subclass Mystacocarida

Mystacocarids are tiny (less than 1 millimetre long) and worm-like and retain what seem to be primitive features, including overt segmentation of the head.

mystacocarid

copepod

Subclass Copepoda

Copepods lack a carapace but have an opulent cephalic shield, which gives them a vaguely tadpole-like appearance; and many have a central, cyclopean eye. Most are planktonic (and are familiar booty of pond-dippers) but some are benthic, and others are committed parasites on fish or invertebrates. Some that are planktonic as adults are parasites as larvae, in gastropods, polychaete annelids, and sometimes echinoderms.

Subclass Branchiura

Branchiurans are one of the crustacean answers to the arachnid ticks: they have flat bodies and cling with their maxillae to fish, choosing a spot where the turbulence is low behind a fin or the gill cover, and sucking blood or body fluids. Between attachments they swim with their thoracopods. About 130 species are known.

branchiuran

Subclass Cirripedia

Barnacles are among the most remarkable of all animals. Most familiar of the four orders is the **Thoracica**, which includes the acorn and goose barnacles. For a long time the acorn barnacles were classified as molluscs; indeed, their carapace has become a sac-like and very molluscan 'mantle' that secretes a calcareous 'shell'. The mobile larvae of acorn barnacles stick themselves by their heads to a rock, ship, or a whale, build their shelly stockade, then feed by waving their seta-fringed ('hairy') legs through the water. Goose barnacles are broadly similar but have a stalk, or **peduncle**.

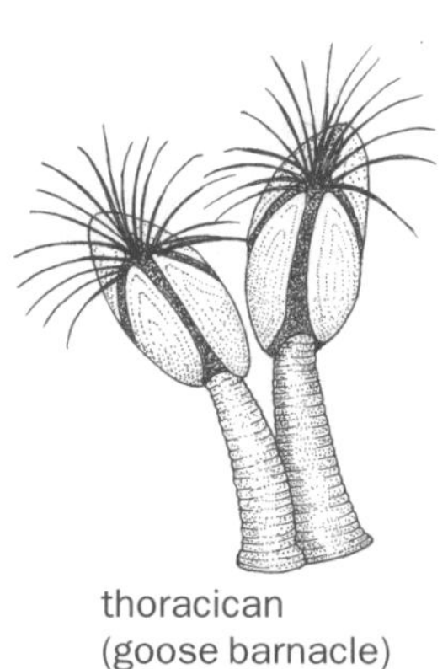

thoracican (goose barnacle)

There are also three further cirripede orders that do not merely stick themselves to rocks or whales but show an increasing tendency to parasitism. Thus the order **Ascothoracica** are parasites of anthozoans and echinoderms; they still have a recognizably crustacean, copepod-like body. Others, in the order **Acrothoracica**, are tiny burrowers, into coral or mollusc shells. But the most extraordinary parasites of all belong to the order **Rhizocephala**, which have effectively reinvented the form of the fungus and parasitize other crustaceans, mostly decapods. *Sacculina carcini* is a famous rhizocephalan, which, when mature, is reduced to a bag of reproductive organs slung beneath the abdomen of a crab, while the rest of its body spreads through the entire host up to the tips of all the claws, like the mycelium of some rot fungus.

acrothoracican barnacle

rhizocephalan barnacle (*Sacculina* on a crab)

But the larvae of cirripedes retain mobility, and some resemble the adult forms of other maxillopods, such as ostracods. This reinforces the notion that these different subclasses are related—and suggests that ostracods may have arisen from the larvae of barnacles that achieved sexual maturity and dropped the sessile stage; that is, by neoteny.

Subclass Tantulocarida

Last and least of the maxillopods, are also parasites as adults, mostly of other, deep-water crustaceans. They are much reduced—they have lost their abdomens and look like blobs—and tiny: less than 0.5 millimetre. They have in their time been placed in various maxillopod groups, including Copepoda and Cirripedia, but were allotted their own subclass in the 1980s.

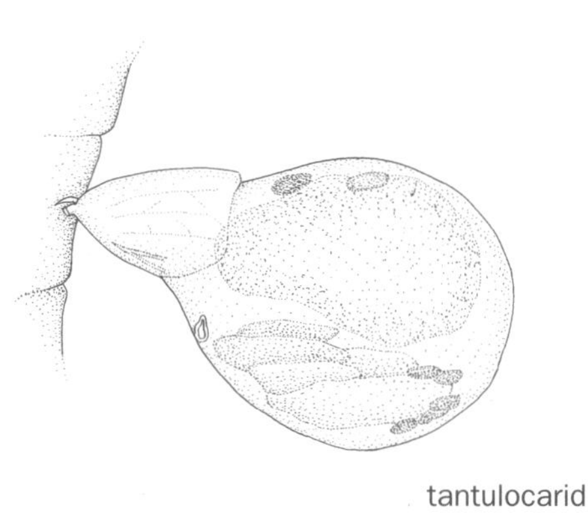

tantulocarid

Crabs and lobsters: class Malacostraca

Most known crustaceans are malacostrocans; and among the malacostracan ranks are all the aristocrats, including some of the largest of all invertebrates such as the giant spider crabs of Japan with their 3-metre leg span (vying with eurypterids as the largest of all arthropods) and the Maine lobster of the United States weighing up to 7 kilograms. Malacostracans are extremely unevenly divided into two subclasses: the **Phyllocarida** are considered to have the more primitive body plan, with 5–8–7 segments plus telson; and the magnificent **Eumalacostraca**, which have a body plan of 5–8–6 plus telson. Malacostracans have compound eyes, which in eumalacostracans are typically, but by no means always, on stalks.

Subclass Phyllocarida

Imagine a more or less shrimp-like creature with the general shape of a terrapin: a big roofed carapace at the front and a thick tapering tail behind. But the phyllocarids are small—mostly 5–15 millimetres long, though with one giant at 4 centimetres. There are only 20 known species in 6 genera and one order, the Leptostraca. All are marine, from the intertidal zone to about 400 metres and most seem to prefer environments with low oxygen; one has now been found around the deep-sea hydrothermal vents of the Galápagos and East Pacific Rise. They differ from the eumalacostracans in their extra abdominal segment and in possessing leaf-like, 'phyllopodous' thoracic appendages. All in all, they are typical of creatures that have not quite hit the jackpot of biological success but survive in a special niche that they exploit particularly well.

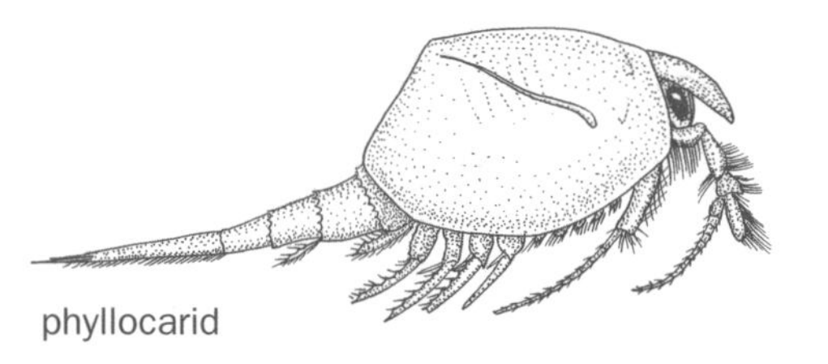
phyllocarid

Subclass Eumalacostraca

This subclass includes 99.9 per cent of all malacostracans, which, of course, means that it includes the bulk of all crustaceans. Again there is a host of shapes and sizes, from lobsters and crabs, long sinuous shrimps as agile as a weasel, to bumbling woodlice and a sprinkling of tick-like parasites. So various are they, so speciose, and of such ecological and economic importance, that I have given them their own tree. Although eumalacostracans are so diverse, however, there are clear synapomorphies. Notably, up to three segments of the thorax—**thoracomeres**—are fused with the head to form a **cephalothorax**; and the appendages of those segments are modified as extra maxillae—maxillipeds—that serve as auxiliary cutlery. This arrangement is beautifully demonstrated, enlarged as if for ease of inspection, by lobsters. The tree and the following account present all four of the eumalacostracan superorders as recognized by Richard and Gary Brusca, but I include only the most significant orders.

Superorder Hoplocarida

The Hoplocarida has one order, **Stomatopoda**, the mantis shrimps. They are slimline, large (2–30 centimetres), and beautifully adapted voracious predators of tropical or subtropical seas, which, like mantises and dragon-fly larvae from the Insecta, seize their prey, including fish, with huge, hinged, raptorial appendages (in this case formed from the second thoracopods). They are ambushers, hiding up in borrows in soft sediments or in crevices. Their long, muscular abdomens and short carapaces allow them to twist in their narrow holes so they can dive in headfirst and turn to face the enemy. About 350 species are known.

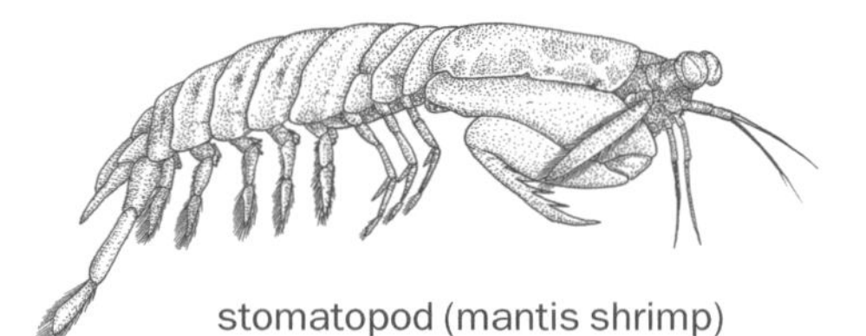
stomatopod (mantis shrimp)

Superorder Syncarida

The Syncarida include two orders—totalling 150 or so known species. Syncarid shrimps are unprepossessing creatures, which, conspicuously, lack a carapace and commonly have reduced appendages. Their reducedness has been ascribed to neoteny, by which juvenile features are retained into adulthood. Syncarids either crawl or swim and live in lakes, streams, and ponds. None is marine.

Superorder Eucarida

The Eucarida include two mighty orders of enormous ecological and economic significance, the Euphausiacea and the Decapoda.

- In the order **Euphausiacea** are the krill. There are only 90 or so species but they occur in huge masses in all ocean environments from the surface down to 5000 metres—often at 1000 individuals (600 grams wet weight) to the cubic metre. Many animals feed on them, including, most famously, the mighty baleen whales. In the Antarctic they exemplify the point that high-latitude ecosystems—in total contrast to the tropics—are based on huge numbers of just a few species. In addition to whales, krill are eaten by crabeater seals, which are among the most common of all large vertebrates, and are in turn preyed on by leopard seals (which also eat krill). In broad terms krill resemble decapod shrimps; but unlike decapods, euphausiaceans lack maxillipeds and carry their thoracic gills external to the carapace. As they swim (and they are all swimmers) the unprotected gills give them a feathery appearance.

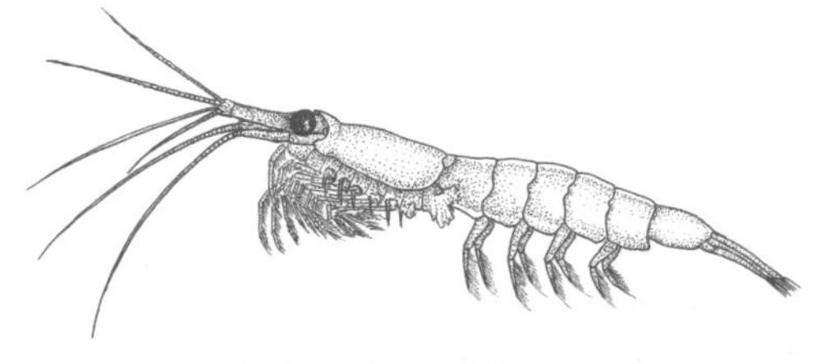
euphausiacean (krill)

- Order **Decapoda** includes some of the glories of the whole animal kingdom, such as the wonderful lobsters and crabs, and also some of the greatest commercial importance, including several groups of broadly similar creatures commonly known as shrimps and prawns. As malacostracans, they have eight segments in the thorax, which bear eight pairs of appendages. But decapods use the first three pairs of these as auxiliary mouthparts or maxillipeds, while the remaining five pairs, **pereopods**, form uniramous or only weakly biramous 'legs', some of which may be modified into claws. Five pairs of legs makes 10 in all and hence the name, 'decapods'. The carapace forms the top of the cephalothorax and extends on either side to protect the gills in the 'branchial chamber'. The gills are attached to the tops of the legs.

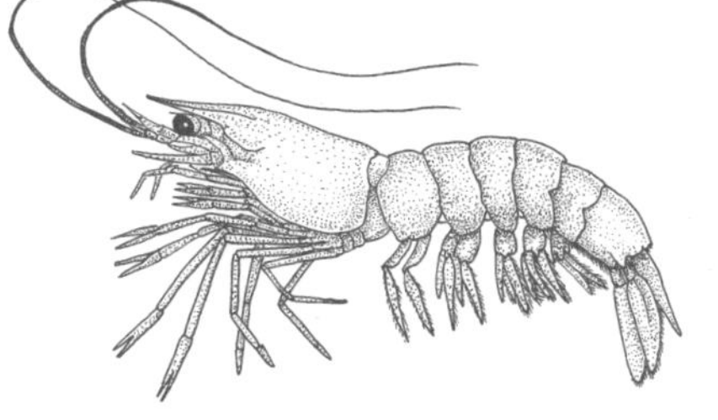
dendrobranchiate shrimp

Two decapod suborders are commonly recognized. The first, the **Dendrobranchiata**, includes 450 species of penaid and sergestid shrimps that grow up to 30 centimetres and are of great commercial importance. They are distinguished by their unique 'dendrobranchiate' gills. The second suborder, **Pleocyemata**, contains all the rest of the decapods; and, to keep things tidy, the pleocyemates are often subdivided into seven infraorders. Three of them are commonly called 'shrimps': the

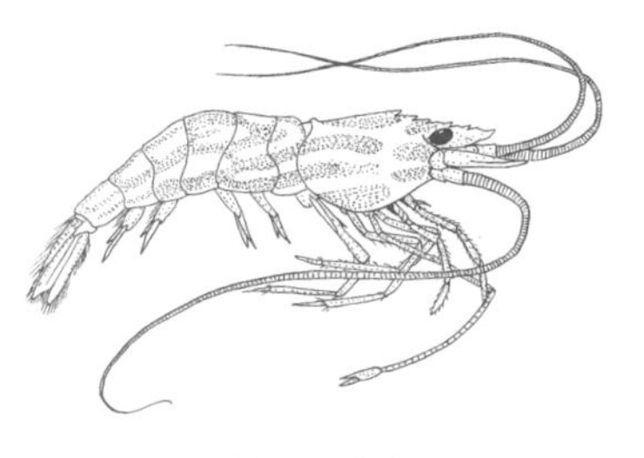

caridean shrimp

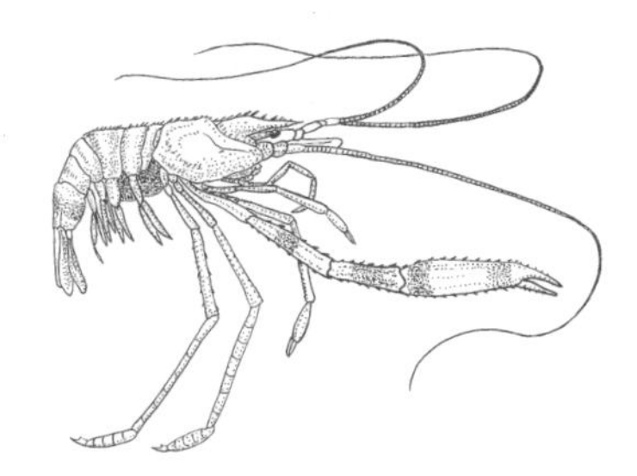

stenopodidean shrimp

thalassinidean shrimp

Caridea, the 2000 or so species of 'caridean shrimps'; the **Stenopodidea**, which often live in coral reefs and include the cleaner shrimps; and the **Thalassinidea**, which are the mud and ghost shrimps.

The fourth infraorder of pleocyemate decapods, **Astacidea**, contains the crayfish and the clawed lobsters. The first three pairs of pereopods are always clawed (**chelate**)—and the first pair forms the great claws. Again we see convergence; there is obvious similarity with scorpion claws although they two groups are at most only distantly related. The astacidean abdomen is big and muscular and terminates in a strong tail fan. It is not obvious (at least to me) why they have such a mighty tail, which must require enormous input of energy to grow and function. Of course, by flexing, the abdomen enables them to retreat rapidly; but perhaps, more generally, it simply provides stability, enabling them to deploy their great claws. Tractors need similar anchorage as they use their fore-loaders. The lobsters are marine but most crayfish live in fresh water—although a few live in damp soil and may excavate extensive and complex burrows. The spiny lobsters or slipper lobsters are in the infraorder **Palinura**. They resemble lobsters but they lack the huge claws. In some, the pereopods are not chelate at all.

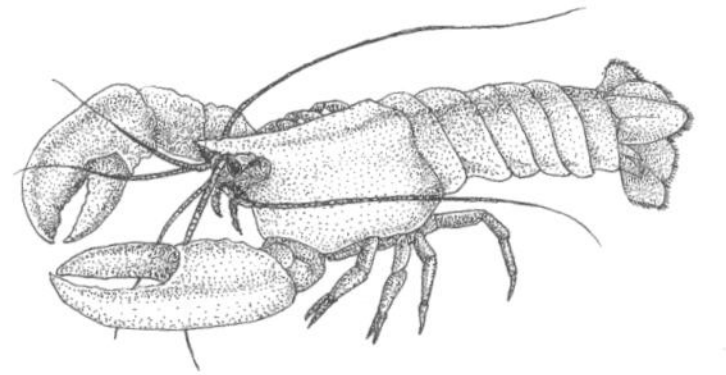

astacidean (crayfish)

palinuran (spiny lobster)

The remaining two infraorders are both 'crabs'. The **Anomura** includes the wonderful, bizarre, hermit crabs, porcelain crabs, and mole and sand crabs, in which the abdomen is either soft and twisted asymmetrically, as in hermit crabs, or short and symmetrical and flexed beneath the thorax, as in porcelain crabs. Hermit crabs, of course, tuck their vulnerable parts into gastropod shells. In these, as in all anomurans, the first pereopods form claws and the second and third are for walking, but the fourth and fifth are much reduced. Robber crabs are giant, semiterrestrial hermit crabs that climb coconut trees

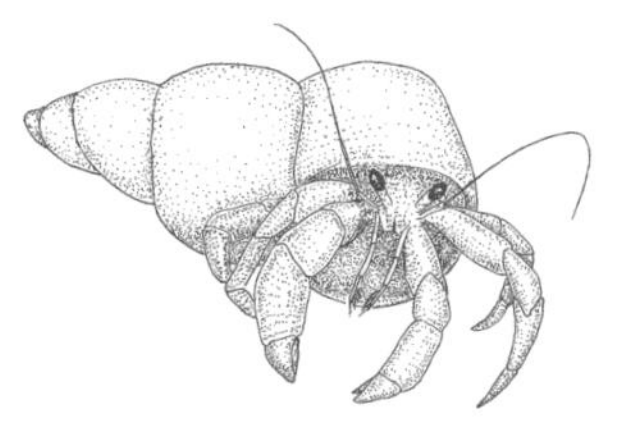

anomuran (hermit crab)

and feed on the nuts. They are among the most extraordinary creatures in nature.

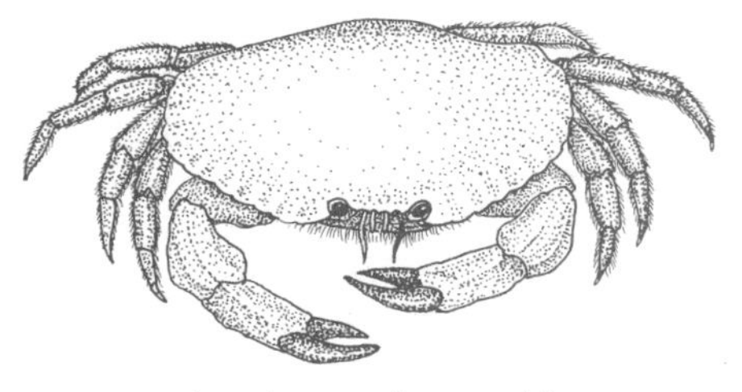
brachyuran (true crab)

Finally, the infraorder **Brachyura** are the 'true' crabs. In them the abdomen is symmetrical but it is much reduced and folded beneath the thorax, which is generally broad and flat. Most are marine but a surprising variety in the tropics are semi-terrestrial, although they depend on the sea, where they lay their eggs and their larvae develop. Crabs are immensely varied and of enormous ecological and economic importance. For example, in Australia crabs are the chief planters and propagators of mangrove trees.

Superorder Peracarida

The Peracarida, the last of the four eumalacostracan superorders, includes about 25 000 species—a significant proportion of all known crustaceans—and are astonishingly varied in form and way of life. They range from krill-like swimmers to crawlers or ticks. In size they range from planktonic types a few millimetres long to benthic crawlers at almost half a metre. They live in land or in the sea as predators, scavengers, parasites, or what you will. Yet the peracarids share a suite of features that suggests they are monophyletic. They all have an extra, hinged joint on the mandible; and they brood their young in marsupial pouches until they finally emerge resembling miniature adults, a mode of life called **direct development** (without a larval stage).

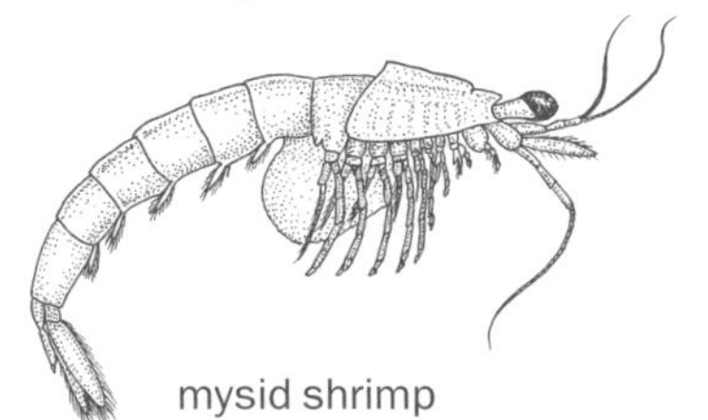
mysid shrimp

The Bruscas list nine peracarid orders in their book *Invertebrates*, seven of them shrimp-like; of these I will single out order **Mysida**, the mysid or opossum shrimps, which superficially resemble krill. Note, again, how commonly the shrimp-like form occurs among different taxa of crustaceans. The remaining two peracarid orders, the Isopoda and Amphipoda, are more varied.

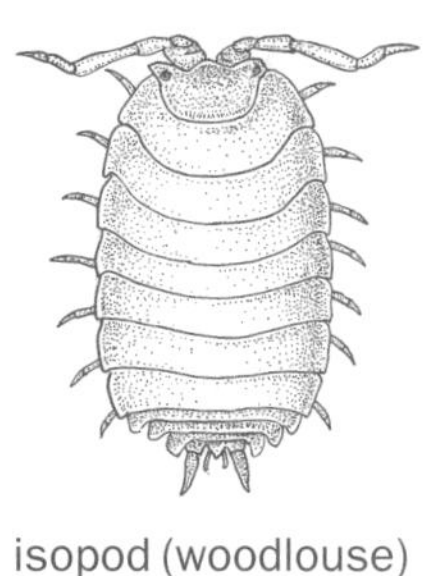
isopod (woodlouse)

- Order **Isopoda** contains about 10 000 species, which include the woodlice (or sowbugs). Woodlice have become the most successful of all the crustaceans that have invaded the land. Their 'direct development' means that they do not need water to breed, as the land crabs do; and they even have 'pseudotracheae', comparable with those of insects and some arachnids, which allow air-breathing. But most isopods are either fresh water or marine, the largest being the huge benthic marine *Bathynomus* at almost half a metre—like a giant, solid woodlouse. Other isopods are partly or wholly parasitic, sucking tissue fluid from fish or other crustaceans.

- Order **Amphipoda** also contains about 10 000 species. Most are vaguely shrimp-like, though without carapaces; but a few are like lice and cling to whales. Some are only 1 millimetre long, while some deep-sea giants reach 25 centimetres. In many marine habitats amphipods form most of the biomass. Some shelter inside jellyfish, where they rear their young, and some apparently eat their jellyfish hosts. Most familiar perhaps, in the suborder **Gammaridea**, are the *Gammarus* 'shrimps', which are known to pond-dippers, and the intertidal beach-hoppers that leap from the kelp disturbed on the shore. Some are even more semiterrestrial, and turn up in moist soil in gardens and greenhouses. There are about 6000 amphipod species, and their classification remains somewhat fluid.

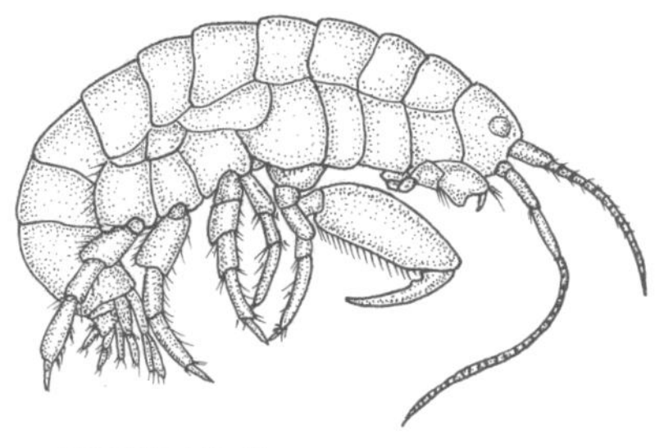

amphipod, *Gammarus*

The Crustacea*, then, is one of the most ancient, ecologically significant, varied, and generally magnificent animal groups of all. The description here is merely an outline, for a complete description would require an entire volume. In the next section I will look at the creatures that may be the crustaceans' descendants: the insects.

10

THE INSECTS

SUBPHYLUM INSECTA

THE INSECTS ARE AMONG the most remarkable creatures that have ever lived. About a third of all *known* animals on Earth today are insects: well over half a million. Indeed about a fifth of all known animals are beetles—members of a single insect order, the Coleoptera. The weevils, who form just one coleopteran family (the Curculionidae), have more known species—around 65 000—than any other *phylum* of non-arthropod animals, apart from the molluscs. Many thousands more would undoubtedly come to light if ever we managed to study tropical forest in depth.

Insects live in every environment this planet has to offer except for the open sea, including hot sulphurous springs and oil seeps; and even in this age of mass extinction many are flourishing as never before, exploiting the novel and sometimes grim environments with which our own species is replacing the pristine world. Thus the termites, of the primitive order Isoptera, are favoured by the destruction of forest that both enhances their food supply—dead wood—and spreads the grassland where many of them build their nests. One albeit extravagant estimate suggests that our planet harbours 750 kilograms of termites for every man, woman, and child. Insects pursue every conceivable lifestyle: as herbivores, scavengers, dung-eaters, parasites, and as fierce (one is tempted to add 'proud') predators—like dragonflies, which capture their prey on the wing like a peregrine falcon.

Insects are also supreme among nectar feeders and hence among plant pollinators. Hummingbirds and many a bat and marsupial, too, are nectar feeders and pollinators; but they are johnnies-come-lately to the saccharine feast. Gymnosperms and insects (probably beetles) pioneered the arts of animal pollination even before flowering plants evolved; but when angiosperms did appear, they and many groups of insects coevolved in a host of intricate symbioses.

Insects can be elaborate lovers, egging each other on with rare and highly specific pheromones or dazzling frills or colours or flashing lights, courting and perhaps copulating for hours or even days at a time; although copulation in many species is taken on the wing and can then be a speedy affair, executed in free fall. Some groups—notably the termites and various Hymenoptera, the bees, wasps, and ants—have taken social life on to a plane of co-operativeness that very few other creatures approach; perhaps only the naked mole rats are comparable, with their reproductive queen and sexually suppressed workers. Once more, however, the insects invented this lifestyle.

All that insects lack among the criteria by which 'success' is often judged, is size, for most are only 0.5 to 2 or 3 centimetres long. Size is restricted (or so, at least, is commonly suggested) by their method of respiration, which limits their oxygen intake: a system of air-tubes called **tracheae** that open into **spiracles** in the surface armour. Some insects—thrips, some beetles, and the parasitic wasps—are almost microscopic. Even so there are relative giants, like the Goliath beetle of equatorial Africa whose males weigh in at 70–100 grams (almost a quarter of a pound) and some 'stick insects', order Phasmida, up to 30 centimetres (a foot) long. Greatest of all by far were some of the ancient dragonflies that approached a metre in body length and in wingspan.

Insects also lack individual 'plasticity' of behaviour: the ability to improvise that we see in particular in many mammals, such as apes, rats, squirrels, dogs, and elephants. Even so, some seem more flexible than is at first apparent (and so a honeybee may bite its way into a flower that is too deep for its tongue); and they make up in behavioural complexity, and in their minute adaptation to circumstance, whatever they may want in mammalian-style intelligence.

As with all animals, insect success is founded in their body plan, which is unfussy and lends itself to many variations. Absolutely characteristically, insect bodies are divided into three clear tagmata. The head is composed of six basic segments carrying four pairs of appendages, which are the **antennae** (second segment), **mandibles** (on the fourth segment, just behind the mouth), **maxillae** (fifth segment, equivalent to the first maxillae of crustaceans), and **labia** (on the sixth segment—commonly a flap-like structure formed by fusion of ancestral second maxillae). Insects also typically have **compound eyes**, which are sometimes enormous but are absent in many groups. The thorax always has three segments, each carrying a pair of walking legs—so there are six legs in all (hence the alternative name of 'hexapod'). The dorsal sclerite of the first thoracic segment forms a shield, the **pronotum**, which in some groups, like the locusts (order Orthoptera), is worn like a stiff Eton collar.

Most living insects have wings—'most', because flight is the true secret of insect success and the ones that evolved wings have far outstripped the ones that have not, ecologically and in diversity (although, as I discuss later, many winged insects—like many birds—have later foregone the power of flight). Insects bear their wings on the middle and last of the thoracic segments (meso- and metathorax), although in beetles the mesothoracic wings are thickened to form shields or **elytra**, which protect the rearmost, flying wings, and in flies and mosquitoes and their relatives (order Diptera) the metathoracic wings are reduced to whirring stabilizers called **halteres**. Wings are extensions of cuticle: the supporting veins that are so helpful in identification are probably derived from tracheae.

Wings, and the mechanisms that enable powered flight, probably evolved only once in insects and all the winged insects form a clade, a subclass of the Insecta, called the Pterygota (from the Greek *pteros* meaning wing or feather). Indeed, pterygotes are the only invertebrates that truly fly although flight evolved at least three times, independently, among vertebrates. But how did insect wings evolve? Here there are two kinds of question: first, what selective forces prompted their evolution —which essentially asks why they evolved; and, second, from what parts of the ancestral body did they develop?

The first of these issues—why insects evolved wings—is like the question sometimes asked of the human eye. The problem in both cases is to imagine the intermediates. That is, we can see that a fully evolved human eye is a wonderful organ whose value is obvious; but it must have evolved in stages—but what use is half an eye? What natural selective forces could have prompted the lens to evolve before there was a retina on which to focus light? But what use is a retina without a lens to direct light upon it? It is very hard to see how two such complex components could have evolved side by side if natural selection works in the way that modern biologists think it does—chance genetic mutations producing small physical changes that just occasionally bring advantage. Similarly, insect flight does not depend simply on the visible wings. It also requires modification of the entire thoracic skeleton, which in advanced insects changes its shape as the wings beat, and an intricate array of muscles and nerves. What use were the wings—flaps of cuticle—before there were muscles to move them? But how did natural selection favour the evolution of muscles before there was anything for them to move?

Such questions are not so devastating as anti-evolutionists like to suppose. Half an eye really is better than no eyes at all. A retina is extremely useful even without a lens to provide fine focus. Even when poorly developed it can distinguish light from dark, and perceive movement. Indeed, a single photoreceptor is useful, let alone a retina. Lenses could have evolved first of all as transparent protectors and developed their focusing power later: at its simplest, after all, this requires nothing more than convexity. Thus as Darwin himself pointed out, we can observe thousands of creatures with far simpler eyes than the human's, down to and including the single eyespots of many protists. It really is very easy not only to imagine the intermediate stages, but to see examples of intermediates still in action.

Similarly, the first insect wings presumably were mere extensions of the cuticle without muscles. We can envisage several possible functions. Perhaps they served for respiration. Perhaps they evolved as heat exchangers, like the ears of an elephant or the sails of a sail-back reptile, or, indeed, like the wings of butterflies, which sunbathe in the morning to get themselves warmed up and active. Probably some early insects jumped (jumping can be a good way to get around), as many creatures demonstrate from beach-hoppers to spiders to kangaroos—and flaps improved their stability and increased their chances of landing upright. Modern grasshoppers demonstrate the principle, although, of course, the wings they use for gliding are not primitive; they are bona fide modern insect wings. We can never know for sure what really happened; but we can certainly envisage various ways in which flaps would be useful even before there were muscles to drive them. With the flaps in place, muscles could evolve later. One ancient and extinct order, the Palaeodictyoptera, does have fixed lobes extending from the **nota** (the dorsal plates of the armour in each thoracic segment) that may be wing precursors.

So how did the flaps evolve? Some authorities suggest they grew as extensions of the body wall: but others have argued that they evolved from the gills of aquatic insect ancestors. Modern crustaceans provide a clue to this.[1] Thus in many crustaceans (crabs, lobsters, shrimps) the gills are feathery extensions from the topmost joints of the legs. There is much evidence, provided not least by Jarmila Kukalová-Peck of Carleton University in Ottawa, that the topmost joints of the ancestral insect leg became integrated into the body wall of modern insects, which would conveniently move the gills on to the body wall. Such structures could then be modified into the aerodynamic or heat-exchanging flaps that we need to hypothesize.

The notion that insect wing-precursors might indeed have evolved from crustacean gills is now supported by intriguing new evidence from Michalis Averof (who is now at the European Molecular Biology Laboratory in Heidelberg, Germany) and Stephen Cohen. They reported in 1997 that genes involved in the development of insect wings are homologous with genes that promote the development of gills on the legs of the brine shrimp *Artemia*—this being an extension of Averof's work at Cambridge with Michael Akam on Hox genes (as described in Section 8). When two organs with structural similarities, occurring in two creatures that are clearly related (insofar as they belong to the same phylum), prove to be fashioned by the same genes, there is a good reason for declaring that the two organs are homologous. It is not particularly surprising that the function and structure of the organ have shifted so radically. This happens all the time in nature: jawbones become ear bones, feet become hands, and so on. Natural selection is opportunistic, and is much more inclined to expropriate, and modify, than to invent.

Finally, the third of the insect tagmata, the abdomen, primitively has 11 segments although in many orders the number is reduced through loss or fusion. Only the most primitive insects (and some embryos) carry true abdominal appendages but many have **cerci** on the last segment of the abdomen, which are generally sensory but in earwigs (Dermaptera) form prehensile callipers.

It is hard to find two entomologists who agree on every detail of insect classification; and there are plenty of authoritative opinions. Insects are of such supreme ecological and economic importance that entomology has emerged as a discrete discipline that has attracted thousands of biologists, including some of the most eminent. The tree shown here is synthesized from two main sources, which in practice agree in all but small details: Richard and Gary Brusca's *Invertebrates* (1990) and N. P. Kristensen's much-acclaimed chapter on 'Phylogeny of extant hexapods' in *The Insects of Australia* (1991). Both the Bruscas and Kristensen divide the Pterygota into 25 orders, although some other authorities offer more. The basic 25 are mentioned here in the text; but eight are minor and the tree shows only 17—all of which are truly significant.

[1] Some zoologists feel that some ancient crustacean was the ancestor of insects; others disagree. But even if the parents of the first 'true' insect was not a crustacean, crustaceans can still illustrate the principle.

A GUIDE TO THE INSECTA

Most living insects either have wings, or are descended from ancestors who had wings and have subsequently abandoned them. That is, most living insects belong to the great clade—we may call it a class—known formally as the **Pterygota**. But there is a significant minority of insects who lack wings not because they have abandoned them, but because they never evolved them in the first place; and these can be called **apterygotes**, meaning 'without wings'.

In precladistic days, apterygotes were commonly placed in a formal grouping, the Apterygota; but as the tree shows, they do not form a clade. There are, in fact, five discrete apterygote lineages—Collembola, Protura, Diplura, Archaeognatha, and Thysanura—and they are all quite distinct, apart from the Collembola and Protura, which clearly enjoy a special relationship. Thus, although the term 'apterygote' remains a useful adjective, the traditional taxon Apterygota is best abandoned. The pattern among insects, therefore, is the same as among mammals and birds. All of those great groups have given rise to several distinct lineages, but in each case only one has come to dominate. Among birds the neornithines represent just one surviving lineage out of eight, while among mammals the therians and the egg-laying monotremes (now represented only by the platypus and echidnas) are again the sole survivors of eight known groups.

All insects have six legs—it is a key synapomorphy; hence the alternative name Hexapoda. But here there is a complication, because different systematists have used the term 'Hexapoda' in different ways. For some, 'Hexapoda' is simply synonymous with 'Insecta'; others call the whole group 'Hexapoda' while reserving the term 'Insecta' for the Pterygota + Thysanura + Archaeognatha. In fact, the Pterygota + Thysanura + Archaeognatha do form a clade, and one thing they have in common is the deployment of mouthparts, which in all three groups stand proud of the face and so are said to be **ectognathous**. By contrast, in the Collembola, Protura, and Diplura the face has grown around the mouthparts so that they appear to be sunk into a pit: the condition described as **entognathous**. But entognathy is not simply a primitive condition, as sometimes supposed, for it now seems that the Diplura on the one hand and the Protura + Collembola on the other, each evolved entognathy independently. Finally, just to complete the complications, Sidnie Manton in the 1970s applied the term 'Insecta' only to the winged insects—so, for her, it was synonymous with 'Pterygota'—and regarded each of the aptyerygote groups as a subclass.

In this book I am following the first, and perhaps the most conventional route: treating Hexapoda and Insecta as synonyms (although this convention actually renders the term 'Hexapoda' redundant). So now let us put nomenclature to one side and look at the creatures themselves, working as usual down the tree from top to bottom.

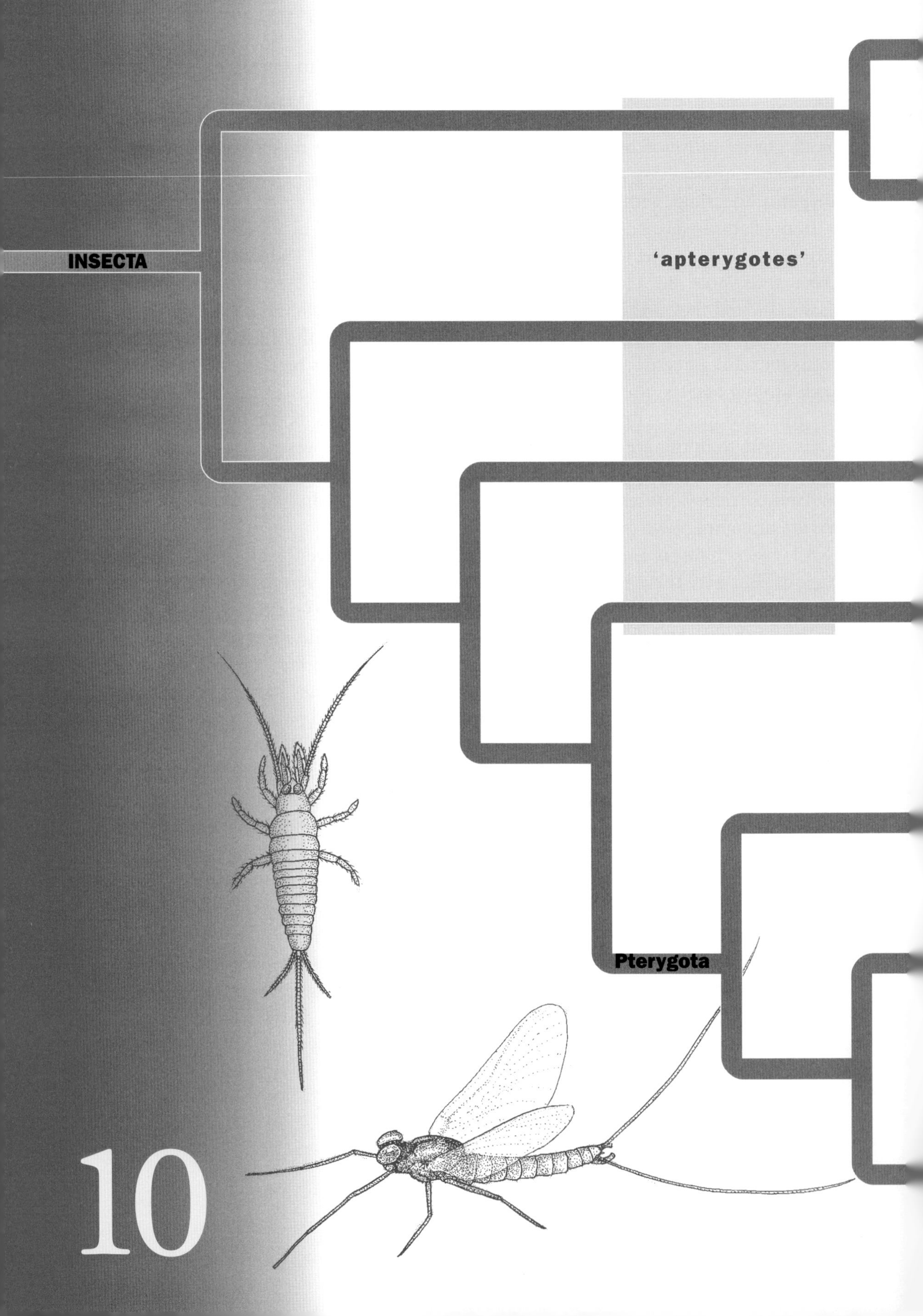
INSECTA
'apterygotes'
Pterygota
10

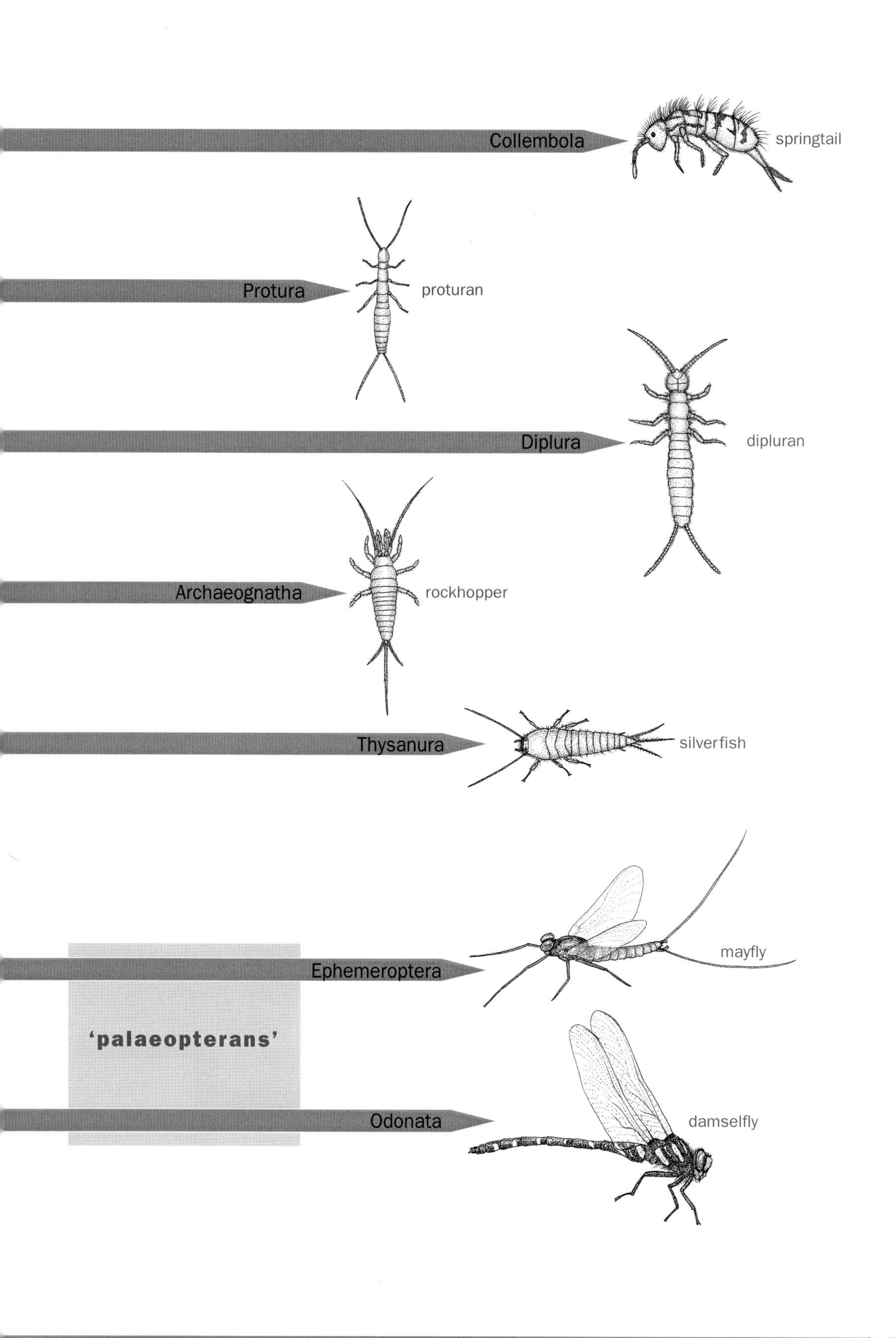

10a · Neoptera

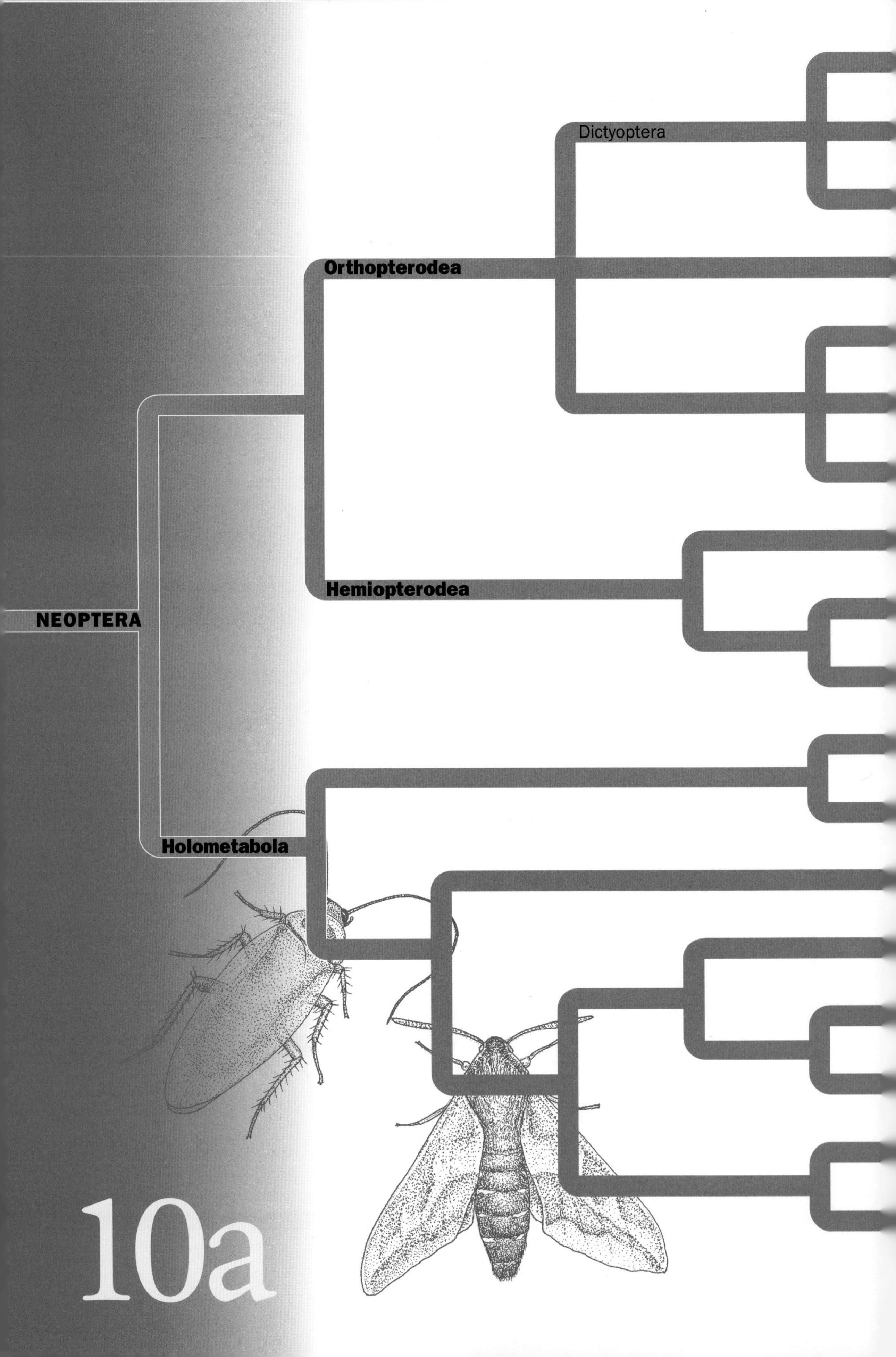
Dictyoptera
Orthopterodea
NEOPTERA
Hemiopterodea
Holometabola
10a

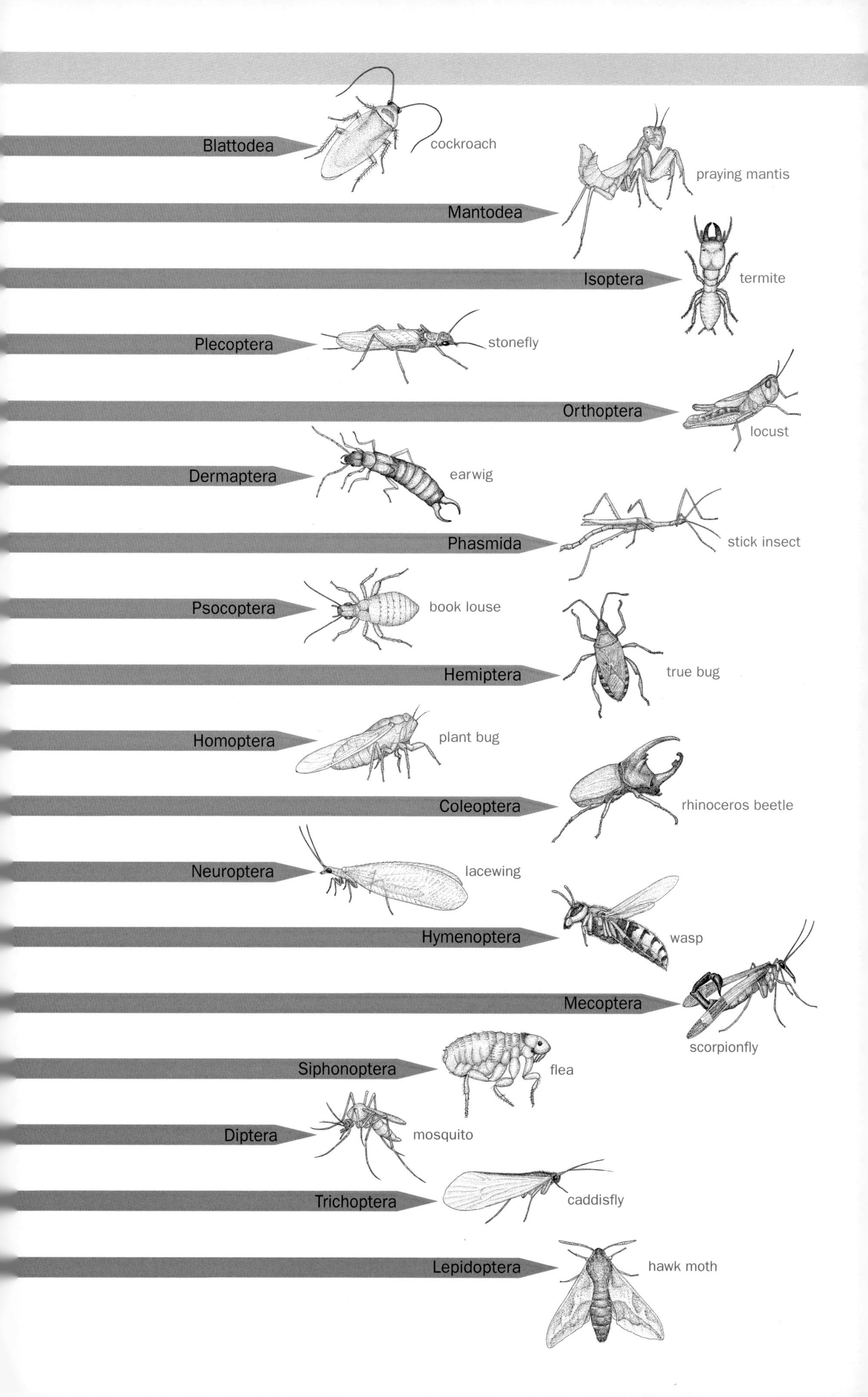
Blattodea
cockroach
praying mantis
Mantodea
Isoptera
termite
Plecoptera
stonefly
Orthoptera
locust
Dermaptera
earwig
Phasmida
stick insect
Psocoptera
book louse
Hemiptera
true bug
Homoptera
plant bug
Coleoptera
rhinoceros beetle
Neuroptera
lacewing
Hymenoptera
wasp
Mecoptera
scorpionfly
Siphonoptera
flea
Diptera
mosquito
Trichoptera
caddisfly
Lepidoptera
hawk moth

INSECTS WITHOUT WINGS: THE 'APTERYGOTE' CLASSES

As the tree shows, living apterygotes divide into five main groupings that separated from each other an extremely long time ago and, logically, deserve the high ranking of class, more or less as Sidnie Manton proposed. Besides, I like the contrast between the ecological and the phylogenetic status: that such lowly creatures should represent such deep phylogenetic branches. Three of them—**Collembola**, **Protura**, and **Diplura**—are entognathous, while two—**Archaeognatha** and **Thysanura**—are ectognathous. We could say that the archaeognaths and the thysanurans appeared *after* the evolution of ectognathous jaws but *before* the evolution of wings. Two of the groups (the Collembola and Protura) do seem to be related, as shown, but the others are separate.

The five wingless classes all practice indirect insemination, as myriapods do: that is, they transmit the sperm in a **spermatophore**. (Flying insects, by contrast, practise copulation.) The apterygote groups also have **direct development** (the young roughly resembling the adults) rather than passing through different specialist stages, as winged insects often do. In general—like myriapods and the humbler arachnids, or indeed the crustacean woodlice—wingless insects are creatures of damp places.

Springtails: class Collembola

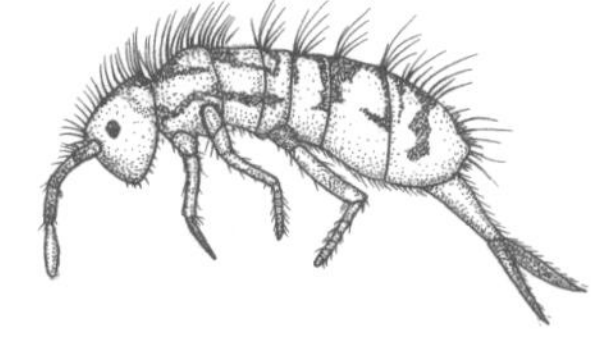
collembolan (springtail)

Springtails at least are well-enough known to have a common name; one that derives from the appendages of the fourth or fifth abdominal, which form a hydraulic spring (and operates by the pressure of haemocoelic fluid). Their abdomens are much shortened, with six or fewer segments; and they often lack tracheae (again, diffusion is enough). Their eyes are peculiar, and indeed are an apomorphy: they are composed of up to 10 'stemmata', each of which seems to be formed from the fusion of compound-eye facets. In other words, they seem to be derived but simplified compound eyes. Their mouthparts are for biting and chewing. About 2000 species are known, in 11 families.

Proturans: class Protura

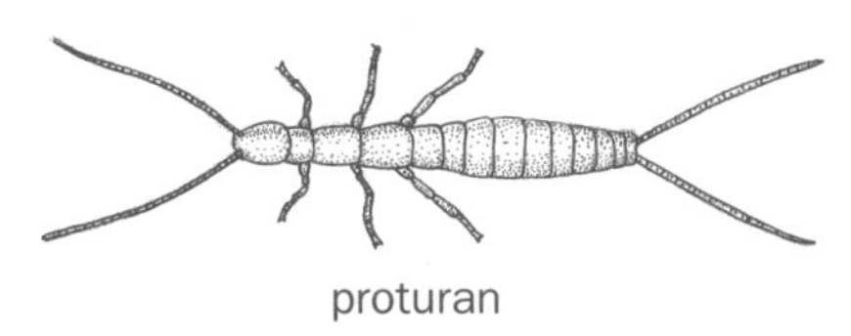
proturan

Proturans are ultimately humble: eyeless, whitish, and never larger than 2 millimetres. They have sucking mouthparts. Their antennae are vestigial but, oddly, they carry their first walking legs aloft to serve as antennae. They are rare, but about 100 species are known.

Diplurans: class Diplura

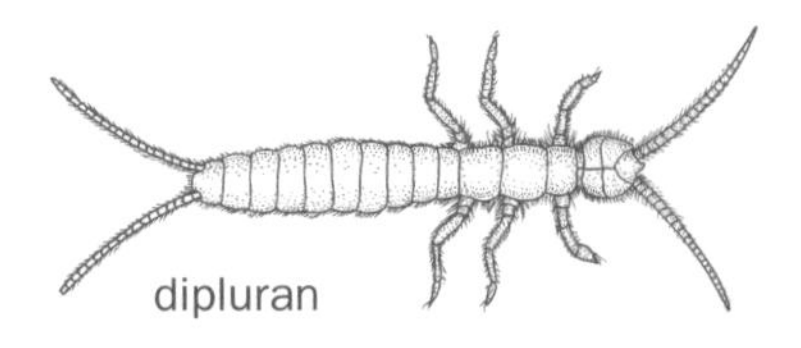
dipluran

Here are more whitish and eyeless creatures, all less than 4 millimetres long. They do, however, have tracheae, and their 10-segment abdomens bear seven pairs of simple appendages called **styli**. Their mouthparts are for chewing. About 100 species are known, in seven families.

Silverfish and rockhoppers: classes Thysanura and Archaeognatha

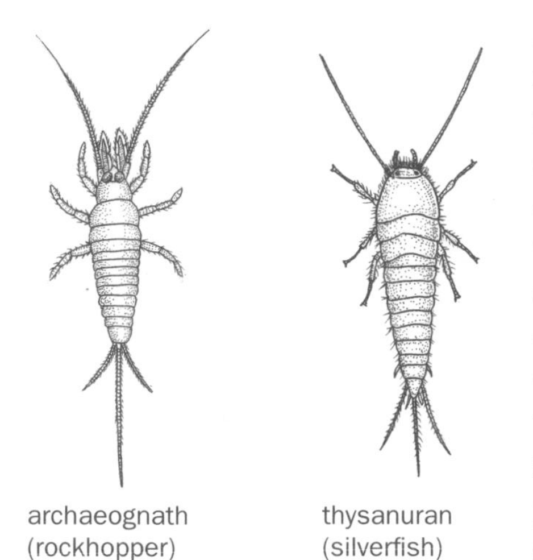
archaeognath (rockhopper) thysanuran (silverfish)

Small creatures again. The thysanurans (silverfish) have much-reduced compound eyes and mandibles for biting and chewing, while the archaeognaths (rockhoppers) have enormous eyes that meet in the midline of the head, and their jaws are specialized picks for chiselling algae and lichen from the rocks. Both groups are distinguished by their lateral abdominal styli (3–7 pairs) and their long, three-pronged tail: a triptych of caudal filaments. Some thysanurans impinge on human beings as minor household pests. The two classes between them contain about 700 named species.

So much for the apterygotes. By far the majority of insect species have wings—or, if they do not, it is because they have abandoned them, like flightless birds. Flight is what makes insects important and interesting. All flying insects belong to a clade—reasonably ranked as a class—the Pterygota.

WINGED INSECTS: CLASS PTERYGOTA

The pterygotes are divided into two great groupings, depending on their mode of flight. Two orders, the Ephemeroptera (mayflies) and the Odonata (dragonflies and damselflies), practise a form of flight, which, so aerodynamic analysis and the fossil record suggest, is primitive. Specifically, they have vertical (dorsoventral) muscles within the thorax that power the wings 'directly', simply by pulling them down. (Note that 'primitive' need not mean 'crude': dragonflies are marvellous fliers, hovering, reversing, taking their insect prey on the wing.) In all other flying insects, like bees and flies, powerful longitudinal muscles move the wings 'indirectly', by distorting the shape of the thorax.

In traditional classification, the Ephemeroptera and Odonata were grouped together to form the infraclass Palaeoptera, while the rest—the majority of well over 20 orders—formed the infraclass Neoptera. In these cladistic days all is not so simple.

The Neoptera do seem to form a clade, and **Neoptera** can remain as a formal taxon. But, again, the erstwhile 'Palaeoptera' is really a loose, paraphyletic assemblage of creatures that do not seem to be closely related, and the term 'palaeopteran' is best treated as an informal adjective. The intricate pattern of the wings of palaeopterans is also primitive, and so is the way they are held at rest: straight out like an aeroplane, or along the abdomen with the dorsal surfaces pressed together. All neopterans can fold their wings in various ways, over their abdomens.

THE PALAEOPTERAN ORDERS

Mayflies: order Ephemeroptera

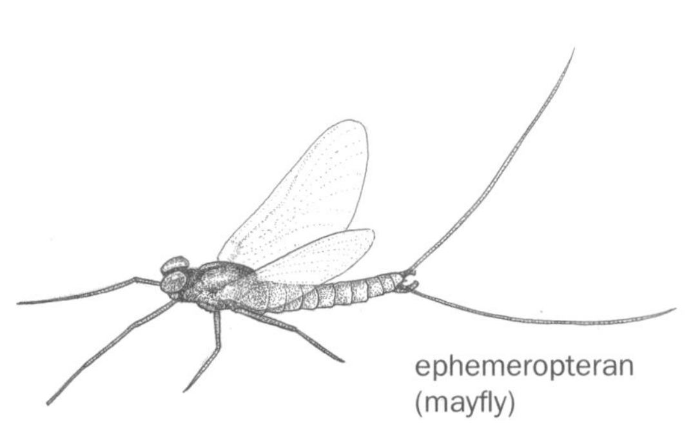
ephemeropteran (mayfly)

Mayflies are queens for a day. The aquatic larvae with their lateral gills and substantial mouthparts dominate the life cycle, while the adults, with only vestigial mouthparts, do not feed at all but live only for as long—a few hours or days—as it takes to find a mate and lay eggs. To hurry things along they sometimes form huge nuptial swarms and copulate on the wing. Uniquely, mayflies have an near-adult stage called a **subimago instar**, which is winged but the wings, like the rest of the body, are encased in an extra membrane. Mayflies have long lateral cerci, usually with a caudal filament in between. About 2100 ephemeropteran species are known.

Dragonflies and damselflies: order Odonata

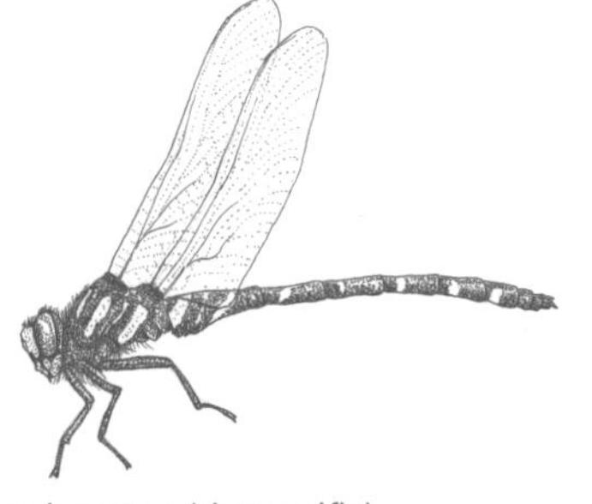
odonatan (damselfly)

Dragonflies and damselflies are glorious aerial predators, like bats or swifts or peregrines. They locate their insect prey with their huge compound eyes. Then they seize it in their legs, which, directed slightly forward, form a basket—the nearest thing in nature that I can think of, to a butterfly net. Then they crush their prey in massive mandibles. The two main odonate lineages, dragonflies and damselflies, demonstrate the two palaeopteran ways of holding the wings at rest: dragonflies straight out, damselflies over the back, with dorsal surfaces together. The nymphs (immature stage—see below) are aquatic, with rectal gills. They, too, are fearsome predators, seizing minnows and other small creatures with labia modified as extending tongs. At metamorphosis, the nymph crawls out of the water and the fabulous **imago** (full adult) emerges straight into the *bel air*. More than 5500 odonate species are known.

Whether the Odonata or the Ephemeroptera are closer to the Neoptera is uncertain. But it has been argued that the Ephemeroptera are different from all other pterygotes because they, uniquely among living types, moult one more time even after they have acquired wings. But others claim to have fossil evidence that some other groups also had wings as subadults. If this is the case, then we would have to argue that subadult wings are a primitive feature of insects, which other groups happen to have lost—independently of each other—but which happens to have been retained in mayflies. In such a case, retention of subadult wings in mayflies would not be sufficient to justify classing them as the sister group of all other winged insects. Some other evidence, however, suggests that the Odonata are sisters to the rest, which would also leave the Ephemeroptera on the outside. In short, the arrangement shown here—with Odonata between Ephemeroptera and Neoptera—is as good as any, at least for want of a better.

SUBCLASS NEOPTERA

The subclass Neoptera is divided into three main groupings, commonly known as divisions, according mainly to the way in which the insects develop: how they progress from egg to adult. Insects, being arthropods, need to moult in order to grow: their armour, once hardened, cannot be stretched. This means that the life of arthropods is perforce divided into separate stages, which in turn means that natural selection can (and does) operate on each stage separately, and often produces young forms that differ enormously from the adults in structure and in way of life. We see this phenomenon spectacularly among crustaceans, whose larvae may bear no obvious resemblance to the adults at all. We also see it among insects. But the extent to which the young differ from the adults varies among the three neopteran groups.

Thus in two of the three divisions, the **Orthopterodea** and the **Hemiopterodea**, development is said to be direct, or **hemimetabolous**. The young forms, commonly but not invariably called **nymphs**, may not closely resemble the adults (although they often do) but they are very obviously insects, with jointed bodies and six legs: all they invariably lack is wings and sexual maturity. Each of the various stages through which the young passes between moults is called an **instar;** the adult is called an **imago**.

But in the third division, the **Holometabola**, development is indirect or **holometabolous**. Characteristically the life cycle—after the egg stage—is broken into three distinct phases. First comes the **larva**, which is dedicated to feeding and growth, and may take any one of a huge variety of forms. Larvae are often worm-like—sometimes with legs, sometimes without, and sometimes with lobopod-like abdominal appendages called **prolegs**. Such larvae have various common names: 'grubs' (Coleoptera), 'maggots' (flies), or 'caterpillars' (Lepidoptera). Yet they can vary tremendously even within orders. Thus fly larvae are maggots, while those of the related mosquitoes are active swimmers. But whatever the shape of the larva, it always contains patches of embryonic tissue called **imaginal discs**; and it is from these that the adult phase is finally formed.

Stage two is the **pupa**. It is superficially inactive and may serve as a dormant stage to help the insect through harsh times, but inside there is turmoil. Much of the larval tissue is broken down—most, indeed, apart from the imaginal discs. These grow and are reorganized to form the organs of the adult—wings, sex organs, compound eyes, and all the rest. The broken-down larval tissues provide the raw materials. Sometimes the pupa is encased in silk to form a cocoon or **chrysalis**. Finally the imago emerges from the pupa. The transitions from larva to pupa and pupa to imago are called **metamorphosis**.

Division Orthopterodea

The orders Blattodea (cockroaches), Mantoda (mantids), and Isoptera (termites) form an obvious grouping, and are sometimes combined within the superorder **Dictyoptera**. The cockroaches may indeed be paraphyletic—ancestral to the termites. These three orders plus the Plecoptera (stoneflies), Orthoptera (locusts), Dermaptera (earwigs), and Phasmida (stick insects) are in turn combined to form the Orthopterodea, which it is convenient to describe simply as a 'division'. The Orthopterodea are generally primitive: hemimetabolous, with two pairs of wings, and biting-chewing mouthparts.

Cockroaches: order Blattodea

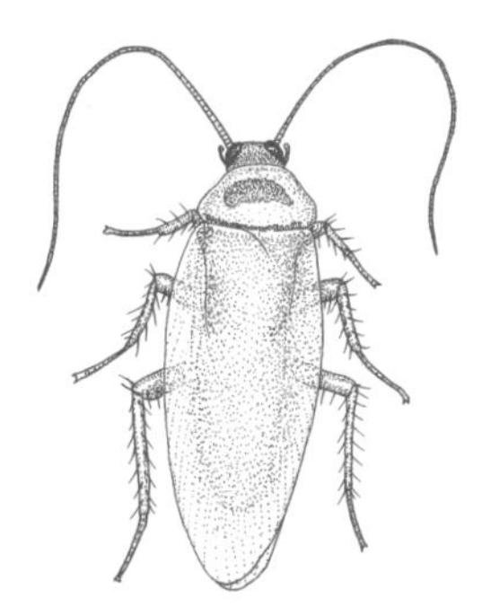

blattodean (cockroach)

Cockroaches have flat bodies with leathery forewings, large fan-like hindwings, and long legs for running. The dorsal sclerite (tergite) of the first segment of the thorax—the **pronotum**—forms a distinct shield in front of the wings. There are distinct cerci near the end of the abdomen. Cockroaches lay their eggs in cases called **oothecae**. Most cockroaches are tropical, but some are temperate. Some live in caves, some in deserts, some in the nests of ants or birds. Some are omnivores and others more restricted; some live on wood and have an intestinal flora to help digest cellulose. Termites are thought to have evolved from cockroach-like ancestors. Some cockroaches are notable and much-feared pests. They are often called 'black beetles' in Britain, though they are not black and look nothing like beetles, while in the US 'roach-phobia' largely replaces the British arachnophobia. But fewer than 40 of the 3700 known species of cockroach live in houses.

Mantids: order Mantodea

As hunters, a mantid is the insects' answer to the chameleon; a static, well-disguised creature that preys on insects and spiders by a lunge—not of the tongue, as in the chameleon, but of the great raptorial front legs. Mantids have a highly mobile

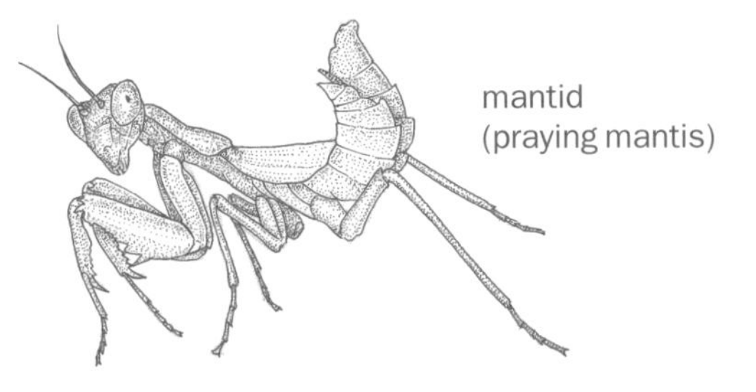

mantid (praying mantis)

periscope of a head with huge compound eyes, set at the end of an elongate, often elaborated, neck-like front part of the thorax (prothorax). Their forewings are thickened, their hindwings membranous. Female mantids are famous for eating their mates, often during copulation. The method is gruesome but economical, and avid Darwinians might suggest that a female who eats her mate ensures thereby insures that he will not inseminate any other females, whose offspring would, in turn, be rivals to her own. Most of the 1800 species of mantid are tropical. A few are temperate.

Termites: order Isoptera

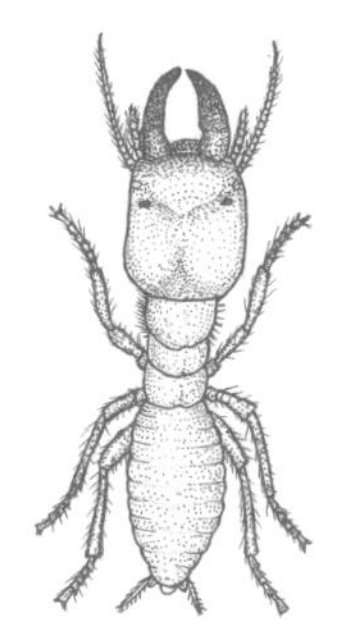

isopteran (termite)

Termites are small, soft-bodied wood-eaters. They, together with the hymenopterans (notably bees, wasps, and ants), invented the extreme form of communal living described as **eusocial**. In eusocial societies only the queen and the specialist males can reproduce while the rest form distinct **castes** of general-purpose workers or protective 'soldiers' that generally are effectively sterile. There are some eusocial mammals, notably a subterranean rodent called the naked mole-rat; but, in them, sterility is reversible, and the worker males and females can become reproductive when the queen dies.

Termite nests are wonderful, sometimes several metres high and with a vast and intricate labyrinth below ground; superbly ventilated in ways that architects are only now beginning to appreciate. All termites harbour microbes in their guts to help digest wood cellulose. More primitive species use flagellate protozoans, while the more advanced favour bacteria. Termites are becoming even more plentiful as tropical forests are felled (simultaneously providing food and space) and the methane generated by the flora and fauna of their guts contributes significantly to global greenhouse warming. About 2000 termite species are known.

Stoneflies: order Plecoptera

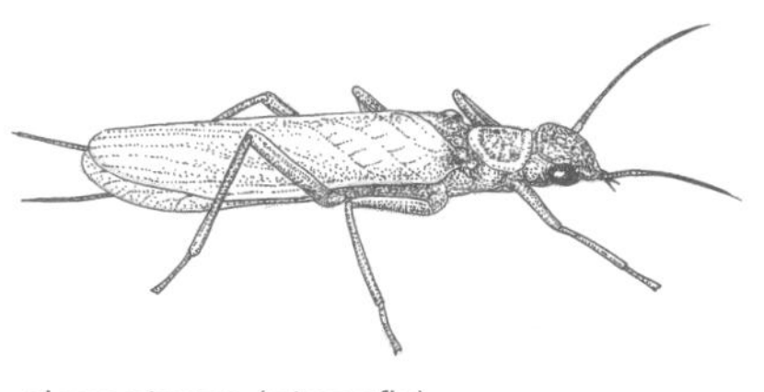

plecopteran (stonefly)

In stoneflies, as in mayflies, the aquatic nymphs dominate the life cycle; in fresh water, they are important predators and also significant prey. The adults have reduced mouthparts and although they generally feed they are short-lived and do not long survive mating. Know them by the primitively veined wings that they fold over their bodies at rest, their long antennae, and—usually—their long articulated cerci. About 1600 plecopteran species are known.

Grasshoppers, locusts, and crickets: order Orthoptera

Orthopteran form is unmistakable: bodies flattened side-to-side; large head; large pronotum like a stiff Eton collar; leathery forewings and fan-like hindwings; often

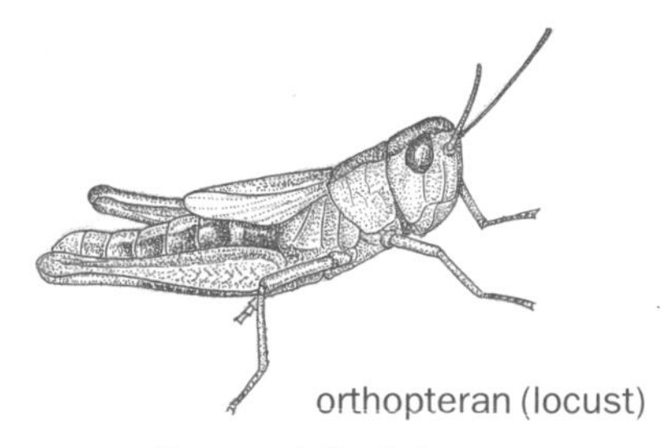
orthopteran (locust)

large rear legs cocked into springs for a hop or a veritable leap. Some of the largest insects are orthopterans—up to 12 centimetres long and 24 centimetres in wingspan. Some are omnivores or even predators but most are herbivores—locusts are prodigious consumers of crops. Some orthopterans **stridulate**, especially males, by rubbing their specially modified forewings together or the hind femur against a special vein of the forewing (but *not* by rubbing their hindlegs together!). Different cricket species have different calls; in many cases this is the only way to tell them apart. (The species that features in all American movies set in the tropics is known to aficionados as 'the Hollywood cricket'.) About 20 000 orthopteran species are known.

Earwigs: order Dermaptera

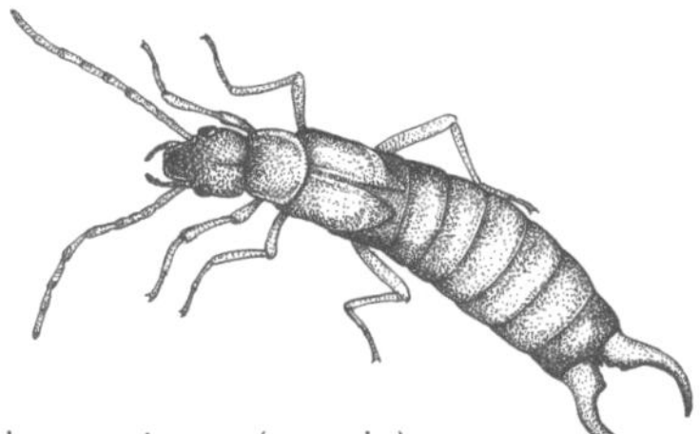
dermapteran (earwig)
Forficula auricularia

Earwigs are mostly tropical, but there are many temperate species; instantly recognized by their heavily sclerotized and prehensile cerci, which serve a variety of purposes, including helping to fold the membranous hindwings (when present) under the small, leathery, protecting forewings. Little is known about earwigs, but most seem to be nocturnal omnivorous scavengers. The females are good mothers; the hemimetabolous offspring are often found gathered around their mothers. About 1100 species of earwig are known.

Stick insects, leaf insects, and their allies: order Phasmida

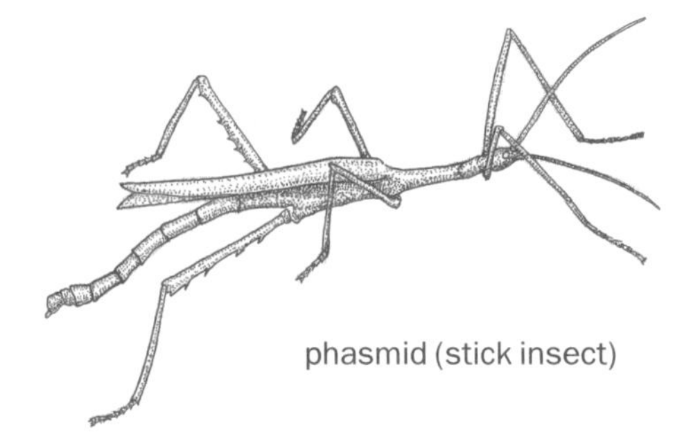
phasmid (stick insect)

Phasmids generally have long bodies like sticks or flat and often frilly bodies like leaves; they are masters of disguise. Like earwigs, they hide fan-like hindwings under short leathery forewings—although these are sometimes lacking. They are herbivores. Although most are under 4 centimetres long some stick insects reach 35 centimetres. More than 2500 phasmid species are known.

Division Hemipterodea

A further seven orders—Psocoptera (book lice), Thysanoptera (thrips), Zoraptera (no common name), Hemiptera (true bugs), Anoplura (sucking lice), Mallophaga (biting lice), and Homoptera ('plant bugs')—form the division Hemipterodea. These creatures have an economical quality—their anatomical features are in general simplified. Thus their antennae are usually short, they lack cerci, and their wings, when present, have reduced venation. Some are fluid feeders, with mouthparts for piercing and sucking like hypodermic syringes. Like the orthopterodeans they develop hemimetabolously but may have one or two pupa-like stages.

The tree shows, and I describe here, only three of the hemipterodean orders—the Psocoptera, Hemiptera, and Homoptera.

Book and bark lice: order Psocoptera

Psocopterans have long and filiform (whip-like) antennae and chewing mouthparts. They are small, but not necessarily minute: 1–10 millimetres long. Like many a snail, psocopterans are generally consumers of microflora—algae and fungi—and live in the moist places where the microflora is found. Because these places include food stores, museum collections, and even the bindings of books, psocopterans can be pests. About 2600 psocopteran species are known.

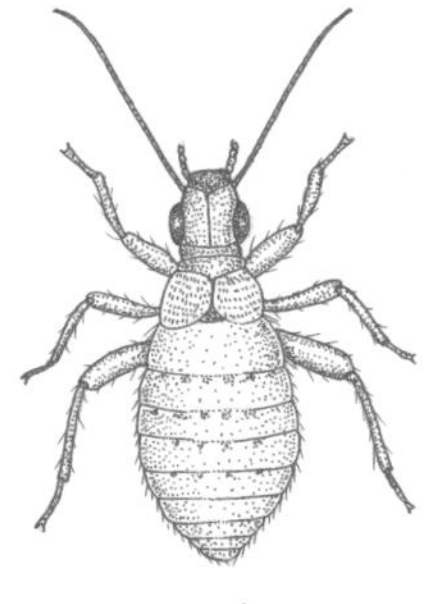
psocopteran (book louse)

True bugs: order Hemiptera

Bugs are varied and ecologically significant creatures with piercing and sucking mouthparts that form an articulated beak. The pronotum forms a conspicuous shield, as in cockroaches; the two pairs of wings lie along the abdomen at rest and the forewings, smaller than the hindwings, are hardened near the body (basally) but membranous towards the ends (distally). Most bugs are herbivores and some of these are pests of huge economic importance; but there are also many predators, and some that are specialized as ectoparasites of vertebrates. Some, like bedbugs, are extremely irritating and some are serious carriers (vectors) of disease, such as the protozoan that causes Chagas' disease, which is the Latin American equivalent of African sleeping sickness (the parasites of which are carried by a dipteran fly). Hemipterans also include some endearing creatures like backswimmers and water boatmen. An enormous order: about 35 000 species are known.

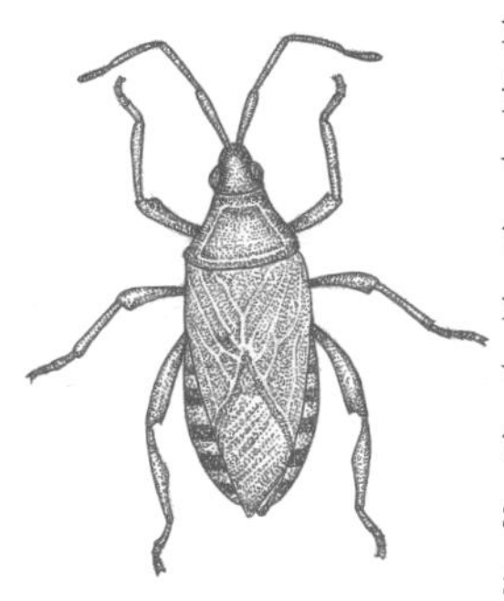
hemipteran (true bug)

Cicadas, aphids, and their allies ('plant bugs'): order Homoptera

'Plant bugs' are characterized by a piercing and sucking beak, two pairs of wings held tent-like over the abdomen, a large pronotum, hindlegs often adapted for jumping, and a body in many protected by waxy secretions. Homopterans, in short, are the plant hoppers, leaf hoppers, tree hoppers, spittlebugs and froghoppers, whiteflies, aphids, cicadas, scale insects and mealy bugs; suckers of plant juices and often major pests, wilting crops and transmitting viruses. Most species exude excess sugars from the anus as

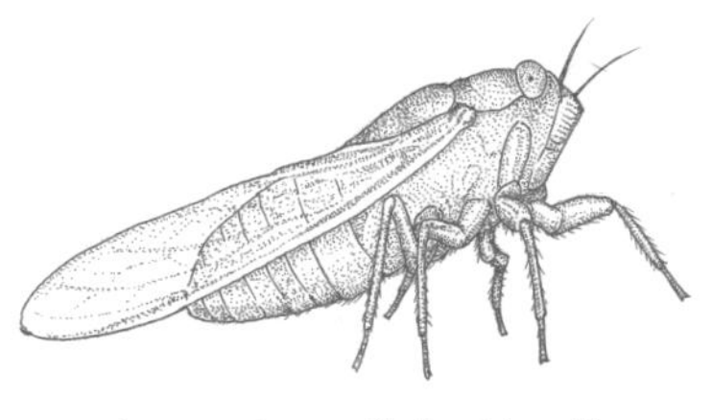
homopteran ('plant bug')

'honeydew'. More than 33 000 species are known from most environments around the world.

Division Holometabola

The eight orders that are commonly grouped within the Holometabola include some of the supreme delights of all nature. These are the insects with 'complete', holometabolous development, whose larvae (bearing buds of the future wings internally) usually live quite different lives from the adults. All eight holometabolous orders demand description.

Beetles: order Coleoptera

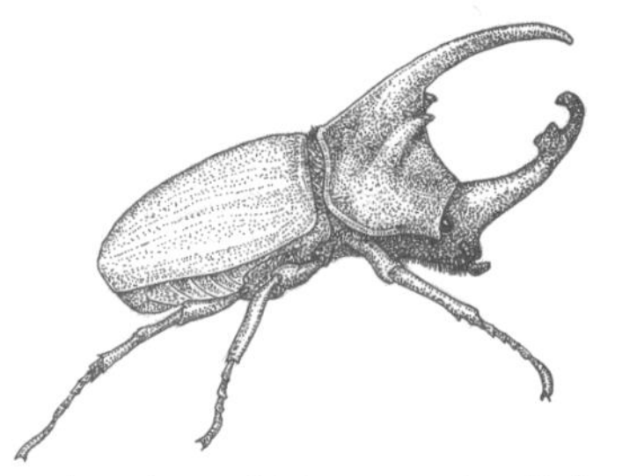
coleopteran (rhinoceros beetle)

Who would guess that a heavily armoured insect body with forewings modified as leathery or horny protective elytra would prove the most versatile of all forms? But almost a third of all known species on Earth are beetles. Sixty-five thousand species are weevils, but there are scores of others, from ground beetles, rove beetles, tiger beetles, long-horned beetles, leaf beetles, click beetles, death-watch beetles, and ladybirds (which Americans erroneously call ladybugs, mistaking the predator for the prey), to the whirligig beetles and fierce diving beetles of fresh water, and a host of amazing specialists such as fireflies and glow-worms. In all, 300 000 species of beetle are known, and there must be hundreds of thousands more.

Lacewings, antlions, and snakeflies: order Neuroptera

Adult neuropterans rest their highly veined wings like a tent over the abdomen. They have mouthparts for biting and chewing, and many species prey on other insects. The young forms are remarkable: the larvae have well-developed legs—and the pupae, too, have free appendages and functional mandibles, and may walk about before the metamorphosis (albeit without feeding). More than 4700 neuropteran species are known.

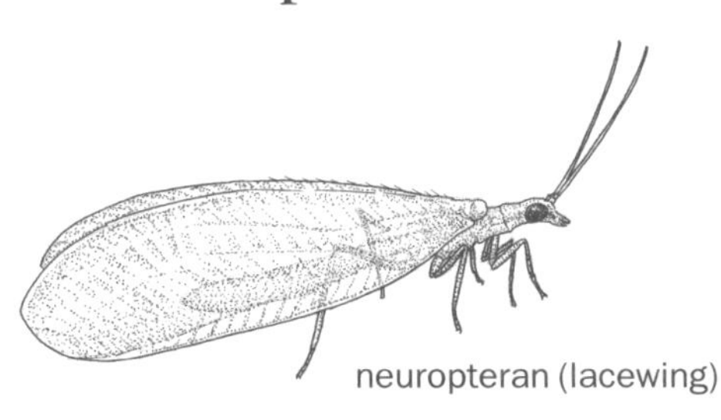
neuropteran (lacewing)

Ants, bees, wasps, and sawflies: order Hymenoptera

In hymenopterans, the smaller, membranous hindwings are connected to the forewings by hooks, to form a single functional unit. Most have biting mouthparts and some, like bees, also have an extended labium to serve as a nectar-lapping tongue. Beetles were probably the first insect pollinators, forming symbiotic relationships with gymnosperms before angiosperms evolved. But the Hymenoptera (especially bees) and the Lepidoptera have turned nectar feeding and pollination into an art form. They have shaped the evolution of flowering plants, and coevolved in their turn. But hymenopterans also feed in many other ways. Among the most remarkable are the many parasitic (ichneumon) wasps, some almost microscopic, that lay their eggs in the bodies of other insects and the larvae eat their way out. They are widely deployed as agents of biological pest control.

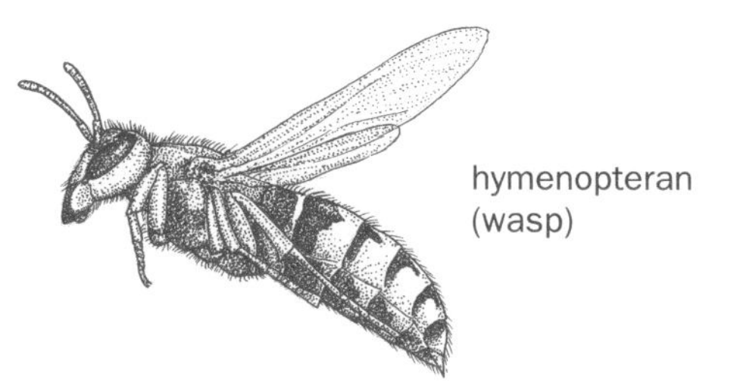
hymenopteran (wasp)

Female hymenopterans are diploid, with the usual two sets of chromosomes—one set inherited from each parent. The males are haploid—with only one set of chromosomes. This means that female hymenopterans inherit only half of their mother's genes, yet inherit *all* of their father's genes. So hymenopteran sisters—workers—have three-quarters of their genes in common and sociobiological theory predicts that this should prompt them to co-operate to an extraordinary extent. So they do. Hymenopterans have reinvented eusocial behaviour (remarkably similar to termites) many different times independently although (unlike termites) there are still many solitary hymenopterans.

There are two suborders of Hymenoptera. The more primitive **Symphyta** are the 'thick-waisted' sawflies and horntails, in which the males and females are similar and always have wings, and the larvae are generally caterpillar-like. The suborder **Apocrita** includes the more advanced ants, bees, and wasps, with a clear constriction ('waist') between the first and second abdominal segments. These are the more social types, often living in nests with distinct castes, and tending the leg-less, grub-like larvae. About 4700 symphytan species and about 125 000 apocritan species are known.

Scorpionflies and their allies: order Mecoptera

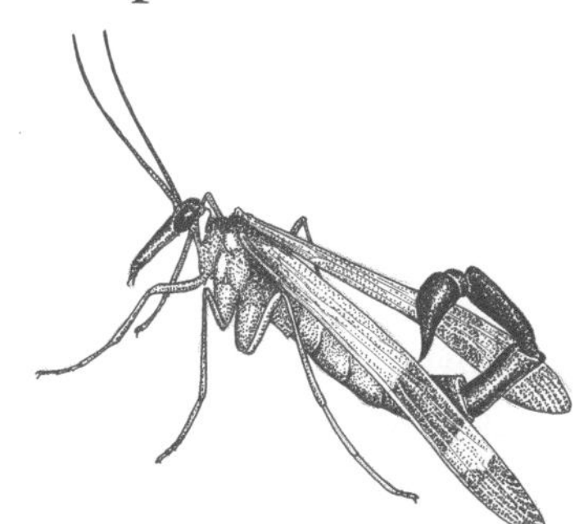
mecopteran (scorpionfly)

Scorpionflies are slender creatures, with narrow membranous wings held at the sides when at rest, long thin legs, and antennae half as long as their bodies. The females have a pair of cerci and the males have prominent and complex genitalia that resemble the sting of a scorpion; hence the name. Most are forest-dwellers, some are nectar feeders, some scavengers, and some prey on other insects. About 500 mecopteran species are known worldwide.

Fleas: order Siphonoptera

Fleas are small ectoparasites (less than 5 millimetres long) of birds and mammals with mouthparts for piercing and sucking, legs for clinging and jumping, and compressed bodies with hard armour that elude easy capture. The antennae are short and

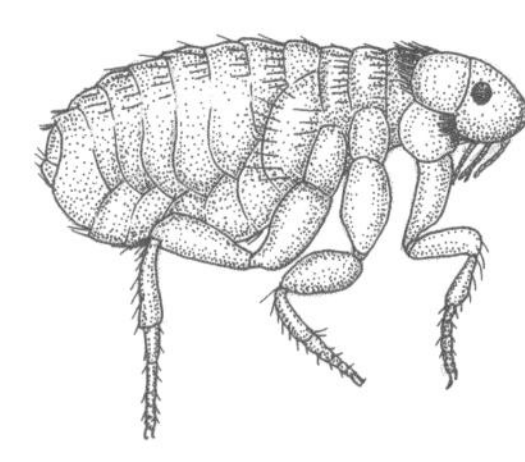
siphonopteran (flea)

lie in deep grooves on the sides of the head; compound eyes are often absent. Like lice, the Siphonoptera have abandoned wings. The larvae feed away from but close to the host—for example, on nest debris—and the pupae are cocooned. Some stick mostly to one host species but others hop happily from species to species. They may carry other pests and diseases from host to host: parasitic worms from dog to dog, or dog to cat; the plague bacterium between rodents, or from rodents to human beings. About 1750 species of flea are known.

Flies, mosquitoes, and gnats: order Diptera

Diptera means 'two wings': only the forewings function as wings; the hindwings are whirring stabilizers called halteres. With their huge compound eyes set on large mobile heads dipterans are brilliant fliers with commensurately excellent vision. They feed in various ways—predation, scavenging, dung-feeding, bloodsucking—and their mouthparts are adapted for sponging, sucking, or lapping, with different parts serving as piercing **stylets** in the bloodsuckers. Many dipteran larvae are maggots without true legs, but others have prolegs or pseudopods for walking, while many are adapted in various ways for life in water.

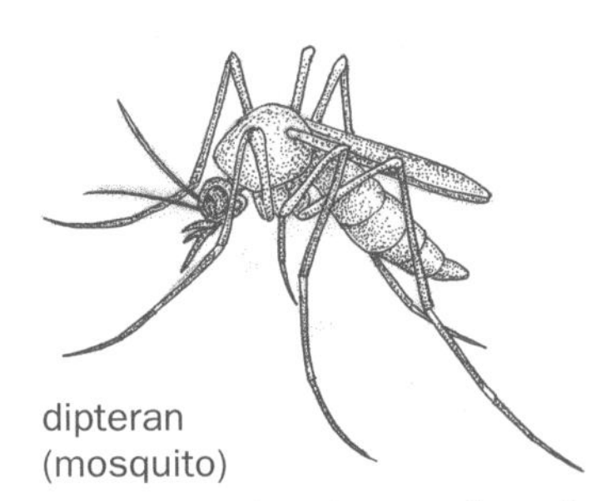
dipteran (mosquito)

Dipterans occur worldwide in a staggering range of environments from hot springs and oil seeps to tundra pools and even shallow sea water. They are the most significant of all insect vectors of disease—from malaria, yellow fever, and African river blindness (mosquitoes) to African sleeping sickness (tsetse flies), sister disease of the Latin American Chagas' disease that is transmitted by bugs. Besides mosquitoes and tsetse flies, dipterans include flesh flies, house flies, stable flies, bee flies, hover flies, picture-winged flies, midges, and crane flies ('daddy-long-legs') whose larvae are the 'leatherjackets' that chew plant roots. About 150 000 dipteran species are known.

Caddisflies: order Trichoptera

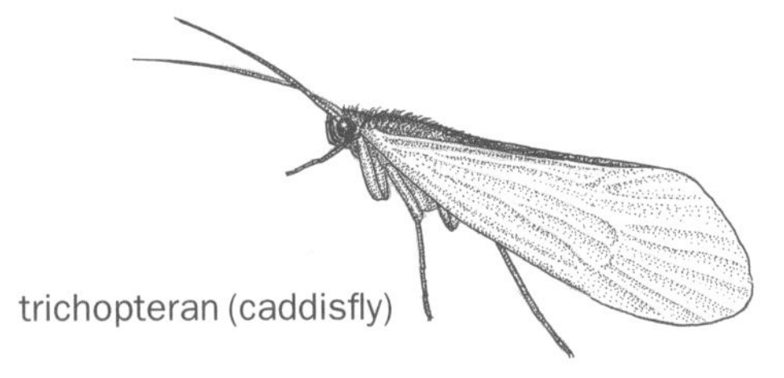
trichopteran (caddisfly)

Adult trichopterans look like small moths, but with bodies covered in hairs rather than scales. The legs are long and slender, and the setaceous (bristly) antennae are as long or longer than the body, and they rest their two pairs of wings over the abdomen like a roof. They take a liquid diet. Caddisfly larvae are more famous than the adults; they build underwater 'houses' or cases out of grains of sand and debris that they bind with silk, and live mainly as herbivorous scavengers. There are about 7000 species of caddisfly.

Butterflies and moths: order Lepidoptera

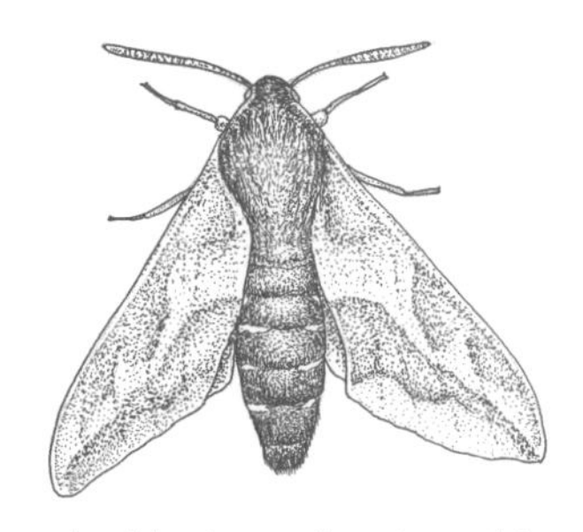
lepidopteran (hawk moth)

The Lepidoptera bring this account of the insects to a suitably climactic finish, for they are truly among the glories of nature. Uniquely, the bodies, heads, and wings of lepidopterans are covered in dense scales, like powder. The adults feed primarily on nectar through a long, sucking **proboscis** formed from the coupled first maxillae, which is coiled when not in use. The caterpillar larvae, with four pairs of prolegs on abdominal segments 3–6, feed mainly on green plants but there are some interesting variations, like those of the large blue butterfly that hitch a ride into an ant nest and preys on their hosts' larvae. About 120 000 lepidopterans are known.

11

SPIDERS, SCORPIONS, MITES, EURYPTERIDS, HORSESHOE CRABS, AND SEA SPIDERS

SUBPHYLUM CHELICERATA AND SUBPHYLUM PYCNOGONIDA

THE SUBPHYLUM Chelicerata includes the classes Arachnida, Xiphosura, and Eurypterida. The Arachnida are all the groups that now live on land—spiders, scorpions, and so on; the Xiphosura are the horseshoe crabs; and the long-extinct Eurypterida superficially resembled scorpions, but sometimes grew as long as a small rowing boat, and were among the most prodigious invertebrates of all time. Eurypterids should be remembered with the kind of awe with which we contemplate dinosaurs. Modern molecular evidence now suggests that the Pycnogonida, the weirdly etiolated 'sea spiders', could be bona fide chelicerates, too. But this does not seem to be unequivocally established, so it seems safest, for the time being, simply to treat them as sisters to the Chelicerata.

About half of all known species of living arachnids are spiders, possessed of a poisoned bite and skilled in silk. But the arachnids also include the much-feared but sadly beleaguered scorpions and the solifugids ('sun spiders' or 'wind spiders'), which sometimes venture into the tropical sun and can indeed run like the wind; plus the familiar harvestmen or 'daddy-long-legs' (a name they share confusingly with spiders of the genus *Pholcus*, and with crane flies); and the mites and ticks, which, as ubiquitous parasites and bearers of disease, are of global significance ecologically and economically (there are, in fact, probably many more species of mite than of spider, but they are less well studied).

Then, too, there is a shortlist of recondite arachnid groups such as the whip scorpions, palpigrades, whip spiders, false scorpions, and ricinuleids—which commonly are small, in some cases seem to be rare, and mostly lurk in dark dank places under stones, in caves, or in leaf litter. Finally, there are several once-successful arachnid orders that failed to pass the Permian–Triassic boundary, of which the spider-like Trigonotarbida are the most significant.

The term 'Arachnida' is applied by convention to the chelicerates that live on land and indeed the arachnids could be the oldest land animals; at least, the oldest known arachnid fossils date from the Silurian, more than 417 million years ago. But the Chelicerata as a whole spent many millions of years in the sea before they became terrestrial. The oldest known chelicerate of all is the marine *Sanctacaris*, dating from the mid-Cambrian. The fossils do not reveal whether *Sanctacaris* had chelicerae, the characteristic chelicerate mouthparts, but it did have other chelicerate features. Among living arachnids, the scorpions are known to have had a long maritime history that persisted well after some of them had ventured on to land—and some of the early marine types were huge: up to a metre in length. Paul Selden of the University of Manchester suggests that the Opiliones, the group that now includes the harvestmen, may also have had a preterrestrial history in the sea. The Xiphosura (one of the two great chelicerate groups that are not classed as arachnids) still swarm in their millions like little clockwork toys along the eastern coasts of North America and Asia. The eurypterids first appeared in the Ordovician more than 450 million years ago, lasted until the Permian, and truly came into their own in the Silurian and Devonian, when, for a time—until the jawed fishes surpassed them—they were among the most formidable predators on Earth; in their time perhaps matched only by the larger ammonites, distant relatives of squids (see Section 7).

Overall, chelicerates have been and are extremely successful and wonderfully varied. About 65 000 species are known, of which about 35 000 are spiders; but further study of mites in particular should enormously expand the list. They extend in size from eurypterids down to mites that can hitch a ride on the hairs of a bee: only such taxa as the mammals, ray-finned fish, reptiles, crustaceans, and molluscs have such a striking range. The chelicerate way of life is equally various but they achieved their ecological success by evolving the arts of predation and most remain as predators—some ambushers, some chasers.

The basic chelicerate feeding technique is to mash up the food with the dagger-like **chelicerae**, generally prodding and macerating—much as one might attack sticky rice with chopsticks. Then they pour digestive enzymes over the resulting mess, and suck out the liquefied tissues. With greater sophistication, false scorpions and the more-advanced spiders pierce the prey with their chelicerae and inject the digestive enzymes, before sucking out the semi-liquefied pabulum. Harvestmen and some mites can ingest solid food. But some chelicerates can be scavengers (like horseshoe crabs and harvestmen), some are herbivores (some mites), and some are parasites (some mites and ticks). Many chelicerates are highly specialized feeders, whereas others are generalists.

But all chelicerates clearly share the same basic body plan. Their bodies are divided into only two clear tagmata. The front tagma, the **prosoma**, is covered or partially covered with a carapace-like shield and has six segments, each of which bears a pair of appendages: the first two are the chelicerae and **pedipalps**, and the last four are the walking legs (with the last pair modified as paddles in many eurypterids). Thus the prosoma accommodates both head and limbs. The rear tagma, the **opisthosoma** or 'abdomen', has 12 segments. In, for example, the eurypterids and scorpions, the opisthosoma is further subdivided into the **mesosoma** and the tail-like **metasoma**; but this is not at all comparable with the three-tagmata body plan of insects.

The chelicerae and especially the pedipalps serve a huge variety of purposes: between them they are to chelicerates what the trunk is to an elephant. The chelicerae —from which the class gets its name—are basically for biting, either stabbing like a sabre-tooth or squeezing like pincers, and (in spiders) injecting poison. Pedipalps serve for sensory perception (for chelicerates have no antennae); communication (as by the waving pedipalps of mating crab spiders); catching prey (in scorpions, false scorpions, and solifugids); and in some groups (including spiders) for conveying sperm. In primitive chelicerates the opisthosoma also carries appendages, as can still be seen in horseshoe crabs. Most modern chelicerates have lost most of the abdominal appendages, but a few have been kept for specialist purposes. Thus they form the peculiar **pectines** on the belly of scorpions and the **spinnerets** of spiders. Primitively, the opisthosoma ends in a **telson**, which in scorpions forms the sting. Chelicerates generally have simple eyes placed medially, but some (such as xiphosurans, eurypterids, and extinct scorpions) also have lateral compound eyes.

The tiniest arachnids like palpigrades and the smallest mites respire simply by exchanging gases through their skin but most chelicerates are endowed with specialist breathing apparatus of various kinds. Horseshoe crabs and the earliest, aquatic ancestors of the scorpions had **book gills** (with respiratory sheets like the pages of a book; each 'sheet' being hollow and filled with blood-like 'haemolymph'). Modern scorpions, spiders, whip scorpions, and others have **book lungs** (a similar structure but adapted for breathing air). Some spiders, false scorpions, solifugids, and others, have tracheae.

In the arachnids that have book lungs the pigment that carries oxygen in the blood is the blue, copper-based haemocyanin. Haemocyanin has a higher affinity for oxygen than the red, iron-based haemoglobin of vertebrates and tends to release oxygen only when the concentration is very low. Hence haemocyanin is more useful for oxygen storage than for rapid release at the tissues. The arachnids that possess haemocyanin, therefore, often show extreme bursts of activity but have remarkably little stamina, and for most of the time are spectacularly inert. Orb web spiders show this mode of life beautifully. Octopuses, which are molluscs, also have haemocyanin and demonstrate much the same mixture of lethargy and freneticism. Arachnids that have tracheae in general do not have haemocyanin, and some of these are very active —including the jumping (salticid) spiders, and some mites and harvestmen.

Finally, we should note the group of twig-like creatures known cavalierly as 'sea spiders' but, more accurately, as Pycnogonida (and sometimes 'pantopods', after the name of the most important pycnogonid order, the Pantopoda). Pycnogonids belong on the shortlist of creatures that make very little impact on the world, often contain few species, but seem anatomically to be so different from anything else that they must be allotted a major taxon of their own. Pycnogonids do not seem to be clearly tagmatized but they do have the chelicerate quota of appendages, the first two pairs of which may well be homologous with the chelicerae and pedipalps. In practice, the Pycnogonida are commonly acknowledged as a separate group, but as the sister taxon of the Chelicerata.

The chelicerate tree and the accompanying notes are based primarily on the ideas of Jeffrey Shultz at the University of Maryland, with additional comments from Paul Selden at the University of Manchester and Bill Shear at Hampden-Sydney College, Virginia. As we will see in the following guide to the group, Jeffrey Shultz in some respects takes a fairly radical view of chelicerate taxonomy, so this is one tree in this book that cannot be called conservative. I have made clear, however, where these ideas diverge from the orthodoxy.

A GUIDE TO THE CHELICERATA

The main features of this tree will not upset most zoologists. Most agree that the Chelicerata is indeed monophyletic, as shown here. The Pycnogonida are presented in their own group, as sister to the subphylum Chelicerata. This may not be correct: for example, pycnogonids *could* be degenerate arachnids. I have given the Pycnogonida subphylum status although they are usually treated simply as a class or an order. After all, if they are sisters to the Chelicerata, then Linnaean consistency demands that they have the same ranking, and the fact that they are recondite creatures with few species should not affect the issue. Molecular studies are currently in progress that should clarify the pycnogonids' position.

Then again, every account I have ever seen of chelicerate phylogeny describes the Xiphosura as the sisters of all the rest. So there are no problems here. That eurypterids are sisters to the Arachnida is also widely accepted; at least, this is a reasonable way to present such evidence as is known. Many traditional accounts place the Xiphosura together with the Eurypterida in a group called Merostomata, and show the Merostomata as sisters to the Arachnida. But modern accounts deny any specific relationship between xiphosurans and eurypterids, because the characters the two have in common seem merely to be primitive. It seems best to abandon the term 'Merostomata' altogether because it does not describe a clade, and—unlike 'protist' or 'microbe'—it seems to serve very little purpose as a general descriptive term.

So far, then, there is little to provoke controversy. But Jeffrey Shultz does break more sharply with tradition in dividing the class Arachnida into two subclasses, the **Dromopoda** and the **Micrura**. As you see, his Dromopoda includes the harvestmen, scorpions, false scorpions, and solifugids; and his Micrura includes the rest, among which the mites (Acari) and spiders (Araneae) are by far the most significant ecologically and numerically. But what many zoologists will find most startling in this classification is the position of the scorpions. Thus the most influential paper on chelicerate classification in the modern age is that of P. Weygoldt and H. F. Paulus of 1979. They show the scorpions as the sister group of all other arachnids, and the eurypterids as sisters of the scorpions + other arachnids. Thus they suggest that scorpions are especially primitive—effectively having an ancestral relationship to all other arachnids—and imply that scorpions have a special relationship with eurypterids. Many other accounts agree that scorpions and eurypterids are particularly close (although, says Jeffrey Shultz, 'Nothing very recent or authoritative').

But Shultz emphatically denies any special relationship between eurypterids and scorpions. To be sure, there is a superficial resemblance: but, he says, the features the two have in common are either convergences or symplesiomorphies. Indeed, he says, the eurypterids have so many primitive characters that it is hard to pin down their position at all. On cladistic grounds alone they could be placed at any one of several places in the arachnid tree—either as a distant relative of harvestmen among the dromopodes, or of spiders and mites among the micrurans. The eurypterids as a whole might even be paraphyletic or even polyphyletic, with different types belonging in different places. In any case, the position they are given here—as primitive sisters to all the rest—is a kind of commonsense compromise. Because the eurypterids all died out so long ago, we cannot expect molecular evidence to sort out this particular problem.

By contrast, says Shultz, scorpions look very primitive on the outside but their internal structures are highly derived. He further suggests that they share many of those derived features with harvestmen, pseudoscorpions, and solifugids, and so groups all these creatures together in the Dromopoda. This is the single most controversial aspect of this tree. Take out the scorpions, and place them on their own branch between the eurypterids and the rest of the arachnids, and the remaining features of Shultz's tree, bar a few details, would not cause too much wrangling. Because scorpions are very much alive, molecular studies of them and suitably chosen arachnid and non-arachnid outgroups should, in time, largely decide whether Shultz is right or wrong. At least, if the molecules show that scorpions have no special relationship with the other 'dromopodes', and, indeed, are different from all other arachnids, then they will suggest he is wrong.

But although the relationships between the main chelicerate groups is still up for discussion, there is relatively little controversy about the authenticity and composition of the groups themselves. Most zoologists agree that a spider is a spider and a palpigrade is a palpigrade. So we should look at the individual groups.

Horseshoe crabs: class Xiphosura

Like the eurypterids, horseshoe crabs may have arisen as long ago as the late Precambrian although the earliest known fossils are Ordovician in age. Three genera (including four species) are still miraculously with us, as if to tell us what early chelicerate life was like: *Limulus* of North America; *Tachypleus* of Southeast Asia; and *Carcinoscorpius* of Malaysia, Thailand, and the Philippines. All live in shallow marine waters, generally preferring clean sandy bottoms, crawling on the surface or

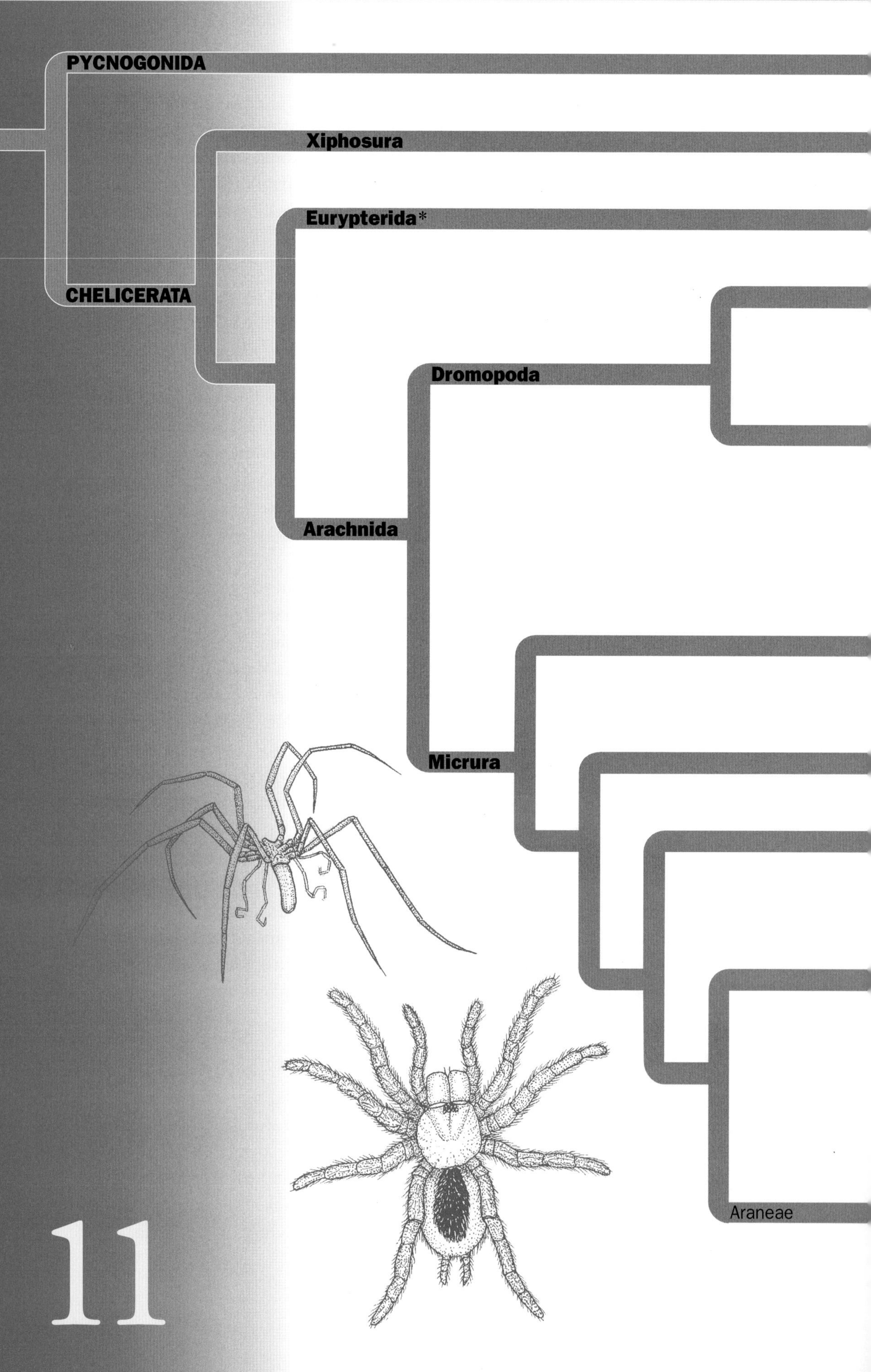
PYCNOGONIDA
Xiphosura
Eurypterida*
CHELICERATA
Dromopoda
Arachnida
Micrura
Araneae
11

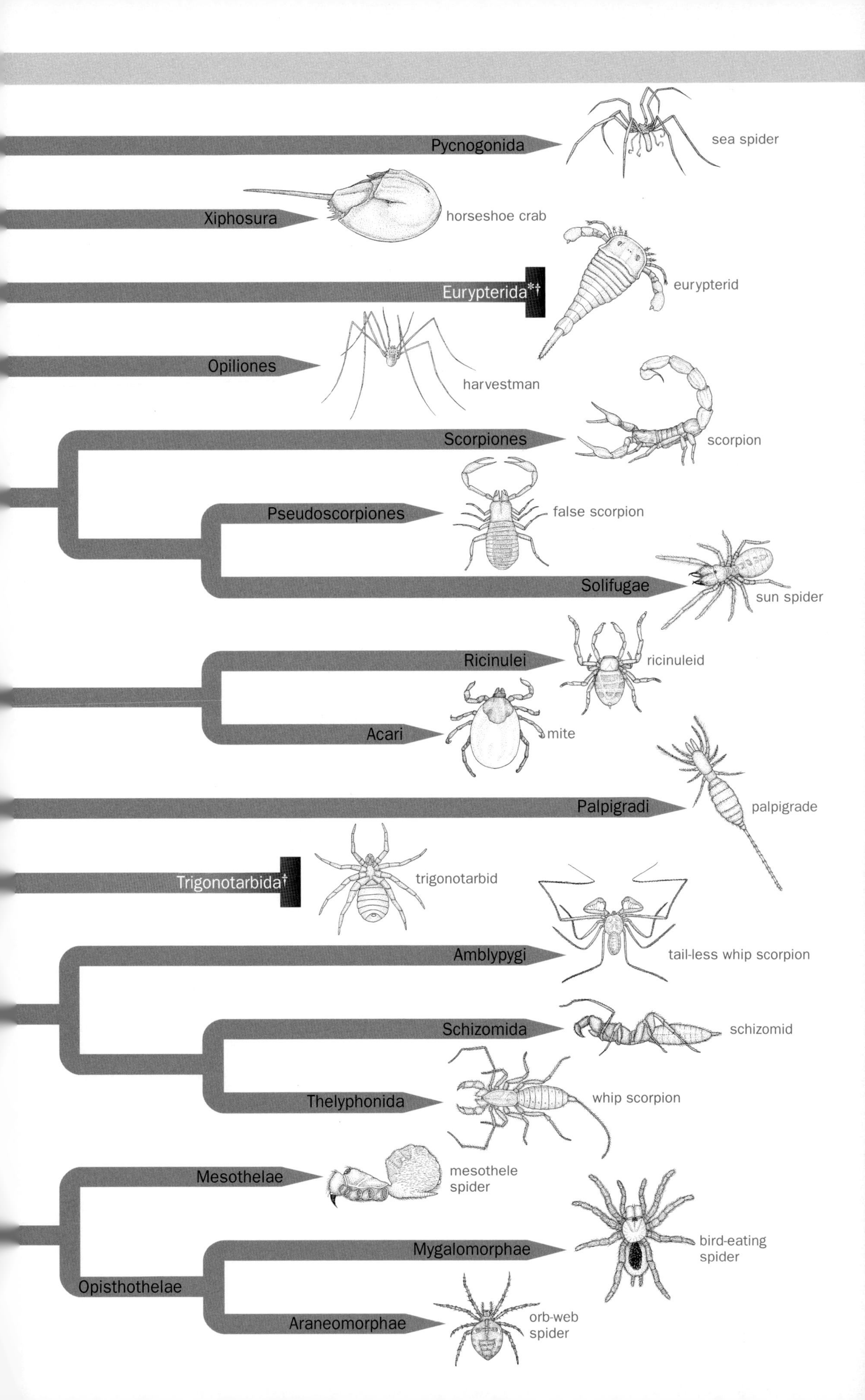

Pycnogonida
sea spider
Xiphosura
horseshoe crab
Eurypterida*†
eurypterid
Opiliones
harvestman
Scorpiones
scorpion
Pseudoscorpiones
false scorpion
Solifugae
sun spider
Ricinulei
ricinuleid
Acari
mite
Palpigradi
palpigrade
Trigonotarbida†
trigonotarbid
Amblypygi
tail-less whip scorpion
Schizomida
schizomid
Thelyphonida
whip scorpion
Mesothelae
mesothele spider
Mygalomorphae
bird-eating spider
Opisthothelae
Araneomorphae
orb-web spider

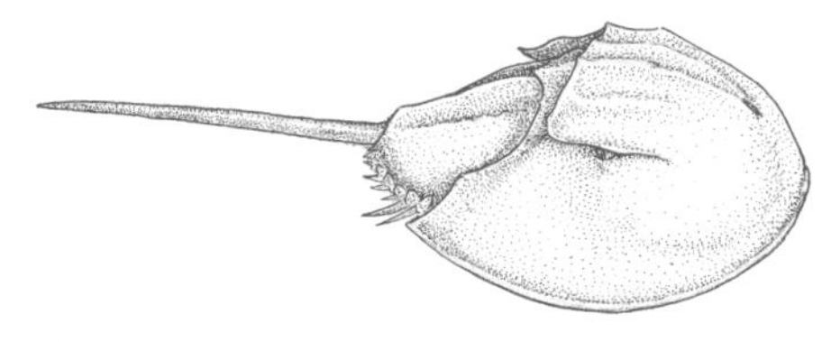
xiphosuran (horseshoe crab)

burrowing just below, and preying on small invertebrates or scavenging. As chelicerates go they are among the more versatile feeders. *Limulus polyphemus*, the common horseshoe crab of the North American Atlantic coasts, has the dubious pleasure of being an extremely popular laboratory animal; the Asian types find their way to the dinner table.

The pedipalps and walking legs of xiphosurans are all chelate (grasping); they gather food by any of these appendages and pass it to the gnathobases along the ventral midline, where it is ground to small bits and then shuffled forwards to the mouth. The sixth prosomal appendage of xiphosurans is modified at the ends for support on soft substrates (whereas that of eurypterids is a paddle). The xiphosuran opisthosoma is quite different from eurypterids, too, being unsegmented and undivided. The xiphosurans are also the only living chelicerates whose opisthosoma bears book gills.

Eurypterids: class Eurypterida*†

The earliest known eurypterid fossils date from the Ordovician period, almost 500 million years ago; but, although there is no evidence, they could be much older than this. They flourished in the Silurian and Devonian and until well into the Permian; and through much of that huge stretch of time (almost 250 million years)—until the fishes finally got into their maritime stroke—they were matched as marine predators only by the largest ammonites and in fresh water had no peer. At least half of eurypterid families included species that were more than 80 centimetres long and some, the pterygotids, reached 2 metres and more.

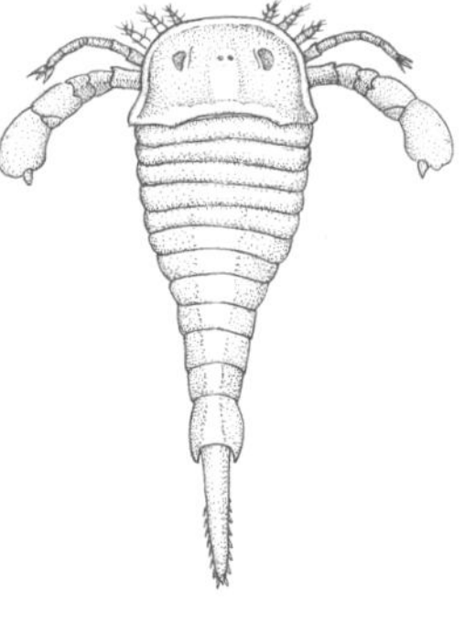
eurypterid

Their varied anatomy suggests a corresponding diversity of lifestyle. They evidently swam to a greater or lesser extent, for in most of them the sixth prosomal appendages (which anatomically is the last pair of walking legs) was paddle-like. Recent work on the anatomy (albeit controversial) suggests that some at least of the eurypterids could move these oars in a figure of eight, thrusting hard in the backward stroke and providing lift and modest thrust in the forestroke, so that they would effectively have 'flown' underwater like a penguin. Some came on land and may have holed up in the hollow trunks of lycopod trees, where their fossil remains have been found.

Feeding styles of eurypterids also probably varied. In some, the chelicerae were extremely reduced whereas in others they were large and grasping. Some swimming eurypterids had a scorpion-like opisthosoma—segmented, and divided into a mesosoma with flaps that covered the gills and a narrow metasoma. In *Carcinosoma*, the metasoma ended with a bent and pointed telson that was very scorpion-like, and may have been fitted with a poison gland (although there is no evidence for this).

Like dinosaurs, eurypterids have a sobering, Ozymandias-like quality; that such magnificent creatures, which flourished, radiated, and often dominated for nearly 250 million years, should finally have gone the way of all flesh. It would be good to have just one survivor to admire.

Arachnids: class Arachnida

Of all the chelicerates, it is the arachnids that have inherited the Earth. They include the most conspicuous and famous—the spiders, scorpions, harvestmen, mites, and ticks—and provide most of the 65 000 or so living chelicerate species that are known. The prosoma of arachnids is wholly or partially covered by a carapace-like shield, while the opisthosoma may be segmented (as in scorpions) or unsegmented (as in most spiders, and ticks); it is undivided in some groups (like spiders) but divided into mesosoma and metasoma in others (like scorpions). Opisthosomal appendages are either lacking, or modified into special organs, such as the pectines of scorpions or the spinnerets of spiders. Apart from some mites and a few spiders that have again taken to the water, arachnids are terrestrial. They breathe by means of tracheae, book lungs, or both. Unlike eurypterids and xiphosurans, arachnids generally lack compound eyes (although these were present in some ancient scorpions); the large simple eyes of spiders protrude from the prosoma like Cat's-eyes in a road. The extinct trigonotarbids nicely show the eyes in an intermediate stage of reduction.

Harvestmen: order Opiliones

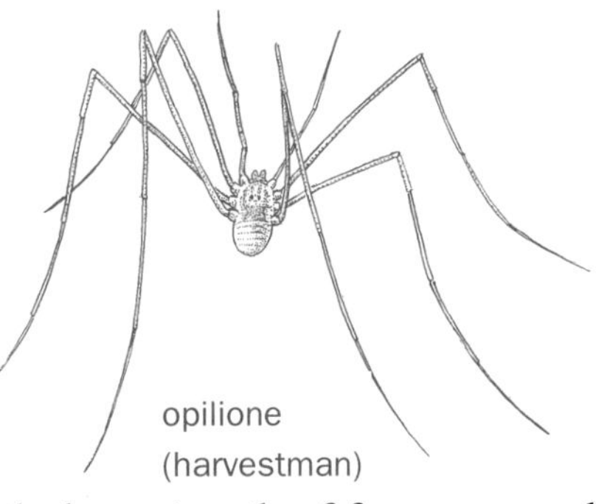

opilione (harvestman)

The Opiliones are another widely familiar group; perhaps their most-conspicuous feature is the fusion of the prosoma and opisthosoma to form a single rounded box. Most of the 5000 or so known species live in tropical South America and South-east Asia, but they also live in damp shaded areas in all climates, including the sub-arctic, and can sometimes be found in droves beneath the arris rails of fences around temperate suburban gardens. 'Harvestmen' is perhaps the best of the English common names, for they do, indeed, mature in the autumn; but 'daddy-long-legs' is appropriate, too, for creatures like this, with bodies usually less than 2 centimetres long and legs up to 10 centimetres (this view of the opiliones is skewed, however, for most of the tropical types have relatively short legs). Harvestmen can be predators, grasping small invertebrates in their pedipalps and chewing with the chelicerae, but they also scavenge on dead animals or even plants. They are among the few arachnids able to ingest small solid particles. They also have a pair of repugnatorial glands, producing secretions containing quinones and phenols to repel attackers.

When reproducing, most arachnids pass the sperm indirectly from male to female: male spiders, for example, eject sperm on to a tiny web built for the purpose on the ground, then gather it in large scoops at the ends of their pedipalps (the distinguishing feature of the males) and inject it into the female genital opening beneath the opisthosoma. Other arachnids cover the eggs as they leave the female's body. But male harvestmen, almost uniquely among arachnids, have a penis for direct copulation. The only other arachnids thus endowed are some mites.

True scorpions: order Scorpiones

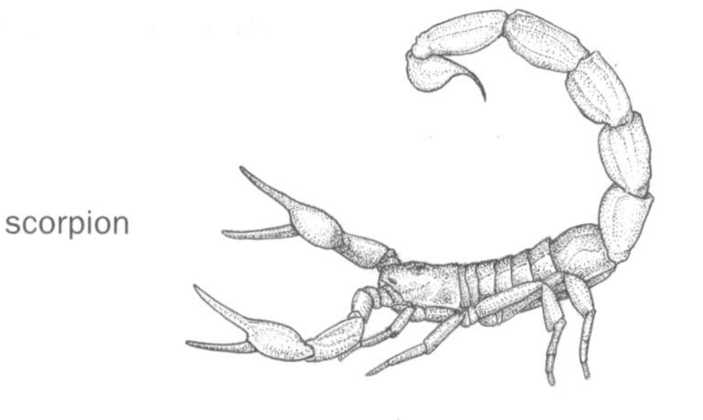
scorpion

Scorpions are the stuff of icons, like snakes and eagles. The pedipalps form great pincers with which to grab the prey and press it to the short chelicerae, which then mash it up. The opisthosoma is clearly divided into two regions: the first, the mesosoma, carries the genital pore, four pairs of openings called **stigmata**, which lead into book lungs, and a pair of pectines underneath. The pectines extend outwards in a V-shape from just behind the genital pore—close to the last walking legs—and are comb-like. The metasoma has no appendages but ends in the telson, which has become a sting. The scorpion applies the sting to its prey by arching its back like a circus contortionist. Scorpions prey mainly on insects, though some prey mainly on other scorpions. The larger ones take lizards and snakes and the venom of one North American species is said to match that of a cobra. Such scorpions have killed people. Some scorpions are 18 centimetres in length—the biggest of all arachnids—but, as noted below, some extinct types that were probably at least in part aquatic grew up to a metre.

Most scorpions are nocturnal; they detect their prey by vibrations, picked up by sensitive hairs called **trichobothria** (not the pectines, the function of which is unknown). They are long-lived (some Australian types are known to survive at least 30 years) and relatively slow-breeding: a life strategy more like that of a tiger than of most invertebrates. Most of the 1200 or so species of living scorpion live either in desert or in rainforest and some are actually arboreal; but some live in the Mediterranean and there is even an imported type in south Essex, near London.

Overall, then, modern scorpions are emphatically terrestrial; but modern research has revealed their aquatic origins and indeed suggests that the Scorpiones as a whole were primarily aquatic. Moreover, ancient aquatic scorpions evidently had plates to protect their gills that are absent in modern scorpions, although they were also present in eurypterids.

Specifically, the oldest known scorpion fossils come from the middle Silurian period, around 425 million years ago. Details of anatomy are largely lacking but it is now clear that the later *Tiphoscorpio*, from the Devonian of New York State, had true gills. In the swamps of the Carboniferous, whose vegetation fossilized into coal, scorpions were probably amphibious. Bill Shear and his colleagues have presented evidence recently of true air-breathing scorpions as old as the early Devonian, around 400 million years ago. But the original and more various aquatic scorpions persisted long after the terrestrial types appeared—indeed until after the Triassic, which ended some 210 million years ago. Some were very big: *Praearcturus gigas* of the early Devonian was a metre long, and *Brontoscorpio* was not much smaller. To be sure these giants may have come on land, but creatures of that size must have returned to the water to moult, for their bodies would have needed support after they had shed their old cuticles, and before the new cuticle hardened off. The biggest living types are weeds by comparison—stressing, perhaps, that the glory days of scorpions, lasting several hundred million years, were spent in the water.

False scorpions: order Pseudoscorpiones

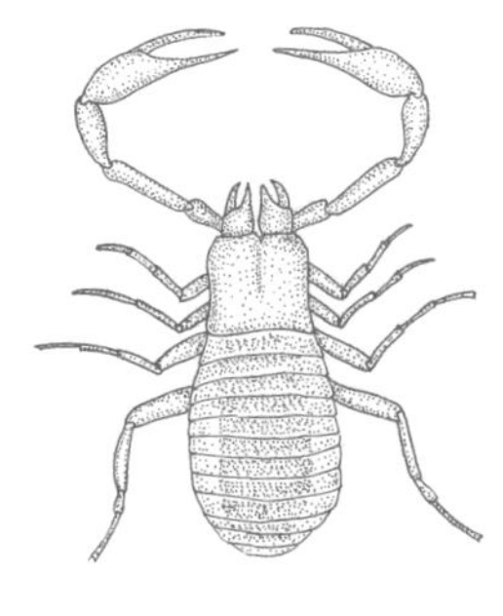
false scorpion

The 2000 or so species of false scorpions are small, the largest being only about 7 millimetres long, but they are successful and formidable. They live worldwide in a great variety of habitats—under stones, in litter, soil, bark, animal nests, and one prefers sandy beaches. They do, indeed, resemble scorpions, with their chelate pedipalps—which also carry poison glands with which to immobilize prey. Typically they grab the prey—a mite, say—with the pedipalps, then rip it with the chelicerae and suck the fluids. Some false scorpions practise 'phoresy': hitching a lift on larger animals, which they grab with their pedipalps. For example, *Chelifer cancroides*, commonly hitches to house flies and hence is a frequent cohabitant of human beings.

Sun spiders or wind spiders: order Solifugae

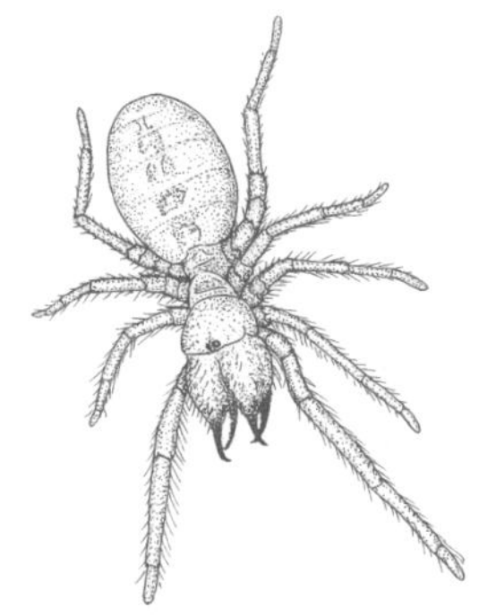
solifugid (sun spider)

The 900 or so species of solifugids, otherwise known as solpugids (Solpugida), are up to 7 centimetres long and look remarkably like spiders; indeed, they are commonly called 'sun spiders' because they often hunt by day, or 'wind spiders' because the males run so bewilderingly fast. But they are clearly not spiders; both the prosoma and opisthosoma are clearly segmented and they lack poison, just ripping their prey apart with enormous chelicerae. Many show a preference for termites or other arthropods but most seem to have broader tastes. Most live in tropical and subtropical deserts in America, Asia, and Africa. Solifugids are generally presented as the sister taxon of the false scorpions.

Ricinuleids: order Ricinulei

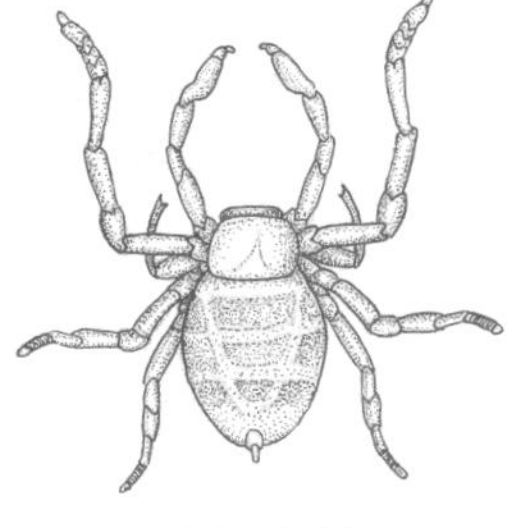
ricinuleid

The 35 or so species of ricinuleids are not prepossessing; they are slow-moving predators that feed on tiny invertebrates in caves and leaf litter in West Africa and tropical America. They are all less than 1 centimetre long. They are, however, probably sisters to the mites, the most numerous and economically most significant of all chelicerates.

Mites, ticks, and chiggers: order Acari

There are fewer known species of mite than of spider; 'only' 30 000. But this must simply reflect their relative inconspicuousness. No one doubts that thousands of species—probably hundreds of thousands—remain to be discovered, many living parasitically or commensally within and around the hordes of other creatures that have still to be revealed in tropical forest. Mites, along with beetles and nematodes, are outstanding among the creatures whose diversity has probably been barely sampled.

Mites are extremely varied in their way of life. Most are terrestrial but some are aquatic, some are predators, some omnivores, some are herbivores, many are parasites, and many are extreme specialists. Because they are so varied, some mite specialists—acarologists—have suggested that Acari is not monophyletic, but is simply an assorted bag of arachnids that have separately adopted small size and compact shape—a form that enables them to squeeze into tiny niches that are unavailable to other creatures, and so lends itself to many different lifestyles. But Jeffrey Shultz points out that variousness *per se* does not imply polyphyly. Successful lineages often radiate prodigiously. Applying cladistic logic, too, he points out that if mites are polyphyletic, then this means that some mites must be more closely related to some non-mites, than they are to other mites. If this is so then where, pray, he asks, are these putative non-mite relatives? There is no convincing answer to this. But if all known mites are more closely related to other mites than any of them are to non-mites, then the mites by definition are monophyletic. Whether they are monophyletic or not, however (but let us assume they are), the mites are divided into three suborders: the Opilioacariformes, Parasitiformes, and Acariformes.

- Mites in the suborder **Opilioacariformes** in general are primitive; ventrally at least they retain segmentation of the opisthoma; and the opisthoma is separated from the prosoma by a transverse groove. The opilioacariforms are omnivores and predators of tropical forest floors and dry temperate habitats.

- Suborder **Parasitiformes** includes free-living and symbiotic (or parasitic) mites and ticks from all over the world. The free-living ones are found in the soil, leaf litter, decaying wood, among mosses, and in the nests of insects and small mammals. Many of them prey on small invertebrates; many of them dwell full time upon other animals (often other arthropods), either when immature or as adults. In some cases the relationship is merely phoretic—the mites are merely hitching a lift—but in others the parasitiform is a true parasite.

The most familiar among the parasitiforms are the ticks, which live as ectoparasites of vertebrates and in one case of beetles, attaching themselves to the skin of their hosts that they slice with smooth chelicerae. They are the largest of all the acarids; engorged on blood some may swell to 2–3 centimetres. The 'hard' ticks (family Ixodidae) cling to reptiles, birds, and mammals for days at a time and sometimes even for weeks, feeding the while. Some of these are vectors of disease, such as *Dermacentor andersoni*, which carries Rocky Mountain spotted fever. The 'soft' ticks (family Argasidae) lack the heavy dorsal shield of the ixodids and feed only briefly, typically for less than an hour at a time, from birds and mammals, especially bats. When not attached to their hosts they hide in cracks and crevices or in the soil. They, too, may carry disease, including tick fever or African relapsing fever borne by *Ornithodoros moubata*.

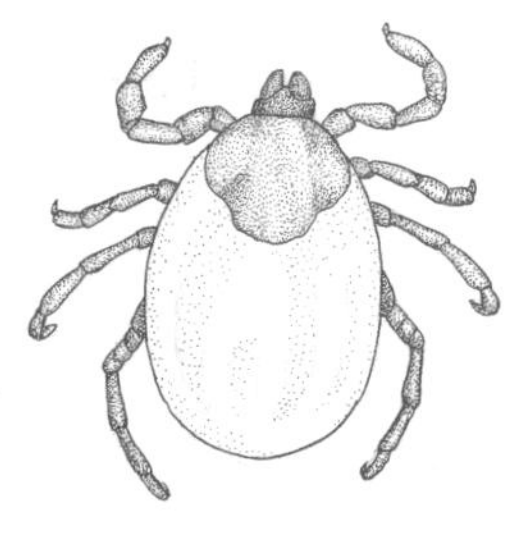
acariform mite

- Suborder **Acariformes** includes most of the species of acarids. Some are known generally as mites, and some as chiggers. Overall, their ways of life are hugely various. Free-living forms on land live in soil, leaf litter, mosses, lichens, fungi, and bark at all altitudes and latitudes; some are herbivorous (including frugivorous or fruit-eating) and some are predators, and many of these can ingest solid as well as liquid food. Some are serious pests of crops while others are used as agents of biological pest control. Others live in land or in the sea, some on algae and some in the ocean depths. A few aquatic forms are 'suspension feeders', filtering out tiny particles. Many others live in association with other creatures, and most of these are true parasites. Hosts include marine and freshwater crustaceans, freshwater insects, marine molluscs, the pulmonary chambers of terrestrial snails and slugs, terrestrial arthropods, the outer surfaces of all groups of terrestrial vertebrates, and the nasal passages of amphibians, birds, and mammals. Many others are phoretic.

Whether you call the plant-feeding mites 'predators' or 'plant parasites' seems largely a matter of taste. Like many of the parasitiforms, the parasitic acariforms can do severe damage. *Demodex cani* causes mange in dogs, while *D. folliculorum* and *D. brevis* share the spoils of the human forehead, the former inhabiting sebaceous glands and the latter preferring hair follicles; a fine example of extreme niche-preference. The human follicle mite apparently does no harm but similar mites cause mange in many animals, wreck the wool of sheep and cause birds to lose their feathers.

Palpigrades: order Palpigradi

Palpigrades are arachnid minimalists. None of the 60 or so known species exceeds 3 millimetres in length and their exiguousness allows them to dispense with organs of circulation and gaseous exchange: they get all the oxygen they need through their tissue-thin and colourless cuticles. Most have been found under rocks and some in caves, although some live on sandy beaches. They are considered rare but, more probably, they are simply overlooked for they have been found here and there in many parts of the world. All in all, palpigrades are strange, cryptic creatures whose evolutionary course has been one of reduction. Frankly, little is known about them.

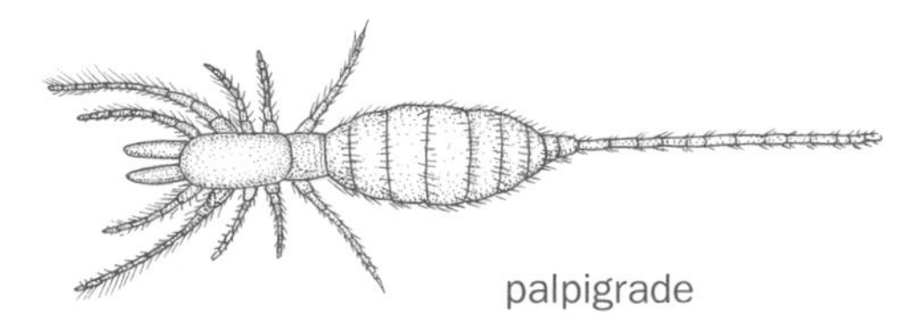
palpigrade

Trigonotarbids: order Trigonotarbida†

trigonotarbid

Superficially, at least, trigonotarbids look like spiders, albeit with clearly segmented abdomens. They were among the first animals known on land—the very earliest fossils, from the west country of England, seem to date from the Silurian period. The best preserved are from the Devonian of Scotland, where they were first described in the 1920s, and from New York: but fossils have been found all over the place, including Argentina and Spain, Czechoslovakia and Germany. The first trigonotarbid fossils known were described by the English geologist and clergyman William Buckland in 1837, the year that Queen Victoria ascended the British throne. Those fossils date from the late Carboniferous when, it seems, trigonotarbids were more important than spiders and presumably filled at least some of the same niches. They are presented here as sisters to the spiders and their closest relatives; and they certainly belong in this position or thereabouts.

Tail-less whip scorpions and whip spiders: order Amblypygi

The 70 or species of amblypygids are bizarre but fearsome beasts up to 3 centimetres long, which look like whip scorpions but resemble spiders internally (though lacking spinnerets and poison glands). Some earlier phylogenists regarded them as the sister group of the spiders, but Jeffrey Shultz denies this. They are creatures of the dark: night hunters that scurry with a sideways gait through the leaf litter and under the bark in many a hot and humid forests, while a few species are cave-dwellers. As in uropygids and schizomids the first walking legs of amblypygids are sensory. But in amblypygids these are impressive feelers indeed—extended up to five times the length of the animal, so a 5-centimetre creature may have 25-centimetre sensors. Thus, like Robert Louis Stevenson's Blind Pew, they feel out their prey. Then, in typical arachnid style, they grab the victim in their pedipalps, tear it open with the chelicerae, and suck out the body fluids.

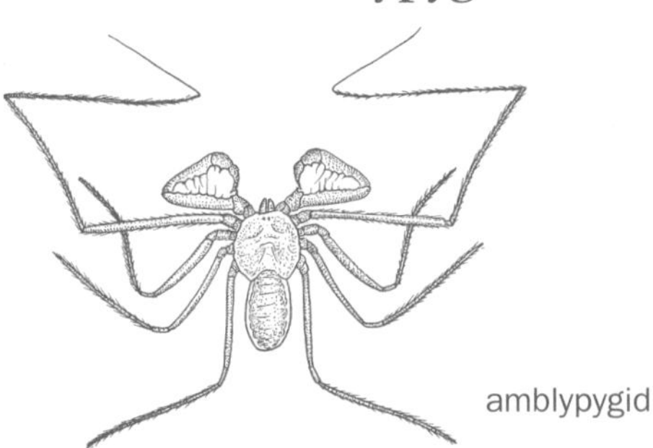
amblypygid

Schizomids: order Schizomida

The order Schizomida is a minor group of about 80 species, none more than a centimetre in length, which live in the common fashion of land arthropods in leaf litter, under stones and in burrows, mostly in tropical and subtropical Asia, Africa, and Americas, though there are a few temperate species. Some class schizomids as miniature uropygids but their prosoma is

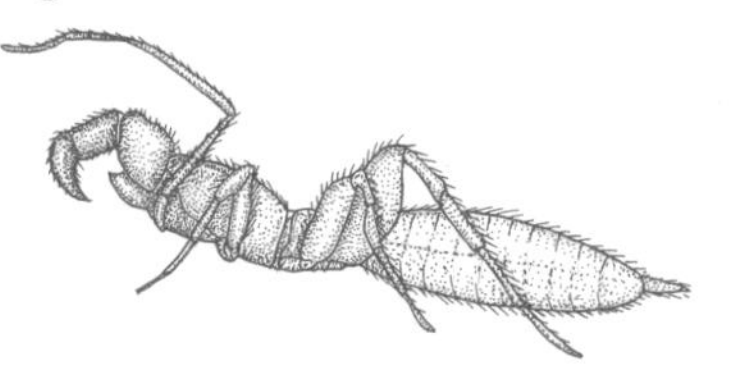
schizomid

divided and the telson is not whip-like. As in uropygids, however, the first walking legs of schizomids are sensory, and they, too, have repugnatorial glands to repel attackers.

Whip scorpions and vinegaroons: order Thelyphonida (or Uropygi)

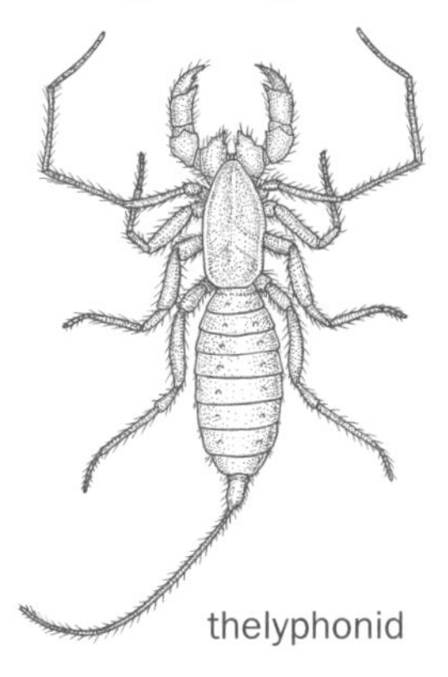

thelyphonid

The 100 or so species of uropygids are middle-sized (1–8 centimetres) and hail mostly from Southeast Asia, although there are a few in the southern USA and South America and some have been introduced into Africa. Their shape is generally scorpion-like; and, like scorpions, they are nocturnal hunters, catching various invertebrates with their pedipalps and passing them to the chelicerae for grinding. They generally lurk in humid places, under rocks and in the leaf litter, but some are adapted to deserts. Thus in general ecology, too, they resemble scorpions. But uropygids have non-scorpion peculiarities, some of which they share with other groups. Thus their first walking legs are adapted as 'feelers' to find their prey. They also (though not quite uniquely) have repugnatorial glands near the anus that produce acid or quinones. Some of these acids are high in acetic acid, which is why some whip scorpions are known as 'vinegaroons'.

Spiders: order Araneae

Spiders are the reason that everyone knows the name 'arachnid': so conspicuous, so various, so ubiquitous, so charming, so sinister, so *clever*. The spinning of the web is one of those animal achievements that take away the breath, like the leap of the dolphin and the fall of the peregrine. The Araneae, however, represent just one order from the entire chelicerate subphylum. Arachnologists have named about 35 000 species of spider and conventionally divide them into two suborders: the Mesothelae and the Opisthothelae.

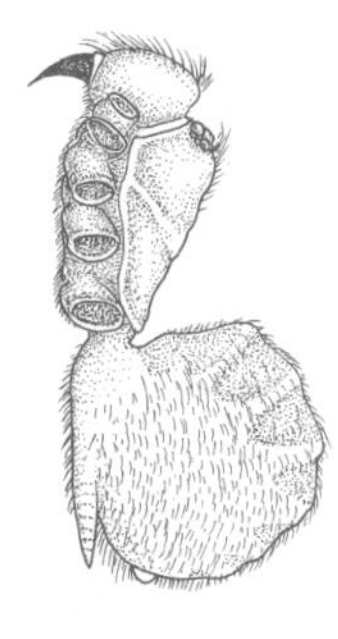

mesothele spider

- Suborder **Mesothelae**, the trap-door spiders: they live in burrows with a lid, a cat-flap with a silken hinge, which they open to dash upon some hapless invertebrate passer-by, and close for privacy. Some of these trap-door spiders are small—1 centimetre or so—but some go up to 3 centimetres. The Mesothelae are not the only trap-door spiders, however; some others, quite unrelated, follow the same technique. Two key anatomical features distinguish the Mesothelae. First, their chelicerae move orthognathously—the spiders stab downwards with them in the manner of a sabre-toothed cat. This technique works only if the prey is held against some solid substrate (or, as in the case of the sabre-tooths's prey, is extremely heavy). Second, and more definitively, the mesothelans look primitive: the opisthosoma still shows clear signs of segmentation.

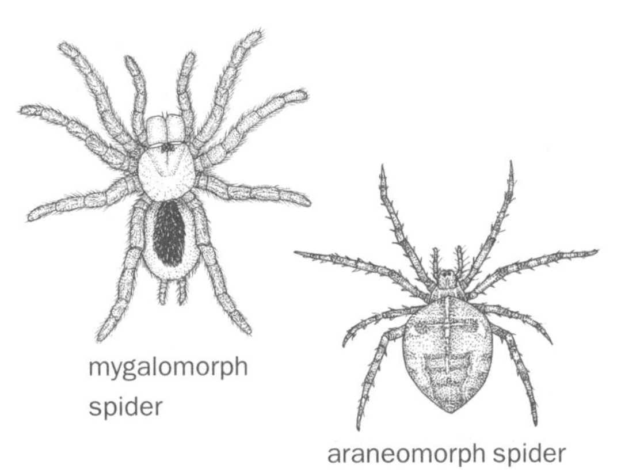

• Most species of spider belong in the second suborder, the **Opisthothelae.** Their opisthosomas are unsegmented, to form the typically polished or furry or velvety bag-like 'abdomen'. The Opisthothelae are further subdivided into two infra-orders. The first, the **Mygalomorphae**, contain spiders that truly inspire awe. From among the 15 or so families I will pick out just three. The Ctenizidae, like the primitive Mesothelae, build tunnels with trap-doors. The Atypidae, such as *Atypus affinis*, are the purse-web spiders. The Theraphosidae include the giants: the bird-eating spiders and the hairy thick-limbed beasts colloquially called 'tarantulas'. In mygalomorphs, the pedipalps are large and serve as additional walking legs. As with the Mesothelae, the chelicerae of mygalomorphs move orthognathously. The second infraorder of Opisthothelae is the **Araneomorphae**. This includes about 75 families of 'typical' spiders, the kind we all know, that hang in their webs over paths and streams or high-step over the carpet, frantic for the dark. In araneomorphs the chelicerae are 'labidognathous', which means they move in and out like pincers. Thus the araneomorphs need no solid substrate against which to rest their prey. They can capture their prey and eat it while in their webs.

What spiders in general share, besides the general suite of chelicerate features, is poison and silk. The poison is a proteolytic (protein-digesting) enzyme that acts as a neurotoxin, which is carried in glands in the prosoma and delivered through a pore at the end of the **fangs**—fangs formed from the second and more distal segment of the chelicerae. (Contrast scorpions, where the poison glands are opisthosomal.) Silk is produced in opisthosomal glands and extruded through the spinnerets, which are derived from opisthosomal appendages. Primitive spiders seem to have had four pairs of spinnerets, which is still the case in some Mesothelae. But most other spiders have three pairs, while some have only two or even one.

The silk itself is one of the wonders of nature. It is a complex protein, composed mainly from the amino acids glycine, alanine, and serine. It is produced in liquid form and polymerizes and hardens as it leaves the body to form a fibre with a complex, intricate structure that is almost as strong as nylon and twice as elastic. Spiders between them have about six main kinds of silk gland that fabricate different grades for different purposes, such as wrapping prey, making cocoons for eggs, forming the main wires or the trap lines of the webs, or providing the 'drag-line' that spiders leave behind them as they walk and enables them to drop from a leaf and hang like a climber from a cliff to escape a predator and resume their journey later. Because of its special properties spider silk is now a target of biotechnologists, who have transferred the genes that produce the tough, elastic protein into microbes, which are then cultured. Fanciful though it may seem, this technique seems to be working. Incidentally, a shirt made of spider silk would be bullet-proof. But it is also so elastic that a bullet would punch the fabric right through the body, and the fact that the shirt itself would not be damaged would probably provide little consolation.

The weaving skills of orb-web spiders, such as those of the family Araneidae, are awesome. Many, like *Araneus*, can build an entire web in less than 30 minutes, and most make a new one every night. To avoid wasting good protein, they eat the old web, and the rate of recycling is prodigious. Scientists have attached radioactive labels to amino acids in silk and shown that after the silk is ingested the labels reappear in new silk within minutes—exuded afresh from the spigots of the spinnerets.

Most familiar spiders seem inveterately solitary. But many are extremely sociable when young and in some species they cluster on their mother's back like cygnets. Even more impressively, 20 tropical species in at least six families are now known to co-operate in building webs, catching prey, and rearing the spiderlings. They may hang their eggs in the web, where they are tended by several different females.

Virtually all spiders are predators, catching their prey by a bewildering variety of techniques: trapping, in many styles of web; fly-fishing with a single thread tipped with glue; throwing a silken net like a gladatorial retiarius; ambushing, often out of burrows and silken funnels; leaping, as in the jumping spiders, Salticidae; chasing, like the wolf spiders, Lycosidae; or arresting their prey by spitting glue, like the remarkable Scytodidae. Once they catch their prey they immobilize or kill it with injections of venom and then consume it in the typical arachnid manner: press it to the chelicerae, which mash it with dentate (tooth-like) projections, then dowse it in digestive enzymes and suck the resulting soup. Filters in the mouth and pharynx prevent the ingestion of food over 1 micron (a thousandth of a millimetre). Most spiders hunt at night except for those who have a particular need to do otherwise, like the Salticidae and Lycosidae. The babies of some species seem to subsist for a time on pollen.

Anyone may suffer from contact with almost any animal if, for example, they happen to be allergic or in other ways particularly sensitive. But only a few animals are in general considered dangerous to humans. This short list includes about 24 spiders, among them the black widow of America, a Brazilian wolf spider, the brown recluse spider, an Australian funnel-web, and some hunting spiders.

A GUIDE TO THE PYCNOGONIDA

There is a village in Sussex in southern England (I forget its name) that claims to be the twin town of Berlin. In much the same way the humble pycnogonids (sea spiders), spindly maritime creatures that seem to have been fashioned from pipe-cleaners, might well be the sister group of the ubiquitous and magnificent chelicerates.

The oldest pycnogonids date back to the Devonian, and a recent classification acknowledges around 1000 living species in 86 genera. They feed on a variety of organisms but their forward-pointing **proboscis** limits their choice. The tagmata of

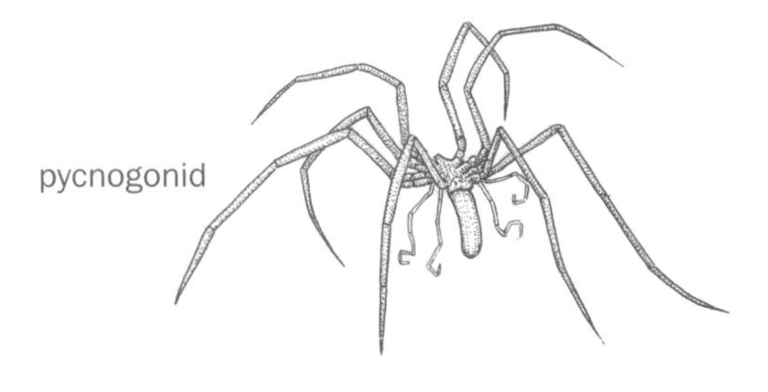

pycnogonids is not obvious, but there is a small, vestigial abdomen. Whether or not the pycnogonids are truly related to the chelicerates largely depends on the relationship between their first two segmental appendages, and the chelicerae and pedipalps of the chelicerates. If these appendages are truly homologous then the two subphyla would indeed be sister groups, and the Pycnogonida and Chelicerata would form a clade, which, so some have suggested, could be called the 'Cheliceriformes'. If pycnogonids and chelicerates are not closely related then, of course, 'Cheliceriformes' would not be a sensible grouping. Molecular studies should soon establish the true relationship; perhaps the pycnogonids will prove to be bona fide if 'degenerate' chelicerates—closer to the arachnids than the xiphosurans are. Again, we will just have to wait and see.

12

STARFISH AND BRITTLE STARS, SEA URCHINS AND SAND DOLLARS, SEA LILIES, SEA DAISIES, AND SEA CUCUMBERS

PHYLUM ECHINODERMATA

THE ECHINODERMS—sea lilies, starfish, sea daisies, brittle stars, sea urchins, sand dollars, and sea cucumbers—are extremely peculiar animals. Their most striking feature, though not immediately evident in some living types, is their **pentaradial symmetry**: their general resemblance not to other animals but to a daisy. Yet this five-sidedness is secondary. The earliest echinoderm-like ancestors were bilaterally symmetrical. Echinoderm larvae today are also bilaterally symmetrical; the radial body develops from just a part of the larva, like a flower growing from a cactus. Yet some modern echinoderms—notably the sea cucumbers or holothurians—have reimposed a form of bilateral symmetry back on to the secondary radial symmetry, as the internal structure still reveals.

Then again, just to keep things complicated, the ancestors of today's echinoderms attached to the substrate by a stalk, and kept both the mouth and anus pointing to the sky; and although the modern-day crinoids—sea lilies and feather stars—retain that ancient form (though the latter have lost their stalks), the other groups have flipped upside down so that the mouth points downwards, as seen in starfish and brittle stars. But the anus of these creatures has migrated so as to face upwards again—though now it is opposite the mouth. (The side with the mouth is called the 'oral' side; and the opposite side is the 'aboral'.) Modern sea cucumbers, however, lie on their sides. In short, you have to look at echinoderms for a very long time, and from a great many angles, to get a sense of how they are put together.

Yet there are still further oddities. Even before echinoderms evolved their radial symmetry they developed a peculiar, internal calcium-based skeleton that develops from the mesodermal cells—the same cell layer as produces the muscles; and in sharp contrast to the ectodermal skeleton of, in particular, arthropods. Note that the skeleton of vertebrates is also mesodermal. But the echinoderm skeleton is compounded not from bones but from elements called **ossicles**, each of which has a typically porous or 'stereomic' structure. In sea urchins, the ossicles are fused to form a **test**—a kind of internal shell—but in sea cucumbers they remain separately embedded as **spicules** in the muscular body wall, and they are somewhere between these extremes in starfish and brittle stars.

Then again—uniquely—echinoderms are equipped with an extraordinary internal system of hydraulics, the **water vascular system**, which is derived from one of the chambers of the coelom. The fluid within the system is different in composition from the surrounding sea water (for example, it contains proteins and has more potassium) although it usually connects to the sea via the **madrepore**, which in starfish and brittle stars at least is on the aboral surface. Among other things, the water vascular system operates the **tube feet**, which protrude to the surface as the soft-fleshed **podia** and perform an array of tasks from attachment and locomotion to respiration and feeding. The tube feet lie in **ambulacral** grooves that run to the mouth and are clear to see beneath the arms of starfish.

Then, as if to leave no animal convention unflouted, echinoderms lack a brain. They have a perfectly good system of nerves, including a ring around the mouth, but no distinct organizing centre. Yet many of them are significant predators (we tend to assume, at least casually, that mobile predators *need* to be brainy).

In addition—although this is not generally listed as a diagnostic feature—the echinoderm surface has a variety of features that protrude like bollards in a city street. These include tubercles and movable spines (*echinos* is Greek for spine and 'derma' means skin) and some also have spines called **pedicellariae**. Pedicellariae have grasping pincers at their tips and seem to have a life of their own; they have their own nerves and muscles and respond reflexly to the outside world, apparently independently of the rest of the animal. It took a long time for zoologists to realize that pedicellariae are part of the animal and not some external parasite.

Strange though they are, however, echinoderms do fit neatly into the grand genealogical tree of animals. Their larvae reveal that they are perfectly good deuterostomes, on the same lineage as the chordates. Indeed, in 1986 Richard Jefferies of the Natural History Museum in London suggested that the echinoderms are sisters to the chordates—that they share a range of synapomorphies. Most zoologists seem to have rejected this idea, but the broad relationship between echinoderms and chordates remains.

Unlikely though it may seem, too, the echinoderms are a huge ecological success—even though they remain exclusively marine, and apparently always have been. They are major players in many maritime contexts and in some cases are fearsome top predators. About 7000 living species in six classes are recognized, but another

13 000 fossil types have been assigned to 20 or more classes; these date right back to the Cambrian, more than 500 million years ago. In short, echinoderms have a highly significant present and an extremely long and impressive past.

The tree shows the living classes with a few representative orders; at least at the class level there is little disagreement on what belongs where, although Claus Nielsen in *Animal Evolution* (1995) prefers to include the sea daisies (Concentricycloidea) among the starfish (Asteroidea) while many modern scholars see them as highly modified echinoideans (sea urchins). I have shown them, conservatively, as one tine of a trichotomy whose other branches include the starfish and the sea urchins. The tree shows how the living groups could be related to each other and also to what seem to be some of the pivotal fossil types. This tree is certainly sensible and tidy—two prime requisites—but as Nielsen points out, there is no universal consensus on the relationships even between living echinoderms, and still less with the extinct types. Time, again, may tell.

A GUIDE TO THE ECHINODERMATA

Classification should follow phylogeny—evolutionary history—and so we should ask: by what routes did the echinoderms come into being? The task as ever is twofold: to provide a plausible scenario—to show what evolutionary routes were actually possible—and to relate the plausible routes to the fossil evidence. With echinoderms, there is a lot to explain; a lot of loose ends to tie up.

The fossils show, first of all, that echinoderms radiated enormously in the Cambrian. Perhaps there were echinoderm-like ancestors even in the Precambrian: these could have been bilateral burrowers that later crawled and swam near the bottom. But the Cambrian, and then the Ordovician, can reasonably be seen as their heyday. All the living classes except the holothurians are known to have appeared by the Ordovician; but all the classes known from the Cambrian died out later in the Palaeozoic. It is clear, too, that Cambrian types had ambulacral plates, showing that they also had the characteristic water vascular system with (presumably) the accompanying tube feet. A key feature that demands explanation is the evolution of pentaradial symmetry. Among the early types some were bilaterally symmetrical, some seem oddly asymmetrical, and some were triradially symmetrical. One modern theory has it that the pentaradiality arose from bilaterality via triradiality. But still the question arises, how? How does an originally bilateral animal become a pentaradial animal?

Up until now, two main hypotheses have contested. The first says that the echinodermal ancestors were essentially like worms, and that in those ancestors the head joined up with the tail to form a kind of lifebelt or doughnut. Then (the idea has it) different sections of the doughnut grew arms; and five happened to be the favoured number (even though some modern types have multiples of five). The second notion says that the first bilaterally symmetrical echinoderm stood upright, like a little fat doll. But this doll had a wide skirt, like a tutu: and this tutu was divided into sections.

12 · PHYLUM ECHINODERMATA

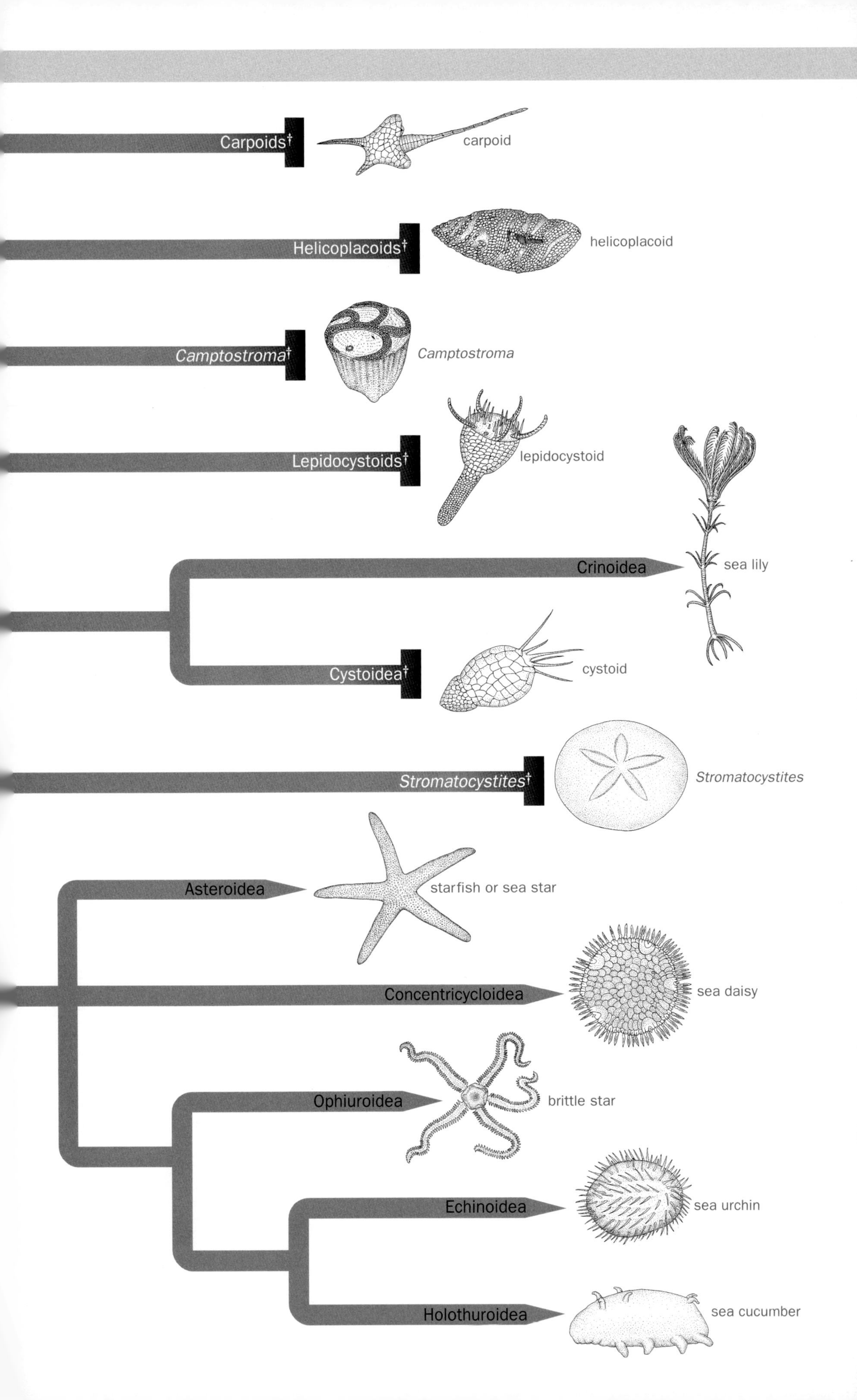

Carpoids†
carpoid
Helicoplacoids†
helicoplacoid
Camptostroma†
Camptostroma
Lepidocystoids†
lepidocystoid
Crinoidea
sea lily
Cystoidea†
cystoid
Stromatocystites†
Stromatocystites
Asteroidea
starfish or sea star
Concentricycloidea
sea daisy
Ophiuroidea
brittle star
Echinoidea
sea urchin
Holothuroidea
sea cucumber

Then the doll became squashed, the head being shoved down into the feet, leaving the tutu sticking out to form five arms.

Some marvellous molecular studies by Greg Wray at the State University of New York at Stony Brook now suggests, however, that both these hypotheses are false. Once again, the required insight is supplied by the homeotic genes, which (broadly speaking) determine what each part of the developing animal ought to be. In fact, Wray focused on the gene known as *engrailed*, which is not itself a Hox gene but accompanies the Hox genes and, like them, is involved in laying out the sequence of appropriate tissues along the long axis of animals, from head to tail (for further details about Hox genes, see Part I, Chapter 5). Wray first 'fished out' (as he says) this gene from brittle stars. He was then able to apply markers to this gene, and show the regions in the developing brittle star in which it is expressed. It transpired that the gene is expressed along each of the arms of the brittle star in the same way, as if each arm were a separate animal with its own axis. In other words, to put the matter crudely, the brittle star is not a worm coiled round with its head joined to its tail: it is like five complete worms joined head to head. The mystery of echinoderm radiality seems solved, but the solution, as so often proves the case in nature, outstrips the imagination.

One aspect of echinoderm development makes Greg Wray's solution particularly satisfying: the way in which the pentaradial subadult develops effectively as an outgrowth of the larva; the rest of the larval tissue is then simply resorbed into this budding adult. In principle, this is not quite so outlandish as it may seem. After all, holometabolous insects (see Section 10) rebuild themselves within the pupa from a few patches of tissue, the 'imaginal discs', using the unwanted larval organs as raw materials. Then again, mammals abandon a sizeable chunk of the fetal tissue at birth—to whit, the placenta. But this mode of development makes it easier to see how the Hox genes have been given the freedom to build a new animal virtually from scratch, from the tissues that were around. This stop–start mode of development may seem strange but it works, and there are many variations on this theme throughout the animal kingdom.

So much for the theory. The historical facts must be gleaned from the huge array of echinoderm fossils. Among them, zoologists have discerned a shortlist that seem to be of particular significance, a selection of which are shown on the tree here: the carpoids and helicoplacoids; the cystoids; the lepidocystoids; and a few individual genera such as *Camptostroma* and *Stromatocystites*.

The **carpoids** date from the early Cambrian. They were bilaterally symmetrical, although somewhat lopsided. They clearly had some echinoderm features, including a skeleton formed from stereomic ossicles, though whether they had a water vascular system is uncertain. They were probably suspension feeders, trapping food in a grooved arm (or **brachiole**) and ferrying it to the mouth. Most zoologists do not feel that carpoids should themselves be classed as echinoderms, largely because they lacked radial symmetry. Instead, car-

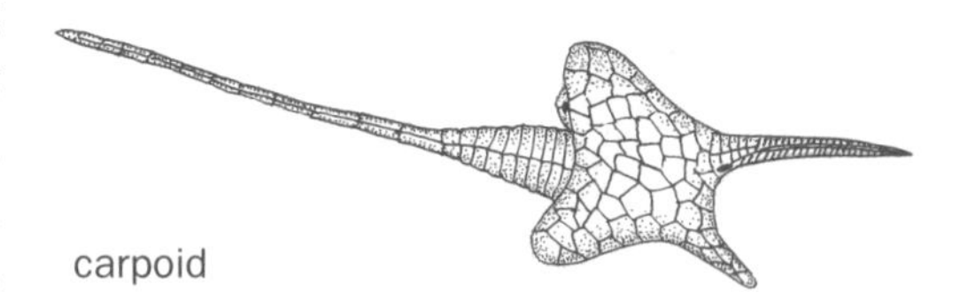

carpoid

poids are often shown as the sister phylum of the echinoderms. It is worth discussing whether carpoids are or are not echinoderms, just as it is worth asking whether the late synapsid reptiles were or were not mammals—but only in so far as such discussion helps us to understand the animals themselves and their relationships. Merely semantic arguments serve little purpose. In any case, carpoids are commonly shown as the sister group of the echinoderms, which is sensible and is the course adopted here.

The **helicoplacoids** were the first of the lineage to show radial symmetry, and hence they are commonly presented as the first true echinoderms, which, again, is the course adopted here. They were not pentaradial, however, but triradial—as reflected in the three open ambulacral grooves, which presumably carried food to the mouth. They clearly did have a water vascular system (which goes along with ambulacral grooves) probably with a madrepore that opened near the mouth; and the mouth was on the side of the body, not the tip.

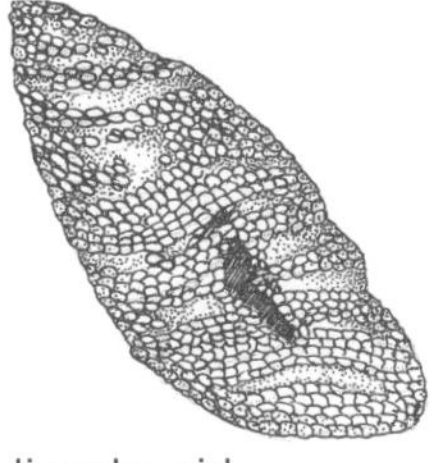
helicoplacoid

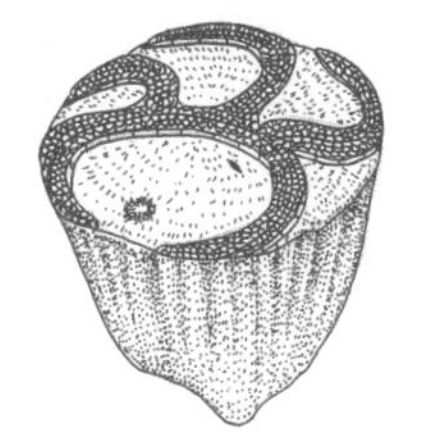
Camptostroma

Camptostroma was one of the earliest known types to attain pentaradial symmetry, with five open ambulacral grooves. *Camptostroma* had a stalk and, like the modern sea lilies, it had both mouth and anus on the upper (oral) surface. *Camptostroma* can plausibly be seen both as a descendant of the helicoplacoids and as the sister (that is, similar to the ancestor) of all later echinoderms: a pivotal creature indeed, if this is truly the case. The **lepidocystoids** seem to be sisters of a clade that includes the crinoids (the modern sea lilies) and the ancient **Cystoidea**, which were like sea lilies. This leaves ***Stromatocystites*** as a plausible sister (a close relative of the presumed common ancestor) of all the remaining living classes.

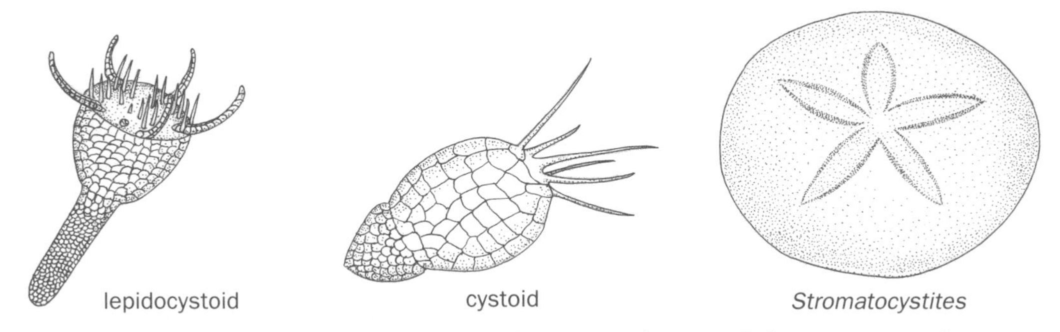
lepidocystoid cystoid *Stromatocystites*

Among the truly modern groups, the sea urchins and the sea cucumbers are thought to be sister groups: the brittle stars have been cited as sisters to sea urchins + sea cucumbers; and the starfish are commonly taken to be sisters to the brittle stars + sea urchins + sea cucumbers. The relationship between the sea daisies (Concentricycloidea), first described as recently as 1986, and the other living classes is not finally established. The discoverers of sea daisies, A. N. Baker, F. W. E. Rowe, and H. E. S. Clark, suggested they are sisters to the starfish; Claus Nielsen places them among the starfish; while many modern scholars regard them as modified echinoideans. I show them here as one tine of a trichotomy; content to do as Joel Cracraft recommends, and be honest about the areas of uncertainty (while at the same time being as tidy as is reasonable).

Finally, some zoologists place the crinoids and cystoids together in their own subphylum, the **Pelmatozoa**; and all other living groups in another subphylum, the **Eleutherozoa**. This is an old idea that has been recently revised. It does emphasize that all the living groups apart from the sea lilies belong in a single clade—one containing about 6400 of the 7000 or so living echinoderms.

Here is a brief overview of the living echinoderm classes.

Sea lilies and feather stars: class Crinoidea

Sea lilies look like flowers, with the oral side upwards like an upside-down starfish (though the gut coils back on itself and the anus points upwards too) and the aboral side (the side opposite the mouth) attached to the substrate by a stalk. But some sea lilies have lost their stalks—and so, too, have feather stars. With the aid of well-developed arm muscles they can walk on the tips of their arms, with the central disc well clear of the substrate; or swim by moving their arms alternately so that at any one time half are up and half are down. Both sea lilies and feather stars feed by trapping suspended particles with their arms and shuffling the nutritious fragments towards the upwards-pointing mouth with their tube feet and cilia. This may be how echinoderm tube feet originated, and hence the whole water vascular system. About 625 living species of crinoid are known.

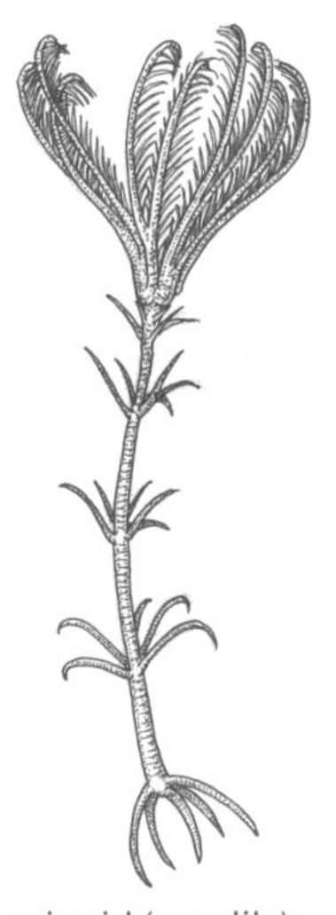
crinoid (sea lily)

Starfish or sea stars: class Asteroidea

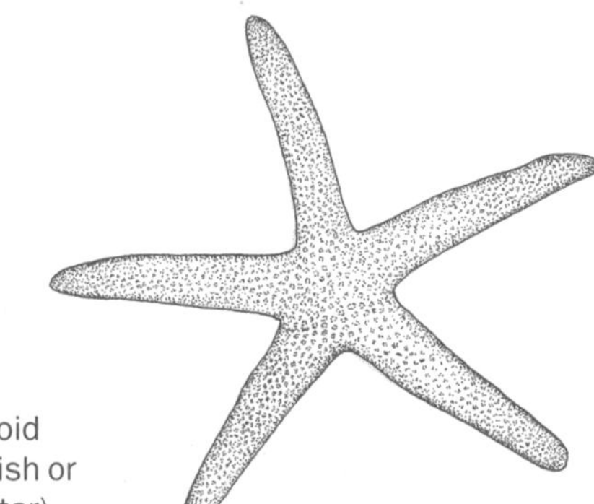
asteroid (starfish or sea star)

Asteroids are the highly successful and influential starfish or (as the Americans say) 'sea stars': there are 1500 or so species in five orders. They have stellate (star-shaped) bodies with five arms or more, which are attached to the central disc without any obvious 'join'. Asteroids exemplify the arts of travelling with tube feet, moving each podium with a power and recovery stroke, shortening, lengthening, sticking, and bending on recovery. The movements are out of phase—there is nothing here to compare with the ordered (metachronal) beating of a millipede's feet—but the overall drag of the body is smooth. Most starfish move slowly but a few species, like *Pycnopodia*, can get a move on.

Most starfish are scavengers or opportunistic predators with an extraordinary way of feeding: they push out or 'evert' their stomach, spread it over the intended food, secrete enzymes, and then suck in the resulting soup. Little escapes them: they can push their stomachs even between the valves of a tight-shut clam, to dissolve the helpless flesh within. Most of the ones that live on the shore between the tides and are seen in rock pools belong to the order Forcipulatida; the famous *Acanthaster* or crown-of-thorns starfish that eats coral polyps and has been doing such damage to Australia's Great Barrier Reef is one of the Spinulosida. Some are strict specialists, like *Solaster stimpsoni* of the northeast Pacific, which feeds on sea cucumbers; while *S. dawsoni* feeds on *S. stimpsoni*. A few starfish are suspension feeders, feeding on plankton or detritus.

Sea daisies: class Concentricycloidea

The sea daisies are diminutive creatures, less than 1 centimetre across, that were discovered as recently as 1988—in deep water, on sunken wood, in New Zealand. Their bodies are discoidal with a ring of spines on the margin, but without radiating arms. Only two species in one genus are known: *Xyloplax turnerae* and *X. medusiformis*. Their guts are remarkably reduced: *X. turnerae* has an incomplete gut with a stomach but no intestine or anus while *X. medusiformis* has no gut at all. They may absorb organic matter, derived from bacteria on the rotten wood, directly through the membrane or **velum** that covers their entire oral (down-facing) surface; and indeed the velum may represent the gut, like a starfish stomach permanently everted.

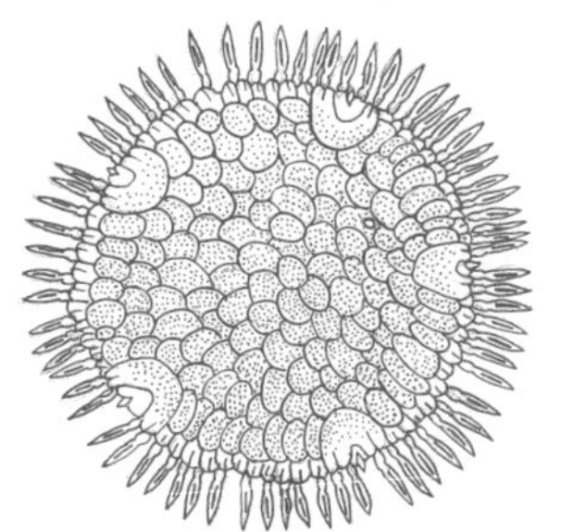

concentricycloid (sea daisy)

Brittle stars and basket stars: class Ophiuroidea

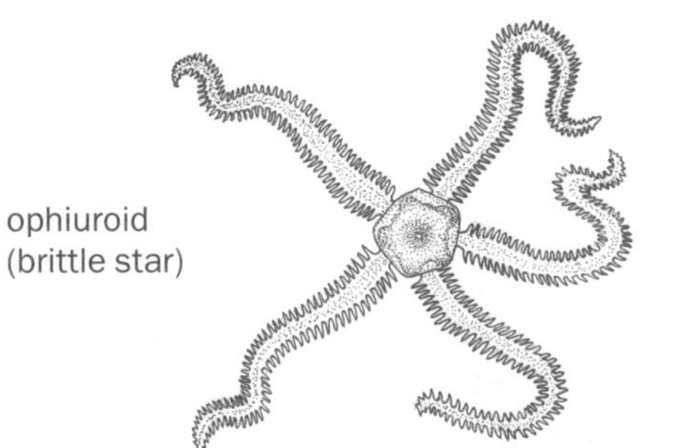

ophiuroid (brittle star)

In ophiuroids—brittle stars and basket stars—the body is again stellate or star-shaped with five arms that in some species are branched. But—in contrast to asteroids—their arms are clearly demarcated from the central disc. Their arms can move freely and powerfully from side to side but not up and down and if you try to bend an arm upwards it breaks off—hence 'brittle star'. By the lateral movements, however, they crawl or cling and in some cases dig. They feed in various ways, from predation to suspension feeding, and some do more than one thing. For example, some wave their arms through the water and trap organic material on mucus, which the podia then roll into a bolus and shuttle to the mouth; and some basket stars coil their arms around larger prey almost like an octopus. All in all they are highly variable—and successful. There are about 2000 species.

Sea urchins and sand dollars: class Echinoidea

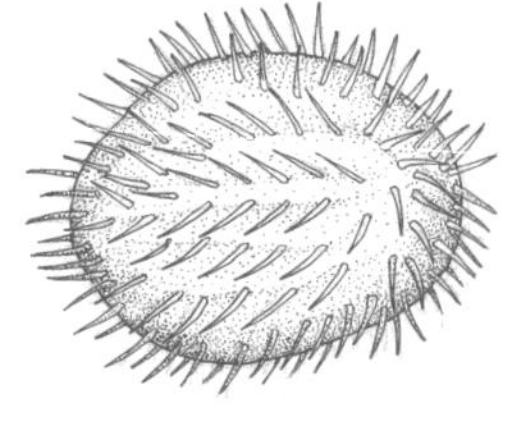
echinoid (sea urchin)

The echinoids are the sea urchins and sand dollars, with bodies shaped like globes ('globose') or discs. Some have modified the radial symmetry to become bilateral again—although, of course, the radial symmetry of echinoderms is itself secondary (as we can see from the bilaterally symmetrical larvae). Characteristically, in echinoids the ossicles are fused to form a complete test, much beloved as seaside souvenirs; and the animal is armoured with movable spines (hence 'sea urchin').

Two subclasses of echinoid are still extant. The Perischoechinoidea, often considered the more primitive, have large, pencil-like spines. Most species are extinct, but about 140 remain. The other subclass, the Euechinoidea, contains about 800 species in four superorders. Two of them—the Diadematacea and the Echinacea—include 'regular' urchins; one, the Gnathostomata, has reverted to bilateral symmetry and includes the sand dollars; and the fourth, the Atelostomata, includes the 'irregular' urchins.

A key feature of sea urchins is an extraordinary skeletal structure in the centre called **Aristotle's lantern**. It has five hard teeth that are moved by an amazing complex of struts and muscles as in some medieval grinding machine. In sand dollars the lantern is much modified, and the teeth reduced, while some echinoids lack the lantern altogether. Echinoids tend to be herbivores (on algae) or feeders on particles although some sea urchins are predators. Sea urchins move both by podia and by movable spines. Many burrow, and some use the teeth of their Aristotle's lantern to scour themselves a hole in solid rock, in which they may become incarcerated. Sand dollars live in or on soft sediments, sometimes on the surface and sometimes burying themselves completely, digging with their movable spines.

Sea cucumbers: class Holothuroidea

holothuroidean (sea cucumber)

Finally, the holothuroideans or sea cucumbers have fleshy bodies: the skeleton is reduced to isolated ossicles embedded in the muscular body wall. They are sausage-shaped and they appear to lie on their sides so that their mouth is at the front. Tentacles surround the mouth, varying in form from finger-like to leaf-like, and in number from 8 to 30. The number and form of the tentacles help to define the three subclasses of sea cucumber found today.

Most sea cucumbers live on the surface of various substrates or burrow into soft sediments, though some lodge in crevices or under rocks and a few are feebly pelagic (swimming). They crawl by podia and muscular movements of the body wall. The standard way to feed is via the mucus-covered tentacles that trap suspended particles, including live plankton, which are then pushed into the mouth; but some use their tentacles to ingest organic material on the substrate.

Sea cucumbers have two remarkable tricks. One is evisceration: the entire digestive tract, sometimes with other organs as well including the gonads, is spewed out through the mouth or anus, depending on species. Experimentally they do this under various stresses although why they do it in nature is unclear, but perhaps the jettisoned guts serve as a decoy, like the ink-cloud of a cephalopod. In any case, the lost organs usually regenerate. Sometimes, too (the second trick), some genera rupture the hindgut to extrude clusters of hollow sticky threads known as **Cuvierian tubules** that wrap around a molester. As I can attest, having once been caught molesting a wild sea cucumber, these are most off-putting.

13

SEA SQUIRTS, LANCELETS, AND VERTEBRATES

PHYLUM CHORDATA

THE PHYLUM CHORDATA includes the most exciting, dramatic, and in many ways accomplished creatures that ever were brought forth by Earthly evolution: in the subphylum Vertebrata are sharks and rays, sturgeons and marlins, lungfish and frogs, turtles and dinosaurs, ostriches and hummingbirds, whales, tigers, bats, squirrels, elephants, human beings—plus several classes of animal that are long-since gone, like the great armoured placoderms, the acanthodians, and a mixed bag of jawless but often armoured 'fishes' colloquially known as 'ostracoderms', and still somewhat mysterious, lamprey-like creatures that started to come to light in the 1980s, the conodonts.

The vertebrates, along with the arthropods and sometimes the molluscs, set the tone of most of the world's ecosystems—for the creatures that sit at the top of the food chains shape the lives of the ones below at least as much as the ones below influence those above. Without vertebrates, life on Earth would have been sadly impoverished these past 500 million or so years. Among other things, there is no good reason to suppose that any other lineage would have produced mammalian-style intelligence. It is highly unlikely, therefore, that any creature would have emerged that was able to appreciate the others. All flowers, as Thomas Gray put the matter, would have bloomed unacknowledged in a conceptual desert.

Yet included, too, in the phylum Chordata—for reasons that seem phylogenetically unarguable—are two other subphyla of unredeemed humility, quite bereft of vertebrate dash and fire. Sisters to the Vertebrata are the Cephalochordata, or lancelets, and best known for *Branchiostoma* (amphioxus); jawless, eyeless, transparent slips of creatures superficially like baby eels. Sisters to the Vertebrata + Cephalochordata are the Urochordata (also called sea squirts, or tunicates because they were once known as Tunicata), often sessile and sometimes brightly coloured, which casual scuba-divers could easily confuse with anemones.

Then, again, the Chordata as a whole has an odd genealogy; a collection of relatives whose portraits would, if this were an ancient household of aristocrats, be hung in some dark and secured gallery. As shown in the master tree of the Animalia in Section 5, the sister group to the Chordata as a whole are the Enteropneusta or acorn worms, such as *Balanoglossus*. In his book *Animal Evolution* (1995), Claus Nielsen unites the Chordata with the Enteropneusta in the supra-grouping he calls **Cyrtotreta**. The cyrtotretes in their turn are linked with the peculiar pterobranchs and—unlikely though it may seem—with the echinoderms.

In short, the chordates, which—in the vertebrates—include the most powerful, most swift, most versatile, and most intelligent creatures that have ever lived, emerged from amongst the ranks of animals that in large part are either sessile or seem to have a hankering to be so and some of which (notably the echinoderms) are without brains, not simply in the loosely pejorative sense but quite literally. Chordates, in short, have strange bedfellows. Nature, it seems, likes to startle.

The chordate tree as a whole is based partly on the ideas of Claus Nielsen, as recorded in his *Animal Evolution*, while the vertebrate section of the tree leans heavily on Michael Benton's *Vertebrate Palaeontology* (second edition, 1997) and the ideas of Axel Meyer of the State University of New York at Stony Brook and Christine Janis of Brown University, Rhode Island.

A GUIDE TO THE CHORDATA

Chordates are revealed as deuterostomes by various plesiomorphic features, as discussed earlier in Part II (Section 5); they include spiral cleavage of the young embryos in those types in which cleavage is complete and visible. Of more interest here are their synapomorphies: the derived features peculiar to the phylum.

First—though they share this feature with the enteropneusts, or acorn worms—the front end of the gut of chordates is generally somewhat enlarged, to form a **pharynx**; and—the key characteristic—a series of **gill slits** perforate the pharynx from the inside (the endodermal cell layer) right through to the outside (the ectodermal cell layer). The gill slits are separated by **gill bars**, which in more complex types like the vertebrates are elaborated into full-blown **gill arches**. In many chordates the gill slits are enclosed within a kind of pouch, formed by an outfolding of the ectoderm—as in tunicates, where the whole front end seems to be held in a loose bag. But in some, like sharks, the slits can be seen in all their glory opening to the surface of the animal. Adult tetrapods (four-legged land vertebrates) lack gill slits; yet the embryos of tetrapods, even of inveterately terrestrial creatures like horses and human beings, still show fleeting signs of them.

It is because both enteropneusts and chordates have gill slits that Nielsen links the two together to form the Cyrtotreta. But the Chordata also share a group of three synapomorphies: features that are absolutely characteristic and diagnostic of themselves, which they do not share with the enteropneusts:

- The body of chordates is stiffened by a rod formed from a cylinder of cells close-packed within a toughened membrane. Most zoologists call this rod the **notochord**, but Nielsen prefers simply to call it a 'chord'. In urochordates (or tunicates) the chord is confined to the rear end, and Nielsen then refers to it as a urochord (where 'uro' implies 'tail'). He reserves the term 'notochord' for the chord of cephalochordates and vertebrates that extends along the whole body; and, indeed, he proposes that the cephalochordates and vertebrates together should be linked to form the 'Notochordata'. In typical vertebrates, of course, the stiff rod of cells in the primitive notochord is replaced by the series of bones that form the **vertebral column**.

- The central nervous system of chordates is a tube along the back—a **dorsal nerve cord**—that is formed by an infolding of the ectoderm.

- At least at some stage in their lives, chordates have bands of muscle along either side, at the rear end, which enable them to wag or undulate their 'tails' and thus drive them along. The muscle fibres are arranged longitudinally—they run along the length of the animal: but commonly the **longitudinal muscles** are divided so that the animal appears to have zigzag bands of muscles running *around* the body—a pattern obvious in fish such as dogfish. In practice, contraction of longitudinal muscles would cause the body to shorten, rather than to bend, unless the body was stiffened by some internal skeleton. In ancestral chordates, the metacoel (a variant on a theme of coelom—see Section 5) may have provided a hydrostatic skeleton; but the chord—or the vertebral column—now does the job.

In practice, these three key chordate features—chord (or notochord), dorsal nerve cord, and longitudinal muscles—probably evolved together. They certainly develop together in the embryo, interdependently.

The chordates are also distinguished by a peculiar organ at the base of the pharynx called the **endostyle**. This consists of bands of ciliated cells that in urochordates and cephalochordates secrete mucus. The mucus traps food and then, with its burden of nutrient, is shuffled by beating cilia in a more or less continuous conveyor into the oesophagus. This mode of **mucociliary feeding** is not aesthetic, but it works. For reasons that I have not been able to discover, however, the cells of the primitive endostyle also trap iodine. In vertebrates, the endostyle is absent, except in the larvae (called ammocoetes) of the jawless lampreys. But close examination shows that many cells of the **thyroid gland** are the same as those of the endostyle. In other words, it seems that in vertebrates the endostyle has evolved into the thyroid, a crucial endocrine gland: and the ability to absorb iodine has become a vital feature, because one of the main hormones of the thyroid, thyroxine, contains iodine as a key component. Here, it seems, is another example of evolutionary opportunism—comparable with the way in which surplus bones in the jaws of ancestral reptiles were commandeered as ear bones in the mammals that descended from them.

Looking at the chordates as a group (and perhaps throwing in the related enteropneusts and pterobranchs, just to round out the picture) we can discern various general evolutionary themes. First of all it is evident that the gill slits served originally for filter feeding: water entered the mouth and was wafted through the gills by cilia

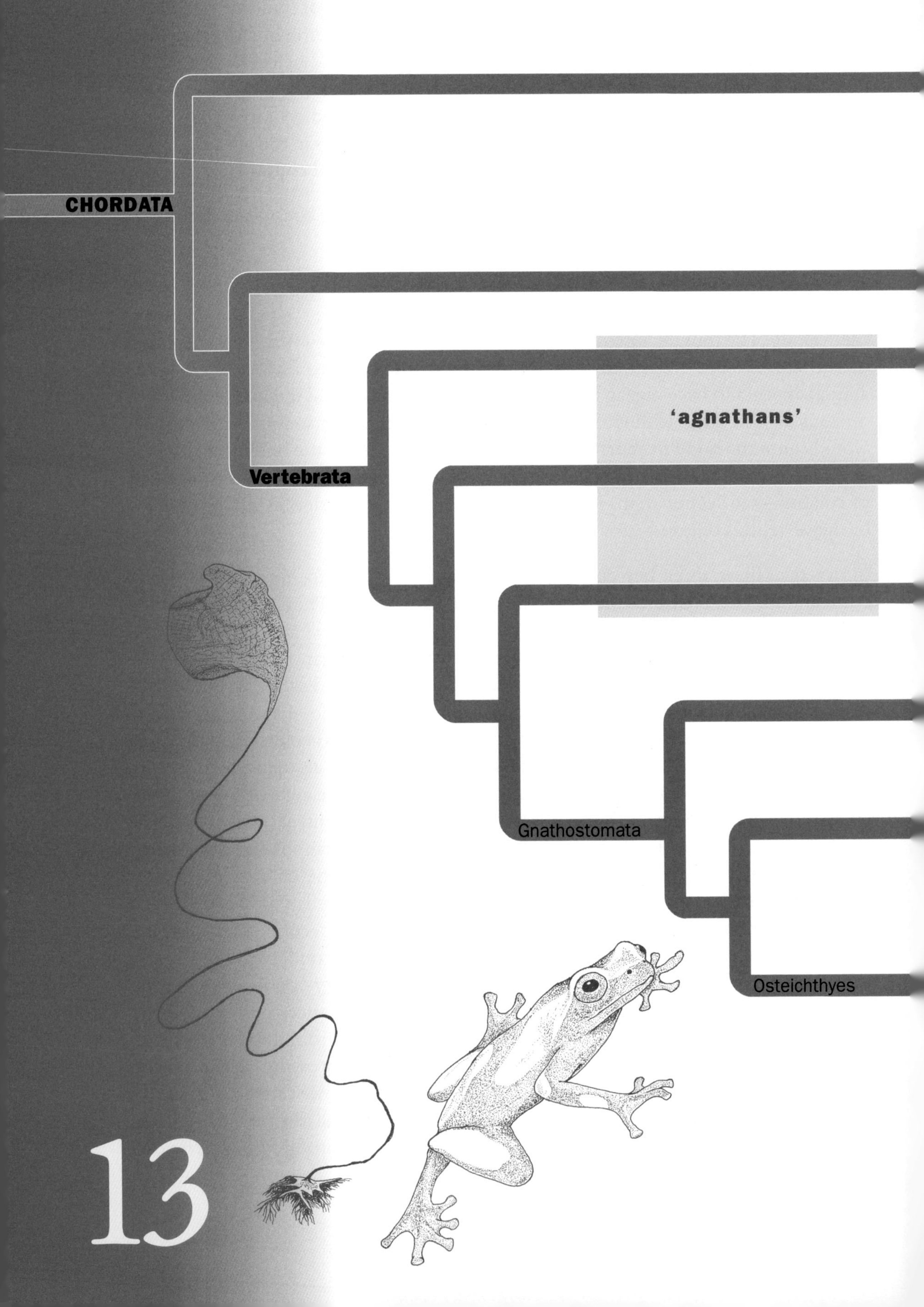
CHORDATA
Vertebrata
'agnathans'
Gnathostomata
Osteichthyes
13

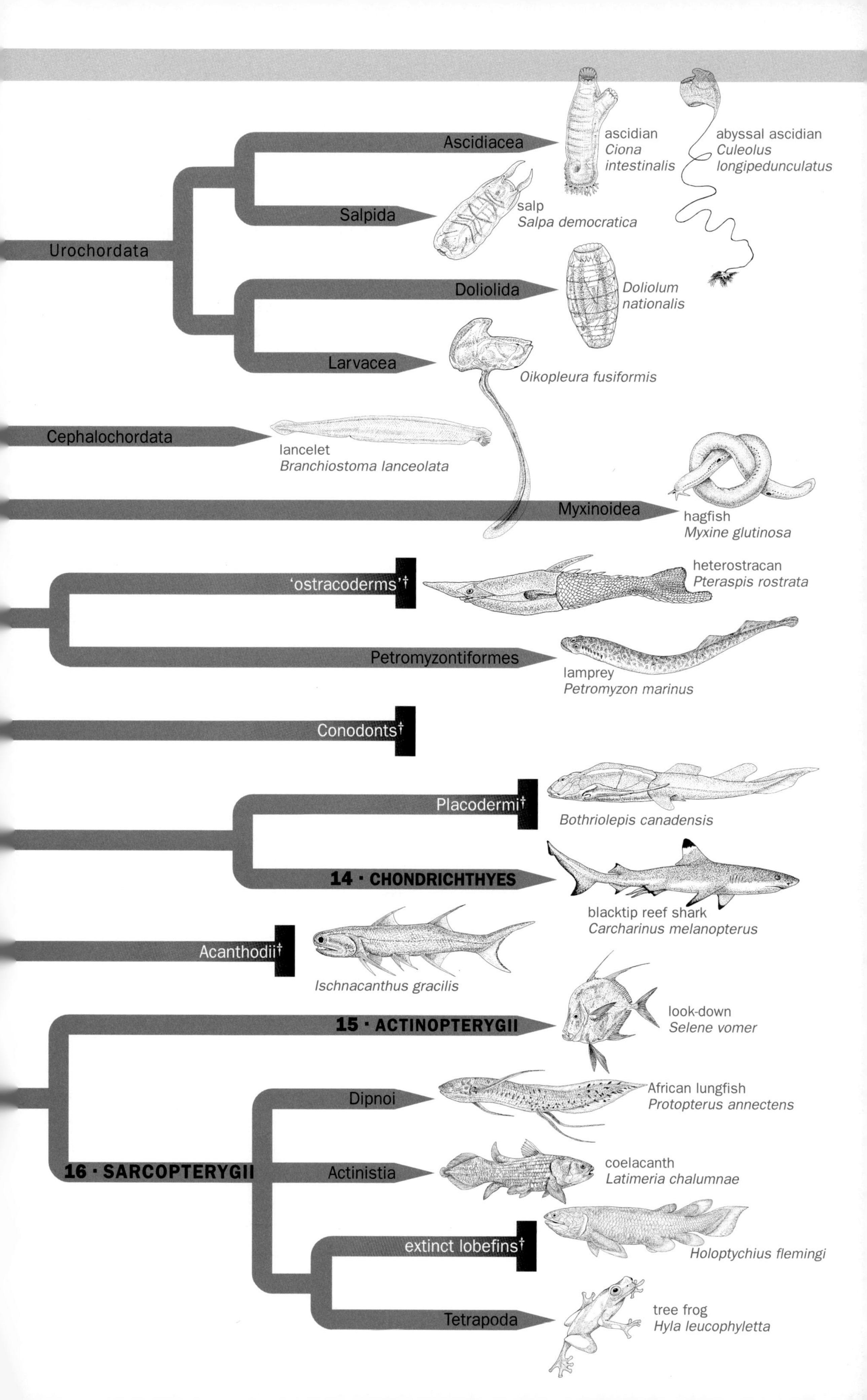
Ascidiacea
ascidian
Ciona
intestinalis
abyssal ascidian
Culeolus
longipedunculatus
Salpida
salp
Salpa democratica
Urochordata
Doliolida
Doliolum
nationalis
Larvacea
Oikopleura fusiformis
Cephalochordata
lancelet
Branchiostoma lanceolata
Myxinoidea
hagfish
Myxine glutinosa
'ostracoderms'†
heterostracan
Pteraspis rostrata
Petromyzontiformes
lamprey
Petromyzon marinus
Conodonts†
Placodermi†
Bothriolepis canadensis
14 · CHONDRICHTHYES
blacktip reef shark
Carcharinus melanopterus
Acanthodii†
Ischnacanthus gracilis
15 · ACTINOPTERYGII
look-down
Selene vomer
Dipnoi
African lungfish
Protopterus annectens
16 · SARCOPTERYGII
Actinistia
coelacanth
Latimeria chalumnae
extinct lobefins†
Holoptychius flemingi
Tetrapoda
tree frog
Hyla leucophyletta

on the gill bars. In other words, this whole pharyngeal apparatus served originally for feeding on small particles, although it probably also always served some function in respiration as the oxygen-rich water was wafted over the thin tissues over the gill bars. In urochordates and cephalochordates, efficiency is enhanced by mucus emanating from the endostyle, which provides the mucociliary mechanism. But the basic system is the same: cilia on the gill bars create the necessary current of water. In later forms—seen in ammocoete larvae of lampreys, and also in fish—the current is driven from the mouth through the gill slits not merely by cilia but by muscular contractions of the pharynx.

At the same time, the function of the pharyngeal apparatus shifted. For vertebrates ceased to be filterers of fine particles, and became macrophagous: eaters of big things. So their gill slits were no longer involved in feeding but instead became focused on respiration—as is clearly seen in modern fish, whose gills are blood-rich and feathery. Note, however, that many modern fish have re-invented filter feeding, and by means of various devices attached to the inside of the gills they do filter their food from the flowing current. Paddlefish, relatives of the sturgeon, do this spectacularly, swimming with their gills expanded like lobster pots,[1] in a technique reminiscent of baleen whales—except, of course, that whales no longer have gill slits through which to shove out the surplus water but instead push it out through the mouth, the way it came in. Just to round off the story, vertebrates no longer needed to secrete vast quantities of mucus in which to entrap their food, and the endostyle was also freed to take on another function—that of the thyroid.

So much for feeding and respiration. At a more general level, the chordates have clearly pursued two paths. First, as seen in most urochordates, some have gone down the sessile route. The adults stick to the rocks like up-market anemones. But urochordates do have mobile larvae, which typically are tadpole-like, whose function is to find new places to live. In short, the typical tunicate lifestyle is a social metaphor: mobility is a childish pursuit, its purpose only to find a place whereon to settle into adult sedentariness. Yet one renegade class of tunicates—the Larvacea—have renounced the staidness of middle age and retain tadpole-like features throughout their lives; exhibiting once more the phenomenon of neoteny. For the other two chordate subphyla, the cephalochordates and the vertebrates, sedentariness is not an option. They glory in their mobility. In fish, the ranks of muscle along the body provide a supremely powerful engine of propulsion—retained in much modified form even in terrestrial animals, as in the flexing body of the cheetah (albeit with the plane of movement shifted from the lateral to the vertical).

The general themes are clear enough: but the historical facts of the case—how the different chordate forms evolved from each other—is less obvious. There have been two main theories. One says that the original chordates were tunicate-like, and that cephalochordates and vertebrates evolved neotenously from the tadpole-like larvae of tunicates, just as the Larvacea seem to have done. The other idea—favoured by Claus Nielsen, and suggested in the tree presented here—is that the tunicates were

[1] The feeding paddlefish is illustrated on page 378.

and are an evolutionary dead end. In other words, the first ancestral chordate was *not* tunicate like: it crawled along the bottom. Thus, in general terms, it had two evolutionary options: to become more committed to the bottom, and become tunicate-like, or to free itself from the bottom, and become a swimmer. Some chose one option, and became the Urochordata, while the other pursued the mobile route, and became the Notochordata, or Cephalochordata + Vertebrata.

Finally, and briefly, there are other ways of classifying the chordates and their relatives. A typical traditional taxonomy is presented in *The Invertebrata* of Borradaile, Eastham, Potts, and Saunders—the famous 'BEPS', the third edition of which, dated 1959, was my standard textbook at university in the early 1960s (as mentioned in Section 3). The authors recognized a grand grouping called the Protochordata, in which they included the single phylum Chordata. They then divided the Chordata into four subphyla. Three of them—the Urochorda, the Cephalochorda, and the Vertebrata—are as described here. But the fourth they called the Hemichorda: and within it they included—as classes—the Enteropneusta, the Pterobranchia, and the long-extinct but once common Graptolita. I mention this traditional (precladistic) system because the term 'Hemichorda' or 'Hemichordata' still crops up. As Claus Nielsen argues, however, 'Hemichordata' is clearly an untidily paraphyletic grouping; and the system that he advocates—to rank the different hemichordate classes as separate phyla, as shown on the tree in Section 5—is clearly preferable. Again, however, it may sometimes be convenient to retain the term 'hemichordate' as a general adjective, to describe the group of phyla that are most closely related to the chordates, but lack the chordate synapomorphies.

So let us look at the three chordate subphyla one by one.

SEA SQUIRTS OR TUNICATES: SUBPHYLUM UROCHORDATA

The typical and best known of the 1250 or so known species of urochordates are like little, brightly coloured bags of jelly stuck to the reef or some such, filtering nutrient from the surroundings by drawing in water through the gaping mouth by cilia, trapping particles of food in mucus, and wafting out the surplus water through the gill slits in the pharynx. The pharynx, with its gill slits, in turn is typically enclosed within a loose **tunic**, formed from ectoderm, like a sleeping bag—hence the alternative name of 'tunicate'. The common name, 'sea squirts', refers to the way the animals eject surplus water. Tunicates find their feeding sites with the aid of mobile, tadpole-like larvae and members of one group, the Larvacea (or appendicularians), neotenously retain the tail of the larva.

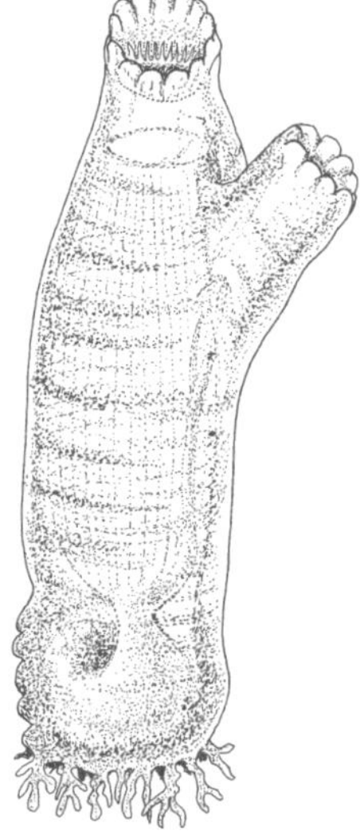

ascidian
Ciona intestinalis

Urochordates are traditionally divided into three classes: the Ascidiacea, the Thaliacea, and the Larvacea. But Nielsen

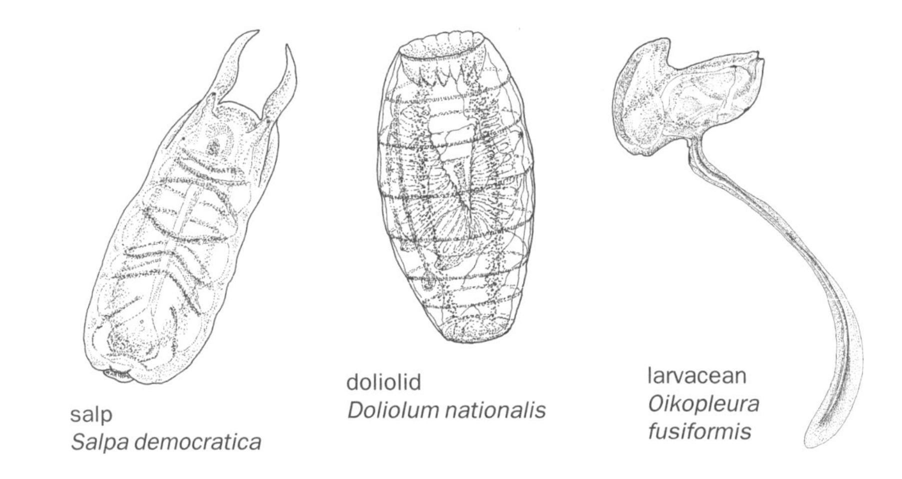

salp
Salpa democratica

doliolid
Doliolum nationalis

larvacean
Oikopleura fusiformis

abyssal ascidian
Culeolus longipedunculatus

points out that the Thaliacea seem to contain two quite separate lineages, which he calls the Salpida and the Doliolida, thus making four classes in all. The ascidians, class **Ascidiacea**, have a thick, tough tunic, and some forms are colonial. The salps, class **Salpida**, have a 'hyaline' (translucent) tunic, which is sometimes brightly coloured. The doliolids, class **Doliolida**, have a thin ectoderm with a thin cuticle, and from time to time they moult large areas of their tunic, which keeps the body free of debris. The larvaceans, class **Larvacea**, secrete materials that form an all-embracing filter, like a house, and pump water through it with their muscular tail to concentrate plankton.

LANCELETS: SUBPHYLUM CEPHALOCHORDATA

There are only about 25 species of cephalochordates, in two families; the commonest is *Branchiostoma*, traditionally called amphioxus, which spends its life half-hidden in coarse sand. A cephalochordate is like a précis of a chordate: an effectively headless, transparent body like a willow leaf, with a notochord running from end to end and rows of muscles either side by which to wiggle and so swim, or bury itself; and a spacious pharynx with an impressive row of parallel gill slits for feeding (which are U-shaped, as is typical of primitive chordates). The common term for cephalochordates as a whole, lancelets, is highly appropriate. Cephalochordates share so many features with vertebrates that the two are obvious sister groups. But amphioxus and its like also have some autapomorphies or special features—including peculiar asymmetry in the juveniles—that show that the cephalochordates as we know them cannot include the ancestor of the vertebrates.

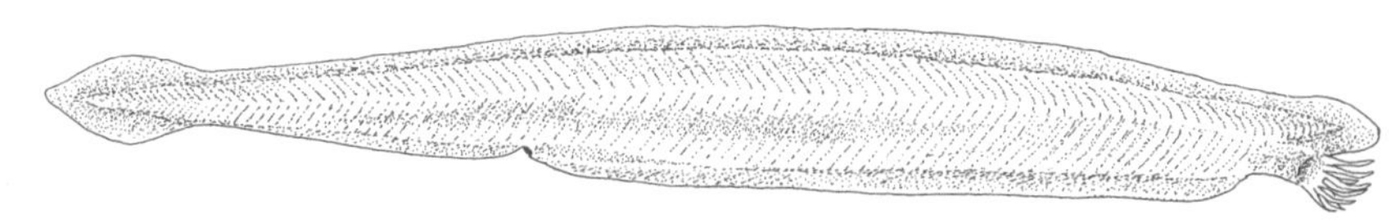

lancelet, *Branchiostoma lanceolata*

VERTEBRATES: SUBPHYLUM VERTEBRATA OR CRANIATA

The vertebrates are most obviously distinguished from other chordates by their bony skeletons. **Bone** has converted the stiff but non-versatile notochord into a non-compressible but nonetheless flexible spine that encloses the dorsal nerve cord, and protects it. In the head, bone encases the brain—hence **Craniata**, the alternative name for the Vertebrata—and without this protection, we may reasonably surmise, the vertebrate brain could never have evolved into the fabulous organ it has so often become. With bone to support them, too, the flaps of skin that serve pre-vertebrate chordates as stabilizers, can evolve into versatile and mobile fins and limbs, acquiring their own musculature and moving independently of the body wall. But, of course, nature is never content.

One evolutionary trend since bony skeletons evolved has been to make them heavier and stronger—there have been heavy, armoured creatures in almost every vertebrate class (except the birds). But another trend has been to lighten the skeleton, so that the skulls of modern fish are often minimalist, with thin, strong plates placed judiciously where they do most good, while bird skeletons are a model of aeronautic economy. Several groups—notably the modern hagfish (Myxinoidea) and lampreys (Petromyzontiformes), the sharks, rays, and chimaeras (class Chondrichthyes) and several groups of ray-finned fish (Actinopterygii)—have abandoned bone entirely, at least in their main skeletons, and make do with rubbery supports of cartilage. But bone is what turned chordates, or some of them, into vertebrates: and it is the vertebrates that have ensured that the chordates are more than also-rans. Indeed, in versatility and ecological impact, vertebrates match arthropods.

The earliest fossils of conodonts—rasping teeth—date from the Cambrian, more than 495 million years ago, and if they are indeed vertebrates (as now seems beyond dispute) then they are the oldest known representatives of the subphylum. Apart from them, the earliest known vertebrates appeared in the Ordovician, the period after the Cambrian; but there could have been much earlier types, which simply have not been found. The earliest types were certainly bony—indeed many quickly acquired heavy armour, and so were bony outside and in. But they lacked jaws; and without jaws they must have fed by various forms of sucking and rasping. The ancient and long-extinct 'ostracoderms', though often impressively armoured, were jawless; but the only remaining jawless vertebrates are the naked, eel-like, hagfish and lampreys. Jawed vertebrates arose from among the ranks of the jawless types and form a distinct clade, the **Gnathostomata**. The jawless types were traditionally classed as the 'Agnatha', but because such a grouping is paraphyletic and highly various, this formal term should be abandoned. But the term **'agnathan'**, spelt informally with a lower-case 'a', can be retained as a useful adjective.

It is difficult to classify the various groups of vertebrates in ways that satisfy the conventions both of cladistics and of Linnaean tradition. At least, the phylogeny itself seems reasonably clear—as clear as in any major taxon—because the vertebrates are well studied and they provide good fossils. But the naming and ranking raise problems. Thus I could simply declare that all the crown groups shown here with formal Graeco-Latin names are classes: the Myxinoidea (hagfish), Petromyzontiformes (lampreys), Placodermi (placoderms), Chondrichthyes (sharks and rays), Acanthodii (acanthodians), Actinopterygii (ray-finned fishes), Dipnoi (lungfish), Actinistia (the coelacanth), and the Tetrapoda (the four-legged land vertebrates). 'Ostracoderms' are recorded informally in the tree because they are such a miscellany, but are treated more fully below; conodonts could be represented as Conodonta; while 'extinct lobefins' are another mixed bag that will be dealt with in more detail in Section 16 (on the Sarcopterygii).

But this simple scheme raises two problems. First, the huge clade of the Osteichthyes ('bony fish') divides cleanly into two subclades: the Actinopterygii and the Sarcopterygii. If we propose that the Actinopterygii should be ranked as a class then, logically, the Sarcopterygii should be a class as well. But the Sarcopterygii embraces at least four major clades: lungfishes, the coelacanth, the mixed bag of extinct lobefins, and the tetrapods. Worse: the tetrapods, in turn, include the creatures generally known as amphibians and reptiles, plus birds and mammals, each of which is traditionally ascribed the rank of class.

So I suggest—in the spirit of Neolinnean Impressionism (see Part I, Chapter 5)—that we should simply wave our arms and agree to be arbitrary. In this book, therefore, I am treating the Actinopterygii as a Linnaean class, and the Sarcopterygii as a major grouping of unspecified ranking that includes a number of classes. Purists will argue that this is messy; but it is no messier than any other possible solution. The phylogenetic history of living creatures is the truth. The trees presented here, based on the best possible data and framed by the rules of cladistics, are hypotheses that attempt to reflect that truth; but the names and the Linnaean rankings are simply an *aide-mémoire* and an exercise in information retrieval, and should be pragmatic. It is inconsistent to treat the Actinopterygii as a class and the Sarcopterygii as a kind of super-superclass, as I have done here; but if we do this while bearing the shortcomings in mind, we can at least communicate with each other, and link the modern literature to that of the past four centuries.

Finally, a brief word on the meaning of 'fish'. For zoologists, 'fish' is a term for a grade: any vertebrate that is basically streamlined and is basically a swimmer, propels itself primarily by undulations of the body, and breathes by gills is colloquially called a fish. In fact, any vertebrate that does not belong among the Tetrapoda is a fish—and this, as you can see, includes creatures from a great many quite distinct lineages. In truth, some fish are far more closely related to tetrapods than to other fish: because the Actinopterygii are the sister group of the Sarcopterygii (the group that includes the tetrapods, lungfish, and lobefins) we can say that the salmon is closer to the horse than it is to the shark. But salmon and sharks can both properly be called 'fish'. The formal term 'Pisces', however, which appears in many traditional texts, seems to have no place in modern classification.

We can now look briefly at the main vertebrate groups, focusing on those that are not treated in more detail in later sections.

VERTEBRATES WITHOUT JAWS: THE AGNATHANS

Two main general groups of ancient, extinct, agnathan fishes are known: the lamprey-like conodonts, and the ostracoderms.

The **conodonts** were first described, from teeth, only in 1983, and at first many doubted whether they were vertebrates at all. But in the mid-1990s remarkable fossils came to light, showing among other things that they had bands of muscles along the sides that are very chordate-like, and large vertebrate-like eyes, and most zoologists now admit them to the vertebrate fold. But they have some puzzling features—not the least being that they lack any sign of gills. Perhaps, though, as Philippe Janvier of the Laboratoire de Paléontologie in Paris has noted, it is just that their gills have so far failed to fossilize. Such clues as there are suggest that the conodonts are fairly close to the gnathostomes, which is how I show them on the tree.

The **'ostracoderms'** were often armoured and they achieved great diversity in the Ordovician and the Devonian. In the past they were all formally classified as Ostracodermi ('bony skin'), but it is now clear that the group we call 'ostracoderms' includes at least five distinct classes (all of which, of course, are long extinct).

The 'ostracoderms'†

The earliest-known 'ostracoderms' are the **Heterostraci** whose front ends were enclosed in enormous bony head shields of extremely variable shape; some indeed (the cyathaspids) were completely encased in bony plates and scales. Yet heterostracans were clearly sophisticated: patterns on well-preserved fossils suggest they were equipped with sense organs that may have enabled them to detect their prey through movements in the water or weak electrical currents, just as various modern fish do.

heterostracan
Pteraspis rostrata

Two 'ostracoderm' classes lack heavy armour. The **Thelodonti** from the late Silurian and early Devonian, some 420–400 million years ago, are not well known although there many odd scales lying around, which are lozenge-shaped and are made from dentine with a pulp cavity, like a tooth. One of the few thelodonts of which we have reasonable knowledge was about 70 millimetres long and flattened, with a broad snout and wide mouth. The **Anaspida**, like the thelodonts, also lacked heavy head shields.

The two remaining classes of 'ostracoderms' were again heavily armoured. The **Galaspida**, of which there are some remarkable fossils from the early Devonian of China, had broad head shields, often with an impressive array of projections ('processes'). Finally, the **Osteostraci** of the late Silurian–early Devonian are thought to be the sisters of the Gnathostomata. They were heavily armoured around the head like heterostracans, and typically their head shields were flattened, curved, and semicircular like the toe of a boot; but they became much more variable as time passed, so that some were bullet-like, some square, some hexagonal, and some had spines pointing either forwards or backwards. The osteostracans were clearly a highly successful group. But creatures who have no jaws would not find it easy to compete with those that have; and the gnathostomes were themselves becoming extremely diverse in the Devonian.

The living agnathans: hagfishes and lampreys

Two groups of agnathans are still with us—but neither, unfortunately, retains anything like the impressive, gothic structures of the 'ostracoderms'. Indeed, both are much 'reduced' to slippery, eel-like forms, which, however, unlike eels, lack paired fins and supporting limb girdles and have skeletons exclusively of cartilage. These two are the hagfishes and the lampreys, the **Myxinoidea** and **Petromyzontiformes**

Because they look similar, both groups used to be placed together in the class Cyclostomata. Indeed, some modern zoologists still argue that Cyclostomata is a perfectly valid taxon—that the similarities between the two are not simply convergences, but true synapomorphies. Most, however, feel that the resemblances are either convergences or plesiomorphies and that the differences in detailed anatomy are profound. Thus lampreys have a nostril on the top of the head that runs into a pouch beneath the brain, while the single nostril of hagfishes connects directly with the pharynx; and lampreys have two semicircular canals on each side of the head while hagfishes have only one each side; and so on. So the two are commonly ascribed these days to quite different lineages, as shown in the tree: the Myxinoidea and the Petromyzontiformes (a name that is not quite so cumbersome as it seems because *Petromyzon* is the name of the best-known genus). Indeed, as the tree suggests, hagfishes emerge as the sister group of all other vertebrates (although they obviously are not the most ancient vertebrates, because their bodies are clearly so reduced and specialized); whereas the lampreys emerge as the sister group of all the gnathostomes. Lampreys, in short, are more closely related to gnathostomes than they are to hagfishes.

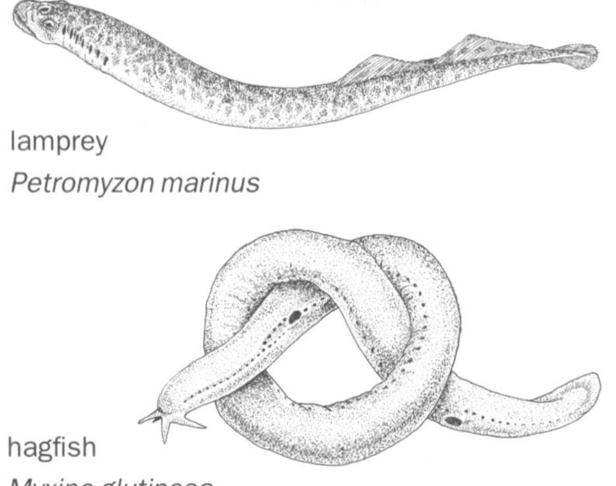

lamprey
Petromyzon marinus

hagfish
Myxine glutinosa

All hagfishes are marine and they live in burrows in soft sediments. Their mouths are surrounded by six tentacles and are equipped with horny, toothed plates that can be pulled in and out to pinch and grasp the worms and rotting cadavers on which they feed. Their method of feeding is bizarre—unique among animals. They tie themselves into a knot and slide the knot forwards until it braces against the flesh, and tears a chunk away.

In contrast, the 30 or so species of lampreys are ectoparasites of other fish to which they attach themselves by mouths shaped like suckers, which, again, are lined with rings of pointed teeth; then rasp away with a toothed, protrusile 'tongue'.

Lampreys spend at least some of their life in fresh water, where they breed; their life cycles are comparable with those of salmon and trout.

But the vertebrates that came storming through after the Silurian, sweeping most of the agnathans aside during the Devonian, were the ones with jaws: the gnathostomes.

VERTEBRATES WITH JAWS: THE GNATHOSTOMATA

Tradition has it that jaws arose from the first three sets of gill arches, each of which, in fish, are compounded from several different bony elements. The general hypothesis has it that some of these elements became incorporated in the lower jaw; some in the upper jaw (which in many fish, even modern ones, is separate from the brain case above); and others became incorporated into the skull itself. There are problems with this. For example, it seems that the bones of the gill arches of gnathostomes are not homologous with the bones of the gill arches of agnathans: gnathostomes evidently replaced the original set with a new set. So there was apparently no direct transition of agnathan gill arches into gnathostome jaws. Neither are there any appropriate fossils to show us the transition. But at least at a commonsense level, the idea is obviously plausible. There certainly seems to be *some* connection between gill arches and jaws.

The tree shows eight lineages of gnathostome. The long-extinct but formidable **Placodermi** (placoderms), are generally supposed to be the sister group of the **Chondrichthyes**—the great clade of cartilaginous fish that now includes the chimaeras, rays, and sharks. These two together are commonly supposed to be the sister group of all the remaining fishes, plus the tetrapods. The **Acanthodii** (acanthodians) are conventionally considered to be the sister group of the Actinopterygii plus Sarcopterygii. Acanthodians are sometimes called 'spiny sharks', and although they emphatically were not sharks they were certainly spiny.

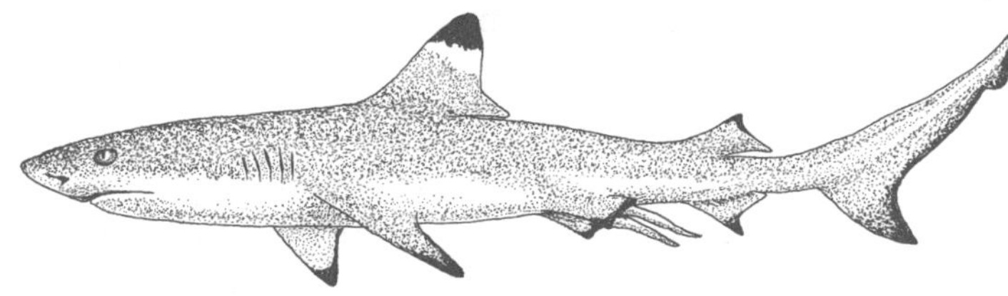
chondrichthyan
(blacktip reef shark)

The **Actinopterygii** are the ray-finned fishes, which are still very much with us—indeed they are easily the most diverse of living vertebrates. Their sister group, the great clade of the **Sarcopterygii**, includes the lungfishes (**Dipnoi**), which are also still with us; the living coelacanth and its extinct relatives, in the class **Actinistia**; a wide range of **extinct lobefins** and **'amphibians'**; and the modern **Tetrapoda**. Because the two great lineages—Actinoptergygii and Sarcopterygii—are sister groups, the

actinopterygian
(look-down)

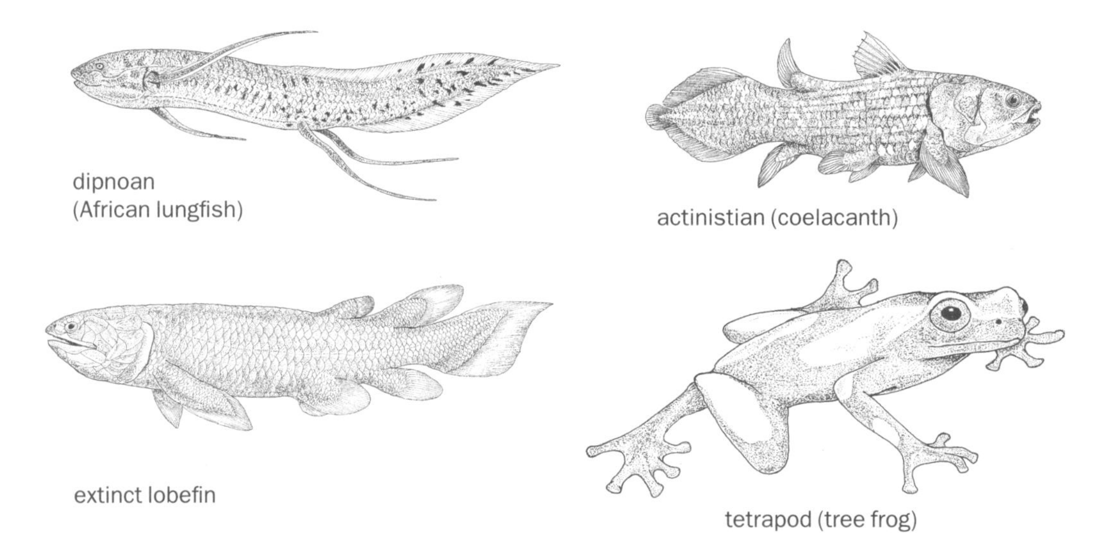

two together also form a clade that is traditionally called **Osteichthyes**. It is good to have a well-established formal name for such an important clade; but, by cladistic convention, the taxon Osteichthyes must include all the descendant groups within it, which means it includes the reptiles, birds, and mammals. 'Osteichthyes' means 'bony fish', so this implies—if we take the term literally—that you, I, tigers, and robins are all bony fish. The term 'Osteichthyes' was, of course, coined in precladistic days, when taxonomists were happy to apply it *only* to bony fish. So here we have yet another clash of Linnaean and cladistic nomenclature. Osteichthyes (in the form of bony fish) are first known from the Devonian, and the distinction between the actinopterygians and the sarcopterygians is evident from very early on.

The Chondrichthyes, Actinopterygii, and the various groups of Sarcopterygii are all described in the following sections, but here is a brief word on the Placodermi and the Acanthodii.

Placoderms: class Placodermi†

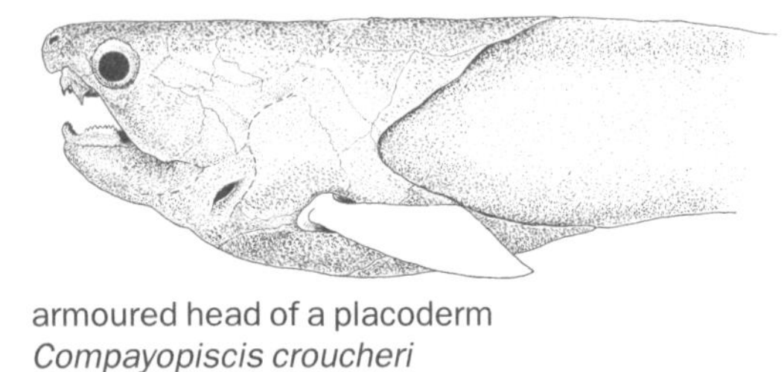
armoured head of a placoderm
Compayopiscis croucheri

Many placoderms look superficially like some of the 'ostracoderms' with whom they shared the Devonian seas because their front ends were typically protected by bony shields and sometimes they were heavily scaled down to their asymmetrical (**heterocercal**) tails. But they had jaws—and that made them far more formidable. Indeed, some of the placoderms from the late Devonian reached 10 metres in length: by far the biggest vertebrates that had lived until that time, and the most fearsome predators of their day. Typically they had paired fins slung from corresponding limb girdles, and many must have been powerful swimmers. In addition, the armour that encased their front ends was technologically advanced: it had a hinge behind the head so that placoderms opened their mouths both by lowering the lower jaw, as we do, and also by lifting the front end of their heads. It has been suggested that some of them they might have fed by scooping their lower jaws into soft mud, as a modern skimmer bird does at the surface of the water, and that it would be easier to do this by raising the front of the head as well as by lowering the jaw. Placoderms became extremely variable, with an enormous range of shapes and sizes: almost 200 genera are known, which are commonly distributed among nine orders. But more than half of them belong to the single order, the **Arthrodira**.

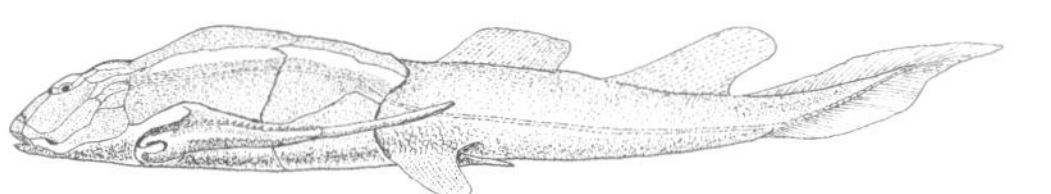

placoderm, *Bothriolepis canadensis*

Exactly where placoderms fit in the phylogenetic scheme has long been controversial (and still is) but the zoologists I have consulted generally favour the notion that they are, or were, sisters to the Chondrichthyes, which is how they are shown in the tree.

Acanthodians: class Acanthodii†

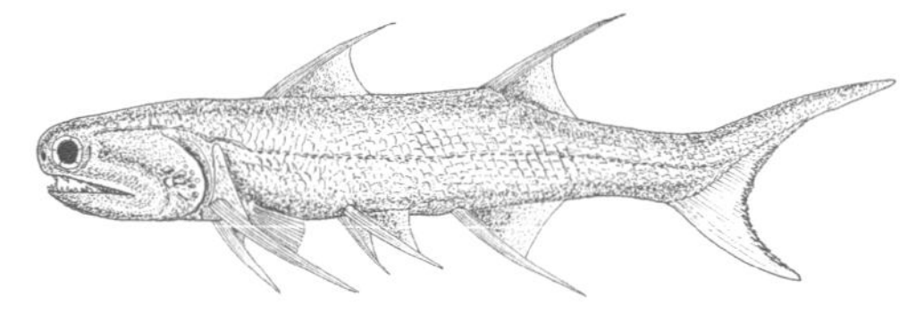

acanthodian, *Ischnacanthus gracilis*

The oldest known acanthodian is Silurian and hence is older than the oldest known placoderm—although, if the tree shown here is correct, the oldest members of the placoderm–chondrichthyes lineage must be at least as old as the oldest acanthodian. (Thus sound classification helps us to predict the kind of fossils we *ought* to find.) Acanthodians were generally small—less than 20 centimetres long—and were very finny and spiny. They had one or two dorsal fins, an anal fin, an asymmetrical tail, and pectoral and pelvic fins that may have been modified into long spines; and the early types had up to six pairs of spines along their bellies. The name 'Acanthodii' relates to these spines. Apart from this, their bodies were covered in small, close-fitting scales made from bone and dentine; apparently the number of scales was fixed throughout life, and the scales grew as a fish grew. Most acanthodians had teeth and presumably were predators—one acanthodian fossil apparently has a smaller fish inside it; but there were some toothless forms that presumably were filter feeders: their gills are equipped with gill rakers. Acanthodians were abundant in the Devonian, and several lines lasted right through until the early Permian, some 280 million years ago.

Enough has been said to introduce the vertebrates. The next nine sections will look at various groups of them in greater or lesser detail. We will begin our survey with the ancient and honourable class of the sharks and rays.

14

SHARKS, RAYS, AND CHIMAERAS

CLASS CHONDRICHTHYES

THREE GROUPS OF FISH form the class Chondrichthyes—the chimaeras, sharks, and rays: and of the three, speaking both phylogenetically and ecologically, the sharks prevail. Although rays can be huge and impressive, on the whole they are amiable mollusc-crushers; and chimaeras, some of which are known as ratfish or rabbitfish, are also crunchers of hard food and for the most part are deep-sea specialists. The sharks, then, are typically the top predators of the group, some of them fearsome even to dolphins, for all the latter's speed and aquatic mastery. Some sharks specialize in catching mammals, such as sealions; many come close inshore to feed and/or to breed; and many are territorial in habit.

Because some sharks have attacked human swimmers, and much more rarely have eaten them, they are widely feared and often hated. We would not be surprised to be attacked by a top predator on land if we wandered noisily into its territory; but sharks seem to be resented in a way that grizzly bears or tigers, say, are not. Most sharks, however, are eaters of other fish; although the largest of all by far—the whale sharks and basking sharks, either of which may be 12 metres or more long—are consumers of plankton, like baleen whales. Whale sharks also take small fishes.

Sharks vary wonderfully in the ways in which they reproduce. Internal fertilization by copulation is usual, with the males clinging to the females by means of 'claspers' around the genital opening. Some, like the bullhead sharks, are **oviparous**, meaning they lay eggs—typically within leathery cases called **mermaid's purses**. Oviparity is generally supposed to be the primitive state. Some, like the spiny dogfish, *Squalus* are **ovoviviparous;** they produce eggs, but retain them within the female's genital tract until they hatch, so that live young are released. But some, like the requiem sharks, are truly **viviparous**. Not only do they produce live young, but they also feed them in the uterus, sometimes producing a kind of 'milk' within the female tract on which the youngsters feed (and a rich milk it can be, too, with up to 13 per cent solids) and sometimes developing a structure that is like a mammalian placenta. Viviparity in different forms seems to have evolved within the sharks and rays independently at least three times. Some species have also evolved a way of feeding their young in the uterus called **oophagy**, which seems somewhat grisly, and has again helped to fire the mythology of sharks: the bigger youngsters in the womb feed on their smaller siblings. It is not aesthetic, maybe, but it works.

'Chondrichthyes' means 'cartilaginous fish', for their skeletons do not contain true bone although the notochord is calcified to form bone-like vertebrae. Because gnathostome fish as a whole seem to have evolved from bony agnathan fish, we must assume that the absence of true bone is a secondary feature. Placoderms are now widely held to be the sister group of the chondrichthyans (as shown in the tree in Section 13) and the placoderms, of course, were positively armoured in bone.

In their lack of bone, the chondrichthyans clearly differ from the bony fish, represented by sarcopterygians (lobefins) and actinopterygians (ray-fins). There are other obvious visible and less-visible differences as well. Thus chondrichthyans lack lungs or a swim-bladder, which in some bony fish enables some air-breathing and/or typically confers neutral buoyancy. Chondrichthyes are generally heavier than water and sink if they do not swim; but their asymmetrical tails provide lift as they beat from side to side and their large, oily livers increase their buoyancy. Externally the gills are visible. Sharks have between five and seven gill slits behind the head; in chimaeras the slits are tucked beneath the cranium; and in rays the mouth and gill slits are beneath. But chondrichthyans lack the bony operculum that protects the gills of actinopterygians. The scales of chondrichthyans are also very different from the neatly overlapping and/or interlocking shingles of actinopterygians: they are **placoid**, like tiny teeth; and, in sharks, the formidable teeth are like a continuation of the placoid scales that cover the whole body. The tails of sharks are typically but not invariably asymmetrical, or heterocercal, with the top lobe much larger than the lower one; whereas in modern bony fish at least (like teleosts) the tail fin (though not the underlying skeleton) is symmetrical or homocercal. A dogfish, commonly dragged aboard in fishermens' nets, is a small shark. No one could mistake it for a cod. They are both 'fish', but they are very different animals.

Chondrichthyes as a whole are various; sharks are generally torpedo-shaped, but in rays the pectoral fins are fused to the body to produce a beautiful form like that of a delta-wing plane that seems to be unparalleled in nature, except, perhaps, to some extent, by some free-swimming nudibranch molluscs. Yet chondrichthyans are much less various than modern actinopterygians and are far less speciose: fewer than 750 sharks, chimaeras, and rays are known against 24 000 ray-fins. But this may be largely because most chondrichthyans are of substantial body size whereas many actinopterygians are minute; and there is less room in the world for viable populations of large animals than for small. In addition, only a few chondrichthyans have evolved to live in fresh water—for example, the stingray *Potamotrygon* in the Amazon—whereas there are thousands of freshwater ray-fins.

In short, actinopterygians have radiated into virtually every aquatic habitat whereas chondrichthyans have in general specialized as predators, usually in a marine environment. Within their maritime, predatory niche, however, chondrichthyans are unsurpassed. Overall, the ecological relationship between chondrichthyans and actinopterygians is rather like that of reptiles and mammals in Pleistocene Australia about 100 000 years ago. The mammals were more varied in form and behaviour, but the reptiles—in the form of giant lizards and snakes, and fast-moving land crocodiles—largely commanded the top predator niche.

Chondrichthyes have not proved easy to classify: molecular studies of living types are not extensive and the fossil record, though wonderful in parts, is extremely patchy. Cartilage does not fossilize well and most extinct types are known only from their teeth. So the tree supplied here should probably be regarded as more speculative than most. It is, however, based on a summary of modern ideas and on the studies of William Bemis, who runs a brilliant undergraduate course on fish biology and systematics at the University of Massachusetts in Amherst. Willy Bemis, in turn, acknowledges his debt to Gary Nelson, formerly of the American Museum of Natural History in New York.

A GUIDE TO THE CHONDRICHTHYES

Note first of all that although zoologists always partition the class Chondrichthyes into three—chimaeras, sharks, and rays—there are, in fact, only two major divisions: between the **Holocephali** (chimaeras), on the one hand; and the **Elasmobranchii** (sharks and rays) on the other.[1] Furthermore, the rays—Rajiformes—in this classification form only one out of nine living elasmobranch orders, and the other eight are all sharks. But some of those sharks—the Squatiniformes and Pristiophoriformes—do look somewhat ray-like, which suggests the continuity between the two kinds of animal. Also note, more generally, that the elasmobranch division of the tree is riddled with polychotomies. Indeed, one four-pronged polychotomy embraces all the living elasmobranch orders except one, the Heterodontiformes, which really do seem to be different from the other living types; and the Squatiniformes, Pristiophoriformes, and Rajiformes form a somewhat speculative trichotomy. In short, the classification of living sharks and rays is not in an advanced state. But the present classification does help us to make sense of a complex group, and this is certainly the first step to resolution.

Chimaeras: subclass Holocephali

Chimaeras are now confined to deep subarctic and Antarctic waters. Their mouths are equipped with 'tooth plates', for crushing hard food, and they have a dorsal spine with a venom sac at the base. The 30 or so living species are classified into just three families: the Callorhynchidae, the ploughnose chimaeras; the Chimaeridae, which sometimes are called 'ratfishes' on account of their whip-like tails, sometimes 'rabbitfishes' on account of their strange-unfish-like heads, or sometimes just 'chimaeras';

[1] Taxonomists have sometimes divided the Elasmobranchii into four large groups, which we may call superorders, three of which include extinct types, and one of which—the 'Selachii'—includes the modern sharks and rays. I mention this only because 'Selachii' appears in many textbooks. I regard it as a redundant term.

14 · CLASS CHONDRICHTHYES

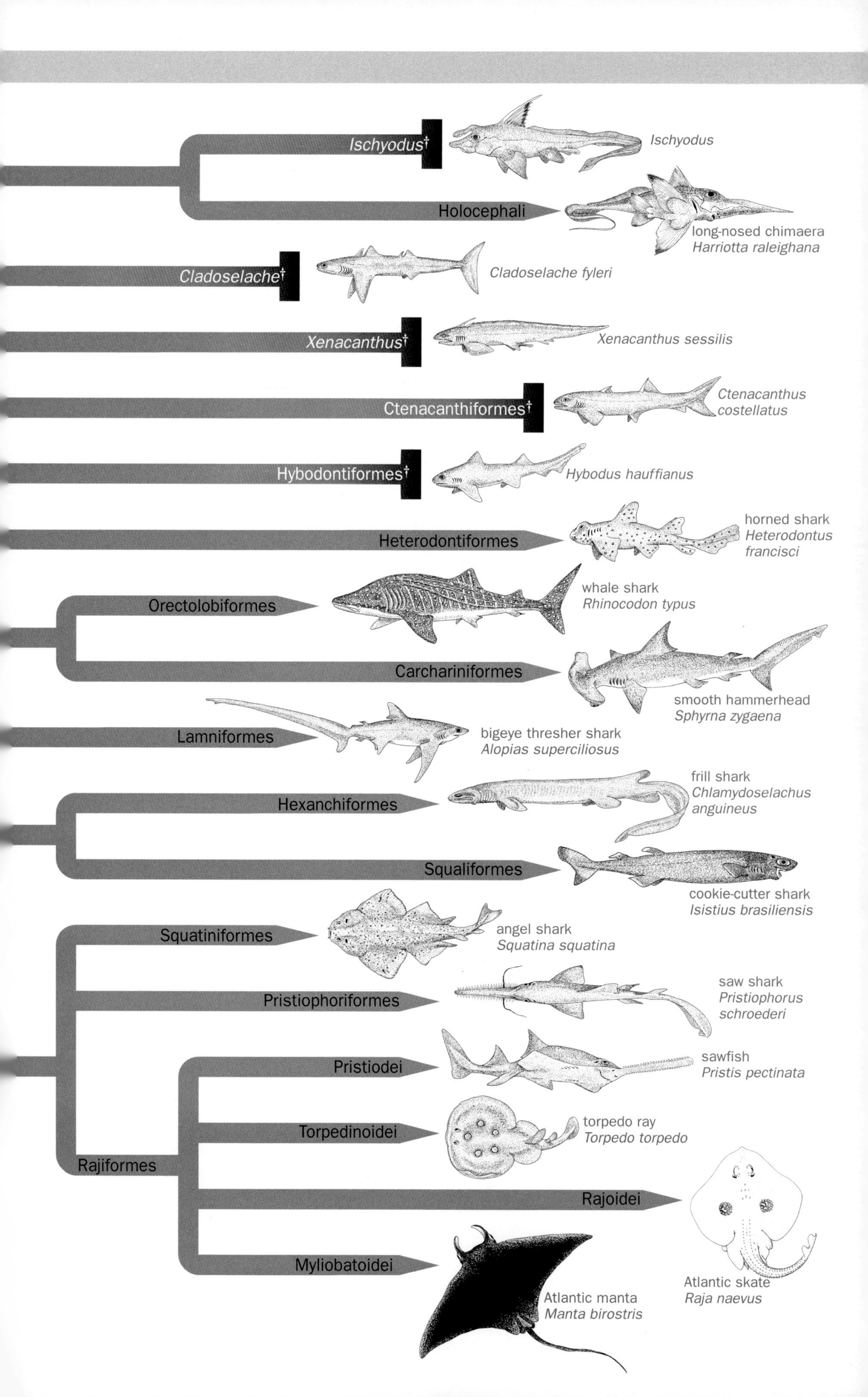
Ischyodus†
Ischyodus
Holocephali
long-nosed chimaera
Harriotta raleighana
Cladoselache†
Cladoselache fyleri
Xenacanthus†
Xenacanthus sessilis
Ctenacanthiformes†
Ctenacanthus costellatus
Hybodontiformes†
Hybodus hauffianus
Heterodontiformes
horned shark
Heterodontus francisci
Orectolobiformes
whale shark
Rhinocodon typus
Carchariniformes
smooth hammerhead
Sphyrna zygaena
Lamniformes
bigeye thresher shark
Alopias superciliosus
Hexanchiformes
frill shark
Chlamydoselachus anguineus
Squaliformes
cookie-cutter shark
Isistius brasiliensis
Squatiniformes
angel shark
Squatina squatina
Pristiophoriformes
saw shark
Pristiophorus schroederi
Pristiodei
sawfish
Pristis pectinata
Torpedinoidei
torpedo ray
Torpedo torpedo
Rajiformes
Rajoidei
Atlantic skate
Raja naevus
Myliobatoidei
Atlantic manta
Manta birostris

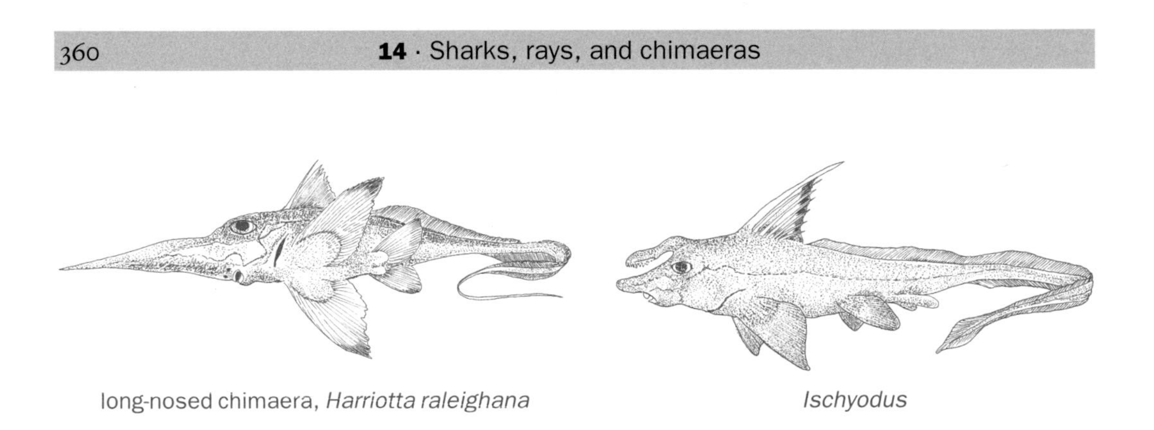

long-nosed chimaera, *Harriotta raleighana*

Ischyodus

and the Rhinochimaeridae, which are the longnose chimaeras. Little is known of their history, for their relationship to most fossils is problematical. But *Ischyodus*, from the Jurassic, is commonly presented as the sister of the living types. Overall, the chimaeras are a strange sidewater in the grand river of evolution.

Sharks and rays: subclass Elasmobranchii

Cladoselache fyleri

There are many fossil elasmobranchs from the later Palaeozoic era—Devonian, Carboniferous, and Permian periods—of which two are commonly singled out. ***Cladoselache***, from the late Devonian, is generally shown as the sister group to all other sharks and rays; it looks generally shark-like (though with a terminal mouth) and a tail that appears symmetrical (homocercal) although the tail skeleton is clearly confined to the upper lobe. But the early types were very variable, sometimes with peculiar ornaments that are lacking in modern species. Thus the curious ***Xenacanthus*** from the Devonian to the Triassic had leaf-like fins, a superficially newt-like body, a tail of the kind known as 'diphycercal', and a long unicorn-like spine behind the head.

Xenacanthus sessilis

In the system presented here modern sharks and rays are all classed as **Euselachii**, which might reasonably be ranked as a superorder. Within the Euselachii there are two extinct orders: the **Ctenacanthiformes** are from the Upper Devonian and the Lower Carboniferous, and the **Hybodontiformes**, which had heavy and convincing spines in front of each of their two dorsal fins, are from the Triassic to the Cretaceous. Some zoologists feel that the hybodont sharks resemble living bullheads, in the order Heterodontiformes. There are nine living orders, as follows.

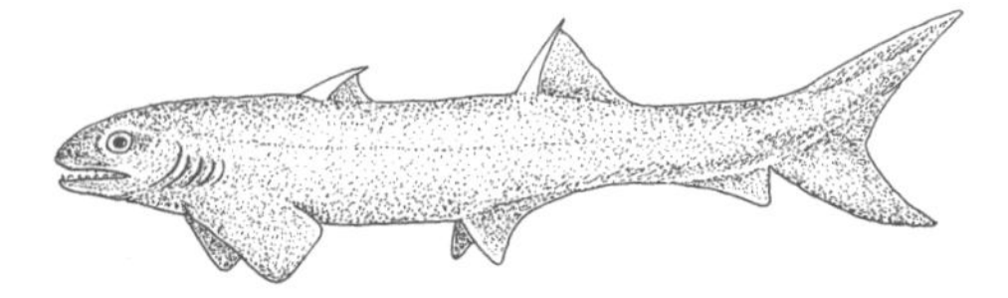

ctenacanthiform, *Ctenacanthus costellatus*

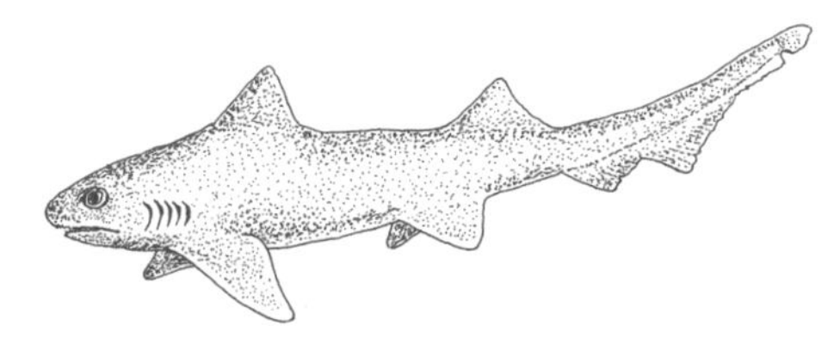

hybodontiform, *Hybodus hauffianus*

Bullhead or Port Jackson sharks: order Heterodontiformes

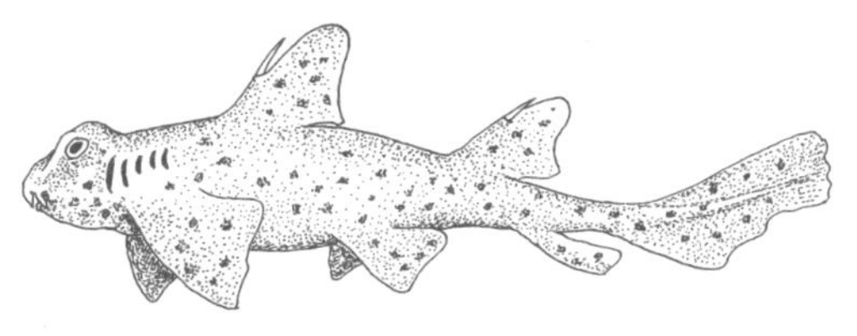
horned shark, *Heterodontus francisci*

The bullhead or Port Jackson sharks in the order Heterodontiformes are shown as the sister of all other living groups: an early diversion of sharkdom. There is just one family, the Heterodontidae, and indeed just one genus, *Heterodontus* with eight species, one of which is *H. portusjacksoni*. Bullheads are small bottom-dwelling sharks that live by crushing molluscs. Like the extinct hybodonts they have a spine on the leading edge of both dorsal fins. They are oviparous, and the egg capsules have a unique, spiral form.

Carpet sharks: order Orectolobiformes

There are only 31 species of orectolobe sharks but they are spread among 14 genera in seven families and are extremely variable. They include the wobbegongs, like the small, sluggish, cryptic (camouflaged) *Orectolobus* of Australia in the family Orectolobidae; the nurse sharks of the family Ginglymostomatidae, which are also sluggish but can be dangerous to bathers because they can be 4.3 metres long and come well into the shallows to breed; and the gigantic whale shark, *Rhinocodon typus*, sole representative of the family Rhinocodontidae. This is the biggest fish of all—often more than 12 metres and sometimes as much as 18 metres long, which is almost 60 feet. But the whale shark has tiny teeth. Like baleen whales, it is a plankton eater, though it also feeds on small fishes. All the orectolobe sharks are oviparous.

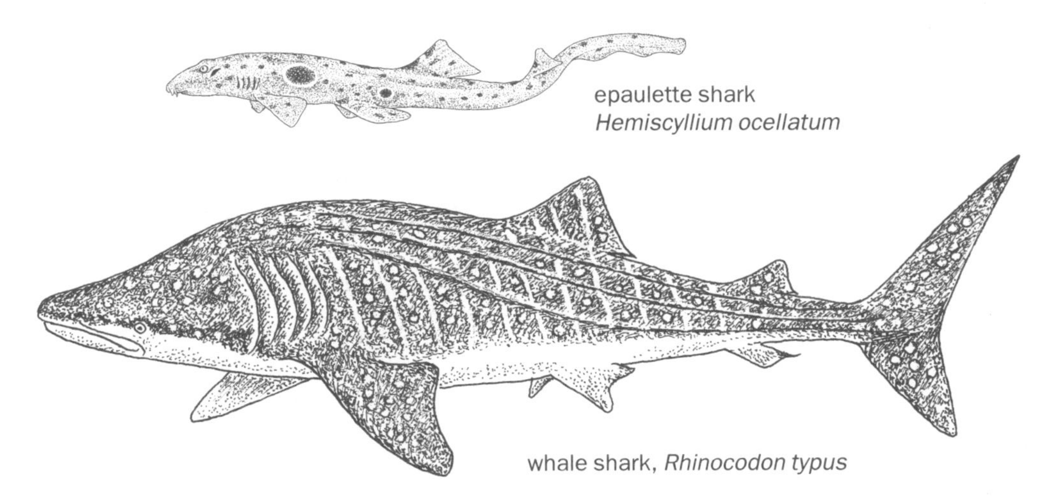
epaulette shark
Hemiscyllium ocellatum

whale shark, *Rhinocodon typus*

Ground sharks: order Carchariniformes

The Carchariniformes, or ground sharks, are shown here as sisters to the Orectolobiformes. Again, it is a very large and enormously various grouping (208 species in 47 genera spread among 7 families). Some ground sharks are oviparous, some ovoviviparous, and some practise true viviparity. Many are large and some are dangerous to human beings. Three of the seven families will serve to illustrate the variety of the Carchariniformes: Scyliorhinidae, Triakidae, and Carcharinidae.

The Scyliorhinidae are the cat sharks (96 species in 15 genera). They are small, sluggish, benthic (bottom-living, in deep water), and oviparous. Among the many species are some that are colloquially known as 'dogfish'. The Triakidae (9 genera, 39 species) are the hound sharks; some small, like *Mustelus canis*, the smooth dogfish, and some medium-sized, like the deep-living *Triakis maculata*, the leopard shark, of the California coast. As you can see, feline and canine common names are interchangeable among sharks, with both cat sharks and hound sharks including dogfish, and hound sharks including leopard sharks. (This is one of many reasons why formal and less-flexible scientific names are a good idea.)

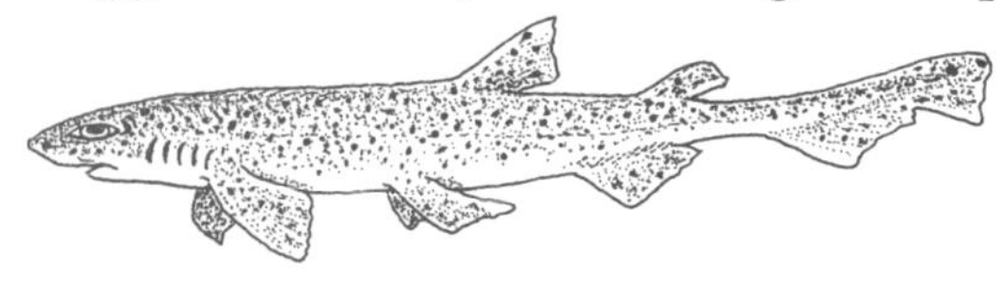
spotted dogfish, *Scyliorhinus canicula*

The Carcharinidae (13 genera, 15 species) contain the somewhat chillingly but aptly termed 'requiem sharks'; they range from medium-sized to extremely large. Some can live in fresh water, and many are dangerous to swimmers, including some of the ones that come up-river and strike where people do not expect to find them. Their upper teeth are typically broader than those of the lower jaw: as Willy Bemis says, they are 'knives versus forks'. The carcharinids are viviparous. Carcharinidae include *Carcharinus*, the whaler sharks, such as *C. leucas*, the bull shark—large and very dangerous close inshore. *Negaprion brevirostris* is the lemon shark: again, large and dangerous. *Galaeocerdo cuvieri* is the tiger shark—an enormous creature up to 7.5 metres (24 feet) long, with a short, broad snout. *Sphyrna* is the genus of the hammer-

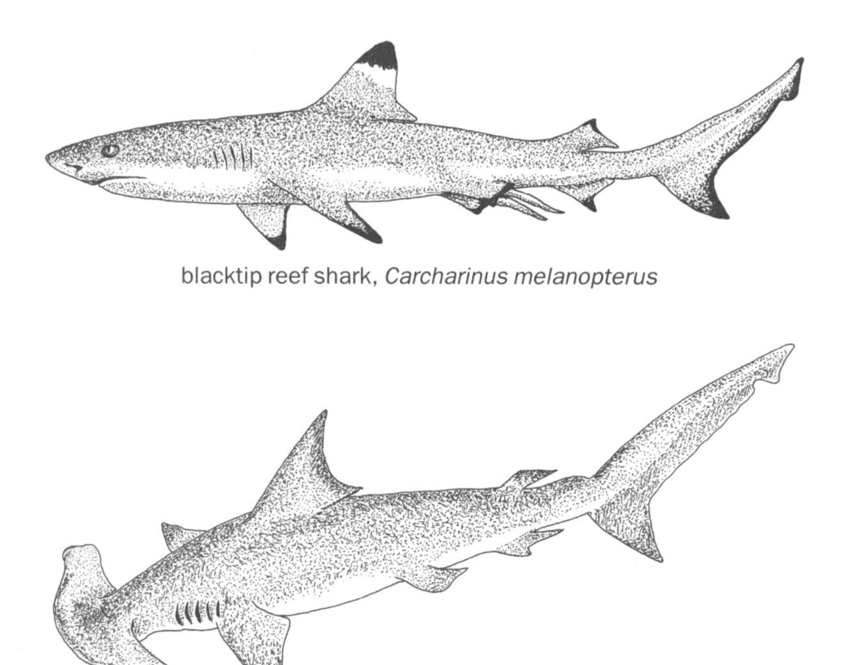
blacktip reef shark, *Carcharinus melanopterus*

smooth hammerhead, *Sphyrna zygaena*

head sharks, which carry their eyes and nostrils at the tips of long 'wings' at the side of the head. *Prionace glauca* is the delightfully elegant and slender blue shark, with its long pectoral fins; pelagic (swimming in open waters) and often basking at the surface; and astonishingly fecund—giving birth to anything from 28 to 54 live young per litter.

Mackerel sharks: order Lamniformes

bigeye thresher shark
Alopias superciliosus

The mackerel sharks include only 16 species, but again they are highly varied, with 10 genera in 7 families. The family Odontaspidae are the sand tiger sharks, with long and slender teeth that typically protrude from their mouths. They are among the groups whose offspring practise oophagy—eating their siblings in the womb. The Megachasmidae include just one species: the filter-feeding *Megachasma*, or megamouth shark, which despite its enormous size (more than 15 metres) was only recently discovered. The wonderful Alopiidae are the thresher sharks; with just one genus, *Alopias*. *Alopias* has a huge tail, the upper lobe long and curved like a sail; and it thrashes its tail at small fish and so stuns them, which makes the eating easier. Cetorhinidae again includes just one species: the enormous basking shark, *Cetorhinus maximus*, up to 15 metres long; enormous but harmless, for *C. maximus* is a filter feeder.

The Lamnidae, for whom the order is named, are the mackerel sharks (5 species,

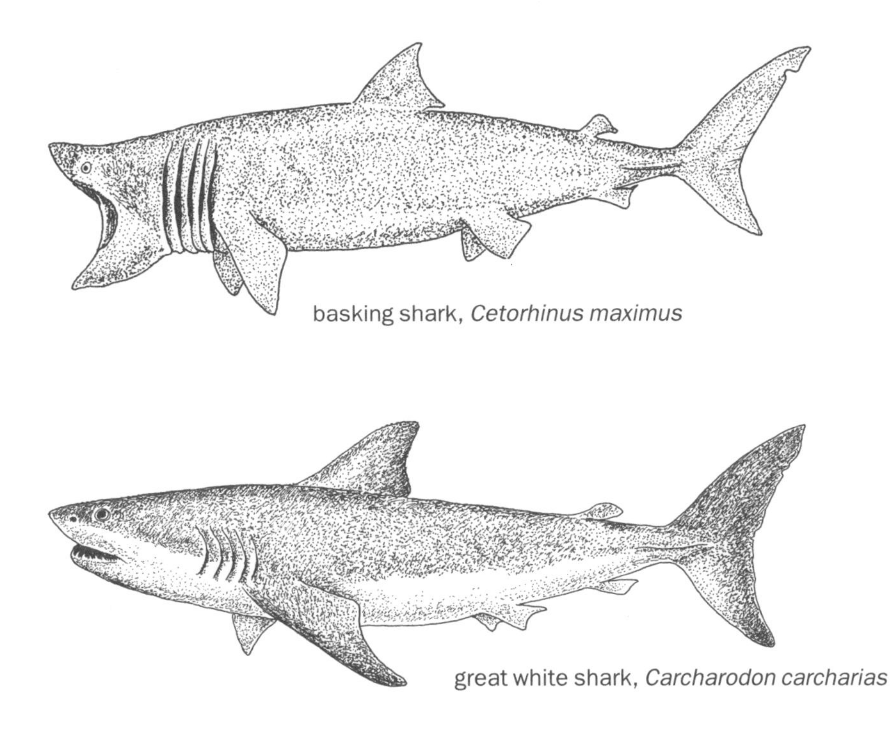
basking shark, *Cetorhinus maximus*

great white shark, *Carcharodon carcharias*

3 genera). They are deep-bodied and heavy, with broad, long, pectoral fins, and the caudal fin is scimitar-shaped like that of a tuna or a mackerel. They have a relatively small number of teeth, but those they have are large or extremely large. Lamnids include *Carcharodon carcharias*, the great white shark up to 8 metres (26 feet) long. It has huge triangular teeth and comes to the California coast to bear its pups and to prey on marine mammals; it has killed bathers and surfers. *Isurus* are the mako sharks—spectacular high-speed swimmers. Their young practise oophagy. *Lamna nasus* is the porbeagle shark, which is sluggish compared with the makos. Porbeagles are also oophagous. Lamnids are 'warm-blooded': like tunas (which are actinopterygians) they have a heat-exchange system that raises their blood temperature above the surroundings, and so speeds their metabolism.

Frill and cow sharks: order Hexanchiformes

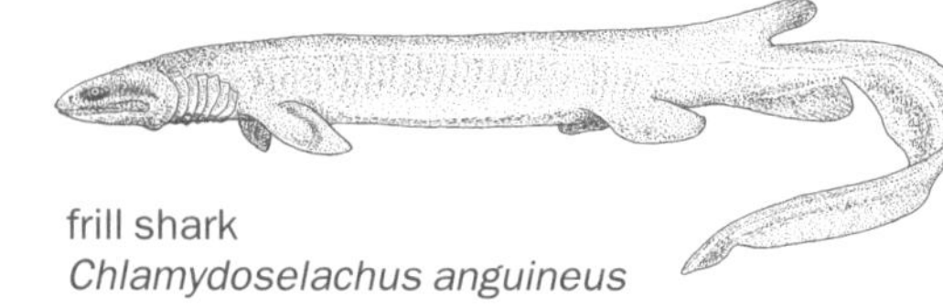

frill shark
Chlamydoselachus anguineus

The Hexanchiformes includes just two families. The Chlamydoselachidae contains only one species, the frill shark, which has a long body, lives near the bottom in deep water, and has six gill slits (instead of the more usual five). The Hexanchidae includes four species, in three genera, of cow sharks; examples are *Hexanchus*, which again has six gill slits and is called the sixgill shark, and *Heptranchius*, the sevengill sharks, with seven gill slits.

Sleepers, cookie-cutters and spiny dogfish: order Squaliformes

The Squaliformes (74 species, 23 genera, 3 families) seem to be sisters to the Hexanthiformes. The sleeper sharks include some extraordinary creatures. The most famous is the Greenland shark. It is the biggest fish of the Arctic at 6.4 metres; but its scientific name, *Somniosus microcephalus*, aptly describes its character. It is sluggish to the point that boats may run into it as it dozes at the surface—and even then

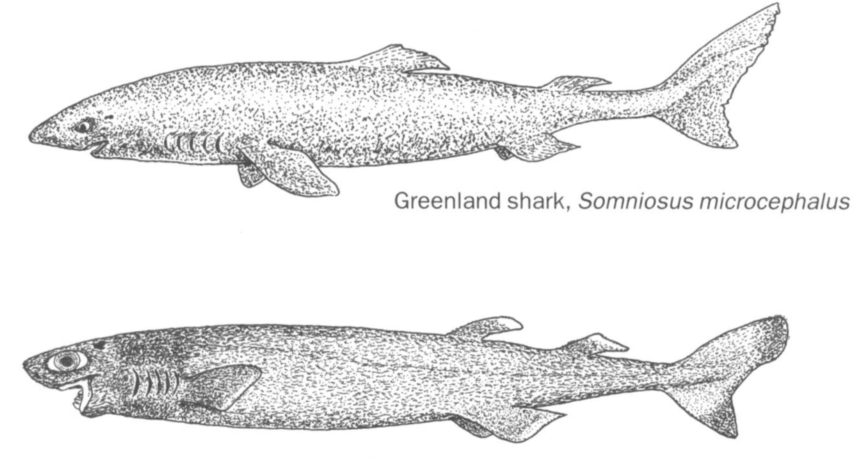

Greenland shark, *Somniosus microcephalus*

cookie-cutter shark, *Isistius brasiliensis*

it may not move away. The related *Isistius* is the cookie-cutter shark of deep water; it has a luminescent belly and bites chunks from larger vertebrates. *Squalus acanthias* is the spiny dogfish or spurdog. It is migratory and abundant along northern coasts, though only in cool waters between 4 and 16 degrees Celsius. *Squalus* is ovoviviparous.

Angel sharks: order Squatiniformes

The Squatiniformes (12 species, 1 genus) are flattened and superficially like rays, Rajiformes. But, unlike rays, their broad pectoral fins are not fused to the head nor to the posterior part of the trunk; their common name of 'angel sharks' is apt. They also swim like sharks rather than rays: that is, they are driven forward by sideways sweeps of the tail, and not by pectoral flying. They live in temperate coastal waters. Squatiniformes are shown in the tree as part of an unresolved trichotomy. That is, they seem to be relatives of the Pristiophoriformes (saw sharks) and the Rajiformes, but the nature of the relationship is unclear.

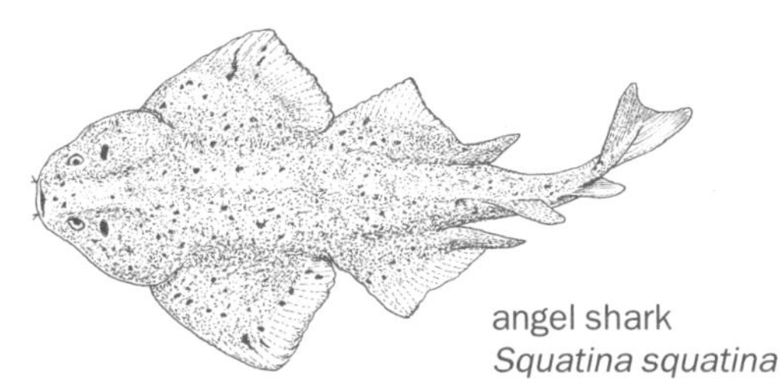

angel shark
Squatina squatina

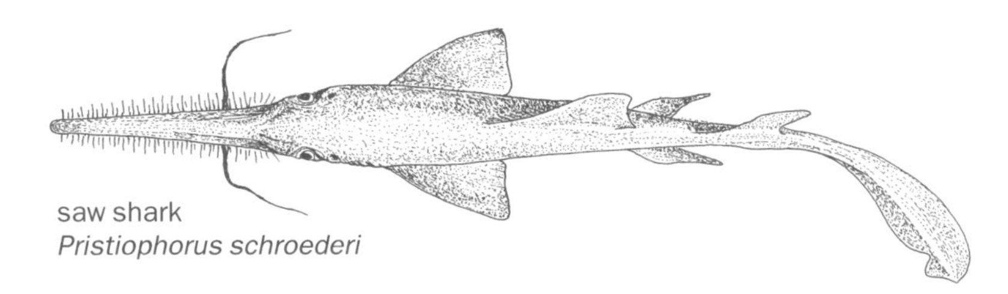

saw shark
Pristiophorus schroederi

Saw sharks: order Pristiophoriformes

There are just five species of saw sharks in two genera, one of them, *Pristiophorus*, with five gill slits and the other, *Pliotrema*, with six. Their outstanding feature, however, is their snout: equipped with a flat blade with teeth that are alternately large and small. They are primarily tropical, from the Indo-Pacific.

Rays, skates, sawfish, and torpedos: order Rajiformes

The fish of the Rajiformes power themselves through the water with great pectoral fins, spread wide on either side and fused along all the trunk and to the head; overall they are flat, with their eyes and spiracles on the upper side and their mouth and gill slits beneath. They all have teeth like pavements, right across the palate, for crushing their shelly prey. Well over half of all living chondrichthyans are Rajiformes—456 species in 52 genera spread among 12 families. This big and diverse order is conveniently subdivided into four suborders: Pristiodei, Torpedinoidei, Rajoidei, and Myliobatoidei.

• The **Pristiodei** are the sawfishes (6 species, 2 genera, 1 family). Like the saw sharks, Pristiophoriformes, their snout carries a blade with teeth, but pristiod teeth are of equal size. *Pristis*, of the western North Atlantic, grows to almost 6 metres. Sawfishes are ovoviviparous.

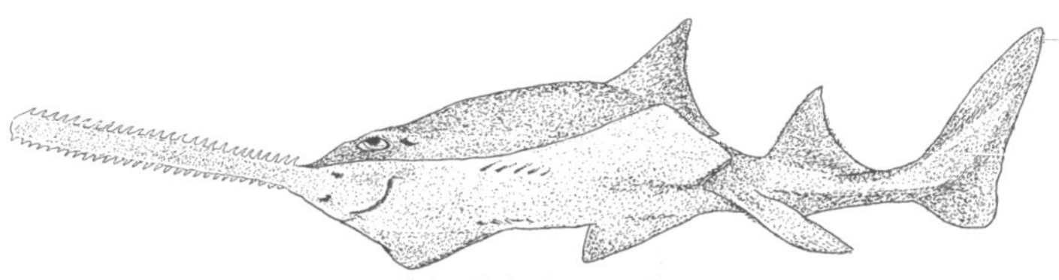

sawfish, *Pristis pectinata*

• The **Torpedinoidei** are the torpedos (38 species, 11 genera, 2 families): the Torpedinidae are the electric rays and the Narcinidae are the lesser electric rays. Like some actinopterygians (notably the electric 'eels'), torpedoes can generate electric charges in their muscles (gill muscles in this case) powerful enough to stun their prey. *Torpedo nobiliana*, the Atlantic torpedo, grows to 1.8 metres.

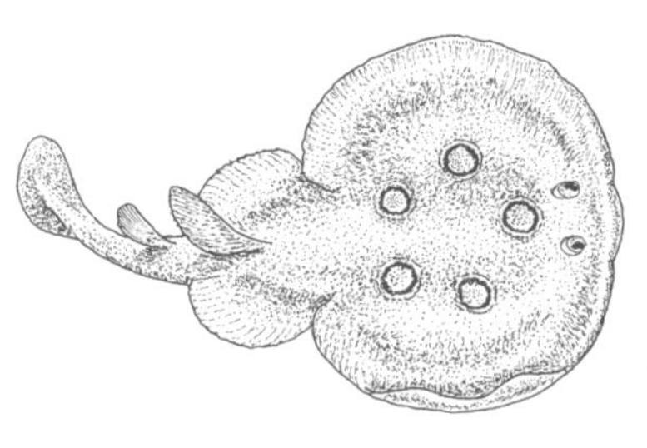

torpedo ray
Torpedo torpedo

• The **Rajoidei** include the guitarfishes, skates, and rays (209 species, 22 genera, 3 families). The Rhinobatidae are the guitarfishes (45 species, 7 genera). They have a triangular, ray-like head but a shark-like tail and trunk. They also have an underslung mouth, like most sharks, and swim like a shark. The Rajidae are the rays (200+ species, 18 genera) and include *Raja*, the skate. Rays are oviparous. Most of the mermaid's purses that are washed up on beaches belong to skates.

• The **Myliobatoidei** are the stingrays, eagle rays, and their kin (158 species, 23 genera, 6 families). The Dasyatidae are the stingrays (70 species, 9 genera), the 'sting' being made from one or two modified dorsal fin spines that are serrated and barbed and have a venom gland at the base. Unlike rays and skates, stingrays and eagle rays are viviparous. *Potamotrygon* is the freshwater stingray of South America. The Myliobatidae are the eagle rays (43 species, 7 genera). The Mobulinae, a subfamily of the Myliobatidae, contain the most magnificent rays of all—the mantas, which may be 6 metres across and weigh 2 tonnes. Mantas, though, like most of the truly enormous chondrichthyans, are filter feeders.

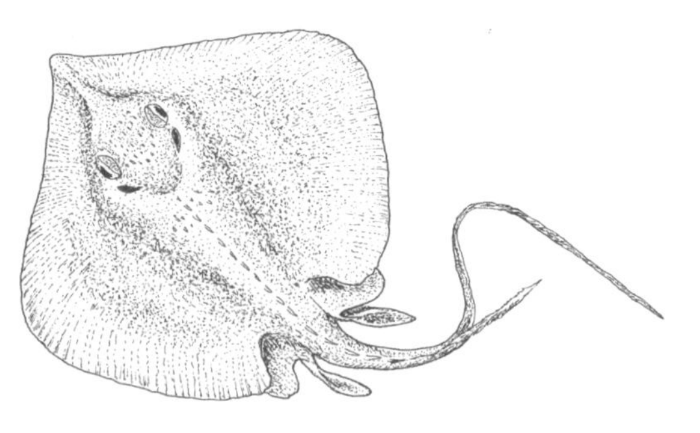

Atlantic stingray, *Dasyatis sabina*

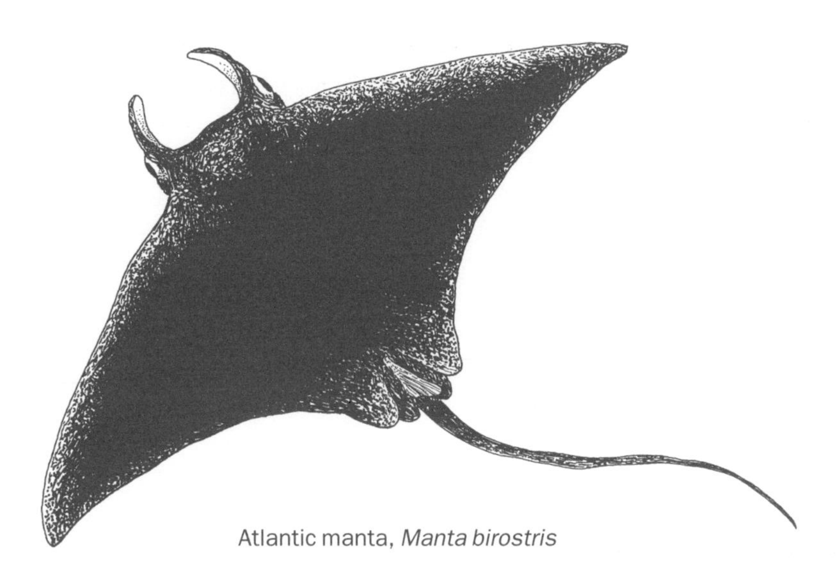

Atlantic manta, *Manta birostris*

Overall, the chondrichthyans are magnificent; they include some of the greatest and most ecologically significant creatures of the modern world. They have received a bad press because some of them are indeed dangerous and also, because they have changed remarkably little over the past few hundred million years or so, people feel they are 'primitive'—primitive in the pejorative, vernacular sense. As we have seen, however, they have some extraordinary adaptations that in some cases are reminiscent of mammals, including 'placentae', 'milk', and 'warm-bloodedness'. Besides, as one shark enthusiast said to me, why change a winning formula?

15

THE RAY-FINNED FISH

CLASS ACTINOPTERYGII

ALMOST HALF OF ALL known species of vertebrates are ray-finned bony fish, or Actinopterygii: more than 20 000 species of them. Some individual families, like the Cichlidae (pronounced 'sick-lid-ee'), which include the famous and economically important *Tilapia*, contain more than 1000 species. Africa's Lake Victoria alone, until recent years, harboured about 300 cichlid species. Yet 20 000 is probably a severe underestimate, for each isolated pool and streamlet may in time evolve its own kinds of fish, and some vast regions—notably in tropical forest—have hardly begun to be systematically explored. The true number of living actinopterygian species could well exceed 40 000.

Fish are constrained by the need to swim so we might expect that most of them would be torpedo-shaped; yet the actinopterygians display a prodigious range of forms. Most are indeed torpedo-like, more or less. But many are flattened dramatically from side to side, which, among other things, enables them to manoeuvre in small spaces—like angel fish, threading through the reeds. Others, known informally as 'flatfish', are equally compressed from side to side but lie on their sides on the bottom. Some, including some catfish and anglerfish, are flattened dorsoventrally and they, too, generally forage near the bottom. Some fish are stiff-bodied whereas others are as lithe as, well, eels. Puffer fish can blow themselves into great round spiny balls, evidently to repel predators; box fish are indeed cuboidal; and sea horses swim upright, driven by a flickering dorsal fin. The tails of sea horses, which in most fish provide propulsion, are prehensile like a spider monkey's and hold the fish fast to weeds; with their long snouts they feed like little vacuum cleaners; and the male sea horses—not the females—brood the babies in a pouch. No fantasist could have invented the sea horse.

Many actinopterygians are adorned with frills and spines of the kind that only an aquatic creature, freed from the constraints of gravity, can truly indulge—like the dragon fish. Fish colours and patterns can beggar belief—although on the coral reef, where many of the most colourful live, it is hard to say if the gaudy is intended to advertise or to conceal, because all is so colourful, and flashes of colour protect by deceiving the eye. The range of size is huge. Some of the gobies are the smallest of all vertebrates, only 7.5 millimetres long and weighing around 5 milligrams, while the beluga *Acipenser huso*, a giant sturgeon that feeds in the Adriatic, Black, and

Caspian seas and breeds in the Volga and the Danube, can be more than 4 metres long and weigh the best part of a tonne; and the Ocean sunfish *Mola mola* averages around 2 metres in length and commonly exceeds a tonne. (A tonne is a million grams, and a gram is a thousand milligrams: so the biggest bony fish weigh 200 million times as much as the smallest. Ray-finned fish, in short, are fabulously various.)

Ray-finned fishes truly illustrate the need for classification. There are so many types, and their fossil history is so rich, that it would be well-nigh impossible even to allude to them if they were not neatly categorized. Fittingly, then, they have attracted some of the most outstanding of all systematists over the past few decades. But they also illustrate the difficulty of devising classifications that are based on genealogy for it has proved—and is still proving—extremely difficult to determine precisely who is related to whom, and so the literature is both rich and controversial. Even some of the commonest and most well-studied groups, like the salmons and trouts, are still difficult to fit into a coherent genealogy.

The tree shown here is informed by several authorities, as outlined in the Acknowledgements. But my principal source has been William Bemis of the University of Massachusetts at Amherst, who, over the past decade, has developed the classification shown here. Willy Bemis in turn acknowledges his debt in particular to Gary Nelson of the American Museum of Natural History.

A GUIDE TO THE ACTINOPTERYGII

As explained in Section 13, five classes of jawed vertebrates are properly called 'fish'; and the different classes have effectively taken over from each other in diversity and ecological importance. The first of them appeared in the Silurian. The placoderms and acanthodians were the early runners, but now share a dubious distinction as the only gnathostome classes to have become totally extinct. The other three fishy classes all began to make their presence felt in the Devonian but although they have all flourished—and are with us still—their histories have followed different patterns. Thus the Chondrichthyes, now represented by the sharks, rays, and chimaeras (Section 14), have enjoyed several radiations but the latest was in the Jurassic when the dinosaurs were in their pomp—so the major living chondrichthyan groups can be traced right back to that time. The lobefins, in the clade of the Sarcopterygii, outstripped the actinopterygians at least until the Carboniferous, but faded in the Mesozoic era and now are reduced to three genera of lungfishes and one coelacanth.

The Actinopterygii are considered to be the sister group of the Sarcopterygii—and both groups are presumed to have arisen from their common ancestor in the Silurian, more than 417 million years ago, although what that ancestor was is unknown. But the actinopterygians got off to a slow start compared with the lobefins, and did not begin to radiate dramatically until the Carboniferous. Since then, however, they have never looked back. They have enjoyed successive radiations, which are still continuing. At least 99 per cent of living actinopterygians belong to the subclass known as the Teleostei; the teleosts did not arise until the Jurassic; and most of the living teleost families arose either in the Cretaceous or the Cenozoic. In short, as John Maisey of the American Museum of Natural History has pointed out, most families of living bony fish are not at all ancient even though fish as a whole extend deep into antiquity; and the Cenozoic era (the past 65 million years), which from a land-based perspective is reasonably called 'The Age of Mammals', might equally well be called 'The Age of Teleosts'.

Actinopterygians are acknowledged to be a true clade, united by at least half a dozen synapomorphies. The chief and most obvious of these—the one that gives the group its name—is the ray fin, which is built around a fan of radials, each of which is a narrow rod of cartilage or bone. This style of fin was and is a great technical innovation: light, collapsible, and mobile. Its design is in sharp contrast to the fin of sarcopterygians, which is built around a single basal bone. It is easy to see how the fin of lobefins could have evolved into a weight-bearing limb; and one or other of the lobefins (though it is not certain which one) undoubtedly is the ancestor of the tetrapods. But fins built like fans do not lend themselves to weight-bearing, and although the loach and the mudskipper and a few others do 'walk' up to a point on stiffened rays of the pectorals, actinopterygians in general have been content to remain as very palpable fish—indeed as the most successful fish of all time.

Other features uniting the actinopterygians include the single dorsal fin (lobefins and sharks generally have two). The scales of ray-finned fish are also special in various ways: for example, they are bedecked with layers of the enamel-like **ganoin**; and the front of each scale has a peg that articulates with a socket in the scale in front. In actinopterygians, too, the pelvic and pectoral fins have characteristically different structures, whereas in lobefins and sharks the two sets of fins are effectively the same.

GRADES AND CLADES OF RAY-FINNED FISHES

Ever since the 1840s zoologists have recognized successive 'grades' within the Actinopterygii; and although it has been fashionable in some biological circles to eschew the concept of 'progress' in evolution there can be no doubt that the types that are traditionally said to be of 'higher' grade arose later in evolution and also are technologically superior to the earlier types when judged according to the objective criteria of engineering. Until recent years taxonomists gave the different grades the status of formal taxa, dividing the ray-finned fishes into 'Chondrostei', 'Holostei', and 'Teleostei'. But modern taxonomists recognize that the old-style 'Chondrostei' and 'Holostei' each include several distinct lineages, and of the three traditional groupings only the Teleostei survives as a formal taxon. All the same, the other two terms still may find some use as informal adjectives, as in 'chondrostean' and 'holostean'.

The grades allude to changes in the composition of the skeleton in general, and in particular in the layout of the bones and muscles to produce animals that in general are lighter but nonetheless as strong as they need to be, are more flexible, and have more mobile skulls and mouths that enable them to feed and respire far more efficiently. In particular, the bones and musculature of the modern teleost's head

15 · CLASS ACTINOPTERYGII

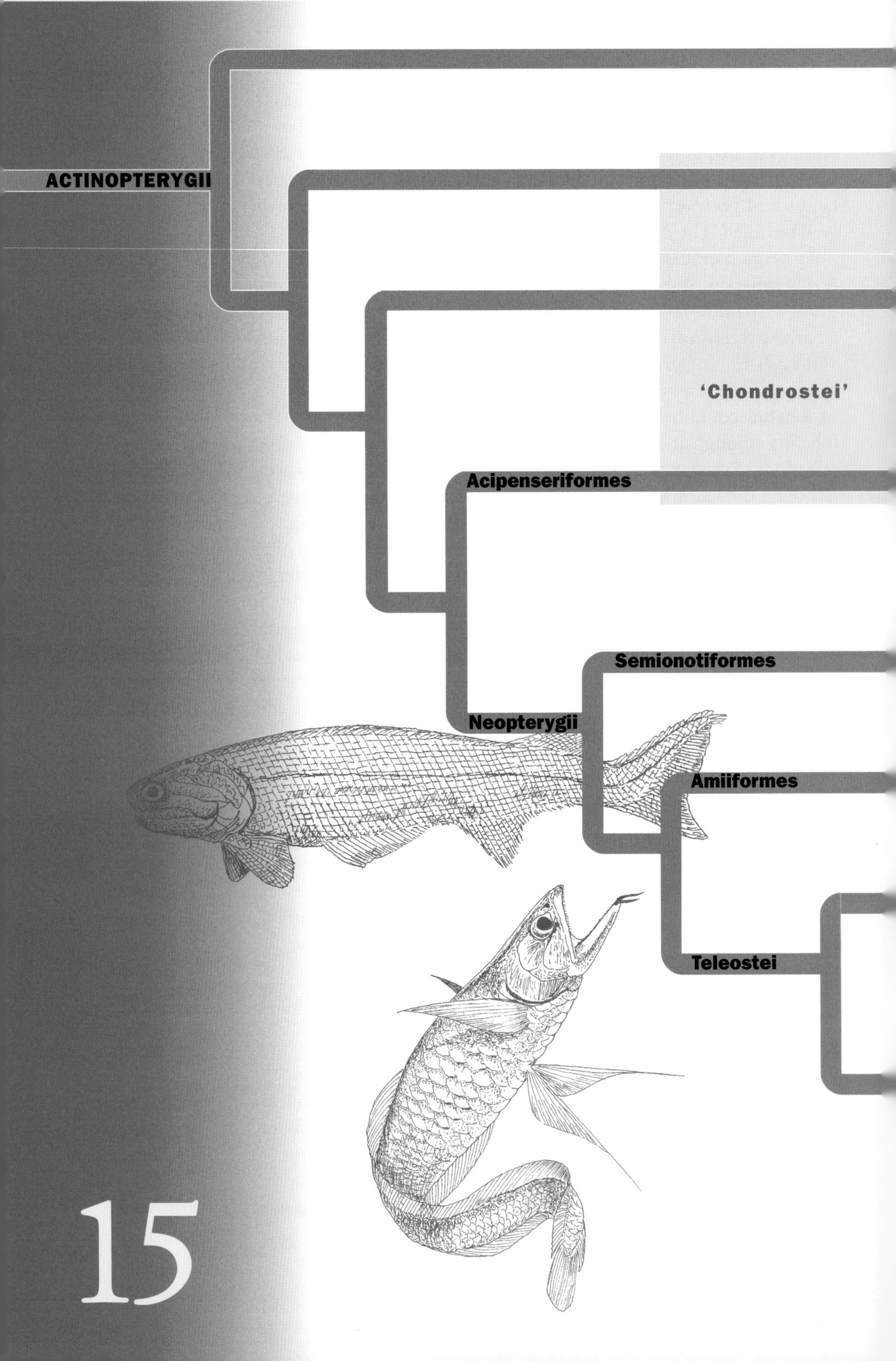

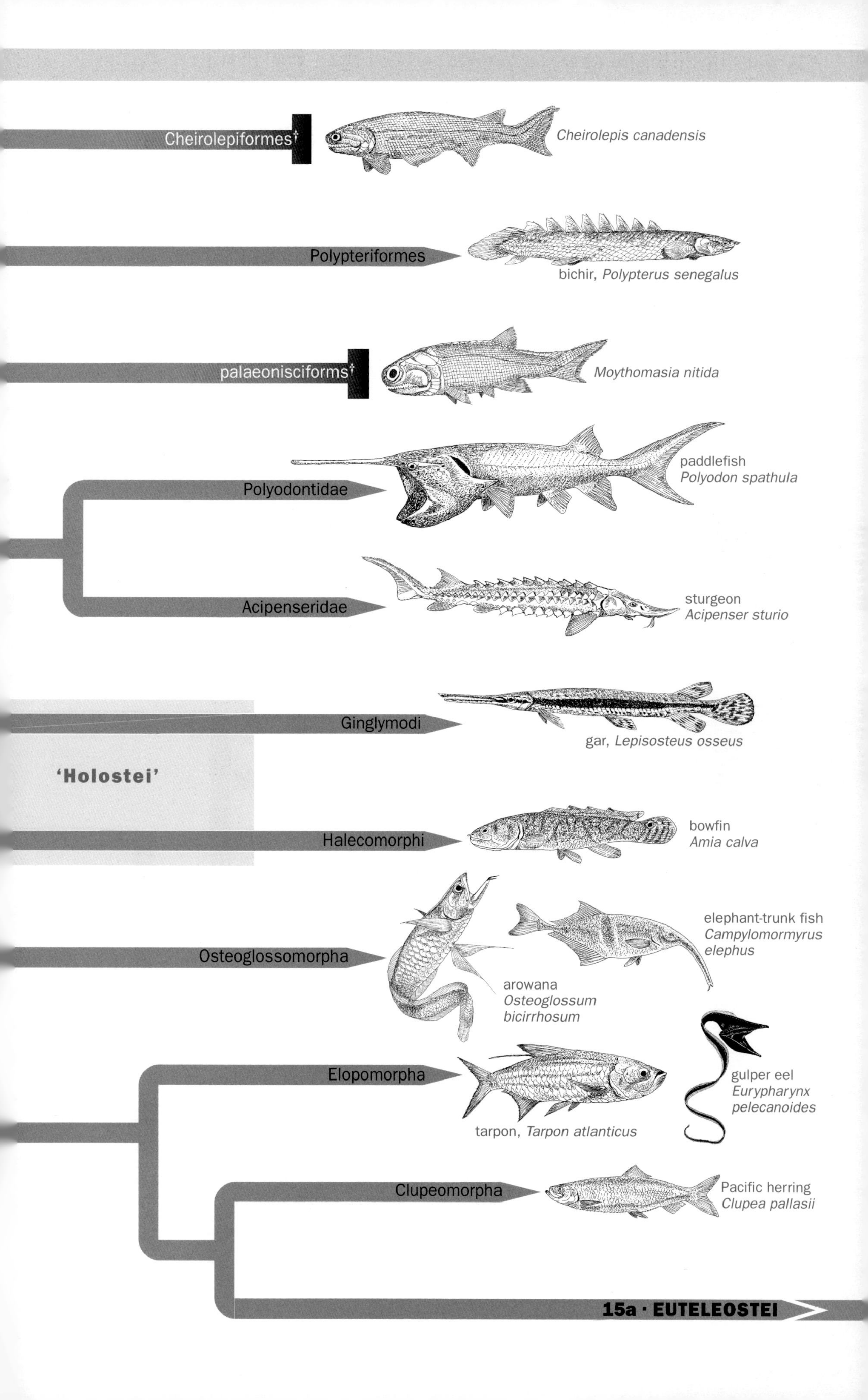

Cheirolepiformes†
Cheirolepis canadensis
Polypteriformes
bichir, Polypterus senegalus
palaeonisciforms†
Moythomasia nitida
Polyodontidae
paddlefish
Polyodon spathula
Acipenseridae
sturgeon
Acipenser sturio
Ginglymodi
gar, Lepisosteus osseus
'Holostei'
Halecomorphi
bowfin
Amia calva
Osteoglossomorpha
elephant-trunk fish
Campylomormyrus
elephus
arowana
Osteoglossum
bicirrhosum
Elopomorpha
gulper eel
Eurypharynx
pelecanoides
tarpon, Tarpon atlanticus
Clupeomorpha
Pacific herring
Clupea pallasii
15a · EUTELEOSTEI

15a · ACTINOPTERYGII · EUTELEOSTEI

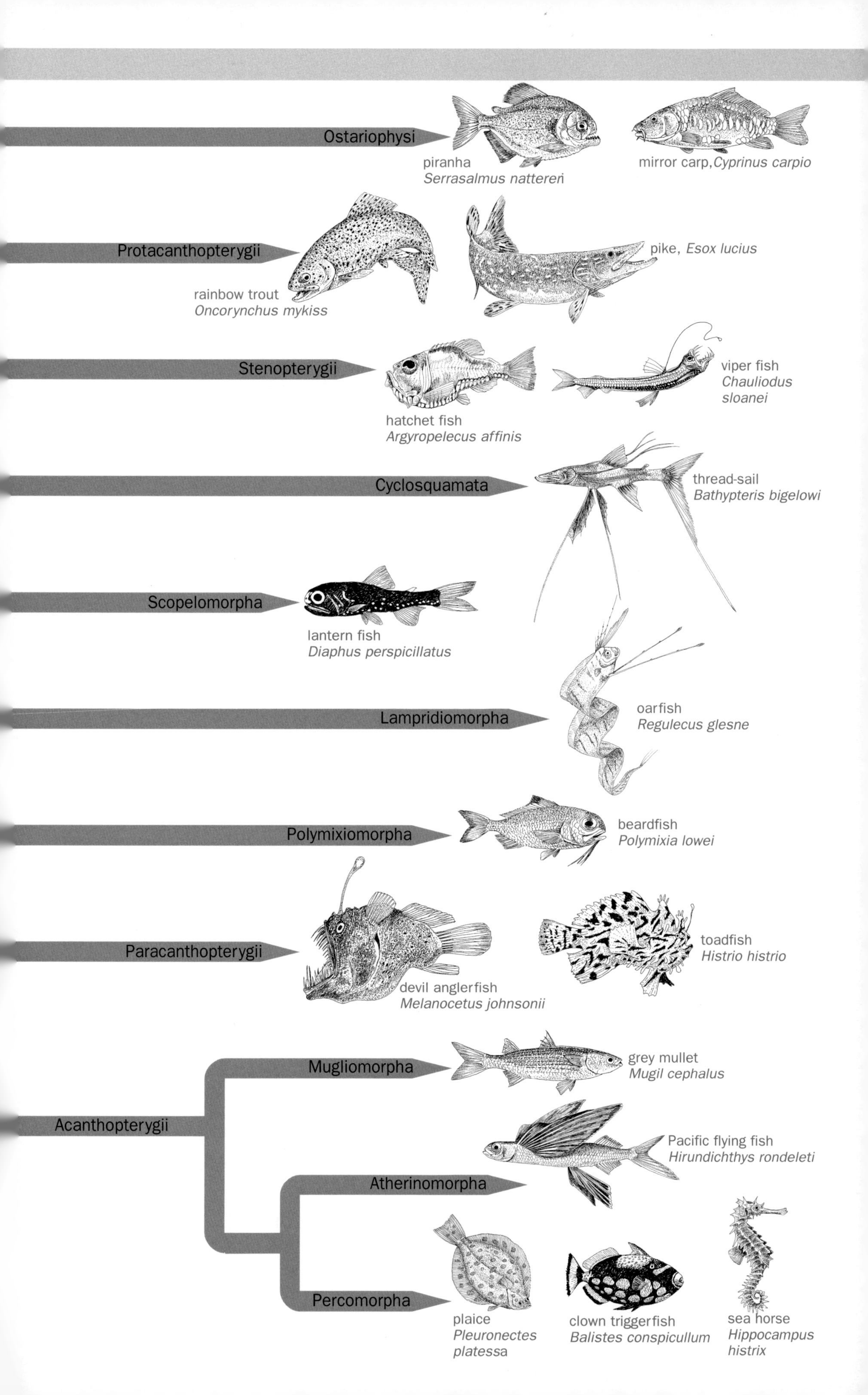
Ostariophysi
piranha
Serrasalmus nattereri
mirror carp, Cyprinus carpio
Protacanthopterygii
pike, Esox lucius
rainbow trout
Oncorynchus mykiss
Stenopterygii
viper fish
Chauliodus
sloanei
hatchet fish
Argyropelecus affinis
Cyclosquamata
thread-sail
Bathypteris bigelowi
Scopelomorpha
lantern fish
Diaphus perspicillatus
Lampridiomorpha
oarfish
Regulecus glesne
Polymixiomorpha
beardfish
Polymixia lowei
Paracanthopterygii
toadfish
Histrio histrio
devil anglerfish
Melanocetus johnsonii
Mugliomorpha
grey mullet
Mugil cephalus
Acanthopterygii
Pacific flying fish
Hirundichthys rondeleti
Atherinomorpha
Percomorpha
plaice
Pleuronectes
platessa
clown triggerfish
Balistes conspicullum
sea horse
Hippocampus
histrix

enable it to create negative pressure inside its mouth, so that it positively sucks in oxygen-rich water, which is then pushed out over the gills; and the species that take small prey use this technique for feeding as well as for respiration. Perhaps of less functional significance, but nonetheless conspicuous, has been a change in the shape of the tail. In sturgeons, which are of the 'chondrostean' grade, the tail is asymmetrical or heterocercal, like a shark's; but most teleosts have a caudal fin shaped in a way that gives the tail a symmetrical or **homocercal** appearance.

The various lineages of the **chondrostean grade** radiated from the Carboniferous to the Triassic, and are so named because their skeletons are or were comprised mainly or exclusively of cartilage. But the loss of bone must be secondary in chondrosteans, being descended from bony ancestors, and, of course, chondrosteans have no special relationship with the Chondrichthyes of Section 14. Most chondrosteans belong to a group of lineages that are loosely called 'palaeonisciforms'; they are known from the Devonian, were rare after the Triassic, and finally went extinct in the Cretaceous. But two chondrostean groups are still with us: the Polypteriformes and the Acipenseriformes; these two together traditionally formed the now-defunct 'Chondrostei'.

Fish of the **holostean grade** emerged and radiated from the Triassic to the Jurassic, and became fairly diverse. Again, they include a mixture of lineages that nowadays include the Semionotiformes or gars (sometimes called garpike) and the bowfins, in the Amiiformes. In traditional taxonomies these two groups are lumped together (to form the 'Holostei') but modern biologists acknowledge that the two have no special relationship, and indeed that the amiiforms are probably sisters to the teleosts. The teleosts themselves began to radiate in the Jurassic and have continued unabated ever since.

In summary, then, the tree shows the Actinopterygyii divided into seven major groups, each of which can be ranked as a subclass. The subclass **Cheirolepiformes**, here represented by the type genus *Cheirolepis*, stands in for the most ancient actinopterygian subclass—the sister of all the rest. Of the 'chondrostean' subclasses, **Polypteriformes** now includes the bichir and reedfish; the extinct **palaeonisciforms** are a mixed bag (and so are written informally, spelt with a lower-case initial 'p'); and the **Acipenseriformes** are now represented by the sturgeons and paddlefish. The remaining three subclasses make up a grouping (there seems little need to give it a formal ranking) called the **Neopterygii**. The first two of these neopterygian subclasses are the old-style holosteans that include the **Semionotiformes** (gars); and the **Amiiformes** (bowfins). The **Teleostei**, the 'modern bony fish', are a large and varied group; it requires a great deal of further subdivision.

I will summarize the principal ray-fin taxa in turn.

Cheirolepis: subclass Cheirolepiformes†

Cheirolepis, known especially from the Middle Devonian of Scotland although similar types are known worldwide, is generally acknowledged as the first bona fide ray-

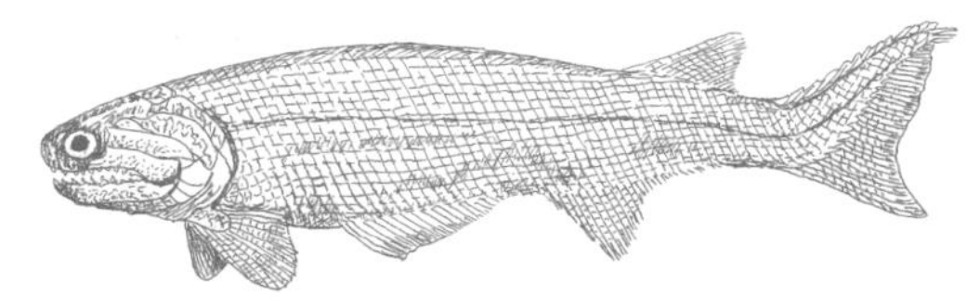
Cheirolepis canadensis

finned fish. *Cheirolepis* itself could conceivably be the common ancestor of all the rest; if it is not, then one of its close relatives presumably was. In either case, it is fittingly shown in the tree as the sister of all the rest. *Cheirolepis* was presumably a fast-moving big-eyed predator, feeding on acanthodians, placoderms, and lobefins. But it was not large—typically around 25 centimetres long—and its body was stiff, driven forward by the powerful tail; a tail that was strongly heterocercal, although the tail fin made it externally almost symmetrical. Neither was *Cheirolepis* particularly manoeuvrable, its pectoral fins lacking the mobility for acute steering.

Cheirolepis had lovely scales—lozenge shaped, arranged in sweeping diagonals, and articulating by pegs and sockets in the region of the tail, with larger ridge scales on the upper edge of the tail cutting the water. The skull is heavy with bony cheeks, which is often interpreted as the original condition of the actinopterygians. Yet the skull is also kinetic, with several quasi-separate units that can move against each other, helping the jaws to produce a tremendous gape. Even so, the skull and mouth are less manoeuvrable and versatile than those of that of later ray-fins.

Bichirs and reedfish: subclass Polypteriformes

Polypteriformes (or the Cladistia) are the most ancient surviving lineage of ray-finned fish. Only two genera remain: *Polypterus*, the bichir, with 10 species; and *Erpetoichthus*, which includes just one species of reedfish. In the shallows of African rivers bichirs are formidable predators, stalking and then lunging at other fish and invertebrates. They are in many ways primitive. The range of movement of the bones that form their jaws and flattened heads is limited; they are 'chondrosteans', with a skeleton of cartilage rather than of bone; their scales are stout and rhomboidal; and their tails are still heterocercal. The dorsal fins of bichirs run the whole length of their bodies and are divided into 'finlets', each with a spine at the front.

When bichirs are young they have external gills, and when they are adult they have gills like other fish—but they also have highly vascularized lungs; so although the adults cannot live on land they can survive in water with a very low oxygen content. In practice, many ray-finned fishes have lungs, or lung-like projections of the gut, but this emphatically does *not* mean that actinopterygians with lungs are especially related to lungfish (Dipnoi) or to tetrapods, any more than those with cartilagenous skeletons are related to the Chondrichthyes. Lungs (and external gills in juveniles) may simply be primitive features of Osteichthyes. In teleosts, the 'lung' becomes the

bichir, *Polypterus senegalus*

swim-bladder, used mainly to control buoyancy (although also with some secondary functions in some groups, as described later).

The palaeonisciforms†

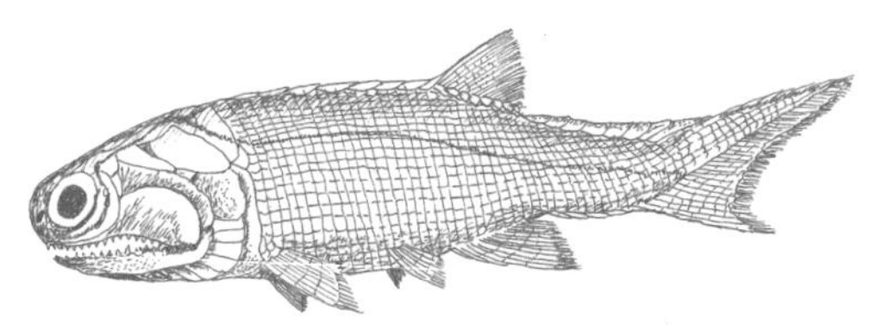
palaeonisciform, *Moythomasia nitida*

The taxon that was once known as the Chondrostei was originally conceived to include the sturgeons and the paddlefish, but as time went by more and more fossil types were included that seemed somewhat similar; in fact, about 200 genera from the Palaeozoic and Mesozoic eras. These fossil types were called palaeonisciforms. Modern cladistic studies, however, group sturgeons and paddlefish together in the Acipenseriformes, which is considered to be a clade; while the palaeonisciforms clearly include several quite different lineages, some of which are indeed close relatives of the Acipenseriformes, and some of which are clearly different although they share some primitive 'chondrostean' features. Here I have lumped all the lineages together informally as 'palaeonisciforms'. Such cavalier treatment is justified only because they are complex, long extinct, and known only to specialists. But in their day they were very significant.

Sturgeons and paddlefish: subclass Acipenseriformes

The Acipenseriformes include two living groups, the sturgeons and the paddlefish. Remarkably, the 20 or so species of sturgeons in the family **Acipenseridae** form the largest living family of osteichthyan fishes outside the Teleostei. They live throughout the Northern Hemisphere and some are among the biggest of all actinopterygians. But there are only two genera of paddlefish, family **Polyodontidae**, one of which lives in the Mississippi and the other in the Yangtze (and similar rivers). Both groups have highly mobile jaws, but they evolved along different lines and quite separately from those of bowfins and teleosts.

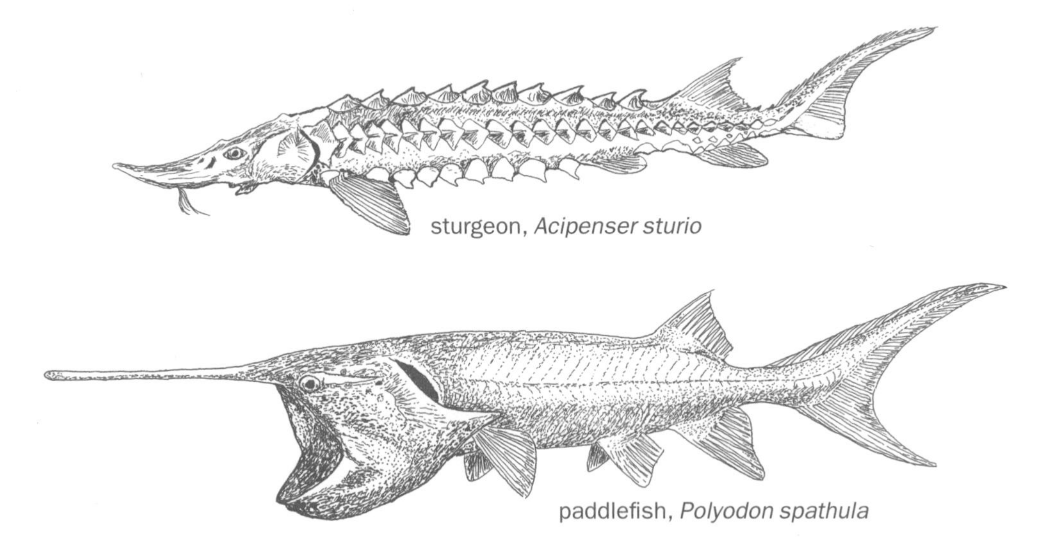
sturgeon, *Acipenser sturio*

paddlefish, *Polyodon spathula*

Sturgeons have poorly ossified skeletons and their scales are reduced to five rows or bony plates; and they are the prime source of caviar (I say 'prime' because paddlefish eggs are also pressed into service). Paddlefish are truly remarkable. They are up to 2 metres long and their long flat snouts—'paddles'—account for up to a third of their total length. The illustration shows how they feed—stretching their gill arches to form a great basket for filtering plankton and small fishes.

THE NEOPTERYGIAN FISH

The neopterygians include the remaining three subclasses: the Semionotiformes and Amiiformes, which are 'holosteans'; and the Teleostei. In neopterygians we see the evolution of what an engineer would call technological refinements: a reduction in the thickness of the scales that offers greater flexibility and lower mass; and a trend towards a homocercal (symmetrical) tail. Also, and crucially, we see increasing mobility of jaws and general ability to control the flow of water into the mouth and out through the gills.

Gars: subclass Semionotiformes

The living Semionotiformes are the gars or garpike. The seven known species form just one family, the Lepisosteidae, within the single order, **Ginglymodi**. Gars are fierce hunters of shallow and weedy fresh water. They are confined these days to North America down to Cuba, although fossils have been found in Europe, Africa, and South America. Like bichirs, gars can breathe air, although their lungs have a different shape and structure. *Lepisosteus* itself, the gar, is 1–4 metres long and lives in warm to temperate, fresh and brackish water, lunging and grasping its prey in the needle teeth that line its long jaws. This very same genus can be traced right back to the Cretaceous—so it can be called a 'living fossil': a slowly evolving lineage that has persisted, though always in low diversity.

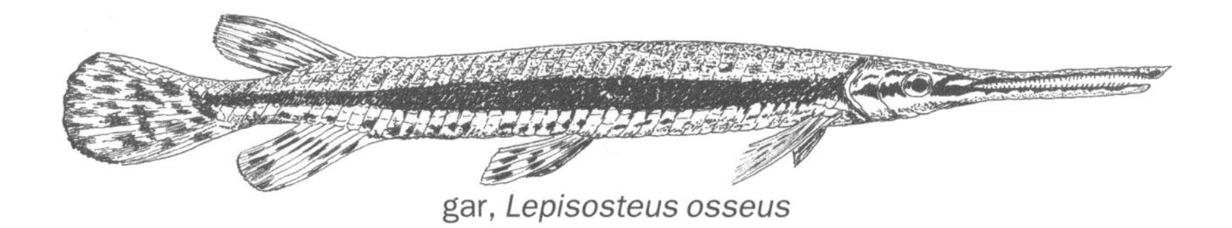

gar, *Lepisosteus osseus*

There are few types of holostean fish today, but there were many more in the past, particularly in the Jurassic. Thus the extinct family Semionotidae, which included about 25 genera such as *Semionotus*, were small and active with nearly symmetrical tails and large dorsal and ventral fins. Semionotids occurred in great diversity in some areas such as the eastern seaboard of North America; often with 10–20 species in the same lake. Apparently entire faunas were wiped out at a time by catastrophic drying but re-evolved when the good times rolled again, like the cichlids of modern African lakes.

Bowfins: subclass Amiiformes

The living Amiiformes are the bowfins. There is one order, the **Halecomorphi**, which contains just one family, the Amiidae. Bowfins are first known from the Jurassic period (about 206–144 million years ago) and are now thought to be the sister group of the teleosts. Bowfins live in fresh water in North America—and we might note in passing that North America harbours some of the world's most phylogenetically significant living fish: sturgeons and paddlefish, gars and bowfins, besides a brilliant range of teleosts. Bowfins are 0.5–1 metre long and prey on a wide range of organisms that may be the same length as themselves.

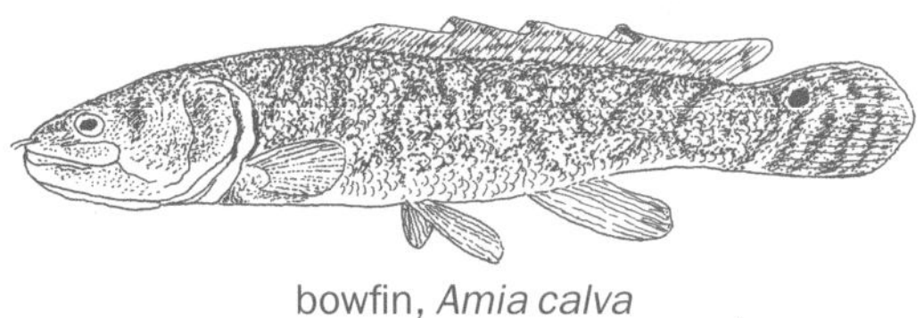

bowfin, *Amia calva*

Although there is only one species of *Amia*, in only one continent, there are many other species from the Cretaceous and Cenozoic, from Europe, Africa, the Middle East, and Asia, as well as North America. The order Halecomorphi includes many species in several families that date back to the Triassic. They are characterized by a specialized jaw joint that involves the symplectic and quadrate bones, which in general gives them a more mobile and versatile mouth than those of more primitive fish. But *Amia*'s jaw is intermediate between that of, say, some palaeonisciforms and that of some teleosts, which have the most mobile and dextrous mouths of all.

'Modern' ray-finned fish: subclass Teleostei

Teleosts are immensely various; they are indeed what most people mean by 'fish'. But they do seem to be monophyletic, which means they all descended from the same ancestor, who apparently lived in the early Mesozoic. Several synapomorphies unite the Teleostei, of which the most obvious are the homocercal tail and the great mobility of the mouth. Importantly, too (although this character is not a synapomorphy, because it is not special to them), teleosts make great and various use of the **swim-bladder**. This is a diversion of the gut that is presumably homologous with the lungs of bichirs on the one hand and lungfishes and tetrapods on the other, but in teleosts is usually a blind sac that can be filled or depleted by secreting or absorbing oxygen from the blood that supplies it. The swim-bladder serves primarily to provide neutral buoyancy, so the fish can stay at whatever depth it chooses without effort; but is also modified in various groups to serve for sound production (as in drums) or to enhance hearing (as in herrings).

Gary Nelson divides the huge array of teleosts among about 40 orders although other biologists at times have defined more than 50. But—mercifully for those trying to get to grips with the group—the 40 orders do fall naturally into four apparently monophyletic 'series': the **Osteoglossomorpha** or bony tongues; the **Elopomorpha**, which include the various eels and tarpons; the **Clupeomorpha**, which are the herrings and their relatives; and the **Euteleostei**, which include all the groups not mentioned so far, from salmon and sun fish to sea horses and minnows. Of these four, the

Euteleostei is by far the largest with at least 30 orders; but Willy Bemis has shown how these orders can be grouped in nine superorders (one of them further divided into three series) , which are shown in the tree and outlined below.

In practice, it is not yet certain whether all the acknowledged teleost orders are monophyletic. Much work still needs to be done, particularly perhaps of a molecular kind. But the classification shown here certainly enables us to make good sense of an extremely difficult group and the work that remains seems to be more a matter of detail than of extreme reorganization.

Bony tongues: the Osteoglossomorpha

The Osteoglossomorpha series contains about 150 living species in just one extant order, the Osteoglossiformes, which probably arose in the late Jurassic and now form two main groups—one containing such glorious creatures as the arowana of South America, and the other including the featherbacks, Old World knifefishes, and the extraordinary members of the Mormyridae, the 100 or so species of elephant-trunk fishes from Africa. What unites them all is their 'bony tongue': a tongue fitted with teeth that shear the prey against the roof of the mouth. But in body form and way of life they are immensely variable. Arowanas look like great muscular eels, and can leap high from the flooded Amazon waters to capture insects and other prey from the foliage overhead. Elephant-trunk fish, by contrast, have long snouts (of many different shapes) that they use to poke around for prey on river bottoms, which they locate via their highly developed system of electrolocation. Many actinopterygyian groups have independently evolved ways of using electricity, both for detection and orientation and even—in the case of electric eels, which are mentioned later—for stunning their prey.

arowana
Osteoglossum bicirrhosum

elephant-trunk fish
Campylomormyrus elephus

Eels, bonefishes, tenpounders, and tarpons: the Elopomorpha

The 650 or so species within the Elopomorpha are again tremendously various in outward physical form: on the one hand there are the tarpons—fast-moving game fish of the kind that excited Ernest Hemingway; and on the other are the eels ('Elopomorpha' means 'eel-form')—although, to be sure, some of them like the conger and moray are also tough and impressive. What unites them all is the peculiar form of their juveniles—a thin, leaf-like larva called a **leptocephalus** (although

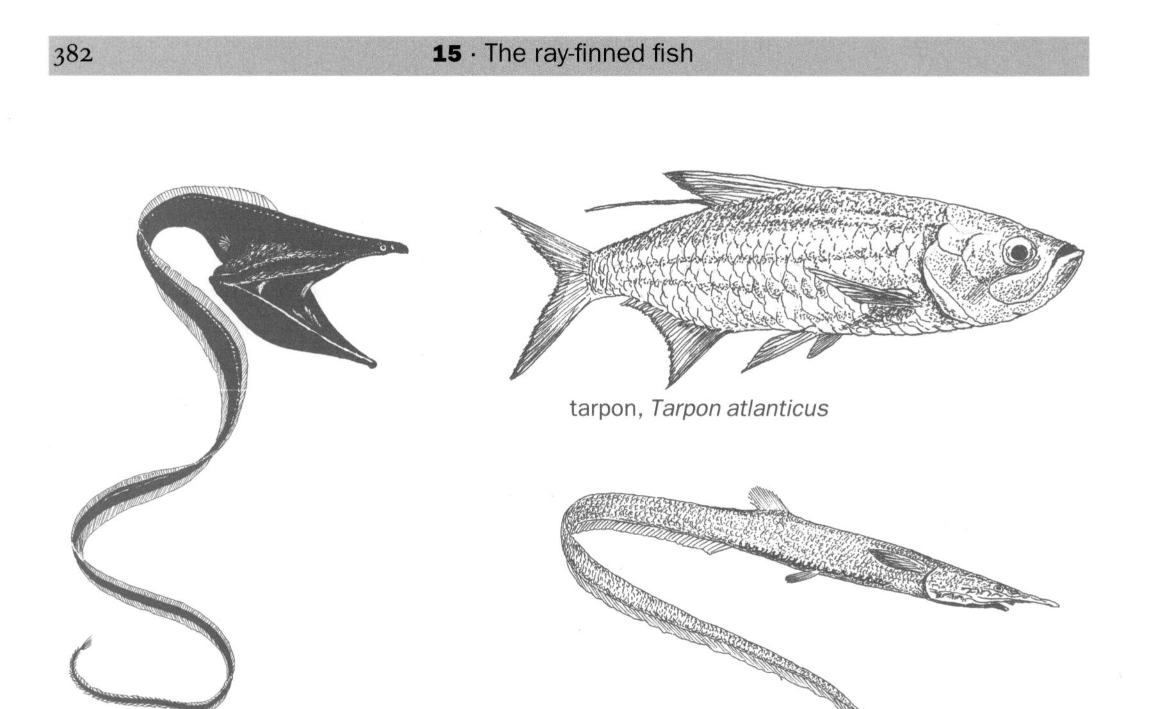
gulper eel, *Eurypharynx pelecanoides*

tarpon, *Tarpon atlanticus*

notacanth
Aldrovandia rostrata

particular larvae of particular elopomorphs may have their own names, as in the 'elver' of the eel). Leptocephalus larvae can migrate long distances before migrating. The Elopomorpha are known from the early Cretaceous, which is not particularly ancient.

Eels are highly modified; besides their obvious elongation they have lost the pelvic girdle and elements of their upper jaw have become fused. Even by the standards of eels, however, the swallowers (saccopharyngoids) are strange. They have lost their scales, their ribs, and their tail fin while their skull (in the words of Michael Benton in *Vertebrate Palaeontology*) is now 'just a huge pair of jaws with a tiny cranium set in front'. In fact, swallower eels are like a set of jaws on a spring; they lie like jacks-in-the-box on the sea bed, ambushing and engulfing passing prey that may be several times their own size.

Herrings and anchovies: the Clupeomorpha

More than 300 living species and 150 fossils dating back to the early Cretaceous are known within the Clupeomorpha. But all the living types at least are contained within a single order the Clupeiformes, which are the remarkable and successful herrings and their relatives. The fined-down bones of herrings are uniquely arranged in a lattice that runs between and even within the muscles, and lends them extraordinary suppleness. But the most obvious synapomorphy that unites the Clupeiformes is the swim-bladder, which extends forwards into the brain case to connect with the inner ear, and thus acts as a sound-chamber. Herrings and anchovies are plankton feeders that form truly vast shoals, and has made them prime targets for industrial fishing.

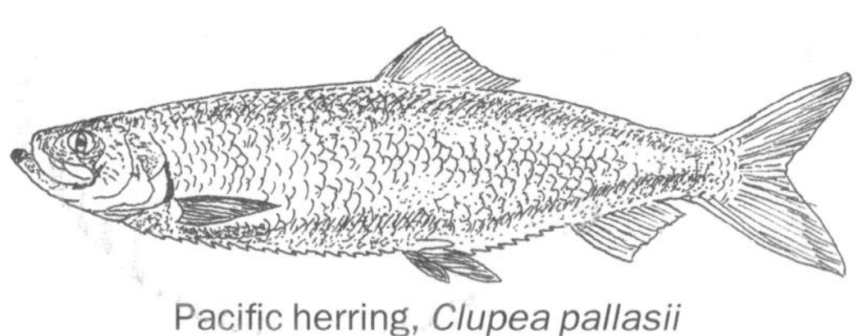
Pacific herring, *Clupea pallasii*

Thoroughly modern ray-finned fish: the Euteleostei

By far the majority of teleosts (about 17 000 living species in 375 families in about 30 orders) belong in the series Euteleostei. Thus the actinopterygians illustrate a principle seen time and time again throughout this book: that any one clade—in this case the Teleostei—is likely at the beginning to essay may different designs, but, typically, just one of these designs takes off and eventually far outstrips all the rest in diversity. Yet the Euteleostei can sensibly be subdivided into a few large groupings. Several modern authorities have divided them into three. Thus the Salmoniformes include the salmon and trout and their relatives, which are generally considered primitive—but also, unfortunately, are probably polyphyletic; the Ostariophysi include the carp, minnows, catfish, and most other freshwater species; and the Neoteleostei include all the rest.

In his classification, however, Willy Bemis further subdivides the neoteleosts into seven superorders, and so the system he advocates, and is presented here, recognizes a total of nine euteleostean superorders. The Ostariophysi is effectively unchanged; the group he calls Protacanthopterygii is essentially the same as the traditional Salmoniformes; and then come the seven neoteleostian superorders. So we finish up with the Ostariophysi, Protacanthopterygii, Stenopterygii, Cyclosquamata, Scopelomorpha, Lampridiomorpha, Polymixiomorpha, Paracanthopterygii, and Acanthopterygii.

- The **Ostariophysi** is a huge clade, with perhaps as many as 10 000 species: almost half of all known teleosts. Bemis includes five extant orders. The Anotophysi include the milkfish. The Cypriniformes include carp, grass carp, zebrafish, dace, chubs, various minnows, and barbs within the single family, Cyprinidae; and the loaches within the Cobitidae. The Characiformes are another huge group of around 1000 mainly freshwater species, including piranhas, neon tetras, freshwater hatchetfish, and the vast grouping of characins. The Gymnotiformes include the electric eels and the knifefishes. And, finally, the 2000 or so species collectively known as catfishes make up the Siluriformes.

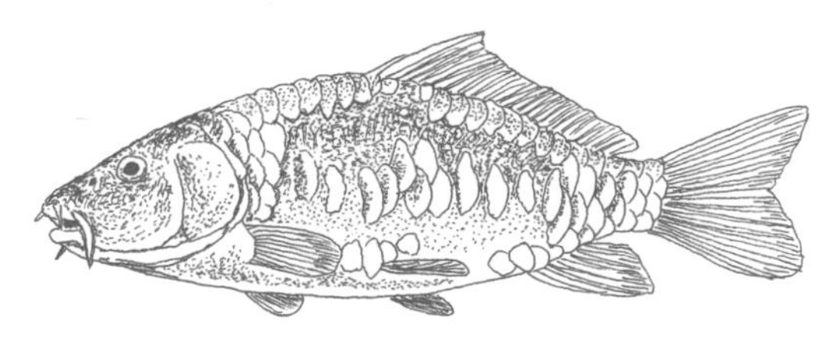

mirror carp *Cyprinus carpio*

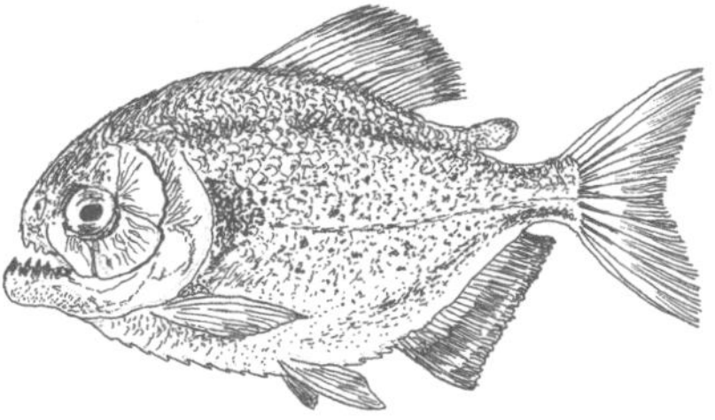

piranha, *Serrasalmus nattereri*

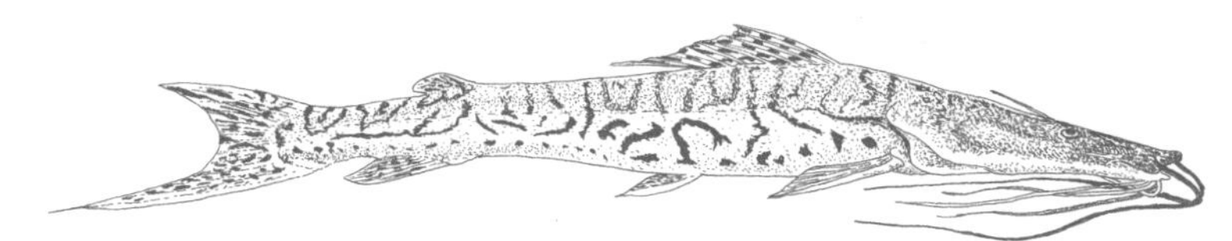

catfish, *Pseudoplatystoma tigrinum*

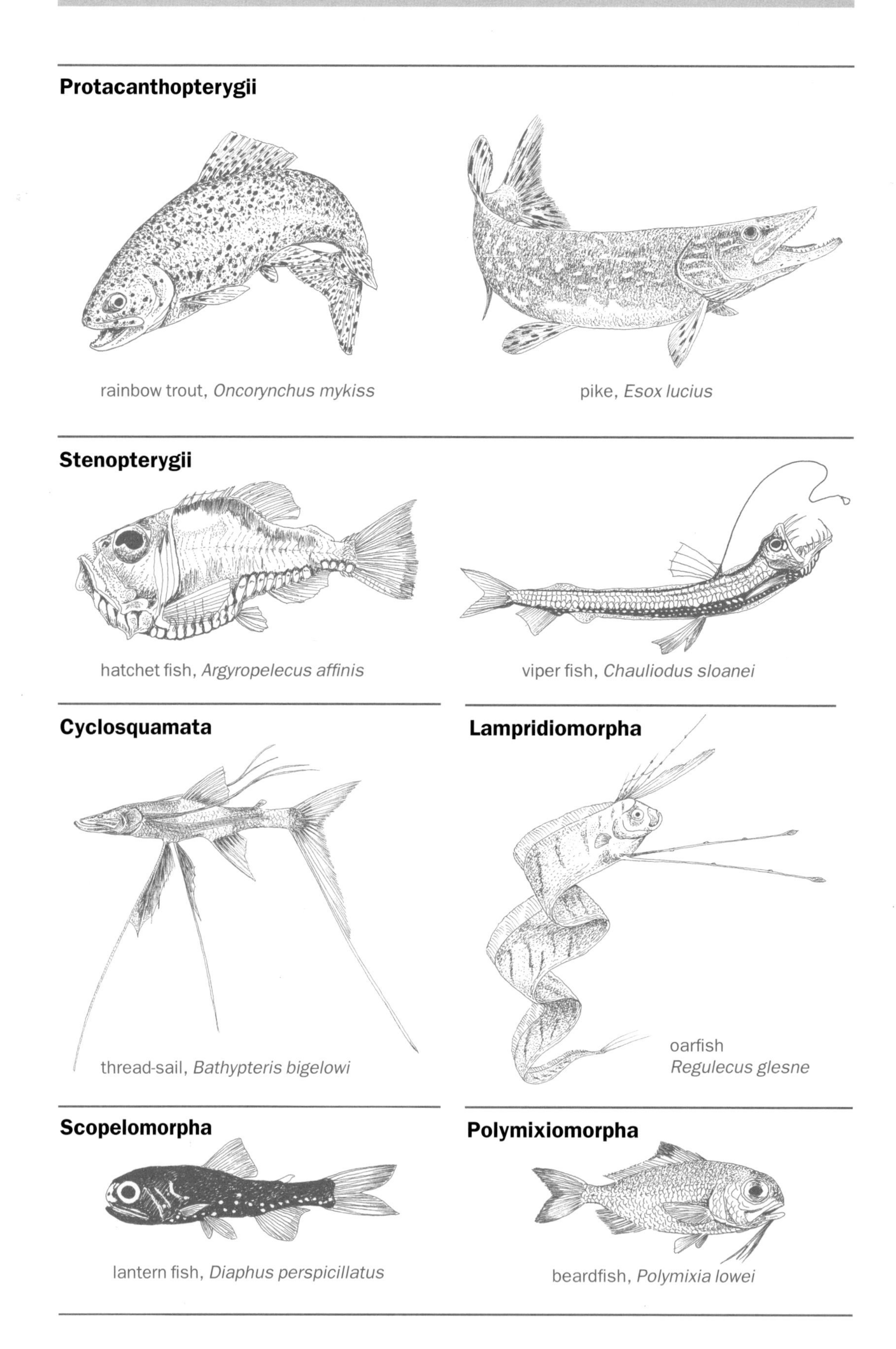
Protacanthopterygii
rainbow trout, *Oncorynchus mykiss*
pike, *Esox lucius*
Stenopterygii
hatchet fish, *Argyropelecus affinis*
viper fish, *Chauliodus sloanei*
Cyclosquamata
thread-sail, *Bathypteris bigelowi*
Lampridiomorpha
oarfish
Regulecus glesne
Scopelomorpha
lantern fish, *Diaphus perspicillatus*
Polymixiomorpha
beardfish, *Polymixia lowei*

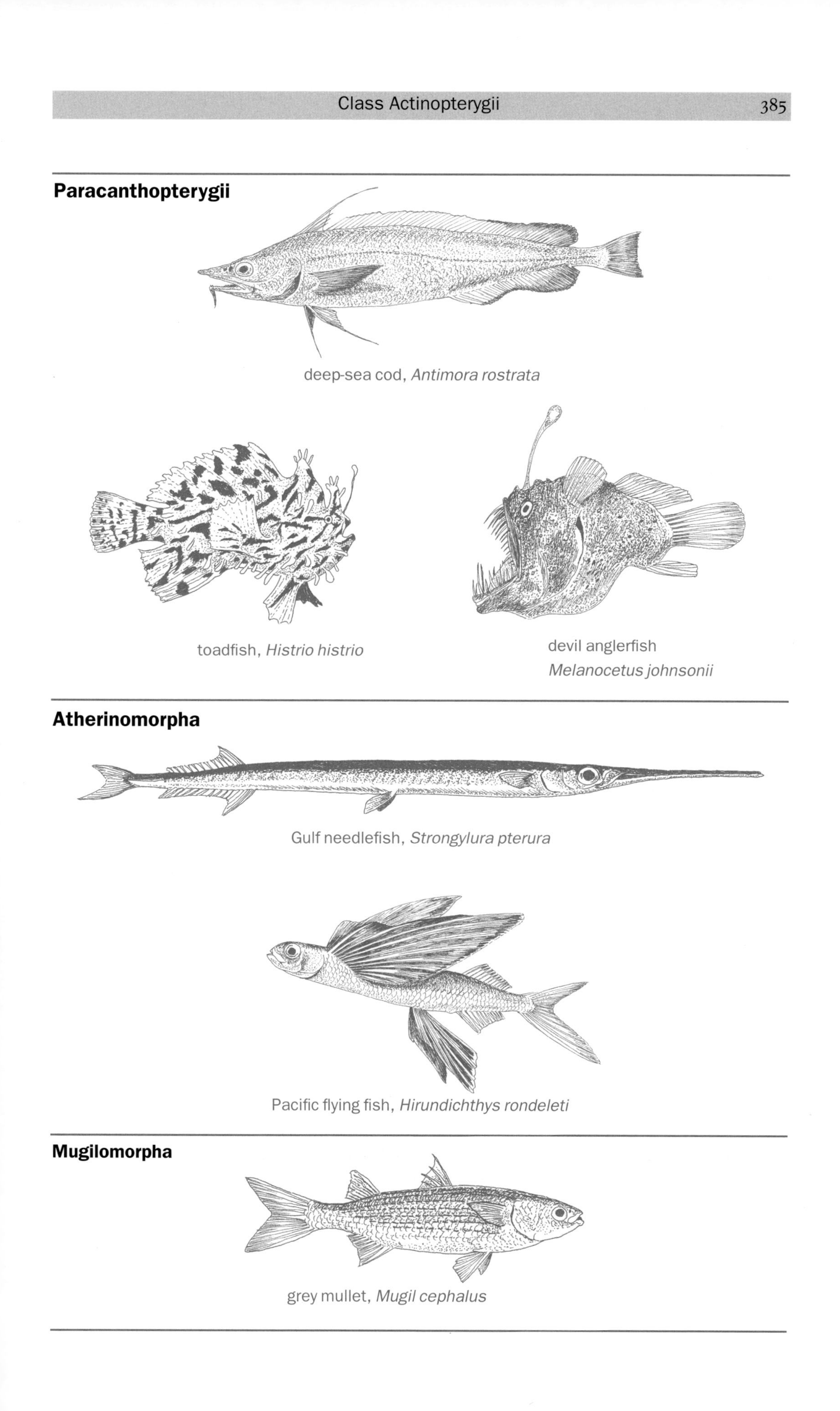

Paracanthopterygii

deep-sea cod, *Antimora rostrata*

toadfish, *Histrio histrio*

devil anglerfish
Melanocetus johnsonii

Atherinomorpha

Gulf needlefish, *Strongylura pterura*

Pacific flying fish, *Hirundichthys rondeleti*

Mugilomorpha

grey mullet, *Mugil cephalus*

Percomorpha

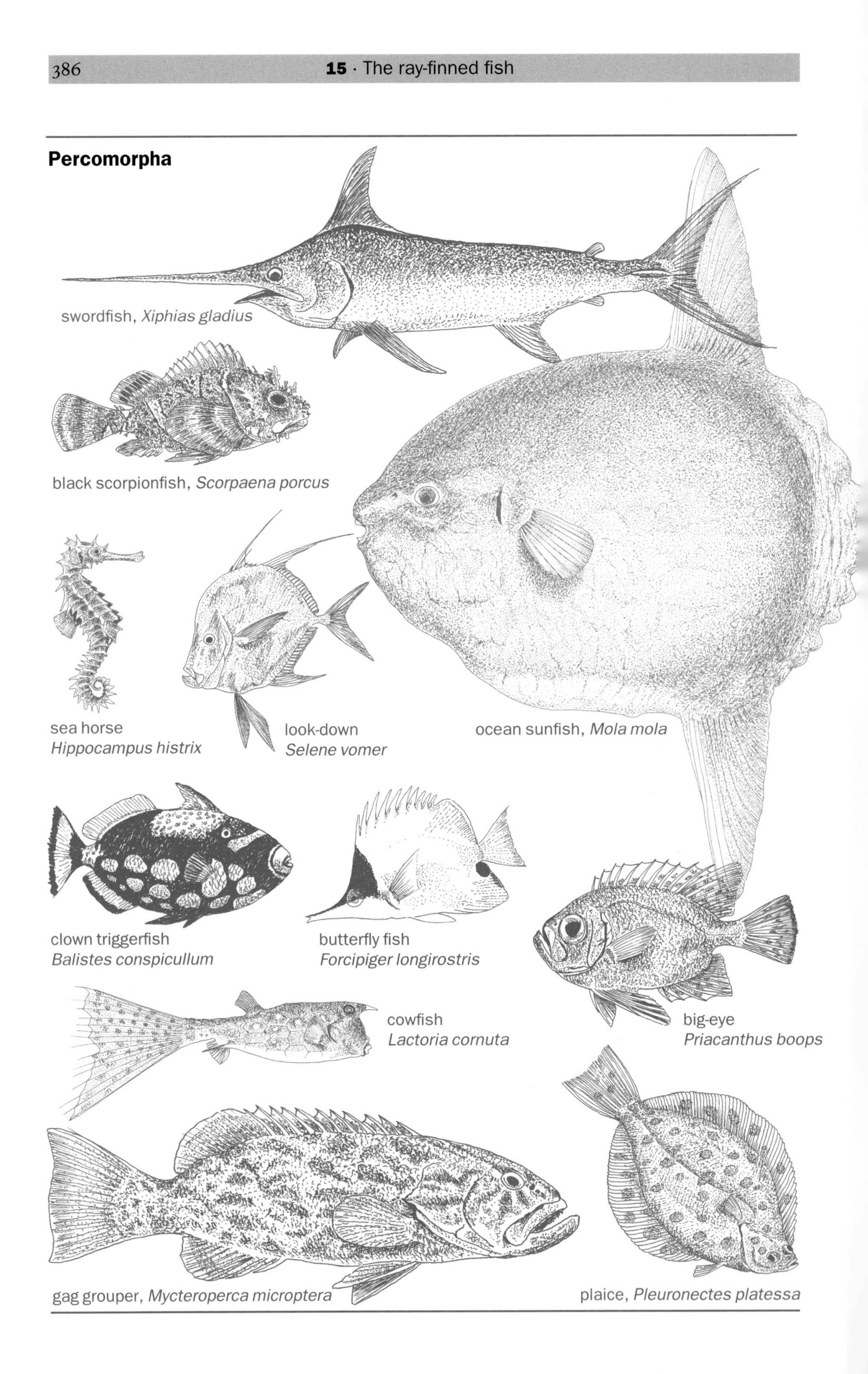

The Ostariophysi are united by a peculiar apparatus called the **Weberian ossicles**. A series of modified ribs and neck vertebrae provide a sound link from the front of the swim-bladder to the ear. The swim-bladder thus acts as a sounding chamber, and sounds impinging on the fish are greatly amplified.

- The **Protacanthopterygii** has three living orders, embracing the salmon, trout, smelts, and pike—most of the traditional Salmoniformes.

- The **Stenopterygii** has two living orders, including the light fishes and the marine hatchet fishes.

- The **Cyclosquamata** has one living order of somewhat obscure fish: the thread-sails.

- The **Scopelomorpha** has one extant order: the lantern fishes, also called myctophiforms, which produce light from organs known as photophores.

- The **Lampridiomorpha** has one living order: the opahs, tube-eyes, ribbonfishes, and oarfishes.

- The **Polymixiomorpha** has one living order: the beardfishes.

- The **Paracanthopterygii** has five living orders: the cods and anglerfishes, and the troutperches, pearlfishes, and toadfishes.

- The **Acanthopterygii**: this is a large and complex group, so called because members have stiff spines. They also have stiff bodies but nonetheless generate tremendous thrust from the tails and some of their members are the quickest of all bony fish and indeed among the quickest of all the creatures in the sea—like the tuna, which can reach 70 kilometres per hour. Willy Bemis divides the acanthopterygians into three 'series'. The **Mugilomorpha** includes the single extant order of the mullets. The **Atherinomorpha**, with three extant orders, includes the killifish, livebearers, silversides, halfbeaks, needlefish, and flying fish. The group is completed by the huge series of the **Percomorpha**, which contains nine orders embracing the sticklebacks, sea horses, scorpionfishes, flatfishes, 'spiny eels', and others. But most of the Percomorpha includes the 7000 or so species of the order Perciformes—perches, cichlids, flatfishes, snappers, grunts, drums, remoras, sunfishes, darters, and many more.

It would be easy to fill the entire book with actinopterygians. They are a hugely successful, diverse, and influential group. But we must pass on to their fellow bony fish, their sisters within the osteichthyes: the great clade of the Sarcopterygii, which includes various more groups of 'fish' plus the tetrapods.

16

LOBEFINS AND TETRAPODS

THE SARCOPTERYGII

THE SUBPHYLUM OF THE Vertebrata (all the backboned animals) culminates in the grand clade of the Osteichthyes, which, in turn, divides into two great sister clades. I described the first of these, class Actinopterygii, in Section 15; the second is the group of classes that together form the Sarcopterygii.

As we saw in the last section, actinopterygian 'limbs' are fins supported by rays—wonderfully mobile and versatile, but effectively committed to remain as fins. So although some actinopterygians venture on to land (such as eels, mudskippers, and climbing perch) and many can breathe air, they are all very obviously fish. In sarcopterygians, however, the limbs are constructed around central struts. Such limbs can serve as fins, and sarcopterygian fishes—informally known as 'lobefins'—were once far more significant than the actinopterygians, although the only sarcopterygian fishes left to us now are the three genera of lungfishes (Dipnoi) and the coelacanth (Actinistia). Much more significant, though, is that in one sarcopterygian lineage the strutted limbs evolved into legs that enabled their owners to move on land. Thus the sarcopterygians gave rise to the only truly terrestrial vertebrates—the members of the clade known as the Tetrapoda: four-legs.

Within the modern Tetrapoda we can see two clear grades, which reflect the evolutionary sequence. First are the tetrapods that spend much or virtually all of their adult life on land but produce naked, unprotected eggs that must be laid in water to avoid desiccation. Thus they need to return to water to reproduce, and inevitably their juveniles—larvae—are aquatic. These creatures thus enjoy a dual life: first living essentially like fish, and then metamorphosing into terrestrial tetrapods. So they are colloquially called 'amphibians' from the Greek *amphi*, meaning 'two', and *bios*, 'life'. In truth many amphibians have devised various ad hoc methods for producing young on land and dispensing with the fish-like larval stage, but in general the duality remains. In the past there have been many lineages of 'amphibians' but today only one remains—the class Lissamphibia, which includes the frogs and toads, salamanders and newts, and the peculiar, tropical, limb-less, earthworm-like caecilians.

But from among the ranks of the 'amphibians' there also evolved a new grade of creatures whose eggs are surrounded by a series of membranes, and also typically by a leathery or stony shell, which can resist desiccation and can be laid on land. One of the protective membranes is called the **amnion** and these creatures are accordingly collectively called amniotes; in fact, they all belong to one lineage, a single clade, which is formally and properly known as the Amniota. As the tree shows, the Amniota is the only remaining part of the larger tetrapod clade of the Reptilomorpha (as discussed later). Amniota includes the miscellany of creatures called 'reptiles', plus the mammals and birds.

In summary, then, the Actinopterygii have remained as fish. Nowadays they are easily the most speciose class of vertebrates, but all of them are fish nonetheless. The Sarcopterygii, by contrast, have undergone two evolutionary and ecological transitions, to produce three grades of creature. The first of these remain as fish and the second two are tetrapods—the semiterrestrial 'amphibians', and the fully terrestrial amniotes (even though some of the amniotes, like whales and ichthyosaurs, have chosen to return to the sea). In the following six sections (17–22) I focus on various subgroupings of the amniotes. In this section, I introduce the first two grades of the sarcopterygians: the sarcopterygian fishes—lungfish and lobefins—and amphibians, the first tetrapods. First of all, what kind of changes were needed to turn lobe-finned fish into amphibian tetrapods?

THE TRANSITION FROM WATER TO LAND

Creatures that contrive to process from life in water to life on land must come to terms with four new exigencies. They must find ways of coping with gravity; they must breathe air; they must avoid desiccation; and they must evolve senses that are appropriate to air rather than water.

For a land vertebrate, the rude and constant assertion of gravity demands that they convert their backbone into an arch, like a suspension bridge, from which to sling the organs of the trunk: and support that bridge with limbs that transmit the weight of the whole animal straight to the ground. Primitively, too, the vertebral column is 'sprung' upwards by the weight of the tail and of the head and neck, pulling down at the ends; the big quadrupedal dinosaurs, like *Diplodocus*, show the principle precisely. But the vertebral column of a fish is merely an unshortenable rod that flexes from side to side. Hold a fish at each end and its body sags. The transition from watery form to land form is made by increasing the amount of bone in the main body—**centrum**—of each vertebra, and elaborating the shape of the vertebrae so that they articulate, and lock if bent backwards. Typically, too, the structure is further strengthened, particularly around the thorax, with ribs of various shapes and attitudes.

The weight of the body now hangs from the arched vertebral column and this weight must be transmitted directly to the ground. The limbs must support the body, and to do this they must engage internally with the pectoral and pelvic girdles, which in turn must be firmly attached to the backbone. But the pectoral girdle of fish is connected to the skull, or at least with the gill arches; and if this arrangement persisted in a land animal the entire skull would vibrate with each step. Pectoral girdle and skull needed to become independent. The problems of the pelvis are different; for the pelvic bones of fish are merely held in the muscular body wall, while a land animal requires a solid basket of bone firmly attached to the backbone. At the distal end of the limbs the animal needs weight-bearing feet, with pads to stand on and spreading toes to distribute the weight.

Air-breathing *per se* was probably not a great problem for the early terrestrial vertebrates. Many actinopterygians have lungs of a sort, and so do the living sarcopterygian fish. After all, water can often become severely depleted in oxygen, particularly when it is warm and still. Probably, indeed, lungs are a primitive feature of osteichthyes. Gills become an embarrassment on land for they function poorly in air and lead to desiccation. Intriguingly, it now seems that some at least of the early tetrapods, such as the late Devonian *Acanthostega* from eastern Greenland, retained their gills, implying that although they had legs, they remained aquatic. As Christine Janis of Brown University, Rhode Island, has commented, 'the gap between fish and tetrapods is narrowing'.

For animals living exclusively on land the main respiratory problem is probably not to acquire oxygen, but to jettison surplus carbon dioxide—which is easy in water because carbon dioxide is so soluble. Even in mammals such as ourselves, with a long evolutionary history on land, the reflexes of breathing are primarily geared not to the level of oxygen in the blood but to the amount of carbon dioxide. It is the build-up of CO_2 that causes us to pant.

Land animals solve the problem of desiccation in various ways but the most straightforward is simply to develop an impervious skin (which many fish have anyway). Intriguingly, modern amphibians like frogs and newts have a highly glandular skin that they keep permanently moist, through which they acquire oxygen and excrete carbon dioxide. Presumably some ancient amphibians also had such a skin but it seems at least likely that the large, more terrestrial amphibians had impervious skin. Loss of gills in terrestrial forms also reduces desiccation and the production of amniotic eggs has been the final refinement.

Finally, terrestrial animals needed to modify their sense organs. In particular, terrestrial vertebrates have abandoned the **lateral-line system**, the sensory system by which fish pick up vibrations from the water, in favour of an ear formed around a vibrating tympanic membrane, or **eardrum**. But some aquatic amphibians have retained the lateral line.

In practice, we see signs of these kinds of changes—or at least of such changes as involve the skeleton and so are visible in fossils—in many different sarcopterygians from around the early Devonian, some 400 million years ago; so we can say that the tetrapods began about that time. As the Devonian progressed and passed into the Carboniferous these early tetrapods became enormously diverse, with many becoming very large—some as big as modern crocodiles—and some becoming more and more terrestrial; many others even returned to the water. These early non-amniotic tetrapods—loosely known as 'amphibians'—dominated until one of their lineages, the Reptilomorpha, gave rise to the amniotes. Once the amniotes began to spread, beginning in the Carboniferous, through the Permian and so into the Triassic, the non-amniotes faded. In fact, most of the early amphibian lineages had become extinct by the middle Permian (around 260 million years ago) and the only ones that survived into the Triassic and beyond (after 248 million years ago) were some fully aquatic types belonging to the temnospondyl clade. Only two groups of Temnospondyli survived beyond the Triassic. One of these groups lived in Australia, and lasted until the Cretaceous; the other gave rise to the Lissamphibia, which are the modern amphibians: frogs, toads, newts, salamanders, and the peculiar caecilians.

The lissamphibians represent a late flowering of the amphibian grade. There are roughly as many known species of living amphibian as there are of mammals—and perhaps as many amphibian genera as there were in the heyday of 'amphibians', in pre-amniotic times. They are of huge ecological significance in wetlands worldwide, in tropical forests, and sometimes in dry areas, both as food for creatures as diverse as storks and bats, and as significant predators. Ask any Australian about the cane toad!

So much for the general background. It is time to look at the creatures themselves, living and extinct. For the tree and the following notes I am indebted in particular to Michael Benton at the University of Bristol; Christine Janis of Brown University, Rhode Island; Gordon Howes; and Axel Meyer of the State University of New York at Stony Brook.

A GUIDE TO THE LOBEFINS, LUNGFISH, AND 'AMPHIBIANS'

A point of nomenclature first of all. The groups shown at the points of the tree are not all of the same rank. The living clades of the **Dipnoi**, **Actinistia**, and **Lissamphibia** can be considered to be 'classes'. But the only other living group—the **Amniota**—comprises a group of three classes, and perhaps is best left without formal ranking. Of the extinct types (which most of them are), ***Ichthyostega*** is the name of a genus, while the **panderichthyids** can be considered to be a family. The rest can take whatever ranking you may feel is appropriate. So now let us work our way down the tree.

LOBEFINS: THE SARCOPTERYGIAN FISHES

Sarcopterygian fishes got into their stroke before the actinopterygians, and for a time in the Devonian they were the dominant fish. Several extinct groups are known but only two of the principal groups are shown here: the **Porolepiformes** and the **Osteolepiformes**. These two, together with some other extinct lobefin lineages, were formerly classed as 'Rhipidistea' but because they are separate clades it now seems preferable either to abandon the term all together, or perhaps use **rhipidistean** as an informal adjective to denote extinct lobefins in general. Formerly, too, all the

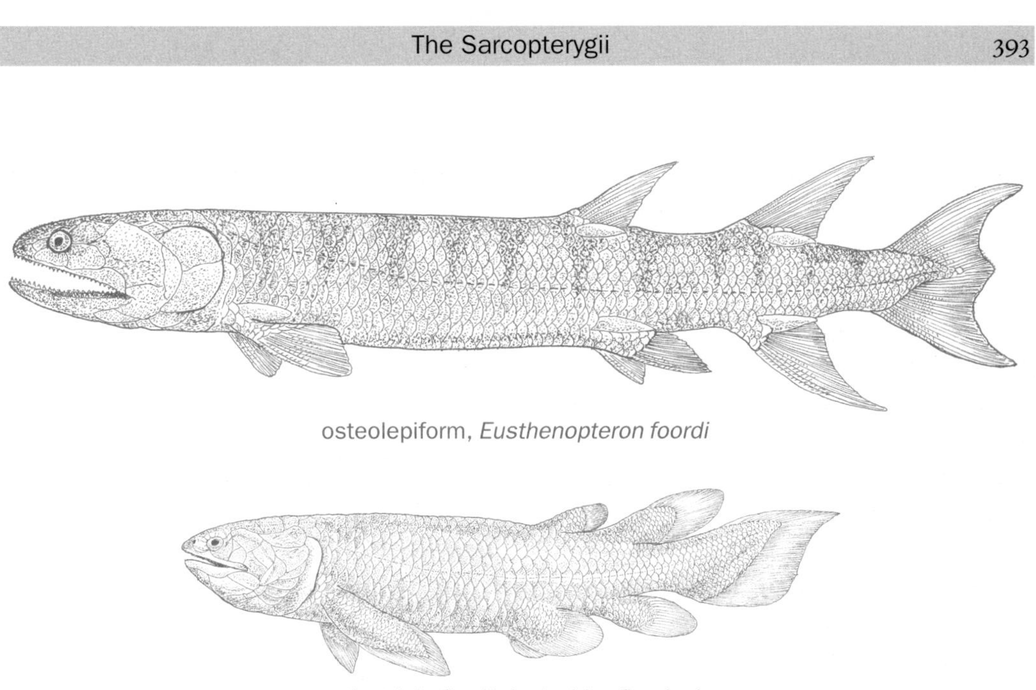
osteolepiform, *Eusthenopteron foordi*

extinct lobefin, *Holoptychius flemingi*

lobefinned fishes apart from the lungfishes (that is, all the extinct types plus the coelacanths) were classed as 'Crossopterygii'. So zoologists traditionally recognized two sarcopterygian classes of fish: the Dipnoi (lungfish) and the Crossopterygii. Again, the formal name Crossopterygii' seems redundant because it clearly embraces several distinct lineages, but the informal **crossopterygian** may be worth retaining.

The glory days of sarcopterygian fishes are long past but two classes remain.

Lungfish: class Dipnoi

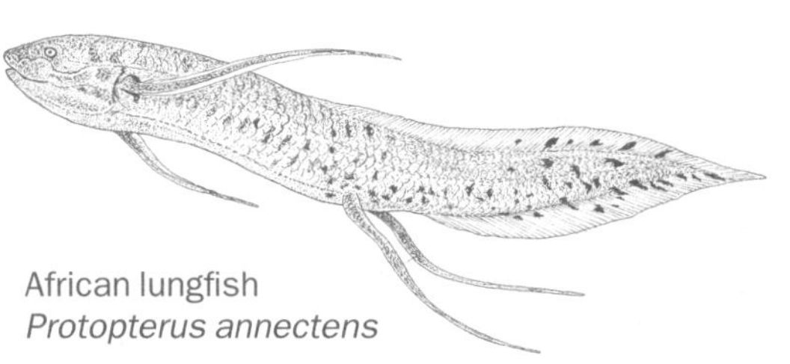
African lungfish
Protopterus annectens

The lungfish are reduced to just three genera today: *Lepidosiren* of South America; *Neoceratodus* of Australia; and the four species of *Protopterus*, which are among the biggest freshwater fishes of Africa. In their native lakes and rivers lungfish can be formidable predators, built as they are like thick moray eels. But although they have gills they also have lungs and can breathe air—and, indeed, can drown; and so they can also live in oxygen-deficient ponds. When the ponds dry up lungfish can 'aestivate': burrowing in to the mud at the bottom, clothing themselves in a mucus cocoon, and waiting for the good times to return.

The coelacanth: class Actinistia

The class Actinistia now contains just one species—the coelacanth, *Latimeria chalumnae*. *Latimeria* was first discovered only in 1938, dragged from deep waters around the

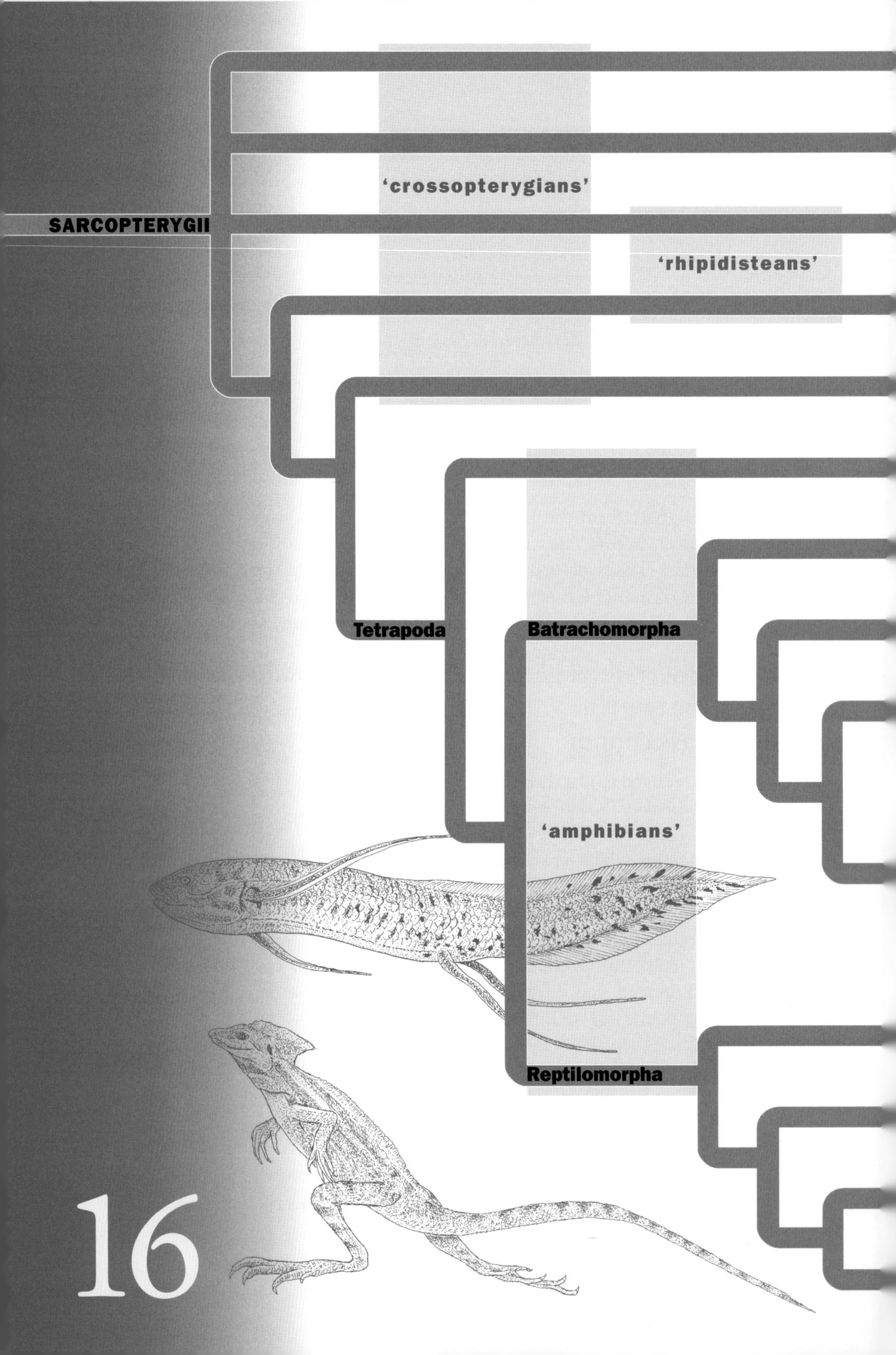

'crossopterygians'
SARCOPTERYGII
'rhipidisteans'
Tetrapoda
Batrachomorpha
'amphibians'
Reptilomorpha
16

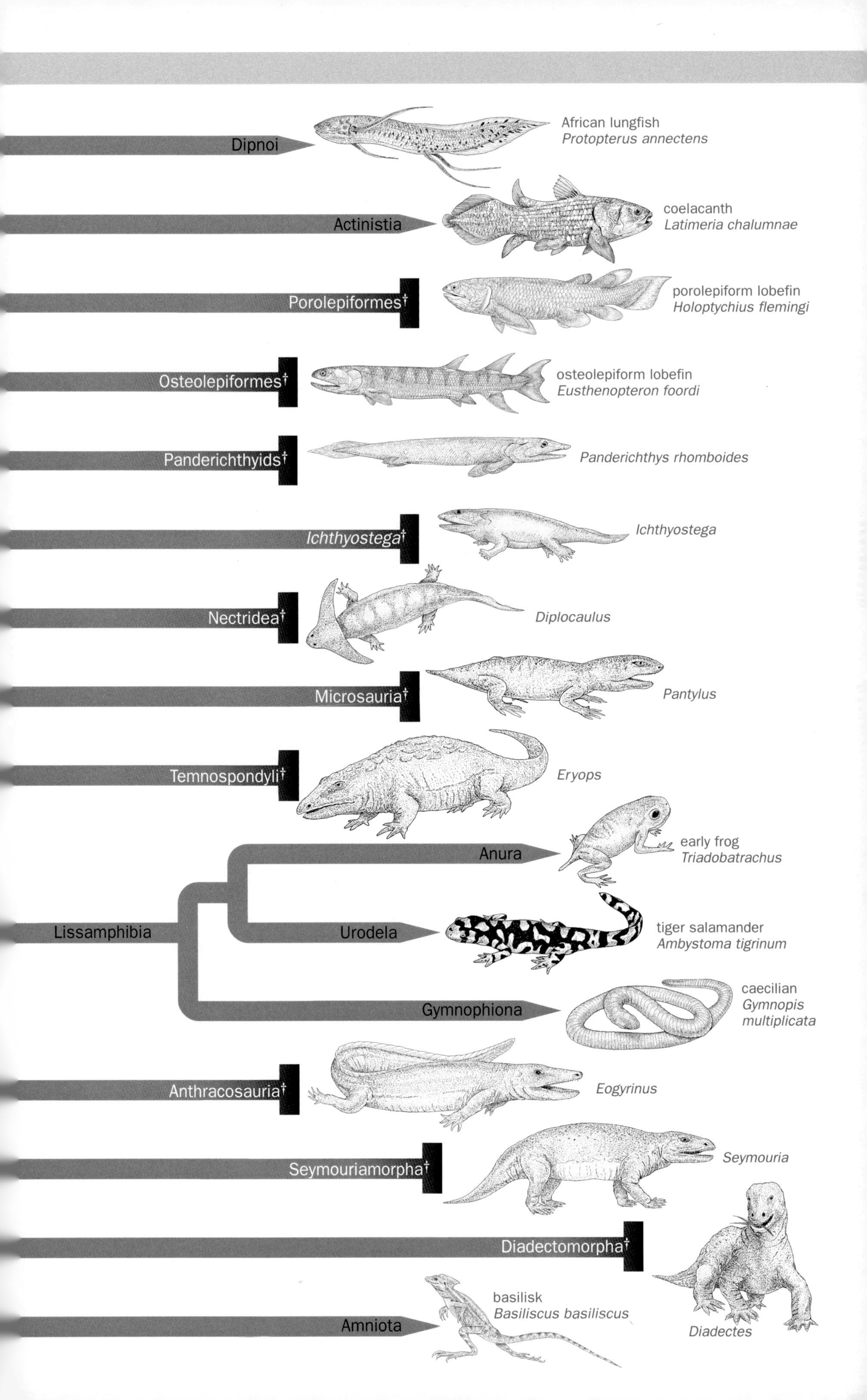

Dipnoi
African lungfish
Protopterus annectens
Actinistia
coelacanth
Latimeria chalumnae
Porolepiformes†
porolepiform lobefin
Holoptychius flemingi
Osteolepiformes†
osteolepiform lobefin
Eusthenopteron foordi
Panderichthyids†
Panderichthys rhomboides
Ichthyostega†
Ichthyostega
Nectridea†
Diplocaulus
Microsauria†
Pantylus
Temnospondyli†
Eryops
Anura
early frog
Triadobatrachus
Lissamphibia
Urodela
tiger salamander
Ambystoma tigrinum
Gymnophiona
caecilian
Gymnopis multiplicata
Anthracosauria†
Eogyrinus
Seymouriamorpha†
Seymouria
Diadectomorpha†
Diadectes
Amniota
basilisk
Basiliscus basiliscus

Comoro Islands near Madagascar. More coelacanths have since been caught from time to time and the wild population is thought to number a few hundred. But until the extraordinary catch of 1938 zoologists believed that the actinistians had gone extinct in the Cretaceous, some time before the dinosaurs.

coelacanth, *Latimeria chalumnae*

The fossil record and all that we know about zoology proclaim that the tetrapods arose from among the ranks of the sarcopterygian fishes. So the question now arises, *which* lineage of lobefin (or lungfish) gave rise to the **tetrapods**? Or, to put the matter in more formal, cladistic language—which sarcopterygian fish is the sister group of the tetrapods?

WHICH LOBEFIN (IF ANY) IS SISTER TO THE TETRAPODS?

In practice, all known groups of sarcopterygian fishes—or, at least, all the groups shown here—have at one time or other been mooted as the possible sister group of the tetrapods. The two living groups—Dipnoi and Actinistia—can, of course, be studied by molecular means, but the extinct types are known only from fossil bones. Some molecular studies suggest that lungfishes are closer to tetrapods, whereas others favour coelacanths. Axel Meyer suggests on the basis of studies of mitochondrial DNA that Dipnoi and Actinistia are sisters to each other, and that some common ancestor of them both is sister to the tetrapods. Mitochondrial DNA mutates relatively rapidly, and is generally used to differentiate between groups that diverged relatively recently. Lungfishes and coelacanths probably diverged around 400 million years ago. Even so, Meyer's studies give clear results. (He point outs, however, that coelacanth material tends to be less than ideal because specimens, dragged occasionally from the depths, have had a tendency to lie in the sun on Madagascan quaysides before any biologist gets to them, which is not an ideal way to begin molecular studies.) Coelacanths, lungfishes, and tetrapods seem to have diverged from each other within a relatively short period—say 20–30 million years—but did so some 400 million years ago; and when groups split within a relatively short time of each other, a long time ago, it is difficult by any molecular means to discover the order of their splitting.

Anatomically, however, several extinct lobefin groups make more convincing sisters to the tetrapods than either of the living groups. Both porolepiforms and osteolepiforms have been proposed but of the two, the osteolepiforms are the leading candidates. Thus the Devonian osteolepiform *Eusthenopteron* already had some of the bones in their limbs that are present in tetrapods: humerus, radius, and ulna in the pectoral fins, plus a few of the typical tetrapod wrist bones; and femur, tibia, and fibula in the pelvic fins, plus a few ankle bones. But *Eusthenopteron* did not use its fins as a tetrapod uses its legs—for example, they pointed backwards—and it could not have walked on land. To become true walking legs, these bony fins needed to evolve additional bones in wrist and ankle, and then acquire proper hands and feet with digits to spread the weight, then the angles and attitudes of the limbs needed to change, and they needed to acquire true elbows and knees so that the feet and hands were brought into four-square contact with the ground.

Of the creatures that can truly be considered to be 'fish', the osteolepiforms do indeed seem closest to the tetrapods. But now a creature has emerged that really does seem to be abandoning fishy ways even though it seems to have been entirely or at least primarily aquatic. This is, or was, *Panderichthys*—here referred to more generally and informally as the 'panderichthyids'. Panderichthyids were crocodile-like, with long flat snouts and eyes on the tops of their heads, which apparently lurked in shallow waters. They, rather than any lobefin, are now regarded as the likely sisters of the tetrapods.

The panderichthyids†

The panderichthyids have prompted a fresh look at the way that tetrapody began. Zoologists have traditionally supposed that tetrapods arose when ancient 'rhipidistean' fishes found themselves stranded in drying ponds, and set out over land to find

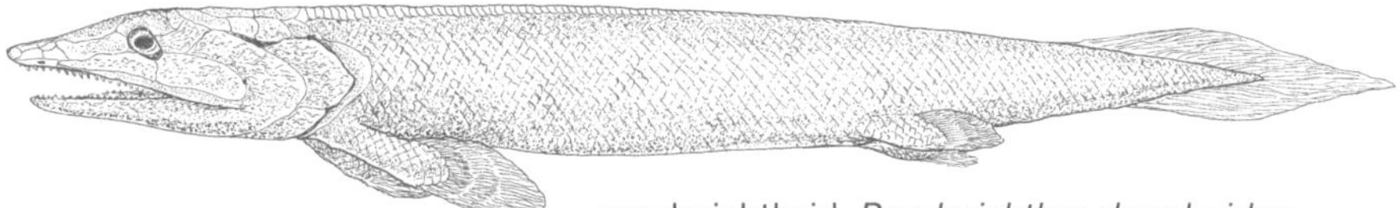

panderichthyid, *Panderichthys rhomboides*

new ones. The panderichthyids suggest a different origin: that a grade of sarcopterygians evolved that hunted by prowling in shallow waters, their eyes half above the water, propelling themselves along by flicking the bottom with limbs that at first were too weak to support the body on land but could propel the creature smartly enough when it was supported by water. In the same way (albeit living at depth) lobsters glide along the seabed perfectly well, propelling their heavy bodies with legs that are far too weak to support them on land. Anyone who has swum in shallow water must know the technique, and very relaxing it is. There could well have been vast stretches of shallow lagoon in the warm Devonian, so there was no shortage of such habitat. Air-breathing would be worthwhile in such an environment because shallow water is liable to be warm and so deficient in oxygen; gases dissolve better in cold water. Thus we can see how air-breathing and walking legs evolved quite independently of one another—so that an animal might well retain gills, as well as lungs, even though it has already gone some way to develop walking legs. Yet although lungs and walking legs evolved independently, they may have done so in the same sets of conditions.

In short, *Panderichthys* provides a truly convincing link between lobefins and tetrapods.

THE MEANING OF 'AMPHIBIAN'

Traditional zoologists recognized a great class that they formally called the 'Amphibia'. As you can see from the tree, this grouping includes a huge variety of creatures (most of them extinct). Also—more to the point—the grouping is paraphyletic, because it has given rise to the amniotes. Because the group is paraphyletic, modern zoologists suggest that the term 'Amphibia' should be dropped altogether, and prefer to call tetrapods that are not amniotes simply 'non-amniotic tetrapods'. Under my own proposed rules of Neolinnaean Impressionism I could call them Amphibia*, with the asterisk acknowledging their paraphyletic status. On balance, however, it seems best simply to drop the formal grouping Amphibia, or Amphibia*, but to retain the term 'amphibian' as an informal term—one that describes a perfectly discernible grade. After all, 'amphibian' is a deal less cumbersome than 'non-amniote tetrapod' and provides a convenient bridge between traditional and modern nomenclature.

If we want to refer formally to the living amphibians (frogs, toads, newts, salamanders, caecilians) we can call them Lissamphibia (because they are a true grade), or lissamphibians (in informal mode). But if it is obvious that we are talking only about living creatures, as we might be doing in an ecological context, then we can simply refer to them as 'amphibians' as naturalists have always done. Note, however, that if we retain the term Amphibia without an asterisk—as the name of the great clade that includes all the amphibians and the amniotes to which they give rise—then Amphibia becomes synonymous with Tetrapoda (which is a true clade, although there seems little point in assigning a formal ranking to it). There is no point in retaining synonyms. Tetrapoda is clearly preferable to Amphibia.

So let us work through the amphibians, from top to bottom.

The first tetrapod?: *Ichthyostega*†

Ichthyostega from the late Devonian is one of several creatures, roughly seal-like in general shape, that might reasonably be regarded as the first tetrapod. It may have belonged to the batrachomorph lineage or it may be the sister of all other amphibians, which is how it is shown here. Like a seal it had supporting front limbs and flipper-like hind limbs; although with its big, snapping-jawed head and short fish-like neck it was nothing like so handsome. The pectoral girdle was clearly separated from the skull—one of the prerequisites of terrestriality. Because *Ichthyostega* was still basically lobefin-like its back must have been weak and inclined to sag under force of gravity. But it was equipped with massive ribs, that were flanged along the hinder edge to form a powerful thoracic box, and perhaps this compensated. Yet *Ichthyostega* clearly spent much of its time in water. It retained the lateral line and had a tail fin and presumably swam with powerful sweeps of the tail. It also had sharp teeth and was

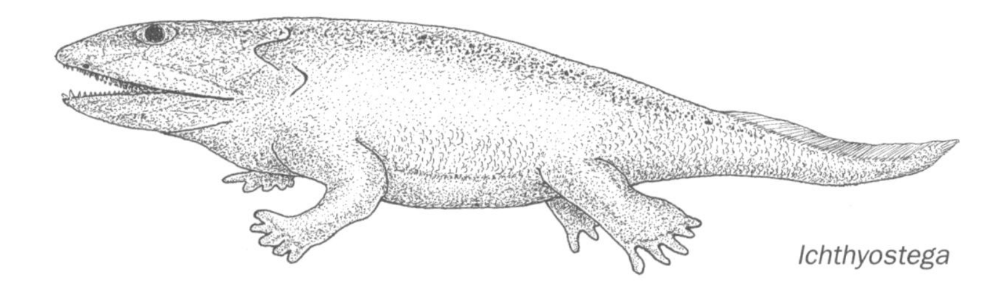

Ichthyostega

probably a fish-eater. The first *Ichthyostega* was found in eastern Greenland in 1932, and others have come to light since.

Incidentally, it has been a matter virtually of dogma for at least 100 years (and probably a lot more) that the first tetrapods had five fingers on each limb; or, as a cladist would say, that **pentadactyly** is a synapomorphy of the tetrapods. But it now transpires that *Ichthyostega* had seven toes on its hind feet, although its front feet are unknown. Another early type, the somewhat more aquatic *Acanthostega*, also known from eastern Greenland, had eight toes on its front feet; and *Tulerpeton*, a Devonian tetrapod known from Russia, had six toes on each foot. So pentadactyly took some time to evolve. Pentadactyly was surely established by the time the reptilomorphs arose, however, so we can say that at least for all amniotes, pentadactyly is a primitive feature (one common both to the amniotes themselves and to the larger group, the reptilomorphs, of which they are a part).

All the remaining tetrapods are either batrachomorphs or reptilomorphs. The tree shows that the amphibians form one great clade, but it is deeply divided into two smaller clades: the **Batrachomorpha**, and the **Reptilomorpha**. Old-fashioned taxonomies divided the ancient amphibians differently—into the Labyrinthodonta on the one hand, and the Lepospondyla on the other. The labyrinthodonts included the larger types, and the enamel of their teeth was folded convolutedly, hence their name. The lepospondyls were smaller, with teeth of simpler structure. It is now clear, however, that these were not natural divisions: it was just that the labyrinthodonts were large and lepospondyls were small, and their other characteristics were simply related to the difference in size. The deep division into batrachomorphs and reptilomorphs, however, does reflect tetrapod phylogeny. The fate of the two taxa has been profoundly different.

THE BATRACHOMORPHA

The batrachomorphs were the more emphatically 'amphibian' of the two great tetrapod clades. *Batrachos* is Greek for 'frog', and naturalists traditionally referred to living amphibians as 'batrachians'. The modern amphibian class, the Lissamphibia, are batrachomorphs. Batrachomorphs flowered during the Devonian and Carboniferous and became immensely various. Many returned to the water—a route many a tetrapod has taken since, from plesiosaurs and ichthyosaurs to whales and manatees. Many lost their legs to become snake-like or eel-like, as the modern caecilians have done. Here I have selected just a few of the ancient types to illustrate something of the phenotypic diversity and the phylogenetic intricacy: the **Nectridea** (nectrids), **Microsauria** (microsaurs), and **Temnospondyli** (temnospondyls)

Nectridea†

Nectrids flourished in the late Carboniferous and early Permian and show how bizarre the ancient amphibians could be. To be sure, many were merely newt-like in

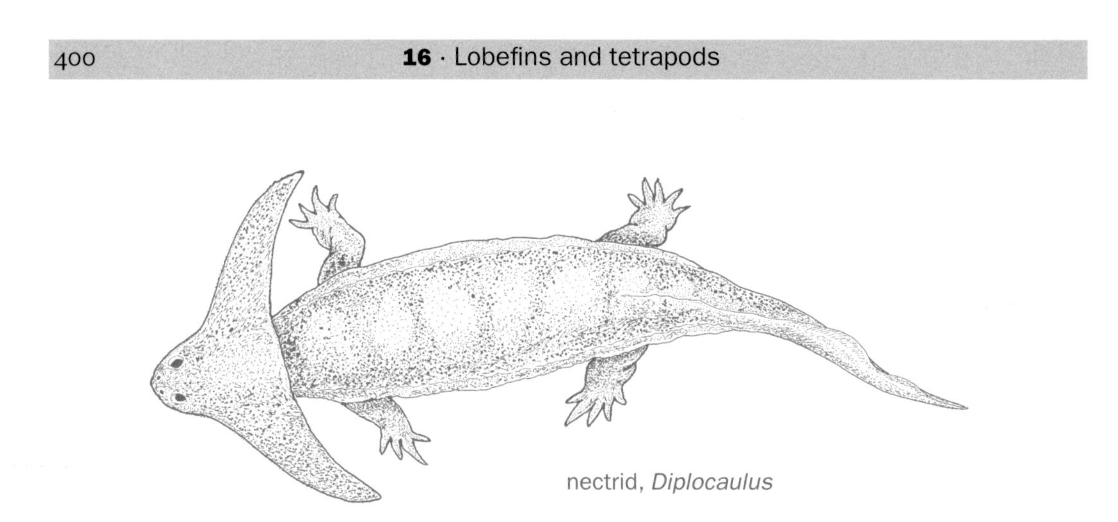

nectrid, *Diplocaulus*

appearance, with long flattened tails for swimming. But *Diplocaulus* and *Diploceraspis* from the early Permian of Oklahoma and Texas had the most extraordinary crescent-shaped skulls, with 'horns' extending backwards over their necks, which Michael Benton compares to a boomerang. These 'horns' grew longer as the animal aged. It has been suggested that this was for hydrodynamic purposes, perhaps giving the animal 'lift' as it leapt off the bottom after its prey. But this looks to me like the kind of feature that develops as a result of sexual selection, like the peacock's tail or the antlers of a stag—or indeed the frills on many modern newts. Female nectrideans might have preferred their males to have boomerang-like heads. (The sexual-selection hypothesis is weakened, however, unless we can show that males had more boomerang-like heads than females.)

Microsauria†

The Microsauria were the largest group of the smaller amphibians (the creatures formerly known as 'lepospondyls') with 11 different families. They thrived in the late Carboniferous and early Permian (which means either side of around 290 million years ago). Some were lizard-like and terrestrial, like the early *Tuditanus* from the late Carboniferous of Ohio. Others, like *Microbrachis*, which lived at around the same time in Czechosloviakia, had evidently returned to the water, and had small limbs and limb girdles. Other microsaurs were apparently burrowers. In short, the microsaurs radiated rapidly and widely (although terms like 'rapidly' are, of course, relative: tens of millions of years were involved).

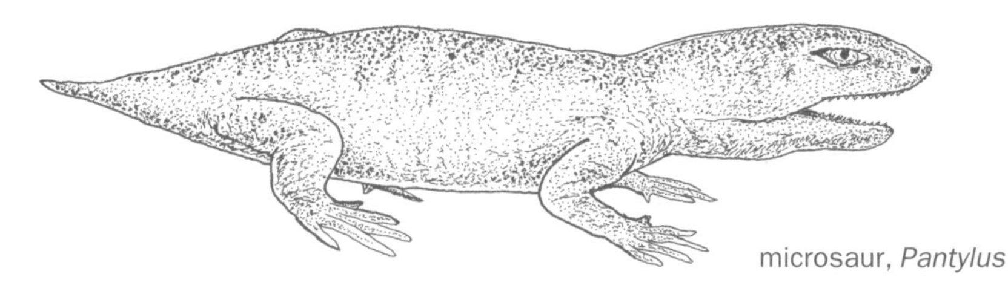

microsaur, *Pantylus*

Temnospondyli†

The temnospondyls were the greatest of all the batrachomorph groups (although we could say they still are the greatest of the amphibians, and many would argue that the

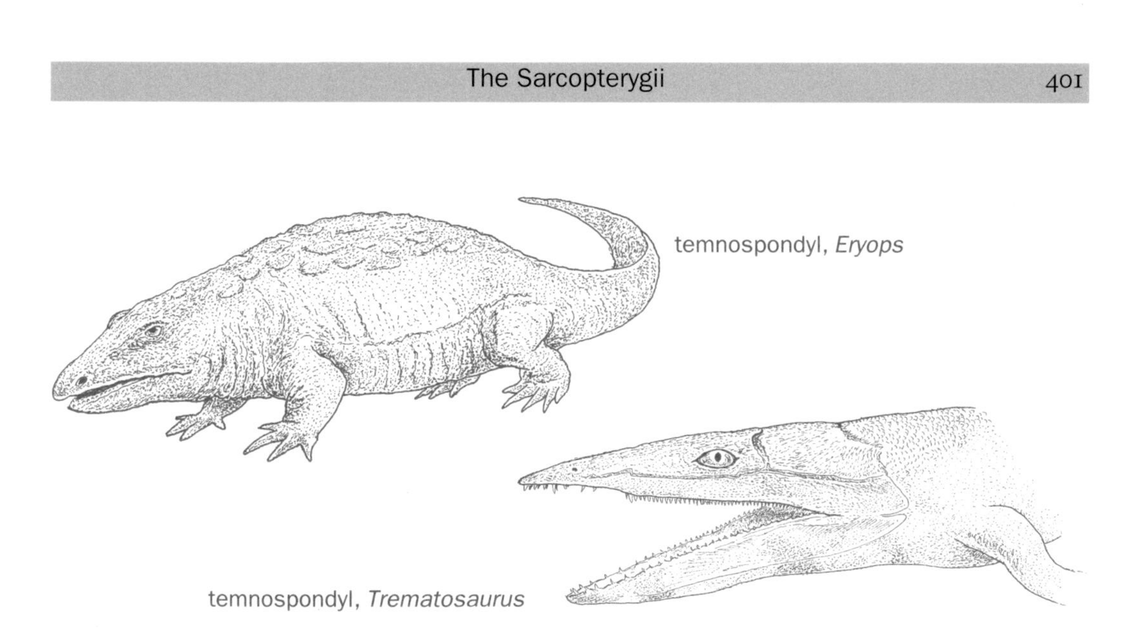

living lissamphibians are temnospondyls, although they are shown here, conservatively, simply as the sister group). The temnospondyls (excluding lissamphibians) lasted from the Permian all the way through to the Cretaceous. Some were emphatically terrestrial, like the heavy-boned *Eryops* from the early Permian of North America, which was 2 metres long and was clearly one of the top predators of its time and place. Other early Permian temnospondyls were fish-eaters, shaped like modern gharials (the slim-nosed crocodilians of Indian rivers). Fifteen families of temnospondyls are known from the Mesozoic—mostly aquatic types with broad flat skulls and reduced limbs. The Mesozoic really belonged to the reptiles, however, and although a few temnospondyls survived almost to the end of that era, most died out in the Triassic. Before the Triassic ended, one of the temnospondyl lineages is presumed to have given rise to the lissamphibians, whose descendants are with us still.

The living batrachomorphs: class Lissamphibia

The three modern amphibian orders look very different: the **Anura**, with about 3500 species of frogs and toads; the **Urodela**, the 360 or so newts and salamanders; and the **Gymnophiona**, sometimes known as the Apoda, which include the 200 or so species of caecilians. Caecilians look for all the world like earthworms or perhaps like leg-less millipedes, with their limb-less, banded bodies, and they are little known because they burrow in tropical soils in pursuit of the creatures they so resemble. But all the lissamphibians share features that suggest their monophyly—including peculiar **pedicellate teeth**, in which the base and crown are separated by a zone of fibrous tissue. The same cladistic analysis that unites the lissamphibians also places them among the temnospondyls. Although the lissamphibians arose in the Triassic, the earliest types are known mainly from scattered fossils in the Jurassic and Cretaceous.

Frogs and toads: order Anura

Modern frogs are supremely adapted for jumping. They have extremely long, muscular hind legs with an elongated ankle to give a 'five-cranked' limb; a front end adapted

early frog
Triadobatrachus

modern frog
Hyla leucophyletta

to absorb the shock of landing; a shortened body—with as few as four vertebrae in the trunk, no ribs, and the posterior vertebrae fused to form a rod called the urostyle, and an elongated pelvis. The earliest known frog is *Triadobatrachus* from the early Triassic of Madagascar and already it shows some of these features: a reduced number of vertebrae (though not as few as four!), reduced ribs, and elongated pelvis. *Viaraella* from early Jurassic South America is down to nine vertebrae and still has traces of ribs, but otherwise seems a very palpable frog. Only some of the 20 or so living anuran families, however, can be traced back to the Mesozoic. Most are known only from the Cenozoic—the past 65 million years. The lissamphibians may be the only amphibians to survive the Mesozoic, but those that did have radiated ebulliently.

Newts and salamanders: order Urodela

Urodeles are less striking; the shape of newts and salamanders is simply that of an archetypal tetrapod. Many, however, have adopted a peculiar way of life. The famous axolotl provides zoology with one of its most obvious examples of neoteny; for although axolotls grow into fair-sized beasts, about 30 centimetres long, they remain as giant larvae all the days of their life, sexually mature but with the feathery, juvenile gills of a larva. It seems that their inveterately aquatic way of life is an adaptation: in the Mexican uplands in which axolotls live there is more to eat in the water than on land. Other, related salamanders are facultatively neotenous: they may retain their larval form but if conditions are right they lose their gills and become like normal adults. Yet other urodeles demonstrate the amphibian propensity for reducing their limbs; like the sirens of the southern USA that retain the forelimbs but have lost the

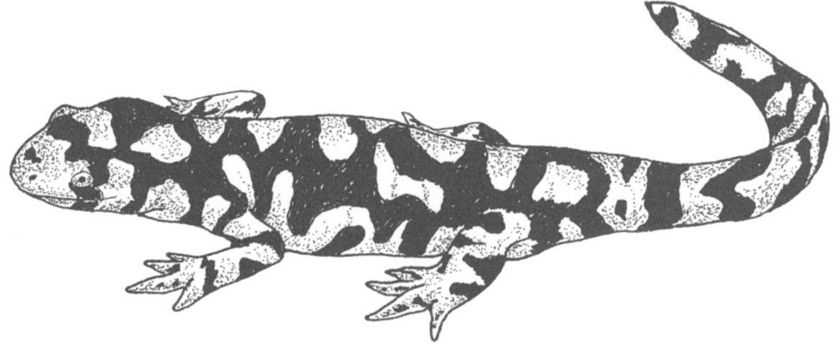
tiger salamander, *Ambystoma tigrinum*

hind ones all together; the eel-like *Amphiuma*, also from the southern USA, whose four tiny legs bear only two toes each; and the eel-like, blind olm, a cave-dweller from the Karst regions of Dalmatia, with only three toes on each diminutive leg.

Caecilians: order Gymnophiona

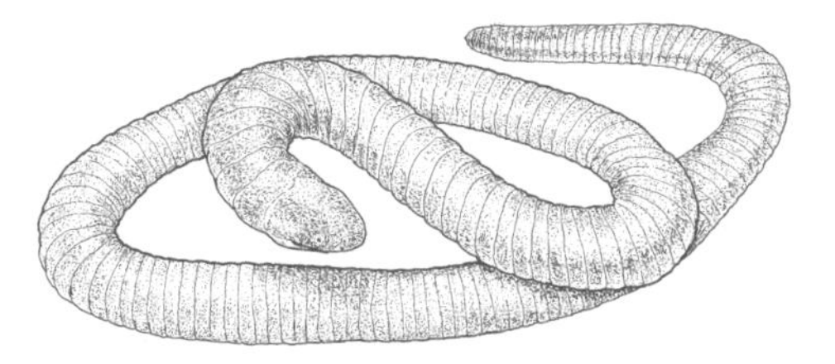

caecilian, *Gymnopis multiplicata*

Finally, the weird, legless caecilians have very little fossil record at all, apart from one recently reported from the early Jurassic of North America and a few scattered vertebrae from the late Cretaceous and Palaeocene of Brazil. Their worm-like bodies result from an elongated trunk, with up to 200 vertebrae; the tail is generally short. But the head is solid like a battering ram, for thrusting through the soil; much like that of a millipede.

The batrachomorphs no longer command all the exploited terrestrial niches as they did in pre-amniote days but the one group that remains to us, the lissamphibians, still contains roughly as many species as there are mammals, and they are of huge ecological significance worldwide. Indeed, the lissamphibians have come into their own in the Tertiary; so although the Tertiary is customarily known as 'The Age of Mammals' it could just as well be called 'The Age of Lissamphibians' (or indeed of songbirds, snakes, teleost fish, butterflies, or flowering plants). Despite the significance of the lissamphibians in the vertebrate fossil record, the tetrapod clade that has dominated the past 250 million years or so is the batrachomorphs' sister group, the Reptilomorpha.

THE REPTILOMORPHA

The reptilomorphs were (and are) the more reptilian of the two great tetrapod clades; sometime in the Carboniferous, they gave rise to the amniotes. Their phylogeny, like that of the batrachomorphs, has been immensely intricate. Some of this intricacy is shown in the next section, on the Reptilia*, and in subsequent sections devoted to mammals and birds. But here it is appropriate just to sketch in the three basal groups that are clearly related to the amniotes. They are the **Anthracosauria** and the **Seymouriamorpha**; and the ones that are thought to be the amniotes' sister group —the Diadectomorpha.

Anthracosauria†

The anthracosaurs—the name means 'coal-lizards'—were apparently of enormous ecological significance. They included about 15 genera of moderate-sized flesh eaters that lived from the early Carboniferous to the late Permian. Some of them were apparently terrestrial, and some seem to have returned to the water.

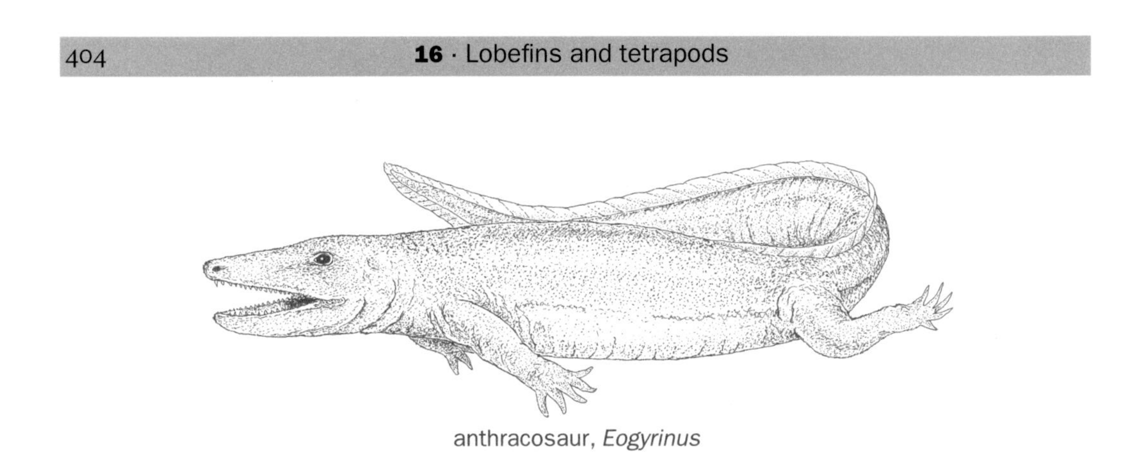

anthracosaur, *Eogyrinus*

Seymouriamorpha†

The seymouriamorphs are of great phylogenetic significance. The group lived in the Permian, and was named after the early Permian *Seymouria*. Here was an active terrestrial animal about 60 centimetres long, which was fairly abundant in the south-western USA. Most importantly, its powerful limbs held its body clearer of the ground than in most of the earlier types: the seymouriamorphs were clearly close to the lineage that gave rise to the reptiles.

seymouriamorph, *Seymouria*

Diadectomorpha†

The diadectomorphs, from the late Carboniferous and early Permian, were perhaps the most significant of all the early Reptilomorpha. They truly seem to be the sister

diadectomorph, *Diadectes*

group (implying ancestors) of the reptiles and hence the rest of the amniotes. *Diadectes* of the western USA was fairly heavily built with short limbs that engaged with massive limb girdles, and heavy vertebrae and ribs. Evidently, it was among the first of all terrestrial vertebrate

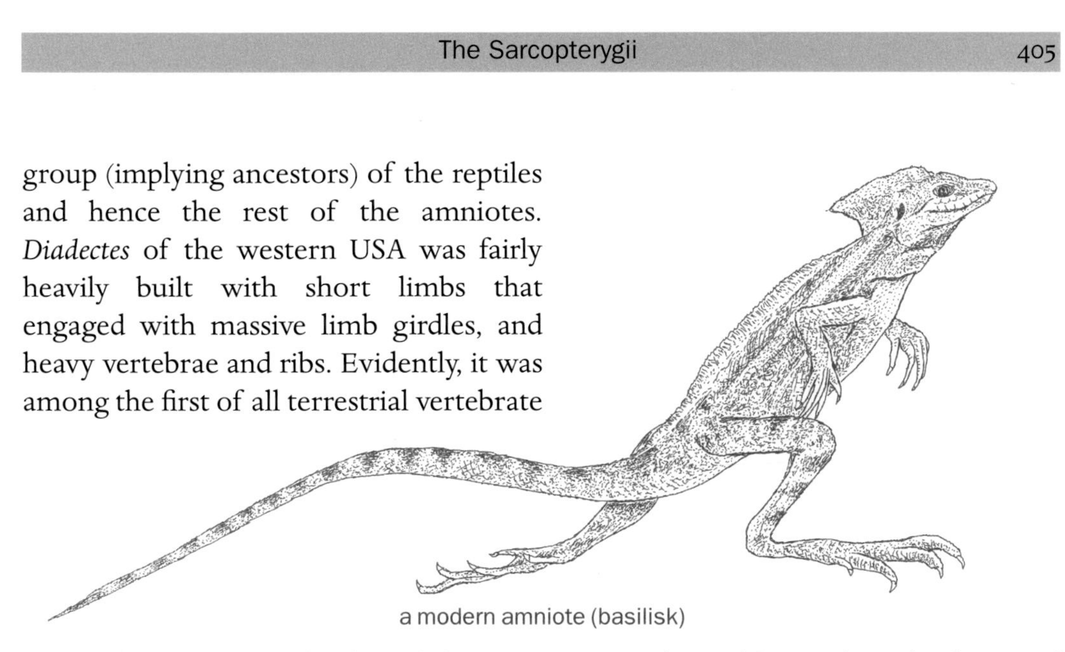

a modern amniote (basilisk)

herbivores: it apparently clipped the vegetation with peg-like teeth at the front and ground it with rows of blunter teeth along the cheeks.

So now we come to the amniotes themselves. In various degrees of detail these occupy the next six sections.

17

THE REPTILES

CLASS REPTILIA*

REPTILIA* IS A HUGE, rambling, ancient, and marvellously various group that embraces some of the most magnificent animals past and present that have ever lived. The reptiles are primarily terrestrial and include the biggest land vertebrates of all time—dinosaurs. The herbivorous *Seismosaurus* of the late Jurassic of New Mexico was the greatest of all. Until 1996 *Tyrannosaurus rex* was accepted as the biggest ever terrestrial predator but now we know of *Carcharodontosaurus saharicus* ('shark-toothed reptile from the Sahara'), which, with an estimated length of 14 metres, was probably even bigger, and with its steak-knife teeth was certainly a carnivore.

Reptiles have essayed an extraordinary variety of ecomorphs[1] and ways of life. On land, there are or have been sprinters (both four-legged and two-legged), rhino-like lumberers and batterers, and a host of burrowers, several lineages of which have lost their legs. Many reptiles flourish in trees. Many different groups have returned to the sea or to fresh water, sometimes partially like the modern marine iguanas of the Galápagos, and sometimes much more committedly—like the Cretaceous mosasaurs, which were giant lizards, or the plesiosaurs, nothosaurs, and placodonts and especially the ichthyosaurs, which were as maritime as any dolphin. Two quite separate lines of reptiles have also developed powered flight, in quite different ways: the extinct but once rampant pterosaurs; and the birds, which phylogenetically are aerial dinosaurs.

Note the asterisk on Reptilia*: this is a paraphyletic grouping, like the Crustacea* (which is also extremely various). Out of the reptilian ranks evolved the clades of the Mammalia and the Aves, which are excluded from Reptilia*. Cladists abhor such groupings and do not admit Reptilia* as a formal taxon, with or without an asterisk. But it seems to me (and to many other biologists) that if we drop the term Reptilia*, then we merely raise more complications. In truth, we cannot define Reptilia* according to acceptable cladistic criteria: reptiles cannot be formally defined by a list of synapomorphies. They are simply the amniotes that lack the special features of mammals (such as the suckling of young) or of birds (such as feathers). To be sure,

[1] 'Ecomorph' refers to a bodily form that is adapted to a particular habitat and way of life. Among mammals, for example, many different groups—deer, antelope, pronghorns, horses—have adopted the ecomorph of the big, hoofed running mammal.

reptile skin tends to be scaly, but this is a symplesiomorphy—scaliness shows up also on the legs of birds (and, we might say, on the tails of rats). Hence reptiles are a grade of creatures rather than a clade.

But Reptilia* does have some of the qualities of a clade. Notably, it does seem that all amniotes (reptiles + mammals + birds) arose from a single clade founder—one of the reptilomorph amphibians that were introduced in the last section. Reptilia*, therefore, represents a neat slice of a clade—the bit that remains after the daughter clades of the Mammalia and Aves have been hived off. So Reptilia* is a tidy grouping, and tidiness is one of the main requirements of a classification. We cannot define Reptilia* in cladistic terms, but we can easily say exactly what we mean by it.

Besides, if we drop the term Reptilia* as cladists demand, then we seem obliged to accept other infelicities. If cladists want to refer to the creatures that most of us call 'reptiles' they should say, 'non-mammalian, non-avian amniotes'. Similarly, most naturalists distinguish instinctively enough between dinosaurs and birds. But because birds seem to have evolved from dinosaurs, cladists must declare that birds *are* dinosaurs and so must speak of *Tyrannosaurus rex*, say, as a 'non-avian dinosaur'. It seems easier just to admit a little paraphyly here and there. Note, finally, that cladists would allow the term Reptilia—without an asterisk—to refer to the entire amniote clade, *including* mammals and birds. But this would imply that people and horses and house sparrows are reptiles; and, of course, 'Reptilia' would be redundant because it would be synonymous with 'Amniota'. For the time being, then, I am sticking with Reptilia*—meaning, 'non-mammalian, non-avian amniotes'.

The term 'Amniota', which defines the whole reptile–mammal–bird clade, derives from the structure of the eggs.[2] The amphibians, in their transition to land, completed only half the task. They continued and developed the air-breathing that had already been essayed by several groups of fish. They developed, for the first time, a skeleton and, in particular, a pair of limb girdles that enabled them to hold their bodies clear of the ground, even though unsupported by water. But they stopped short of terrestrial reproduction. They continued to produce eggs that were and are naked and hence are prey to desiccation. Apart from a few inventive types like the gastric brooding frog, which swallows its eggs to keep them moist, amphibians must return to the water to breed and their newborn young are aquatic larvae: tadpoles.

The reptiles completed the transition to the land: and they did so by providing each egg with what Michael Benton of the University of Bristol calls a 'private pond'. Reptile eggs are subdivided by a series of membranes and the whole is surrounded by a semipermeable shell that permits the flow of gases but prevents the escape of water. The shell is usually calcareous but is sometimes leathery, as in marine turtles, snakes, and some lizards. The novel internal membranes are the **chorion**, which surrounds the embryo and yolk sac; the **allantois**, which is involved in respiration and stores waste materials—and hence expands as development proceeds; and the

[2] Of course, most modern mammals bear live young—as do some reptiles, including ichthyosaurs whose babies were born at sea and would otherwise have drowned. But egg-laying mammals still exist and the details of mammalian pregnancy still betray their eggy associations.

amnion, which lies within the chorion and surrounds the embryo. Because reptiles invest more in their eggs than amphibians do, and protect them better, they generally lay fewer of them. Birds and mammals retain the basic structures of the reptilian egg and the clade Amniota is, of course, named after the amnion.

Living Reptilia* now include about 6500 or so species in four main groups: 270 or so species of turtles, tortoises, and terrapins in the Chelonia: more than 3700 lizards, 2300 snakes, and 140 amphisbaenids in the Squamata; the few remaining species of tuatara in the Sphenodontida, all now confined to islands around New Zealand; and the 22 species of alligators, crocodiles, and gharials in the Crocodylia. So although we tend to think of reptiles as also-rans they still include many more living species than mammals do, and they are ecologically significant from one end of the globe to the other, particularly, but by no means exclusively, in the tropics. Wherever food is sparse or space is limited—notably in deserts and on islands—reptiles tend to have the advantage over mammals, because the living kinds are poikilothermic, which means they do not officiously maintain a body temperature significantly above that of their surroundings, and so need far less food, weight for weight. Thus reptiles even today are sometimes the top herbivores, like the marine and land iguanas of the Galápagos and the giant tortoises of various island groups; and even more often they are top predators, like the Komodo dragon of Indonesia and the goanna of Australia (both of which are monitor lizards), the crocodilians of tropical rivers and some coasts, and, in some contexts, snakes.

As the tree shows, however, the reptiles of today are a fitful relict of glories past. Survivorship has been extremely spotty. In fact, of the 22 major lineages represented in the tree just six have survived and two of those—mammals and birds—are not conventionally recognized as reptiles at all. Most of the major reptilian lineages have gone. The once huge lineage of synapsids survives only in the form of mammals. The Anapsida live on as chelonians but various other 'parareptilian' anapsid groups, such as the marine mesosaurs[3] (not to be confused with the mosasaurs—see below), have long disappeared. We can properly regret that none of the huge diapsid marine reptiles is still with us—the placodonts, ichthyosaurs, nothosaurs, and plesiosaurs (unless a plesiosaur lives on in Scotland's Loch Ness—an extremely long shot!). The lepidosaurs have survived as sphenodonts and squamates (lizards, including amphisbaenids, and snakes); but they, too, have lost some magnificent representatives, including the fearsome maritime Cretaceous mosasaurs that were relatives of the living monitor lizards, which now include the Komodo dragon and the goanna. The crocodiles and birds are all that is left from the huge grouping of the Archosauromorpha, which gave rise to the ubiquitous Triassic rhynchosaurs and then to the archosaurs, such as the dinosaurs.

Over the 300 million years plus of reptilian history there must have been at least 100 species for every one alive now—so the total inventory of reptilian species past and present can hardly be less than half a million, and is probably far more. We

[3] At least, some zoologists include the mesosaurs among the 'parareptiles' but some put them outside the Anapsida altogether. Such early creatures are often difficult to place.

surely have a right to feel regret. It really is a pity that the magnificent class of reptiles has been wiped so conclusively. All the more reason to be grateful for the ones that have survived.

Reptiles are so ancient and have been so various that we must trim their genealogical tree somewhat in order to make sense of it; in its unpruned state it is a thicket that can still bewilder even the specialists. So the present tree shows all the living groups, and all the major divisions of the extinct types plus a few minor outgroups just to acknowledge the complexity that I have left out, and to indicate some of the range of body forms. The taxonomy is firmly rooted in the ideas of Michael Benton, with whom I have had excellent conversations over many years and whose *Vertebrate Palaeontology* (1997) is a classic; but I lay claim to the mistakes. Reptilian palaeontology also attracts many fine scholars so the ground is constantly shifting, and no tree can claim to be definitive.

A GUIDE TO THE REPTILIA*

The class Reptilia* is deeply divided into two great subclades, the **Synapsida** and the **Sauropsida**. The Sauropsida is in turn divided deeply into the **Anapsida** and the **Diapsida**. In effect, then, we have three great groups: the Synapsida, the Anapsida, and the Diapsida.

The terms Synapsida, Anapsida, and Diapsida refer to the various holes—**fenestrations**—in the side of the skull, or the lack of them. In the Anapsida the skull is an uninterrupted box of bone; in the Synapsida there is a fenestra in the lower temple; and in the Diapsida there are two such fenestrae, one in the lower position (as in Synapsida) and another higher up on the temple. These temporal fenestrae lighten the skull and, more importantly, provide extra surfaces for muscle attachment and spaces into which contracting muscles can expand. They are derived features and I would say that in engineering terms they certainly represent a functional improvement.

Note, however, that various marine diapsids—the placodonts, ichthyosaurs, nothosaurs, and plesiosaurs—have evolved yet another form of fenestration, known as **euryapsid**. Their skulls have only one fenestra, but it is positioned high on the temple, like the upper fenestra of a diapsid. All the euryapsid groups are aquatic. Because of these striking similarities, many taxonomists have invoked a fourth subclass, the 'Euryapsida'. But the euryapsid state now seems merely to be derived from the diapsid condition. Furthermore, euryapsid fenestration seems to have evolved at least twice, independently: once within the ichthyosaurs, and again within the placodonts + nothosaurs + plesiosaurs. Thus the formal subclass Euryapsida is no longer recognized although the informal adjective 'euryapsid' is still useful.

The very first reptiles of all, dating from the Carboniferous, were anapsid; so the anapsid state is primitive. The animals included in the bona fide subclass Anapsida have retained the anapsid condition. But because this character is symplesiomorhic for Reptilia* as a whole, it cannot define the subclass. In fact, the members of the Anapsida are recognized as a true clade because they possess other, quite separate synapomorphies (although the details of these need not delay us). As the tree shows, the Anapsida and Diapsida are regarded as sister clades—which together form the Sauropsida. The Synapsida is then seen as the sister of the Sauropsida.

Specifically, the first known unequivocal reptile was *Hylonomus* from the late Carboniferous (around 310 million years ago) of Nova Scotia. It was a small, lizard-like creature—about 20 centimetres long including tail—with a small head, light-boned skull, and small sharp teeth suggesting a diet of invertebrates. The skull from the slightly later *Paleothyris* is better known than that of *Hylonomus* and it clearly had a group of jaw muscles (the pterygoideus) that the amphibians do not have, which presumably enhanced the range of movement. Reptiles, in short, are more adept feeders than amphibians.

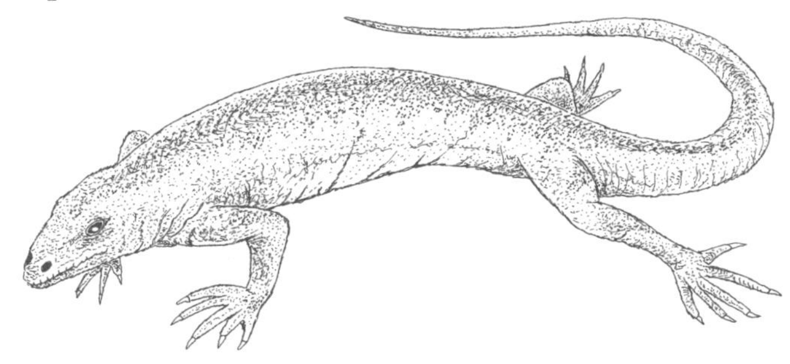

the first known reptile, *Hylonomus*

Ideally, it would be good to examine fossil eggs from these earliest reptiles, to gauge whether they really were amniotic; but in fact the oldest putative reptile egg is from the early Permian of Texas, dated at 270 million years ago. So we have to judge the reptilian status of the earliest types from their bones: for example, reptile heads tend to be much smaller, relative to their bodies, than are those of ancient amphibians, and the bones of the back of the reptile skull are particularly reduced. *Hylonomus* and *Paleothyris* meet these reptilian criteria. Because they are Carboniferous, it is a reasonable guess that amniotes (meaning reptiles) first appeared in the earliest Carboniferous, some time after 350 million years ago.

The different groups of reptiles then emerged in a series of waves—first one major taxon dominated, then another, then another. The synapsid reptiles peaked early, in the Permian, and the synapsid lineage is still with us—in the form of the mammals. But in the Mesozoic—often known as 'The Age of Reptiles'—the diapsids dominated. The herbivorous rhynchosaurs prevailed in the Triassic, and then, from the Triassic to the end of the Cretaceous, the world's continents belonged to the dinosaurs. Several of the once-dominant groups seem to have crashed in mass extinctions, perhaps occasioned simply by a change of climate (in turn perhaps brought about by the sundering and clash of continents as they drifted around the globe) and sometimes, it seems, caused by some 'catastrophe'—including collision with asteroids. The biggest mass extinction in the history of the world took place at the end of the Permian, around 248 million years ago, and marked the boundary between the

17 · CLASS REPTILIA*

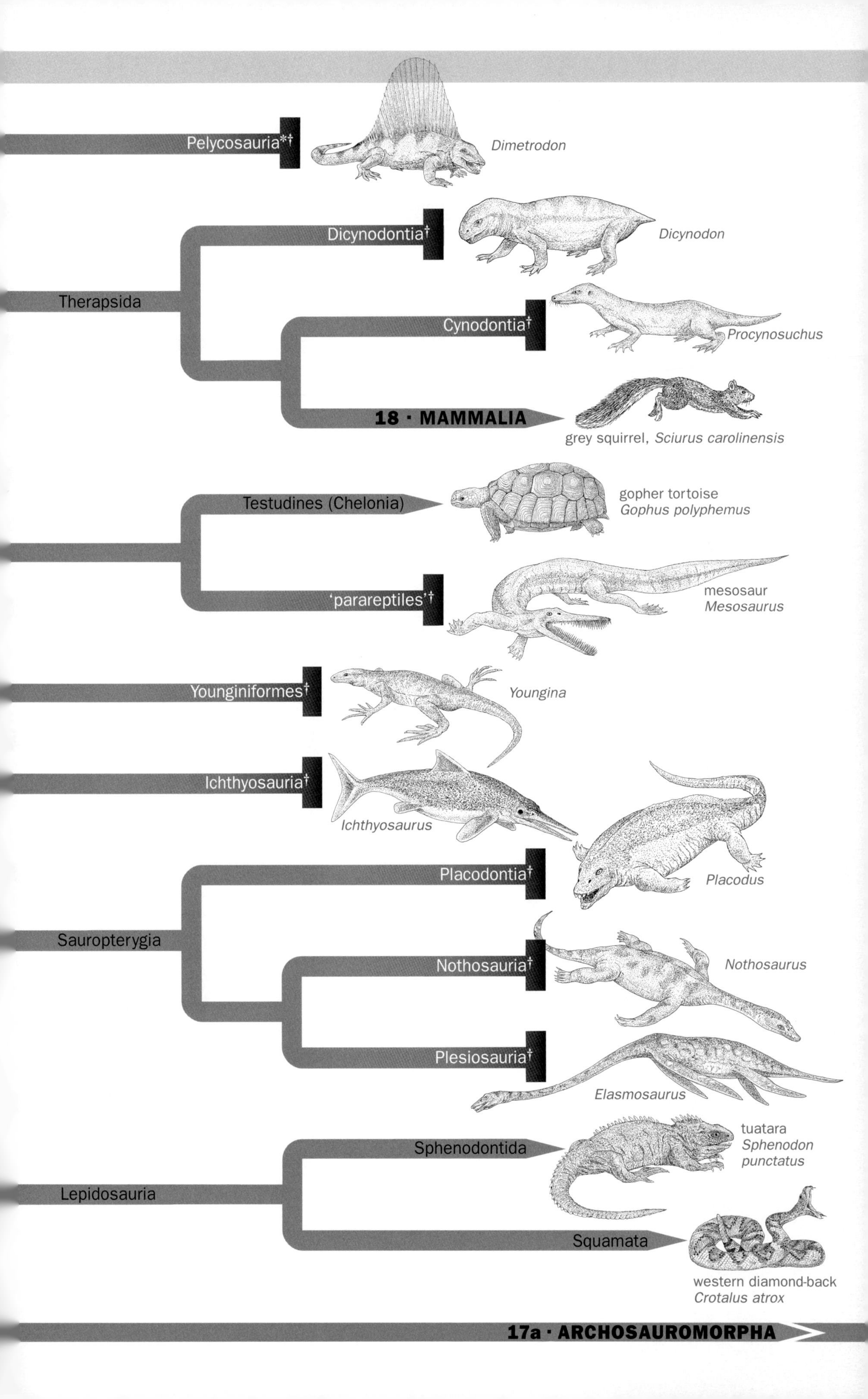

Pelycosauria*†
Dimetrodon
Dicynodontia†
Dicynodon
Therapsida
Cynodontia†
Procynosuchus
18 · MAMMALIA
grey squirrel, Sciurus carolinensis
Testudines (Chelonia)
gopher tortoise
Gophus polyphemus
'parareptiles'†
mesosaur
Mesosaurus
Younginiformes†
Youngina
Ichthyosauria†
Ichthyosaurus
Placodontia†
Placodus
Sauropterygia
Nothosauria†
Nothosaurus
Plesiosauria†
Elasmosaurus
Sphenodontida
tuatara
Sphenodon
punctatus
Lepidosauria
Squamata
western diamond-back
Crotalus atrox
17a · ARCHOSAUROMORPHA

Archosauromorpha
Archosauria
Ornithodira
Dinosauria
17a

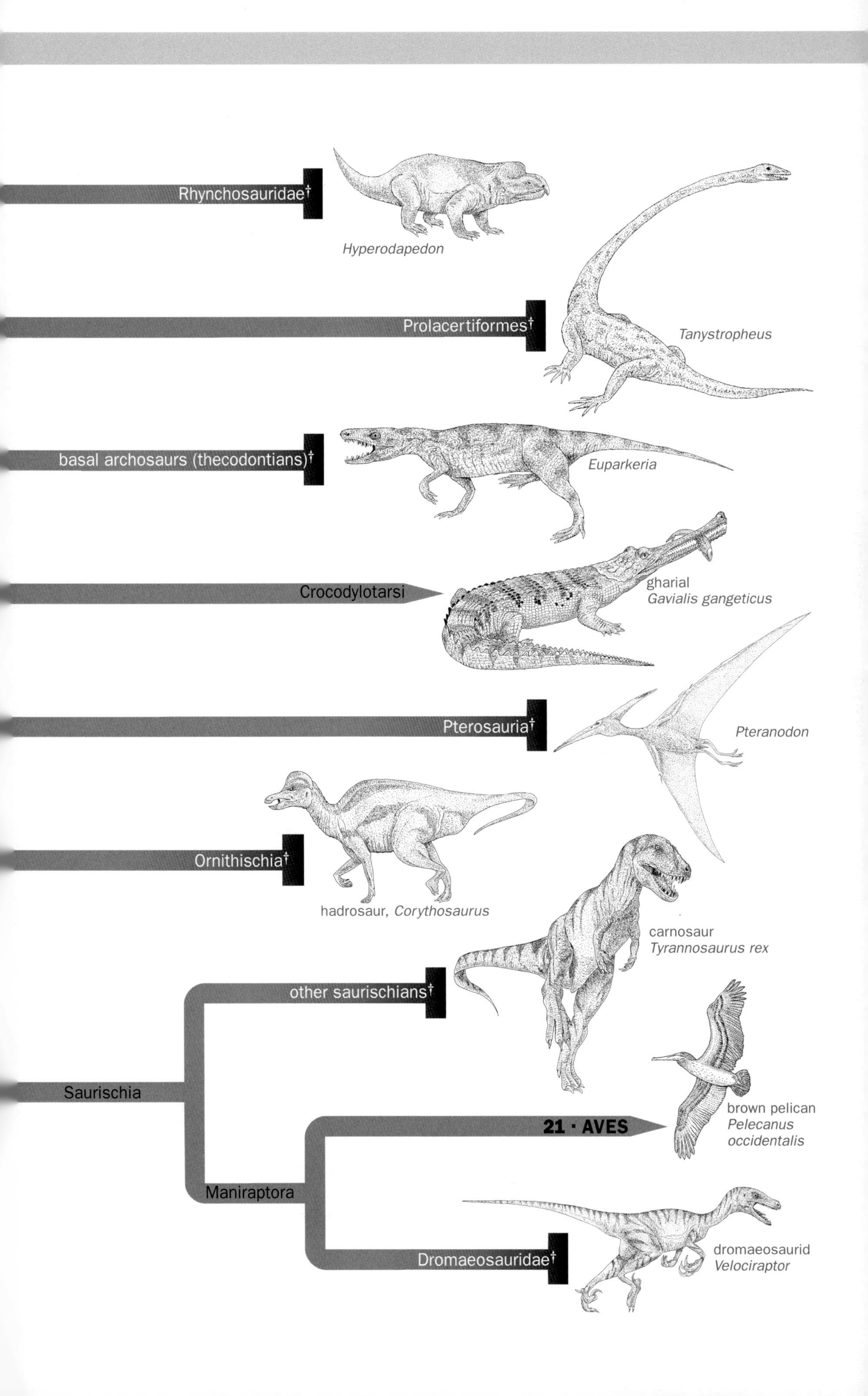

Rhynchosauridae†
Hyperodapedon
Prolacertiformes†
Tanystropheus
basal archosaurs (thecodontians)†
Euparkeria
Crocodylotarsi
gharial
Gavialis gangeticus
Pterosauria†
Pteranodon
Ornithischia†
hadrosaur, Corythosaurus
carnosaur
Tyrannosaurus rex
other saurischians†
Saurischia
21 · AVES
brown pelican
Pelecanus
occidentalis
Maniraptora
Dromaeosauridae†
dromaeosaurid
Velociraptor

Palaeozoic and the Mesozoic eras. The fossil record also suggests yet another mass extinction at the end of the Triassic, which may have wiped out the rhynchosaurs. The most famous of the mass extinctions was at the end of the Cretaceous, some 65 million years ago: that put paid to the dinosaurs.

THE SYNAPSIDA

The Synapsida as a whole include the **Mammalia**. If we want to refer to the synapsids of reptilian grade, then it seems to me reasonable simply to call them 'synapsid reptiles'. Some vertebrate zoologists object to this expression, but I cannot see why. Synapsid defines the clade, and 'reptile' defines the grade. What could be simpler? As the tree shows, the clade Synapsida divides into two subclades: the **Pelycosauria*** and the **Therapsida**.

Sail-backs and their relatives: the Pelycosauria*†

The six families of the pelycosaurs were really the first reptiles to make a serious ecological impact. They emerged in the late Carboniferous and dominated the early Permian and between them account for 70 per cent of all reptilian genera known from that time. At least three groups of pelycosaurs—apparently independently—developed remarkable 'sails' on their backs, formed from skin and supported by vast extensions of the neural spines of the vertebrae. The function of these sails is uncertain but they apparently had a rich blood supply, and probably helped to regulate body temperature—extremely efficiently, so simple physics suggests. *Edaphosaurus* was a herbivorous sail-back reptile; *Dimetrodon*—shown on the tree—was a carnivore. But most pelycosaurs were sail-less. Pelycosauria* should be seen as a paraphyletic group, for the Therapsida, the mammal-like reptiles, arose from among their ranks in the late Permian.

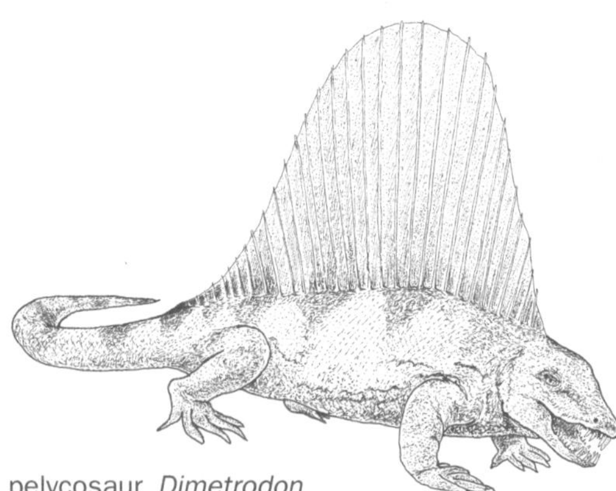

pelycosaur, *Dimetrodon*

Mammal-like reptiles: the Therapsida

Therapsida is a clade and it includes the mammals—the only remaining members of the clade Synapsida (therapsids of reptilian grade lasted only until the Triassic). The Therapsida split into several subgroups of which just two are shown on the tree: the **Dicynodontia**, and the **Cynodontia**. The dicynodonts radiated into more than 70 known genera, some of which were leading carnivores while others were the dominant herbivores of the late Permian. Some were large, 3 metres and more. *Kannemeyria*

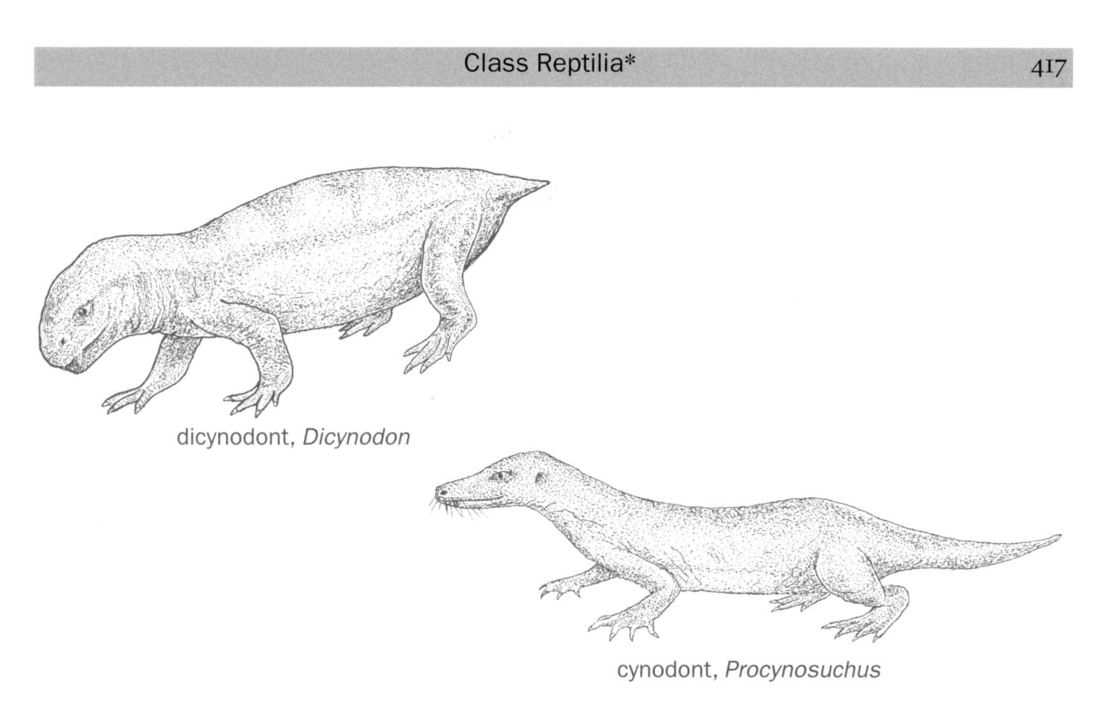

dicynodont, *Dicynodon*

cynodont, *Procynosuchus*

had a narrow-pointed snout and a crest on its head, vast ribs and heavy limbs. The gorgonopsans were formidable carnivores with skulls a metre long. One of them, *Arctognathus*, had huge canines like a sabre-tooth cat and could open its jaws to 90 degrees.

The clade of the Cynodontia includes the mammals, cynodonts of reptilian grade becoming more and more mammal-like as the Permian gave way to the Triassic. Specifically, cynodonts began to show the widely flaring **zygomatic arches** of the skull ('cheek bones') that accommodate large jaw muscles; enlargement of the **dentary bone** of the lower jaw, which characteristically remains as the only bone in the mammalian lower jaw; and the step-by-step development of the typical mammalian **secondary palate**. The *Procynosuchus* from the late Permian of South Africa is a typical early cynodont; some say it was an insectivore whereas others feel it was an otter-like fish-eater. From such beginnings, creatures that are well on the way to being mammals had emerged by the mid-Triassic, around 235 million years ago. Note, therefore, how astonishingly ancient the mammals are. They did not truly flourish until after the dinosaurs disappeared, 65 million years ago; but they made their first appearance *before* the first dinosaurs!

So much, then, for the Synapsida: an early emerging group with some extremely bizarre representatives, including the sail-backs; but the clade nonetheless that includes our own species. The second great lineage of Reptilia* is the Sauropsida, which in turn splits into two subclades: the Anapsida and Diapsida.

THE ANAPSIDA

Tortoises, turtles, and terrapins are the only living anapsid reptiles; they form the clade of the **Testudines**, or Chelonia.

Tortoises, turtles and terrapins: order Testudines (or Chelonia)

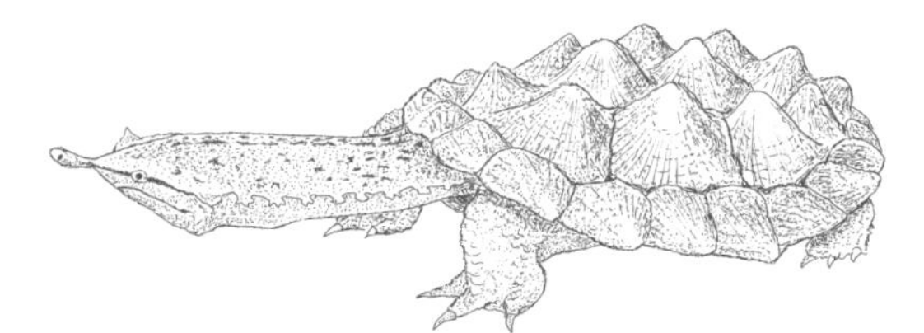
pleurodiran (matamata), *Chelus fimbriatus*

cryptodiran (gopher tortoise), *Gophus polyphemus*

The first testudines, dating from the late Triassic, could not withdraw their heads into their shells; but the later types, from the Jurassic to the present time, are able to do so. The more primitive types, in the suborder **Pleurodira**, kink their necks sideways to withdraw their heads—as do the living snake-necks and matamatas. Those in the **Cryptodira** bunch their necks in a vertical 'S', as do most of the familiar tortoises. Chelonians are still a remarkably successful and widespread group, even though some of the giant island types have suffered greatly in recent centuries from the spread of human beings and their domestic animals, including some giant tortoises from Madagascar that are now extinct. The biggest of all non-marine turtles was *Stupendemys* from Pliocene Venezuela, whose upper shell (carapace) was 2.2 metres long. Perhaps the most bizarre were the horned tortoises, mainly from Pleistocene Australia, whose spiky skulls were around half a metre across.

Mesosaurs and their allies: the 'parareptiles'†

The Anapsida also includes a group of about half a dozen families that were once assumed to be polyphyletic and were bunched together more or less purely for convenience as 'parareptiles'. Among them are the **Mesosauria**[1] from the early Permian of South America and South Africa, which were the first reptiles truly to return to the water. They were only about a metre long, but had long bodies and necks and a very long flat-sided tail that was obviously used in swimming; and long thin jaws with interlocking teeth that seem to have been used as strainers, like the teeth of a

mesosaur, *Mesosaurus*

[1] See footnote on page 409.

modern crab-eater seal or the baleen of a whale. The point, though, is that modern work shows the relationship of mesosaurs and the other 'parareptiles' to each other and to the Testudines (Chelonia), and so the Anapsida as a whole has emerged as a true clade. As already noted, the character that gives the group its name—the anapsid skull—cannot define the group because it is merely a primitive reptilian feature.

THE DIAPSIDA

The largest and phylogenetically most intricate reptilian clade is the Diapsida. It includes all the living reptiles apart from the tortoises and turtles, plus dinosaurs, which in turn gave rise to the birds, plus the flying reptiles, plus the ancient and once-dominant herbivores known as the rhynchosaurs, plus the ichthyosaurs, plesiosaurs, and some other groups of aquatic reptiles whose skulls have the derived, euryapsid form, plus a miscellany of long-gone stem types and other phylogenetic twiglets that are too numerous and recondite to include here. In the beginning—late Carboniferous—the fossil record shows a few stem creatures that were clearly setting the diapsid trend. By the Permian we can discern two clear subclades: the Younginiformes and the Neodiapsida.

Palaeontologists have shuffled the **Younginiformes** around the reptilian tree in recent years, presumably because their appearance is so generalized and lizard-like, and they are difficult to place. Until recently they were regarded as the sister group of the Lepidosauria, the group that includes the modern lizards and the sphenodonts. But a lizard-like form is merely primitive and now younginiforms are thought to be the sister group of all the remaining Diapsida. They are certainly ancient enough for such a role, for they emerged in the Permian and indeed were wiped out by the mass extinction that ended the Permian and ushered in the Triassic. Some younginiforms were terrestrial, and some were aquatic. *Youngina* was typical: a late Permian lizard-like insectivore and carnivore about 30–40 centimetres long, with a short neck and long limbs.

younginiform
Youngina

It is the other diapsid branch, the **Neodiapsida**, that really carries the action. As the tree shows, it again divides into two great clades: the **Lepidosauromorpha** and the **Archosauromorpha**. The Lepidosauromorpha in turn divides into three clades whose exact relationship is unknown, so they must be shown as an unresolved trichotomy: the **Ichthyosauria**; the **Sauropterygia**, which includes the plesiosaurs; and the **Lepidosauria**.

THE LEPIDOSAUROMORPHA

Four great lineages of neodiapsids are aquatic, which mainly means marine. The Ichthyosauria form one clade by themselves, while the other three groups—Placodontia, Nothosauria, and Plesiosauria—together form a second clade, the Sauropterygia. All of them have euryapsid skulls, although they are now considered to be a derived subgrouping of the Diapsida, the lower of the original two temporal fenestrae having been lost. The euryapsid character seems to have arisen independently at least twice and perhaps three times: in the placodonts, the ichthyosaurs, and the nothosaurs + plesiosaurs, who do seem to share a common ancestor and so form a clade together.

Ichthyosaurs: the Ichthyosauria†

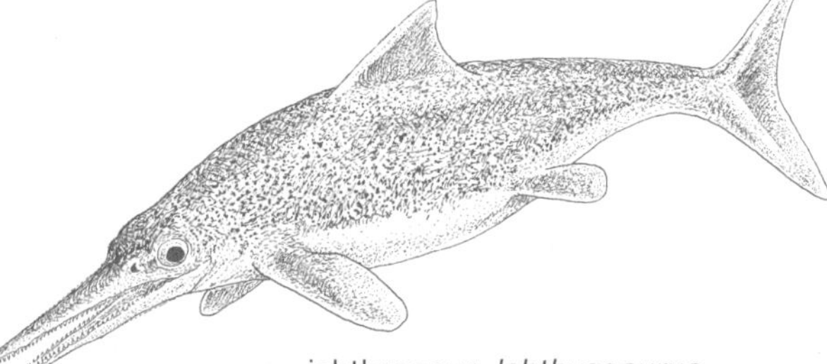

ichthyosaur, *Ichthyosaurus*

The ichthyosaurs were wonderful creatures: their name means 'fish-lizard' but in truth they were like reptilian porpoises. They were supremely streamlined—no neck, paddle-limbs, and a heterocercal tail like a fish's. They also had big eyes, reinforced with bone to resist the pressure of deep water, and nostrils placed well back from the tip of the snout. Peg-like teeth reveal they had a fishy diet. They look as if they should have succeeded and indeed they did, remaining very similar from the time of their first appearance in the early Triassic, to the end of the Mesozoic; but they seemed to reach their pinnacle of size—around 15 metres—in the late Triassic. They bore live young: several famous fossils seem to represent mother and baby in the moment of birth. The babies apparently were born tail-first, like a dolphin's, which is an obvious precaution in an aquatic animal that breathes air.

Placodonts, nothosaurs, and plesiosaurs: the Sauropterygia†

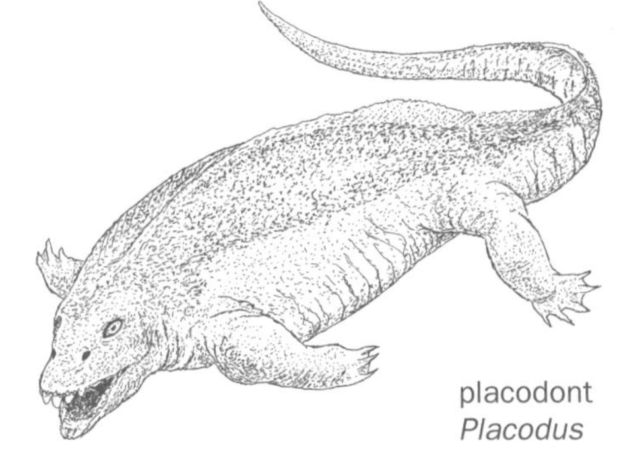

placodont
Placodus

The big heavy bodies of the **Placodontia** hardly look built for the water at all but their remains are prominent in shallow marine beds from the middle Triassic of Central Europe. The plate-like teeth on their palates, from which they take their name, suggest a diet of molluscs, which they presumably prised loose with their spatula-shaped (spatulate) incisors.

nothosaur, *Nothosaurus*

The **Nothosauria** were elongated animals with small heads, long necks, long tails that they used for propulsion, and paddle-like limbs. Their peg-like teeth again suggest a fishy diet. They ranged in size from about 20 centimetres up to 4 metres and are best known from the middle Triassic of Central Europe. *Nothosaurus* was a typical representative.

The very first **Plesiosauria** appeared at the end of the Triassic, and seem to be descendants of the nothosaurs. They are generally larger than nothosaurs (typically 2–14 metres long) and drove themselves along with powerful paddle-like limbs that attached to heavily reinforced limb girdles. The swimming action seems to have generated lift and, therefore, hydrodynamically speaking, would have been a kind of 'underwater flying', comparable with that of modern-day marine turtles or penguins. There were several discrete groups of plesiosaurs, some with long necks and some with short. It has often been suggested that the Loch Ness Monster is a plesiosaur, which would probably be the case in the highly unlikely event that it exists at all. The long-necked forms presumably swam slowly and perhaps hunted fish by ambush. The **Pliosauria** are a subgroup of the plesiosaurs that had long heavy skulls, usually with short necks, and presumably swam faster, so they could have run down their prey. Pliosaurs were up to 12 metres long and may have fed on other plesiosaurs and ichthyosaurs.

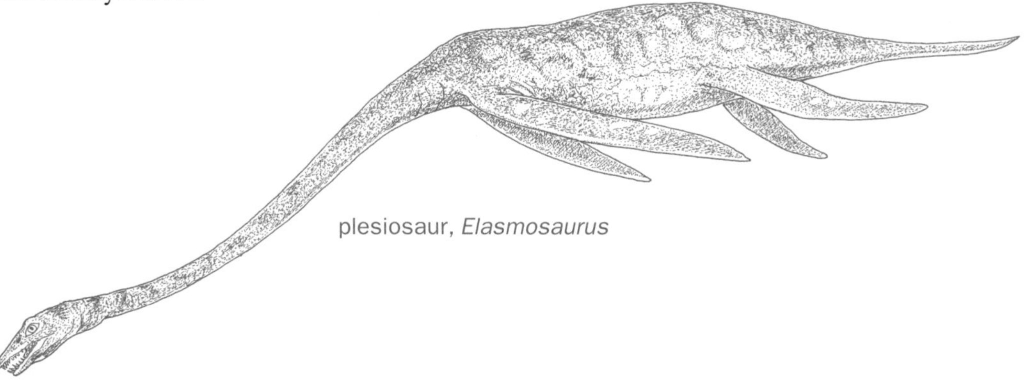
plesiosaur, *Elasmosaurus*

Superorder Lepidosauria

The third great branch of the Lepidosauromorpha are the Lepidosauria, many of which are still thriving as snakes, lizards, and sphenodonts (or tuataras). The lepidosaurs again divide neatly into two main groups: the **Sphenodontida** and the **Squamata**.

Tuataras and their relatives: order Sphenodontida

The sphenodonts are lizard-like in general form, but they lack the special lizard features including the peculiarly flexible skull. Neither do they shed their skin. Only a few are left to us: the tuataras which are now confined to a few New Zealand islands, having been swept from the mainland after human beings arrived in about the tenth century AD. But sphenodonts were once widespread; for example, their fossils have turned up in England.

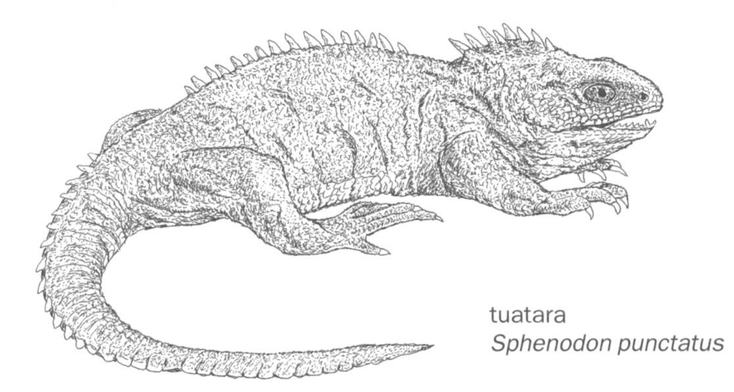

tuatara
Sphenodon punctatus

Snakes, lizards, and amphisbaenids: order Squamata

The order Squamata includes the snakes, the peculiar burrowing amphisbaenids, and the creatures that are commonly called lizards. Squamates as a whole have two outstanding characteristics. First, they shed their skin at intervals, to emerge clean and shiny.[1] Second, they have extraordinary jointed skulls and jaws that lend them both flexibility and strength; so they can on the one hand get their jaws around far larger prey than they otherwise could, and on the other can grip more powerfully and with more even pressure.

The modern groups are known first from the late Jurassic (about 160 million years ago) but the oldest known squamate fossils are mid-Jurassic in age—and we must assume that the order as a whole first appeared before then. The conventional division of squamates into snakes, lizards, and amphisbaenids does not do justice to the underlying phylogenetic reality. In truth, there are six subgroups, which are often ranked as infraorders. One, the Serpentes, includes the snakes; the Amphisbaenia are, of course, the amphisbaenids; while the other four—Gekkota, Iguania, Scincomorpha, and Anguimorpha—are all commonly called 'lizards'.

Traditional classifications generally combine the four lizard groups together in a suborder called Lacertilia, which is generally taken to include the amphisbaenids as well; but because the different groups of lizards are not necessarily related more

[1] Of course, other creatures shed their skins as well; indeed, most household dust is flakes of human skin. But most creatures constantly shed a little at a time. Squamates shed the whole at once, like a tight-fitting vest. Perhaps we should be grateful that we do not do this. The results would be eerie.

closely to each other than they are to the snakes, Lacertilia is a paraphyletic grouping (and the term 'Lacertilia' seems redundant). The relationship between the six infra-orders is not firmly established, but the Scincomorpha and the Anguimorpha seem to be most derived, and are sister groups; the Amphisbaenia and Serpentes may be closely related, and together emerge as the sister group of the Scincomorpha and Anguimorpha; the Gekkota are the sister group of those four together; and the Iguania are the sister group of all the rest. The sphenodonts are the sister group of the Squamata as a whole.

• Snakes, **Serpentes**, are known from the early Cretaceous. Constricting types appeared first, which kill their prey by suffocation. Venomous types are not known until the late Eocene. Snakes diversified mainly in the Tertiary period—the past 65 million years—so that their radiation ran parallel with that of the mammals, which form their main prey. Thus they demonstrate once more, lest there should be any doubt on the matter, that the groups we choose to recognize as 'dominant' do not have things all their own way. Like the birds, teleost fish, flowering plants, and insect groups such as the butterflies and hymenopterans, the snakes demonstrate that the Tertiary was a flourishing time for a great many groups, and was not simply 'The Age of Mammals'. The biggest living snakes are 6–7-metre-long constrictors, but the biggest of all lived in the Palaeocene of North Africa, and seemed to have reached about 9 metres.

western diamond-back, *Crotalus atrox*

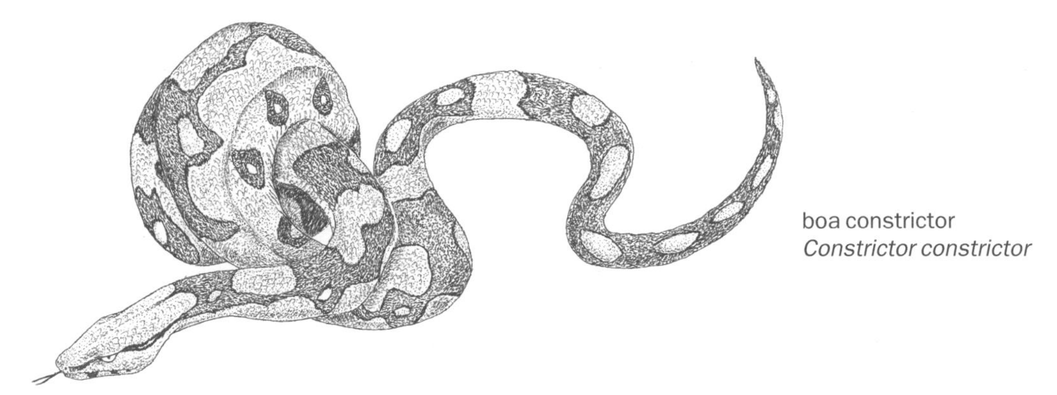

boa constrictor
Constrictor constrictor

amphisbaenid
Amphisbaena

• The **Amphisbaenia**, the amphisbaenids, are peculiar creatures adapted to burrowing—the reptilian equivalent of the amphibian caecilians. Many (though not all) have lost their legs while their heads are reduced to battering rams with which to drive through the soil. Like the caecilians they are little known because they are rarely seen, but about 140 species are known.

- The **Gekkota** are the gekkos, known for their remarkable ability to walk up walls or even across the ceiling. Many a home in Pakistan or India harbours a family of gekkos behind the pictures on the walls, dashing out at intervals to snatch inattentive insects.

Malayan house gekko
Gekko gecko

basilisk, *Basiliscus basiliscus*

- The **Iguania** include the iguanas of the New World (one of which is the astonishing basilisk, which can run on water), the agamid lizards of the Old World, and the chameleons.

- The **Scincomorpha** include the skinks and the lacertid lizards of Europe, like the creatures that run over the rocks in the Mediterranean, and Britain's common and sand lizards.

- The **Anguimorpha** are the most diverse of the lizards. Living forms include the monitors (family Varanidae), such as the modern Komodo dragon; the anguids, which are mostly limbless, like Europe's 'slow worm'; the venomous gila monster, and others. There are also three extinct families of marine anguimorphs in the late Cretaceous, of which the **mosasaurs** were the most spectacular. Mosasaurs were heavy-skulled fish eaters, ranging in length from 3 to 10 metres.

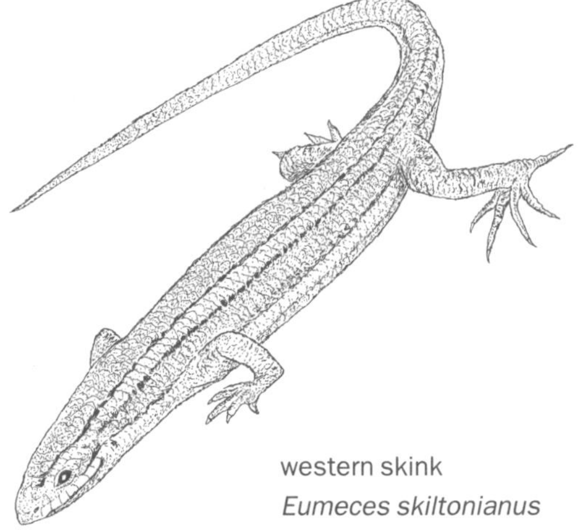

western skink
Eumeces skiltonianus

For sheer spectacle, grandeur, and long-lastingness, however, no group of creatures of any kind has ever approached the archosauromorphs.

THE ARCHOSAUROMORPHA

As a glance at the tree shows, the Archosauromorpha includes some of the most remarkable reptiles that have ever lived—the rhynchosaurs, dinosaurs, crocodilians, and the flying pterosaurs—and also gave rise to the birds, which are modified dinosaurs. A few other earlier types are worth mentioning as well, just to sketch in some of the outlying diversity.

The rhynchosaurs (family **Rhynchosauridae**) are shown here as the sister group of all the other archosauromorphs. They were the dominant herbivores of the Triassic, in many places accounting for more than half of all the skeletons found. They were bulky creatures with extraordinary skulls that looked triangular when viewed from the top and were wider at the base than they were long; and their lower jaw, opening and closing via a simple hinge without any to-and-fro motion, slotted into the upper jaw like the blade of a penknife, in what is known as the 'precision shear system'. Rhynchosaurs seemed adapted to eat tough vegetation, like 'seed- ferns'; and the hind feet of at least some of them, like *Hyperodapedon*, had huge claws as if for digging tubers. Rhynchosaurs did not survive the Triassic.

rhynchosaur, *Hyperodapedon*

The **Prolacertiformes** arose in the late Permian, radiated in the Triassic, and are worth mentioning for two reasons. First, they were extraordinary. At least, some were simply vaguely lizard-like but others developed long, stiff necks. In *Tanystropheus* of Central Europe the neck was more than twice as long as the trunk; and it must have been inflexible because it contained only 9–12 vertebrae, although long spines from those vertebrae suggest attachment of powerful muscles. How *Tanystropheus* lived is unclear, though its teeth suggest a diet of flesh and its limbs suggest a marine existence; so perhaps it lived in the shallows, darting its long neck after fish. The Prolacertiformes demand mention mainly as the sister group of the **Archosauria**, the most striking reptiles of all.

prolacertiform, *Tanystropheus*

THE GREAT CLADE OF THE ARCHOSAURIA

As the tree shows, most of the Archosauria divide neatly into two great lineages: the **Crocodylotarsi** on the one hand, and the **Ornithodira** on the other. But there is also a group of early archosaurs, or **basal archosaurs**, which have not yet assumed the characteristic features of the crocodylotarsans or the ornithodirans. Pre-cladistic palaeontologists commonly grouped these basal archosaurs together into the group Thecodontia. But the 'thecodontians' simply represent an early radiation of primitive archosaurs—a loose assemblage of creatures, one of whose members gave rise to the crocodylotarsians, and another to the first ornithodiran. I like traditional terms, but in this case it seems safest (least confusing) to drop the formal 'Thecodontia*' (a paraphyletic grouping) and also the adjective, 'thecodontian', that goes with it. 'Basal archosaur' is a neutral term that describes the creatures in question precisely: the first, primitive archosaurs, including the ancestors of the Crocodylotarsi and the Ornithodira.

The basal archosaurs†

The Triassic was dominated mainly by rhynchosaurs and synapsids but the primitive, basal archosaurs (previously known as **thecodontians**) were also a significant presence. Indeed, as the Triassic wore on, basal archosaurs took over the carnivorous niches of the synapsids.

Most basal archosaurs were small by archosaur standards. *Proterosuchus* from South Africa was typical: a slender beast, about 1.5 metres long, whose long teeth showed it to be an efficient predator, presumably of small synapsids. Like the synapsids that were its contemporaries, and like modern salamanders and lizards, *Proterosuchus* clearly had a sprawling gait, with upper arms and thighs held outwards,

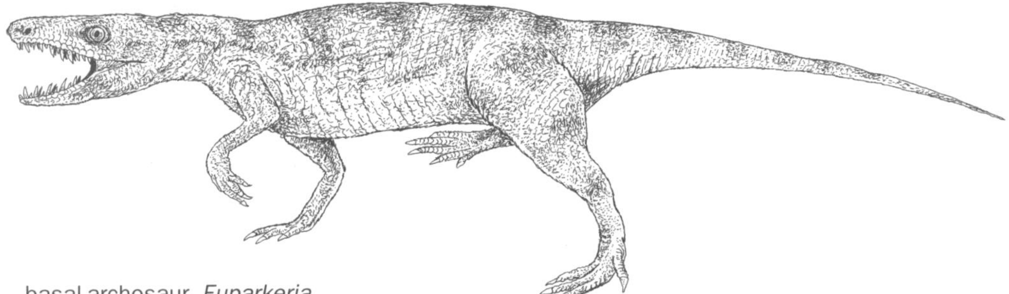

basal archosaur, *Euparkeria*

and knees permanently flexed. But it had typical archosaur characteristics: flattened, rather than rounded teeth; an extra opening (fenestra) in the skull between nostril and eye socket; and an additional knob-like protrusion of the femur called the fourth trochanter. So *Proterosuchus* was not an exciting creature in itself, but it was among the first of a distinguished lineage. *Euparkeria*, a better-known basal archosaur, was clearly more advanced. It was only 0.5 metres long but it had long limbs with flexible ankles and may have been able to walk on all fours or, when it chose, bipedally. The teeth of *Euparkeria* were set individually in sockets as in all later archosaurs. It was a predator.

CROCODILES, ANCIENT AND MODERN: THE CROCODYLOTARSI

The living crocodilians all look much of a muchness: typically sprawling on river banks (and sometimes on the seashore), half in and half out of the water, feeding either on fish (as the gharials do) or grabbing passing mammals (as crocodiles tend to do). But the sprawl of the modern crocs is not a primitive feature. They can raise themselves on their albeit short legs when they choose—and their ancestry shows that their sprawl is a secondary adaptation to their present way of life.

The Crocodylotarsi arose in the Triassic and they radiated in the middle and late Triassic into many forms. Many of the early types were bipedal. The chief crocodylotarsan predators of the Triassic were the rauisuchians (family Rauisuchidae), which were sometimes up to 7 metres long, and clearly terrestrial. Their general shape was croc-like, with a long heavy tail, but their jaws were shorter—not obviously adapted for fish—and they apparently held their legs beneath them like pillars, in the manner of dinosaurs or, indeed, of modern mammals.

All the same, many early crocodylotarsans did essay the modern crocodilian form and way of life. *Parasuchus* of Triassic India, of the primitive family Phytosauridae, was distinctly gharial-like with its long jaws and low-slung body—but although its nostrils were raised to poke above the water as in a modern crocodile, they were near the eyes rather than the tip of the nose. Although *Parasuchus* was clearly a fish-eater, fossilized stomach contents show that it also ate land animals, such as rhynchosaurs and prolacertiforms, which it presumably dragged into the water like a modern croc. Although the first crocodylotarsans were predatory (the stem types of every new reptilian lineage seem invariably to be predators!), some herbivorous types evolved later. The **aetosaurs** (family Stagonolepididae) were the first known archosaur herbivores. Their body shape again was generally croc-like, but stocky; and instead of the long croc jaws with their gripping rows of teeth they had short heads and a snout that went up at the front like a shovel, presumably for digging roots.

But this broad crocodylotarsan lineage is now represented by just 22 species, which are arranged in three families: the Gavialidae, which are the fish-eating gavials (or gharials); the Crocodylidae, or crocodiles; and the Alligatoridae, or alligators. These three families are closely related, and are all placed within the suborder Eusuchia ('true crocodiles') within the order **Crocodylia**. The Crocodylia first arose in the Jurassic but the modern eusuchians are latecomers, which did not appear until the late Cretaceous. Unlike all earlier crocodylotarsans modern crocodylians have a full secondary palate, allowing them to feed and breathe at the same time as a

gharial
Gavialis gangeticus

mammal can do; very handy for creatures which, like crocodiles, live by dragging larger prey, like buffalo, into the water. But the crocodilian palate is not homologous with the mammalian palate. The eusuchians still have a significant presence nowadays but they have an even more glorious past, for in the late Cretaceous and the Tertiary they spread throughout Europe north to Sweden, and in North America up to Canada, and provided dozens of species in the tropics and subtropics. The crocodiles again demonstrate that the Tertiary does not belong exclusively to the mammals.

The other great group of archosaurs, sisters to the Crocodylotarsi, were the Ornithodira. These include some of the finest creatures that have ever lived: the pterosaurs, dinosaurs, and birds.

PTEROSAURS, DINOSAURS, AND BIRDS: THE ORNITHODIRA

The Ornithodira, the second main lineage of the archosaurs, divides deeply into two great clades: the **Pterosauria**, and the **Dinosauria**. Out of the Dinosauria emerged the birds, which are conventionally placed in the class **Aves**. Hence the Ornithodira contains three great ecomorphs: pterosaurs, dinosaurs, and birds.

Flying reptiles: Pterosauria†

The amniotes have invented true, powered flight three times—in bats, in birds, and in pterosaurs. Many others have invented gliding, including snakes, lizards, weigeltisaurs, various possums and squirrels, and so on. The pterosaurs first arose in the late Triassic and through the Jurassic most of them remained as small (crow-sized) but ecologically significant fish-eaters. But they truly diversified in the Cretaceous. Thus, *Rhamphorhynchus* had long, widely spaced teeth apparently adapted for piercing and holding fish. It may well have hunted like a modern skimmer, a relative of gulls and terns, which flies just above the surface with its elongated lower bill trailing in the water. Other types had lost all or most of their teeth but may have skimmed anyway, shovelling their prey straight into capacious throats. *Pterodaustro* had hundreds of flexible teeth in each jaw that equipped it to catch plankton. Others were more land-based. *Dimorphodon* had short teeth and was perhaps an insect eater. *Pteranodon* from Kansas was huge, with a wingspan of 5–8 metres, and a head equipped with a remarkable 'weather vane' at the back that gave it a total length of around 1.8 metres, longer than the body. *Quetzalcoatlus* of Texas was even bigger; indeed its estimated wingspan of

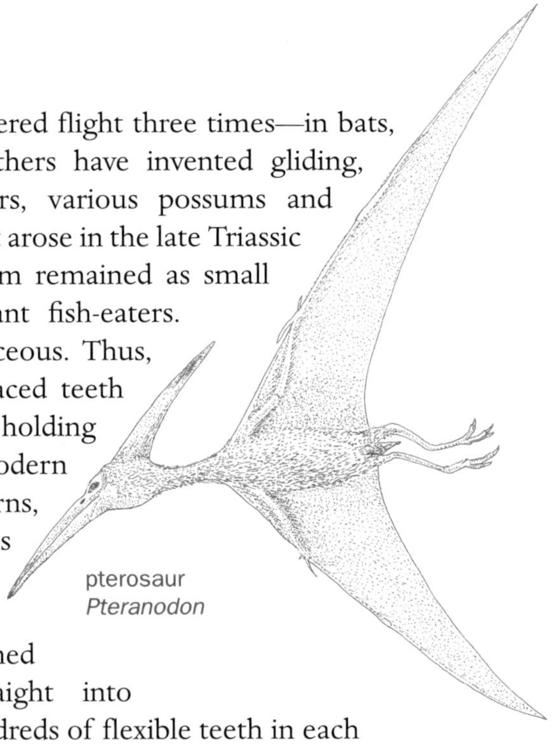

pterosaur
Pteranodon

11–15 metres makes it by far the biggest flying animal of all time; 30 per cent larger than a Mark I Spitfire, the great British fighter plane of World War II.

The wings of pterosaurs contrast intriguingly with those both of bats and of birds. In pterosaur wings a membrane extends from a vastly elongated fourth finger, probably back to the top of the pelvis, giving a long, narrow outline like a gull; and toes one, two, and three remained poking out and presumably functional (perhaps enabling the pterosaur to clamber through the trees) at the bend of the wing. Their bodies were short and the bones of the pelvis were fused, as in birds: strength being the premium. Whether pterosaurs scrambled about on all fours when they landed, as bats are inclined to do, or waddled or dashed along bipedally like birds, is still discussed. They do seem to have maintained a high body temperature, however, like bats and birds. At least, fossil imprints show that they had hair. It would have been good if at least one pterosaur had survived the mass extinction at the end of the Cretaceous. They were clearly fascinating creatures.

Even more stunning, however, were the Dinosauria.

Dinosaurs: superorder Dinosauria†

Many palaeontologists have argued that the dinosaurs were a polyphyletic group: a mixed bag of unrelated creatures whose resemblances were largely superficial. Michael Benton and others treat them as a clade, however, which is how I represent them here. The synapomorphies that unite them include a femur that articulates in a true ball and socket with the pelvis, with the ball standing out at an angle from the shaft of the thighbone as in mammals: an arrangement that brings the hind leg right under the animal so that, as in mammals, it supports the body weight like a pillar. This is a mark of true terrestriality. In many other reptiles (including lizards) the upper limbs are still sprawled out sideways from the body, and permanently bent at the elbows and knees.

For convenience we can accord the clade of Dinosauria the rank of superorder—for this enables us to give the two great dinosaurian subdivisions the rank of order. These orders are the **Saurischia** and the **Ornithischia**. 'Saurischia' means 'lizard-like pelvis': the pubic bone points forward, so the pelvis as a whole looks triangular when seen from the side. 'Ornithischia' means 'bird-like pelvis': the pubic bone is bent backwards, towards the tail, so that the pelvis seen from the side looks long and flat as in a modern bird. The resemblance is only superficial, however—for the class Aves arose from within the Saurischia.

'Lizard-hipped' or saurischian dinosaurs: order Saurischia†

Saurischian dinosaurs are grouped between two suborders, the Theropoda and the Sauropodomorpha. The **Theropoda** include various subgroups such as the basal **Ceratosauria**; for example, the Jurassic *Ceratosaurus*—a heavy-tailed, bipedal creature, which, in overall shape, matches most people's image of a generalized dinosaur. The carnivorous **Carnosauria**, includes such luminaries as *Tyrannosaurus rex* and the

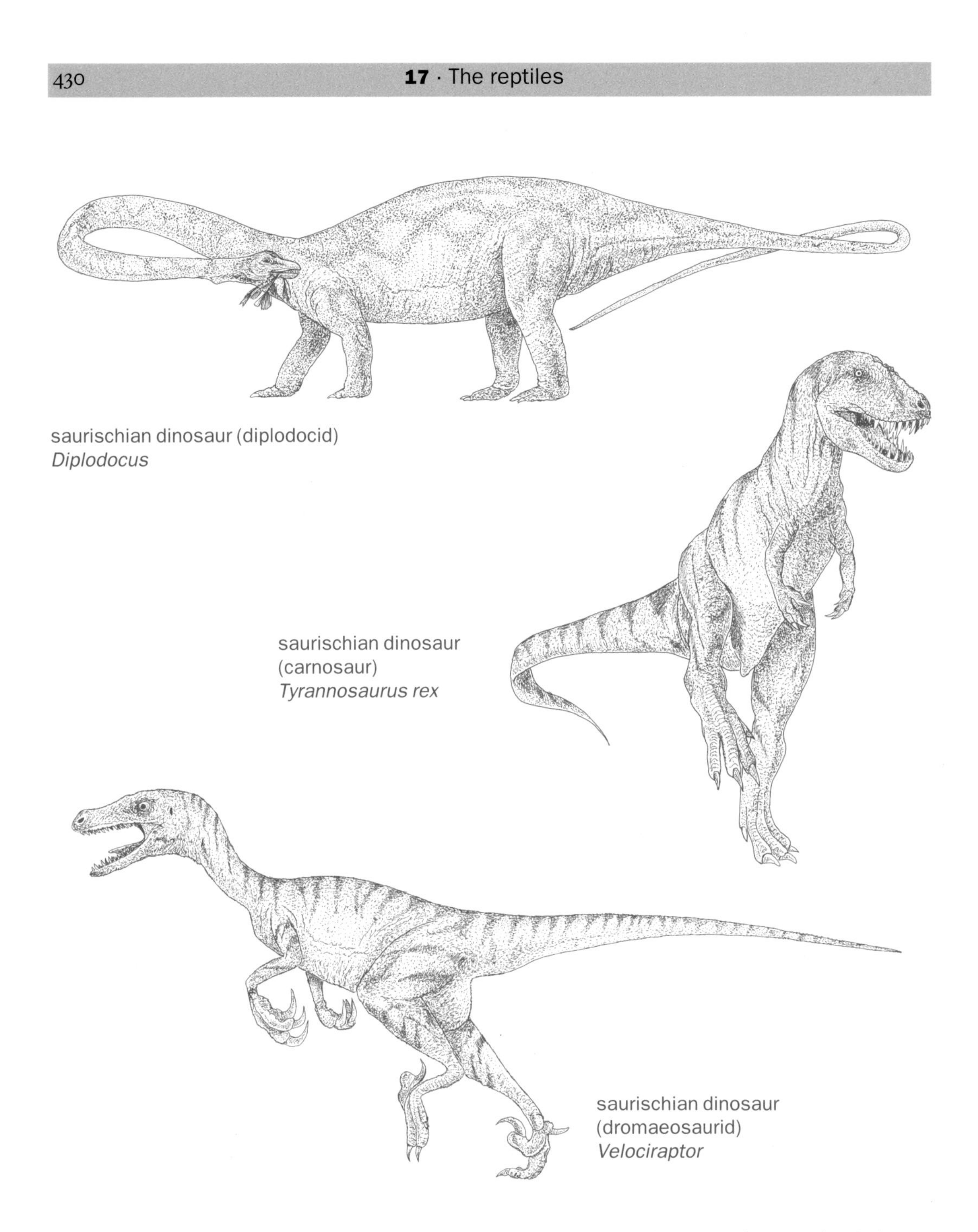

saurischian dinosaur (diplodocid)
Diplodocus

saurischian dinosaur
(carnosaur)
Tyrannosaurus rex

saurischian dinosaur
(dromaeosaurid)
Velociraptor

gracile, presumably swift-moving members of the Ornithomimidae. A third thero-pod grouping, the **Maniraptora**, includes the curious egg-eating oviraptors (which means 'egg-seizing') in the family Oviraptoridae. The maniraptors also include the **Dromaeosauridae**—the family that seems to be the sister group of the Aves. In short, the birds emerge from out of the middle of the dinosaur clade. If we apply Neolinnaean rankings we can see the birds as scions of the maniraptor infraorder of the saurischian order of the superorder Dinosauria.

The **Sauropodomorpha** includes the Prosauropoda and the Sauropoda—the most famous examples of the latter being the huge herbivorous brachiosaurs of the Brachiosauridae, and *Diplodocus* of the Diplodocidae.

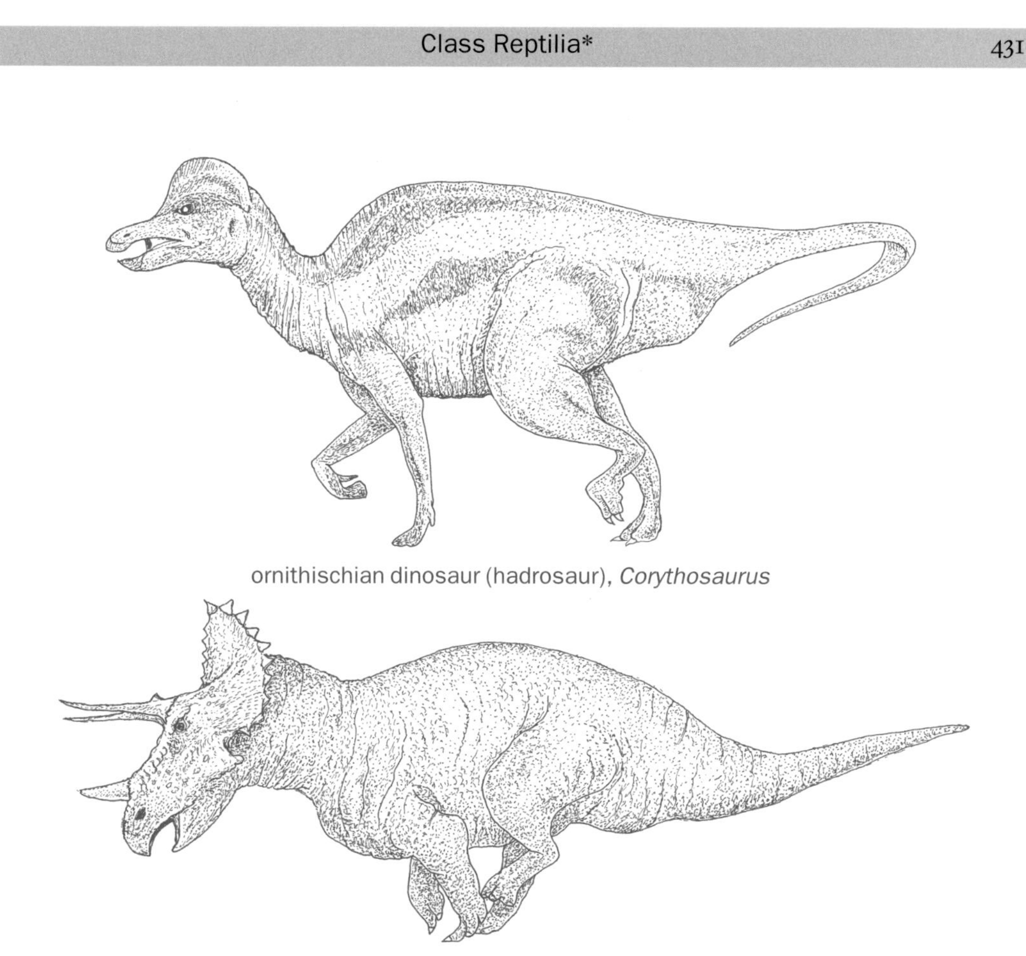

ornithischian dinosaur (hadrosaur), *Corythosaurus*

ornithischian dinosaur (ceratopsid), *Triceratops*

'Bird-hipped' or ornithischian dinosaurs: order Ornithischia†

The Ornithischia, those with the superficially bird-like pelvises, also include two major subgroups, the Cerapoda and the Thyreophora. The **Cerapoda** include the iguanodonts, such as the famous *Iguanodon*; the hadrosaurs, which are the 'duck-billed' dinosaurs; and the ceratopsids with their great horned faces, such as *Triceratops*. The **Thyreophora** include the ankylosaurs and the stegosaurs, with their great spiky tails and rows of vertical plates aligned along their backs—perhaps for defence and perhaps for temperature control.

It would, of course, be easy to devote an entire book to the dinosaurs, but many others have done this already. In a book of phylogeny we might simply note that, wondrous though they were, dinosaurs represent just one phyletic branchlet in the huge reptilian bush; but one that happened to succeed, more emphatically than all the rest put together, for 160 million years (225–65 million years ago). It is sad and salutary to reflect that the only archosaurs left to us are the birds and the crocodylians.

In the next five sections I look at the two new clades—and grades—of creature to which the reptiles gave rise: the mammals, which phylogenetically speaking are synapsids; and the birds, which are diapsids and, more specifically, emerged from among the ranks of the dinosaurs. First the mammals.

18

THE MAMMALS
CLASS MAMMALIA

MAMMALS ARE NOT tremendously speciose—G. E. Corbet and J. E. Hill list 4327 living species in the third edition of their classic *A World List of Mammalian Species* (1991)—but animals of large body size inevitably generate fewer species than small ones because the world cannot contain so many viable populations, and groups that are now extremely low in species are known to have included a great many more in the past. Thus there are only five species of living rhinoceros but the fossil record of the past 50 million years or so shows at least 200—and they include the biggest land mammal of all times, which was *Paraceratherium* (also called *Indricotherium* or *Baluchitherium*) from the Oligocene of Baluchistan (a region of Pakistan). Only two species of elephants are left to us but the fossil record reveals at least 150. The first mammals (or, at least, what most zoologists would call mammals) arose in the late Triassic, around 210 million years ago, and we may reasonably guess that over that time there have been at least 100 mammalian species for every one that exists today. In short, the total inventory of mammals that have ever lived is probably in excess of 400 000—a number that compares with the total catalogue of known, living beetles.

The 4300 or so living species, however, include many that invite superlatives. The baleen whales are the biggest animals that have ever lived—outweighing even the mightiest dinosaurs. No land animal has ever run faster than a cheetah; none has had more stamina than a dog or a horse. Several separate lineages of mammal have produced creatures of enormous behavioural versatility and inventiveness, suggesting great underlying intelligence: pigs, squirrels, porpoises, dogs, and many primates, including ourselves. Taken all in all mammals are by far the most various of vertebrates, in body form and in way of life. The standard shape for a small mammal is that of a shrew or rat, with body low to the ground and a leg at each corner. But the bigger quadrupeds have adopted a huge array of forms shown, for example, by leopards, rhinos, camels, deer, and elephants. And there is also a host of specialist creatures with specialist shapes—kangaroos, spider monkeys, human beings, whales, bats, and flying squirrels.

With such variety, behavioural versatility, and general size and strength, it is hardly surprising that mammals sit at or near the top of food chains in many and perhaps in most environments on Earth. To be sure, fish have a greater impact in the sea, birds still dominate the air, and reptiles are often the dominant forms on islands. But mammals, overall, have been the kingpins of the Cenozoic, setting the tone of ecosystems: the 'keystone' species that largely determined what else lived and survived.

Because mammals are now so successful, because they are so bright and lively, and because the fossil record shows that they began to come truly into their own after the dinosaurs disappeared around 65 million years ago, zoologists and folklorists have often misrepresented their history and their significance. Thus we have been told that mammals, being obviously superior, pushed aside dinosaurs—creatures that, so folklore also has it, were inveterate dullards whose prodigious bulk far outstripped their wanting intellects. If anything, however, the fossil record suggests the precise opposite. First, of course, it is becoming more and more evident that many dinosaurs were not dullards—that they often had complex social lives and took good care of their young, in the manner of modern birds. But, second, it seems that the lineage of synapsid reptiles that led to the mammals and the diapsid lineage that produced the dinosaurs started to radiate at much the same time; and that in so far as the two were in competition, the diapsid dinosaur lineage triumphed.

Thus the lineage of synapsids that led eventually to the mammals is clearly discernible in the late Permian, around 250 million years ago—well before the dinosaurs were apparent. By the Triassic, the weasel-to-dog-sized synapsids of the kind known as cynodonts had already become distinctly mammal-like and most palaeontologists feel that the morganucodonts (pronounced 'morgan-oo-co-donts') of the late Triassic should be classed as true mammals—though this was in an age when dinosaurs had yet to reach their full flowering, and still had more than 150 million years to run. Yet through all the time that mammals lived alongside dinosaurs they were ecological also-rans, rarely larger than a polecat and living what we may suppose was a marginal existence close to the ground or in trees, and probably largely at night. What killed the dinosaurs is still unknown, but a dramatic change of climate triggered by asteroids is the leading suggestion. Without the aid of such disaster the mammals might be skulking in the byways still.

In fact it seems more difficult to explain how the mammals managed to evolve in the first place—for they were the only members of the synapsid lineage to persist once the dinosaurs and their diapsid relatives had achieved such scope and dominance in the Mesozoic. My own suggestion (which I have not seen spelled out in quite this form) is that their initial and continued success lay in their **homoiothermy**. We need not assume that the earliest mammals were as positively or constantly 'warm-blooded' as dogs or people are today; perhaps their general body temperature was lower (as in modern marsupials such as kangaroos) and perhaps they allowed their temperature to fluctuate and often to cool (as do modern-day edentates such as sloths). But we may assume that they were able from an early stage to boost their temperature well above the ambient, not simply by running around, as 'poikilotherms' like reptiles or insects may do, but simply by burning energy for no other purpose than the supply of heat. Thus they were able to forage at night, and hence escaped competition from the generally diurnal diapsids. In the same way, it seems, modern bats are forced to be nocturnal because when they fly by day they are picked off by birds of prey (who themselves belong to the diapsid dinosaur lineage).

A great deal points to such a view. First, although we obviously know them only from their fossil bones it seems that even the late cynodonts had fur. Thus the skulls of Triassic cynodonts like *Thrinaxodon* had small pits around the snout, which some zoologists have suggested resemble those in modern mammals, and carry nerves and blood vessels to the whiskers. So, they suggest, *Thrinaxodon* and its kind might have had whiskers too—and whiskers are modified hair. The prime function of hair is to conserve heat. It is expensive to grow, and would serve little purpose if the body was not being warmed from the inside. This is speculation; but it makes sense.

Less-direct but even more conspicuous evidence for early warm-bloodedness is provided by the steady evolution of the lower jaw and teeth from the reptilian state of the cynodonts to that of the true mammals. Reptilian mandibles (lower jaws) are compounded from several bones, but those of mammals (after the middle Jurassic) include only one, the **dentary**;[1] a one-boned structure seems to be stronger. The lone dentary bone in the mammalian lower jaw has in turn evolved a new process, the condylar, to engage with the skull. Cynodonts and mammals have also evolved a new and powerful muscle on the outside of the dentary, the **masseter**, which closes the lower jaw. This muscle also allows the jaw to move sideways, as well as up and down like a simple hinge. The pattern of tooth wear suggests that cynodonts did not in practice move their jaws sideways, except perhaps the latest ones. But mammals do. At the same time the teeth of mammals are in general more specialized than those of reptiles, and their surfaces **occlude**—meaning they mesh together to make a highly efficient grinding surface, like the well-dressed stones of a flourmill. Perhaps because the teeth are so finely tuned to each other, mammals conventionally grow just two sets in their lives—the 'milk' or **deciduous teeth**, and then the adult set; whereas reptiles conventionally lose and replace teeth throughout their lives. Finally, as if for good measure, cynodonts evolved a full **secondary palate** between the mouth and the chamber of the nose. Originally this palate may have served simply to strengthen the tube-like muzzle; but, as a bonus, it enables mammals to chew and breathe at the same time, which reptiles that lack palates cannot do.

Taken all in all, then, the mammalian feeding apparatus is stronger than in reptiles, and formidably efficient; processing the food more thoroughly before swallowing. It needs to be efficient because mammals are homoiotherms and homoiotherms need a great deal more energy than poikilotherms (such that modern mammals typically need 10 times as much food as a modern reptile of equivalent weight). Thus the price of homoiothermy is high; but if it offers a route to survival—through feeding at night while the opposition chills out—then natural selection will favour it.

Much seems to follow from the notion that mammals first evolved as hairy,

[1] One of the bones of the ancestral reptilian jaw, the articular, now forms the **malleus** bone of the mammalian middle ear—one of three bones that conduct sound from the ear drum to the inner ear. Another mammalian middle-ear bone, the **incus**, derives from the quadrate bone in the reptilian upper jaw; while the third, the **stapes**, is the same in all tetrapods, including reptiles.

quasi-warm-blooded night feeders. First, night-time animals are liable to make a great deal of use of scent; and mammals are, on the whole, smelly. Skin glands—often serving as scent glands—are typical: they may be homologous (at least loosely) with the skin glands of living amphibians (and for the most part are lost in living reptiles and birds). Fur is a useful aid to scent: exudates spread on body hair waft into the wind more efficiently. Mammals are best known, indeed are identified as mammals, by their production of milk, by which they suckle their young. The most primitive living mammals, the monotreme platypus and echidnas, still lay eggs, as if to remind us of their reptilian ancestry; but they suckle their young nonetheless, even though they have not evolved nipples, and their babies must lick their nourishment from their mother's surface. Textbooks commonly suggest that milk glands evolved from sweat glands, but lactation is universal in mammals while sweating (at least for the general purpose of cooling) is not. I think it is much more likely that glands for milk and sweat both evolved from scent glands—and that mammalian smelliness evolved first. Modern humans seek to banish body smells but our ancestors presumably did not; and it seems to me no accident that adult body hair in humans sprouts from the areas that are most odoriferous, as if the early mammalian 'desire' to advertise by scent was with us still, despite our modern fastidiousness.

Finally, the postcranial skeletons of mammals show various special features. Thus, primitively, reptile limbs extend sideways from the body, as in the sprawl of a modern lizard, but mammalian limbs are positioned beneath the body; to enable this, the joints of the limb girdles have been re-angled and the joints of the limbs reshaped. Mammals thus carry the weight of their body directly over the legs, which in big animals offers great mechanical advantage. Furthermore, when a lizard runs it must bend its trunk from side to side, which makes it hard to fill the lungs efficiently, but a mammal needs no such contortion and can breathe perfectly well while running.

Whatever their origins, whatever the forces that drove their evolution, mammals in a world stripped of dinosaurs have proved famously successful: fliers, swimmers, runners, burrowers, and, above all, thinkers, mammals have produced them all in virtually every plausible habitat from pole to pole and from the skies to the deepest ocean beds. The tree shows all the living orders, and the principal lineages of those that are now extinct. It is based mainly on conversations with and papers by Michael Novacek, director of the American Museum of Natural History in New York; André Wyss of the University of California at Santa Barbara; and Christine Janis at Brown University, Rhode Island. I have also borrowed from John G. Fleagle's fine book, *Primate Adaptation and Evolution* (second edition, 1999).

A GUIDE TO THE MAMMALIA

In traditional classifications of mammals, including the seminal text of the great American palaeontologist George Gaylord Simpson (1902–84) of 1945, the class Mammalia was divided into two great subclasses, the Prototheria and the Theria.

The Prototheria contains just one living group, the order Monotremata, which nowadays is represented only by the duck-billed platypus and two species (in two genera) of echidna; creatures that lay eggs, and keep their new-hatched young in a pouch. The **Theria** contains two great groups, which are often ranked as 'infraclasses'. The first, the **Metatheria**, contains the single order of the Marsupialia; and, indeed, the metatherians might as well simply be called marsupials. The marsupials give birth to live young that are still at a very early stage of development, and many of them (but by no means all) keep their babies in a pouch. The second group of therians, the **Eutheria**, produce live young, which in general are more advanced at birth than those of marsupials and in some cases (like foals) are alert and active virtually from birth.

This traditional classification has now been much refined. The 'Theria' still stands up, and can be regarded as a subclass. In any case it is a true clade, for the marsupials and the eutherians are clearly related, and indeed there are fossils from the early Cretaceous (around 120 million years ago) of both Americas that make very plausible common ancestors. Thus we can say that the marsupials and the eutherians both came into being at around that time.

But the 'Prototheria' can no longer be seen as a formal taxon. At least eight distinct non-therian mammalian lineages are known from the Mesozoic. Of these, the monotremes are still with us, while the best known of the extinct groups are the morganucodonts and multituberculates. Most of the extinct lineages are known only from a few teeth and jaws—but these are sufficient to show both that they were indeed mammalian and that each was distinct from the others, although they must have looked much of a muchness for they all seem to have been small with the generalized mammalian shape that is commonly described as 'shrew-like'. Zoologists traditionally assumed that that all of the extinct types were egg-layers but some, like the Multituberculata, might well have invented viviparity (which may have arisen among mammals more than once, just as it has among reptiles). In any case, 'Prototheria' clearly is not a tidy clade—although 'prototherian' with a small 'p' can be used colloquially to mean 'a (loosely defined) mammal that is not a therian'.

Clearly, too, the bearing of live young is not one of the distinctive features of the mammalian group as a whole. Most mammalian lineages in the history of the world were probably egg-layers. The emphatically viviparous therians are just one lineage among eight or more—just as all modern birds represent just one among at least eight distinct lineages of Aves. To get a feel for mammaldom as a whole, we should look briefly at the non-therians.

THE MARGINS OF MAMMALDOM: THE NON-THERIANS

Right on the margins of mammaldom—some would like to include them within the Mammalia but some would not—are the morganucodonts, the **Morganucodonta**, which arose in the late Triassic in China and Europe and persisted in many parts of the world until the late Jurassic. The genus after whom the group is named is

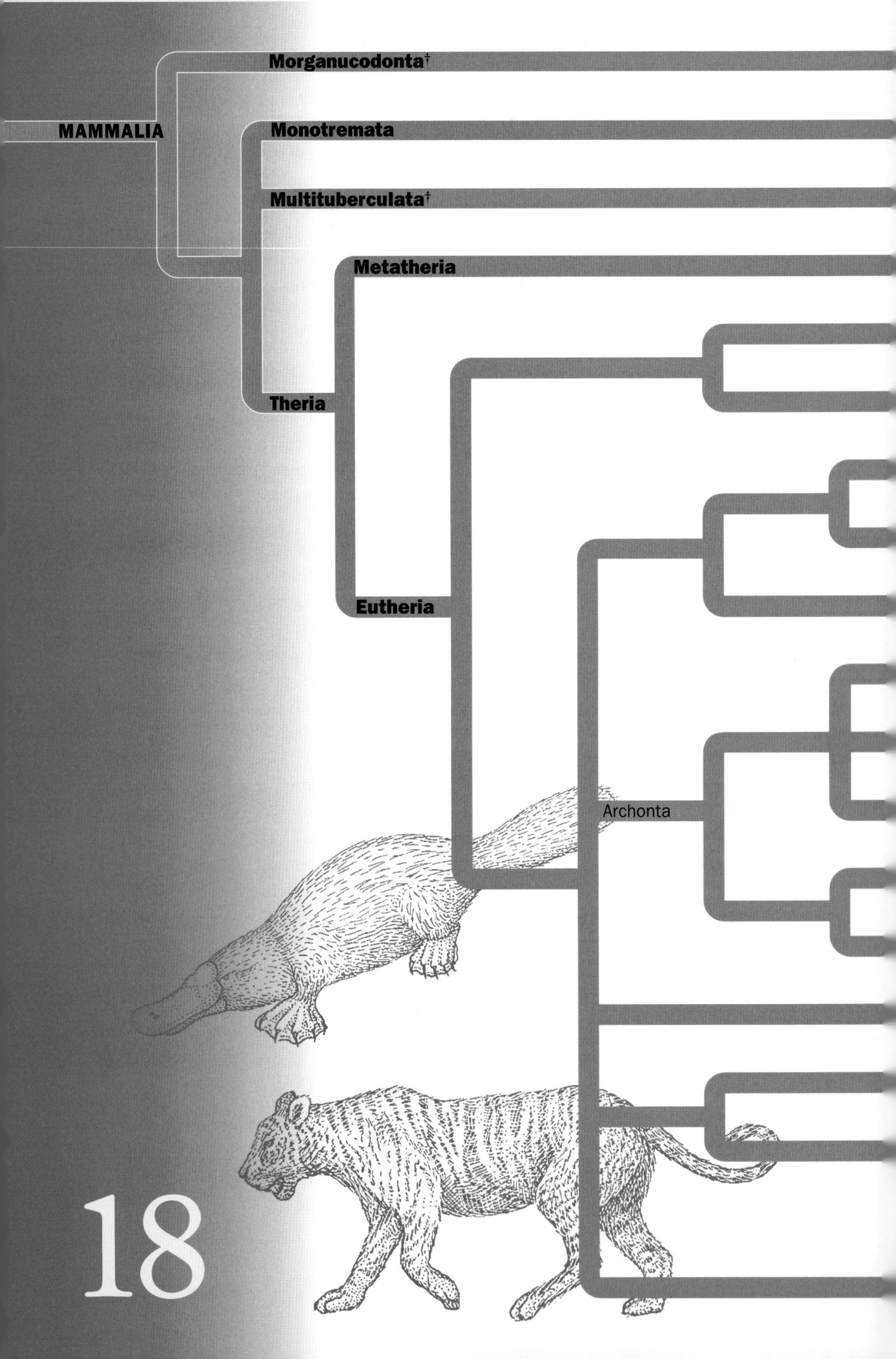
Morganucodonta†
MAMMALIA
Monotremata
Multituberculata†
Metatheria
Theria
Eutheria
Archonta
18

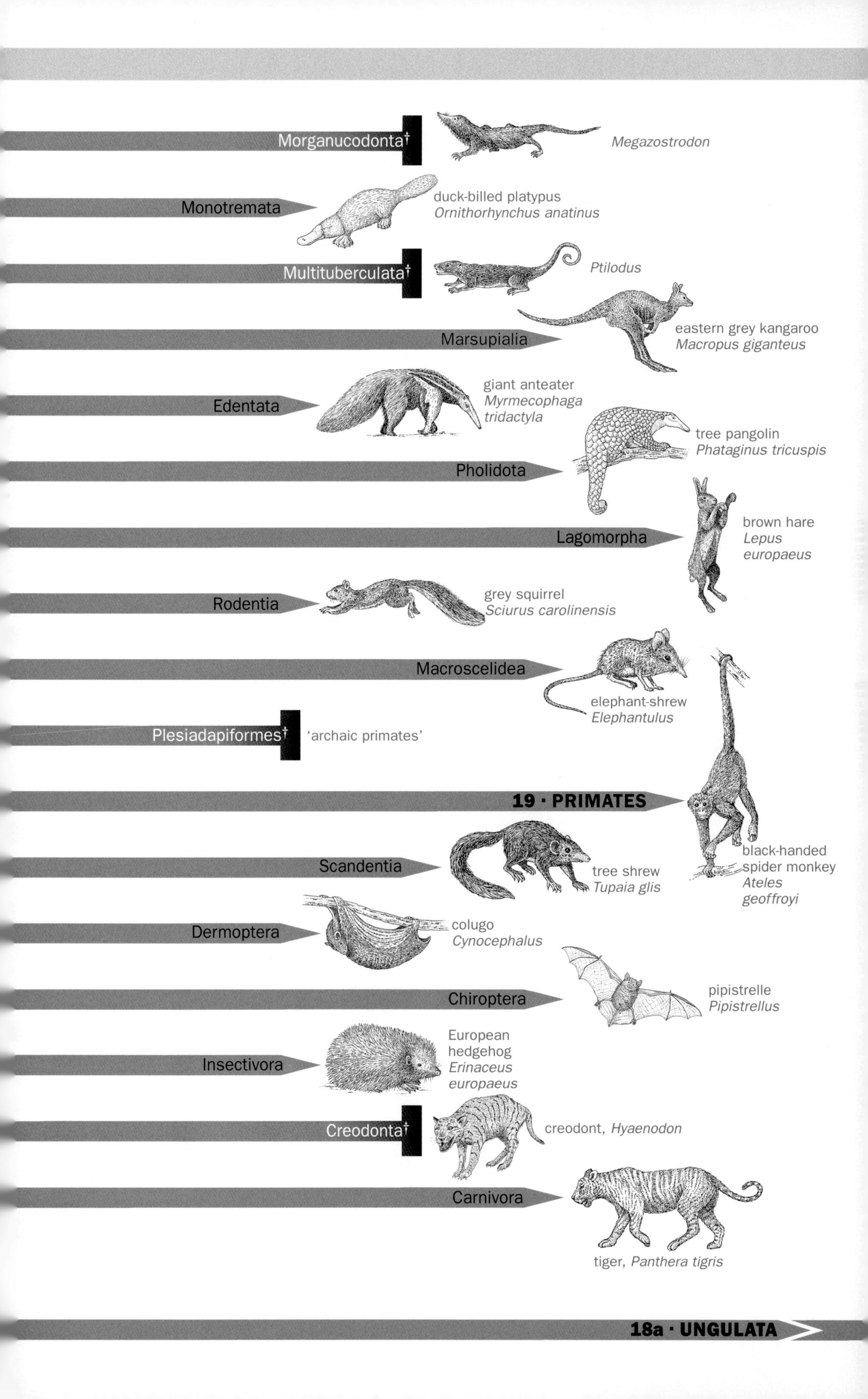

Morganucodonta†
Megazostrodon
Monotremata
duck-billed platypus
Ornithorhynchus anatinus
Multituberculata†
Ptilodus
Marsupialia
eastern grey kangaroo
Macropus giganteus
Edentata
giant anteater
Myrmecophaga tridactyla
Pholidota
tree pangolin
Phataginus tricuspis
Lagomorpha
brown hare
Lepus europaeus
Rodentia
grey squirrel
Sciurus carolinensis
Macroscelidea
elephant-shrew
Elephantulus
Plesiadapiformes†
'archaic primates'
19 · PRIMATES
black-handed spider monkey
Ateles geoffroyi
Scandentia
tree shrew
Tupaia glis
Dermoptera
colugo
Cynocephalus
Chiroptera
pipistrelle
Pipistrellus
Insectivora
European hedgehog
Erinaceus europaeus
Creodonta†
creodont, Hyaenodon
Carnivora
tiger, Panthera tigris
18a · UNGULATA

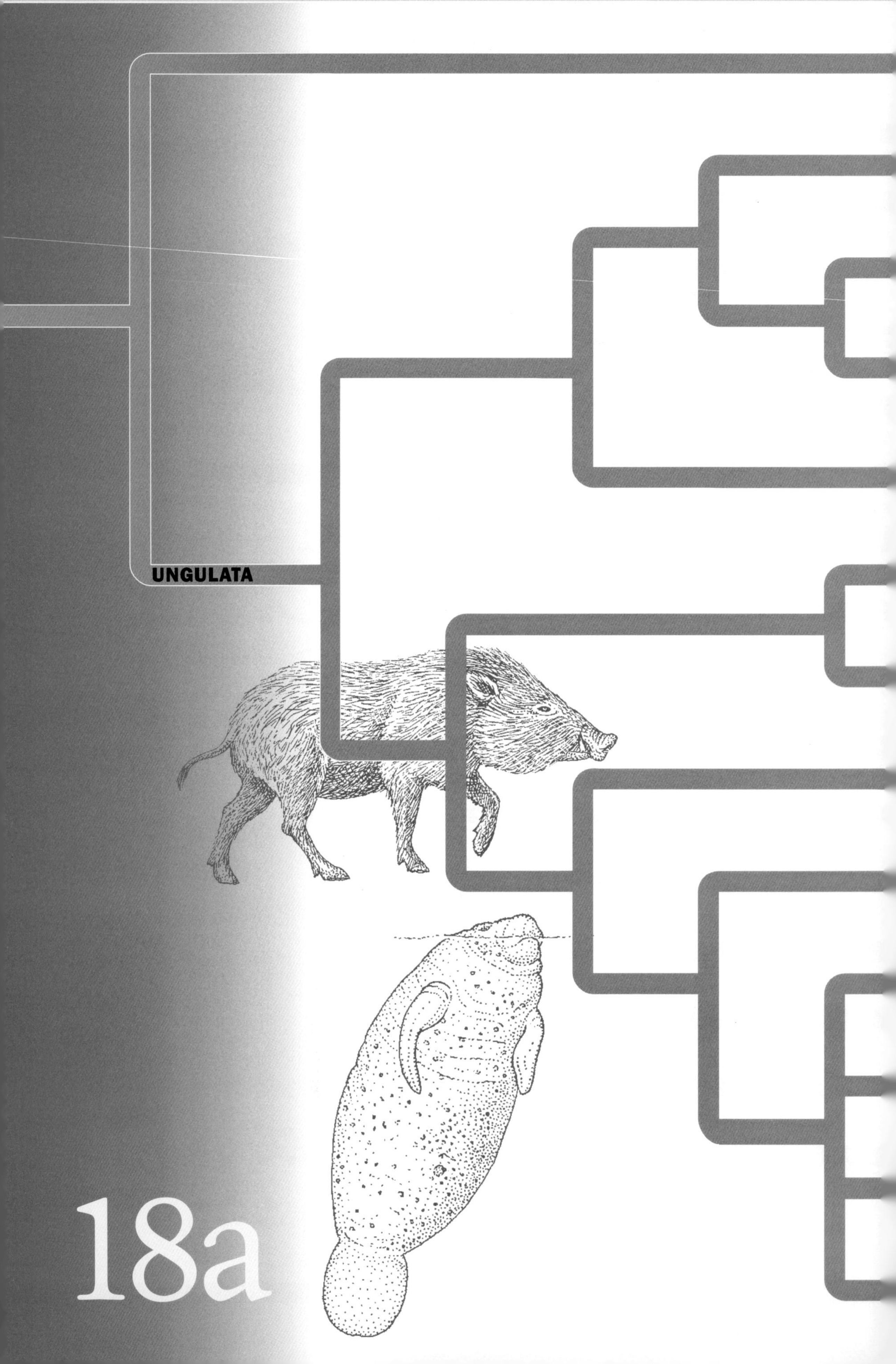
UNGULATA
18a

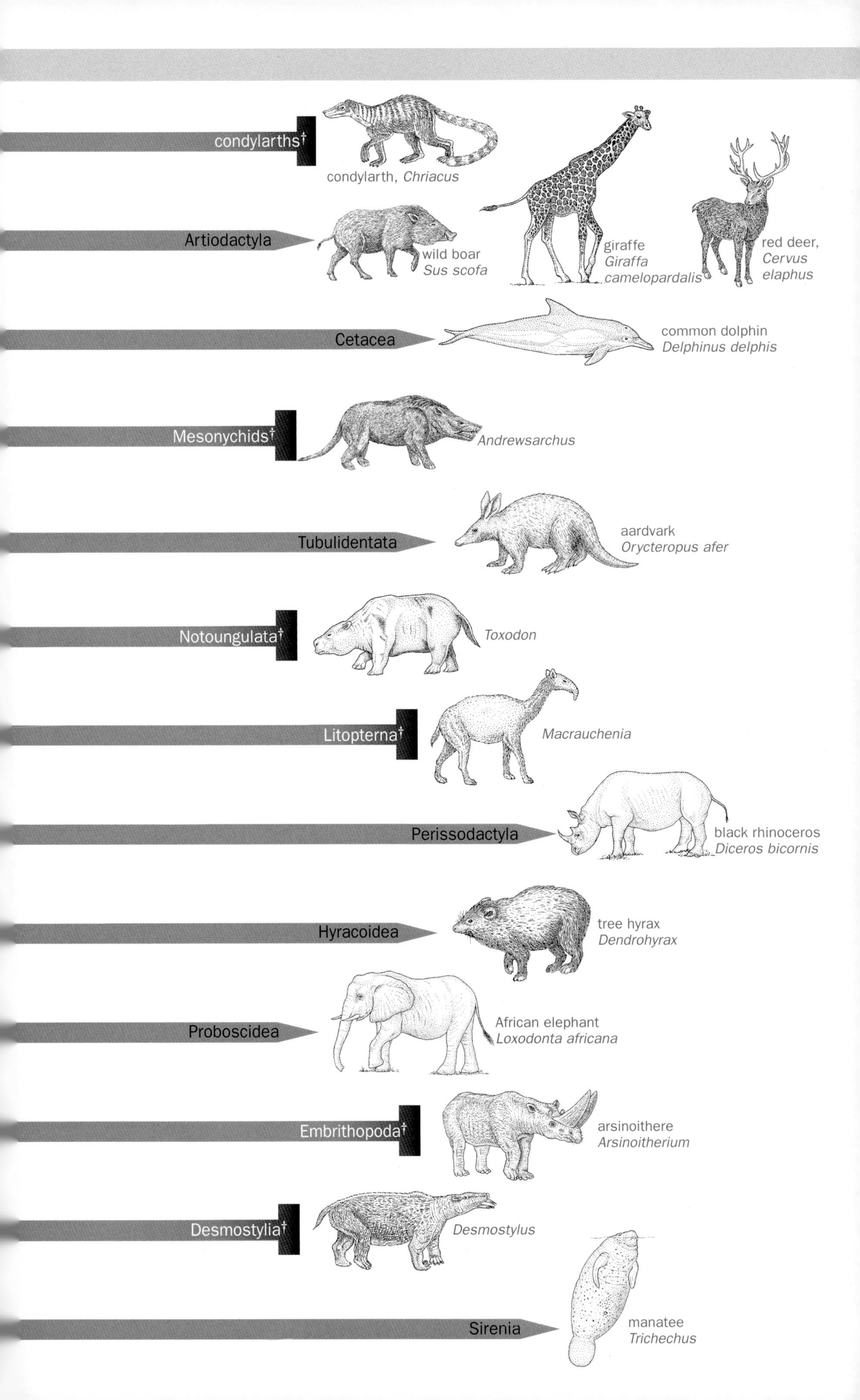
condylarths†
condylarth, *Chriacus*
Artiodactyla
wild boar
Sus scofa
giraffe
Giraffa camelopardalis
red deer,
Cervus elaphus
Cetacea
common dolphin
Delphinus delphis
Mesonychids†
Andrewsarchus
Tubulidentata
aardvark
Orycteropus afer
Notoungulata†
Toxodon
Litopterna†
Macrauchenia
Perissodactyla
black rhinoceros
Diceros bicornis
Hyracoidea
tree hyrax
Dendrohyrax
Proboscidea
African elephant
Loxodonta africana
Embrithopoda†
arsinoithere
Arsinoitherium
Desmostylia†
Desmostylus
Sirenia
manatee
Trichechus

Morganucodon (also known as *Eozostrodon*) and most of its skeleton apart from the head is unknown; but more is known about the skeleton of the related *Megazostrodon* from South Africa and it seems that both creatures moved and looked like true mammals—in fact, like modern shrews. But their skulls and teeth do have some distinct and presumably primitive features: for example, the post-dentary bones of the reptilian lower jaw are still attached to the dentary, and have not been integrated into the middle ear. But morganucodonts probably had hair and suckled their young, as all living mammals do.

morganucodont
Megazostrodon

The order **Monotremata** are still with us—the platypus (family Ornithorhynchidae) and two echidnas (Tachyglossidae) of Australasia—but they probably arose in the mid-Jurassic, although the oldest known fossil is a tooth from the Cretaceous. They have all the diagnostic mammalian features (once we put aside the idea that mammals are supposed to bear live young).

But the largest group of mammals in the Mesozoic were the **Multituberculata**, which survived until the late Eocene—well into the Cenozoic. They were generally rodent-like omnivores and by the standards of their times were quite variable. Thus *Ptilodus* was like a squirrel and, like a squirrel, could turn its hind feet backwards so that it could descend a tree head first. But—more like a kinkajou than a squirrel—it had a prehensile tail, for grasping branches. Multituberculates also had teeth with cusps (little prominences) in rows—sometimes long rows; the individual teeth, then, were like combs. Whether they laid eggs or bore live young is uncertain, but they had narrow pelvises, which suggests that they well have given birth to small, very immature young, as modern marsupials do.

monotreme (duck-billed platypus)
Ornithorhynchus anatinus

Neither is it clear whether the multituberculates were closer to the therians than the monotremes are, or were sisters to the monotremes + therians. The Polish palaeontologist Zofie Kielan-Jaworowska, who works on multituberculates, feels they were more primitive than monotremes because the cochlea of the inner ear is uncoiled, like a reptile's, whereas the monotreme cochlea has half a turn and the therian cochlea makes two-and-a-half turns. But evidence from the shoulder girdles places the multituberculates closer to the therians. Andy Wyss says he is prejudiced towards evidence from the

multituberculate, *Ptilodus*

head, which perhaps *needs* to be more conservative than the postcranial skeleton. Molecules cannot help in this discussion because the multituberculates are long gone. The tree accompanying this section shows monotremes, multituberculates, and therians as a trichotomy—the safe and sensible course when there is room for doubt.

Finally, some zoologists feel that because the mammals clearly grade fairly smoothly into the cynodont reptiles, it is very difficult to define the limits of the clade Mammalia. Accordingly some prefer to confine the term Mammalia to living mammals and refer to the extinct 'prototherians' merely as 'mammaliform'. If, however, the extinct multituberculates were closer to the therians than the monotremes are, then this would make the present-day Mammalia polyphyletic (because by definition it would then be supposed to exclude the multituberculates, even though they would be part of any clade that stretched from monotremes to therians). Other zoologists are content to define Mammalia as shown here, with the extinct prototherians, like the morganucodonts and multituberculates, snugly on the inside of the class.

Although 'prototherians' were the commonest mammals of the Mesozoic, the clade that has dominated more and more since the Cretaceous is that of the Theria. The therians have made the Cenozoic into 'The Age of Mammals'.

BEARERS OF LIVE YOUNG: THE THERIA

The term 'therian' alludes not to the bearing of live young *per se* but to the particular way of doing it: before birth the embryo is fed in the uterus, or womb, via a **placenta**. Marsupials and eutherians *both* have placentas; it is quite wrong to refer to the eutherians exclusively as 'the placental mammals', which implies that marsupials are non-placental. There is, however, a difference in emphasis between the two groups. In both, the young are raised first in the uterus, and then, after birth, by suckling. In both groups, too, birth is a somewhat movable feast—so that sometimes the fetus develops in the uterus until it is quite advanced (as in horses), which takes much of the burden away from the suckling phase; and sometimes the fetus is born while still very immature, so that lactation must provide almost all of the nutrition from conception to self-feeding. In general, though, the balance in marsupials is tipped heavily towards birth at an early stage, while in eutherians the placental stage typically contributes far more (although bears, for example, give birth to very immature babies).

To be sure, the embryo in some marsupials spends a long time in the womb, so that the pregnancy may seem to be prolonged. But in such marsupials (for example, kangaroos and wallabies) development in the womb simply enters a period of **diapause**, meaning that the young embryo stays quiescent, sometimes for months at a time. When growth in the uterus does finally resume, it commonly lasts only a few days; but during this brief phase, the embryo is fed via a placenta, just as in eutherian mammals (although in some marsupials the placenta is simpler in structure than the eutherian placenta). After birth, some marsupial species leave their young in nests or carry them on their backs, but most of the living types, like the kangaroos, wombats, and koalas, and the carnivorous Tasmanian devils and quolls, carry their babies constantly in pouches on the belly, where they are more or less permanently locked on to the teat or teats.

Although marsupials and eutherians are so different, they clearly shared a common ancestor. They form the therian lineage *together*, and so this is a true clade: there is much less phylogenetic difference between them than there is between, say, the morganucodonts and the multituberculates, even though the morganucodonts and multituberculates may look more similar (simply because their general body form is more generally primitive). Their common ancestor apparently lived around 120 million years ago, deep in the Cretaceous. This means, of course, that the marsupials and eutherians themselves both appeared at that time—and hence that they subsequently lived alongside the dinosaurs for more than 50 million years. Indeed several of the modern eutherian orders, including the rodents, lagomorphs (now represented by rabbits and hares), and the primates, as well as the primitive edentates, almost certainly appeared long before the dinosaurs disappeared. I like the notion that the world's first squirrel-like primates would have scolded the questing heads of Cretaceous dinosaur herbivores as they probed the canopy for succulent leaves.

MARSUPIALS: THE METATHERIA

All except one of the 282 species of extant marsupials (order **Marsupialia**) live only in Australia and New Guinea or in South America; and the other—one of the American opossums of the family Didelphidae—lives in North America. In fact, there are a few marsupials elsewhere, including possums in New Zealand and wombats on the Pennine hills of England, but both were taken to those places by human beings (that is, 'introduced').

Marsupials owe the peculiarities of their natural distribution to their own migrations and to the fact that continents do not stay still; indeed over time they have slithered athletically over the surface of the globe. Thus marsupials seem to have originated during the early Cretaceous in the land that nowadays forms North America—or that, at least, is where the oldest marsupial fossils have been found. But in the early Cretaceous all the land in the world was collected into two great continents: **Laurasia** to the north and **Gondwana** to the south. Present-day North America was simply a part of Laurasia, together with present-day Greenland and Eurasia; while Gondwana contained the land masses that now form South America, Africa, Madagascar, India, Australia, New Guinea, New Zealand, and Antarctica. Somehow or other—either via some transient landbridge that is long since gone, or by 'rafting', which means floating across on some mat of vegetation—marsupials managed to reach Gondwana. At least, the fossil record now seems to tell us that they reached Australia via South America and Antarctica. (There is an alternative theory —that marsupials reached Australia via Europe and Asia—but this now seems to be defunct. Some marsupial fossils have been found in Europe and Asia but these, together with some in Africa, seem to be North American types that came to Eurasia via Greenland during the Eocene.)

Gondwana had already begun to break up before the Cretaceous, way back in the Jurassic; although Australia did not finally break free from Antarctica until the Eocene, while South America remained attached to Antarctica until the early Miocene, around 25 million years ago. After the break up, Antarctica remained in the deep south (as deep as it is possible to be) while the other fragments drifted north. Africa made contact with Asia via the Middle East (and intermittently via Gibraltar, for sea level has sunk from time to time and left the Strait of Gibraltar high and dry) while India rammed emphatically into Asia, rumpling the land before it as it went to form the Himalayas. But Australia has remained an island—an 'island continent'—sublimely isolated for the past 40 million years. From time to time, however, Australia has been joined to New Guinea (they are on the same fragment of tectonic plate) and the two separated most recently less than 10 000 years ago, as the last Ice Age ended. South America remained as an island from about 25 million years ago until about 3.5 million years ago (the mid-Pliocene), when it finally made contact with North America via the isthmus of Panama. This meeting allowed the **Great Interchange**, when many American animals from the north flooded south—mainly to move towards the Equator at a time of global cooling—and somewhat fewer migrated from the south to the north.

Of the Gondwanan continents, only South America, Australia, and New Guinea have native marsupials (no fully terrestrial vertebrates survive on Antarctica). The two groups of marsupials on Australia and South America evolved in total isolation from each other and in virtual isolation from the rest of the world; for the two island continents have not been part of the same land mass since the Cretaceous. The marsupials of today are extremely variable, and many zoologists, including me, question whether they should really all be crammed into the same order; whether, in fact, this does not simply reflect northern chauvinism. Despite the obvious anatomical and ecological variety, however, molecular evidence suggests—unsurprisingly in view of their history—that the greatest division within the marsupials is between the South American and the Australian representatives. Incidentally, the one North American species, the lone opossum, is not simply a leftover descendant of the first-ever marsupial. It migrated into North America from South America some time after the two continents made contact in the Pliocene, and the oldest known opossum fossils in North America date from the Pleistocene. The Australian possums are quite different from the North American opossums, and not at all related. The possums belong to the phalanger family, the Phalangeridae.

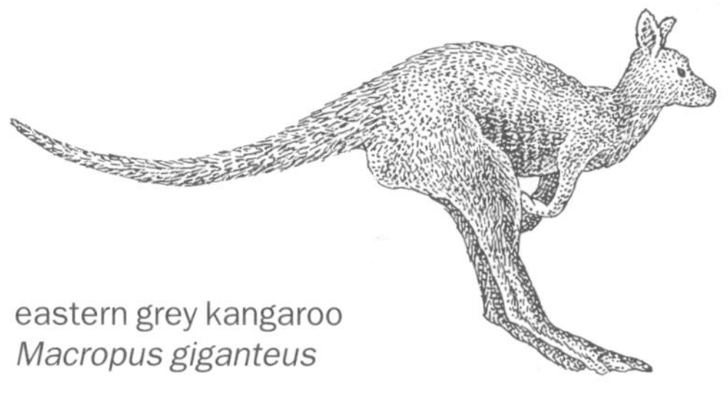
eastern grey kangaroo
Macropus giganteus

The extent to which living marsupials parallel the forms of eutherian mammals is extraordinary, and is reflected in the names that the Europeans first gave to them—'marsupial moles', 'marsupial mice', 'marsupial cats', and so on; the latter two, incidentally, both being placed in the same family, the Dasyuridae. At least until the

1930s, too, there was a dog-like form known as the thylacine wolf (now usually called simply the thylacine); and the yapok of the Andes is like an otter. In addition there are gliding forms (called gliders), some like huge guinea-pigs (wombats), and forms with no parallel in the eutherian world—like the remarkable koalas that somehow contrive to subsist on the fearsomely toxic leaves of eucalyptus and, above all, the macropods (kangaroos and wallabies) whose remarkably efficient leaping locomotion is unique among all large animals that have ever lived.

Yet there have been many other, quite different, and equally remarkable marsupial forms in the past: a one-toed, short-faced giant kangaroo nearly 3 metres high; the rhinoceros-sized diprotodonts of Australia, like giant wombats, which survived until long after the Aborigines arrived and perhaps gave rise to the bunyip myth; the so-called 'marsupial lion', *Thylacoleo*, which was the size of a leopard, had sabre-teeth fashioned not from its canines but from its incisors, remarkable, huge, shearing cheek teeth (molars), and, even more remarkable, prominently clawed thumbs; *Thylacosmilus* of South America, for all the world like a sabre-tooth cat; and the many and varied borhyaenids of South America, which were mostly small and ferret-like (as most carnivores are) though some were more like dogs and others like bears.

Marsupials deserve much more discussion but we should move on to the Eutherians.

THE EUTHERIA

No one doubts that among the eutherian mammals, the **Edentata**—sloths, armadillos, and anteaters—are different from the rest; and some, including Michael Novacek, regard the **Pholidota** (pangolins) as sisters to the rest. Accordingly, the present tree shows the edentates + pangolins as sisters to all the rest.

But relationships within and among the non-edentate eutherians are somewhat more difficult to pin down. At least—with a few exceptions as outlined below—it has not generally proved too difficult to decide which order each non-edentate eutherian should be placed in, and most zoologists would be happy with the ones shown on the tree, which includes all 16 living orders, and half a dozen of the orders that have already gone extinct. But it is not easy to decide the exact relationships between the different orders. The reason is the same as with modern birds: that most of the main orders of modern mammals arose within a few million years of each other in the Palaeocene, between 65 and 56 million years ago (soon after the dinosaurs disappeared), and it is difficult to work out the order of evolutionary events that took place so long ago but followed hard upon each others' heels. Furthermore, the modern molecular investigations that should be shedding light on the problems thrown up by traditional anatomy are doing so only in part: sometimes they support traditional ideas, and sometimes, at least for the time being, they seem to be adding to the confusion.

The tree does to some extent show relationships between different orders, although many of the relationships presented here are controversial; for example, many doubt whether rodents and lagomorphs (rabbits and hares) are really closely related, or have simply come to resemble each other through convergence. But, as you can see, all the 24 non-edentate eutherian orders represented on the tree are grouped into five lineages, which are shown forming a five-pronged polychotomy. In other words, the tree makes no statement about the order in which the five grand lineages emerged; but it does assert that the big five all shared a common ancestor, which was a sister to the edentates + pangolins; and also hypothesizes that the orders that are grouped on any one of the prongs are more closely related to the others on the prong than to those on other prongs.

This is a cautious treatment: some others have suggested that lagomorphs and rodents, together with the macroscelideans (elephant-shrews), should be seen as sisters to all the rest; and that, within 'the rest', the primates, plesiadapiforms ('archaic primates'), scandentians (tree shrews), demopterans (colugos), and chiropterans form one natural lineage while the insectivores, carnivores, and all the hoofed orders (artiodactyls, perissodactyls, proboscideans, and so on) form another. But there does not seem to be enough evidence to regroup the five-pronged polychotomy with any confidence.

So let us look more closely at the main mammalian orders, starting with the pair that are clearly different from all the rest.

Anteaters, armadillos, sloths, and the pangolins: orders Edentata (or Xenarthra) and Pholidota

The edentates (also called the xenarthrans), order **Edentata**, are extremely diverse in form but are generally considered monophyletic. They are definitely peculiar and, for example, have an extra point of articulation between each successive pair of vertebrae—hence the name Xenarthra, meaning 'foreign joint'. They are South American; along with the marsupials (and, as we will see later, with various endemic but extinct groups of hoofed animals) they demonstrate that when animals evolve in isolation they are prone to produce forms that are quite unlike anything that evolves elsewhere. Existing xenarthrans include the giant anteater and three smaller anteaters, the two genera of sloths (three two-toed species and two three-toed), and the 20 species in eight genera of armadillos—one of which, the common long-nosed armadillo, is also doing well in the southern United States, having migrated up from the south in the Great Interchange. But some of the edentates from the fairly recent past were far more impressive. The glyptodonts were like the mammals' answer to the tortoise. They had a round 'shell' like a golf-ball formed from dermal bones, but some were as big as a small family car. The giant sloths were also huge—some would have weighed a couple of tonnes—and were immensely

giant anteater, *Myrmecophaga tridactyla*

diverse and successful. They spread throughout North America once contact was made with South America. Giant sloths are commonly depicted standing on their hind legs and reaching up into the trees with their huge slashing claws, and they are commonly called 'ground sloths'. But their hind legs are typically curved and their feet are clawed like grappling hooks and I believe that they climbed. I know this is an eccentric view, but I really don't see why not. Grizzly bears climb, after all—and polar bears will also climb trees in zoos, at least in those few zoos that give them something more interesting to look at than concrete. Some giant sloths were much bigger than grizzly bears, but then bears are positively agile. I am not suggesting that giant sloths would have leapt around like squirrels, any more than tree sloths do.

tree pangolin
Phataginus tricuspis

The tree shows (following Michael Novacek) the edentates as sisters to the **Pholidota** (pangolins), which include a single family of scaly-skinned insect eaters from Africa and Southeast Asia; and the two together as sisters to all other eutherians. Certainly, much evidence suggests that the edentates broke from the rest soon after the eutherians as a whole split from the marsupials.

Rabbits, rodents, and elephant shrews: orders Lagomorpha, Rodentia, and Macroscelidea

The proposed affinity of these three orders—Lagomorpha, Rodentia, and Macroscelidea—is highly controversial. First, there is the matter of whether the **Lagomorpha**—rabbits, hares, and pikas—are or are not related to the **Rodentia**, which include the rats and mice, the squirrels, and the guinea-pigs and porcupines and between them account for about a quarter of all species of living mammals. The clear similarities between the two groups have prompted many zoologists, including George Gaylord Simpson, to place them together within the superorder Glires. But there are also obvious and less-obvious differences: for example, the fact that rodents have just two gnawing teeth

brown hare
Lepus europaeus

grey squirrel, *Sciurus carolinensis*

in the front upper jaw whereas the lagomorphs have four, with the second pair tucked behind the front pair for support. Some new molecular studies also suggest that the rodents and lagomorphs have nothing much to do with each other. But Michael Novacek presents the rodents and lagomorphs as sisters, while Andy Wyss opines that the superorder Glires is as 'well supported' as any of the superordinal groupings of mammals. So that is how they are presented here.

Yet new trouble has arisen even within the Rodentia, with Dan Graur of Tel Aviv University and his colleagues presenting molecular evidence that the guinea-pig is not a rodent at all. To be sure, the rodent suborder known as the hystricimorphs, which includes guinea-pigs and porcupines, has always been thought to represent a deep branch within the Rodentia. But few have ever doubted that the Rodentia as a whole are a true clade. Graur's work is still controversial: far more evidence is needed, both from other rodents and from non-rodents. If he is right, then the Rodentia will have to be broken up into at least two orders, albeit showing remarkable parallels between them. Yet if he is right, says Wyss, those who strive to do phylogeny by anatomy 'might as well pack up'.

The elephant-shrews, order **Macroscelidea**—12 quaint but enigmatic African species of shrew-like creatures with long noses—have been shuffled all around the Mammalia. Once they were placed within the Insectivora, alongside the 'true' shrews and moles; and the Insectivora, in turn, has commonly been linked to the Scandentia (the tree shrews) and hence to the Primates. But it has long been clear that the Macroscelidea, Insectivora, and Scandentia are different lineages, and only the Scandentia is now thought to have any close relationship with the Primates. Whether the Macroscelidea will retain their putative niche alongside the putative Glires remains to be seen.

elephant shrew
Elephantulus

Primates, 'archaic primates', tree shrews, colugos, and bats: orders Primates, Plesiadapiformes†, Scandentia, Dermoptera, and Chiroptera

The Primates (lemurs, bushbabies and lorises, tarsiers, monkeys, and apes), Plesiadapiformes ('archaic primates'), Scandentia (tree shrews), Dermoptera (colugos or 'flying lemurs'), and Chiroptera (bats) have commonly been linked together in a grouping sometimes called **Archonta**; and this arrangement in general seems to stand up. But there has been controversy in the past few years about the monophyly or otherwise of the bats, and their relationship to the primates.

microbat (pipistrelle), *Pipistrellus*

Thus the **Chiroptera** are commonly and uncontroversally divided into two suborders, the **Megachiroptera** (often colloquially called 'megabats') and the **Microchiroptera** ('microbats'). The former are the fruit eaters, like the flying foxes; and the latter include the rest, embracing all those groups that catch insects on the wing with the aid of echolocation (which the megabats do not possess). The megabats are generally supposed to be ancestral to the microbats, with the two together forming a true clade.

megabat (Franquet's fruit bat) *Epomops franqueti*

Some zoologists of late, however, notably John Pettigrew of the University of Queensland, argue that megabats and microbats are not closely related. Pettigrew agrees that there are close similarities between the two, including the structure of the wings. But, he says, there are also striking differences, not least in the fine structure of the brain. And details of brain structure, he suggests, tell us more about phylogeny than does wing structure, because wing structure directly influences flying efficiency and hence survival and so is exposed to rigorous natural selection, and thus is much more likely to evolve along similar lines in different lineages in response to the common environmental pressures. Details of the brain, on the other hand, do not have such obvious functional significance, are not so exposed to selection, and hence are less likely to converge in different lineages. Pettigrew argues, indeed, that megabats are closest to colugos, whose brains are similar in details, and that megabats, colugos, and primates form a clade; effectively he sees megabats as 'flying primates'.

But others argue that megabats and microbats seem to have very different brains because microbat brains are highly derived: microbats practice echolocation, and megabats do not. The corollary is that megabat brains seem to be more like the brains of colugos and primates than those of microbats, simply because megabat, colugo, and primate brains share more primitive features (because they all lack the special features associated with echolocation). Molecular evidence also seems on balance to support the traditional belief in bat monophyly, with megabats and microbats as sisters; and the arrangement shown here, with the bats as a whole as sisters to the colugos (**Dermoptera**), and those two together as sisters to primates + tree shrews (**Scandentia**), would probably attract the most support.

common tree shrew, *Tupaia glis*

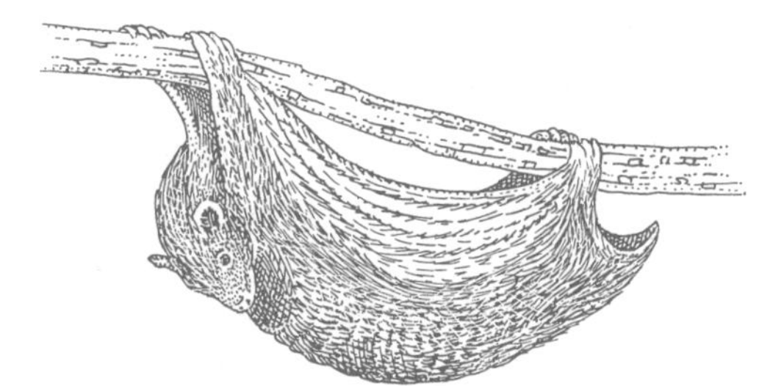

colugo, *Cynocephalus*

The **Plesiadapiformes** are—or were—a diverse group of primate-like mammals, roughly resembling prosimians, that lived in the Palaeocene and early Eocene of North America and Europe and, it now seems, in Asia. They have often been called 'archaic primates' and treated as a suborder of the Primates. But recent evidence suggests that the plesiadapiforms are no more closely related to primates than the tree-shrews are—or the colugos—so I show them here as a separate order. You may feel that they are not worth mentioning, because they were small and long gone. But in their day the plesiadapiforms were extremely successful. In some fossil assemblages they are the commonest mammals, evidently filling the niches that now are taken over by the primates and rodents.

primate (black-handed spider monkey)

I look at our own group, the **Primates**, in more detail in Section 19.

Shrews, moles, and hedgehogs: order Insectivora

European hedgehog, *Erinaceus europaeus*

The modern **Insectivora** include the shrews, moles, and hedgehogs; and they have caused mammalian taxonomists no end of trouble. Taxonomically speaking, the problem seems to be twofold. First, there is the problem of symplesiomorphy, for insectivores retain the general body shape of primitive mammals with few derived additions and so they have tended to be grouped with other mammals with generalized body features. Then there is a problem with convergence, for insectivores have small teeth with pointed cusps, adapted to puncture the exoskeletons of insects; but so do other insect eaters that are not, in fact, closely related. For these reasons, insectivores in the past have at times been linked with the tree shrews (Scandentia) and through them to the primates and bats; and with the elephant shrews (Macroscelidea). It has commonly been assumed, too, that because modern insectivores retain so many primitive features that the order Insectivora must truly be ancient, and closely related to the ancestors of all other mammals. But this idea also seems to have been dropped. Retention of primitive features—yes: for the primitive mammalian shape is ideal for a committed insect eater. Special relationship to ancient, ancestral mammals—no.

So it seems safest simply to show the Insectivora as an order in their own right, difficult to place, but with no obvious relationship to any other living order and with no good reason to assume outstanding antiquity. This is how they are shown in the tree, pending further evidence.

The committed meat eaters: orders Creodonta† and Carnivora

The terrestrial mammalian meat eaters that have arisen in the Cenozoic (after the demise of the dinosaurs) all belong to three orders. The first to produce any large carnivores were the Condylarthra, which are a primitive and diffuse group, difficult to classify, and are shown here (for simplicity) as sisters to all the hoofed mammals, the ungulates (see below). The condylarths flourished in the early Palaeocene, and are long gone.

The other emphatically carnivorous orders are the **Carnivora**, which are still with us; and their sister group, the **Creodonta**, which finally went extinct in the late

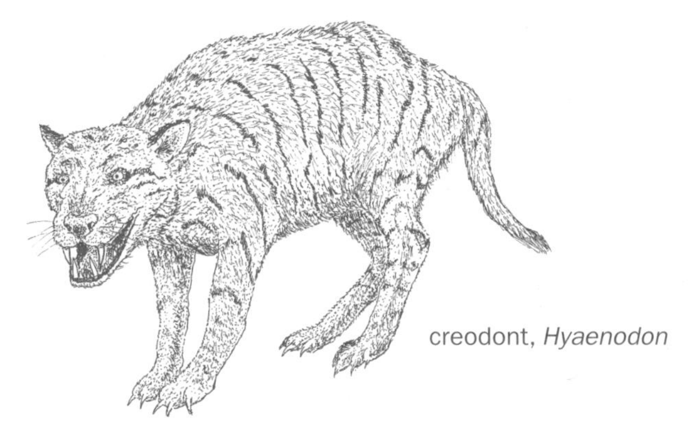

creodont, *Hyaenodon*

Miocene. Of these two orders, the earlier known is the Carnivora:[2] weasel-like creatures from the early Palaeocene. The earliest known creodonts date from the late Palaeocene. The creodonts, however, seemed to be the first to get fully into their stride: that is, the carnivores remained weasel-like until late middle Eocene, while the creodonts produced some large predators by the late Palaeocene (and they replaced the earlier condylarths). In short, although the Carnivora may be the earlier group, the Creodonta were the first of the two to fill the niche of the large predator.

The creodonts were highly successful. They produced two distinct lineages of larger meat eaters. First came the cat-like **oxyaenids**, which were the world's predominant meat eaters until the end of the middle Eocene, when they died out. Then in the late Eocene the remaining creodonts gave rise to a second wave of large meat eaters: the fox-like and wolf-like hyaenodonts. The **hyaenodonts** were prominent in North America in the late Eocene and Oligocene, but then died out in North America. But they persisted as important predators in Africa and Asia until the late Miocene.

In the Palaeocene, the Carnivora split into a cat-like lineage ('feliform') and a dog-like lineage ('caniform'). The feliforms gave rise to the modern group of families known as the **Feloidea**, a superfamily that includes the Felidae (cats), Viverridae

[2] Here we find a somewhat irritating confusion of terms. The word 'carnivore' with a small 'c', and the adjective 'carnivorous', really means 'any creature that eats meat'. A *Tyrannosaurus* or a shark may legitimately be referred to as 'carnivores'. But the specific order Carnivora is a true mammalian taxon; a bona fide clade. The trouble is that if we want to refer to members of the Carnivora colloquially, then we have to call them 'carnivores', just as members of the Rodentia are called 'rodents'.

(civets), the Herpestidae (mongooses), and the Hyaenidae (hyaenas), plus the extinct, very cat-like Nimravidae. The caniforms have given rise to two modern superfamilies: the **Canoidea**, which includes only the dog family, Canidae; and the **Arctoidea**, which include the Ursidae (bears), Procyonidae (raccoons and coatis), and the Mustelidae (weasels, otters, badgers, wolverines). Modern molecular studies suggest that the giant panda is essentially a bear (Ursidae), while the red panda belongs with the Procyonidae. The Feloidea initially arose as an Old World group; the Canoidea were initially New World; and the Arctoidea were initially Old World, but made many incursions into the New World. The procyonids almost all belong to the New World, except for the red panda, which is Asian.

felid (tiger), *Panthera tigris*

The modern **Pinnipedia**—the seals, sealions, and walruses—are conventionally separated off as a suborder of the Carnivora, although with an obvious relationship with the Arctoidea. Traditionally they were assumed to be monophyletic, but in recent decades some zoologists have suggested that they are really diphyletic, with the sealions and walruses related to the bears, and the seals to the mustelids. But as outlined in Chapter 3 in Part I, cladistic studies by Andy Wyss now suggest that the pinnipeds are indeed monophyletic—probably with the walruses closer to the seals than to the sealions; and that pinnipeds as a whole are sisters to the bears; the ursids.

So now we come to the grand assemblage of **Ungulata**, and the groups that have descended from them.

'UNGULATES': HOOFED MAMMALS AND THEIR DERIVATIVES

Simplified classifications commonly placed all the hoofed animals in a single order, the 'Ungulata', from the Latin *ungula* meaning 'hoof'. But now it is clear that the differences between the various groups of ungulates are huge and of ancient origin, so the creatures that were traditionally grouped in the single order 'Ungulata' are now divided among four orders (at least of living types). The **Artiodactyla** are the even-toed ('cloven-hoofed') creatures, which range from cattle and deer through pigs and hippopotamuses to camels. The **Perissodactyla** bear their weight on the middle toe and now include the horses, rhinos, and tapirs. The **Proboscidea**, now represented by just two species of elephant, were also represented in the past by a host of mammoths, mastodonts, gomphotheres, and deinotheres. Ever since Richard Owen in the nineteenth century, the hyraxes, order **Hyracoidea**, have also been included as honorary ungulates. Hyraxes feature in the Bible as 'cavies'. Superficially they resemble big guinea-pigs, but Owen perceived their ungulate affiliation, and suggested that they are specifically related to the perissodactyls (although a powerful lobby since has seen closer affinities with the proboscideans). A zoological friend of mine who has kept hyraxes says they 'behave just like bad-tempered Shetland ponies, and are about as smart'.

The traditional order of the 'Ungulata' also included a range of hoofed animals that are now extinct, and do not belong in any of the living ungulate orders. These, too, are now apportioned among several different orders. Among these extinct groups are the **Embrithopoda** of Oligocene Africa, which included the remarkable, superficially rhino-like arsinoitheres, who bore great paired horns on their noses side by side like giant rifle sights. Four of the extinct ungulate orders were peculiar to South America, the best known being the **Notoungulata** and the **Litopterna**. Both were highly variable, and both include creatures that converged remarkably with ungulates from other orders. Thus the notoungulate *Toxodon* was like a hornless rhino, while *Macrauchenia*, a litoptern, resembled a long-necked camel.

It is clear that many other orders, living and extinct, are related to that central core of traditional 'ungulates'; and that most of these others do not have hooves and are not necessarily herbivorous, as most of the traditional ungulates are. Thus it seems that the whales, **Cetacea**, which, of course, have no feet at all (no hind limbs, and the forelimbs modified as flippers), arose from among the ranks of the ungulates. The **Sirenia** (manatees and dugongs) also have clear ungulate origins, their closest living relatives being the elephants. The extinct **Desmostylia,** also related to the sirenians and proboscideans, were again semi-aquatic and hoofless. Even the bizarre termite-eating aardvark of Africa, order **Tubilidentata** (somewhat anteater-like, somewhat pig-like—indescribable, really), seems to belong among the ungulates, although this is not absolutely certain.

We tend to assume that ungulates are specialist herbivores, like cows, horses, and elephants. But even some familiar ungulates defy this dietary preconception, for pigs are omnivores. Whales are among the most carnivorous of creatures on Earth and the **condylarths**, which we have seen included the world's first big terrestrial mammalian carnivores, are also ungulates. (George Gaylord Simpson also suggested that the Ungulata had a special relationship with the Carnivora and Creodonta, placing them all in a 'cohort', which he called Ferungulata. But this idea has been dropped.)

In short, 'Ungulata' has lost its traditional meaning. The group can no longer be seen as a single order of (predominantly) herbivorous animals with hooves. But 'Ungulata' remains as a clade nonetheless—a monophyletic lineage of the Eutheria (which we could call a 'cohort', but let's stick to 'lineage') that includes seven living orders (Artiodactyla, Cetacea, Tubulidentata, Perissodactyla, Hyracoidea, Proboscidea, and Sirenia) plus numerous extinct orders, six of which are shown in the tree. Let us follow through the ungulate section of the tree from top to bottom.

The condylarths†

At the top of the ungulate clade, shown as sister of all the true ungulates and their immediate relatives, are the condylarths. But the condylarths are shown in this sister-group position only for convenience, for the so-called 'Condylarthra' is a catch-all grouping that clearly includes five or six probably separate lineages from among the confusing mass of large and medium-sized mammals that radiated rapidly in the

Palaeocene, once the cork of the dinosaurs had been removed from the pan-global ecological bottle. Condylarths demonstrate the problem posed by all 'stem-group' creatures: that it is very difficult, going on impossible, to say who is related to whom, or which lineages were destined to die out, or which were to evolve into artiodactyls, or elephants, and so on. Among the best known of the condylarths were the arctocyonids, which roughly resembled raccoons but had broad molars for crushing plants; precisely the kinds of generalized-looking creatures whose relationships are hardest to pin down.

condylarth, *Chriacus*

Cattle, deer, pigs, and their relatives: order Artiodactyla

wild boar, *Sus scofa*

Then we come to the 'true' ungulates. Appropriately, the first to be listed are the artiodactyls, which are among the most significant of all animals; they are hugely speciose, and include some of the strongest and swiftest, the most intelligent, the most beautiful, and the most ecologically successful (when measured, for example, in terms of numbers and crude biomass). Broadly speaking there are three great groups: the non-ruminants, the 'pseudo-ruminants', and the ruminants. The non-ruminants are the pigs, peccaries (New World pigs), and the hippopotamuses. The 'pseudo-ruminants' are the camels, which also embrace the llama, alpaca, guanaco, and vicuna. They practise a form of rumination that differs in physiological details from that of the 'true' ruminants.

The 'true' ruminants are among the most accomplished of all herbivores: browsers eat the leaves of trees and bushes while grazers focus on grass. In either case, they digest the hard plant material very efficiently in a vast fermentation chamber—a modified stomach called the **rumen**—that is packed with bacteria. The bacteria, among other things, break down the cellulose cell walls into materials that provide the animal with energy. Six families of ruminants are commonly recognized: the Tragulidae, Moschidae, Cervidae, Giraffidae, Antilocapridae, and Bovidae.

The Tragulidae are the chevrotains, or 'mouse-deer': four species of small, duiker-like creatures from the forests of Central Africa and Southeast Asia. The Moschidae include the three species of musk deer from East Asia. Moschids resemble deer but lack antlers, and the upper canines of the male form conspicuous tusks (which is also true of some deer). The Cervidae include the 'true' deer—34 known species. Worldwide they are immensely successful. They include the vast northern herds of caribou and reindeer but also flourish in the tropics, and range in size from the tiny pudus of South America to the giant moose (or elk) of northern North America and Europe.

red deer
Cervus elaphus

giraffe
Giraffa camelopardalis

The Giraffidae are closely related to the Cervidae. The giraffid family now includes just two species—the giraffe and the okapi—but there have been many more giraffids in the past, some of which had antlers, as if to demonstrate their cervid affiliations. The Antilocapridae are the pronghorns, once widespread but now reduced to one species; they look antelope-like, but their horns are forked and they shed and renew the outer sheath throughout life.

The Bovidae are the mightiest of all the artiodactyl families, and among the most-influential land mammals of all—for truly they are makers of landscape. Five subfamilies are widely recognized (and, in this case, 'subfamily' is a useful ranking). The Bovinae are the cattle and 'spiral-horned' antelope: 23 living species in 8 genera—yak, bison, buffalo, gaur, the anoas, plus the nyala, eland, kudu, bongo, bushbuck, and so on. The Cephalophinae are the duikers, of which there are 17 species in 2 genera. The Hippotraginae include the 24 species of grazing antelopes—waterbuck, topi, the gnus, impala, oryx, addax, and the rest. The Antilopinae—the gazelles and dwarf antelopes—are the most speciose of all large mammals, with 30 species. Finally, a somewhat mixed bag of 'goat antelopes' form the Caprinae. The 26 species include the saiga 'antelope' of the Asian steppes, the chamois and its relatives, the musk ox, and the world's 17 species of goats and sheep. By whatever yardstick they are measured, the bovids are truly among the all-time successes of mammalian life.

African buffalo
Synceros caffer

Whales, dolphins, and porpoises: order Cetacea

It now seems that the closest living relatives of the artiodactyls are the cetaceans, the whales, dolphins, and porpoises. Here the two orders are presented as sister groups (albeit with the extinct mesonychids intervening, of whom more later); but, in fact, modern molecular studies show the cetaceans as a branch emerging from within the artiodactyls. Cetaceans are 'ungulates' whose feet have long been reduced to flippers. Neither, of course, are they herbivores; indeed as devourers of krill (which are crustaceans), fish, and squid (the standard fare of sperm whales) whales, dolphins, and porpoises are among the most carnivorous creatures of all. As large, full-time deep seafarers they don't have much option.

common dolphin
Delphinus delphis

The mesonychids†

The mesonychids, here shown as the immediate sisters of the Cetacea, were a variable group who appeared in the Palaeocene, and were indeed among the first big mammals to emerge after the dinosaurs had gone. (In fact, the mesonychids—Mesonychiidae—represent just one family among the order called the Acreodi; but every palaeontologist speaks of 'mesonychids', and I have rarely seen the term 'Acreodi', so it seems more sensible to use the term that is most current.) Mesonychids had hooves; but they also had great flesh-seizing teeth, which reveal them as carnivores. Among the most famous mesonychid was *Andrewsarchus*, commonly depicted as an irredeemably unprepossessing beast built like a heavy tiger—although with mini-hooves rather than claws—and a huge head with long, crocodile-like jaws; indeed, *Andrewsarchus* was the biggest and most fearsomely jawed mammalian carnivore ever, with a body 4 metres long and a 1-metre skull (which nonetheless housed a tiny brain, by modern mammalian standards). But the first whales had comparably long, toothed jaws; and present-day whales generally have big long heads, even though baleen whales have lost their teeth; and some mesonychid that was very like *Andrewsarchus* was probably the cetacean ancestor, becoming by stages more and more committed to the water.

mesonychid, *Andrewsarchus*

The aardvark: order Tubulidentata

The order Tubulidentata, shown on the tree as the sisters of the artiodactyls + mesonychids + cetaceans, includes just one living species: the aardvark, or 'earth-pig', which is indeed pig-like but lives by tearing open the nests of termites with its formidable claws and scooping out the inhabitants with its long, serpentine

aardvark, *Orycteropus afer*

tongue. The notion that aardvarks belong among the ungulates is based on molecular studies, which, some say, contradict anatomical studies. Some zoologists place the aardvark with the edentates and pangolins—though others would say that this is simply another case of convergence.

Four extinct South American ungulate groups: orders Notoungulata†, Litopterna†, Astrapotheria†, and Pyrotheria†

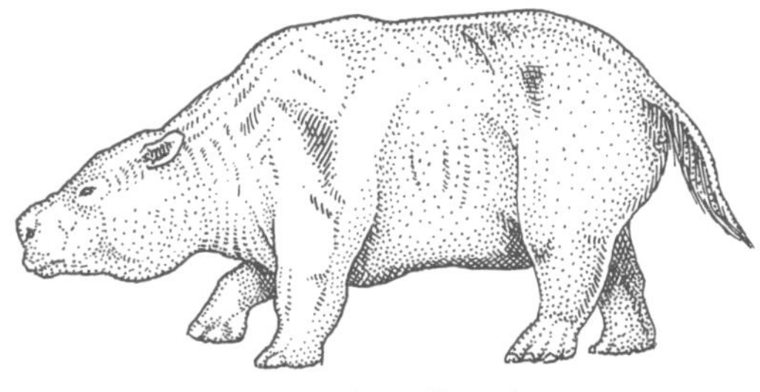
notoungulate, *Toxodon*

The Notoungulata and Litopterna have been mentioned briefly above; together with the Astrapotheria and Pyrotheria they formed a quartet of extinct ungulate orders unique to South America. How they got there is uncertain, but it is hard to explain the provenance of South American mammals in general unless we assume that at some time in the Palaeocene, land links (or closely spaced chains of islands) were established between South America and North America, and between North America and Eurasia. In such a case, the South American ungulate orders would have had Laurasian origins. Modern cladistic studies suggest that the South American ungulates were probably related to the perissodactyls, but they developed entirely along their own lines. Some seemed to die out after South America joined (again) with North America in the Pliocene. Perhaps they failed to compete with the invaders from the north, which included camelids (ancestors of the modern llamas and their kin), peccaries (related to the Old World pigs, though in a different family), deer, horses, gomphotheres, and tapirs. But perhaps they missed out simply because the world was cooling at the time, and the savannah that they preferred—grassland with trees—was changing to prairie or pampas—grassland without trees. The invaders (so this scenario has it) were already adapted to more or less treeless grassland. But some of the South American ungulates, such as *Macrauchenia*, survived until the first incursions of human beings around 11 000 years ago and we may properly regret that they are not with us still.

litoptern, *Macrauchenia*

Horses, rhinos, tapirs, and their extinct relatives: order Perissodactyla

The modern Perissodactyla include the horses, rhinos, and tapirs. But in the past this order also included the chalicotheres, which were huge creatures with horse-like heads and great sloth-like front limbs fitted with peculiar bifurcated claws like toasting forks, which they presumably used to slash vegetation. Cladistic studies suggest that the most primitive of the perissodactyls were the brontotheres that radiated in the middle Eocene into a range of tapir-sized creatures but later produced huge rhino-like creatures with knobbly faces.

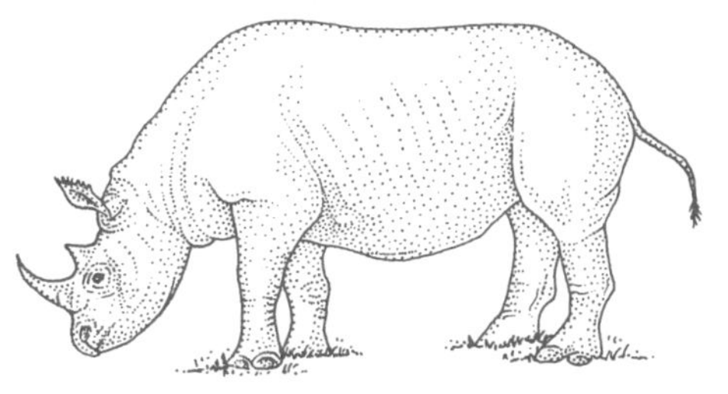
black rhinoceros, *Diceros bicornis*

Hyraxes: order Hyracoidea

So to the enigmatic hyraxes, the Hyracoidea. They are, as mentioned, guinea-pig like. Today there are eight species in three genera. Rock and bush hyraxes live in colonies in and around rocks on the edges of deserts in Africa, from Namibia to Ethiopia and Egypt, while tree hyraxes are more solitary. Although Richard Owen first suggested that they are related to perissodactyls—and the earliest known horse is called *Hyracotherium* because Owen mistook it for a hyrax ('Eohippus' is a colloquial name, but much to be preferred). Later scholars felt that hyraxes are probably closer to proboscideans. In recent years Don Prothero and his colleagues in California have suggested that Owen was right after all: that hyraxes do belong with perissodactyls. But emerging molecular studies apparently favour the proboscidean connection. Here I show the hyraxes interposed between the perissodactyls and the proboscideans (plus a group of other comparably weighty creatures).

tree hyrax, *Dendrohyrax*

Proboscideans, arsinoitheres, and sirenians: orders Proboscidea, Embrithopoda†, Desmostylia†, and Sirenia

Finally, the tree (which follows Novacek) shows the Proboscidea as part of a polychotomy that also includes the extinct Embrithopoda and Desmostylia, and the extant Sirenia. Like all the mammalian orders they are ancient groups and they evidently emerged within a short time of each other, so the order of their emergence is, as American palaeontologists tend to say, 'too close to call'.

The **Proboscidea**, now represented by just two species, are known to have given rise to at least 150 species since they

African elephant, *Loxodonta africana*

embrithopod, *Arsinoitherium*

desmostylian, *Desmostylus*

first appeared in the Eocene, and they have had a tremendous impact on the whole ecology not only in Africa and Asia where they live today but also throughout Europe and both Americas. For example, the presence or absence of elephants largely determines whether forests do or do not survive, and, if so, what kind of forest prevails. The **Embrithopoda**, as mentioned, were the giant, cumbersome, late Eocene arsinoitheres of Africa. The **Desmostylia** were heavy, stout-limbed herbivores that were apparently semi-aquatic; and the **Sirenia**, now represented by the manatees and the dugong (or 'sea-cow'), are gentle herbivores of coasts and rivers. Cetaceans, desmostylians, sirenians—and we might add the artiodactyl hippos; the 'ungulates' seem to have an affinity for water.

manatee
Trichechus

In summary, the traditionally acknowledged mammalian orders—at least of living types—taken individually do seem to stand up reasonably well to analysis. That is, the traditional orders do seem on the whole to be true clades, although there could still be problems with rodents in particular. The relationships between orders is a much more difficult issue, although the arrangement shown here certainly makes sense and would enjoy a great deal of support. The stem groups, such as the condylarths, continue to give trouble and perhaps always will. They lived a long time ago; the fossil record is sparse; and creatures that have not yet had time to evolve a great many clearly 'derived' and special features are innately difficult to fit into phylogenetic niches.

Albeit mainly for chauvinistic reasons, it now seems appropriate to look in more detail at our own mammalian order: the Primates.

19

LEMURS, LORISES, TARSIERS, MONKEYS, AND APES

ORDER PRIMATES

Living primates include the lemurs, lorises, and bushbabies, the tarsiers of Southeast Asia, the monkeys of the New World and Old World, and the gibbons and great apes among whom we must number ourselves. They all unmistakably are of a kind—all have round brainy skulls with high foreheads, forward-looking eyes providing stereoscopic vision, mobile arms and dextrous hands—yet zoologists find the primate order very hard to define.

Primates do have big brains relative to their body weight compared with most other mammals, but this is a matter of degree. Their eyes do face forwards, but so do those of some other mammals. The orbits of their eyes are completely surrounded by bone, but they are not unique in this. Primates have only two mammary glands, on the chest; but, then, so do sirenians and elephants. Their fingers and toes are tipped with sensitive pads, but so are those of tree shrews. Their joints are highly mobile: always their ankles, and sometimes their shoulders, but squirrels have mobile ankles too. Their hands and often their feet are dextrous, but so are those of squirrels (up to a point); in some primates the thumb is fully opposable, but not in all. They have nails instead of claws, which does seem to be a uniquely primate feature, though it is not much by which to define an entire grouping, and some do have claws.

In short, most of the features that seem so characteristic and so unmistakable do not seem to qualify unequivocally as synapomorphies in the manner, say, of the hoofs of artiodactyls. Many are symplesiomorphies: primitive features of mammaldom. Yet when you see a lemur, a bushbaby, a tarsier, a monkey, or an ape you can *feel* their relatedness; and some people indignantly deny our own relationship to apes precisely because the fact is so obvious. Gestalt is not a reliable guide to true relationship; but, with primates, gestalt is hard to improve upon.

Yet this apparent lack of obvious, particular specialties is the strength of primates. In the orders that are easiest to characterize the basic mammalian features have evolved into specialist instruments, or sometimes been lost all together: the wing of the bat, the pillar-like legs of the elephant, the flippers and toothlessness of the baleen whales, the extreme reduction of the equid foot. Primates, by contrast, seem to have lost virtually nothing of the basic mammalian inheritance. They retain the characters of Mesozoic mammals, down to and including the five toes and five fingers (that is, the pentadactyl limbs), which are a general amniote feature. But having retained these characters, they have built on them. The generalized tetrapod form, refined but not recast, gives primates an overall versatility approached only by creatures like bears, raccoons, or squirrels but in practice surpassing all of them.

Because they are so physically versatile, and because they have refined their dexterity, their physical co-ordination, and their depth of vision, primates have been able to become the supreme arborealists. In practice, the arboreal life and the refinements of structure and ability feed on each other. Natural selection has favoured dexterity and the senses and nervous system that go with it because primates live in trees; such features would not be favoured if primates fossicked among the leaf-litter, like hedgehogs. But, in turn, the more dextrous and agile the primates became, the more appropriate the arboreal option became.

Yet many primates have taken to life on the ground, largely if not entirely. The giant extinct lemur known as *Megaladapis*, which went extinct soon after human beings first arrived on Madagascar in the first centuries AD, was probably largely terrestrial. Many Old World monkeys have come to live mainly or virtually entirely on the ground, including the baboons. Among apes, gorillas are primarily terrestrial; and so, of course, is the most extraordinary of apes, *Homo*. But each of them has brought to terrestrial life the versatility and refinements that were honed in the trees. Ground-living primates are very different from mammals that have always been terrestrial, such as dogs and pigs.

As a group, primates are probably surprisingly ancient. Indeed, along with marsupials, edentates, and rodents, they are among the few extant orders that may well have originated in the Cretaceous. In truth, the oldest true primate fossils (like lemurs and tarsiers) date only from the early Eocene, around 55 million years ago. Primate-like fossils that are older than that have turned out not to be primates at all, but plesiadapiforms; and, as discussed in Section 18, although the Plesiadapiformes have been called 'archaic primates' they now seem to be a separate order.

But there are many good reasons for believing that the oldest known fossils of true primates do not represent the first primates. Thus, there are roughly 200 primate species today; and, if the oldest known fossil types are more than 50 million years old, then we can calculate (using ecological rules of thumb that are not sure-fire but at least are sensible) that several thousand primate species must have existed since that time. But only around 250 fossil primate species are known so far—only a few per cent of the number that must have lived. In general, too, all palaeontologists know that the oldest fossils they find of any one particular creature are unlikely to represent the earliest examples of that creature. Fossilization is usually a rare event so that no animal is liable to form fossils until it is already common and widespread; and no group is liable to be common and widespread until it has been around a long time.

Putting all such thoughts together, some primatologists such as Robert Martin of the University of Zurich now conclude that the earliest primates must be at least 30 per cent older than the oldest known fossils; and 55 million years ago plus 30 per cent is more than 70 millions of years ago. If the first true primates were indeed that ancient, then they would have been contemporaries of the late dinosaurs. It is good to think of them up in the trees, chattering their alarm as some great reptilian head poked in among the foliage.

The tree here is based on various texts, including two papers in *Nature* by Robert Martin, plus one by David Dean and Eric Delson; and Michael Benton's *Vertebrate Palaeontology* (1995). The section on New World monkeys is taken from Alfred Rosenberger's treatment in *The Cambridge Encyclopedia of Human Evolution* (1992) and the section on Old World monkeys is based on Eric Delson's essay in the same volume. I have also been guided by the excellent *World List of Mammalian Species* of G. B. Corbet and J. E. Hill (1991). This section has also been informed by conversations with Ian Tattersall at the American Museum of Natural History, New York; Chris Stringer of London's Natural History Museum; and Bernard Wood, then at the University of Liverpool but now at George Washington University, Washington DC.

The tree shows all 11 (though some might say 10) families of living primates, plus a few key lineages and outstanding fossils from the past—the latter serving both to introduce the extinct types that seem most significant and also helping to put the living groups into context. Some of the ancient lineages are represented by single genera (*Proconsul*, *Sivapithecus*, *Dryopithecus*, and *Australopithecus*); and I have also included all the living genera of anthropoids—monkeys and apes. In general shape the tree slavishly reflects the opinions of leading specialists but it cannot represent consensus, because there is none.

A GUIDE TO THE PRIMATES

As the tree shows, by reading from top to bottom, the Primates are conveniently divided into three grades: the **'prosimians'**, including the living lemurs, lorises, bush-babies, and tarsiers, plus the extinct adapids and omomyids; the monkeys, both Old World and New World; and the apes, including humans. In precladistic days, taxonomists recognized two distinct suborders: the Prosimii, which included all prosimians; and the Anthropoidea, which included monkeys and apes. But as the tree shows, the Prosimii includes two deeply divided lineages and is at best paraphyletic—hence the formal term should be dropped; but 'prosimian' with a lower-case 'p' remains a useful adjective to apply to the grade. **Anthropoidea** is a clade, however, and the term can stand (although the informal adjective 'simian' can be used instead).

Until recent years, palaeoprimatologists also admitted a fourth grade of primates whose fossils are known from the Palaeocene (and obviously existed far back into the Cretaceous). These were known formally as the Plesiadapiformes and

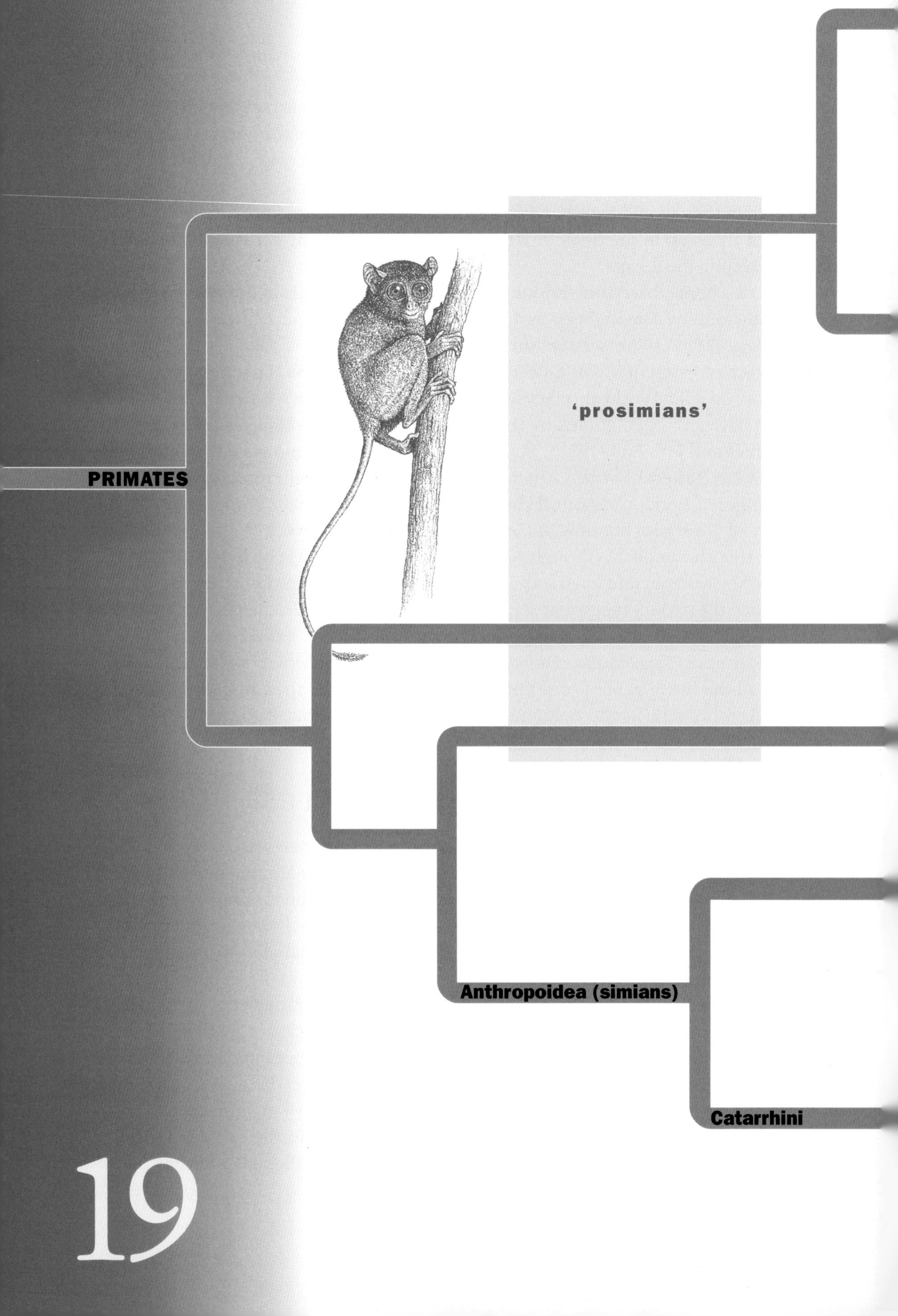
'prosimians'
PRIMATES
Anthropoidea (simians)
Catarrhini
19

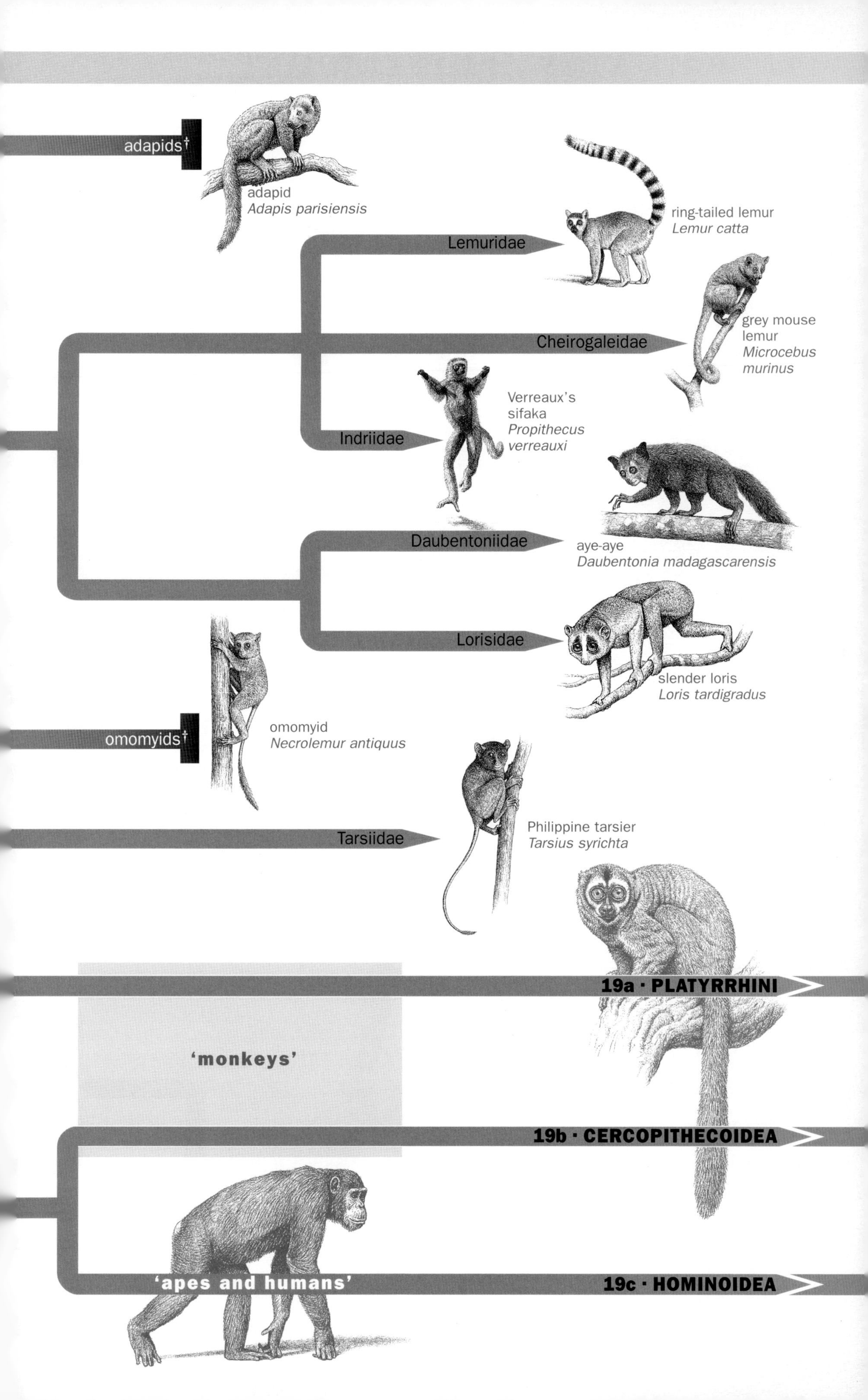
adapids†
adapid
Adapis parisiensis
ring-tailed lemur
Lemur catta
Lemuridae
grey mouse
lemur
Microcebus
murinus
Cheirogaleidae
Verreaux's
sifaka
Propithecus
verreauxi
Indriidae
Daubentoniidae
aye-aye
Daubentonia madagascarensis
Lorisidae
slender loris
Loris tardigradus
omomyid
Necrolemur antiquus
omomyids†
Philippine tarsier
Tarsius syrichta
Tarsiidae
19a · PLATYRRHINI
'monkeys'
19b · CERCOPITHECOIDEA
'apes and humans'
19c · HOMINOIDEA

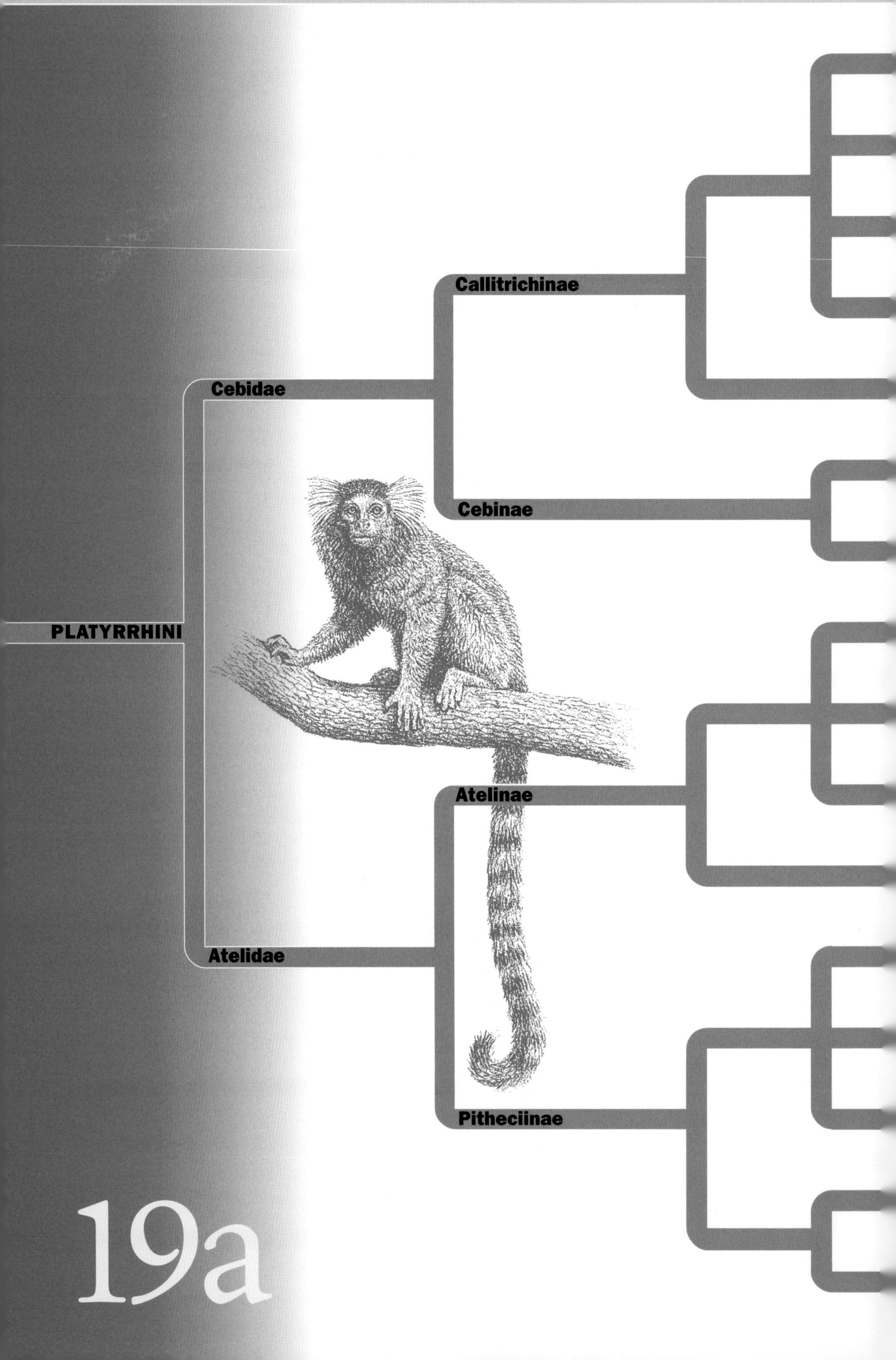
Callitrichinae
Cebidae
Cebinae
PLATYRRHINI
Atelinae
Atelidae
Pitheciinae
19a

Callithrix
Cebuella
Leontopithecus
Saguinus
Callimico
Cebus
Saimiri
Ateles
Brachyteles
Lagothrix
Alouatta
Pithecia
Chiropetes
Cacajoa
Aotus
Callicebus
common marmoset
Callithrix jacchus
white-faced capuchin
Cebus capucinus
black-handed
spider monkey
Ateles geoffroyi
red howler monkey
Alouatta seniculus
white-faced saki
Pithecia pithecia
owl monkey
Aotus trivirgatus

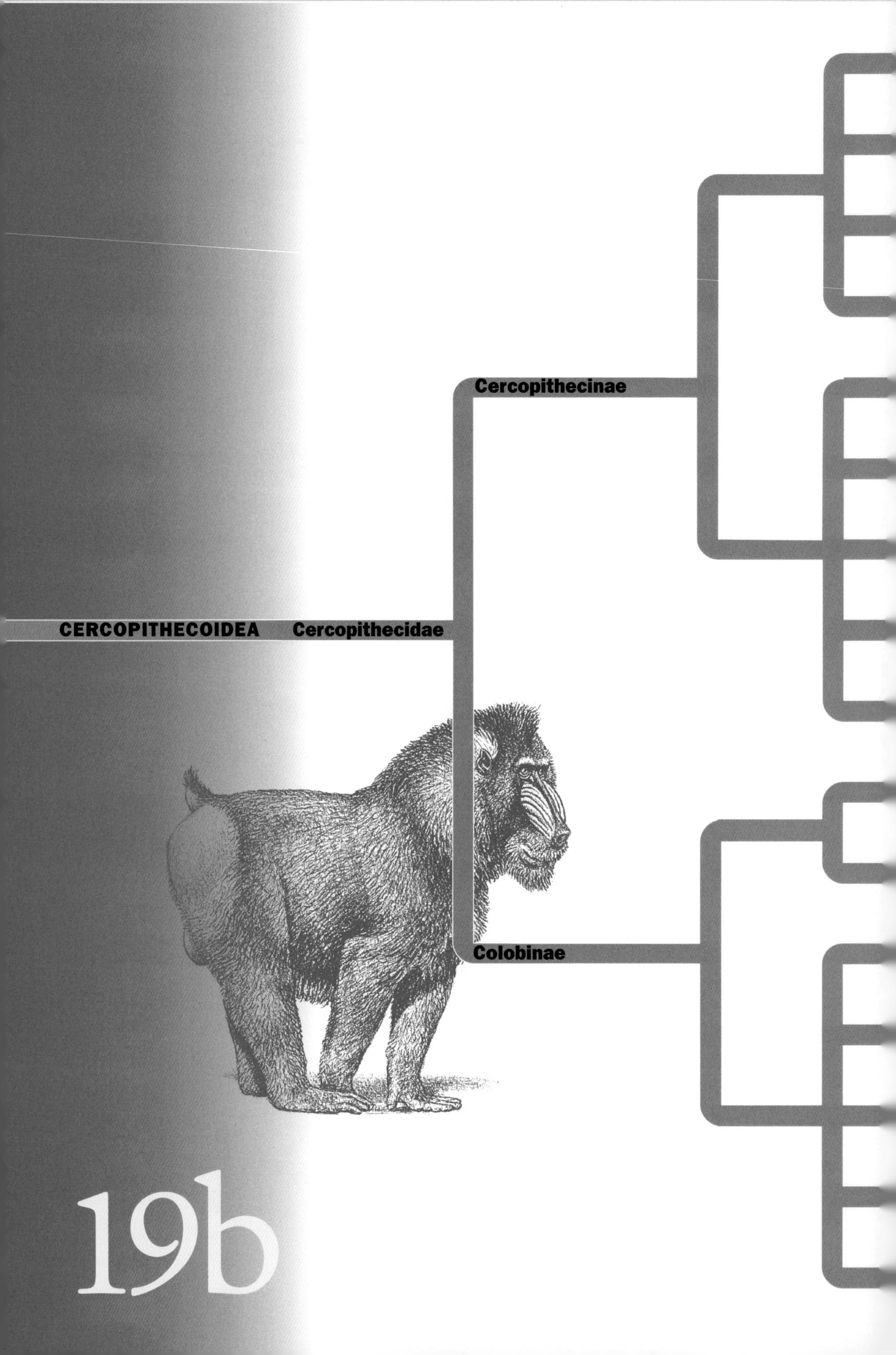
Cercopithecinae
CERCOPITHECOIDEA
Cercopithecidae
Colobinae
19b

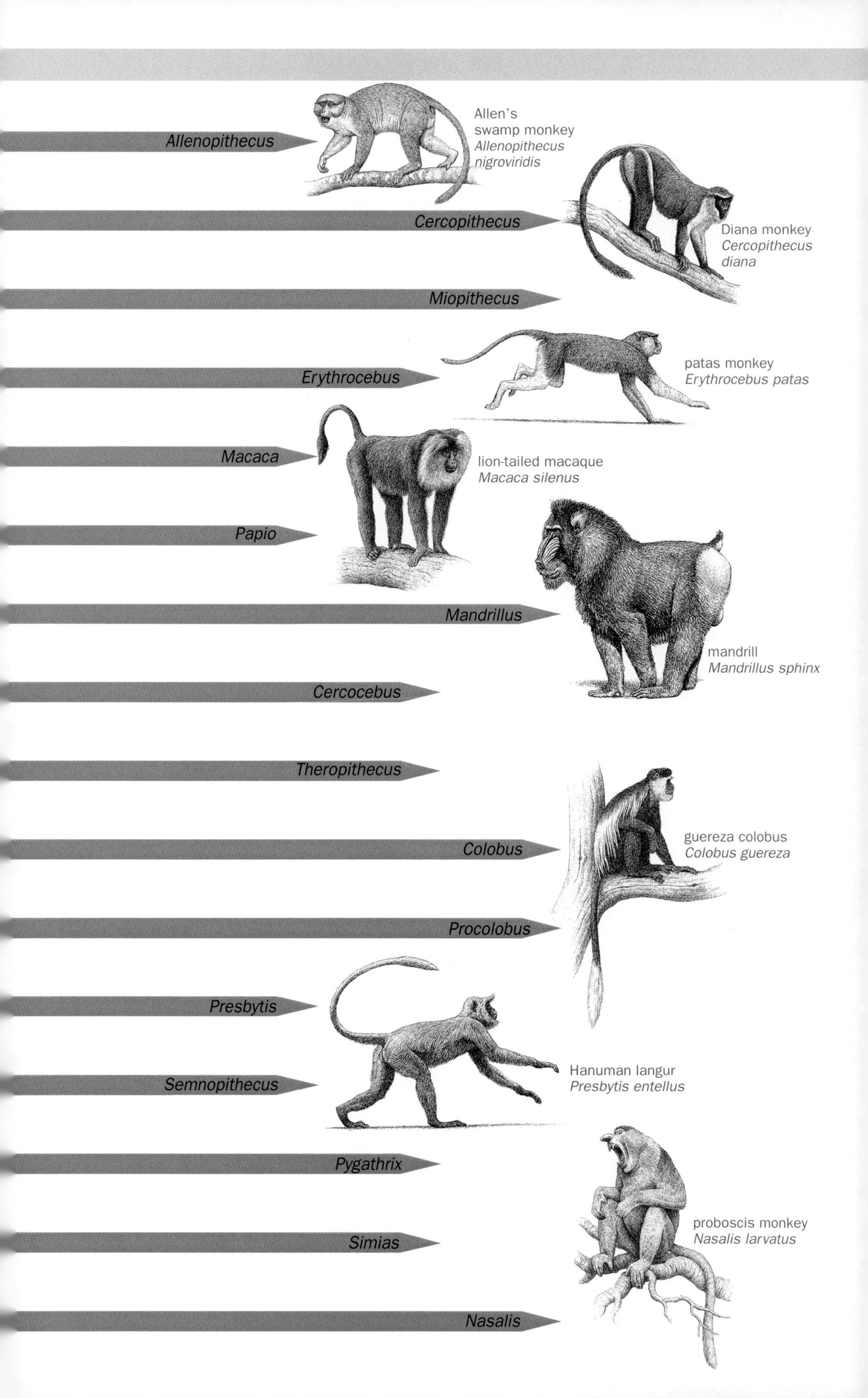
Allenopithecus
Allen's
swamp monkey
Allenopithecus
nigroviridis
Cercopithecus
Diana monkey
Cercopithecus
diana
Miopithecus
Erythrocebus
patas monkey
Erythrocebus patas
Macaca
lion-tailed macaque
Macaca silenus
Papio
Mandrillus
mandrill
Mandrillus sphinx
Cercocebus
Theropithecus
Colobus
guereza colobus
Colobus guereza
Procolobus
Presbytis
Semnopithecus
Hanuman langur
Presbytis entellus
Pygathrix
Simias
proboscis monkey
Nasalis larvatus
Nasalis

HOMINOIDEA
Hylobatidae
Hominidae/Pongidae
19c

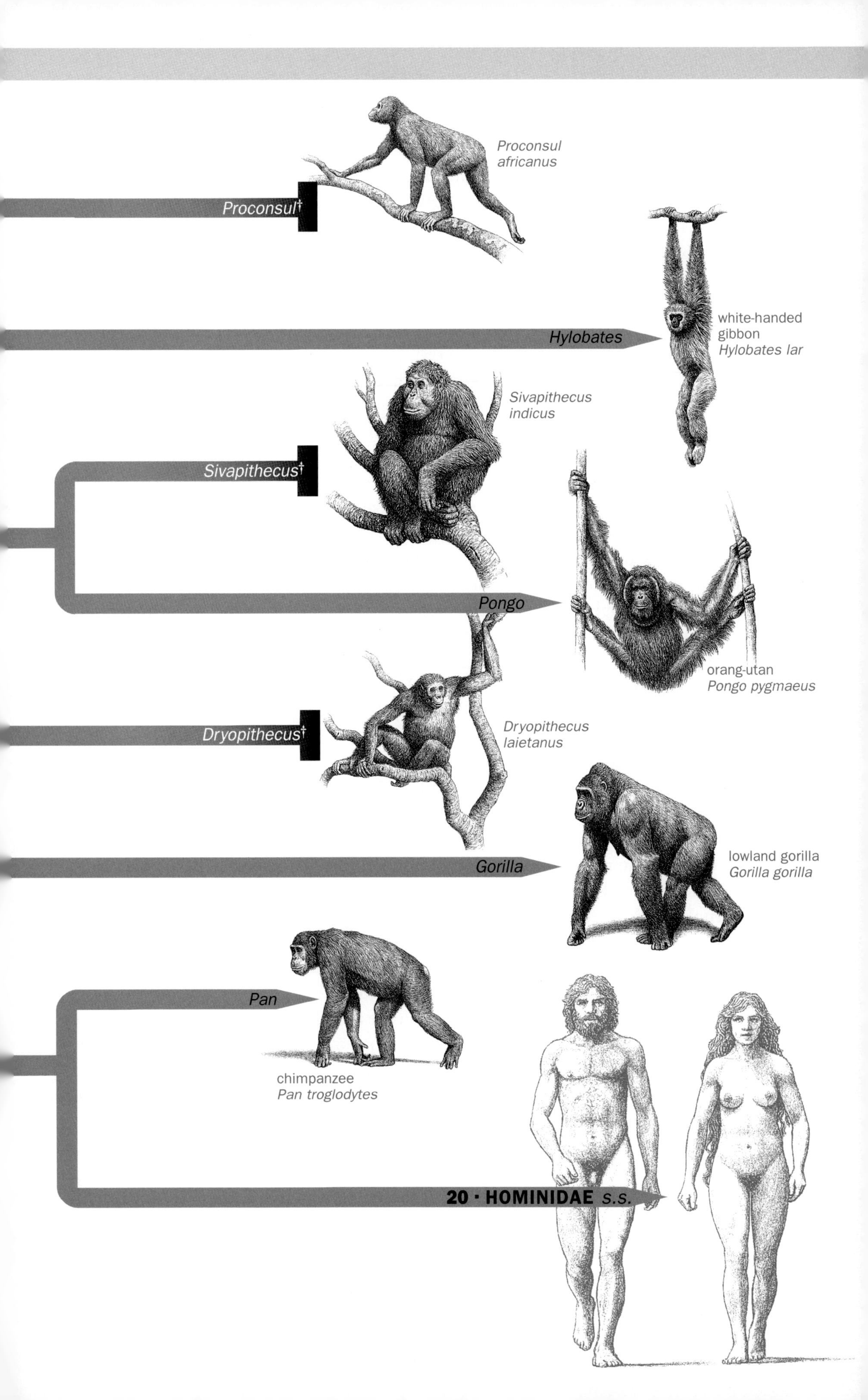
Proconsul africanus
Proconsul†
Hylobates
white-handed gibbon Hylobates lar
Sivapithecus indicus
Sivapithecus†
Pongo
orang-utan Pongo pygmaeus
Dryopithecus†
Dryopithecus laietanus
Gorilla
lowland gorilla Gorilla gorilla
Pan
chimpanzee Pan troglodytes
20 · HOMINIDAE s.s.

informally as 'archaic primates'. But Robert Martin says that the Plesiadapiformes are probably related to the colugos (Dermoptera) rather than to the Primates. I mention the Plesiadapiformes here only because they feature in many traditional and recent texts; but because they probably were not in fact 'archaic primates' they will not be mentioned again.

In general, as we move down the tree from prosimian to simian we can observe several evolutionary trends. Thus prosimians tend to have longer snouts than monkeys and apes (although baboons in particular have re-acquired a long, dog-like face). Also, prosimians tend to rely more on a sense of smell than simians do; but the reduction of the snout in simians allows the eyes to come fully to the front, giving total overlap of the two fields of vision—hence unequivocal stereoscopy.

There is also a nutritional trend. The most primitive primates are insectivorous (like the tree shrew, in the sister order Scandentia), while monkeys and apes move towards fructivory (fruit-eating) and folivory (leaf-eating). But degrees of omnivory are common throughout the primates and baboons, chimpanzees, and, of course, human beings take a significant amount of meat. The callitrichids—tamarins and marmosets—feed on insects and tree gum, but do not seem to be primitive. Apparently they are simply re-evolved the dietary habits of prosimians and scandentians.

Primates also show various locomotory trends. Thus the earliest primates were quadrupeds, with legs of equal length, narrow chests, and long bodies and pelvises. Some monkeys have retained this general form; baboons, for example, are mainly ground-living but some are arboreal and run along the upper sides of stout branches. Some lemurs, bushbabies, and tarsiers have acquired long hind legs that enable them to make tremendous leaps. By contrast, some monkeys—such as the spider monkeys of South America—and apes have short hind legs and long arms; and some, notably the spider monkeys, gibbons, and the much-heavier orang-utan and chimpanzees, have developed the art of brachiation, swinging arm-to-arm. These animals have accordingly acquired tremendous mobility of the shoulders, a barrel-like chest, shorter bodies, and a more massive, basin-like pelvis that supports the viscera when the trunk is held vertically. Human beings brought these powerful arms and chests with them when they came down to earth, and put them to other purposes: and developed the bent hind legs of the ape into what might be construed as the most energy-efficient agents of bipedalism of any animal.

Finally, but crucially, as we move down the tree the brains become larger—both relative to body weight, and in absolute terms; and with the enlarged brain goes greater social complexity. So among the monkeys and apes we find some that are monogamous (like gibbons), some that are polygynous ('many wives') with a single dominant male (as in gorillas), and some that are polygynous but with several breeding males in a group (as in chimpanzees). In general, sexual dimorphism (difference between the sexes) increases as polygyny increases. The ultra-versatile human beings have some difficulty deciding which form of family life comes most naturally, although they do seem to shade towards monogamy: up to a point.

Let us look at the components of the tree in more detail.

THE PROSIMIANS

The prosimians include the modern lemurs, lorises, and tarsiers, and the extinct adapids and omomyids. They are all obviously primates, with their big, generally forward-facing eyes, their flat nails, their sensitive padded finger tips, their two mammaries on the chest, and so on. But they also have more obviously general and primitive mammalian features: longish snouts, with heavy reliance on the sense of smell, and a brain in which the centres of olfaction outweigh those of vision. Because monkeys and apes are 'simians', 'prosimian' is not a bad descriptive term.

But 'prosimian' is *only* a descriptive term; not a phylogenetic one. As the tree shows, the deepest division within the whole primate order runs right through the middle of the prosimians. On one branch, most primatologists agree, are the lemurs and lorises, while the tarsiers are on the other. The tarsiers—so most but not all primatologists maintain—are in turn related to the monkeys and apes, leaving the lemurs and lorises out on their own. The deep division between the lemur–loris group and the tarsier group clearly happened a long time ago—probably in the Cretaceous, before the dinosaurs had disappeared, although the earliest true lemur and tarsier fossils date only from the Eocene, around 50 million years ago.

The oldest known unequivocal primate fossils of any kind date from the Palaeocene and mainly belong to two distinct groups, both long extinct and traditionally labelled Adapidae and Omomyidae. Palaeoprimatologists were once content to place the **adapids** and **omomyids** together, effectively as 'stem primates', but in recent years there have been two major refinements. First, many primatologists now believe that the adapids are close to the lemur–loris lineage, and that omomyids belong on the tarsier–anthropoid line; so this is how the two are shown here. Second, it now seems that the erstwhile Adapidae represents at least two distinct lineages, one

adapid, *Adapis parisiensis*

omomyid, *Necrolemur antiquus*

of which is probably the sister of the lemurs and the other(s) of which is (are) more distant. It is still convenient to talk in a general way about 'adapids' but because they do seem to be polyphyletic I am simply writing them informally into the tree, spelt with a lower-case 'a'. The formal family Omomyidae may or may not survive further scrutiny, but it seems prudent in any case simply to refer to them informally as omomyids.

The living prosimians are included in five families of lemurs + lorises; and the single family of the tarsiers.

The five families of lemurs and lorises

Three prosimian families—the Lemuridae, Cheirogaleidae, and Indriidae—are commonly called 'lemurs', and all of them are endemic (confined) to Madagascar. We should give thanks for island continents: without Madagascar, the lemur lineage (and others) would not have survived; without Australia, kangaroos and koalas would not have evolved at all.

• The **Lemuridae** are what Corbet and Hill call 'large lemurs'. About 11 species are known, but they are mostly forest dwellers and so there must be some out there we do not yet know about; and there is little agreement on how to divide up the genus *Lepilemur*. Be that as it may: *Lemur* is the genus of *L. catta*, the beautiful and famous ring-tailed lemur; *Petterus* includes such lovelies as the black and the brown lemurs; *Hapalemur* are the gentle lemurs; *Varecia* are the big and vociferous ruffed lemurs (the colony of which at Jersey Zoo is worth crossing the Channel for); and *Lepilemur* includes the weasel-lemur and the sportive lemur.

ring-tailed lemur, *Lemur catta*

• The **Cheirogaleidae** are the mouse-lemurs and dwarf lemurs of which seven species are known, all forest dwellers. Four genera are commonly acknowledged: *Microcebus* (embracing *Mirza*), are the mouse-lemurs; *Cheirogaleus* are the dwarf lemurs; *Allocebus*, the hairy-eared dwarf lemur; and *Phaner* is the fork-marked lemur.

• The **Indriidae** are the 'leaping lemurs'; and leap they very definitely do. *Avahi* is the woolly lemur; *Propithecus* includes three species of sifaka; and *Indri* is the indri.

Then there are two sister families—the Daubentoniidae and Lorisidae—separate from but related to the lemurs:

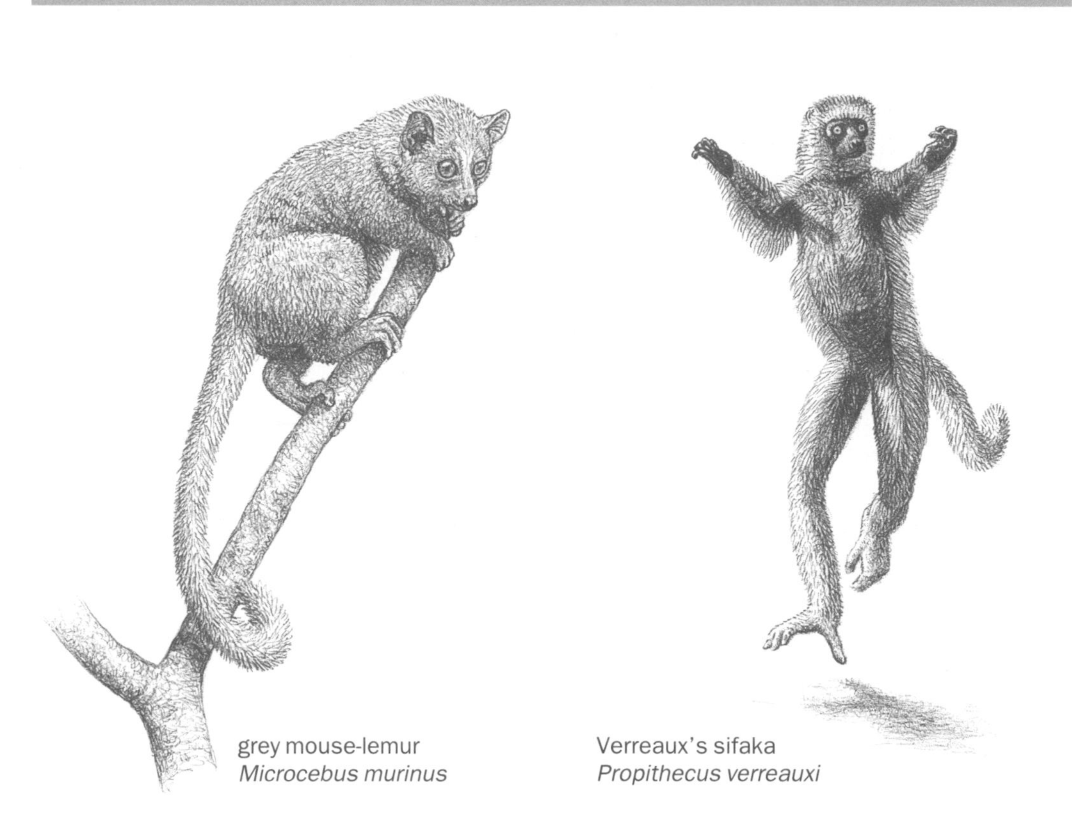

grey mouse-lemur
Microcebus murinus

Verreaux's sifaka
Propithecus verreauxi

• The **Daubentoniidae** is the family of the single species *Daubentonia madagascariensis*, the amazing aye-aye. This is the creature with the huge eyes and a thin, elongated, long-clawed middle finger with which it prises insects from beneath the bark. The combination of stare and probe disconcerts some Malagasy people who see the aye-aye as an evil spirit. I have seen aye-ayes only once, in Jersey Zoo, and they are surprisingly large—as big as a domestic cat. But their mien is of whimsy. They are sweet things. How could they seem evil?

aye-aye, *Daubentonia madagascarensis*

• The **Lorisidae** comprises the only living members of the lemur–loris clade that do not live in Madagascar; indeed they are denizens of India, Southeast Asia, and Africa. Corbet and Hill list eight genera. *Loris* includes a single species, the slender loris, from South India and Sri Lanka. *Nycticebus* includes the slow loris and pygmy slow

slender loris
Loris tardigradus

loris from Southeast Asia. *Perodicticus* is the potto from the forests of Central Africa and Kenya. *Arctocebus* is the angwantibo from along the Rivers Niger and Zaire. Then come four genera of bushbabies, all from Central, West, and East Africa: *Galago* and *Galagoides* (bushbabies), *Otolemur* (greater bushbabies), and *Euoticus* (needle-clawed bushbabies). Until recently, these four genera were taken to include about 10 species. But over the past few years Simon Bearder of Oxford Brookes University has been studying bushbabies by flashlight in West Africa (for they are nocturnal) and in places where he expected to find a couple of species has typically found half a dozen or so. On the one hand, these elusive creatures seem simply to be under-investigated; and on the other, many species of nocturnal or subterranean creatures (including various groups of owls and mice) look very similar to other species because they distinguish members of their own kind not by sight but by sound or smell. Thus the animals themselves can tell who is who, but humans—even specialist biologists—sometimes cannot. Bearder estimates that there could in truth be at least 40 species of bushbaby in Africa, and recent molecular studies suggest that they should probably be divided into at least three families. Because their forest home is being felled, however, some at least of the species will almost certainly have disappeared before biologists have a chance even to take note of them.

These, then, are the living genera of the lemur–loris clade: at least as known so far. But there remains another group of prosimians that is quite separate from the lemurs + lorises: the tarsiers, in the family **Tarsiidae**.

Tarsiers: family Tarsiidae

Tarsiers look like little bushbabies; but beneath the surface there are clear differences, suggesting that the tarsier and the loris lineages separated as far back as the Mesozoic. The features that the two groups share are primate symplesiomorphies or convergences, including the catlike faces and the leaping hind legs. There are just four species in the genus *Tarsier*, all living in Indonesia and the Philippines. But, in the light of Simon Bearder's experience with bushbabies, it would be surprising if the forests did not harbour more.

Philippine tarsier
Tarsius syrichta

MONKEYS AND APES: THE SIMIANS (ANTHROPOIDEA)

More than half of all the recognized families of living primates are prosimian: six out of 10 or 11. Ecologically, however, and in number of species, the prosimians have generally been losing out to the simians, the monkeys and apes, ever since the latter first appeared—again, probably in the Eocene. Monkeys and apes, relative to the prosimians (and, indeed, to all other mammals), have bigger brains both in absolute and in relative terms, shorter faces (which means shorter snouts), and more intricate social lives.

Although the monkeys first almost certainly arose far back in the Eocene, the oldest known unequivocal anthropoid fossils are from the next epoch, the early Oligocene, which began about 34 million years ago. One of the best known of these is *Aegyptopithecus,* from an area rich in mammalian fossils on the eastern edge of the Sahara Desert in Egypt, the Fayum region. In the days of *Aegyptopithecus*, though, the Fayum was lush, wet forest; and fossils found alongside those of *Aegyptopithecus* include the elongated, needle-like footbones of lilytrotters, which enable them to walk with perfect confidence across the floating leaves of waterlilies. Here, though, I will focus on the living anthropoids, beginning with the two families of New World monkeys.

New World monkeys: families Cebidae and Atelidae

Between them, with more than 60 species, the two families of New World monkeys account for about 30 per cent of all living primates. The New World monkeys form the infraorder of the **Platyrrhini**, meaning 'flat noses', with nostrils that point outwards; as opposed to the Old World monkeys or **Catarrhini**, which have straighter noses and forward-pointing nostrils. There seem to have been roughly as many genera of platyrrhines as of catarrhines or of prosimians since they first appeared, probably around 40 million years ago.

At first sight New World monkeys seem less varied than Old World monkeys, because they are all, and always have been, arboreal; whereas Old World monkeys have often ventured to the ground, and spread themselves through the grassland. But the forest of South America is huge and intricate, offering a host of niches, and platyrrhines have radiated accordingly. Thus, closer examination shows that the New World monkeys are more varied in lifestyle than the Old Worlders. In diet, they range from the insectivorous and gum-eating tamarins and marmosets, to the fruit-eating uakaris and the leaf-eating howlers. And whereas the standard platyrrhine locomotion is as quadrupedal climbers, as demonstrated by squirrel monkeys and capuchins, they have also evolved into squirrel-scamperers, like the tamarins and marmosets, while the spider monkeys brachiate like gibbons—though assisted by a wonderful prehensile tail, which truly serves them as a fifth limb. The related woolly monkeys also brachiate, but to a lesser extent.

Platyrrhine classification has not proved easy, not least because there has been very little help from the fossil record, which consists only of a few hundred

common marmoset
Callithrix jacchus

white-faced capuchin
Cebus capucinus

fragments. Basically, there are five obvious groupings. First, the marmosets (*Callithrix* and *Cebuella*) are clearly related to the tamarins (*Leontopithecus* and *Saguinus*); and Goeldi's monkey (*Callimico*), though somewhat different, seems to belong with them as well. Marmosets and tamarins have small and unconvoluted brains, claws instead of nails on their fingers and toes (apart from the great toe), and uncomplicated molars. They rely heavily on their sense of smell and invariably produce non-identical twins (whereas most monkeys and apes have singletons). Small brains, claws, and simple teeth look like primitive features—reminiscent even of tree shrews. Yet many modern primatologists feel that these are derived features—secondary adaptations that go with small size. Being small, they are able to exploit the region just below the forest canopy, feeding on insects and gums. Claws are better than nails as the branches are too thick to grasp in any case, and, in practice, as Alfred Rosenberger comments, they probably use their feet as grappling hooks. In locomotion (though not in diet) tamarins and marmosets are the primate answer to the squirrel.

The second obvious platyrrhine grouping includes the squirrel monkeys (*Saimiri*) and the capuchins (*Cebus*). These are the archetypal organ-grinder's monkeys, with big brains (relative to their body weight) and rounded brain cases, foreshortened faces and close-set eyes. Europeans can see marvellous chirruping flocks of squirrel monkeys at Appenheul primate centre in the Netherlands, where the primates of many species run semi-wild among the visitors.

The fruit-eating spider monkeys (*Ateles*), the woolly spider monkeys (*Brachyteles*), and the woolly monkeys (*Lagothrix*) form a third obvious grouping. The leaf-eating howler monkeys (*Alouatta*) have often been separated off into their own subfamily—

red howler monkey, *Alouatta seniculus*

black-handed
spider monkey
Ateles geoffroyi

their peculiar throats, adapted to their wonderful calls, make them seem very different. But, says Rosenberger, the howler is beyond doubt a 'divergent' spider monkey. Spider monkeys are among nature's greatest acrobats, while woolly monkeys are among the primates' greatest charmers. Those who know them well (you can see them at the excellent woolly monkey sanctuary at Looe, in Cornwall) suggest they are the most intelligent of monkeys.

The sakis (*Pithecia*), uakaris (*Cacajao*), and bearded sakis (*Chiropetes*) together form a fourth grouping. These are strange animals; the red uakari, for example, has a naked brow and crown, coloured pink to scarlet. Some live in deepest Amazonia where the river floods over vast areas for part of the year, deep enough to cover big forest trees, so the fish and the river dolphins swim among the branches. They have powerful incisors and canines for skinning thick-leaved fruit, and grinders for crushing seeds, and also eat toxic leaves. Oddly, uakaris are at times the most terrestrial of the platyrrhines, for they come to the ground to forage for seeds and seedlings when the floodwaters recede.

Finally, the titis (*Callicebus*) and owl monkeys or douroucoulis (*Aotus*) are little fruit eaters that go around in small family groups, the titis by day and the owl monkeys by night, as their name and big round eyes imply.

So how do these five groupings relate to each other? And what, then, should be the formal classification? Traditionally the New World monkeys have been divided into two families: the Callitrichidae included the marmosets, tamarins, and Goeldi's monkey, while the Cebidae included all the rest. This was logical, simple, and probably is what most vertebrate zoologists who are not primate specialists would still maintain.

But as Alfred Rosenberger comments, platyrrhine classification is still 'in a state of flux'. Thus many primatologists over the years have drawn a clear distinction between Goeldi's monkey and the other callitrichids; thus, for instance, P. Hershkovitz

owl monkey
Aotus trivirgatus

white-faced saki
Pithecia pithecia

in 1977 put Goeldi's into its own family, the Callimiconidae (and so divided the New World monkeys as a whole into three families—Cebidae, Callitrichidae, and Callimiconidae). But other primatologists have emphasized other distinctions. Thus S. M. Ford in 1986 divided the Cebidae into two: the capuchins, squirrel monkeys, titis, and owl monkeys he left in the Cebidae; while spider monkeys, the woolly spider monkey, woolly monkeys, and howler monkeys, plus the sakis, bearded sakis, and uakaris, he hived off into a new family, the Atelidae. Callitrichidae he left in its original form—to include Goeldi's, as well as tamarins and marmosets.

But the classification favoured here is the one proposed by Rosenberger, as presented in *The Cambridge Encyclopedia of Human Evolution* (1992). As the tree shows, Rosenberger retains the **Atelidae**—but this grouping no longer includes the capuchins and squirrel monkeys. These two, he feels, are closer to the marmosets, tamarins, and Goeldi's. So all five—capuchins, squirrel monkeys, marmosets, tamarins, and Goeldi's—now form a newly cast **Cebidae**. The family Callitrichidae has been dropped—or, rather, it has become a subfamily, the Callitrichinae, within the Cebidae.

Rosenberger bases his classification on cladistic analysis of esoteric features, which need not delay us. Clearly, it contains intriguing features. It suggests, for example, that the general resemblance between, say, capuchin monkeys and woolly monkeys is either due to convergence or is simply primitive; the semiprehensile tails of capuchins and the fully prehensile tails of howler and spider monkeys certainly represent convergence. The 'prosimian'-like qualities of the tamarins and marmosets, including their claws and their emphasis on olfaction, are seen as derived features. The position of *Callimico*—with some monkey-like features, and some marmoset-like features—thus becomes clearer; it is indeed a kind of half-way creature (and is sometimes popularly called 'Goeldi's monkey' and sometimes 'Goeldi's marmoset'). But, as you can see from the tree, Rosenberger does subdivide his newly defined Cebidae into two subfamilies—the **Cebinae** and **Callitrichinae**—and further subdivides the Callitrichinae into two tribes.

Within his newly defined Atelidae, Rosenberger groups the spider monkeys, woolly spider monkey, woolly monkeys, and howlers into one subfamily, the **Atelinae**; although, again, he further subdivides the Atelinae into two tribes, with the spider monkeys, woolly spider monkey, and woolly monkeys in one, and the howler monkeys by themselves in another. The sakis, bearded sakis, uakaris, owl monkeys, and titis then form a second subfamily, the **Pitheciinae**. But although this subfamily is again subdivided into two tribes—sakis, bearded sakis, and uakaris in one; owl monkeys and titis in the other—it's intriguing that he does group these superficially different monkeys together. Superficially owl monkeys and titis more closely resemble capuchins and squirrel monkeys, and traditionally were grouped with them.

Taken all in all, Rosenberger's classification of New World monkeys seems the most satisfying so far. But whether it will prove to be the last word remains to be seen.

Old World monkeys: family Cercopithecidae

The Old World monkeys now live in Africa, India, and Southeast Asia. Just one species still has a foothold in Europe—the Barbary macaque, *Macaca sylvanus*, commonly if misleadingly known as the Barbary ape (for macaques have very short tails), which lives as a tourist attraction on the Rock of Gibralter; but many fossil Old World monkeys are known from Europe. The Barbary macaque and the Japanese macaque, *M. fuscata*, are the only primates apart from human beings to live outside the tropics and subtropics (Japanese macaques have often been filmed looking miserable in the snow, and sometimes taking refuge from it in Japan's hot springs). Altogether there are about 80 cercopithecid species. Thus more than two-thirds of all primates—more than 140 species out of 200 or so—are monkeys, either Old World or New World.

The Cercopithecidae seem to divide comfortably enough into two subfamilies, the **Cercopithecinae** and the **Colobinae**. The cercopithecines have cheek pouches (or buccal sacs) in which they store food and they will eat almost anything, although they tend to focus on fruit. Their molars are somewhat non-specialized with low, rounded cusps; their incisors are relatively large, and the lower ones lack enamel on the inner surfaces—an arrangement that ensures constant sharpening, as in a rodent. Cercopithecines can be arboreal but many are terrestrial. Their eyes are close together, and their noses tend to be long, especially in baboons. The colobines, by contrast, are leaf eaters, with stomachs specialized to ferment vegetation in bulk—as in a ruminant.

Allen's swamp monkey, *Allenopithecus nigroviridis*

Diana monkey
Cercopithecus diana

lion-tailed macaque, *Macaca silenus*

patas monkey
Erythrocebus patas

Their molars carry high, sharp cusps while their incisors are smaller, relative to body size, than those of cercopithecines. Colobines are almost all arboreal, with small or almost-absent thumbs. Their eyes are wide apart, and their faces are short.

Eric Delson divides the Cercopithecinae into two tribes. The first includes four genera. *Allenopithecus* is Allen's swamp monkey. *Cercopithecus* is the large genus of guenons, living throughout sub-Saharan Africa both in forest and savannah, and including well-known representatives like the vervet, which is a subspecies of the savannah monkey, the Diana monkey; and De Brazza's monkey. *Miopithecus* is the talapoin of Gabon; and *Erythrocebus* is the patas monkey of East Africa. The second tribe of cercopithecines contains five genera. *Macaca* we have met: 16 or so species of macaque from Asia and North Africa (with a presence in Europe). *Cercocebus* are the mangabeys, from Africa. *Papio* are the savannah baboons—commonly divided into

mandrill, *Mandrillus sphinx*

guereza colobus *Colobus guereza*

five species, but in truth probably once forming a continuum across Africa and into Arabia. *Mandrillus* are the mandrill and drill—forest baboons, a kind of super-*Papio*. *Theropithecus* is the gelada from the grasslands of Ethiopia.

Delson also divides the Colobinae into two tribes. The first includes just two genera: *Colobus* monkeys from West, Central, and East Africa include some of the handsomest of all monkeys with their silky hair and their long, bushy-tipped besom-like tails; and *Procolobus* is the olive colobus of West Africa. Finally, five more genera compose the second tribe. *Pygathrix* includes the snub-nosed monkeys of northern Vietnam and China and the Douc langur of Vietnam, Laos, and Cambodia, *Semno-*

Hanuman langur, *Presbytis entellus*

proboscis monkey, *Nasalis larvatus*

pithecus includes more langurs, and *Simias* is the pig-tailed langur from the Mentawai Islands of Sumatra. *Nasalis* is the lugubrious proboscis monkey of Borneo, a leaf eater; and *Presbytis* are the leaf monkeys of India and Southeast Asia, known as surelis.

The Old World monkeys (**Cercopithecoidea**) are without doubt the sister group of the **Hominoidea**: the apes. Both groups may be ranked as superfamilies.

Apes: Hominoidea

Charles Darwin (in *The Descent of Man*, 1871) first established beyond reasonable doubt that human beings are descended from apes, and suggested furthermore that our species arose in Africa and that our closest living relatives are the living African apes—the chimpanzees of the genus *Pan* and the gorillas, *Gorilla*. Darwin is now considered to be absolutely correct on all three counts, as depicted in the tree. Furthermore, modern molecular studies from the late 1960s onwards show that human beings shared a common ancestor with chimps and gorillas only about 5–7 million years ago—with much evidence favouring the later date. But many palaeoanthropologists, who came after Darwin but were working before the molecular evidence became available, seemed anxious to separate our species as far as possible from the living apes and suggested that our own lineage may have evolved separately from the 'great apes' (the African apes plus the orang-utan) as long ago as the Oligocene, more than 25 million years ago. Our own lineage they called the Hominidae; and all great apes were bundled together into the Pongidae.

In some accounts before the 1970s the early Miocene *Proconsul*, from East Africa, was presented as an early representative of the human lineage; and the late Miocene *Sivapithecus*, from Asia, was billed as a later member of the same lineage. In short, a lineage was depicted stretching from *Proconsul* through *Sivapithecus* to *Homo*. It is because these two extinct creatures have featured so heavily in twentieth-century palaeontological literature that I am mentioning them here. But as you can see from the tree, their perceived status has changed. *Proconsul* is now seen as a generalized, primitive early ape, which can reasonably be presented as the sister of all later apes; and later specimens of *Sivapithecus* reveal it as a close relative of the modern orangutans, of the genus *Pongo*. I also include *Dryopithecus*, known from the middle to late Miocene of Europe, as a putative fossil relative of the modern African apes. But *Dryopithecus* was just one among several types of large ape living during the middle to late Miocene between about 15 and 8 million years ago, and there is as yet no consensus about the relationships of these various forms to the great apes of today.

Proconsul africanus

Sivapithecus indicus

Dryopithecus laietanus

With those fossils now put in their place, we can focus on the living families. The **Hylobatidae** of Southeast Asia include the nine living species of gibbons, all in the genus *Hylobates*, one of which, *H. syndactylus*, is commonly known as the siamang. Gibbons are arch brachiators, swinging through the trees with heart-stopping bravura; although they must fail now and again because many wild skeletons have been found with healed fractures. Gibbons are monogamous—and the males and females are very similar.

Now we come to an area of dispute. Traditionally, as we have seen, the human beings of the genus *Homo* and their immediate extinct relatives such as *Australopithecus* were placed in the family Hominidae, and the great apes (all except the gibbons) were thrust into the Pongidae. But recent molecular studies show a huge genetic similarity between chimps (*Pan*) and modern human beings; about 98 per cent of the DNA of *Homo* and *Pan* is more or less the same. *Homo* and *Gorilla* are also very similar. Many feel, therefore, that there is no good case for placing them in separate families. But others argue that family-level divisions are in any case somewhat arbitrary, and that

white-handed gibbon
Hylobates lar

Australopitheus afarensis *Homo sapiens*

on grounds apart from DNA human beings are clearly very different from chimps, and indeed have created a new ecology that makes them qualitatively different in their way of life, and their impact, from all other animals. Mere genetics, these traditionalists say, is not the *only* basis for classification. According to this argument, it is important only to draw the tree correctly, to show the true phylogeny; but the rankings accorded to each clade are devised on commonsensical grounds, for convenience. Traditionalists, then, would keep the Pongidae / Hominidae distinction.

I confess I am torn. If primatologists as a whole decide to place all great apes in the family Hominidae and drop the term Pongidae, so be it. (Incidentally, the term Hominidae was coined first, by Linnaeus, and the rules of nomenclature dictate that the later name, Pongidae, is the one to jettison.) If the majority elects to reserve Hominidae for human beings and the australopithecines, fair enough. Or, if they decide to do what is tentatively shown here—Hominidae reserved for the modern Africans and Pongidae for orang-utans and *Sivapithecus*—then I am happy with that too. I am trying merely to make the point that there are various opinions, and you may come across any one of them in the literature. I discuss this nomenclature problem further in Section 20.

Anyway, somehow or other four living genera have to be divided among the **Pongidae/Hominidae**. *Pongo* includes the two subspecies of orang-utan, the Sumatran and Bornean; the two are somewhat different (there is even a small chromosome difference) although they readily interbreed in zoos without apparent disadvantage. *Pan* includes two species of chimpanzee. *Pan troglodytes* is the 'common' chimpanzee and *Pan paniscus* was traditionally called the 'pygmy chimpanzee', but most primatologists these days favour the alternative name 'bonobo' for the latter; indeed, the term

chimpanzee, *Pan troglodytes*

orang-utan, *Pongo pygmaeus*

lowland gorilla, *Gorilla gorilla*

'pygmy chimpanzee' can be considered defunct. *Gorilla* is traditionally taken to include just one species, *G. gorilla*, but with three races—the Western lowland gorilla, the Eastern lowland, and the mountain gorillas (the ones made famous by Dian Fossey); but some now feel that the different groups should be acknowledged as separate species.

Whether chimpanzees or gorillas are more closely related to human beings is not known for certain and I could legitimately have shown gorillas, chimpanzees, and humans as a trichotomy. But on balance, by a short head, molecular studies seem to support the general anatomical impression that chimps are closer; hence the present arrangement.

Human beings and their immediate (extinct) relatives are discussed in the next section.

20

HUMAN BEINGS AND OUR IMMEDIATE RELATIVES

FAMILY HOMINIDAE *s.s.*

THE '*s.s.*' IN THE TITLE is short for *sensu stricto*, which is Latin for 'in the strict sense'; as opposed to 'Hominidae *s.l.*, which means *sensu lato*, 'in the broad sense'. *Sensu stricto* implies that 'Hominidae' refers only to the genus *Homo*—our own genus—and its immediately related genera, all of whom, as it happens, are extinct. The extinct hominids are *Ardipithecus*, which is the most ancient known hominid genus; *Australopithecus* ('the australopithecines'), which probably evolved from *Ardipithecus*; and *Paranthropus* ('the paranthropines'), which is essentially a heavy-jawed ('robust') version of *Australopithecus*. There may be other hominid genera (*sensu stricto*) that we don't know about yet, and some of the four that are already established should, perhaps, be subdivided further, to form yet more genera. In particular, *Paranthropus* should perhaps be divided into *Paranthropus* and *Zinjanthropus*. But *Ardipithecus*, *Australopithecus*, *Paranthropus*, and *Homo* are the widely acknowledged quartet that appeared on Earth, chronologically, in that order. 'Hominidae *sensu lato*' would also include *Pan* (the chimpanzees), *Gorilla*, perhaps *Pongo* (the orang-utan), and some extinct types such as *Dryopithecus*.

Of course, this is all very messy, and possibly even tedious, but there really are two current schools of thought, as outlined in Section 19. Some palaeoanthropologists declare that chimpanzees, gorillas, and human beings are so similar, genetically, that they should all be placed in the same family; and that this family must be called Hominidae, rather than Pongidae, because the term 'Hominidae' was coined first and must take precedence. Some would also include the somewhat more distant orang-utans within that newly defined Hominidae *sensu lato*, and some would not. Other palaeoanthropologists, equally distinguished but more traditional, point out that *although* humans are so close genetically to chimps and gorillas, we have nonetheless developed a very different body form and a totally novel way of life. Indeed, we could argue that our ecology is so different from all other creatures, including chimps, that we virtually form a new kingdom. Therefore, say the traditionalists,

Hominidae should include only *Homo*, and our immediate, extinct relatives: and this is Hominidae *sensu stricto*.

Besides, the traditionalists argue, when the huge existing literature of palaeoanthropology refers to 'hominids', it invariably means 'members of the Hominidae *sensu stricto*'. 'Hominid', in almost all the published books and papers, emphatically is *not* intended to include the African apes. But if we now broaden the term 'Hominidae' to include chimps and gorillas then they become hominids too, and all the traditional literature will become confusing. We could say, 'Well, that's the price we have to pay for accuracy!'. But this argument is less convincing than some might feel. To be sure, the phylogenetic tree should reflect the literal truth as far as it can be known, and should emphasize the genetic propinquity of chimpanzees, gorillas, and human beings. But the Neolinnaean names that we apply to the various clades and subclades are there for convenience—to ease understanding and communication. Because, therefore, it would create huge confusion to redefine 'hominid', and to rename the Pongidae (or part of it), it is best to stay as we are.

On balance, I side with the traditionalists. To redefine 'hominid' would cause more trouble than it is worth. Some might argue on ethical grounds that we might treat chimpanzees better if we acknowledged them as hominids—be less inclined to use them to test vaccines, for example. But I reject such argument. We should treat other animals with respect because they are our fellow creatures, not simply because we perceive them to be closely related to ourselves. Surely no one would argue (would they?) that orang-utans can be treated worse than chimps, simply because they are more distant from ourselves. Then, again, no infrangible principle of systematics is being compromised if we retain the traditional Hominidae and Pongidae, because readers of this book know that Neolinnaean Impressionist taxonomy is intended simply to carve up the trees into convenient chunks—for ease of thought and communication: an exercise in librarianship. To this end, Neolinnaean Impressionism admits paraphyletic groups, like Crustacea* and Reptilia*; and, by the same token, it accepts that taxa are in practice defined by grade, as well as by clade. Humans do belong on the same narrow clade as chimpanzees and gorillas, but beyond any doubt at all we have founded a new grade. It makes perfect sense, therefore, to define Hominidae narrowly. From now on, in this section, when I say 'Hominidae' or 'hominid', I mean '*sensu stricto*'.

There are many anatomical differences between our own genus, *Homo*, and the ape genera, epitomized by *Pan*. But three sets of differences are outstanding:

First, we (*Homo*) are the most proficient bipedal creatures that have ever lived. We stand and walk perfectly upright and yet are superbly balanced, not because we are counterweighted fore and aft like a bipedal dinosaur with its prodigious neck and tail that act as a tightrope-walker's pole, but because our nervous systems are so beautifully co-ordinated. In the same way, modern fighter planes are aerodynamically unstable—compared, say, with a World War I biplane—but are kept aloft by responsive on-board computers. Accordingly, we hold our spines vertically, with the skull perched on top like the knob on a mace; the place on the skull that articulates with the top of the spine, the **foramen magnum**, is well beneath the brain case. Our long, muscular legs form upright columns beneath the vertical trunk. Chimpanzees, by contrast, move quadrupedally when on the ground, with the spine nearer to the horizontal. The foramen magnum is not directly beneath the skull, as in a human, but further towards the back, so that the head is carried in front of the near-horizontal spine. Chimpanzee hind legs are short and bent. They can walk bipedally, but only at a waddle. No chimp could ever compete in the bipedal stakes with Carl Lewis, or indeed with any of us. There is evidence that some at least of the known australopithecines—notably *A. afarensis*, which is one of the oldest—also walked upright; not exactly in a human way, but certainly more like a human than a chimpanzee. This is one reason for including it in the Hominidae *s.s.*

Second, human beings have enormous brains both in absolute terms, and relative to body size. Thus the brain of a chimpanzee is around 400 millilitres in volume, which is big by most mammalian standards; but our own brains are an average 1450 ml in size. As a rule of thumb, fossil hominids have tended to be admitted to the genus *Homo* (as opposed to *Australopithecus*) if their brains exceed around 700 ml. The brains of the ancient *Ardipithecus*, and of most australopithecines and paranthropines, are more chimpish in size.

Third—although this is a minor character compared with the first two—human beings have much flatter faces than apes, and much smaller canine teeth. Thus modern chimps and gorillas have huge, dog-like canines that they use in display, whereas human beings have smaller teeth that are useful for eating, and serve as extra incisors. Modern apes also have huge brow ridges while we do not. But some ancient species of *Homo*, including *H. erectus* and *H. neanderthalensis*, did have prominent brows; so brow ridges (as opposed to prominent faces) are not characteristically apish. *Ardipithecus*, *Australopithecus*, and *Paranthropus* also had smaller canine teeth and generally flatter faces than chimpanzees.

It is not yet 100 per cent clear that human beings are closer to chimpanzees than they are to gorillas. Gross anatomy suggests that this is so—we look more like chimps than gorillas—but we know that appearances can be deceptive. The molecular data seem to suggest that chimps are slightly nearer to us than gorillas are, but the evidence is marginal. All the same, most anthropologists tend to assume that we are closest to chimps; in other words, *Pan* is the sister genus of the hominids.

As mentioned in the last section, Darwin was the first to state clearly, as a discrete scientific hypothesis, that chimps and humans are close relatives. But many biologists, as well as many laypeople, were reluctant to accept this, even as recently as the 1960s. What shook the palaeontological world to its core were the DNA studies that suggested that chimpanzees and humans had in fact diverged only about 3 million years ago. Some biologists were simply appalled by this news, much as many were appalled in 1859 when Darwin suggested that all living things had evolved from some primeval common ancestor. Others pointed out that apes and humans simply could not have gone their separate ways quite so recently. After all, *Australopithecus africanus* was already known by the 1960s—the first specimen was found in 1925—and was a bona fide hominid; yet this creature seems to have lived almost 3 million years ago. So hominids must have split from apes some time earlier than that. But the molecular evidence continued to accumulate in the 1970s and 80s. Combined molecular and fossil evidence now convinces most palaeoanthropologists that apes—meaning the direct ancestors of *Pan*—split from the first hominids somewhere between 7 and 5 million years ago, in the late Miocene.

The fact that chimpanzees and gorillas are African, plus the modern fossil and molecular evidence, suggest that the split between apes and hominids did indeed take place in Africa, which is what Darwin proposed in *Descent of Man*. (He was right about most things.) In fact, as we will see, *most* of the really big transitions in hominid evolution took place in Africa: all four of the known and acknowledged hominid genera arose in Africa, including *Homo*; and so did our own particular species, *Homo sapiens*.

What prompted hominids to diverge from the apes—to walk so skilfully on two legs, and evolve such enormous brains? And which came first? Did the first humans learn to walk on two legs before they evolved big brains, or were they brainy before they were upright?

WHY AND HOW DID HUMAN BEINGS EVOLVE?

First of all, we know that in general the temperature of the whole world has been dropping over the past 40 million years,[1] and from time to time the cooling has occurred in bursts. One such burst occurred around 5 million years ago, and another about 2.5 million years ago; and the cooling was exacerbated in southern and eastern Africa as the shuffling of continental plates caused the land to rise, so that Nairobi and Johannesburg are now a couple of thousand metres above sea level. When the temperature goes down, rainfall also declines and trees, which in general prefer the climate warm and moist, tend to give way to grass. Thus around 5 million years ago the vast tropical forests of southern and eastern Africa began to turn into open woodland—that is, woodland in which the trees are spaced and there are large gaps in the canopy, as in a medieval English forest (like Robin Hood's Sherwood). Around 2.5 million years ago the open woodland began to give way to wooded savannah, of the kind that still covers much of Africa. Pockets of dense forest remained, but largely only on mountains and along river valleys.

So it seems eminently likely that the first hominids—first *Ardipithecus*, then *Australopithecus*—were adapted to open woodland. Phylogenetically they were hominids—their fossil remains show their relationship to ourselves—but in brain size they were far closer to the apes. Ecologically indeed they were like woodland apes; the apish equivalent of baboons. But around 2.5 million years ago, when the open woodland declined further into wooded savannah, some of those australopithecines grew taller, and began to develop bigger brains, and so gave rise to the first *Homo*.

[1] The reasons for this, and indeed the background to human evolution, are explained in my own book, *The Day Before Yesterday* (1995), which in the USA is titled *The Time Before History*.

The question, 'Did upright walking evolve first, or did brains evolve first?', has given rise to much palaeoanthropological *angst* and infighting. Late Victorian and early twentieth-century anthropologists wanted to believe—indeed, they more or less took it as read—that early hominids evolved big brains before they evolved bipedalism and upright bodies. Those early scholars liked to believe, in fact, that human evolution had been 'brain led'. This seemed to give supremacy to the brain, which was considered very right and proper. Thus they envisaged that our earliest hominid ancestors must have had big heads on apish bodies—or that their heads combined huge rounded brain cases, like ours, with protruding, apish jaws. In 1912, a still-unidentified hoaxer fostered this conceit by burying the dyed skull of a modern man (from an Anglo-Saxon burial) alongside the dyed and doctored jawbone of an orang-utan at a gravel-pit at Piltdown in Sussex, and leading British palaeoanthropologists accepted this concoction as our direct ancestor—the famous, or infamous, 'Piltdown Man'. The fraud was not finally detected until the 1950s (through chemical analysis of the dye and of the bone) although some biologists had long since smelled a large rat. The fraud succeeded, however (perhaps more than its perpetrator intended), because it fulfilled the preconceptions of the day. Large cranium on apish jaw was what the scientists expected, and large cranium with apish jaw was what they found.

Yet the fossil that first showed the lie of this idea was found in 1925, at a lime pit near Taung (or Taungs as it was then called) in northern Cape Province, South Africa, and correctly described as a hominid by the anatomist Raymond Dart (1893–1988). The fossil was part of a skull. Its face and teeth were clearly hominid in shape, but its cranium was small. So here was an early hominid—a putative human ancestor of *Homo*—with a small brain and a human face. Dart called his new hominid *Australopithecus africanus*, meaning 'the southern ape from Africa'. The name *Australopithecus* is unfortunate: 'australo-' sounds like Australia; 'pithecus' means ape, although Dart wanted to emphasize his fossil's humanity; and 'australo-' is Latin while 'pithecus' is Greek, which annoyed the classical purists. (Dart himself was born in Australia, which also annoyed the purists.) But if *Australopithecus*, with a human face and a small brain, was the direct ancestor of human beings, then Piltdown Man, with an apish face and a large brain, could not be. The discoverers of Piltdown preferred the fake, however, and until the 1950s, when the fakery was finally revealed, *A. africanus*, the real thing, was given a rough ride. Since Dart's discovery of the Taung skull, other remains of *Australopithecus africanus* have been found in South Africa. It has proved difficult to date them precisely but all are probably between 3 and 2 million years old.

The Taung skull alone did not show whether *A. africanus* walked upright. So it still did not settle the big question of human evolution—did a big brain come first, or upright walking come first? This was answered by a fossilized early hominid skeleton found by the American palaeoanthropologist Don Johanson and his team in 1974 at Hadar in Ethiopia, which they later called *Australopithecus afarensis*. The skeleton—nicknamed 'Lucy'—included part of a skull, and a remarkable amount of the postcranial skeleton, including part of the arm bones, the pelvis, and the leg bones. These bones, and many others found later, showed beyond reasonable doubt that *A. afarensis* walked upright. Later, in 1978 and 1979, at Laetoli in Tanzania, Mary Leakey's team found 3.6-million-year-old footprint trails, supposedly left by *A. afarensis*, showing that australopithecines walked very much as we do now. *Australopithecus afarensis* lived between 3.8 and 3 million years ago, and is presumed to be the immediate ancestor of *A. africanus*. So now the matter is clear: walking upright, in virtually the same way as we do, clearly preceded big brains. In short, human evolution was not 'brain led'. Our ancestors led with their feet.

So when, why, and how did the brain develop? As explained above, the world's temperature took a sudden dip about 2.5 million years ago, and the vegetation accordingly became harsher. The australopithecines, already well established in open woodlands, apparently responded in two ways. One australopithecine lineage—or, more probably, several lineages (at least one in southern Africa, and at least one in eastern Africa)—developed larger, more heavily boned jaws, with huge millstone teeth for grinding. These 'robust' types are generally lumped together in the genus *Paranthropus*—although, as we will see, *Paranthropus* should probably be divided into two genera, each perhaps including two species. Another, separate group of australopithecines retained their small jaws and became brainier. By about 2.3 million years ago these smaller-jawed, more 'gracile' types had brain volumes of around 700 millilitres, which, albeit arbitrarily, is commonly taken as the threshold of *Homo*. Elisabeth Vrba, palaeoanthropologist at Yale University, has commented that '*Paranthropus* tried to chew his way out of trouble; and *Homo* thought his way out of trouble'. As things have turned out, the *Homo* strategy has prevailed.

There is, however, a world of difference between the 700-millilitre brain of the first *Homo*, and the 1450-millilitre brain of humans today; yet this twofold increase was achieved within about 2 million years. To biologists brought up in the traditional belief that significant evolutionary change requires a huge expanse of time, this is a startling rate of increase. So how and why did it happen?

HOW DID WE BECOME SO BRAINY?

We could wave our arms as many have done in the past and suggest that big brains and corresponding intelligence are *bound* to be advantageous, and so are bound to be favoured by natural selection; but this argument will not do. Brains are expensive organs; our own brain takes 20 per cent of all our metabolic energy (whether we are cogitating or resting—and so, unfortunately, we cannot get slim just by thinking hard). Unless we use our intelligence for purposes that enhance survival and reproductive success, and unless such gains outweigh the obvious costs, then natural selection will *not* favour braininess. Besides, almost all lineages of mammals have developed bigger brains over the past 65 million years or so, since the dinosaurs disappeared—the exceptions are specialists like koalas, which live on toxic leaves. But no other lineage has developed anything like the ratio of brain to body size that *Homo sapiens* now enjoys. So we cannot simply wave our arms. We have to explain why the advantages of extreme braininess outweigh the costs in our particular lineage but not, apparently, in other mammalian lineages; and how, given that that must be the case, our brains had the wherewithal to increase so rapidly over the past 2 million years.

In truth, there seem to be three main plausible hypotheses. The first is based in the simplest form of natural selection, which has been called 'survival selection'—in which creatures are favoured that are best able to overcome life's day-to-day vicissitudes; the second is rooted in Darwin's idea of sexual selection—the notion that the characters of an animal are in part shaped by its need to attract mates; and the third is founded in the notion of 'social selection', which is simply another variant of natural selection, and proposes that animals are likely to survive best when they are able to harmonize and compete effectively in their social groups. We can look at these three ideas in turn.

BRAINS FOR SURVIVAL

A larger brain would not enhance survival unless the increase could be immediately translated into action that brought some pay-off. Thus (as I speculate in *Day Before Yesterday*) a codfish that was suddenly endowed with the brain of Jane Austen would be at no advantage at all, because it would have no physical means of writing down its thoughts, and no one to appreciate them if it did. Such a marvellously endowed fish would undoubtedly be selected *against*. Its attributes would be costly but would bring no reward.

The first ground-living hominids descended from tree-living, largely fruit-eating apes. In the trees, and as fructivores, hominid ancestors had evolved gripping, dextrous hands—and extremely mobile shoulders. This latter point is rarely emphasized but when you think about it, the all-round flexibility of our forelimbs is rare among mammals, yet vital to our success. Cats, for example, supremely agile creatures that they are, cannot whirl their arms through virtually all points of a sphere as we are able to do. Little old ladies in the aerobics class may whirl their arms somewhat creakily but at least they can do it. A cat, more lithe than Rudolf Nureyev, cannot do this. And consider: the dexterity of the hand would largely be wasted if the shoulders were not so mobile. If a horse had the hands of a master tailor it still could not thread a needle, because it could not bring the two hands into the correct relative positions.

Primitively, tetrapods are quadrupedal: all four legs are used for walking. Among the minority that have explored bipedalism some, like *Tyrannosaurus rex*, have abandoned the forelimbs virtually entirely: *T. rex* is a pair of jaws on legs; its forelimbs are vestigial. Birds have turned their forelimbs into specialist wings (or sometimes turned the wings into paddles, as in penguins); but if they abandon flight (whether in air or under water) the wings again become vestigial, as in ostriches and kiwis. Some mammals can walk or stand on two legs up to a point, and do use their forelimbs as 'hands'; but most of these part-time bipedalists remain primarily quadrupedal, and still must use their hands for locomotion. This is true of bears, squirrels, and of most monkeys and gorillas, which are quite manual, but are still basically four-legged.

But when humans came to the ground and became bipedal, they did not simply abandon their forelimbs like *T. rex*, or develop limbs with a single, dedicated purpose, like birds. Instead, they began to employ the dexterity and mobility that had evolved for life in the trees for new purposes, and, furthermore, for a *variety* of new purposes. Once freed from the chore and strain of locomotion our hands and arms became general all-purpose tools and weapons. Among mammals, the only organ of comparable versatility and mobility is the elephant's trunk.

Even before our hominid ancestors evolved the mobility and strength and arm of, say, Achilles, or the dexterity of Picasso, their newly liberated arms and hands would have been useful. Carrying is useful: it opens whole new possibilities for food gathering and child care. Toolmaking is, of course, useful. Chimpanzees make various tools from twigs and leaves, as Jane Goodall discovered; and some chimps, it is now known, even work in stone (although, as forest animals, they work more generally with plant material). Stones at their crudest are clubs for smashing bones and nuts and releasing the good things that lie within, and with only minimal chipping develop sharp edges for slashing and skinning.

Much under-emphasized, too, is the human skill with missiles. Some chimps can throw well; but humans, who are better balanced, can throw better. The missile offers quite new strategic possibilities. Human beings are among the very few animals that can kill or maim at a distance; and—apart from the very occasional chimp—we are the only large mammals that can do this. The only other animals with even a semblance of such ability are archer fish (which capture flies with squirts of water), spitting cobras, and a few spitting and web-throwing spiders; although electric eels also stun or kill at a distance, without recourse to projectiles. Creatures that can kill at a distance can attack large prey without running the gauntlets of hoofs and horns, and delay the attacks even of the most dangerous predators. With a stone, David felled Goliath. With spears, modern-day pygmies bring down forest elephants. The missile, like the stone tool, did not simply make the lives of our ancestors easier. It afforded them an ecological strategy quite distinct from anything that had gone before. Tools and missiles turned us into a new kind of animal; the kind that can kill an animal as big as itself, or bigger, without personal risk.

An animal that can make tools would surely benefit from making tools that were even better; and one that could hunt large prey and outface large predators would surely benefit from superior strategies. It seems to me, then, that any increase in brain (with commensurate increase in intelligence) would bring immediate rewards in such a creature, and would therefore be favoured by natural selection. But once an already dextrous animal developed a larger brain, natural selection would favour the evolution of even greater dexterity—to make better use of the new mental ability. So our ancestors developed a feedback loop of the kind that in many contexts is known to have produced rapid and tremendously precise evolutionary adaptations. The principle is that of 'coevolution'. Darwin noted it in the context of orchids and the moths that pollinate them: the flowers of the orchid become adapted to the proboscis of the moth, which in turn adapts more finely to the peculiarities of the orchid, and so on, until we find two fantastical creatures, each beautifully adapted to the other but each reliant on the other. In hominids we see coadaptation of two different organs in the same animal: brain adapts to hand and then hand to brain, and so on.

This, then, is one route by which our ancestors might have evolved their enormous brains so rapidly: the first hominids already had dextrous hands; big brains enabled them to use those hands more efficiently; this provided impetus for further brain development; and so on. That is the traditional kind of explanation. But Geoffrey Miller, now at University College London, has proposed a quite different mechanism—based on Darwin's idea of evolution by the mechanism of sexual selection. Darwin pointed out that animals cannot reproduce unless they find mates; for many animals, this requires attracting a mate. So it is that male spiders, fish, birds, mammals, and what you will put on mating displays, to show the females how fine they are.

BRAINS FOR SEX

The qualities that male animals display to attract their mates are often wild and fanciful: the plumes of the bird of paradise, the peacock's tail. Yet such qualities are not simply fatuous. Their bearers are dandies but not popinjays, and the female preference is not mere whimsy. For example, as Bill Hamilton of the University of Oxford has pointed out, birds with bright feathers are more likely to be free from parasites than those with dowdy feathers; a colourful, costly display is a certificate of health. Whatever the underlying cause, the male's proclivity for display, and the female's predilection for the males that display most gloriously, set up another feedback loop; a kind recognized by the great mathematician-biologist R. A. Fisher and known as 'Fisher's runaway'. Peacocks with big tails mate with females that have a preference for big tails. The sons of this mating inherit the genes that will give them big tails, and the daughters inherit a preference for big tails. Thus the two characters—the ability to grow big tails and the preference for big tails—coevolve, reinforcing each other and becoming more exaggerated generation by generation.

Miller suggests that early male hominids used their big brains just as peacocks use their big tails: to impress potential mates. In such a context, the brain is not used simply to utilitarian ends, any more than a peacock seeks to impress its mate by superior powers of flight or an extra keen eye for nutritious seeds. The display is deliberately flippant, to show that the actor has strength or intelligence to spare. Thus male hominids, over time, developed vocal skills, wit, and artistry to impress their potential mates; and the more they did so, the greater their mating success. Miller suggests that great poetry and painting, from the caves of Lascaux to the National Gallery of London, are primarily exercises in sexual display.

Finally, we could develop larger brains to help us get on with our fellow humans. Speaking morally—or from the simple standpoint of survival—nothing is more important.

BRAINS FOR SOCIALITY

Human beings are tremendously social animals. Indeed, our great intelligence would bring severely limited benefit unless we shared our thoughts with others, so that human beings operate together a great collective intelligence—another huge innovation. Matt Ridley argues cogently in his recent *Origins of Virtue* (1996) that human beings rely for their survival on the division of labour between individuals to a degree that is matched only by the eusocial insects, like bees and termites. Indeed, we depend on each other for hour-by-hour survival; and we are able to work together as closely as we do because we have such finely honed social skills. These skills prove remarkably complex once analysed, and again require enormous brain power. As Robin Dunbar of the University of Liverpool has pointed out, primates with the largest brains have the most intricate social systems.

So here we have three plausible routes that could have prompted the enormous brain growth of human beings between 3 and 1 million years ago; an unprecedented, and, of course, unrepeated burst that, among other things, illustrates the potential that may lie within dynasties of creatures, perhaps waiting for the opportunity to become manifest. The three routes, of course, are not mutually exclusive, and I feel that all must have played their part in developing the human brain. Sexual selection could encourage the brain to evolve rapidly, despite the cost, just as it has encouraged the evolution of the costly peacock's tail: indeed, to a large extent this evolutionary route succeeds *because* the organ in question is extravagant. On the other hand, it is hard to see how the sexual selective mechanism could have come into play unless early humans were already brainy; and it is easy to see how the hand–brain feedback loop could have operated to set the process in train. Both mechanisms involve some measure of coevolution: hand with brain in one case, male display with female preference in the other. Put both together (and there is no reason whatever why they should not work side by side) and we see why the human brain is so adept in the serious skills of living—hunting, gardening, craftsmanship—and also why we have such a taste for whimsy and apparent excess. Increasing social intricacy also encourages yet further co-operation of hand and brain. For example, a tool-using society works more efficiently if people share the tasks. The people who make the best spears are not necessarily the best hunters. Ideally, the artificers and the hunters should evolve enough social skill to work together.

In short, we can never know for certain why our ancestors developed such enormous brains so rapidly. But there are some very plausible hypotheses, rooted in the fossil evidence, in our knowledge of living human societies, and in evolutionary theory that is well founded (and to a large extent testable and tested); and this, perhaps, is the closest we can hope to come to an understanding of long-gone biological history. So let us return to the creatures themselves, and what the fossil record can tell us.

The following is based largely on the research I carried out for my book on human evolution, *The Day Before Yesterday* (1995), updated to take account of the new fossils found since publication. Sources for *Day Before Yesterday* are detailed in the Bibliography.

A GUIDE TO THE HOMINIDAE

A modern taxonomic tree, put together on cladistic principles, has two agenda: it provides a classification, of course, but also—because it is based on phylogeny—it is an attempt to summarize evolutionary history. It does not represent the absolute, unequivocal truth, because the unequivocal truth cannot be known. It does represent a brave attempt (preferably the best possible attempt) to get as near to the truth as possible, with the evidence available. But the evidence of hominid history is particularly meagre (although it is getting better all the time), and so leaves wide scope for interpretation, and hence for argument. Besides (so some observers suggest) palaeoanthropologists seem to be particularly individualistic, and the field is shot through with rivalries, so that many contrasting hypotheses find fierce advocates. In short, there is no broad consensus; and even if there were, there could be no guarantee that the consensus was true. The Universe, and history, are as they are; they do not adjust themselves to fit what scientists happen to agree is the case.

The tree presented here, like all trees, is on the one hand a hypothesis—a hypothetical summary of hominid ancestry and relationships; and on the other it is a compromise, between many different and rival trees in many different publications. It may be wrong but it is at least sensible. It shows all the known and recognized hominid species except the newly discovered *Australopithecus bahrelghazali* from Chad, which is too little known to be included with confidence. *Pan* is shown as the sister group of the hominids; *Ardipithecus ramidus* is presented as the sister group of all the other hominids; *Australopithecus anamensis* is shown as the sister group of all the other australopithecines, plus *Paranthropus* and *Homo*; and *Australopithecus afarensis* is shown with a pivotal role, as the possible sister group (implying ancestor) of all later hominids. The trichotomy in the middle of the tree—with *Australopithecus africanus* and *A. garhi* on one tine, the paranthropines on another, and *Homo* on the third—is ugly but honest. The relationships between these various lineages is simply unclear.

By reading the tree from top to bottom, however, we can get at least some idea of the evolutionary route that led to us.

The first hominid: *Ardipithecus ramidus*†

Ardipithecus ramidus is one of the latest hominids to be described—by Tim White from the University of California at Berkeley and his colleagues in 1994. But it is the oldest known certain hominid: the fragments that are known include key diagnostic body parts. The remains were found in 1992–3, lying on fossilized volcanic ash (tuff) in the Afar region of Ethiopia, at Aramis near the Awash River, and are dated at 4.4

20

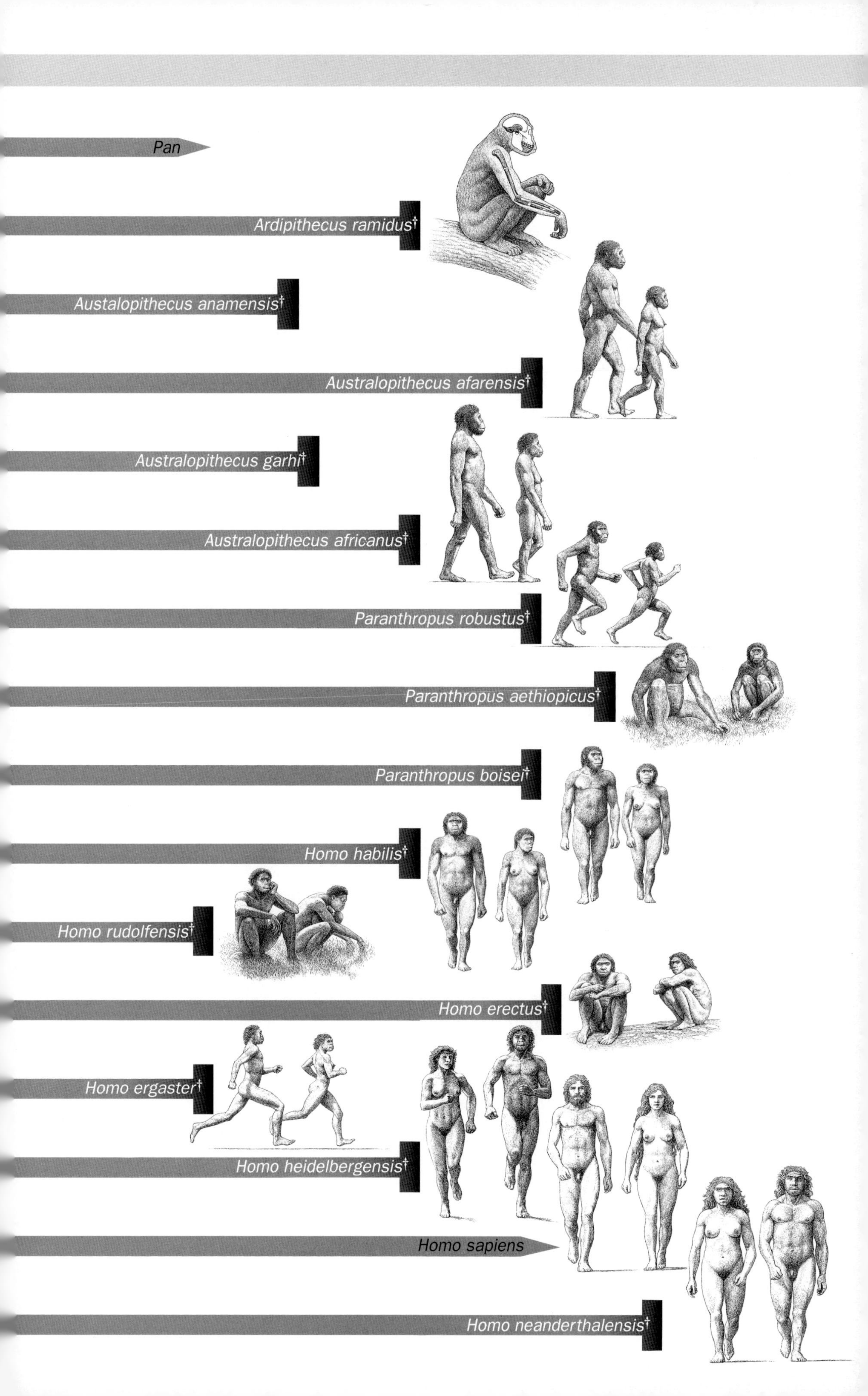

Pan
Ardipithecus ramidus†
Austalopithecus anamensis†
Australopithecus afarensis†
Australopithecus garhi†
Australopithecus africanus†
Paranthropus robustus†
Paranthropus aethiopicus†
Paranthropus boisei†
Homo habilis†
Homo rudolfensis†
Homo erectus†
Homo ergaster†
Homo heidelbergensis†
Homo sapiens
Homo neanderthalensis†

Ardipithecus ramidus

million years ago. At first White and his colleagues called their newly discovered hominid *Australopithecus ramidus*, after the australopithecines that were known already, but later decided that it was different enough to warrant a different genus. '*Ardi*' means 'ground' or 'floor' while the specific epithet '*ramidus*' comes from the Afar word 'ramid' meaning 'root'.

The Aramis fossils between them revealed that *ramidus* had precisely the combination of characters that palaeontologists were hoping to find and indeed were expecting: more ape-like than *A. afarensis*, less apish than a chimp, and with qualities of both. The fragments provided a complete set of teeth, including one deciduous or 'milk' tooth, plus the bases of two skulls and, most unusually, pieces of all three bones from a left arm. The teeth reveal an animal that really does seem to be half-way between an ape and *A. afarensis*. The canines are still prominent but, says White, they are 'low and blunt' compared with a chimp's. In addition, the bottom canines are worn as if they were used for eating rather than display. The enamel of the canines and other teeth is thin like a chimp's, not thick like a human's; it is not as thick even as that of *A. afarensis*. The milk tooth was a first, lower milk molar—a tooth that has always proved particularly instructive in deciding who among the apes and hominids is related to whom. In *ramidus*, this tooth is narrow like a chimp's, not wide like a human's. According to White and his colleagues it is 'far closer to that of a chimpanzee than to any known hominid'.

The arm bones are broken so it is impossible to say whether in life they would have been long like a chimp's or shorter like the known hominids. But the bits that do remain show a 'mosaic' of chimpish and hominid details. Unfortunately the key issue—whether *A. ramidus* walked upright like a human or more shambling like an ape—is not yet resolvable. The remains include no fragments of bones or feet. But the shape of the back of the skull does suggest uprightness—the foramen magnum is certainly further forward than in a chimp. So *A. ramidus* does seem to have been well on the way to bipedalism. It is at least a plausible 'link' between australopithecines and the apes.

One intriguing hypothesis suggests that *A. ramidus* may actually predate the split of chimpanzees and hominids—that indeed it may be a close relative of the common ancestor of them both. This is certainly plausible. *Ardipithecus* is more like a human than it is like a chimp but it is a mistake to think that chimps are in all respects more primitive, and therefore older, just because they are more obviously ape-like. Chimps have many derived features that probably evolved after the split with the hominids; and the common ancestor of us and chimps may well have been at least as much like us as like them. Here, however, I am taking a cautious view—though following the broad censensus; and am assuming that *Ardipithecus* arose after the split, and that it is indeed at the base of the hominid (*s.s.*) lineage.

It seems likely on present evidence, therefore, that *Ardipithecus* gave rise to *Australopithecus* and that between them they spanned 2 million years—as great a period as *Homo* itself.

The australopithecines: *Australopithecus*† and *Paranthropus*†

Within the australopithecines we see two evolutionary trends. Some remained 'gracile'—meaning they had relatively light bones, with modest teeth and jaws; and some became 'robust', developing heavy jaws and teeth suitable for grinding coarse vegetation. The gracile policy seems to have worked better. All the australopithecines have long disappeared but whereas the robusts have died without issue, one or other of the graciles presumably gave rise to *Homo*.

The oldest known species of *Australopithecus* is *A. anamensis*. Fossils were first found in the 1960s, but it was confirmed as a distinct species only in the mid-1990s, when Meave Leakey and her colleagues found more fossils in northern Kenya. All the *A. anamensis* specimens so far date from around 4.2 to 3.8 million years ago. In general, *A. anamensis* resembles *A. afarensis* (which will now be described in more detail).

Until *A. anamensis* came on the scene, the oldest known australopithecine was *Australopithecus afarensis*. The first specimen of this species—'Lucy'—was discovered in 1974 at Afar in Ethiopia. 'Lucy' herself lived a little over 3 million years ago but it now seems that the *afarensis* species first appeared about 3.8 or 3.7 million years ago and lasted until about 3 million years ago—so 'Lucy' was a late representative. The species lived throughout East Africa, and, for an australopithecine, lasted for a long time.

Australopithecus afarensis

Australopithecus afarensis, like *Ardipithecus ramidus*, shows a nice mosaic of humanish and apish qualities, though leaning more towards the human. Its human-like qualities include the straight, stilt-like legs, which we now know allowed a true bipedal gait; and when it stood tall it was 1–1.5 metres high. Its arms were shortish relative to its legs, though not as relatively short as a modern human's. But its brain was small—an apish 400–500 millilitres in volume. It also retained at least some of the apish prominence in the canine teeth. In general, *A. afarensis* was a slim-boned creature—a gracile skeleton—with a modest jaw and teeth, evidently adapted for an omnivorous diet.

The next oldest australopithecine in East Africa that we know about was *Australopithecus garhi*. Part of a skeleton of this australopithecine was reported from Bouri in Ethiopia only in 1999; its discoverers—the Ethiopian palaeoanthropologist Berhane Asfaw and his colleagues—dated the finds about 2.5 million years old. They also found stone tools of the same age nearby, and suggest that *A. garhi* made those tools. If so, then we would have to bury forever the idea that human beings—genus *Homo*—can be defined as creatures that make stone tools. Moreover, Asfaw and his team found animal bones near the *A. garhi* fossils that were scarred as if they had been butchered, which suggests, first, that *A. garhi* was a serious meat-eater and, second, that it used tools for this purpose.

Australopithecus africanus—the species that Raymond Dart identified and named in 1925—was a contemporary of *A. garhi*, but lived in southern rather than eastern Africa, from around 3 to nearly 2 million years ago. *Australopithecus africanus* stood about 1.1–1.4 metres high. Its canine teeth were less prominent than those of *A. afarensis*—another apparent shift towards the human condition. As outlined above, it was not until the Piltdown hoax was finally exposed in the 1950s that everyone felt able to acknowledge the true significance of *A. africanus*: perhaps as our own ancestor or at least, as a modern cladist would say, as the sister of *Homo*.

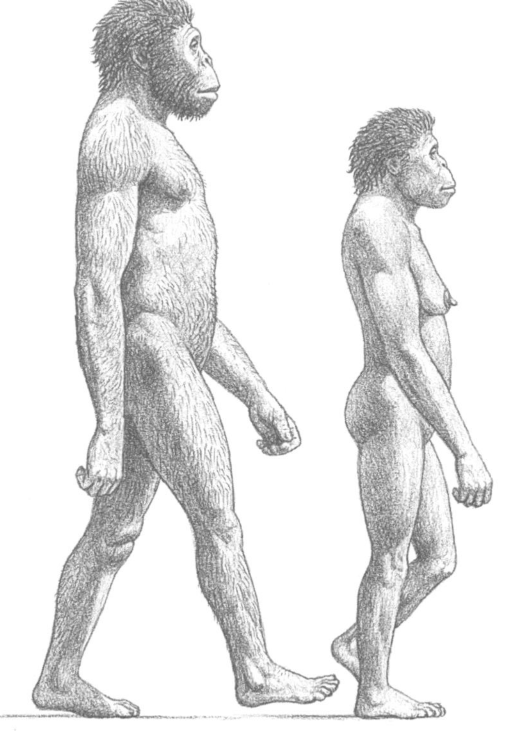

Australopithecus africanus

Finally, yet another gracile australopithecine—*A. bahrelghazali*—was described recently from Chad. It is dated at 3.5–3.0 million years ago and because it lived about 2500 kilometres (1800 miles) west of the East African australopithecines (namely *A. afarensis*) that are known from that time, it shows that early australopithecines were far more widespread than has been appreciated. Too little is known of *A. bahrelghazali* to venture its inclusion in the tree.

At the time of *A. africanus* the world's climate cooled yet again; the African woodland gave way to savannah, and the coarse vegetation that goes with it, and at least one australopithecine lineage developed big teeth and heavy jaws to house them in, evidently shifting from an omnivorous to a more vegetarian diet, and adapting to the tough plants that now prevailed. This lineage gave rise to the genus now generally known as *Paranthropus*. These creatures are also still commonly referred to by the old soubriquet of 'robust australopithecines' although they should now, of course, be

Paranthropus robustus

called 'paranthropines'. In fact, there were two distinct groups who *may* represent quite different lineages and should, if they do, have different generic names.

For the time being, however, most palaeoanthropologists recognize *Paranthropus robustus* who lived in southern Africa from about 2 to 1 million years ago; again a shortish creature, around 1.1–1.3 metres tall, but heavier than the gracile australopithecines, and with a brain size of around 530 millilitres—generally bigger than in the graciles (but not relative to body weight). The second group of paranthropines lived in East Africa, and, it seems, also in Malawi. Two species are recognized. The best known and later kind was *Paranthropus boisei,* who was first discovered by Mary Leakey in 1959 and named *Zinjanthropus* or 'Nutcracker Man' by her husband Louis. The species is estimated to have lived from about 2.3 to around 1.2 million years ago. If it is finally decided that the East African paranthropines are not related to *P. robustus*, then they will need a new generic name, and *Zinjanthropus* has precedence. *Paranthropus boisei*—or *Zinjanthropus boisei*—was extremely heavily built, stood about 1.2–1.4

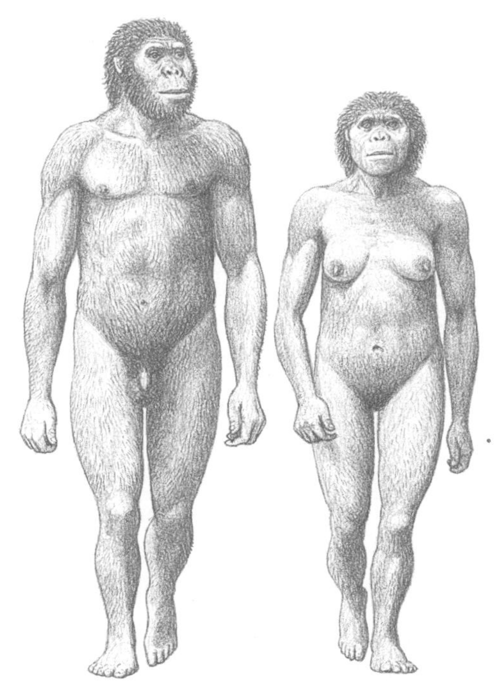

Paranthropus boisei

Paranthropus aethiopicus

metres tall, and had a smallish brain of around 410–530 millilitres. Immediately preceding *P. boisei* in East Africa was a similar but less-extreme type who is given separate species status, as *P. aethiopicus*. A remarkable Kenyan fossil of this paranthropine, called the 'Black Skull', was found by Alan Walker in 1985. The species lived from around 2.6 to 2.3 million years ago.

Another group of australopithecines responded differently to the new, harsher conditions of around 2.5 million years ago. This lineage retained the gracile body shape of *A. afarensis* and *A. africanus* but over time became taller and brainier. Thus emerged the genus *Homo*.

The first people: *Homo habilis*† and *Homo rudolfensis*†

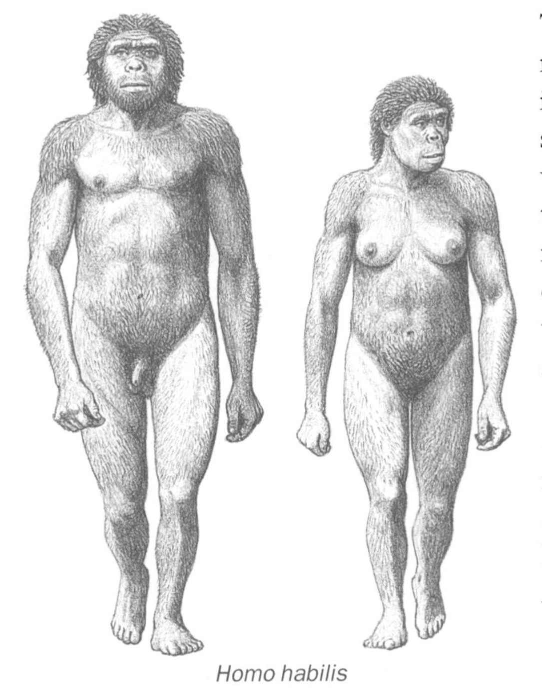

Homo habilis

The earliest known hominid that can reasonably be ascribed to the genus *Homo* is *Homo habilis* from eastern and possibly southern Africa. The biggest *H. habilis* were a little taller than the australopithecines, though still only about 1–1.5 metres high; but their brains, at around 600–800 millilitres in volume, were distinctly larger, although still about half the size of the modern human brain (range 1200–1700 ml). Louis Leakey, John Napier, and Phillip Tobias first discovered and named *H. habilis* in 1964, but met with scepticism. Some felt that the new hominid was not different enough from *Australopithecus* to justify being placed in the genus *Homo*, while others said that it was just another *H. erectus*. But Bernard

Homo rudolfensis

Wood, now at George Washington University in Washington DC, says that *H. habilis* is certainly different both from *Australopithecus* and from *Homo erectus*. He also suggests that the 'habilines' seem to form two distinct species: *H. habilis* and *H. rudolfensis* (*rudolfensis* was named by another specialist after Lake Rudolf, now known as Lake Turkana). *Homo habilis* and *H. rudolfensis* apparently lasted only from about 2.3 to 1.6 million years ago. They seem not to have left Africa.

The habilines have several features that make them extremely interesting. For example, their remains are associated with crude stone tools, which, until the recent finds of *A. garhi*, made them the earliest known hominid toolmakers—hence the name '*habilis*', meaning 'handy'. Phillip Tobias suggests that the inside of the cranium of *H. habilis/rudolfensis* shows that the habiline brain had the beginnings of a Broca's area—the region that, in modern humans, is associated with speech. So did habilines also have the beginnings of language? Many doubt this; but the idea is certainly interesting.

Tall and upright people: *Homo erectus*† and *Homo ergaster*†

One of the habilines—Bernard Wood suggests it was *H. rudolfensis* rather than *H. habilis*—gave rise to *Homo erectus* ('upright man'), who first is known from about 1.8 million years ago. Again, there seems to have been more than one species (a second one is called *H. ergaster*, or 'working man'), and it is simpler to refer to them collectively as '*erectus*-grade' people. In height they compared with modern *Homo*; their brains ranged from around 750 to 1250 millilitres, so there was some overlap with the larger habilines at the lower end and with small-brained modern humans at the higher end. *Erectus*-grade people were probably the first hominids to migrate out of Africa, possibly as long ago as 1.6 million years. They reached the furthest reaches of Asia; indeed the first *H. erectus* to come to light was found by Eugène Dubois at Trinil in Java in 1890 (although he called his discovery *Pithecanthropus*). There are reports of *H. erectus* surviving in China and Java as late as 100 000 years ago, but some palaeo-

Homo erectus

Homo ergaster

anthropologists include these people in the more-modern *H. heidelbergensis*—or say that these late forms belong in another species. It is yet another area of controversy.

Erectus-grade people, or their stone tools, have been found on various islands of Southeast Asia—so how did they get there? In some cases there is no problem: Java and Sumatra were once part of the mainland, and became cut off as the sea level rose (most recently, at the end of the last Ice Age). Now, however, Australian and Indonesian scientists have found stone tools on the island of Flores that they propose were made by *erectus* people, some 840 000 years ago. But Flores is surrounded by deep water and could never have been attached to the mainland. To reach Flores, those ancient people must have undertaken at least two sea voyages, one of 25 kilometres. Is it conceivable that *erectus* people built boats so long ago? How else could they have crossed the open ocean?

Many *erectus*-grade fossils have now been found in Africa. According to Bernard Wood, *H. erectus* has specialized features that indicate that it could not have been our direct ancestor, as is traditionally supposed; it was an evolutionary dead end—albeit a highly successful one. For example, the skull of *H. erectus* slopes inwards at the back in a sharp shelf, as ours does not. Wood believes that *H. ergaster* is the more convincing human ancestor: it stayed at home (in Africa) and evolved to give rise to the hominids that have been known as 'archaic *Homo sapiens*' but which many palaeoanthropologists now call *Homo heidelbergensis* (named after the German city that in the past has yielded some fine 'archaic' fossils): *H. heidelbergensis* is the ancestor both of the Neanderthals and of ourselves.

'Archaic' people: *Homo heidelbergensis*† and *Homo neanderthalensis*†

A certain amount of taxonomic confusion arises at this point. The first 'archaic' humans probably arose in Africa at least 400 000 years ago and possibly as much as 800 000 years ago (if palaeoanthropologists agree that new finds from Gran Dolina in Spain are indeed 'archaic'). In general, the adjective 'archaic' seems reasonably apt. These people did have significantly bigger brains that those of the *erectus* grade and, at around 1100–1400 millilitres, the biggest ventured well into the range of modern humans. But the 'archaics' also had primitive or apparently primitive features, including prominent brow ridges and low, sloping skulls. They would certainly have been conspicuous in the average bus queue.

Confusion arises because people of 'archaic' grade are known from virtually all over the Old World (they never reached the Americas) and they varied considerably in detail in different places and at different times. For this reason, palaeoanthropologists have coined all kinds of different specific names for the different 'archaics', such as *Homo rhodesiensis* for 'Rhodesia Man', *Homo daliensis* for the form found in China, and so on. Nobody really knows, however, whether these populations were truly isolated from each other, in which case it would be reasonable to ascribe them to different species, or whether they more or less formed a continuum, as modern human beings do. The division shown here is a compromise. Most of the creatures traditionally called 'archaic *Homo sapiens*' are ascribed to *H. heidelbergensis*. The one traditional 'archaic', with its own specific status, is *H. neanderthalensis*, Neanderthal Man. The first Neanderthal that truly attracted attention was found in

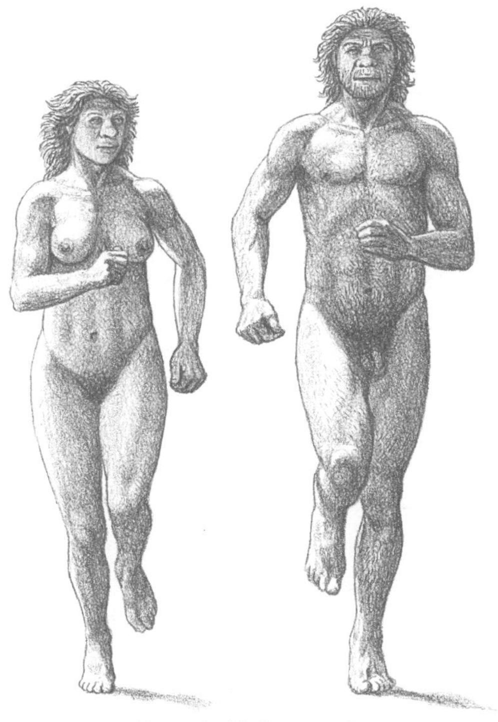

Homo heidelbergensis

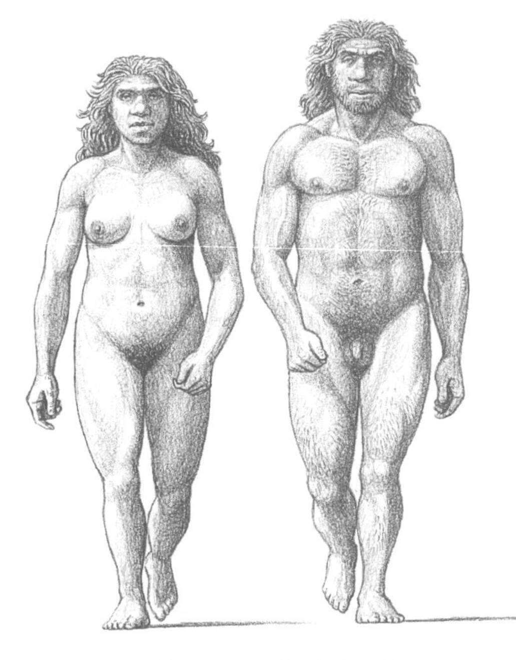
Homo neanderthalensis

the Neander Valley near Düsseldorf in Germany in 1856 and the bones were at various times said to belong to a deformed lunatic and/or to a Cossack soldier. Only later were they recognized as an ancient type of human being.

The relationship between Neanderthals and the 'archaics' (*H. heidelbergensis*), and indeed with *H. sapiens*, remains uncertain. Some regard Neanderthals as an extreme form of 'archaics', whereas others have treated them as a subspecies of modern human beings (that is, *H. sapiens neanderthalensis*). The tree here suggests that *H. neanderthalensis* is a separate species, and is the sister group of *H. sapiens*, both species arising together from among the ranks of *H. heidelbergensis*. This may or may not be the case, but it is reasonable (and tidy).

Neanderthals had big brains—the biggest, at 1750 millilitres, exceeded even the largest known modern human brain—but they also had huge brow ridges, wide flat noses, receding chins and foreheads, big teeth, and, above all, very heavy bones with large areas for muscle attachment. Physically, then, the Neanderthals were extremely powerful, but their mental ability is hard to judge. It seems that they buried their dead, and there is some (disputed) evidence that they did so with ceremony, which suggests some cultural attainment. But they evidently made the same kinds of tools throughout their existence so clearly they were not innovators; and whether they had anything resembling speech is again controversial. Neanderthals are known from the Middle East and through all of Europe, but the most extreme types lived towards the end of their time on Earth, after 90 000 years ago, in northern Europe. It has often reasonably been suggested, therefore, that their extreme physical development was an adaptation to cold. But the fossil record seems to show a gradual evolution of Neanderthal-type features from less-extreme types over a period of 300 000 or so years.

Neanderthals survived until about 30 000 years ago. Fully modern human beings—true *H. sapiens*, just like us—arrived in Europe about 40 000 years ago, so the two species evidently lived side by side for about 10 000 years. Why Neanderthals eventually died out is unknown. Perhaps our own ancestors exploited the territory more efficiently. They may simply have been skilful hunters or more astute gatherers of plants—but I feel that there was probably more to it than this. I suggest that whereas Neanderthals were never more than hunters and gatherers, modern people began from an early stage to manage their environment.

Archaeologists traditionally argue that human beings first began farming around 10 000 years ago in the Middle East, in the 'Neolithic Revolution'—which, after all, is what the archaeological evidence seems to suggest. But I argued in *Neanderthals, Bandits, and Farmers* (1998) that by Neolithic times, people had long since learned to control the movement of animals—for example, by the judicious use of bushfires, in the manner of modern Australian Aborigines—and to protect and propagate the plants that were most useful to them. In other words, our own Palaeolithic ancestors were not mere hunters and gatherers: they were game managers and quasi-horticulturalists—a combination that I have called 'proto-farming'. Present-day hunting and gathering people usually turn out to be 'proto-farmers' in this sense, once anthropologists look closely. Thus *H. sapiens* possibly outcompeted *H. neanderthalensis* by superior game and land management. Or perhaps the increasingly invasive activities of *H. sapiens* began, in the end, to interfere directly with the more passive efforts of the Neanderthals.

modern people, *Homo sapiens*

On the other hand, it is at least possible that our own ancestors, with their superior brain power and technology, simply hunted Neanderthals and wiped them out. Genocide has been common among modern human beings. Perhaps it is an ancient proclivity.

Palaeoanthropologists still discuss whether and to what extent modern humans interbred with Neanderthals before the latter disappeared, and how many specific Neanderthal genes are now circulating through the modern human population. As related in Part I (Chapter 2), some Neanderthal fossils have miraculously yielded fragments of DNA, which suggest that there was little interbreeding between them and *H. sapiens*. But this is not conclusive.

Modern people: *Homo sapiens*

Our own species, *Homo sapiens*, apparently first arose about 100 000 years ago—or maybe somewhat earlier—again in Africa (although there is also evidence from the Middle East). Early moderns had light bones, flat high foreheads, domed skulls housing brains of around 1450 millilitres, and small teeth and jaws. People of precisely that description are known to have reached western Europe about 40 000 years ago. The first Australians arrived from Southeast Asia at least 40 000 years ago but, more likely, as long as 60 000 years ago. They could have made their journey only with some sort of seacraft—an extraordinary feat. Asian people crossed into North America later, at least by 15 000 years ago, and spread south to Tierra del Fuego over the following 2000 years or so, evidently killing most of the big species of wild mammal along the way in an episode known as 'the Pleistocene Overkill'. The first Americans arrived not by sea but by crossing a wide area of land called Beringia that links Alaska and Siberia during ice ages, when the sea level may fall by almost 200 metres.

HOW MUCH DO WE REALLY KNOW?

In general, most modern palaeoanthropologists envisage that the major groups of hominids—*Ardipithecus*, the gracile and robust australopithecines, the genus *Homo*, the *erectus*-grade people, the 'archaics', and modern *Homo sapiens*—all originated in Africa. Darwin surmised that hominids must have arisen in Africa, though even he could not have envisaged just how great the African contribution really is. The conventional view, then, has it that some of the hominid lineages—*H. erectus*, the 'archaics', and modern *H. sapiens*—migrated out of Africa, in a series of diaspora.

But some palaeoanthropologists tell the history of human evolution somewhat differently. A few, for example, argue that *Homo*—in the form of *Homo erectus*—emerged from Africa only once, and then evolved in different sites into many different forms, which grew into the modern races. Most, however (including me), consider this scenario is a little eccentric. I believe that even the three-migration model—first *erectus*, then *heidelbergensis*, then *sapiens*—is probably too conservative. It seems to me far more likely that from the time of *H. erectus* onwards populations of hominids probably migrated out of Africa many times—and migrated back again: perhaps killing other populations that they met along the way, and perhaps mating with them, and probably both. All we can gather from the fossils, after all, are the tiniest clues to what was probably an extremely rich picture; as if we had found just a wing tip or two and a few grenades from the battles of World War II.

More and more fossil evidence, too, tends to push the significant events in human prehistory back into time. As we have seen, the earliest stone tools (from eastern Africa) now seem to date back to at least 2.5 million years ago, probably made by an australopithecine. The shaft of a spear has turned up in Germany. It is balanced as finely as a modern javelin, but it is a million years old and was presumably made by an *erectus*. And, as we have seen, *H. erectus* might have had the capability of building some sort of water craft by 840 000 years ago.

The four hominid genera that are now widely acknowledged contained, between them, at least 15 species, as shown on the tree. Perhaps the fossils that are known already between them represent more than 15 species; some—taxonomic 'splitters'—would put the figure as high as 20, and some even higher. Probably, too, there have been hominid species in the past that we simply do not know about at all. Nonetheless, we probably do know roughly the number of hominids that have lived in the past. During most of the past 3 million years, and until about 30 000 years ago when the last of the Neanderthals disappeared, there have been several different hominids at any one time. Just over a million years ago, for example, *Paranthropus robustus* was still living in South Africa, the even more robust *Paranthropus boisei* was doing well in East Africa, *Homo ergaster* was getting into his stride in Africa, and *Homo erectus* had spread through Asia to China. Robert Foley of the University of Cambridge has argued on ecological grounds that Africa and Eurasia were unlikely to have supported more than five hominid species at any one time; if that is so, then we seem to know at least a fair proportion of them.

This, then, is a lightning overview of our own family; but our discussion of animal phylogeny should not end with human beings. Human beings are mammals, and mammals were not the most recent major clade of vertebrates to evolve. The class Aves is newer than the class Mammalia. Proceeding chronologically, then, we should now look at the birds.

21

THE BIRDS
CLASS AVES

SOME GROUPS OF CREATURES are hard to pin down—'What *are* they exactly?'—but birds seem unmistakable. Like mammals, they are warm-blooded, or homoiothermic; they have feathers; and their whole bodies are beautifully adapted for flight. Thus the skeleton is reduced and fused for strength: the ribs and sternum (breastbone) form a light and compact box and the breastbone of all modern birds (except in the flightless ratites like the ostrich and its kin) juts forwards in a deep 'keel' that provides anchorage for the huge flight muscles. The bones of the hand are struts for hanging feathers from, and the pelvis and tail effectively form one unit. Indeed, the bony tail has almost gone—reduced to a stumpy **pygostyle**; what we call the 'tail' of the bird is just extended feathers. Individual bones, especially in a small bird, seem almost weightless; reduced to a minimalist web of internal struts, a triumph of mechanical engineering. Birds lack teeth—a bony and horny beak serves just as well, sometimes serrated to improve the grip, as in mergansers. The air rushes through the lungs and into air sacs beyond, reducing weight and speeding the exchange of oxygen. The skull is rounded, housing a brain that is geared to perception and the extraordinary co-ordination that flight requires. All in all, it seems, birds are unmistakable: homoiothermy, feathers, and the adaptations of flight carry the day.

For more than a century, too, the evolutionary history of birds seemed straightforward. Ever since biologists first began to feel that evolution is the general way of things, they have agreed that birds have descended from reptiles. Modern birds have many overtly reptilian features, including scaly legs. More specifically, the highly accomplished amateurs of the early nineteenth century soon realized that birds are diapsids; and Thomas Henry Huxley, the great friend and defender of Darwin, was among those who argued that their ancestors are to be found among the dinosaurs. Although some biologists still try to link them with the crocodiles, and others look for an ancestor among the most primitive archosaurs or for a forebear that is not an archosaur at all, most are content to place the bird lineage—the monophyletic class Aves—firmly among the carnivorous, theropod dinosaurs. For committed cladists, birds *are* dinosaurs. When they want to refer to *Tyrannosaurus rex* and its ilk, they call them 'non-avian dinosaurs', to distinguish them from ducks and sparrows.

The link between birds and reptiles in general (though not specifically with dinosaurs) was confirmed beyond all reasonable doubt after 1861 when palaeontologists in Bavaria described an almost complete skeleton of *Archaeopteryx*. An even better skeleton turned up in 1877 and four others (and a single feather) have been found since. *Archaeopteryx* seemed wonderfully to combine the features of reptiles and modern birds. It clearly had feathers, for their imprints are miraculously preserved in the rock that surrounds the fossil bones. It had unmistakably bird-like wings—although it still had three functional claws at the bend of each wing. It also retained a long reptilian tail, and reptilian teeth. It dated from the Jurassic, around 150 million years ago—when, of course, dinosaurs were still at their very considerable peak.

For a long time, then, and until a few years ago, it seemed very easy to define what a bird is and to see (at least in broad outline) how the modern group evolved, and the creatures they evolved from. But evolutionary biologists seem to have fallen once again, as they all too often have in the past, into the philosophical traps known as **orthogenesis** and **teleology**. Orthogenesis is a view of evolutionary history that shows lineages of creatures changing in a straight line, without deviation from some presumed ancestral state to the modern condition. Thus, human evolution has often been presented as a straight race from shambling, small-brained apedom to the glorious, upright, cerebral paragons we are today; but as Section 20 describes, hominid evolution has taken many turns (including the paranthropines in the early days) and our own particular triumph was by no means pre-ordained. Teleology (from the Greek *teleos*, meaning 'end') implies that evolution follows a plan—that a lineage of creatures 'knows' in advance what it is going to evolve into. Thus in orthogenetic and teleological vein, biologists have tended to depict the evolution of birds as an unwavering and apparently prescribed progression from the weakly flapping *Archaeopteryx* to the falcon and the albatross—with just a few eccentrics, like the penguin, abandoning flight along the way in favour of some tempting alternative lifestyle.

But although there must have been a lineage of birds that led from an *Archaeopteryx*-like ancestor to the modern peregrine, this is by no means the only course that the class of Aves has pursued. In fact, the history of birds raises difficulties at every turn. Most fundamentally, it is not quite as easy as it once seemed to define what a bird is—or not, at least, when extinct types are taken into account; and although some descendants of *Archaeopteryx* (or some similar creature) must have pursued the aerial route to modern bird-dom, others emphatically did not. We should look beyond the neornithine subclass that happens to survive, to the whole avian assemblage; and when we do that, the concept of 'bird' becomes a little harder to describe than at first it seems.

THE PROBLEM OF FEATHERS

Feathers, until recently, seemed to be the feature that everyone could agree upon. If a creature has or had feathers, then it is or was a bird. No argument. But in 1998, Ji Qiang and his colleagues from China, Canada, and the USA reported that they had found partial skeletons from two turkey-sized theropod dinosaurs in the Liaoning province of China, dating from the late Jurassic or early Cretaceous. These two dinosaurs are very interesting, for although there is nothing to suggest that they could fly, they very definitely had feathers. Qiang and his colleagues called their new finds *Protarchaeopteryx robusta* and *Caudipteryx zoui*.

No one supposes that these newly found feathered dinosaurs are the ancestors of *Archaeopteryx*. Were it so, it would be an extraordinary coincidence. Besides, they could be from the late Jurassic, and so are too young. But they must be related to *Archaeopteryx*, for it does seem likely that feathers evolved only once. So it seems that *Archaeopteryx*—which is classified as the oldest known member of the Aves—belongs to a larger group of feathered dinosaurs, which included *Protarchaeopteryx* and *Caudipteryx*. Unless we want to increase the scope of the class Aves to include such creatures, we have to say that feathers no longer define the Aves. They can no longer be taken as a synapomorphy of Aves. They are, indeed, a primitive feature, shared with some other dinosaurs. Of course, if we consider only the living birds, then such niceties need not delay us. But the specialists have to rethink.

The new Chinese dinosaurs also prompt us to reconsider the origin of feathers—why they evolved in the first place. There have been three main hypotheses: insulation, flight, and display, which mainly means sexual display.

The insulation idea at first sight seems strong. As we saw in the case of mammals, there are several sound reasons for becoming warm-blooded, even though the cost of maintaining body heat is very high. Some biologists still argue that some dinosaurs were homoiothermic, and the smaller ones at least would have benefited from insulation. Modern birds surely could not fly as they do unless they had the kind of high metabolic rate associated with homoiothermy (although this must be seen only as a bonus; natural selection does not look ahead, and could not have favoured homoiothermy simply because, at some time in the future, it might favour flight!). The new Chinese finds are in line with the insulation idea; or, at least, they show that feathers are not always associated with flight, which leaves insulation as a reasonable alternative.

On the other hand, as Luis Chiappe of the American Museum of Natural History in New York has argued, it now seems that some bona fide birds dating from the Cretaceous were not fully homoiothermic; as discussed in more detail later, this can be inferred from the fine structure of their bones. These birds were much younger than *Archaeopteryx*, and if they were not fully homoiothermic, then neither, perhaps, was *Archaeopteryx*. But if feathers evolved before their owners had evolved warm-bloodedness, then this suggests that they first evolved for reasons other than insulation: like, for example, flight. But then we come back to the Chinese dinosaurs, which had feathers but clearly did not fly. We can, of course, envisage some compromise. For example, some modern mammals are less homoiothermic than others: marsupials in general maintain a lower body temperature than eutherians do, and edentates (armadillos, sloths, and their kind) allow their body temperature to track the ambient temperature, at least to some extent. Partial homoiothermy is, therefore, a known and established strategy; and feathers would assist this. Even so, the hypothesis that says that feathers means warm-bloodedness does not seem to work as neatly as might be supposed; neither does the idea that feathers means flight; and neither, it seems, does the notion that feathers equals birds—unless we define 'birds' more broadly than seems sensible.

The notion that I believe does stand up is the one that has been least discussed—that of sexual selection. Sexual selection in general was underestimated for at least a century after Darwin first proposed it in 1871 (in *The Descent of Man and Selection in Relation to Sex*). Biologists seemed to think that the mere attraction of mates—showing off—was far too frivolous to be a serious driving force in evolution. Yet a little modern thinking shows that nothing is more serious, for a creature that does not find a mate does not pass on its genes; whereas, contrariwise, a gene that enhances an animal's ability to find a mate much increases its own chances of being passed on. The peacock is the world's exemplar: no creature invests more obviously in sexual display. But all creatures must invest the time and trouble needed to attract a mate. At least, those that do not die without issue, and their genes die with them.

Modern reptiles are wizards of sexual display. With their frills and flashes many would do credit to a peacock. Many mince and bob like a bird of paradise. Feathers seem to be derived from scales. What better device by which to show off: with big flashy scales, serrated and barbed to diffract the light. The features of feathers that now seem to us to be caprice—the elusive colours, the shimmers, the crests, the vibrations—were perhaps the prime movers that drove feathers to evolve in the first place. Once the quills and plumes were in place, then they could be adapted to the more mundane business of insulation, or indeed for flight. But you only have to watch a modern reptile as it struts before its intended mate, to see how its relatives might have benefited from feathers without any reference to insulation and no thought at all of flight.

Once the feather had developed, however, it was pre-adapted for other uses. Reflecting the sun and trapping the air, it could easily become an excellent agent of insulation—brought to supreme heights in the breast of the modern swan or the eider. It could also become the supreme instrument of flight: light, strong, stiff but flexible, and with a wide surface area. Perfect. Bird flight is not built around flaps of skin, as in pterosaurs and bats, but around feathers. But although modern bird flight is wonderful, and its advantages are obvious, in its earliest days it must have been crude, inefficient, and painful. So how did it evolve in the first place?

THE PROBLEM OF FLIGHT

Archaeopteryx clearly flew. It had feathers, and the curve and asymmetry of the pinions show beyond reasonable doubt that they were used for flight. But how did it fly, and how well? Was it a powered flier, driven by flapping wings, or merely a glider like many a squirrel? Some say it could not have flown because it lacked the anatomy: in particular, it did not have the deeply keeled breastbone that anchors the flight muscles in modern birds. But others point out that bats fly very well without such refinements: they attach their power-stroke muscles elsewhere, and so might *Archaeopteryx*. Most now agree that it was a genuine flapper, albeit a weak one. With its long tail and its paper-aeroplane shape it would have been highly stable—like the kind of plane now used for training novices. Modern fighter planes, by contrast, are far less stable although much faster and more manoeuvrable, and need supremely gifted pilots and fast-reaction computers to stay in the air. By the same token, modern sparrows and other such birds are much less stable than an *Archaeopteryx*, but far more manoeuvrable; and they, too, rely on rapid responses and fine co-ordination to stay aloft.

However it flew, flapper or glider, its flight must have been weak. So what good was it? Was it favoured by selection? How could flight have evolved in the first place if, in its earliest days, it was inevitably feeble? Some suggest that the precursors of *Archaeopteryx* clambered around the forest canopy—aided by the wing claws that *Archaeopteryx* itself retains—and came to glide from tree to tree, like flying squirrels, snakes, and frogs; then extended the glide with more and more convincing flaps. But further studies show that the place *Archaeopteryx* is known to have lived—by lagoons at Solnhofen in Germany—had sparse vegetation, no more than 3 metres high. This lends weight to a second much-favoured hypothesis: that *Archaeopteryx* ran along the ground like other small, carnivorous, theropod dinosaurs—such as the dromaeosaurids, *Protarchaeopteryx*, and *Caudipteryx*—and may have leapt into the air to catch flying insects.

I feel that both kinds of ideas are too specific. Many kinds of animals benefit from jumping, in all sorts of ways. When domestic cats or dogs of all kinds—dingos, wolves, foxes—hunt rodents and insects (which they are all fond of doing) they leap high in the air and come down forefeet first; approaching behind the line of sight of the prey, and leaving no direct escape route. Conversely, modern pheasants rise vertically from the forest floor to escape predators and to roost. Jumping in kangaroos is thought to have evolved in forests, as proto-macropods negotiated the undergrowth. Flying fish take to the air to avoid predators and to save energy—air is less viscous than water and the ride through the wave-level wind is free. Grasshoppers jump for safety and to avoid obstacles and also control their leaps with rudimentary flight. More broadly, our own ancestors—*Australopithecus*—evidently evolved in low, open woodland, which, once we make a suitable adjustment for scale, seems not so very different from the terrain of the crow-sized *Archaeopteryx*. For the hominids, as perhaps for *Archaeopteryx*, the scattered vegetation added another dimension to their otherwise terrestrial existence, and one that came in very handy.

Putting all these thoughts together, I see *Archaeopteryx* living a little like an *Australopithecus* and a little like a modern ground hornbill. That is, it was a predatory generalist of open 'woodland' that itself was all too easily preyed upon. Jumping helped it to avoid obstacles, as in a grasshopper or primordial kangaroo, to leap to safety like a pheasant, and to attack from unexpected angles like a wolf catching lemmings. Theologians who felt uncomfortable with Darwin's idea that evolution always proceeded step by step used to ask, 'What use is half an eye?', to which the response is that *any* vision for a creature that does not live in total darkness is better than none at all, and any improvement on the mere ability to distinguish light from dark is liable to bring rewards, so in practice it is easy to envisage a graded progression from total insensibility to the sensitivity of Vincent van Gogh. By the same token, a leap has a thousand applications in appropriate territory. If arm flaps can extend the creature's range so much the better; and so we may proceed by increments to the grace and power of the fulmar and the peregrine.

Two further thoughts. First, flight is costly both in energy and in neural ability, and brings hazards of its own, including that of cross-winds. We need not be surprised, then, that natural selection has sometimes favoured the loss of flight—just as it has often favoured the loss of eyes—effectively allowing birds to return to theropod status. So it is that many birds, both among the Aves in general and among the Neornithes in particular, have abandoned the ability to fly in favour of something simpler. In short, just because we think that flight is wonderful, that does not mean that natural selection will always favour those who practise it over those who do not. The same goes, in our own lineage, for intelligence.

Second, as Luis Chiappe again emphasizes, taxa should be defined by their phylogenetic relationships, *not* by their way of life. The birds are a clade: collectively they form the monophyletic class Aves. The most conspicuous feature of birds is that they fly. But that does not mean that members of the clade Aves *have* to fly, and we need not be surprised if they do not. After all, says Chiappe, we do not demand that members of the clade Mammalia should adopt one particular lifestyle. We are not surprised by mammals that have reinvented the form of the fish, as whales have done, or by mammals that fly, as bats do. So why should we feel that birds need be more restricted? Why shouldn't birds radiate as freely as mammals have done? Well, in fact, birds have not proved as versatile as mammals; and, indeed, the surviving subclass of birds (the Neornithes) and most of the extinct subclasses are all mainly fliers, and are much of a muchness. But two known subclasses were clearly not fliers and in one, represented by the extinct genus *Mononykus*, the forelimbs are reduced to extraordinary little pegs whose function can only be guessed. *Mononykus* reminds us indeed of the basic lessons: not to assume that a particular taxon has to pursue a particular way of life, because all taxa in principle might try anything at all; and not to assume that evolution proceeds orthogenetically, or indeed towards any particular pre-envisioned goal.

A GUIDE TO THE AVES

The tree shows the eight known subclasses of the Aves, as envisaged by Luis Chiappe, although some are represented only by single genera: that is, the Aves extends from *Archaeoptyeryx*, which is traditionally held to represent the oldest true birds, right through to the modern Neornithes (they have their own tree in Section 22)—what most people mean by 'bird'; but which, in truth, are the only surviving fragment of

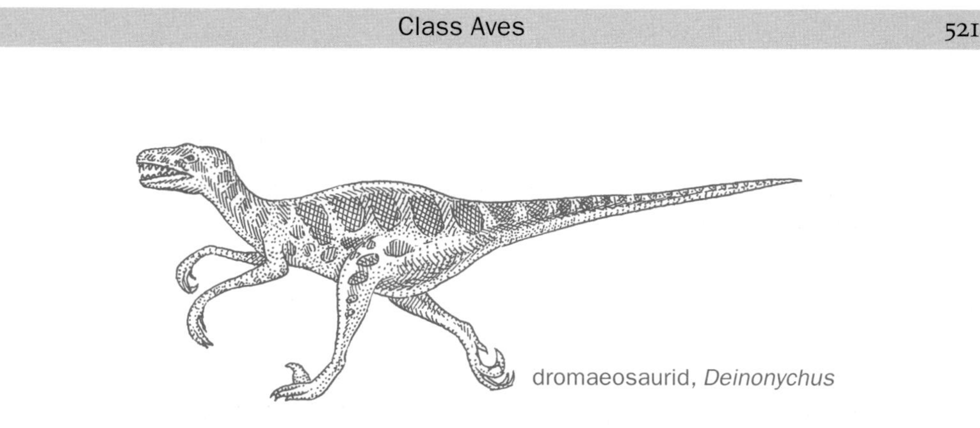
dromaeosaurid, *Deinonychus*

bird-dom. Also shown is the family of carnivorous theropod dinosaurs that has traditionally been recognized as the sister group of the Aves: the family **Dromaeosauridae**, best known in the form of *Deinonychus*. *Protarchaeopteryx* and *Caudipteryx* should probably be inserted between *Deinonychus* and *Archaeopteryx*.

Deinonychus (representing the dromaeosaurids) was a remarkable creature from early Cretaceous Montana. It was up to 3 metres long and 1 metre high, carried its body and long tail stiffly and horizontally, and had wicked, curved, serrated teeth like steak knives. *Deinonychus* was a very definite theropod dinosaur but it had bird-like features: well-developed forearms with hands that were nearly half the length of the arm; wrists that look very avian; bird-like proportions in its legs including a short femur; and four toes. The second toe ended in a powerful claw that the animal held clear of the ground as it ran, and was presumably used for disembowelling: *Deinonychus* means 'terrible claw'. It is at least intriguing that many modern birds fight and kill with a downward strike of the foot, like secretary birds and fighting cocks and ostriches.

Phylogenetically speaking, however, the Dromaeosauridae were just one of about six known theropod families that are long extinct. Their sisters, the Aves, have proved more durable—one subclass, the Neornithes, has survived into the Tertiary and to the present day.

THE EIGHT AVIAN SUBCLASSES

Archaeopteryx†

As we have seen, the first *Archaeopteryx* remains that were recognized as birds were described from the Solnhofen marshes (as once they were) in Bavaria, in 1861. There was an earlier specimen—found in Haarlem in 1855—but it was mistaken for a pterodactyl.

Archaeopteryx

Archaeopteryx beautifully combined primitive reptilian features with those of a modern bird. It had feathers, and it probably had at least a weak flapping flight, as described above. Because it had feathers and flew, *Archaeopteryx* is always regarded

21 · CLASS AVES

21

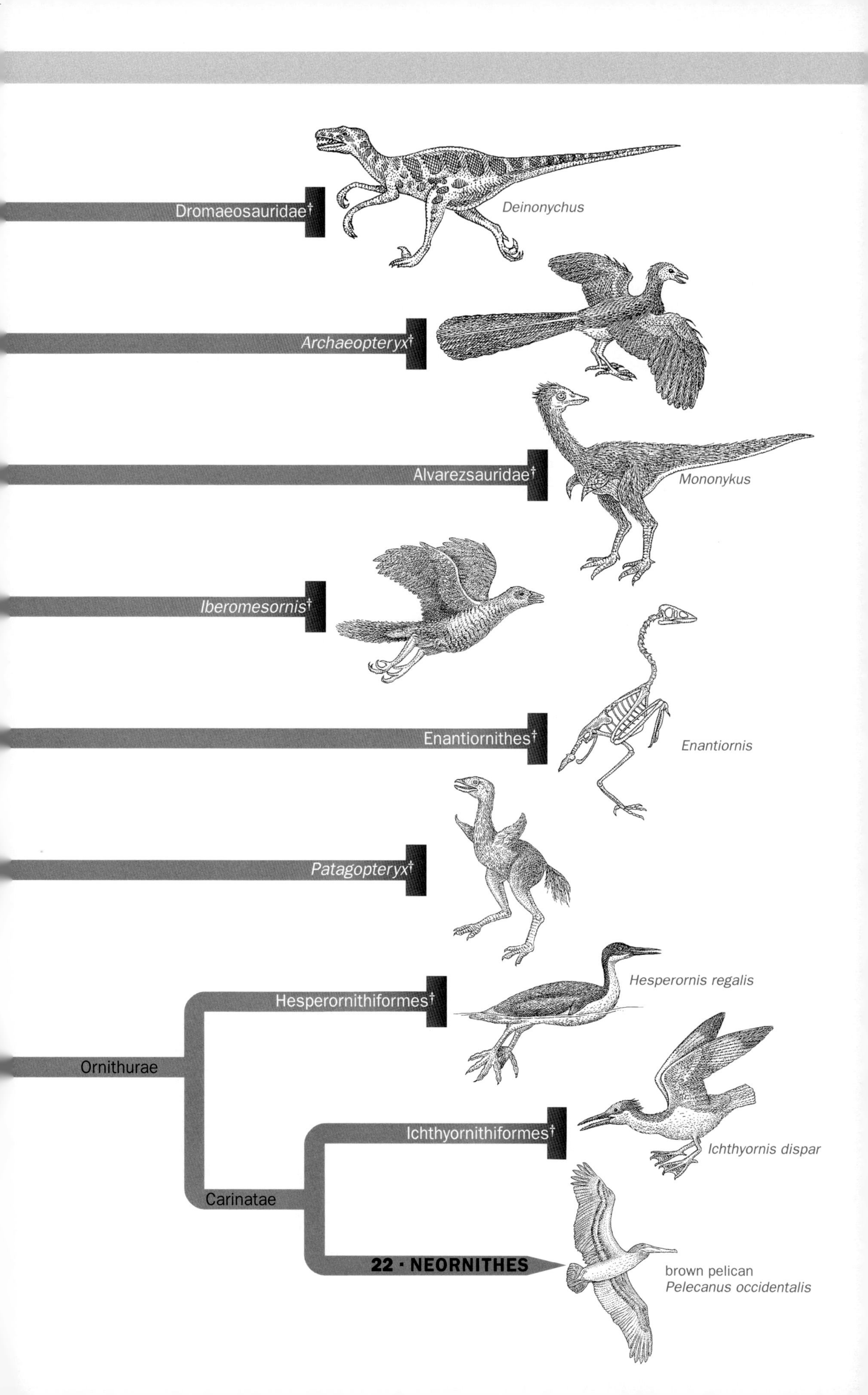
Dromaeosauridae†
Deinonychus
Archaeopteryx†
Alvarezsauridae†
Mononykus
Iberomesornis†
Enantiornithes†
Enantiornis
Patagopteryx†
Hesperornithiformes†
Hesperornis regalis
Ornithurae
Ichthyornithiformes†
Ichthyornis dispar
Carinatae
22 · NEORNITHES
brown pelican
Pelecanus occidentalis

as the oldest known bird, although the newly found feathered dinosaurs from China could yet challenge that status if taxonomists feel that the boundaries of the Aves should be extended. Old-fashioned taxonomists were wont to say that *Archaeopteryx* was the ancestor of all later birds, but as discussed in Part I (Chapter 3), it really is most unlikely that one or other of the known skeletons was, in fact, the particular ancestor. It is much safer to suggest that *Archaeopteryx* and all other birds shared a common ancestor that was itself a bird—and which, in fact, was probably very like *Archaeopteryx* as the sister group of all other birds, leaving open the option that it *might* be their ancestor.

The avians that succeeded *Archaeopteryx* through the rest of the Mesozoic are only just coming to light; the number of known Cretaceous taxa doubled in the 1990s, and revelations are emerging by the month not least from central Asia and South America. But the fossils are still fragmented and confusing. Some are more obviously bird-like than *Archaeopteryx* whereas others—superficially at least—are much less so. Without the rigour of cladistic analysis, it would not be possible even to begin to sort out the fossils or even, in cases like *Mononykus*, to recognize their avian provenance. In general, though, the known fossils show that by the end of the early Cretaceous and during the late Cretaceous, avians varied in size and lifestyle from flightless runners and foot-propelled divers to waders and tree-dwellers.

Mononykus: Alvarezsauridae†

The second major grouping of the Aves shown on the tree is at present represented by a few genera that have been grouped in a single family, the Alvarezsauridae. Some, like *Alvarezsaurus*, come from South America. But the first to be discovered, and the most peculiar, is from the late Cretaceous of the Gobi Desert of central Asia: the turkey-sized *Mononykus*.

alvarezsaurid *Mononykus*

Mononykus (the name means 'one claw') to the untrained eye seems nothing like a bird. It was flightless, with long hind limbs that as it ran it swung mainly from the hip like a dinosaur, and not so much from the knee like a modern bird. But then we might expect the hindquarters to be reptilian—as in *Archaeopteryx*. It is the forelimbs that make *Mononykus* so strange: they were reduced to little struts—very strong, and with fused bones like a bird, but absolutely inappropriate for flight. Indeed it is not clear what they were used for. Some have suggested that they were used for digging, though Luis Chiappe says he finds this most unlikely. Perhaps they served like the claspers of a shark, to help the male keep hold during mating.

But a complete *Mononykus* skull has now been found, and details, not least in the orbit of the eye, reveal the avian affiliation. Of course, such peculiarities could have evolved independently; but common ancestry provides a more likely, more 'parsimonious', explanation. By all the accepted principles of modern taxonomic practice *Mononykus* has to be fitted within class Aves, and the tree shows how it seems to fit best: as the sister group of all later birds.

Iberomesornis†

The third suggested subclass also lacks a formal name, and again is known only from a few fossils, this time from the early Cretaceous of Spain. The only genus is the sparrow-sized *Iberomesornis*. Its wings and shoulders suggest that it was a better flier than *Archaeopteryx*: the forearm (ulna) is longer than the upper arm (humerus) as in a modern bird, which improves the leverage and efficiency, while the coracoid bone of the shoulder is more strut-like than in *Archaeopteryx* and so provided stronger support. The feet of *Iberomesornis* are fully adapted to perching in trees—another feature of modern birds that clearly evolved very early!

Iberomesornis

With a date of about 125 million years ago, the known *Iberomesornis* fossils are older than the known *Mononykus* remains but this does not mean that its lineage is older: the fossil record is far too spotty to make such a judgement. Cladistic analysis of the creatures' anatomy suggests that the *Mononykus* line appeared first, as shown in the tree.

Enantiornithes†

Many kinds of Enantiornithes are known from all over the world, dating from the early Cretaceous more than 130 million years ago until almost the end of the Mesozoic (the fossils suggest that they died out before 65 million years ago). All in all, the Enantiornithes form a very significant slice of avian history and, I suggest, of Cretaceous ecology. Enantiornithine fossils have been known for almost a century but they have come to prominence only after new discoveries in the 1970s and their significance as a huge, diverse, and discrete subclass of Mesozoic avians has been recognized only since

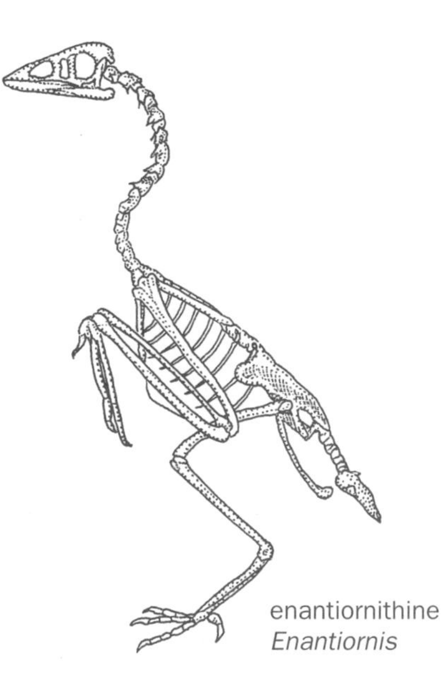
enantiornithine
Enantiornis

1981. Until the 1980s they were placed among the non-avian theropods or squeezed into modern groups.

All of the enantiornithines look as if they could fly reasonably well. Some were long legged and lived by the shore and others were aquatic, but most probably lived in land and perched in trees. The early enantiornithines were tiny and had teeth, like *Cathayornis* and *Sinornis* from China; but the wings of the later *Enantiornis* and others spanned a metre or more and some of the later kinds, like *Gobipteryx*, had no teeth. Because the sister group of the modern birds, the Ichthyornithiformes, seems to have had teeth, we must conclude that avians have lost their teeth more than once. But there is nothing strange in this: after all, many different groups of non-avian dinosaurs also lost their teeth.

The fine structure (histology) of enantiornithine bones is most intriguing. In modern birds and nearly modern ones like the Hesperornithiformes, the bones are richly served by blood vessels and the pattern of growth can be seen to be rapid, continuous, and uninterrupted. This is the kind of pattern you expect when growth is not significantly delayed by the seasons—and is the pattern found in warm-blooded animals for whom the weather outside is less of a problem. But the bones of enantiornithines are almost devoid of blood vessels, and clearly grew slowly and seasonally. Indeed, they apparently grew for several seasons after hatching, while the skeletons of most modern birds reach adult size in a year. Surprisingly, then, the enantiornithines—and presumably the other avians that preceded them—do not seem to have been true homoiotherms in the modern sense: Luis Chiappe suggests that their metabolism was intermediate between that of, say, a modern lizard and a modern bird. It seems, then, that true homoiothermy, a quality so characteristic of modern birds, was apparently a late avian invention. This kind of finding must also cast doubt on the suggestion that other dinosaurs were homoiothermic.

Patagopteryx†

Here is yet another bird—again different enough from the rest to warrant its own subclass, but as yet without a subclass title—that demonstrates that avians do not have to be fliers; that reversion to dinosaur two-legged running is an ever-open option. The only known genus of this group is the hen-sized *Patagopteryx* from Patagonia. Again, it was originally ascribed to the flightless ratites—which are modern birds, in the subclass Neornithes. But *Patagopteryx* lacks many of the special, 'derived' features of modern birds, or, indeed, of the **Ornithurae** as a whole, which, as the tree shows, include the Neornithes + Ichthyornithiformes +

Patagopteryx

Hesperornithiformes. Not only is *Patagopteryx* not a ratite, then: it is nowhere near being a ratite. It is, in fact, the sister group of the Ornithurae.

Hesperornithiformes†

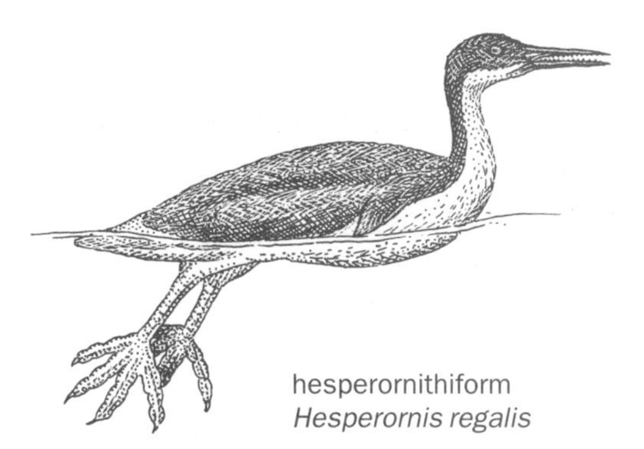

hesperornithiform
Hesperornis regalis

The Hesperornithiformes are among best known of the Mesozoic birds. They had teeth, and they were flightless, heavy-boned, fish-eating divers that drove themselves through the water with their feet (like diving ducks and grebes but unlike cormorants or penguins, which 'fly' under water). Their feet were probably lobed like modern grebes, and they have often been regarded as primitive grebes or loons, or sometimes even as aquatic palaeognaths: that is, as relatives of the ratites. But the bulk of evidence makes them the sister group of the **Carinatae**, the group that includes both the modern birds and the Ichthyornithiformes. The Hesperornithiformes are known mainly from the late Cretaceous of eastern Europe and western Asia.

Ichthyornithiformes†

ichthyornithiform, *Ichthyornis dispar*

The late Cretaceous Ichthyornithiformes were tern-sized flyers from the western interior of North America, and have been known as fossils since the nineteenth century. They are the sister group of the modern birds, but they still have their reptilian teeth. Again, however, they have in the past been linked erroneously with modern birds; in particular, with the Charadriiformes—the waders, gulls, and auks.

Neornithes

So to the modern birds, the Neornithes: the one surviving avian subclass out of the eight that are known. This pattern is seen again and again throughout nature: a successful new clade radiates, throws up many different forms and body plans, of which in the end only one prevails. But, again, we find that the one group that survives radiates to become as diverse in form and ways of life as all the previous groups put together. To be sure, there is nothing like *Mononykus* among the modern birds. But there are plenty of running, flightless kinds, re-exploring their theropod inheritance.

neornithine
(brown pelican)

The Neornithines are indeed the moderns but they, too, were well represented in the Mesozoic: ducks, fowl, gulls, and albatrosses all have late Cretaceous relatives. Indeed, it would have been good to have been a Cretaceous ornithologist, and to have seen the true array of birddom in full pomp; all those and the pterosaurs too. But so long as we have the neornithines to admire, we have little to complain about. They are the subject of the next section.

22

THE MODERN BIRDS

SUBCLASS NEORNITHES

ONLY ONE OUT OF the eight known subclasses of avians is around today: the Neornithes. Indeed, only the Neornithes (neornithines) seem to have survived into the Cenozoic—the past 65 million years. When most of us speak of 'birds', therefore, we mean 'neornithines'.

The single subclass of the neornithines has re-explored almost all the ecological niches and ways of life that, in the late Mesozoic, were essayed by the Aves as a whole; all, that is, except the niche exploited by the wingless *Mononykus*, which was clearly doing things that no biologist has yet pinned down and no living creature may emulate. Almost 10 000 living species are known (with new ones coming to light regularly) and they seem to fall into 27 orders—of which just one, the Passeriformes or perching birds, contains about three-fifths of all living species. All in all the neornithines in the Cenozoic have been just as diverse as were the Aves as a whole in the Cretaceous. Most of the Cenozoic types have been fliers, like the majority today, fully exploiting their wonderful inheritance of wings and feathers. But some have reverted to the traditional dinosaur mode and re-emerged as flightless, bipedal runners. Some have exploited a third option, as divers, propelling themselves either by their feet, like a grebe (or a *Hesperornis* in the Cretaceous), or using their wings to 'fly' under water. In this last case they may either retain the power of aerial flight (like most but not all cormorants and auks) or lose it (like some auks and cormorants, and all penguins).

Flight is a wonderful escape mechanism, and opens entire new niches for feeding and nesting. It turns two dimensions into three and thus transforms the structure of the world. Furthermore, it is easier to move through air than through water or over land. Many birds—even small ones—fly regularly from continent to continent or even from one world's end to the other. Thus for many a bird from many a family, the entire planet becomes the habitat; or, rather, flying birds compound their habitats through space and seasons—from a marsh here, a group of trees there, an eave or an estuary somewhere else. Swallows, for example, can spend their lives in perpetual warmth, exploiting the long summer days of the temperate north and whiling away the northern winters in the tropics. Others, like the Arctic tern, seek unending light and commute virtually from Pole to Pole.

But flight also exacts an enormous price. It requires commitment from the entire anatomy and a huge input of energy, and it can be dangerous—flying birds on islands may be blown out to sea. So natural selection has often favoured flightlessness, and within the neornithine orders there are flightless pigeons (including the dodo and solitaire), flightless cormorants and auks, many flightless ducks and geese, flightless ibises (including species from Rodrigues Island and Hawaii), an almost flightless parrot (the kakapo from New Zealand), and so on. Added to which are the flightless orders of ratites and penguins. We would know of many more flightless species as well, were it not that they live mainly on islands where they have proved vulnerable in recent centuries. Some have been wiped out by sailors and human colonists: birds like the dodo, solitaire, and great auk, plus the many extinct moas of New Zealand and the elephant birds of Madagascar.

Throughout the Cenozoic both the neornithines and the mammals (the other homoiothermic tetrapods) have explored both flight and terrestriality. Birds seem to have won the contest in the air, for although there are more than 800 species of bat (about a fifth of all living mammals), most bats are nocturnal. One reason for this, it seems, is the competition from birds; when bats are occasionally forced by hunger to fly by day, they are picked off within hours by hawks. Mammals, on the other hand, have clearly won the battle on the ground. There are, though, some big and successful ground-living continental birds, including the ostriches of Africa and the bustards of Africa and Eurasia; and in the not-so-distant past there have been some huge neornithine terrestrial predators like the phorusrhacoids of South America with beaks like eagles and heads as big as horses. But most flightless birds live on islands or island continents like Australia (or as South America was until the Pliocene) where they have largely been free from serious mammalian competition. If mammals had succumbed at the end of the Mesozoic, as the dinosaurs did, we may reasonably assume that flightless birds would now fill most of the terrestrial niches that the mammals do now. There would be huge, striding herds of elephantine birds where now there are bison and antelopes, and formidable bands of phorusrhacoid-like predators where now we have lions and wolves. In short, the presence of dinosaurs prevented mammals from realizing their potential in the Mesozoic while in the Cenozoic, mammals are preventing birds—the direct scions of the dinosaurs—from realizing theirs.

Variable though the neornithines are, however, no biologist seriously doubts these days that they are a true monophyletic group; a clade. They share some primitive features with the extinct Hesperornithiformes and the Ichthyornithiformes, such as the pygostyle (short, fused tail) and fused pelvis; but they also have special apomorphies of their own, including the **toothless beak**. Although there are some loose ends—for example, no one yet seems entirely clear whether New World vultures are really vultures, or are modified storks—the different kinds of neornithine seem easy enough to define and hence to arrange into discrete orders: a duck is a duck and a pigeon is a pigeon. But the relationships between the orders have been much harder to pin down, not least because the modern orders seem to have radiated within a short time of each other, early in the Cenozoic; and it is notoriously difficult to distinguish phylogenetic events that occurred close together, but a long time ago.

The tree in this section, which presents the 27 neornithine orders and attempts to show the relationships between them, is based on the ideas of Joel Cracraft at the American Museum of Natural History in New York—one of just a few bird taxonomists who are bringing strict cladistic principles to bear on the problems of bird phylogeny. As you see, however, the tree is full of polychotomies. Present data enable us to see broad relationships between groups (or 'series') of orders, but does not yet enable us to decide the precise sequence in which the different orders appeared.

A GUIDE TO THE NEORNITHES

The phylogeny and hence the modern taxonomy of birds—here meaning neornithines—has not proved easy to unravel. The fossils are extremely patchy and give less help than they might. The morphology of living birds is, of course, the main guide but bird evolution is rife with convergence, not least because flying and diving impose such restraints on form. Birds, too, like any other creature, are prone to rapid divergence.

The great taxonomic study of recent years has been by Charles Sibley of Yale University and Jon Ahlquist at Ohio University, who used the technique of DNA hybridization (Chapter 2) to reclassify all birds, and provided many fresh insights. But their favoured technique does have several drawbacks. In particular, it reveals the overall similarity or distance between the DNA of different species, but seems innately unable to distinguish between similarities that are merely primitive and those that are derived; yet only the latter reflect special relationship. So the work of Sibley and Ahlquist stands as a great pioneer study, whose findings must influence bird taxonomy in the decades to come. They propose many hypotheses that demand to be tested. But their conclusions so far cannot be taken as definitive. We need cladistic studies based both on morphology and on features of mitochondrial and nuclear DNA. Such studies are now taking place at the American Museum of Natural History under Joel Cracraft, and at the Smithsonian Institution in Washington DC under Michael Braun. Early in the twenty-first century we should be able to describe the phylogeny of modern birds in detail and with confidence; but not yet.

The present classification by Joel Cracraft, summarized on the following pages, has much to support it, but there are still some ancient puzzles to be resolved—as is reflected (though not entirely encapsulated) in the polychotomies. Furthermore, I have been unable to persuade him to commit himself on the positions of the Psittaciformes (parrots); Cuculiformes (cuckoos, roadrunners, and touracos); and the Trogoniformes (trogons). Here, I have slotted them in roughly according to the hypotheses of Charles Sibley; that is, they belong to the large clade of neognaths, extending on the cladogram from grebes to passerines, and roughly in the region of swifts, owls, kingfishers, and woodpeckers. Clearly it is unsatisfactory to mix the ideas of different authorities who have used different methods, but it seems better to me to put everything roughly in place pro tem than to leave groups hanging untidily. The first requirement, after all, is to be explicit. Truth remains the ambition but

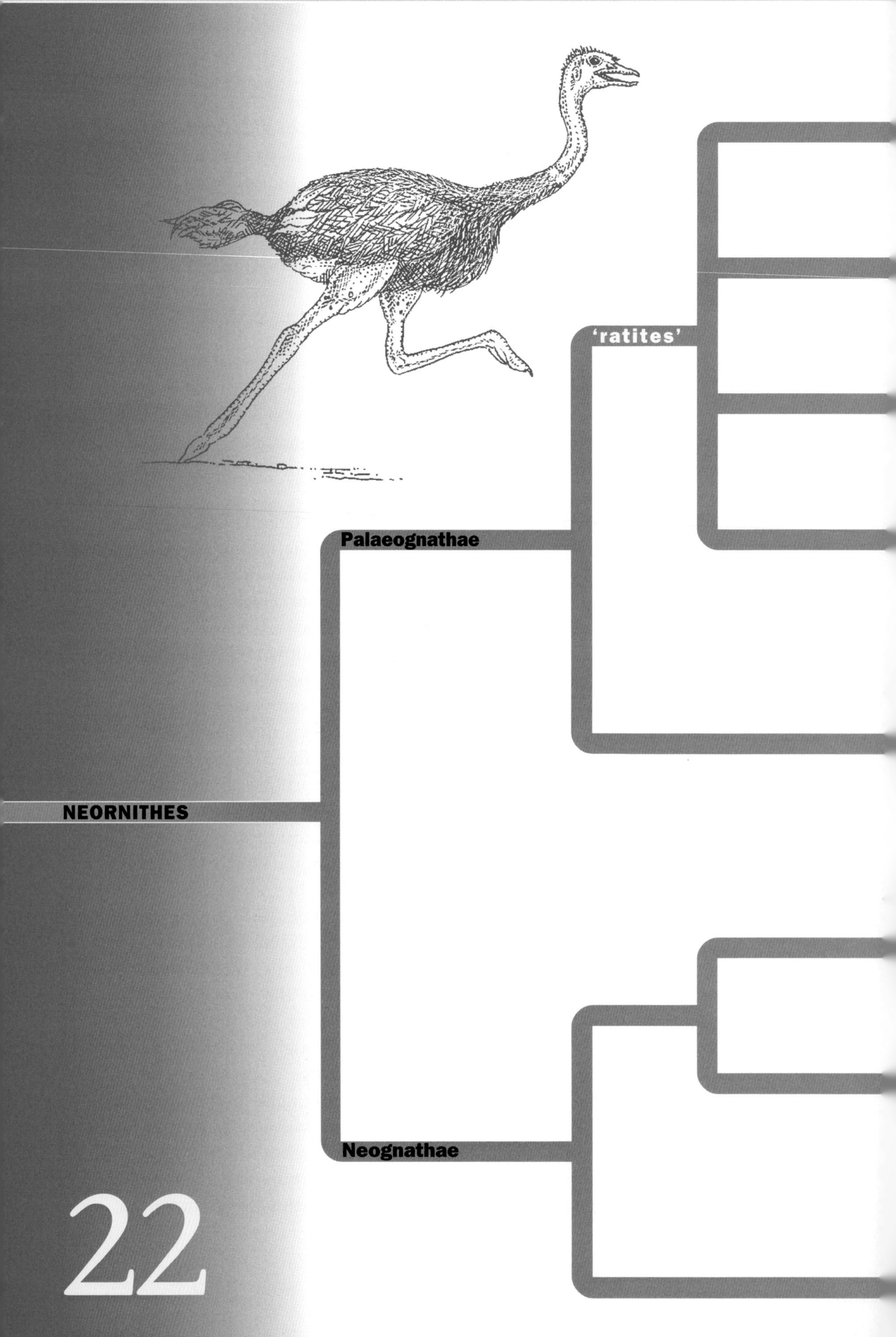
'ratites'
Palaeognathae
NEORNITHES
Neognathae
22

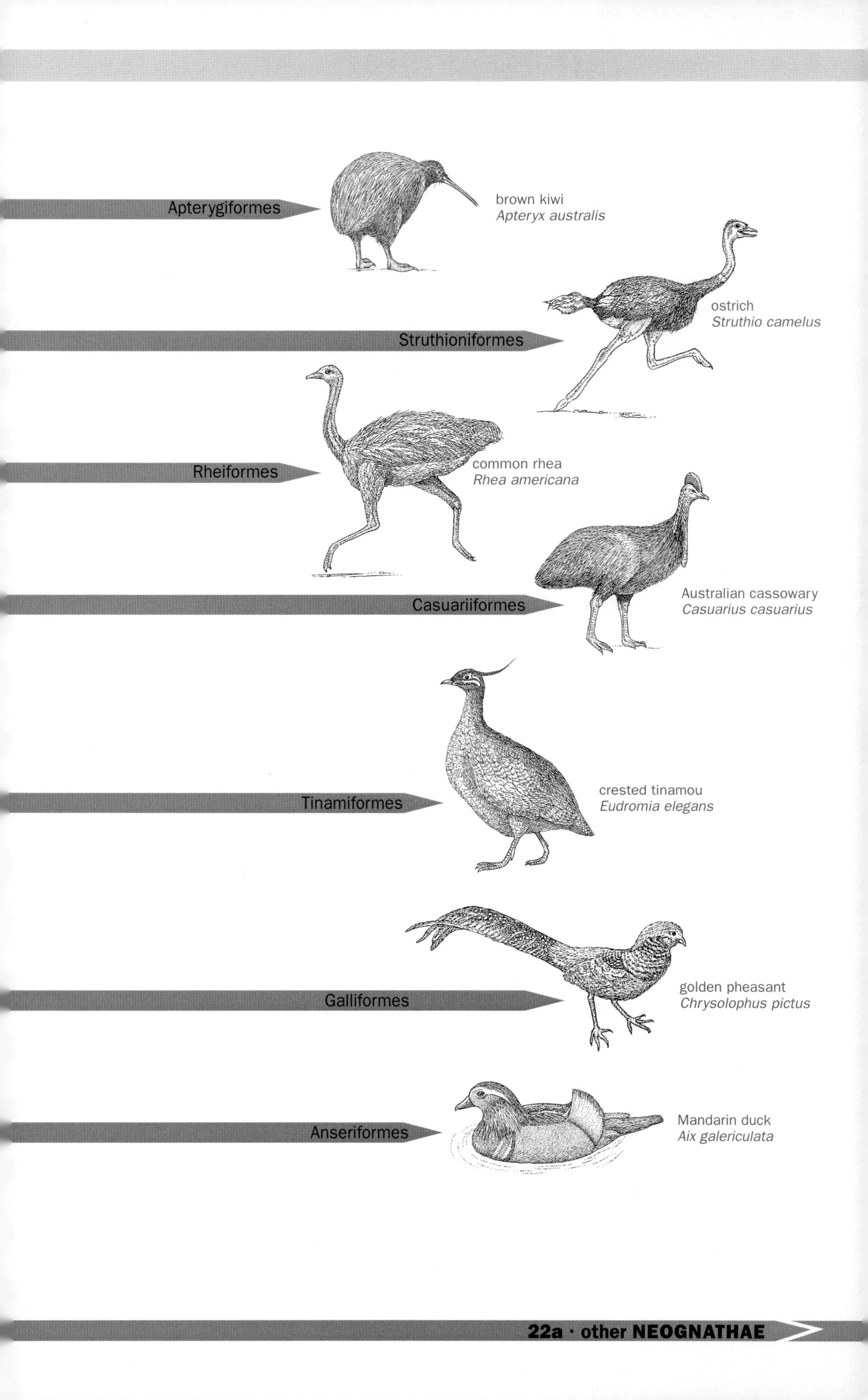
Apterygiformes
brown kiwi
Apteryx australis
ostrich
Struthio camelus
Struthioniformes
Rheiformes
common rhea
Rhea americana
Casuariiformes
Australian cassowary
Casuarius casuarius
Tinamiformes
crested tinamou
Eudromia elegans
Galliformes
golden pheasant
Chrysolophus pictus
Anseriformes
Mandarin duck
Aix galericulata
22a · other **NEOGNATHAE**

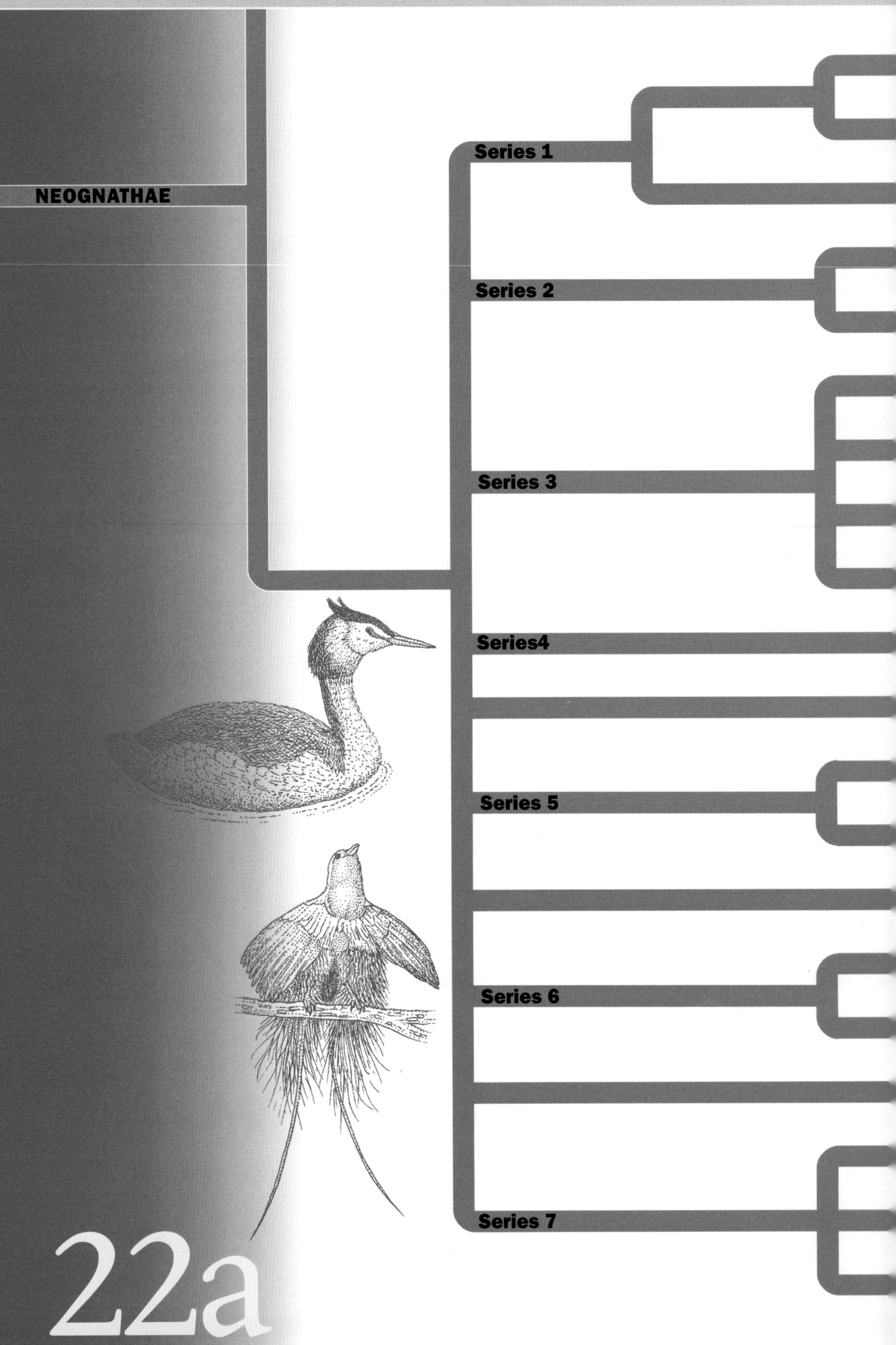
NEOGNATHAE
Series 1
Series 2
Series 3
Series4
Series 5
Series 6
Series 7
22a

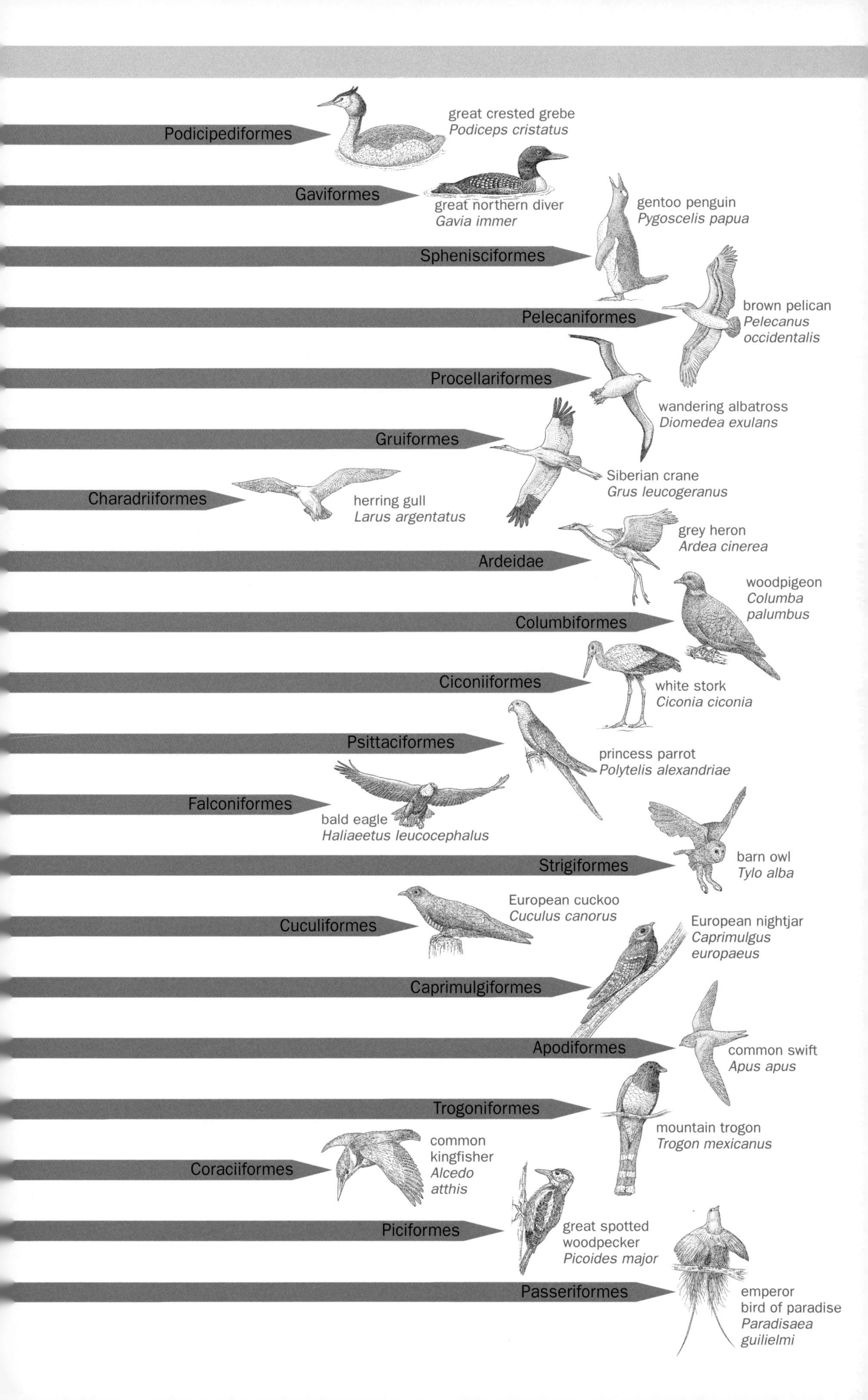
Podicipediformes
great crested grebe
Podiceps cristatus
Gaviformes
great northern diver
Gavia immer
gentoo penguin
Pygoscelis papua
Sphenisciformes
Pelecaniformes
brown pelican
Pelecanus
occidentalis
Procellariformes
wandering albatross
Diomedea exulans
Gruiformes
Siberian crane
Grus leucogeranus
Charadriiformes
herring gull
Larus argentatus
grey heron
Ardea cinerea
Ardeidae
woodpigeon
Columba
palumbus
Columbiformes
Ciconiiformes
white stork
Ciconia ciconia
Psittaciformes
princess parrot
Polytelis alexandriae
Falconiformes
bald eagle
Haliaeetus leucocephalus
Strigiformes
barn owl
Tylo alba
European cuckoo
Cuculus canorus
Cuculiformes
European nightjar
Caprimulgus
europaeus
Caprimulgiformes
Apodiformes
common swift
Apus apus
Trogoniformes
mountain trogon
Trogon mexicanus
common
kingfisher
Alcedo
atthis
Coraciiformes
Piciformes
great spotted
woodpecker
Picoides major
Passeriformes
emperor
bird of paradise
Paradisaea
guilielmi

cannot be attained without the facts; and, of course, truth and certainty are by no means the same thing.

So, conscious that we are looking at work in progress, let us explore the tree. The neornithines as a whole are divided into two clear groupings: the **Palaeognathae**, which embrace the ratites and the tinamous, and the **Neognathae**, which include all the rest. The two are distinguished by the form of their palate—which does seem valid even though the features that are commonly cited to distinguish the two are, in fact, a mixture of both primitive and derived.

THE PALAEOGNATHAE

The **ratites** are flightless, their wings reduced to vestiges, and they lack the deep keel on the breastbone to which the flight muscles would attach; but because members of the neornithine sister group (that is, the Ichthyornithiformes—see Section 21) do have a keel, we can assume that its loss in the ratites is secondary. The greatest of the ratites were the moas of New Zealand and the elephant birds of Madagascar, but they were driven to extinction in historical times—the former by the Maoris from the

tenth century AD onwards, and the latter by the first Malagasy in the first few centuries AD. So we are left with just four main orders of ratite with only a few species in each, which, between them, span the Gondwanan continents of the Southern Hemisphere. The kiwis of New Zealand form the order **Apterygiformes**; the ostriches of Africa (and formerly also of Arabia) are the **Struthioniformes**; the rheas of South America form the **Rheiformes**; and the emus of Australia plus the cassowaries of northern Australia and New Guinea make up the **Casuariiformes**. Perhaps the ostriches, rheas, emus, and cassowaries should not be split among three orders; perhaps one or two could reflect the important differences. But when in doubt it is probably better to split than lump because it is easier to re-lump groups that have been over-split than to re-split those that have been cavalierly bundled together.

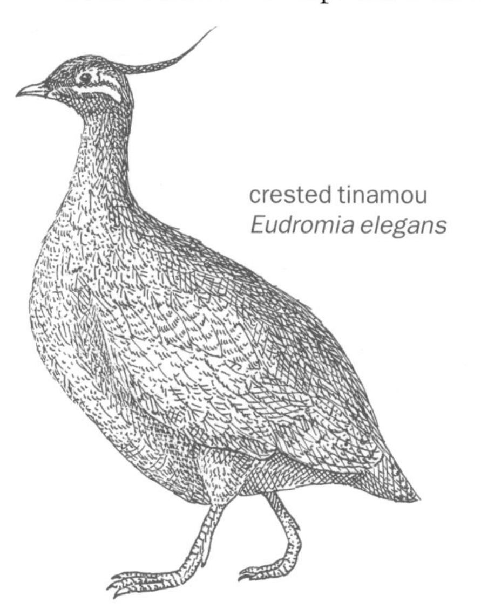

crested tinamou
Eudromia elegans

The tinamous certainly form a discrete order, the **Tinamiformes**, which are sisters to all the ratites. Tinamous are fast-flying ground nesters from South America; very like grouse in form and way of life but absolutely unrelated to them.

THE NEOGNATHAE

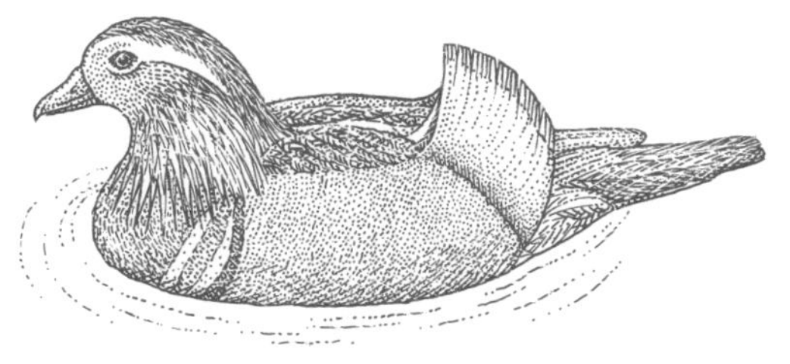

Mandarin duck, *Aix galericulata*

The Neognathae also divide clearly into two main groups. The first comprises the **Galliformes,** fowl, and the **Anseriformes**, waterfowl. At first the relationship between the two seems surprising: chickens and pheasant seem nothing like ducks, geese, and swans. But the Anseriformes also includes the turkey-sized screamers from South America that wade and swim as other anseriformes do but have feet that are only cursorily webbed and have distinctly pheasant-like beaks and faces. The Galliformes include the megapodes, curassows, guans, grouse, chickens, pheasants, peafowl, guineafowl, turkeys—and possibly the curious South American hoatzin, whose newly hatched young retain two claws on the bend of the wing,

golden pheasant
Chrysolophus pictus

Archaeopteryx-like, although these are lost as the bird matures. Overall, it is a pleasant thought that the birds of the barnyard—chickens and turkeys, ducks and geese—provide a nice lesson in bird phylogeny, representing between them the primitive wing of the neognaths, the sister group of all the rest.

Joel Cracraft is content for the time being to divide the remaining neognaths into 17 orders, most of which correspond with the traditional bird orders apart from the Ardeidae, the heron family, which he has separated from the storks and flamingos (Ciconiiformes) but are not yet widely acknowledged as a formal order in their own right. These 17 seem to cluster in seven main series (series being an informal but phylogenetically valid grouping), each containing between one and four orders; but, as noted above, Cracraft's published series omit the trogons, parrots, and cuckoos, which I have roughly sketched in. The overall polychotomy is not desirable; phylogeny more probably proceeded as a series of dichotomies that should be reflected in a finished tree. Sibley and Ahlquist do provide a much more finished phylogeny, but Joel urges caution: present evidence, he says, permits only an interim statement, in which the polychotomies delineate areas of ignorance or perhaps more optimistically signify 'work in progress'. Even as it stands, however, the taxonomy is full of intriguing insights—and of unresolved puzzles. I will first look at the Cracraft's seven groups one by one, and then deal briefly with the three that got away: the parrots, the cuckoos, and the trogons.

Series 1: grebes, divers, and penguins

Cracraft's Series 1 contains the **Podicipediformes** (grebes), the **Gaviformes** (divers), and the **Sphenisciformes** (penguins). In some earlier taxonomies the grebes have been

great crested grebe, *Podiceps cristatus*

great northern diver, *Gavia immer*

gentoo penguin *Pygoscelis papua*

linked with the ducks on morphological grounds, while the penguin order has often been placed effectively on its own: a neognath right enough, but distinct from all the others. But penguins merely show once again how easy it is to be deceived by divergence. Penguins do look very different; but their DNA shows their affinity to the other somewhat primitive water birds, grebes and divers.

Series 2: pelicans and albatrosses

Series 2 somewhat controversially links the **Pelecaniformes** with the **Procellariformes**. The former are a large and varied group including pelicans, tropicbirds, cormorants, gannets, boobies, frigatebirds, and anhingers (or darters or 'snakebirds'—cormorant-like diving birds from Florida and elsewhere that swim with their bodies submerged so only their heads and necks are above the water). The Procellariformes are distinguished by their 'tube-noses'—their nostrils extending along their upper beaks in tubes, like short drinking straws. The Procellariformes include the albatrosses, fulmars, shearwaters, storm petrels, and diving petrels. Fulmars, riding the winds on the cliff face, are superfically gull-like; it is good to think of them as miniature versions of the albatross, the loneliest and most romantic of birds.

brown pelican
Pelicanus occidentalis

wandering albatross
Diomedea exulans

Series 3: cranes, rails, gulls, herons, and pigeons

Series 3 links four orders. The **Gruiformes** include the cranes and rails (such as the moorhen and coot) and a variety of recondite creatures like the mesites and hemipodes, limpkins and bustards. The **Charadriiformes** include the waders such as jacanas, snipes, oystercatchers, plovers, lapwings, sandpipers, stilts, avocets, phalaropes, and stone curlews; the gulls, skuas, terns, and skimmers; and auks like the guillemots, razorbill, and puffins. The **Columbiformes** are the pigeons and doves

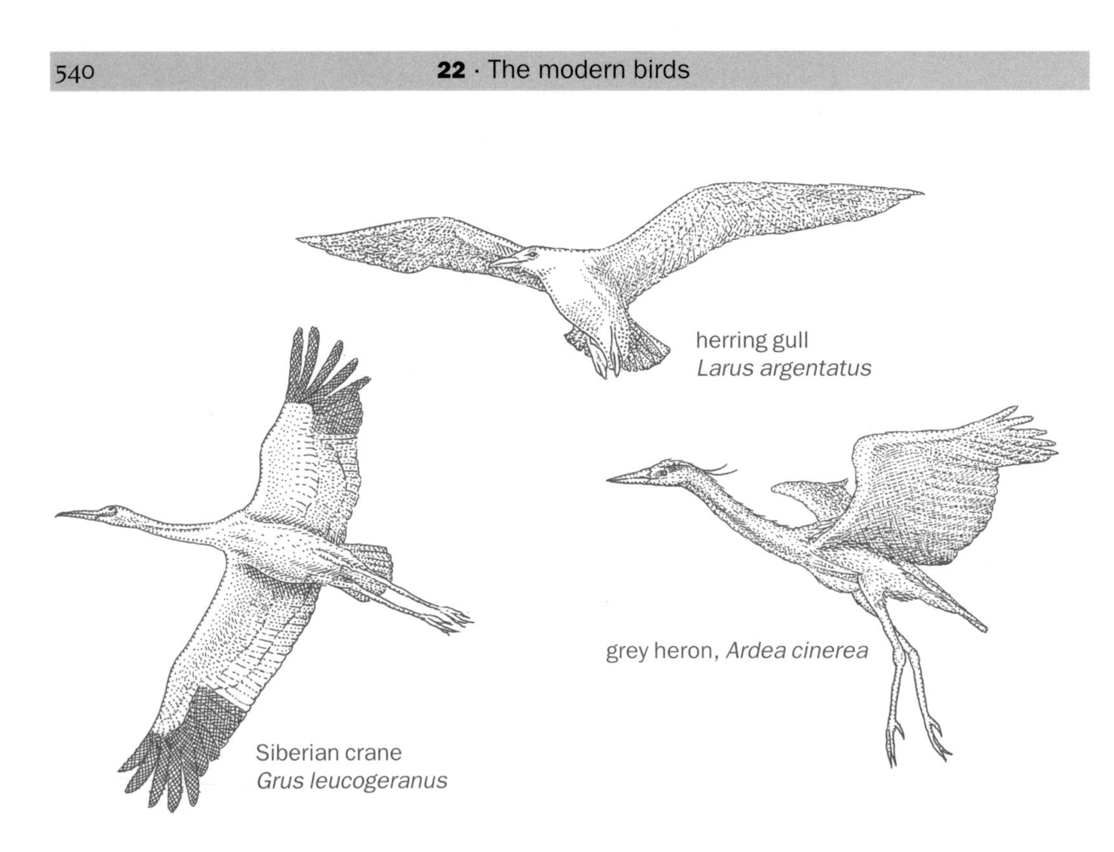

herring gull
Larus argentatus

grey heron, *Ardea cinerea*

Siberian crane
Grus leucogeranus

—but also include the dodos and solitaires, and the sandgrouse. If pigeons in their feral form were not so common we surely would rate them more highly. They include some of the world's most endangered and beautiful birds, and some of the most powerful fliers. The final members of this series are the herons, bitterns, and egrets: family **Ardeidae**. Traditionally herons and their kind have generally been grouped on grounds of general morphology with the storks, in the Ciconiiformes. Again the molecules tell a different story.

woodpigeon,
Columba palumbus

Series 4: storks and allies

Series 4 contains just one order: the **Ciconiiformes**. Traditionally this group contains the boatbill, the whalehead, and the hammerhead; the ubiquitous and varied storks; ibises and spoonbills; and the flamingos.

white stork, *Ciconia ciconia*

Series 5: diurnal birds of prey, and owls

Series 5 contains the diurnal birds of prey, order **Falconiformes**; and the primarily nocturnal birds of prey, the owls or **Strigiformes**. Here there is still plenty of scope for controversy. Traditional ornithologists defined Falconiformes widely, to include the Old World vultures, hawks, eagles, kites, and harriers; the ubiquitous and beautiful osprey; the falcons and caracaras; plus the American vultures (including the condors) and the secretarybird.

bald eagle, *Haliaeetus leucocephalus* barn owl, *Tylo alba*

Many taxonomists have tried to break up this group, arguing that many of the similarities are due to convergence; after all, we might expect birds that live as swooping raptors (from the Latin for 'seizing') to have talons and hooked beaks whatever their provenance, just as we would expect carnivorous mammals from whatever order to have enlarged canines. Many ornithologists have suggested that the New World vultures and condors are allies of the storks—and when you get close to a condor you can see what they mean. Sibley and Ahlquist's DNA hybridization studies also link New World birds of prey with storks. But current DNA studies at the American Museum of Natural History, for the time being at least, seem to re-affirm the traditional view—condors and their like remain Falconiformes. This discussion is not over yet. Some ornithologists, too, have maintained that owls are not monophyletic and that they have nothing to do with diurnal birds of prey—but, again, the DNA studies seem to show differently. Strigiformes does seem at present to be a clade, and does seem to emerge as sister group of the Falconiformes. Joel Cracraft rejects attempts to link the owls with the nightjars of the order Caprimulgiformes for this, he says, would require some very unparsimonious special pleading.

Series 6: nightjars and swifts

Series 6 intriguingly links the **Caprimulgiformes** with the **Apodiformes**. The former include a strange group of crepuscular or nocturnal birds known for their huge gaping mouths: the oilbird, frogmouths, potoos, owlet frogmouths, and the nightjars. The Apodiformes at first sight could hardly be more different: they include the swifts and the hummingbirds. But a swift seen close to is not unlike a nightjar; both catch insects on the wing in wide-open mouths, and both have long tapering wings. All in all, swifts provide a wonderful example both of divergence and of convergence. They are very different in appearance and way of life from their closest living relatives, the hummingbirds. On the other hand, however, they do resemble swallows—both with their forked tails, often flying together to take insects on the wing. But swallows are perching birds—passerines. Thus does a knowledge of phylogeny open whole new vistas of appreciation.

European nightjar
Caprimulgus europaeus

Series 7: kingfishers, woodpeckers, and perching birds

Series 7 includes most living species of birds, in three sizeable orders: the **Coraciiformes**, which are the kingfishers, bee-eaters, rollers, hoopoes, and hornbills; **Piciformes**, which are the woodpeckers, barbets, honeyguides, jacamars, and toucans; and the huge order of the **Passeriformes**, which are the perching birds. The Passeriformes include about 60 per cent of all living species. Familiar to everyone in the temperate north are the tits, sparrows, finches, pipits and wagtails, swallows, martins, larks, wrens, warblers, thrushes, starlings, dippers, wrens, jays and crows—including the largest passerine of all, the raven. Twitchers look out for waxwings, tree creepers, nuthatches, and shrikes. North Americans all know their version of the blackbird

common kingfisher
Alcedo atthis

great spotted woodpecker
Picoides major

emperor bird of paradise
Paradisaea guilielmi

(quite unrelated to the European blackbird), plus grackles, cowbirds, and mockingbirds. Passerines throughout the tropics and neotropics include some of the most lavish of all creatures, showing the lengths to which animals may be driven by sexual selection: cocks-of-the-rock and umbrellabirds, lyrebirds, and birds of paradise.

Here, then, is the old familiar pattern: one group among many expands until they outnumber all the rest. Yet the speciosity of the passerines should not be confused with 'dominance'. Though they are, of course, immensely varied, they all essentially continue to exploit the niche of the small percher. They are excellent at what they do—and the fossil record seems to show them outcompeting the coraciiforms, which seemed to have commandeered the niche of the small diurnal flier at least until the Miocene. But the passerines are so various largely because they are predominantly small; and for small animals the world is a bigger place, which provides more niches. In short, passerines are numerous in the way that rodents are numerous among mammals: extremely diverse and conspicuous, but occupying niches that offer no challenge to the larger and inevitably less various fellows.

The groups missing from Joel Cracraft's original treatment—the Psittaciformes, Cuculiformes, and Trogoniformes—are all quite distinctive but, obviously, hard to place phylogenetically.

Parrots: order Psittaciformes

The parrots, order Psittaciformes, are extremely various, but nonetheless distinctive. There are at least 315 known species, including parrots, parakeets, cockatoos, cockateels, conures, lories, lorikeets, macaws, amazons, lovebirds, and budgerigars; some stumpy-tailed (like the popular African grey and cockatoos) but most with long tails (like the amazons, macaws, and budgerigars); and ranging in size from pygmy parakeets from Papua at 9 centimetres to the glorious blue hyacinthine macaws of South America at some 102 centimetres long. Yet you cannot mistake psittaciforms, with their big heads and short necks, their great hooked beaks with the nostrils set neatly at the top in a 'cere' (sometimes naked as in budgies but often feathered), and their powerful, gripping feet with two toes pointing forwards and two backwards in the manner known as 'yoke-toed'. They now live throughout the tropics (though they are not well represented in Africa) but in former times they were more widespread, for although the palaeontological record is not good there are Miocene fossils from France and North America almost into Canada. Even in recent times their range was far larger, for in the nineteenth century the Carolina parakeet flourished up to North Dakota and New York, and the last was killed only in the 1920s, in the Florida Everglades.

princess parrot, *Polytelis alexandriae*

Parrots are fairly uniform in their habits: most are mainly seed eaters, and most nest in unlined holes in trees. But a few have broken the mould, such as the pygmy parakeets, which nest in termite mounds, and the grey-breasted parakeet of Argentina, which builds a huge communal nest (where it is sometimes joined by other species such as wood ducks and opossums), while the kaka parrot of New Zealand is a grub eater and its relative, the kea, is a notorious attacker of sheep. New Zealand also harbours the kakapo, the only known flightless parrot (though it has been known to glide up to 82 metres). As a group, parrots are among the most endangered of animals: around 200 of the 300 or so species are thought to be threatened in the wild.

Touracos, cuckoos, and roadrunners: order Cuculiformes

European cuckoo, *Cuculus canorus*

Order Cuculiformes includes two families: the 19 known species of touraco in the Musophagidae (which means 'banana eater') and the cuckoos, roadrunners, and coucals in the Cuculidae. The touracos are large-ish—around 38–63 centimetres long—with long tails and, typically, a crested head. They live in pairs or small family parties deep in the African forest where they run along the branches like squirrels (although they do fly as well). Like cuckoos (and like parrots) the feet of touracos are yoke-toed (two toes forwards and two backwards) but, like ospreys and owls, they can move the outer toe either forwards *or* backwards.

The cuckoos, in the family Cuculidae, are best known for their 'social parasitism': laying their eggs in other birds' nests and allowing the foster parents to raise them. People often wonder why the foster parents put up with this (don't they notice those great ugly fledglings?). But some recent studies suggest they may have little choice, because the big, menacing mother cuckoo remains on hand to make sure that her young are properly cared for. In fact, cuckoos are not the only social parasites among birds (others include honeyguides and weaver finches) and some cuckoo species do build their own nests. The road-runners of the American south-west, immortalized in cartoons, sprint at a timed 23 miles (37 kilometres) per hour on long, stout legs and kill small snakes and lizards for food by pounding them with their heavy bills. The related Malayan ground cuckoo is about 60 centimetres long.

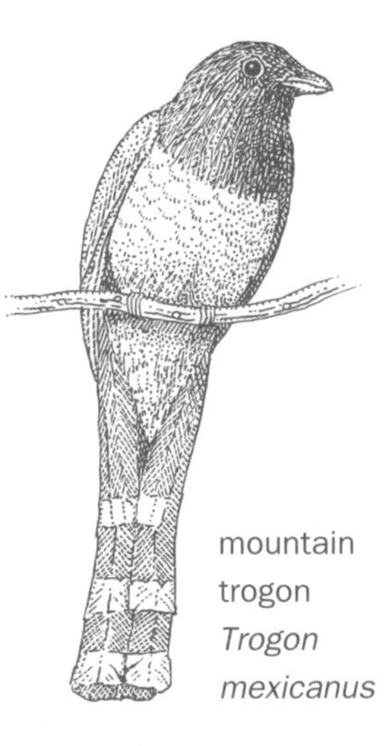
mountain trogon *Trogon mexicanus*

Trogons: order Trogoniformes

Finally, the trogons, order Trogoniformes, include one family of 34 species from the Americas, sub-Saharan Africa (three species), and Asia from India to the Philippines. They are beautiful. We may speak of convergent evolution with the quite unrelated birds of paradise; especially the male, breeding quetzal of Central America with its 60-centimetre-long, trailing tail feathers that the Aztecs and Mayas harvested without killing the bird (for trogon skins are quite remarkably thin and weak, and their feathers fall easily). Again, trogons are yoke-toed, but the inner (second) toe is the one that points backwards (whereas in parrots and cuckoos it's the outer, fourth, toe that has shifted). Trogons mostly eat insects, which they catch in flight.

This completes our survey of the animal kingdom. The last three sections are devoted to plants.

23

THE PLANTS
KINGDOM PLANTAE

THERE ARE THREE GREAT kingdoms of autotrophic mega-eukaryotes: the red seaweeds; the brown seaweeds; and the plants, which include the green algae. Of these three only plants have successfully invaded the land, and only some of them have done so (the ones known as the embryophytes) for the green algae have almost all remained aquatic. The 250 000 or so living species of land plant are the most conspicuous terrestrial creatures of all: liverworts, hornworts, mosses, clubmosses, *Selaginella*, quillworts, psilophytes, horsetails, ferns, cycads, ginkgos, conifers, and the host of flowering plants—they *are* the forests, the grasslands, the tundra. Indeed, as autotrophs (which they achieve by photosynthesis), plants are inevitably at the root of terrestrial food chains, which means that they provide the principal support for all the others. Without help from fungi, however, plants themselves may never have become established on land in the first place.

Yet there is still some disagreement even in the highest botanic circles as to what a plant actually *is*. Some still prefer to regard the green algae as 'protists', and accord them their own kingdom (or some lesser grouping) like the fungi and the red and brown seaweeds. Yet green algae and land plants have much in common—enough to leave no doubt that the land plants' ancestor was a green alga. Land plants and green algae both have the same kind of chlorophyll, with carotenoids as accessory pigments in photosynthesis. Both deposit starch inside the chloroplasts, whereas red and brown seaweeds store the products of photosynthesis outside the chloroplasts. Cellulose is the main component of cell walls in land plants, and is also important in some cell walls of green algae. And so on: the biochemical and ultrastructural similarities are clear, and there is no reason to doubt that they are synapomorphies, and therefore that green algae and land plants form a true clade. Molecular evidence is now reinforcing this impression. Thus it is absolutely justified phylogenetically to place green algae and land plants in the same kingdom—the Plantae, which is how they are treated in this account.

Within the Plantae thus defined we can discern a series of grades, from alga to flowering plant. Between the different grades there is a series of transitions—or, rather, two series of transitions running side by side: one very obvious, which I will call 'ecomorphic', alluding both to way of life and to general structure; and one more recondite but no less important, which I will call 'cryptic'.

ECOMORPHIC TRANSITIONS WITHIN THE PLANTS

Single-celled to many-celled

The obvious transitions are ecological and structural, which means ecomorphic. Clearly, the first plants of all were unicellular algae, and among them various lineages independently developed **multicellularity**. Thus both the green seaweeds of the Ulvophyceae, and the plants of the great clade of the Embryophyta, must each have evolved from *different* protist ancestors. To put the matter another way: the clade of the Ulvophyceae, and the clade that includes the Charales, the Coleochaetales, and the Embryophyta (I know of no formal name for this clade) both *include* members that are protist in grade (formed from single cells or a small group of cells) and some that are true, multicellular organisms. In principle, after all, multicellularity is not a terribly difficult trick to pull, for at least at its most basic level it merely involves cell division (asexual reproduction) without cell separation. The individual cells within the clusters thus formed evolve to become dependent on one another to various degrees, and then began to specialize for different functions, until, finally, we see the emergence of true tissues and the extreme co-operation between cells that is characteristic of truly multicellular organisms such as bracken and oak trees.

Water to land

The second great transition was from water—fresh water—to land; a transition taken to completion by only one of the plant lineages. In some ways this transition was ecologically simple. Light is more available on land: the phototrophic zone in theory runs from the surface of the ground to the top of the atmosphere, while in water the light fades rapidly after the top few metres. The vital gases, oxygen for respiration and carbon dioxide for nutrient, flow freely. Other vital nutrients such as nitrogen are abundant. But there are problems on land: to stay in the light and air while avoiding desiccation.

All land plants (from liverworts upwards) solve this problem at least in part by developing a waxy layer, the **cuticle**, over all or most of their aerial surfaces. But although this reduces the loss of water it also restricts the essential exchange of gases; to compensate, therefore, all land plants (apart from liverworts) have apertures in their leaves and often in their stems called **stomata**, meaning 'mouths'. Like real mouths, the stomata can close; and they do so when water loss is excessive.

Liverworts, hornworts, and mosses are the most elementary of living land plants. Until recently they were commonly placed together in the 'division' known as 'Bryophyta'; but nowadays these are acknowledged as three separate lineages and the term 'bryophyte' is used informally—to describe the grade. The formal term 'Bryophyta' lives on, however (at least among some botanists), as the taxonomic name of the mosses. For the most part, liverworts, hornworts, and mosses solve the gas–water problem by remaining small and living in damp places: the same device adopted by animals such as woodlice and tardigrades (water bears). Like tardigrades, too, many mosses are able to resist desiccation, often to a quite extraordinary degree.

Two final adaptations are crucial. First, the sporophytes of land plants (the term 'sporophyte' is explained later) develop via true **embryos**, which in their early stages depend on the gametophyte for succour and protection. So the entire clade of land plants can be called the Embryophyta[1] or, informally, the embryophytes. Second, land plants evolved the **spore**: an agent of dispersal, borne by the wind, tiny, and with only minimal reserves of nourishment—but waterproof, which is the crucial innovation. The leathery or calcified shell of the reptile is the comparable device.

Plumbing: the vascular plants

Mosses have strands of specialist cells in their stems called **tracheary elements**, which carry water and nutrients from one part of the organism to another. In plants of 'higher' grade, these elements have evolved into more specialized conducting vessels known as **tracheids**, and such plants are said to be **vascular**. Vessels are needed for the rapid transport of water and minerals up from the ground and of the products of photosynthesis down from the leaves, so plants cannot grow large unless they are vascular. The clade of the Tracheophyta—which nowadays includes the psilophytes, horsetails, ferns, cycads, gingkos, conifers, gnetophytes, and angiosperms (flowering plants)—have the most derived and efficient conducting vessels and they include the biggest plants of all: giant angiosperms like some of the eucalypts and the high-storey trees of tropical forests, and huge conifers like the redwoods and Douglas fir. In the extinct rhyniophytes and zosterophyllophytes, and in the clade of the Lycophyta, the vessels in general are somewhat less highly developed. But lycophytes can be large, too: at least, extinct types included significant trees up to 40 metres tall. Fossils of giant eurypterids and early reptiles have been found in the hollowed stems of those ancient, lycophytic trees.

In effect, vascular plants act as wicks, drawing a surging 'transpiration stream' of water up from the ground and through the conducting vessels, ultimately to evaporate from the leaves. Thus they regulate the water content of the soil, and so affect the entire climate and the nature of the terrain. Grass may thus act to keep the soil drier than it would otherwise be, as shown beautifully by the great flat lands of northern Eurasia at the end of the last Ice Age, when the dry grassland gave way to wet, boggy moss and the great herds of mammoths that grazed on the grass disappeared. The traditional notion is that the shift in climate caused this change, and that the mammoths suffered as a result. Now some biologists tell the story the other way around. They suggest that the mammoths sustained the grass by grazing it and that they were driven to extinction, not by the change in vegetation, but by late Palaeolithic human hunters, who were flooding back into northern Eurasia as the Ice Age ended, around 10 000 years ago. When the mammoths had been wiped out, the tracheophyte grass became rank and was ousted by non-vascular moss; and so the

[1] Some biologists do not accept that any of the algal grade plants should be included within the kingdom Plantae. For them, therefore, 'Plantae' is synonymous with 'Embryophyta'. But a classification that is truly based on phylogeny should include green algae, so that Embryophyta becomes a clade (we might call it a subkingdom) within the Plantae.

dry and grassy steppe gave way to dank, cold tundra. At the other end of the scale, however, the tall trees of the tropical forest provide shade that reduces the loss of water by evaporation, and retains the waters of the rainy season that would otherwise simply run way. Thus tropical trees are ameliorative; they maintain the richest biota on Earth in territory that, but for them, would in many cases be desert punctuated by flood.

Spores to seeds

Among the tracheophytes we see three further transitions, all involving reproduction and dispersal. Thus clubmosses, ferns, and horsetails reproduce and disperse themselves by spores, just as bryophytes do. But—and here is the first transition—some of these spore-bearers, such as the clubmosses, are **homosporous**, producing only one kind of spore; and some, like *Selaginella*, are **heterosporous**, producing large **megaspores** and small **microspores**, which correspond to female and male. Then—the second transition—the cycads, ginkgos, conifers, flowering plants, plus various extinct sister groups, produce more elaborate structures for reproduction and dispersal, replete with their own food stores, known as **seeds**.

Ferns, horsetails, and psilophytes were traditionally classed formally as the Pteridophyta; but these groups, too, like the various 'bryophyte' taxa, are now treated as quite separate lineages. The informal term 'pteridophyte' can be retained, however, to describe the grade of vascular plants that reproduce by spores.

Naked seeds to protected seeds

We see one final transition among the seed-bearers. In some—the group that traditionally were called 'gymnosperms'—the unfertilized egg or ovule born by the female organs is naked. At least, it is protected in folds of tissue, and kept moist, but the invading pollen interacts with the ovule directly. In the more derived seed-bearers (the angiosperms), however, the ovule is completely cocooned in layers of tissue and the pollen can make contact with it only by shooting out a long pollen tube that works its way through a canal that runs between the enveloping cells. 'Gymnosperm', like 'Pteridophyta', no longer persists as a formal taxonomic term. It is a paraphyletic grouping that includes the cycads, ginkgos, and conifers, and is useful only informally, to describe the grade of plants with naked seeds. But 'angiosperm' (or Angiospermae) does persist as a formal taxon—the only traditional name that I remember from the 1950s to survive for formal purposes. But, as we will see in the next section, the traditional division of angiosperms into 'monocots' and 'dicots' no longer seems as straightforward as it once did.

CRYPTIC TRANSITIONS

Sex

The other, parallel series of transitions, though no less significant, is far less obvious. Thus the first eukaryotic cells of all presumably had only one set of chromosomes, and hence only one set of genes, and were called 'haploid'; those earliest types may or may not have practised sex (although I would guess that eukaryotes in general were truly sexual from a very early stage, and that sexlessness is generally secondary). Exactly how sex arose need not delay us but once it had arisen its advantages quickly became obvious. In the short term, it provides instant variability in the offspring, which reduces the chances of epidemic disease; and in the longer term, through constant recombination of genes, it greatly enhances the possibilities of evolution. Through recombination, too, sex enables the lineages to jettison harmful mutant genes. Sex, of course, involves the fusion of two cells. Hence immediately after such fusion the two haploid cells form a 'diploid' cell: one with two sets of genes.

Thus, in any organism that practises true sex, a haploid 'generation' of cells alternates with a diploid generation. That is simply a logical consequence of sex. The question now is, which generation is the more conspicuous? In animals such as ourselves the question seems rather silly. The cells of our bodies are diploid. The haploid generation of cells are simply the gametes—highly specialized eggs and sperm produced ad hoc for the purposes of forming a diploid zygote (a zygote being the initial, one-celled embryo). But in some other animals, including aphids, haploid 'gametes' may grow parthenogenetically to produce a generation of haploid offspring that in some cases are free-living organisms in their own right. Then the true meaning of the expression 'alternation of generations' becomes apparent.

Alternation of generations

In plants, **alternation of generations** has become a key feature.[2] Thus in some single-celled algae the usual state is haploid; but two haploid cells may fuse together as if they were gametes to form a diploid zygote, which then promptly undergoes meiosis or 'reduction division' to produce daughter cells, each of which is haploid again. In these algae, then, haploidy is the usual state, with short bursts of diploidy immediately after sexual fusion. In plants more complex than single-celled algae, however, the diploid zygote does not revert immediately to haploidy. Instead, it grows into a whole new organism. Thus in *Ulva*, the sea lettuce, the zygote multiplies to form a **thallus**, the familiar flat, green structure that, when stranded, drapes over rocks and piers. Special regions of the thallus then produce haploid spores, which promptly grow up to form more sea-lettuce thalli, which look exactly like their diploid parent, even though the cells of these new thalli are haploid. These haploid thalli then produce gametes, which fuse to produce zygotes, which grow into diploid thalli again. The diploid thallus is called the **sporophyte** generation, and the haploid thallus is the **gametophyte** generation. Diploid sporophyte thus alternates with haploid gametophyte, and in the case of the sea lettuce you cannot tell, without the aid of a powerful microscope, which kind of generation any particular thallus belongs to.

[2] Note that alternation of generations in plants is quite different from alternation of generations in cnidarians. In the latter (at least in some groups), a jellyfish-like medusa alternates with an anemone-like polyp. But in cnidarians there is no concomitant change of ploidy (chromosome number). Medusae and polyps are both diploid; only the gametes are haploid.

Gametophyte to sporophyte

But in some plants, including all land plants, the gametophyte and sporophyte generations look very different—and their roles change from grade to grade. Thus in mosses and other bryophytes the gametophyte generation dominates. The mosses we see around us on walls and tree trunks are haploid. But the little lamp-post-like structures they throw up from time to time, which contain and distribute their spores, are the diploid sporophytes. The sporophyte is relatively short-lived and less conspicuous; but it has the more complex structure.

In vascular plants, however, the sporophyte predominates. A frond of a fern is a diploid sporophyte; and the haploid gametophyte is generally an inconspicuous and short-lived thallus clinging to the ground like a liverwort (though in the epiphytic stags-horn ferns, the gametophyte is big and permanent, clinging conspicuously to the trunk of the host tree as if to form a 'base' for the hanging, sporophytic fronds). Seed plants reduce the gametophyte still further: it simply become part of the seed. In angiosperms, the gametophyte is reduced to just a few cells.

Thus at this cryptic level we first see haploidy interspersed with short bursts of diploidy, brought about by sexual fusion—with uninterrupted, asexual haploidy emerging here and there as a variation. Then we see a transition to true alternation of generations, sometimes with no clear visible distinction between the haploid and diploid phases; then to alternation of generations with the haploid gametophytes clearly dominant; then to alternation with the diploid sporophytes dominant; and, finally, to a point where the gametophyte has been suppressed almost entirely, so that its existence is revealed only by the most diligent probing.

Furthermore, there is a transition in the gametophytes of plants in which the sporophyte dominates. Thus the uniform spores of homosporous plants, like club-mosses and horsetails, typically develop into only one kind of gametophyte, which bears both female sex organs (**archegonia**) and male (**antheridia**). But the megaspores and microspores of heterosporous plants, such as *Selaginella*, germinate to produce two different kinds of gametophyte, one bearing only archegonia and the other bearing only antheridia.

This, then, is the clade—the kingdom—of the plants: green algae at the base, dividing into several lineages; one lineage developing into land plants, or Embryophyta; and the latter evolving through a series of grades through 'bryophytes', spore-bearing 'lycophytes', and 'pteridophytes', to seed-bearing plants with naked seeds ('gymnosperms'), and eventually to flowering plants (angiosperms).

The phylogenetic tree is based primarily on a paper by Michael J. Donoghue of Harvard University, Massachusetts, as shown in the *Annals of the Missouri Botanical Garden* (1994). But the tree also incorporates ideas from a paper by Paul Kenrick and Peter Crane in *Nature*, and by Kenrick and Crane's excellent *The Origin and Early Diversification of Land Plants* (1997). It is also informed by the fifth edition of *Biology of*

Plants by Peter Raven, Ray Evert, and Susan Eichhorn (1992). Botanical nomenclature is extremely varied (different botanists favour different names for the same groups) and the names here are taken in the main from *Biology of Plants*: it seems a good idea to be consistent with a good, standard textbook.

A GUIDE TO THE PLANTAE

The phylogenetic tree as presented here is regrettably uneven. Some groups are named in formal Latin (as in Sphenophyta), some in the vernacular (as in conifers), and some with endings that suggest high-ranking taxa (as in Pterophyta); while others have family or other endings (as in Chlorophyceae or the generic name, *Selaginella*). The tree also includes some groups that could be polyphyletic, including the micromonads, the rhyniophytes, the bennettitaleans (also known as the cycadeoids or Cycadeoidophyta), and the progymnosperms. The widespread polychotomies are also regrettable.

The fact is, however, that plant classification is in a particularly turbulent state. All phylogenetic trees represent work in progress but in this case the work is somewhat frenetic. The many extinct taxa must be integrated with the living groups but the data from the two are very different. Thus the systematics of living plants is now being greatly enriched by molecular studies, while many of the extinct groups have no living descendants and their structure and relationships must be inferred exclusively from fossils; new finds and re-examination of old specimens are constantly changing the picture.

So, for example, the early group of land plants long known as rhyniophytes now seems to be polyphyletic (so I have abandoned the formal 'Rhyniophyta'). Worse: the apparently well-established rhyniophyte genus known as *Cooksonia* now also seems to be polyphyletic—with some 'cooksonias' apparently closer to clubmosses than to other rhyniophytes. Then there is the problem we have encountered throughout this book: that some of the biologists who are now in the vanguard of systematic studies are least interested in the niceties of Linnaean ranking and nomenclature and, for example, are happy to allow the name of a single genus to stand in for an entire order or class. As they point out, this seems eminently reasonable if the group concerned is known only from a single genus—as, for example, the only living member of the Selaginellales is *Selaginella*. But the resulting classification looks messy, at least by the grand eighteenth-century standards of Linnaeus. On the other hand, in the present state of flux, it does seem silly to invent elaborate Graeco-Latinisms that are likely to have a limited lifespan. Overall, then, I can only regret the infelicities but ask you not to shoot the messenger. Such felicity as might be imposed at this stage would largely be wishful thinking.

The tree does, however, perform its principal task. It includes all the high-ranking taxa of living plants with the outstanding extinct groups painted in where they seem to fit. These extinct taxa include the rhyniophytes, zosterophyllophytes, and trimerophytes, which are outstanding among the early land plants with vessels;

'green algae'
PLANTAE
'bryophytes'
Embryophyta
Lycophyta
23

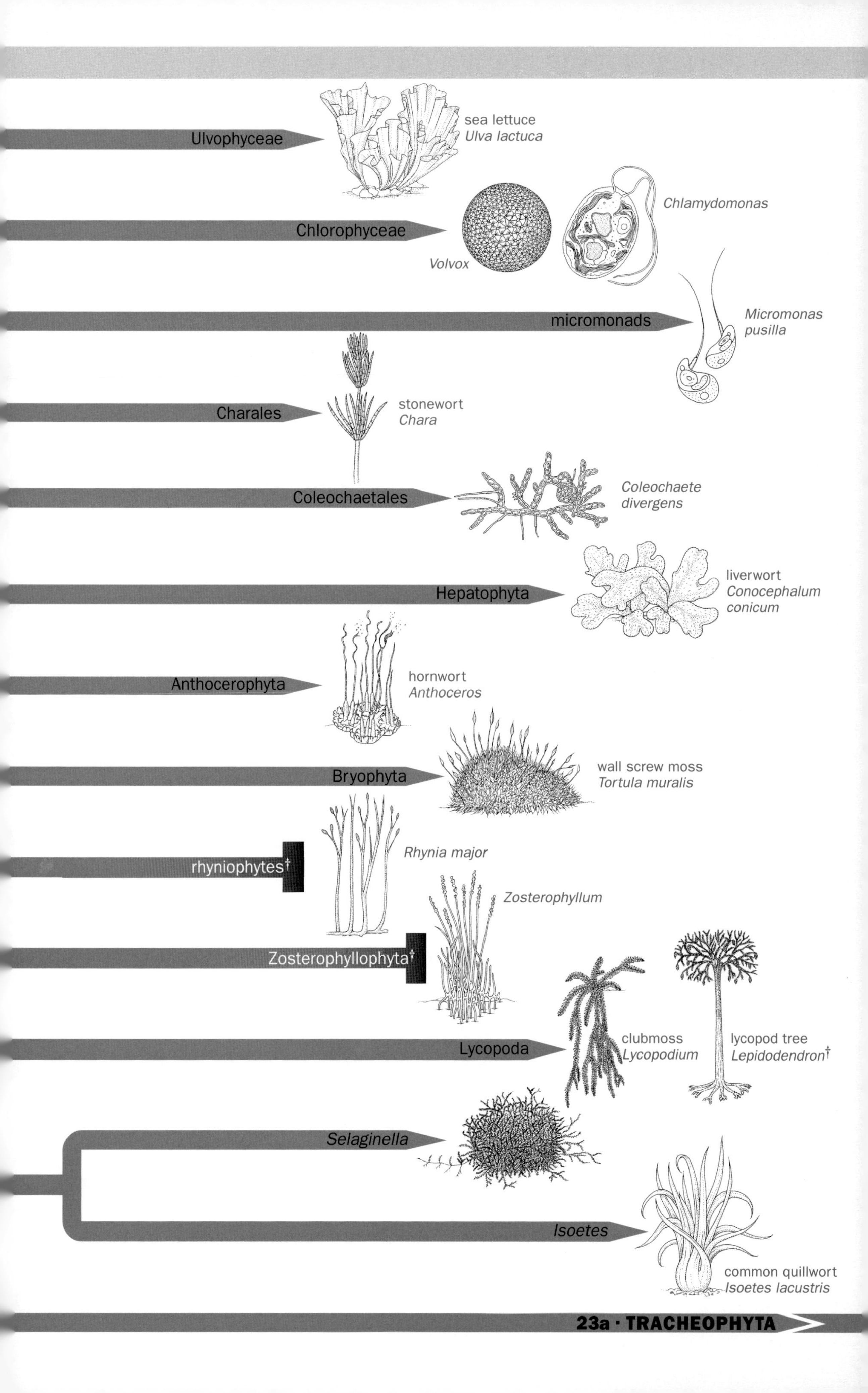
Ulvophyceae
sea lettuce
Ulva lactuca
Chlorophyceae
Volvox
Chlamydomonas
micromonads
Micromonas pusilla
Charales
stonewort
Chara
Coleochaetales
Coleochaete divergens
Hepatophyta
liverwort
Conocephalum conicum
Anthocerophyta
hornwort
Anthoceros
Bryophyta
wall screw moss
Tortula muralis
rhyniophytes†
Rhynia major
Zosterophyllum
Zosterophyllophyta†
Lycopoda
clubmoss
Lycopodium
lycopod tree
Lepidodendron†
Selaginella
Isoetes
common quillwort
Isoetes lacustris
23a · TRACHEOPHYTA

TRACHEOPHYTA
'pteridophytes'
'gymnosperms'
Anthophyta
23a

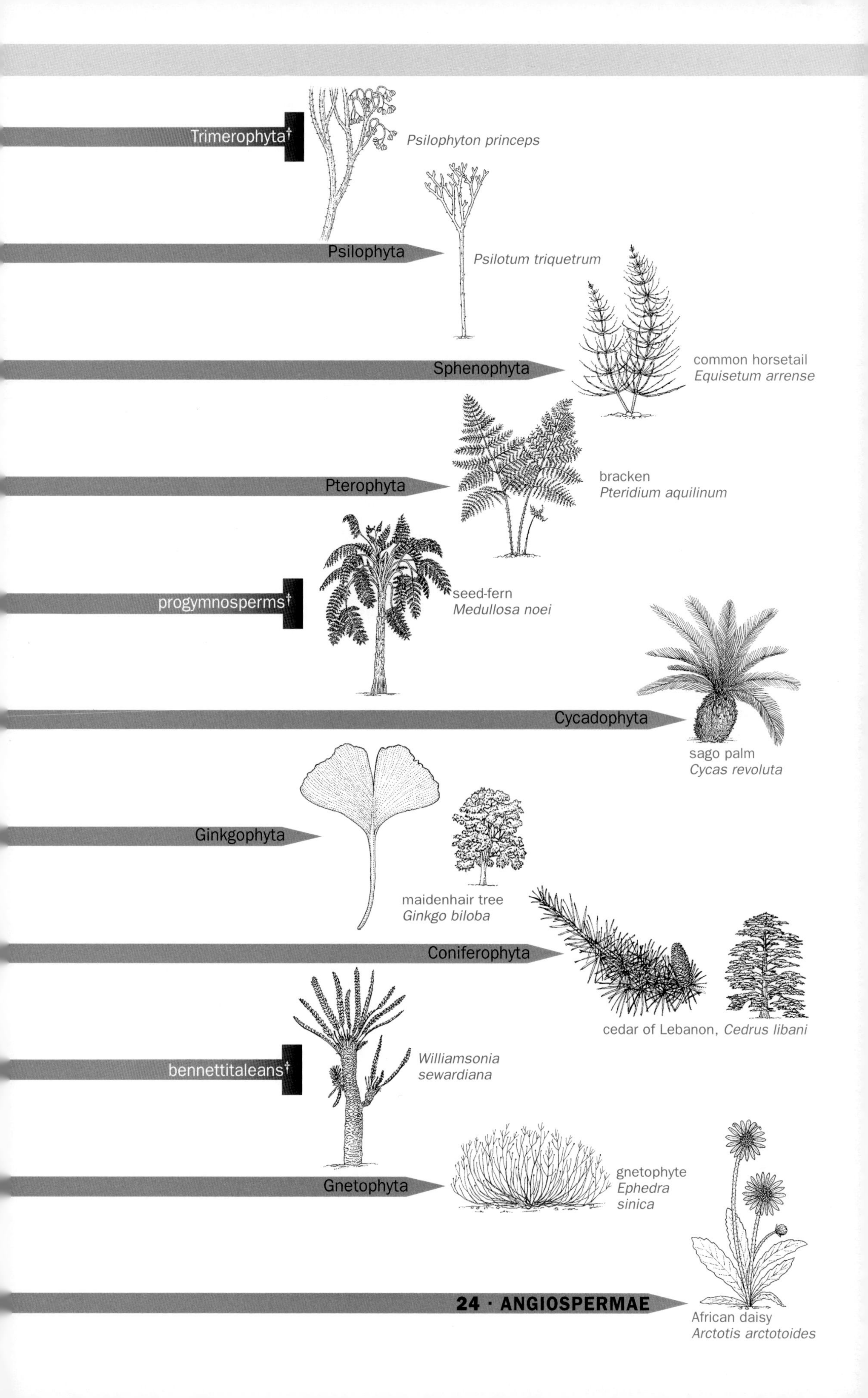
Trimerophyta†
Psilophyton princeps
Psilophyta
Psilotum triquetrum
Sphenophyta
common horsetail
Equisetum arrense
Pterophyta
bracken
Pteridium aquilinum
progymnosperms†
seed-fern
Medullosa noei
Cycadophyta
sago palm
Cycas revoluta
Ginkgophyta
maidenhair tree
Ginkgo biloba
Coniferophyta
cedar of Lebanon, Cedrus libani
bennettitaleans†
Williamsonia
sewardiana
Gnetophyta
gnetophyte
Ephedra
sinica
24 · ANGIOSPERMAE
African daisy
Arctotis arctotoides

the magnificent lycophyte trees that formed great forests in the Carboniferous but are close relatives of the modern clubmosses; the progymnosperms, which apparently include the ancestors of all the living seed plants; and the bennettitaleans, which are not close relatives of the living cycads as their alternative name (cycadeoids) suggests, but seem to be allies of the modern gnetophytes and angiosperms.

Notice the trichotomy at the base (left) of the tree. The topmost branch comprises two great groups of algae: the **Ulvophyceae**, which include the green seaweeds, such as the sea lettuce; and the **Chlorophycaceae**, such as the famous *Chlamydomonas* and *Volvox* of fresh water. The lowermost branch includes the land plants (sometimes collectively called **Embryophyta**, because their development includes a true embryonic phase), plus two more groups of algae, the **Charales** and the **Coleochaetales**, one of whose extinct representatives is thought to include the common ancestor of the land plants. In the middle, awkward and uncertain, sits a swatch of unicellular algae that collectively are traditionally known as 'Micromonadophyta' but, because they may well be polyphyletic, they are rendered here simply as **micromonads**. The micromonads may look out of place, but as Brent Mishler of the University of California at Berkeley comments in the *Annals of the Missouri Botanical Garden* (1994), phylogenetically they do seem to sit between the two great outer branches.

The informal names shown linking some groups of lineages represent the traditional 'divisions' that were widely recognized until recent decades, and still appear in textbooks: **'green algae'** includes all the groups from the green seaweeds to the Coleochaetales; **'bryophytes'** includes liverworts, hornworts, and mosses; the group term **'lycophytes'** is used differently by different authors but certainly includes clubmosses and their extinct arboreal relatives, as here; and the **'pteridophytes'** include the psilophytes, horsetails, and ferns. These terms generally describe grades rather than clades but are still useful colloquially.

So let us look briefly at the different groups, in sequence. The grade of the green algae includes more than 7000 species that are immensely variable. Many are unicellular whereas others, like the fabulous *Codium magnum* of Mexico, one of the Ulvophyceae, are wider than a ping-pong bat and taller than a two-storey house. As we have seen, 'algae' include at least five distinct groupings (though probably more, if the micromonads are polyphyletic) spread across all three principal plant lineages. I begin with the top lineage on the tree: the Ulvophyceae and the Chlorophyceae.

Green algae: the Ulvophyceae and Chlorophyceae

One outstanding feature that unites these two great groups and reveals their difference from the land plants and their allies is their method of mitosis; that is, the process by which their chromosomes divide during cell division. Thus in land plants and their immediate algal relatives the membranes surrounding the nucleus disappear as the chromosomes divide, and reform again when mitosis is over. In ulvophyceans and chlorophyceans, however, the nuclear membranes remain intact throughout. This, and other significant differences of detail, suggest a profound distinction between the ulvophyceans and the chlorophyceans and the rest—a phylogenetic parting of the ways deep in plant history.

The **Ulvophyceae** include what most of us call 'green seaweeds'. There are many freshwater green algae, but the ulvophyceans form the only green algal group that is primarily marine. Most are multicellular. In many, the cells lie end to end to form filaments. In *Cladophora*, for example, the filaments form dense mats both in fresh and sea water. Each cell contains many nuclei in the manner known as coenocytic or **syncytial**. The freshwater *Ulothrix zonata* of cold fresh water is also filamentous but its cells have only one nucleus. In *Ulva*, the cells are arranged in two layers to produce a large, flat **thallus**, held to the substrate by a **holdfast**. The common name, sea lettuce, is extremely apt, for *Ulva*, once the tide has receded, looks just like wet lettuce left out like laundry on the rocks.

sea lettuce
Ulva lactuca

The **Chlorophyceae** contain most of the species of green algae. They include a range of unicellular types, of which some have flagella and some do not. Some are colonial in habit; some motile (capable of moving around) and some not. Some are truly multicellular and, of these, some form filaments and some form flat sheets of cells called **parenchyma**. The chlorophyceans mainly live in fresh water but a few unicellular types live in plankton around the coasts, while some make a showing on land, and live in snow or in soil or on wood.

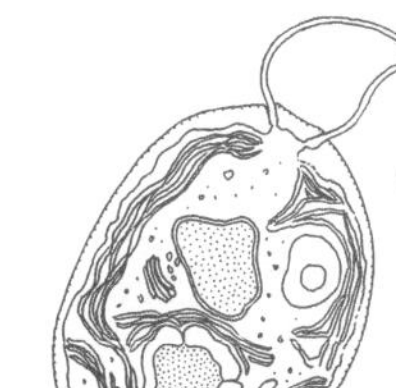

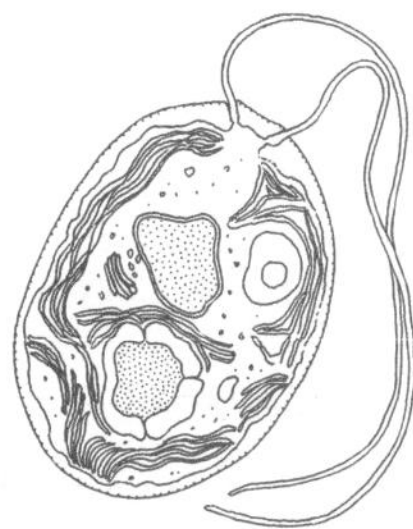

unicellular chlorophycean
Chlamydomonas

Famous among the Chlorophyceae—they feature in all courses of botany—is a series of algae of increasing complexity that extends from *Chlamydomonas* through *Gonium*, *Pandorina*, *Eudorina* to *Volvox*. *Chlamydomonas* is a common, freshwater 'protist', whose single, commonly pear-shaped cells are fitted at the front with two flagella that draw it rapidly but jerkily through the water. Each *Chlamydomonas* cell has a photosensitive eyespot containing the pigment rhodopsin—homologous with the rhodopsin in our own eyes and, apparently, with that of dinoflagellates. Clearly, rhodopsin is a very ancient molecule. The eyespots guide *Chlamydomonas* into light of suitable intensity. When stressed, *Chlamydomonas* cells lose their flagella and their walls become gelatinous, but they recover their verve when conditions improve. They reproduce both asexually and sexually: the latter involving fusion of different mating types.

But in *Gonium*, *Pandorina*, *Eudorina*, and *Volvox*, cells that are virtually identical to *Chlamydomonas* are formed into colonies that function as autonomous organisms; and in the most complex genera such as *Eudorina* different groups of cells take on different functions—so that between them they divide the 'labour' just as the different tissues do in a 'true' organism

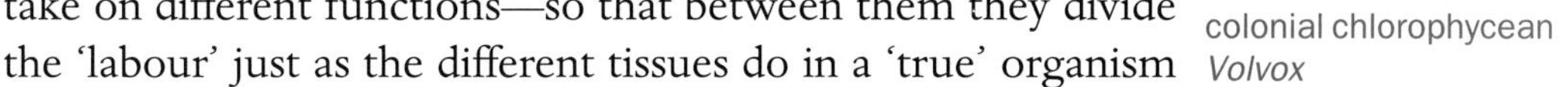

colonial chlorophycean
Volvox

like ourselves. In *Gonium*, the colonies contain 4, 8, 16, or 32 cells, loosely held together in a gelatinous matrix to form a shield-shaped formation, with all flagella beating to drive the colony along. Each cell in the colony can divide to produce a new colony. In *Pandorina*, 16 or 32 cells form a spherical or egg-shaped group that corkscrews through the water, but it has a front and a back, with bigger eyespots at the front. Again, all the cells can reproduce. The colonies are bigger in *Eudorina*—32, 64, or 128 cells; and in this organism, some cells are incapable of reproduction, so here is the start of specialization. *Volvox* is the climactic form: 500 to 60 000 *Chlamydomonas*-like cells form a hollow ball that spins clockwise through the water, and only a minority of cells partake in reproduction. *Volvox* has hence crossed the divide between 'colony' and 'organism'.

Other unicellular chlorophyceans, like *Chlorella*, lack flagella and eyespots; basically such forms are just small green floating spheres. In *Hydrodictyon*, the cells, each with many nuclei, form a reticulum like a fishnet stocking. Others are filamentous—like *Fritschiella*, which grows on land and has superficial resemblances to land plants, including a 'stem' lying close to the substrate (which might be a moist wall) and a 'rhizoid' extending into it. In short, the Chlorophyceae, like the Ulvophyceae, are highly various.

The micromonads

The micromonads are a mixed, possibly polyphyletic but perhaps monophyletic group of unicellular 'flagellates'—archetypical 'protista'. Because they may be polyphyletic, it does not seem wise to persist with their formal family name of Micromonadophyceae. Their precise relationship to the Ulvophyceae and Chlorophyceae, on the one hand, and to the Charales, Coleochaetales, and land plants, on the other, is unclear but molecular studies show them intermediate between the two. Their intermediate phylogenetic position emphasizes, however, that multicellularity among the Ulvophyceae and multicellularity among the embryophytes occurred quite independently.

micromonad
Micromonas pusilla

The Charales and Coleochaetales

The last two groups of algae on the tree, the Charales and Coleochaetales, belong with the lineage that includes the land plants. Indeed, some ancient type among them was the land plants' ancestor. In traditional taxonomies these two groups are commonly combined as 'Charophyceae'.

The **Charales** are the stoneworts: quaint little plants of which there are about 250 living species, found in fresh or brackish water. Some have heavily calcified walls, and so they

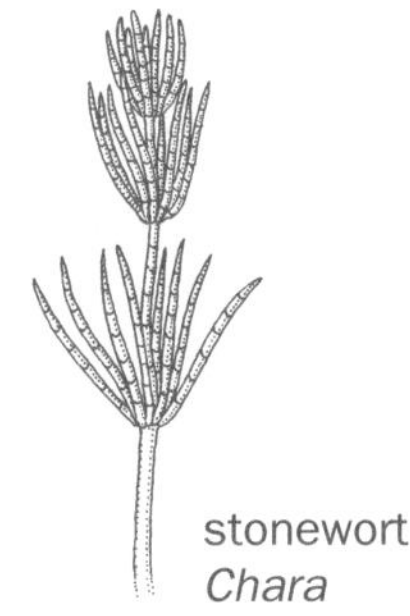

stonewort
Chara

are well represented as fossils. Their structures are complex. Like land plants, they grow by multiplication of cells near the tips (apical growth), and their filamentous 'stems' consist of a series of internodes divided by nodes, from which grow whorls of short branches. They reproduce sexually via eggs and sperm, the latter being produced in remarkably complex antheridia.

Some genera of the **Coleochaetales** take the form of branched filaments, but in others the cells form discs. *Coleochaete* grows on the underwater surfaces of water plants. Among the ranks of the Coleochaetales is thought to be the ancestor of the land plants: the suggestive similarities include the multilayered antheridia, and details of biochemistry and cell division. The living Coleochaetales have their own specializations, however, which rule them out as direct ancestors of the land plants (the actual ancestor is presumably long since extinct).

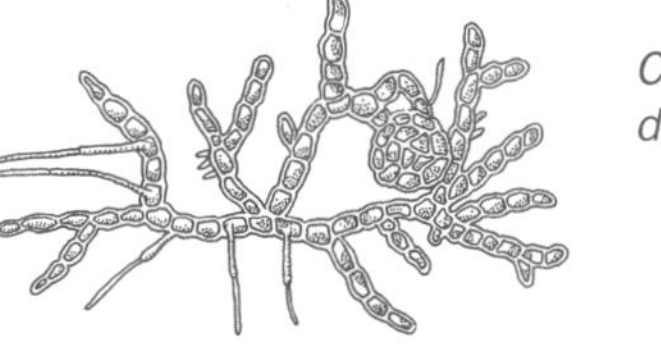
Coleochaete divergens

THE LAND PLANTS: EMBRYOPHYTA

The first land plants were presumably algae, resembling coleochaetaleans, which ventured on to land in the late Ordovician, about 450 million years ago. Even within the Ordovician there are signs of advancement, including bits of cuticle and spores and elongated cells resembling tracheary elements—the columns of specialist cells that conduct water through the plant; and, of course, even the earliest fossils we know about are most unlikely to represent the first organisms of their kind. The first apparent fossils of non-algal land plants are Silurian, from about 430 million years ago. Intriguingly, it now seems that the very first land plants formed mycorrhizal associations with fungi, living symbiotically in their roots; and perhaps—who knows?—their terrestrial algal precursors also lived symbiotically with fungi, so they would qualify as primitive lichens. One way and another, mycorrhizal fungi clearly played a crucial part in the passage on to land.

Pioneers: the 'bryophytes'

The most elementary of the living land plants—informally called 'bryophytes'—include three groups: the **Hepatophyta** (also known as the Marchantiopsida), the liverworts; the **Anthocerophyta** (or Anthocerotopsida), the hornworts; and the **Bryophyta** (also called Bryopsida[3]), the mosses. Phylogenetically the three are separate groups, but they share various general (primitive) features. In all three the gametophyte generation is the more conspicuous and persistent. All of them anchor themselves to the ground via short threads of cells called **rhizoids**—comparable more with a seaweed holdfast than with a tracheophyte root because it is not their special task to absorb water and nutrients. Unlike many algae, bryophytes produce gametes that are

[3] Marchantiopsida, Anthocerotopsida, and Bryopsida is the nomenclature adopted by Paul Kenrick and Peter Crane.

quite distinct—truly motile sperm and eggs. But the eggs are held firmly by special flask-shaped organs in the female gametophyte, and the sperm (perhaps initially washed to the female by rain) must swim up to the egg through a fluid-filled protective tube of cells. After fertilization, the zygote grows for a time as a true, dependent embryo. Thus the mode of reproduction of bryophytes is intermediate between the primitive, essentially protist way of doing things, and the method typical of 'higher' land plants.

This much the three bryophyte taxa have in common. Beyond this, they are clearly very different.

Liverworts: the Hepatophyta (or Marchantiopsida)

The 6000 or so species of liverworts are the lowliest of the true land plants, or at least of those that live today. They should be seen as the sister group of all other land plants, as shown in the tree. They lack one of the typical features of other land plants—stomata in their leaves. But they do have the principal feature of other land plants: the young zygote, when it starts to develop at the beginning of the diploid or sporophyte generation, forms a true embryo, which means it is dependent on the parent gametophyte for nourishment. Protection of the young organism at its most vulnerable stage seems to be the key to a successful life on land: in just the same way, the amniotes, the first true land vertebrates represented by the reptiles, birds, and mammals, protect their embryos in eggs equipped with shells and additional membranes.

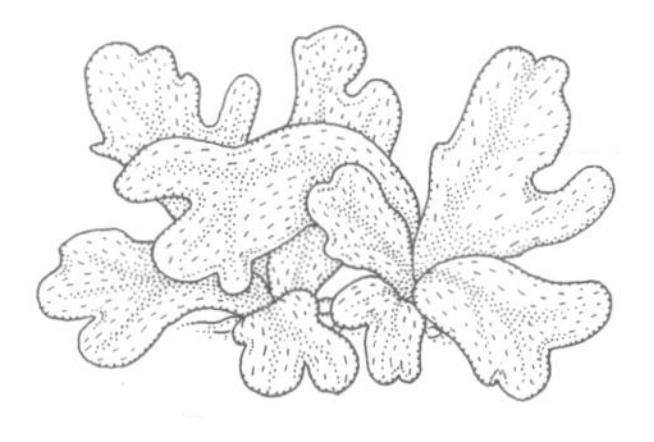
liverwort *Conocephalum conicum*

In some liverworts (the minority), the plant consists of thick flat thalli, like a small but thick seaweed: *Marchantia* is typical. Most—at least 4000 species—are leafy, like mosses. Unlike mosses, however, their leaves lack a thickened midrib, and they are typically arranged in two main rows with a third row of smaller leaves underneath. Moss leaves tend to be of equal size and arranged in spirals.

Hornworts: the Anthocerophyta (or Anthocerotopsida)

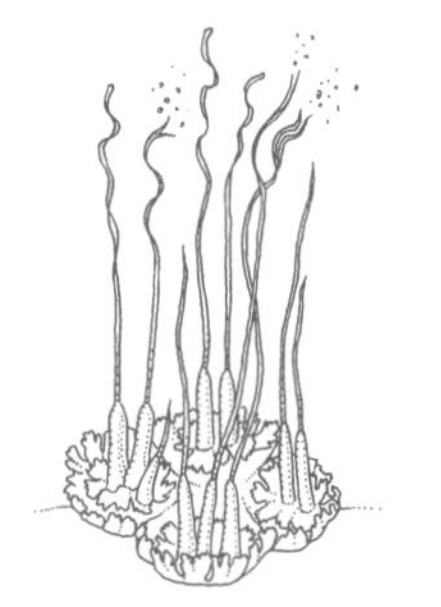
hornwort, *Anthoceros*

The hornworts are a quaint, small group of plants—just 100 species or so in half a dozen genera of which the best known is *Anthoceros*. Superficially hornworts resemble liverworts, with fleshy thalli often arranged in a rosette, the size of a penny. Although the liverworts do seem to emerge as the sisters of the rest, the hornworts have several odd characters in common with green algae—including one large chloroplast per cell, instead of the multiplicity typical of most plant cells. On the other hand, they have stomata, like all the other land plants except the liverworts. Their thalli are riddled with spaces, as in liverworts; but whereas the liverwort spaces are full of air, those of hornworts are full of mucilage within which lurk cyanobacteria of the genus *Nostoc*—a handy symbiosis because *Nostoc* is a 'fixer' of nitrogen, converting nitrogen gas from the atmosphere into soluble nitrogen compounds that help to nourish the plant.

The principal hornwort plant is a gametophyte, as in other bryophytes. But hornworts often send out numerous sporophytes on the same plant, which are green and may grow tall; hence the antlered appearance and the name.

Mosses: the Bryophyta (or Bryopsida)

Mosses are the most successful bryophytes whether assessed by number of species (around 9500) or by ecological impact, which is huge. Thus they occur worldwide from the tropics to Antarctica, mostly in the wet but sometimes in desert and a few at the seaside (although they are never truly marine), while the peat mosses of the genus *Sphagnum* occupy at least 1 per cent of the world's surface—equivalent in area to half the United States! Yet, like lichens, mosses are highly sensitive to air pollution (particularly sulphur dioxide).

There is also a short catalogue of plants that are commonly called 'mosses' but in reality are no such thing. These include 'reindeer moss', which is a lichen; 'club-moss', which is a lycophyte; 'Spanish moss', which is a flowering plant (related to the pineapple); and 'sea moss' and 'Irish moss', which are algae. Real mosses are of three main lineages: the **Bryidae**, which are the 'true' mosses; the **Sphagnidae**, the peat mosses; and the **Andreaeidae**, the granite mosses.

'True' mosses: the Bryidae

These are 'typical' moss: the main plant, the gametophyte, is from 0.5 millimetre to 50 centimetres tall and with leaves, just one cell thick, generally arranged in spirals. Many have conducting cells in their stems called **hyroids**, evidently homologous with the tracheids and other conducting vessels of tracheophytes. Others also have **sieve elements** around the water-conducting cells, similar to the phloem of 'higher' plants. Mosses reproduce by sperm produced in male antheridia, and carried to the female archaegonia by drops of water and sometimes by insects—in which case they may travel great distances.

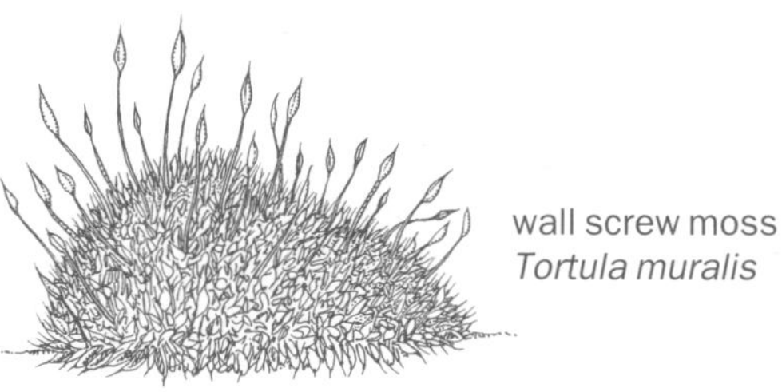

wall screw moss
Tortula muralis

Bryids commonly adopt one of two kinds of habit: 'cushion', with upright gametophytes with little branching, and usually with sporophytes at the tips; or 'feathery', with highly branched, creeping stems and sporophytes born laterally. In both cases the capsules of the **sporangia** release their spores—up to 50 million per capsule!—in a manner highly characteristic of the group: the top comes off like a lid, leaving behind a jagged circle of 'teeth' called a **peristome**. Mosses also reproduce asexually by fragmentation, or by producing little multicellular propagules known as **gemmae**.

Peat mosses: the Sphagnidae

All the 350 or so species of sphagnids belong in just one genus, *Sphagnum*, which, in ecological terms, is surely among the most successful of all genera. Sphagnids are quite distinct from bryids and the two groups clearly diverged a long time ago. Their gametophyte stems are highly branched—often five branches per node—with more branches towards the tips, so they tend to have mop-like heads. Mature plants lack rhizoids and their leaves lack midribs.

Sphagnid leaves, looking bright green or reddish *en masse*, have a most peculiar structure: live cells surround dead cells that are hollow and hold water, such that sphagnum moss can hold 20 times its own weight of water. This is a wonderful adaptation to the boggy ground in which the moss lives, and which it helps to perpetuate. *Sphagnum* accordingly has been used as a dressing for wounds (notably during World War I). Cotton dressings have taken over only because they look cleaner—they hold only four to six times their own weight of water. *Sphagnum* sporophytes are spherical or nearly so, and are red to blackish brown. As is the way with sporangia they scatter the spores by drying and hence distorting and then splitting, but sphagnum does this in spectacular fashion. It bursts with an audible click, throwing the spores far and wide.

The peat bogs formed by sphagnum moss maintain their vastness partly by repelling invaders. The moss produces hydrogen ions that creates high acidity—a pH of 4, which is most unusual in unpolluted environments.

Granite mosses: the Andreaeidae

There are just two genera of granite mosses. *Andreaea* includes about 100 species that form small blackish-green or reddish-brown tufts on rocks, often on granite, on mountains, and high up in the Arctic. Peculiarly, *Andreaea* disperses its spores through splits in the sporangium capsule. *Andreaeobryum* was discovered in Alaska only in 1976. Its capsules split to the apex when it scatters its spores.

PLANTS WITH VESSELS: THE RHYNIOPHYTES, LYCOPHYTA, AND TRACHEOPHYTA

From among the ranks of the bryophytes emerged the grade—and clade—of plants that are equipped with vessels. The tree shows all the plants that have vessels in one great clade, and divides that clade into three smaller clades: the rhyniophytes, Lycophyta, and Tracheophyta. The shape of the tree as shown here is derived directly from Paul Kenrick and Peter Crane but the nomenclature is somewhat novel. Because I am not a professional plant taxonomist you may well feel that I am being presumptuous in applying names in a perhaps unprecedented fashion. So let me explain the logic of it.

First—to begin at the end—many modern botanists apply the term 'Tracheophyta' to all the groups that have true vessels—namely the psilophytes, sphenophytes (horsetails), pterophytes (ferns), cycads, ginkgos, conifers, gnetophytes, and angiosperms (flowering plants), all of which are with us still; plus the mixed bag of progymnosperms and bennettitaleans, which are extinct. Many, however, exclude the extinct Trimerophyta from the Tracheophyta, and show them as the sister group of the rest. But Peter Crane suggests (in a personal letter) that the relationship of the trimerophytes to the other tracheophyte groups is not fully resolved, and, for the time being at least, they should be shown as part of a polychotomy, as here. If they are shown in this way, then the term Tracheophyta must include them. In any case, no one seems to doubt these days that the trimerophytes do share a common ancestor with the other tracheophytes. In all respects, then, it seems perfectly justified to apply the formal clade name Tracheophyta, as shown here, to include the trimerophytes as well as all the rest.

There is, however, a wide and varied group of plants that have vessels, but the vessels do not have the highly derived form of the true tracheophytes. These include lycopods (clubmosses and extinct relatives), the living *Selaginella* and *Isoetes* (quillworts), and the extinct zosterophylls. So far as I know the term Lycophyta has not been applied to all of these groups together, as shown here. Some apply the term exclusively to the Lycopoda, so that 'lycophyte' and 'lycopod' become synonymous. Peter Raven and his colleagues in *Biology of Plants* include the Lycopoda, *Selaginella*, and *Isoetes* within the Lycophyta. But if the zosterophylls are the sister group of the lycopods + *Selaginella* + *Isoetes*, as shown here (and Peter Crane suggests), then it is neat, logical, and cladistically justified to include them, too, within the Lycophyta. This also gives us the tidy division into Lycophyta and Tracheophyta. We certainly should not bend the facts to fit the nomenclature; but it is perfectly acceptable to readjust existing nomenclature to fit what are perceived to be the facts. So that is what I have done.

Lycopods, zosterophylls, *Selaginella*, and the quillworts all have small leaves of the kind known as **microphylls**, which are formed simply as extensions of the stem. The leaves of most other tracheophytes, like ferns and flowering plants, are **macrophylls**, probably formed from a group of lateral branches that have grown together, or 'anastomosed', to form a flat **lamina**. But the possession of microphylls should not be a defining feature of the Lycophyta. Microphylly should probably be seen simply as a primitive feature (a symplesiomorphy) and not as a common, derived feature (a synapomorphy).

The rhyniophytes, as shown here, are commonly assumed to be the sister group of all the more derived plants with vessels, both the Lycophyta and the Tracheophyta. As mentioned earlier, I am calling them 'rhyniophytes' and not 'Rhyniophyta' because they are probably polyphyletic. In traditional classifications, rhyniophytes, zosterophylls, and trimerophytes are commonly grouped together as 'protracheophytes'. But I have dropped this term as the three groups clearly belong on different branches of the tree, as shown here.

Rhyniophytes†

The polyphyletic rhyniophytes take their name from the village of Rhynie in Scotland, near the site where the first examples were found. The oldest date from the

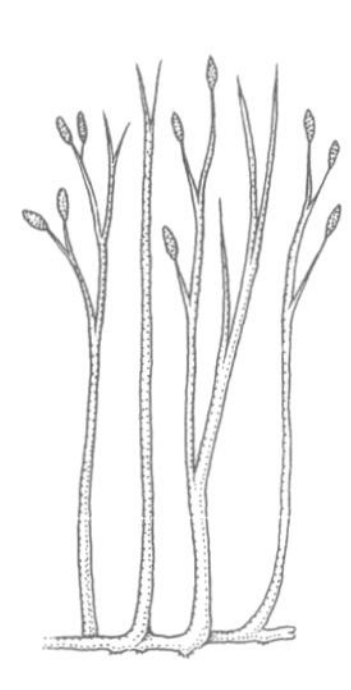
rhyniophyte
Rhynia major

Silurian, more than 420 million years ago, and by the mid-Devonian, about 380 million years ago, they were gone. The oldest of all the rhyniophytes that we know about was *Cooksonia*,[4] whose green, forked, match-like heads poked about 6.5 centimetres above the mud. *Cooksonia* was extinct by the middle Devonian. The best-known rhyniophyte genus is the larger *Rhynia*, with stems about 20 centimetres long, like forked asparagus. The rhyniophytes looked mightily primitive, bearing their spore-filled sporangia at the tips of their leafless, bifurcated branches, but the internal structure of their stems was like that of many vascular plants of today.

Zosterophylls: the Zosterophyllophyta†

The zosterophylls are also Devonian. They, too, were leafless and bifurcated and only the tops of their stems had stomata—so the bottoms of their stems may have been buried in mud. Superficially they resemble the modern *Zostera*, the maritime flowering plant known colloquially as 'sea grass' (although it is not a grass), which is the food of dugongs and the marine iguanas of Galápagos. Unlike rhyniophytes, zosterophylls bear their spherical or kidney-shaped sporangia on short stalks, laterally. The zosterophylls are quite distinct from the rhyniophytes and seem to be the sister group to (and include the ancestor of) the lycopods, *Selaginella*, and the quillworts.

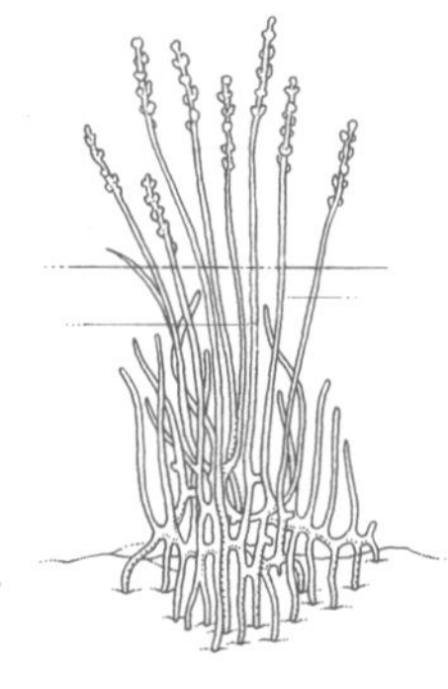
zosterophyll
Zosterophyllum

THE LYCOPHYTA

Clubmosses and their extinct relatives: the Lycopoda

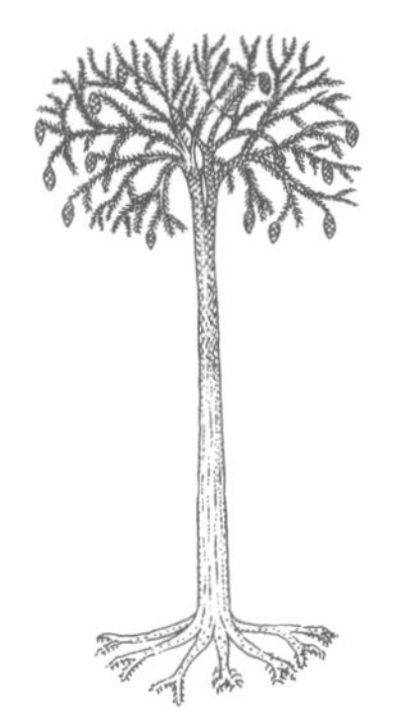
lycopod tree
Lepidodendron

Lycopoda, as defined here, includes the living clubmosses and their tropical relatives, plus their immediate extinct relatives. The lycopods, thus defined, evidently began in the Devonian, probably as descendants of the zosterophylls. At least half a dozen orders are known and although the living lycopods are of humble appearance, many of the ancient types were magnificant trees up to 40 metres tall, although still with the simple, dichotomously branched structure of the earliest vascularized plants. Lycopod trees dominated the Carboniferous, and their carbonized remains form much of our coal. Most had become extinct by the end of the Permian. That none has survived is a

[4] But note the comment above—that *Cooksonia* is probably polyphyletic. Indeed, some plants ascribed to the *Cooksonia* genus are related to other rhyniophytes while others may be more closely related to clubmosses.

great aesthetic loss—even one survivor would have provided a wonderful glimpse of the past, as the *Ginkgo* does.

The living clubmosses and their tropical relatives, in the family **Lycopodiaceae**, are clearly not mosses; the superficially moss-like plant is a sporophyte, while 'real' mosses are gametophytes. Most of the 400 or so known, living species were formerly called *Lycopodium* but this catch-all genus is now being subdivided and probably represents 20 genera or more. Most lycopods are tropical, and most of the tropical kinds belong in the genus *Phlegmarius*. The tropical types are epiphytic, and usually hard to see: the familiar clubmosses of temperate latitudes are denizens of the forest floor, and are conspicuous because they remain green in winter.

clubmoss, *Lycopodium*

Lycopods are homosporous: the spores, when they germinate, give rise to bisexual gametophytes, bearing both female archegonia and male antheridia. The gametophytes may be lobed, green, and photosynthetic, or subterranean, non-photosynthetic, and mycorrhizal; and the archegonia and antheridia take 6–15 years to mature. Lycopods produce biflagellated sperm, which must swim to the archegonium and down its neck to the egg within.

Selaginella

Selaginella is the only genus within the family Selaginellaceae, which is the only family in the order Selaginelles; so it seems reasonable to drop the family and ordinal names and simply refer to all the selaginellas as *Selaginella*. The genus does, however, include 700 known species. Most live in moist places but a few are at home in deserts where they may be dormant in the driest parts of the year, like the 'resurrection plant', *S. lepidophylla* from Texas, New Mexico, and Mexico. Selaginellas commonly look like little ferns although their leaves are microphylls like lycopods, rather than macrophylls, as in true ferns. Unlike lycopods, however, *Selaginella* is heterosporous: its megaspores and microspores germinate to form separate female and male gametophytes. Again, *Selaginella* needs water for reproduction as its sperm must swim to the eggs in their archegonia. In *Selaginella*, 'conquest' of the land is incomplete.

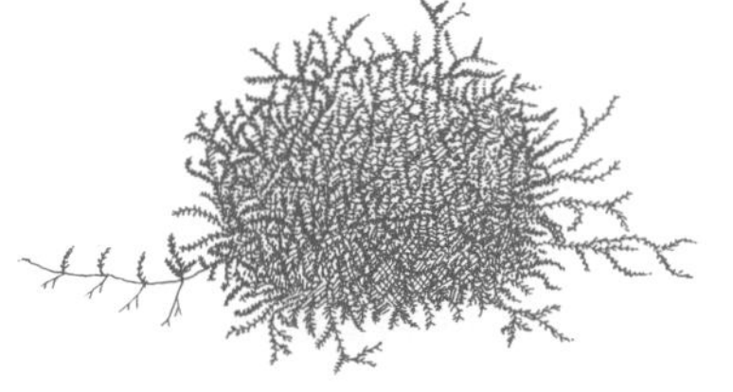
Selaginella

Quillworts: *Isoetes*

Here is yet another major taxon with only one living genus: *Isoetes*, the quillworts. They are curious plants, with short, fleshy, underground corms with quill-like micro-

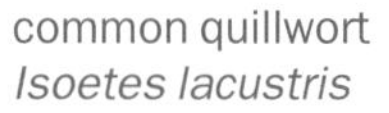
common quillwort
Isoetes lacustris

phylls above and roots below. They grow in water, or in pools that dry up from time to time.

THE TRACHEOPHYTA

Trimerophytes: the Trimerophyta†

The trimerophytes lived only during the Devonian, from about 395 to 375 million years ago. They, too, have the typically leafless, dichotomous look of the other 'protracheophytes' but they were probably up to a metre high—much bigger than rhyniophytes or zosterophylls, and more complex. Some botanists have suggested that trimerophytes arose directly from rhyniophytes, and that they are the sister group of all the tracheophytes—that would give them a pivotal position, as the 'missing link'. But their relationship to the other tracheophytes is unclear and so they should be shown, as here, simply as part of a polychotomy.

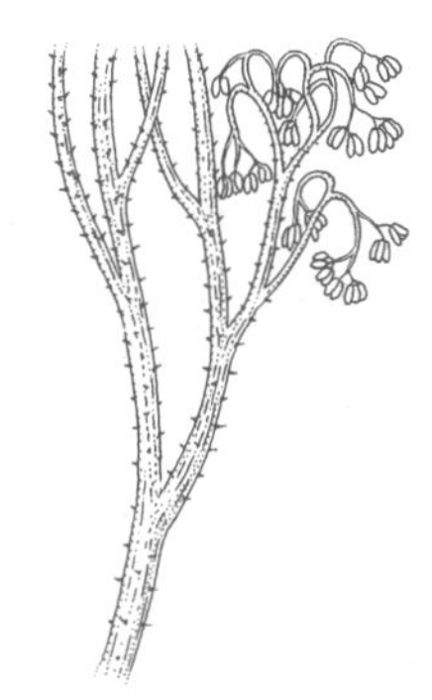

trimerophyte
Psilophyton princeps

THE PTERIDOPHYTE GRADE

Psilophytes: the Psilophyta

psilophyte
Psilotum triquetrum

The psilophytes bring us to the plants of pteridophyte grade: truly vascular plants with highly derived vessels, but reproducing by spores. The psilophytes, however, are a most peculiar group: lacking both roots and leaves they resemble rhyniophytes. The only living genera are *Psilotum*, a common greenhouse weed in Florida and other southern US states, Hawaii and Puerto Rico; and *Tmesipteris*, which grows as an epiphyte in Australia, New Zealand, and elsewhere in the South Pacific. Both produce their spores on lateral branches, which germinate underground to form a subterranean gametophyte that is nourished by a symbiotic fungus. The sporophyte—the visible plant—then grows out from the gametophyte by a 'foot', which later detaches itself.

Horsetails: the Sphenophyta

Wherever it is damp, by streams and at the edge of woods, you are liable to find horsetails. In my own native London I remember them teeming on railway banks (although they seem less common now), looking marvellously primitive and exotic even in that humdrum context. They date from the Devonian and, like the lycophytes, in the late Devonian and Carboniferous they produced great trees known as calamites, up to 18 metres

common horsetail
Equisetum arrense

high and with trunks nearly half a metre thick. They reached their climax of abundance and diversity in the late Carboniferous, around 300 million years ago. Now only one genus remains, *Equisetum*, the horsetail. But *Equisetum* is more or less identical with *Equisetites* known from 300 million years ago, so *Equisetum*/*Equisetites* might reasonably be seen as the oldest surviving genus in the world.

Horsetail stems look vaguely bamboo-like, with sharply demarcated nodes where the needle-like leaves—probably reduced macrophylls—sprout in whorls. The stem between the nodes, the internodes, are ribbed and reinforced with silica, so horsetails are traditionally used for cleaning pots and are also known as 'scouring rushes'. They are homosporous, and bear their spores in sporangia along the margins of umbrella-like 'sporangiophores'. Sometimes the reproductive stems that carry these sporangiophores are creamy white, almost devoid of chlorophyll. The spores germinate to form green, free-living gametophytes about the size of a pinhead, which flourish on nutrient-rich mud. The gametophytes, which are either bisexual or all male, mature in 3–5 weeks, and, again, the multiflagellated sperm need water to reach the eggs: freedom from water is incomplete.

Ferns: the Pterophyta

Ferns, with about 11 000 living species, are the biggest and most diverse group of non-flowering plants. Most species live in the tropics—so that the modestly sized island of Costa Rica has around 1000 species, while the whole of the USA and Canada has a mere 380. But many temperate communities are dominated by ferns, like the ubiquitous bracken, even if the variety is low. About a third of all tropical ferns grow as epiphytes on trees. Most of the most familiar types have divided, feathery leaves but, in some, the leaves are big and 'entire'. *Lygodium* is a climber with extended leaf stalks 20 metres and more in length. Many are easily big enough to qualify as trees—such as *Cyathea*, sometimes more than 24 metres tall with leaves more than 5 metres long. Even with the biggest ones, however, there is no secondary growth of the kind that year-by-year builds up the trunks of pines and oaks. The stems of *Cyathea* look 30 centimetres thick but most of the thickness is brought about by a sheath of fibrous roots.

Ferns fall into two broad grades. In the more primitive, **eusporangiate** types the spore-bearing sporangium forms from a whole layer of cells; but in the more derived **leptosporangiate** types the sporangia develop from single cells. Most living ferns are homosporous—and, indeed, only two orders of water ferns, and various extinct types, are heterosporous. Various orders of ferns are traditionally recognized, of which I will discuss just five.

The two eusporangiate orders: the Ophioglossales and Marattiales

There are only two eusporangiate fern orders—the Ophioglossales and the Marattiales.

The order **Ophioglossales** includes just three living genera of which two are widespread in north temperature regions: *Botrychium*, the grape ferns, and *Ophioglossum*, the adder's tongues. Their gametophytes are subterranean and tuberous. *Ophioglossum reticulatum* has 1260 chromosomes—more than any other organism known.

Ferns in the order **Marattiales** are tropical, and date back to the Carboniferous. Now there are about 200 species in six genera. Extinct types include the tree fern *Psaronius*.

The largest leptosporangiate order: the Filicales

bracken
Pteridium aquilinum

Most ferns are leptosporangiate and most of them—about 10 500 species in 320 genera spread among 35 families—belong to the order Filicales. Filicales are homosporous and bear their spores in sporangia that commonly occur in clusters called **sori**, which are often coloured—yellow, brown, or blackish—and arranged in lines, dots, or broad patches. In many, the young sori are covered by specialized outgrowths known as **indusia**. The spores germinate to produce free-living, bisexual, heart-shaped gametophytes, which grow in moist places, including the damp exterior of flowerpots. Reproduction again requires water, for the multiflagellated sperm must swim to the eggs.

Water ferns: orders Marsileales and Salviniales

The leptosporangiates also include two orders of water ferns, both heterosporous—they are the only living ferns that are. But in other ways the two groups are very different, and almost certainly evolved independently from different terrestrial ancestors.

Order **Marsileales** contains just three genera, including *Marsilea* with about 50 species. It grows in mud, often with leaves floating on the water and resembling a four-leaved clover. *Marsilea's* drought-resistant sporocarps remain viable after a century of dry storage.

Order **Salviniales** contain just two genera, both of which are small and floating. *Salvinia* has three leaves, born in a whorl on a floating rhizome. Two of the leaves are undivided, about 2 centimetres long, and covered with hairs that maintain bouyancy; but the third is highly dissected like a mass of whitish roots and hangs down into the water—although it is clearly a leaf and not a root because it bears sporangia. *Azolla* is a tiny fern that floats among the paddy fields of China and Southeast Asia and is of enormous economic importance. It has pouches in its duckweed-like leaves that house the cyanobacterium *Anabaena azollae*. *Anabaena* is a fixer of nitrogen. Thus, in traditional rice fields, *Azolla* with its cargo of *Anabaena* is a prime source of fertility.

THE SEED PLANTS

The seed is one of nature's great inventions, equivalent to the cleidoic (closed) egg of the reptiles; protecting the embryo—the young sporophyte—from desiccation and thereby meeting the final demand of terrestriality. All seed-bearing plants are heterosporous: they have microsporangia, which produce male microspores; and female megasporangia, which produce female megaspores. But in seed plants the megaspore gives rise to a megagametophyte that is highly reduced—and is retained in the megaspore that gave rise to it; and the megaspore, in turn, is retained within the megasporangium; the whole structure is called the **ovule**. Thus the whole, vulnerable, haploid gametophyte generation on the female side is concertinaed and packaged. The first known seed-like structures appeared in the late Devonian, some 370 million years ago.

On the male side, the microspore is comparably chaperoned by parent cells in a package called **pollen**, which is carried to the waiting ovule by wind or by some animal—beetle, bee, hummingbird, 'honey possum', bat, what you will. In ginkgos and cycads, the pollen produces sperm as in the seedless plants, but the sperm is not released except when the pollen has already been brought to the ovule by a **pollen tube**, which grows out of the pollen and through the protective tissues that surround the female gamete. Most seed plants dispense with motile sperm altogether; fertilization is achieved simply via a pollen tube, with the male gamete reduced effectively to a nucleus. Thus in all the seed plants, including ginkgos and cycads, fertilization is achieved by a carefully orchestrated liaison of female and male gametes, precisely comparable with the copulation that is usual in land animals. In neither case do the gametes need any watery intermediary. Hence, for reproductive purposes, seed plants enjoy total freedom from water.

The seed itself is a fertilized ovule, together with the extra protective layers provided by the mother plant—the **seed coat**. In most modern plants, the embryo (the young sporophyte) develops within the seed before dispersal; a refinement that perhaps was promoted by natural selection during the Permian, when climates were extreme and sometimes cold and dry. The seed, in short, is a brilliant device involving a whole series of innovations; yet it seems that nature has invented it not once, but at least twice. The ultimate ancestors of all seedy plants, perhaps, were the trimerophytes. In the Devonian, the trimerophytes gave rise to a more derived group called the **progymnosperms**. These were spore-bearers, like the trimerophytes themselves. But in their vegetative parts, and notably in the structure of their conducting vessels, the progymnosperms resembled modern conifers, giving them a decidedly modern aspect and suggesting that they were indeed the ancestors of later groups.

In the Devonian, too, apparently in two quite separate evolutionary events, the progymnosperms gave rise to two quite different kinds of seed-bearing plants. The first of these, now totally extinct, were the seed-ferns (once called the Pteridospermophyta). The seed-ferns flourished through the Carboniferous and lasted into the Triassic. Emphatically they were *not* ferns, although they had fern-like leaves. Neither were they the ancestors of the 'true' seed plants. In short, the seeds of seed-ferns and of true seed plants provide yet another, wonderful example of convergence.

seed-fern
Medullosa noei

The true seed plants apparently arose as a clade—a single evolutionary event—from a quite separate group of progymnosperms. Within this clade, there are six lineages, of which five are still with us and one is extinct. The living types are the cycads, **Cycadophyta** or Cycadales; the ginkgo, **Ginkgophyta**; the conifers, **Coniferophyta**; the peculiar gnetophytes, **Gnetophyta**; and **Angiospermae**, or angiosperms, which are the flowering plants. The one extinct division—perhaps polyphyletic—is the cycad-like **Cycadeoidophyta**, otherwise known as the **bennettitaleans**. Cycads, the ginkgo, and the conifers may be referred to informally as **'gymnosperms'.** As the tree shows, they do not form a clade, but 'gymnosperm' is a useful descriptive grade-name; it means 'naked seed' and it denotes those seed plants whose ovules are not totally encased within layers of protective cells, as is the case in angiosperms.

Just as botanists cannot be sure of the order in which the psilophytes, horse-tails, ferns, and progymnosperms arose from the trimerophytes (assuming they did), so they remain uncertain of the sequence in which the seed-ferns, cycads, ginkgos, and conifers arose from the progymnosperms. Thus we have another polychotomy. The relationship of the bennettitaleans, gnetophytes, and angiosperms is also best shown as a trichotomy for the time being. Of course, such arrangements are not ideal; but they are honest, and at least define the areas of certainty and uncertainty (or so we may hope). Let us now look briefly at the true seed plants, group by group:

Cycads: the Cycadophyta

Some cycads resemble palms whereas others have the look of giant pineapples, their elongate or rounded trunks patterned by the bases of fallen leaves, their living leaves bunched in a plume at the top. But they are neither palms nor pineapples. Cycads are cycads. They appeared at least 320 million years ago, in the Carboniferous, and flourished in the Mesozoic alongside the superficially similar cycadeoidophytes or bennettitaleans. So the Mesozoic was not just 'The Age of Dinosaurs'; it was 'The Age of Dinosaurs, Cycads, and Bennettitaleans'. Now there are about 140 species of cycad in 11 genera; most of them are large, and some grow to more than 18 metres. Many are toxic, but some have an edible pith known as sago; hence the common name of 'sago palm'.

sago palm
Cycas revoluta

Cycads still reproduce by means of mobile spermatozoa, though these do not have to rough it across country like those of mosses or ferns, but are decorously introduced to the ovum via a pollen tube. The pollen itself seems generally to be spread by insects, commonly beetles (and, in particular, nowadays, by weevils). Indeed, beetles may have been the first insect pollinators, way back in the Carboniferous, and cycads may well have been the first plants to be insect-pollinated. By comparison with the cycads and their attendant beetles, the consortia of flowering plants with bees, butterflies, and moths are johnnies-come-lately.

Ginkgos: the Ginkgophyta

Only one of the Ginkgophyta survives: *Ginkgo biloba*, the maidenhair tree. The genus, *Ginkgo*, has changed little since deep in the Cretaceous, 80 million years ago; and other ginkgophytes date back to the early Permian, 280 million years ago. As a wild tree the modern ginkgo seems virtually extinct but it flourishes in parks worldwide, growing up to 30 metres, and producing a fine show of golden autumn colour, before the leaves drop off. Ginkgos also do particularly well in cities, as they are fairly resistant to air pollution. Males and females form separate plants. The ovules of the females form fleshy, edible fruits, which Chinese people eat (and gather in the proper season in New York's Central Park). Interestingly, the ovules may not be fertilized until after they have been shed.

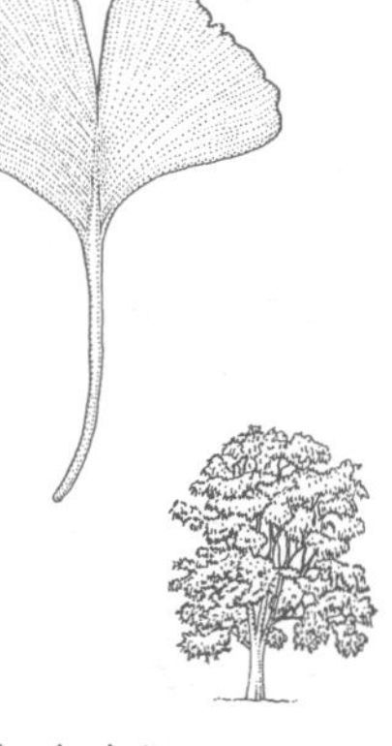
maidenhair tree
Ginkgo biloba

Conifers: the Coniferophyta

The conifers form by far the largest group of living 'gymnosperms', with 550 or so species in 50 genera. Of course they are far less speciose than the angiosperms, which now include around 235 000 species, and far more uniform in appearance. But they are enormously successful, forming huge, often virtually monocultural forests over vast areas of the globe. The world's tallest living tree is a conifer—*Sequoia sempervirens*, the giant redwood, reaching 117 metres, which is well over 350 feet; although Australia's 'mountain ash', which is not an 'ash' at all but a eucalypt and is the tallest living flowering plant, runs it a close second.

Conifers are known at least since the late Carboniferous, around 300 million years ago; prominent among the early, primitive types were the Cordaites, which also contributed enormously to the great coal-forming forests. Many of the modern conifers have features that seem adapted to drought, perhaps reflecting their diversification during the cold, dry Permian, which followed the Carboniferous.

Just a few genera account for most of the modern conifers. Outstanding are the 90-odd species of *Pinus*, dominating huge areas of Eurasia and North America and widely cultivated in the Southern Hemisphere. The arrangement of their leaves is unique among living conifers: they are produced on short side branches whose growth is 'determinate', which means predetermined. The longest lived of all trees is a pine—the bristlecone, *Pinus longaeva*, whose individual leaves remain active for up to 45 years. But because they keep their leaves for such a long time (usually 2–4 years), pines are very susceptible to pollution.

The striking and distinctive feature of the conifers, however, is none of these

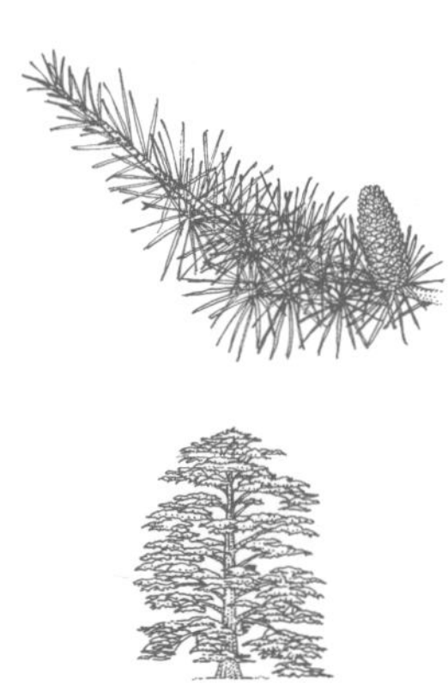

cedar of Lebanon
Cedrus libani

things. It is the **cone**, and the seeds it contains. Cones are of two kinds, male and female; and although both are usually produced on the same tree, pollination is usually effected by male cones from a different tree (so that the trees are generally 'outcrossed'). The female cones commonly take 2 years to mature. The ovules are pollinated in the spring of year one and the mature seeds, winged to assist wind dispersal, are shed in the autumn of year two. The seed is a combination of two sporophytic generations—the seed coat and the embryo; and also includes the gametophyte generation, which provides a food reserve for the embryo. In some species, as in the lodgepole pine *Pinus contorta*, the scales of the cone separate to release the seeds only after subjection to extreme heat or, in other words, after the tree is scorched by forest fire. Many other trees, including many Australian eucalypts, are similarly fire-dependent.

Other notable genera among the conifers are *Abies*, the firs; *Picea*, the spruces; *Tsuga*, the hemlocks; *Pseudotsuga*, the Douglas firs; *Cupressus*, the cypresses; *Juniperus*, the junipers; *Taxodium*, the bald cypresses; the family Taxaceae, the yews, which bear their fruit in fleshy red cups called **arils**; *Sequoia*, the redwoods; and *Metasequoia*, the dawn redwood, which was the most abundant conifer in west and northern North America from the Cretaceous to the Miocene (90–15 million years ago) and was thought to be extinct, but was rediscovered alive and well in the Chinese province of Sichuan in 1944. *Metasequoia* now flourishes in botanic gardens including Kew, in London, and Missouri, at St Louis. The outstanding conifers of the Southern Hemisphere are the araucarias. They are of enormous economic importance, and are familiar to northerners in the form of *Araucaria araucana*, the monkey-puzzle tree, pride of many a suburban garden.

THE ANTHOPHYTA

Bennettitaleans: the Cycadeoidophyta†

Superficially the cycadeoidophytes, or benettitaleans, resemble living cycads but the details of their reproduction are different. They were very much part of the scene in the Jurassic and Cretaceous periods; it is a pity that none are with us today. The group may be polyphyletic.

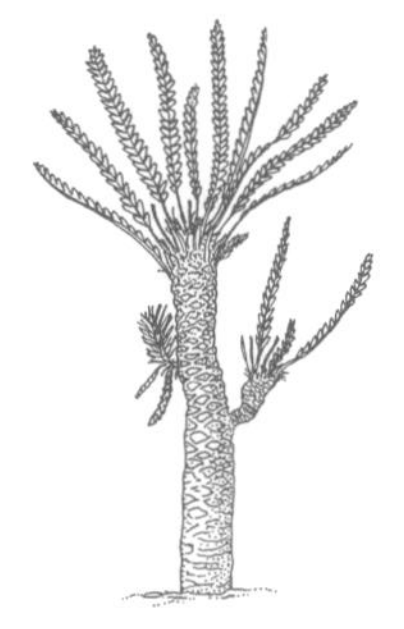

bennettitalean
Williamsonia sewardiana

Gnetophytes: the Gnetophyta

The gnetophytes have much in common with the angiosperms including the minor fact that they all produce nectar and are visited by insects, which presumably facilitate their pollination. Although this and other similarities have often been ascribed to parallel evolution, it is now generally agreed that the gnetophytes are truly the sister group of the angiosperms. The first ever angiosperm probably sprang from the ranks of the gnetophytes, though not from among the living types.

The living gnetophytes contain only about 70 species in three very different but clearly related genera, and a remarkable lot they are. *Gnetum* includes about 30 species of trees and climbing vines with large leathery leaves that closely resemble dicots. They are found throughout the moist tropics. *Ephedra* includes about 35 species, most of which are highly branched shrubs, whose small, inconspicuous, scaly leaves superficially resemble those of horsetails. *Ephedra* species mostly live in deserts. *Welwitschia* is one of the oddest of all plants. Most of it is buried in sandy soil, while the part that shows above ground is a massive, woody, concave disc with cone-bearing branches around the margins, and just two strap-like and remarkably dead-looking leaves; although these leaves split end to end so the plant resembles a bunch of wrack or a pile of abandoned ticker-tape. *Welwitschia* lives in the coastal desert of southwestern Africa in Angola, Namibia, and South Africa.

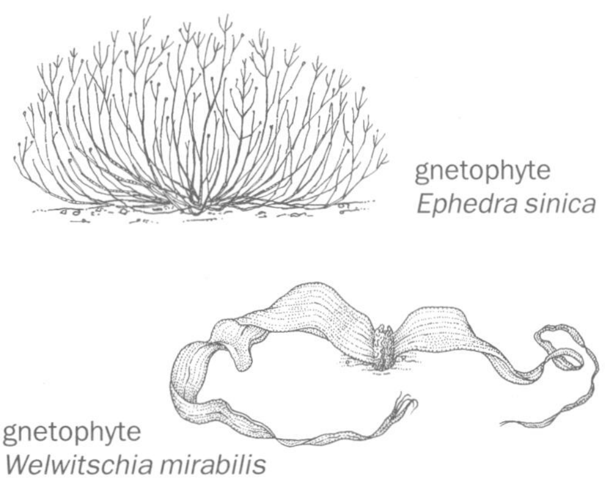
gnetophyte *Ephedra sinica*

gnetophyte *Welwitschia mirabilis*

Flowering plants: Angiospermae

angiosperm (African daisy)

So to the angiosperms. They first appear in the fossil record in the early Cretaceous, around 130 million years ago, although they were already various by then and their origins must date from the Jurassic. They cannot claim to have dominated every terrestrial landscape, for conifers cover vast areas, tree ferns dominate some tropical mountain tops, and *Sphagnum* moss creates one of the biggest of all the world's biotopes within the northern tundra. Overall, however, angiosperms clearly dominate the world's land masses, accounting for most of the vast tropical rainforests and the deciduous forests of higher latitudes, all of the world's grasslands, and a vast variety of other niches in between. They are also by far the most various of plants, with 235 000 species—more than 90 per cent of all plant species. Clearly, they deserve a section to themselves.

24

THE FLOWERING PLANTS

CLASS ANGIOSPERMAE

THE ANGIOSPERMS ARE THE flowering plants and unless you live in some fairly unusual place—an ocean-going yacht, the tundra, a cloud-mountain, the heart of some coniferous forest—then you will be surrounded by them. They have been steadily increasing their influence upon the world since the early Cretaceous, at least 130 million years ago, occupying all but the most recondite niches on land and in fresh water. They have even ventured back into the sea, or at least to the edges of it; like the glassworts *Salicornia*, and the salt-marsh grass *Spartina*, the many species of trees that make up mangrove swamps, and 'eel-grass', *Zostera*, which actually flowers under water and feeds a host of maritime creatures from manatees to the marine iguanas of Galápagos.

The angiosperm range of size and form is stupendous: from Australia's 'mountain ash' (in fact, a eucalypt), which vies with the coniferous redwoods for the title of the world's tallest plant; to the tiniest duckweeds of the genus *Wolffia*, reduced to floating leaves less than a millimetre across. Duckweeds are monocots: their family, Lemnaceae, is closely related to that of the arum lilies, Araceae. The smallest of all flowers are pinhead-sized or less. The biggest is that of the Indonesian parasitic plant *Rafflesia*. Its vegetative parts are pared to a fungus-like 'mycelium' that ramifies through the cambium (growth tissue) of its host, while its flowers, up to a metre across, smell of carrion to attract obliging swarms of pollinating flies. *Rafflesia* is one of about 3000 species of flowering plants that have partly or entirely abandoned photosynthesis and/or the acquisition of nutrients through their own roots and live as parasites off other angiosperms. Many others, like the broomrape and mistletoe, are common wayside plants.

Evolutionarily speaking, the flowering plants seem to be latecomers. The oldest unequivocal angiosperm fossils are early Cretaceous—pollen from about 127 million years ago, and some small flowers from Melbourne from around 120 million years ago that resemble modern pepper in the family Piperaceae. Yet many botanists feel that flowering plants must be much older than this and there are occasional reports of Jurassic and even Triassic angiosperm remains. Many point out, too, that the earliest unequivocally angiosperm pollen resembles that of gymnosperms and that anything much older would simply be undistinguishable, so even if older angiosperm pollen were found it would be misdiagnosed. Some molecular evidence also seems to suggest that the two great groups of angiosperms that have been traditionally recognized, the monocots and dicots, diverged more than 200 million years ago, which would mean that the angiosperms as a whole must be older than this.

But other botanists point out that the bennettitaleans and the gnetophytes, the groups most closely related to the angiosperms, are known only from the Triassic around 225 million years ago, and they must be older than the angiosperms. The molecular evidence that suggests extreme antiquity is so far out of line with all other data that it must surely be erroneous; and the alleged pre-Cretaceous angiosperm fossils really are equivocal. Taken all in all, then, there seems little good reason to suppose that angiosperms are in fact older than the early Cretaceous. Indeed we can reasonably suggest that they probably did arise only about 140–130 million years ago, and then radiated rapidly. The fossil record shows that many living families and even some living genera had appeared by 90 million years ago. Many botanists, too, have doubted whether angiosperms are monophyletic. But modern evidence proclaims that they are a true clade, demarcated by a suite of special characteristics.

WHAT MAKES AN ANGIOSPERM AN ANGIOSPERM?

Five characteristics are outstanding. First, and the most obvious feature of angiosperms is the flower itself. Each flower in principle consists of four 'whorls' of sexual and auxiliary structures. On the outside are the supporting **sepals**, which may be green or may themselves be coloured like petals. Inside the sepals are the **petals**, commonly brightly coloured to attract insects (although the patterns visible to us may not be what attracts the insects—for insects tend to be sensitive to ultraviolet light and to them many flowers appear 'ultraviolet coloured'). Then come the male components, collectively known as the **androecium**, which includes a highly variable number of **stamens**. Each stamen in a modern angiosperm has two parts, like a standard-lamp; an **anther** at the top, which produces and then releases the **pollen**, and the supporting **filament**. In the centre of the flower are the female parts of the flower, collectively called the **gynoecium** and consisting of one or several carpels. Each **carpel** has an **ovary** at the base, containing the ovule (the egg cell plus its nutritive tissue); and is surmounted by a **style** topped by the **stigma**, which is specialized to receive the pollen. There are, however, endless variations on this basic theme of 'flower': many, for example, are unisexed, and many lack sepals and/or petals.

Exactly how flowers arose in evolutionary history remains unclear. The flowers of modern flowers are pollinated either by wind or by animals—mainly insects, birds, bats, or sometimes other mammals. Many wind-pollinated flowers are obviously highly derived, like those of grasses; and so are many insect-pollinated flowers, from orchids to snapdragons to daisies. But some wind-pollinated flowers are clearly primitive, like those of peppers;[1] while primitive insect-pollinated flowers include those of

[1] 'Peppers' in this context means the 'true' pepper—members of the Piperaceae, not the 'sweet peppers' or the chillies, which are fruits of Solanaceae.

buttercups, water lilies, and magnolias, whose flowers are wonderfully showy but simple in design, with the different parts of the flowers separate from one another, and arranged very obviously in a spiral, which is one of nature's primitive forms (and note, in passing, that the most primitive forms are often the most beautiful). 'Primitive' in this biological context means 'close to the form of the common ancestor'. If the angiosperms are truly monophyletic, then this means that all flowering plants share a common ancestor. So was that common ancestor small and wind-pollinated, like a pepper, or big and showy like a magnolia? It could not have been both, simultaneously. The two designs are incompatible.

In fact, botanists have generally subscribed to the notion that insect pollination provided the initial spur to angiosperm evolution, and that wind pollination is secondary among flowering plants. Insect pollination offers the huge advantage of precision: an insect-pollinated plant can afford to produce relatively modest amounts of pollen, which is then carried exactly where it is needed. Hence insect-pollinated plants can find mates even when the individuals and populations are widely scattered; and this enhances the survival of small populations in niches that may be rare and far apart, which, in turn, encourages the evolution of a wide range of species. The very first insect pollinators were probably beetles, the descendants of whom still pollinate living cycads; the butterflies and bees that are now the principal insect pollinators of angiosperms probably came on the scene later . Evolutionarily speaking, the angiosperms probably rode on the backs of the cycads and the bennettitaleans, which over many millions of years had coevolved a mutually beneficial arrangement with a variety of insects. The first angiosperms, siren-like, lured those fickle insects away (or some of them, at least).

But the most fundamental evolutionary question was posed by Charles Darwin and is still unanswered: what is the relationship between the reproductive parts of the flower, and those of other plants? How, if at all, does the spiral-like form of the magnolia relate to that of the gymnospermous cone? What, if any, are the homologies between the bright-coloured flower, evolved to attract insects, and the comparable lures of cycads and bennettitaleans? Many classical botanists have sought to find out, but their conclusions remain equivocal. Molecular studies may resolve the issues. It should be possible to identify the individual genes that code for the different parts of flowers and comparable structures in other plants: the same kinds of studies that are helping to illuminate the homologies or otherwise between insect and crustacean mouthparts. Meanwhile, the extent to which the different parts of flowers are homologous with gymnosperm cones, or are merely analogous, remains almost as uncertain as in Jane Austen's day. That is not a criticism of the investigators; it simply reflects the extreme difficulty of the task.

The second great unifying feature of angiosperms is that the **ovule** is completely enclosed within layers of tissue supplied by the parent plant. Fertilization is achieved via a **pollen tube**, which grows out of the pollen like a hypha of a fungus, and burrows through the style to the ovule, with its sperm nuclei inside.

Third, more than any other plant, angiosperms emphasize the **sporophyte**.

Indeed the male gametophyte is reduced to just three cells, which between them form the pollen: two sperm cells and one that forms the pollen tube. The female gametophyte is reduced to just seven cells. One of these seven is the egg cell. Another of the cells—of which more will be said in the next paragraph—retains two nuclei; so the seven cells contains eight nuclei between them. These seven gametophyte cells form the crucial component of the ovule, the **embryo sac**. The nuclei of these various gametophyte cells are all, of course, haploid. Each carries only one set of chromosomes.

Fourth, we see in angiosperms the peculiar phenomenon of **double fertilization** —although in truth this is also shared by the gnetophytes, which are the oldest living relatives of angiosperms. Perhaps it was practised by the bennettitaleans as well, in which case we could say that double fertilization is a synapomorphy of all the anthophytes (see tree in Section 23). Looked at from an animal perspective, double fertilization seems most extraordinary. It means exactly what it says. One of the two sperm nuclei in the pollen fuses with the egg cell to begin the new embryo. The other sperm nucleus from the pollen fuses with the gametophyte cell that contains the two nuclei, thus forming a cell with three sets of chromosomes in it: two from the female gametophyte and one from the visiting male gametophyte. The resulting 'triploid' cell multiplies to form the **endosperm**, which provides the nutritive tissue for the developing embryo.

The mechanism by which double fertilization provides both embryo and endosperm superbly illustrates the opportunism of natural selection; the way in which a desirable structure (in this case the endosperm) may be cobbled together from cells that seem simply to be lying around—rather as two of the mammalian earbones were fashioned out of spare bones from the reptile jaw and its surrounds. But why should evolution have worked in such a way? Why was it necessary to arrange this secondary fusion of male and female cells—and to provide a special female cell containing two nuclei—to produce an endospermal mass? I suspect this has to do with the way in which genes are switched on and off, and the need on the one hand to produce tissue that divides in a fairly disorganized fashion to provide a nutritive mass, and on the other to draw a clear genetic distinction between such multiplicative tissue and the main plant, lest it take over like a cancer. Be that as it may, double fertilization is a key feature of the angiosperms and, apparently, of their fellow anthophytes.

Finally, angiosperms have evolved some unique vegetative structures—in particular, the special **sieve cells** of the phloem that conduct nutrients through the plant more efficiently than the less-specialized conducting cells of other tracheophytes.

In addition to these five diagnostic characters there are a few more, which, although they may occur in other groups, are in general more common in angiosperms. Thus, many angiosperms are deciduous—a skill that may have evolved in the tropics to protect them against drought, and now serves even more conspicuously to reduce their vulnerability in high-latitude winters. Only a few conifers have adopted deciduousness, such as the larch and bald cypresses. Many angiosperms manage to complete their entire life cycle in a year, or in some cases even less; these are the annuals. This again gives a degree of flexibility that opens many entirely new horizons, and enables flowering plants to overwinter in the form of seeds that sometimes may germinate in the following season and at other times may wait around for many years before springing back to life. Many wayside 'weeds' are such rapid-growing opportunists. Very few tracheophytes apart from angiosperms have essayed the herbaceous form—virtually woodless, with the plant supported solely by the hydrostatic pressure of the cells. But angiospermous herbs abound—their woodlessness allowing them to grow extremely rapidly, with all the benefits that may thereby accrue. More generally, angiosperms have the most diverse of habits: they grow as lianas, vines, trees, creepers, and they display the most various devices for surviving bad times—a host of tubers, rhizomes, bulbs, and what you will.

Then, again, angiosperms produce some of the most robust of seeds, which survive to be widely spread and may last through successions of bad winters and the most prolonged of droughts. The toughness of their seeds and the many devices for dispersal by wind, water, and animals, combined with their precise methods of pollination, ensure that angiosperm reproduction is the most efficient of any plants. Finally, angiosperms are the greatest of the eukaryote pharmacologists, producing an enormous variety of chemical agents, which, in particular, defend them against diseases and herbivores. Incidentally, the qualities of herbaceousness—quick growth, and highly efficient pollination and seed dispersal (in space and time)—are supremely demonstrated by the composites (the daisy family), as described in the next section.

Overall, we simply have to concede that the angiosperms are brilliant. It is no longer fashionable to espouse the notion of evolutionary 'progress', partly because the Victorians tended to conflate 'progress' both with 'destiny' and with moral ascendancy. But if we perceive the flower as a machine for bringing pollen and ovum together, and for dispersing the seed that results from their union, then we have to acknowledge that this particular machine is extremely efficient. An engineer, assessing the tasks objectively and the solutions that the angiosperm brings to bear on them, would surely acknowledge their superiority. Certainly the angiosperms have flourished, spread, and speciated spectacularly during the past 50 million years—since the early Eocene—as the world has grown steadily cooler and in many ways more stressful. But as observed in *Ecclesiastes*, the race is not always to the swift nor the fight to the strong, and the angiosperms have not had things all their own way. Non-flowering plants flourish, too, and in many contexts they outgun the angiosperms. Mosses, ferns, and conifers still prevail over vast landscapes, while the brown and red seaweeds, autotrophic mega-eukaryotes that are very different from plants, remain unsurpassed in the oceans.

The classification of angiosperms is at an interesting stage. Botanists began serious classification even before Linnaeus, and many ideas dating from the early nineteenth century are still current and, indeed, may stand forever. After all, there were many fine botanists in earlier centuries and if modern methods overturned all previous scholarship then the methods would be suspect and surely would be disregarded. Nonetheless, the methods of cladistics and new data, both from molecular studies and from fossils, are now suggesting that the traditional classifications that prevail in textbooks and popular guides do need rethinking, sometimes radically. Unfortunately for me, however, trying to provide a classification that is tidy and comprehensive, and yet is truly based on phylogeny, the studies remain very much 'in progress'. Traditional classifications provide the tidiness and comprehensiveness; but modern studies increasingly question their assumptions without yet offering anything comparably tidy (or widely accepted) in exchange.

So in this guide I have gone for compromise. The tree is based on cladistic studies by Peter Crane, of the Field Museum, University of Chicago (as in 'The Fossil History of the Monocotyledons', by Patrick Herendeen and Peter Crane in *Monocotyledons: Systematics and Evolution*, edited by P. J. Rudall and others, 1995; and a paper in *Nature* by Peter Crane, Else Marie Friis and Kaj Raunsgaard Pederson, 1995)—full details of these references are given in the Sources.

But I have also tried to link these new ideas with traditional accounts that do at least provide a tidy overview of all the flowering plants, and are the rock on which any new classification must be founded. I feel it is important, too, to savour the contrast between the new and the traditional, so as to enhance appreciation of both. So I have also drawn on the traditional taxonomy in *Flowering Plants of the World* (1978) edited by V. H. Heywood. This is among my most treasured reference books—a classic, which deserves to be reissued as it stands. But it was written in the 1970s, just before cladistical ideas and molecular data truly began to come on line; and many of the groupings it recognizes will surely be rethought. What follows, then, is a somewhat hybrid classification, reflecting the state of transition.

A GUIDE TO THE ANGIOSPERMAE

Traditionally botanists have divided flowering plants clearly into two subclasses: the Dicotyledoneae and the Monocotyledoneae. The monocots include the grasses, lilies, onions, irises, palms, bromeliads (pineapples and their relatives), and so on; and the dicots, as traditionally defined, included the rest—oak trees, daisies, peas, cabbages, and so on. Botanically the division is based on the number of seed-leaves, or cotyledons: two, generally, in dicots, and one in monocots. The great English naturalist John Ray (1627–1705) first suggested this division in 1703, and Antoine-Laurent de Jussieu (1748–1836), one of the most influential of all plant taxonomists, also emphasized the fundamental importance of the cotyledons. Most botanists today take it for granted that flowering plants divide cleanly into dicots and monocots. This is the division that Vernon Heywood favoured. Dicots are generally listed first, and then the monocots.

For most purposes this straightforward split into dicots and monocots works well enough. It is certainly tidy. But as shown in the tree, cladistic studies by Peter Crane and others now suggest that this simple division into dicots and monocots does not reflect true evolutionary relationships and should not, therefore, form the basis of a classification that aspires to be 'natural'. In fact the monocots *do* form a true clade—but not simply as the sister group of all the dicots. The monocot clade emerges from among the ranks of dicots, just as the bird clade arises from among the dinosaurs. It is the case, however, that most of the living dicots do form a clade, which Crane calls the 'eudicots'. Only the primitive ones—such as the water lilies, peppers, and magnolias—are hived off. In all, Crane's tree acknowledges six main groupings of living angiosperms, while tradition acknowledges only two.

A general word about the tree is called for. Peter Crane and his immediate colleagues are among the many modern systematists who do not care to spend time adjusting their nomenclature to fit convention. That is, Linnaean (and strict Hennigian!) convention requires that groupings of equivalent rank should have the same name ending. So the names of plant superorders conventionally end in '-idae' (not to be confused with the *families* of animals); plant orders end in '-ales'; and the families should end in '-aceae'. You will see, however, that Crane's six main divisions (which we might call subclasses, although the ranking might equally well be left unspecified) have various endings: Nymphae*ales*, Piper*ales*, Monocots, Aristolochi*aceae*, Woody Magnol*iids* (implying short for Magnoliidae), and Eudicots. At the same time, the groups on the right-hand side of the tree (which we might acknowledge as superorders—or simply leave unspecified) sometimes have names that seem more appropriate to families, and sometimes have names that are indeed conventionally considered superordinal.

Many traditionalists might feel that this way of naming is rough and ready. But many moderns just feel that life is too short to spend too much time on linguistic convention. For my part, I feel that it might be worth trying to tidy up the names when there is wider agreement on what the classification ought actually to be; but, until then, playing with names that at present are widely understood would simply be confusing. So for the time being at least, I favour the rough and ready.

The point, though, is that modern, cladistic studies suggest that many of the conventional groupings do not in reality have the evolutionary significance that traditional classification suggests. Thus, although the monocots do emerge as a true clade, they can no longer be seen as one subclass out of two. Instead, they represent one major division among six—and the other five would all traditionally have been called 'dicots'.

Heywood divided the dicots into six superorders: Magnoliidae, Hamamelidae, Caryophyllidae, Dilleniidae, Rosidae, and Asteridae. Four of the six major divisions that Crane defines were originally fitted into the traditional Magnoliidae; the **Nymphaeales** (water lilies and their immediate relatives); the **Piperales** (peppers and relatives); the **Aristolochiaceae** (aristolochias and relatives); and the **Woody Magnoliids**, which traditionally formed part of the order Magnoliales. The Woody Magnoliids, of course, include the genus *Magnolia*, from which Magnoliales and Magnoliidae take their names. Crane's remaining two major subdivisions effectively reflect tradition, for the **Monocots** remain as a natural group (a true clade) and the **Eudicots** include *most* of the traditional dicots.

24 · CLASS ANGIOSPERMAE

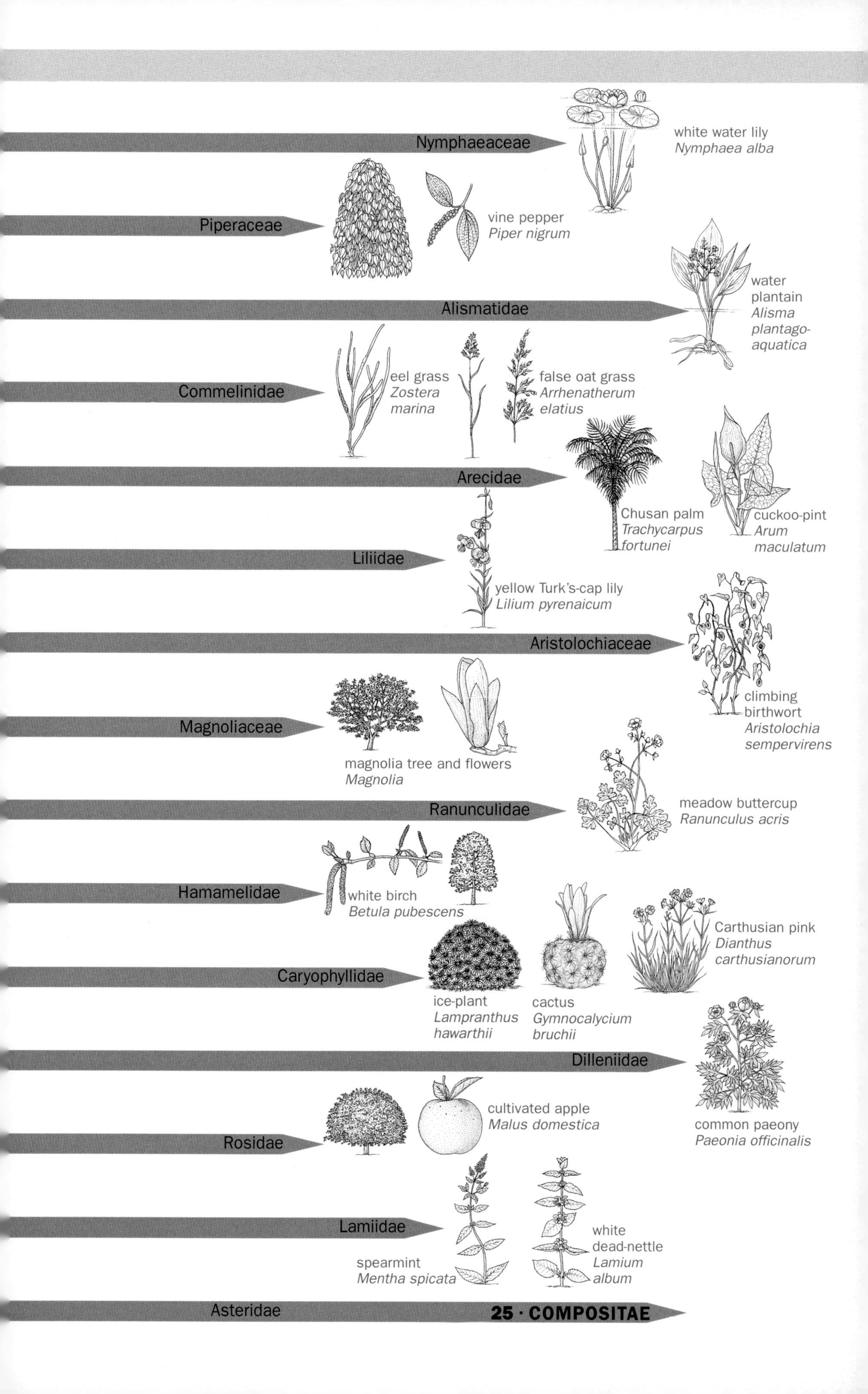
Nymphaeaceae
white water lily
Nymphaea alba
Piperaceae
vine pepper
Piper nigrum
Alismatidae
water
plantain
Alisma
plantago-
aquatica
Commelinidae
eel grass
Zostera
marina
false oat grass
Arrhenatherum
elatius
Arecidae
Chusan palm
Trachycarpus
fortunei
cuckoo-pint
Arum
maculatum
Liliidae
yellow Turk's-cap lily
Lilium pyrenaicum
Aristolochiaceae
climbing
birthwort
Aristolochia
sempervirens
Magnoliaceae
magnolia tree and flowers
Magnolia
Ranunculidae
meadow buttercup
Ranunculus acris
Hamamelidae
white birch
Betula pubescens
Caryophyllidae
ice-plant
Lampranthus
hawarthii
cactus
Gymnocalycium
bruchii
Carthusian pink
Dianthus
carthusianorum
Dilleniidae
common paeony
Paeonia officinalis
Rosidae
cultivated apple
Malus domestica
Lamiidae
spearmint
Mentha spicata
white
dead-nettle
Lamium
album
Asteridae
25 · COMPOSITAE

It is not surprising, however, and is in many ways salutary, that the traditional Magnoliidae should be the traditional group that has been most broken up. For the magnoliids include all the most primitive angiosperms. Yet, as we have seen, very little unites the primitive types except their primitiveness: peppers are small-flowered and wind-pollinated, while magnolias and water lilies are fulsomely flowered and insect-pollinated, and on the face of things could hardly be more different. But then, in traditional classifications, primitiveness *per se* (or perceived primitiveness) was commonly considered to provide sufficient grounds to group different organisms together. Thus many basic-looking but otherwise very different mammals have at various times been bundled together as 'insectivores'. We have seen throughout this book that when new, major groups arise they radiate—take many forms. Generally, most of the first-emerging types go by the board; and so at least half a dozen major groupings of early mammals have disappeared, and seven out of eight known major groupings of birds have gone (leaving only the neornithines) and most of the early arthropods, as seen in the Burgess Shale, failed to survive the Cambrian. But, it seems, six early branches of the angiosperms have survived. Two of them, the monocots and the eudicots, have become highly derived while the other four have retained many primitive features. If phylogeny is to be our guide, however, we should not simply bundle the more primitive groups together.

That, then, in outline, is the contrast between the traditional classification and the modern, and the reason that lies behind difference. The following survey of the angiosperms works as usual from top to bottom.

Water lilies and their relatives: the Nymphaeales

The Nymphaeales are the water lilies and their immediate relatives. The **Nymphaeaceae** is the principal family. As the tree shows, the water lilies now emerge as the sister group of all other angiosperms: the group which, at least in the general plan of their flowers, are presumed to have deviated least from the first ancestral angiosperm.

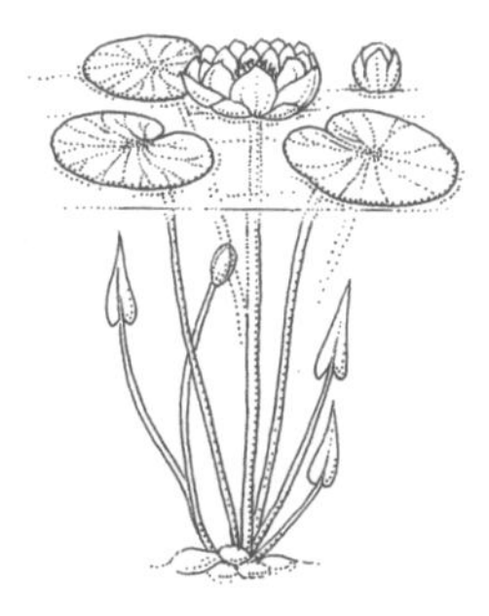

white water lily
Nymphaea alba

Peppers and their relatives: the Piperales

vine pepper
Piper nigrum

The Piperales represent the other group of angiosperms that is presumed to be primitive: small-flowered, and wind-pollinated. The **Piperaceae** are the peppers (the source of the condiment; not the sweet peppers, which are Solanaceae). The related **Sauruaceae** (not shown on the tree) are a small group from East Asia and North America, which are sometimes known as lizard tails and include some garden ornamentals and others that provide folk recipes.

The Monocots

The Monocots remain much as traditionally defined. They include about a fifth (22 per cent) of all known living angiosperms. The four superorders shown here are as defined by Heywood. The **Alismatidae** in Heywood's treatment include four orders that between them contain 16 families. Outstanding among these families are the Alismataceae (water plantains, which give the whole superorder its name); Potamogetonaceae (pondweeds); and Zosteraceae (the 'eel grasses'). Overall, the Alismatidae does not look like a natural group; just a mixed bag of monocotyledenous aquatics.

water plantain
Alisma plantago-aquatica

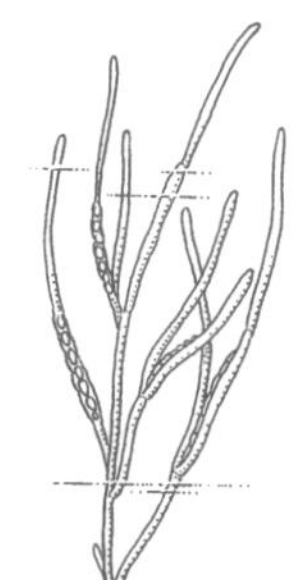

eel grass
Zostera marina

pineapple
Ananas comosus

false oat grass
Arrhenatherum elatius

The **Commelinidae** in Heywood's classification includes nine orders with 20 families, several of which are of enormous ecological and/or economic importance. Among them are the Poaceae or Gramineae (the grasses, which, of course, include the cereals); Juncaceae (rushes); the Cyperaceae (sedges); Bromeliaceae (pineapples and a host of epiphytes such as 'Spanish moss', which drapes the trees in the wetlands of the American south); Musaceae (bananas); and Zingiberaceae (which gives us ginger, cardamon, and turmeric).

Chusan palm
Trachycarpus fortunei

cuckoo-pint
Arum maculatum

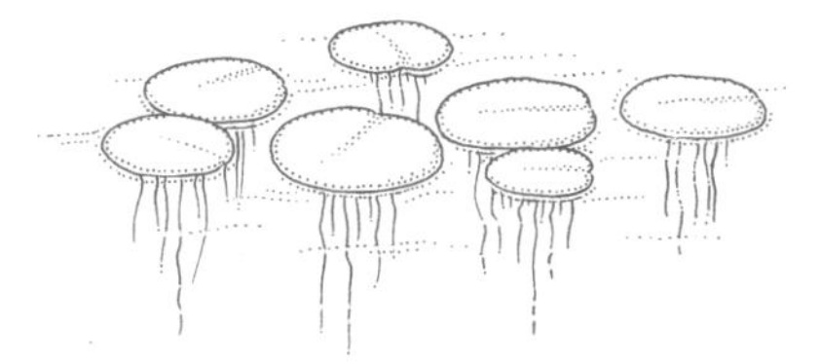

greater duckweed, *Spirodela polyrhiza*

The **Arecidae** include just four orders with only five families—but, again, some outstanding ones. Chief among the latter are the Arecaceae, or Palmae (the palms); Lemnaceae—the duckweeds that I mentioned in the introduction; and Araceae (the aroids, which include the extraordinary cuckoo-pint and its like).

Finally, the **Liliidae** includes just two orders, embracing 16 families, and again including some extremely important families. Among them are the Pontederiaceae, which include the water hyacinth, one of the outstanding weed plants of tropical waterways; Iridaceae (irises); Liliaceae (the lilies, and in some classifications also the onions); Amaryllidaceae (daffodils);

yellow Turk's-cap lily
Lilium pyrenaicum

Agavaceae (agaves, which include sisal); the extraordinary Xanthorrhoeaceae ('grass trees', which look like small palm trees with a tuft of grass at the top); Dioscoreaceae (which include temperate wayside blooms like black briony and also include yams, which are so important in the tropics); and Orchidaceae (the orchids, which include around 18 000 species and a good many hybrids).

The oldest known monocots date from the early Cretaceous. Because they tend to be herbaceous we can expect that the monocot record is liable to be particularly bad; but if this is indeed when they arose, and if the angiosperms as a whole also arose at around this time, then clearly the monocots represent an early division of flowering plants. Like the angiosperms as a whole they diversified in the mid-Cretaceous. Families like the Musaceae (bananas) and Arecaceae (palms) were established by the late Cretaceous, 80 million years ago, and many of the modern types were extant by the early Cenozoic.

Aristolochias and their relatives: the Aristolochiaceae

The Aristolochiaceae includes more than 600 species in seven genera. Many are twining lianas, and some are grown as house plants. *Aristolochia macrophylla* is the Dutchman's pipe, *A. ornicephala* is the bird's head, and *A. grandiflora* is the pelican flower. They are widespread, throughout the tropics and temperate regions, but not Australia. They have never been easy to classify. Heywood puts the Aristolochiaceae in the Magnoliidae: but the features they have in common with the magnolias may merely be primitive. Many botanists have noted similarities with the monocots. As you see, Crane classifies them as the sister group of the monocots: the link between the monocots and the traditional 'dicots'.

climbing birthwort
Aristolochia sempervirens

Magnolias: the Woody Magnoliids

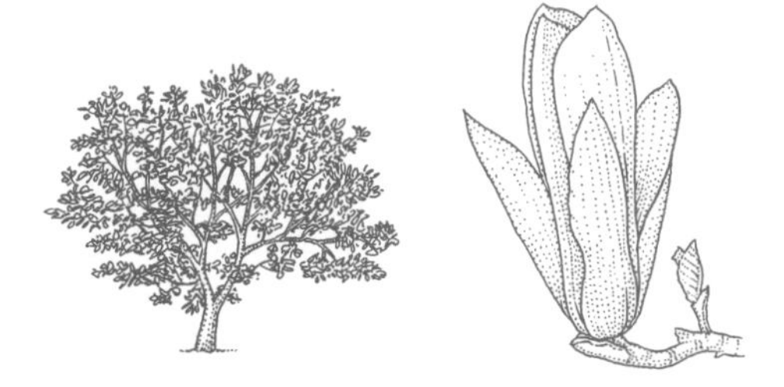
magnolia tree and flower, *Magnolia*

The informally named Woody Magnoliids include the traditional **Magnoliaceae**, best known for the magnolias (*Magnolia*) and the tulip tree (*Liriodendron*). Their big, spiral-formed flowers clearly resemble those of water lilies. They have commonly been considered to be the most primitive of all angiosperms but now they have lost that status and emerge as the sister group of the eudicots. That is to say: we can assume that the ancestors of the eudicots resembled a magnolia.

The Eudicots

Finally, the Eudicots include at least 70 per cent of all flowering plants. The superorders shown here do not quite correspond to Heywood's. Thus Heywood did not nominate the Ranunculidae or the Lamiidae. The former—the buttercups and their relatives—he included as an order (Ranunculales) within the Magnoliidae, and the latter—the mints, teak, and their like—he treated simply as the Lamiales, within the superorder Asteridae. Apart from this, the broad outlines are similar, and I am following the traditional listing.

Ranunculidae

The Ranunculidae include the buttercup family, Ranunculaceae, which, like the magnolias, has clearly separated floral parts and a basically primitive structure. Besides the buttercup, *Ranunculus*, the family includes garden favourites such as the *Helleborus*, *Aquilegia*, *Delphinium*, *Anemone*, *Clematis*, and *Nigella*.

meadow buttercup
Ranunculus acris

Hamamelidae

The Hamamelidae include a wide range of tree families such as the Platanaceae (planes); Hamamelidaceae (witch hazel and sweet gums); Betulaceae (which include the birches, alders, and hornbeams); and Fagaceae (beeches, oaks, and sweet chestnuts).

white birch, *Betula pubescens*

Caryophyllidae

The Caryophyllidae are a very mixed bunch phenotypically. Traditionally they include the Cactaceae (cacti); Aizoacae (ice-plants and mesembryanthemums); Caryophyllaceae (carnations); Nyctaginaceae (bougainvilleas); Amaranthaceae (which

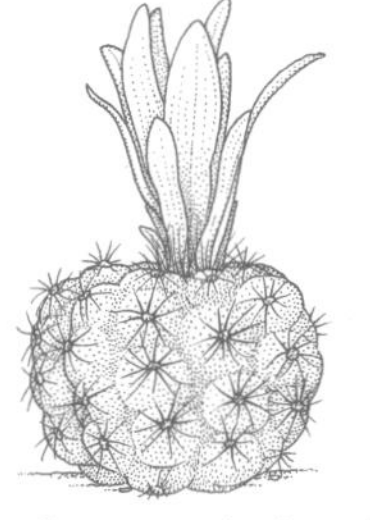
cactus, *Gymnocalycium bruchii*

ice-plant
Lampranthus hawarthii

Carthusian pink
Dianthus carthusianorum

includes love-lies-bleeding and the important Andean grain crop amaranth); Chenopodiaceae (sugar beet and spinach, Good-King-Henry and fat hen, and another traditional South American staple, quinoa); Batidaceae (saltwort); Polygonaceae (buckwheat, rhubarb, and sorrels); and Plumbaginaceae (lavenders).

Dilleniidae

common paeony
Paeonia officinalis

The Dilleniidae include the Paeoniaceae (paeonies); Theaceae (tea and camellias); Guttiferae (mangosteen); Tiliaceae (limes and jute); Sterculiaceae (cocoa); Bombaceae (baobab and balsa); the marvellous family of the Malvaceae (cotton, mallows, hollyhocks, and 'lady's fingers', or 'bindhi'); Ulmaceae (elms and hackberries); Moraceae (figs, hemp, and mulberry); Urticaceae (stinging nettles); Lecythidaceae (Brazil nuts); Violaceae (violets); Passifloraceae (passion flowers); Caricaceae (pawpaws); Begoniaceae (begonias); Cucurbitaceae (cucumbers, marrows, melons, squashes, gourds, pumpkins, and their like); Salicaceae (willows, aspens and poplars); Capparaceae (capers); Cruciferae (cabbage, turnips and mustard); Ericaceae (heathers, azaleas, and rhododendrons); Sapotaceae (gutta percha, the toffee-like sapodillas, and chicle—the tree that yields chewing-gum); and Primulaceae (primroses).

Rosidae

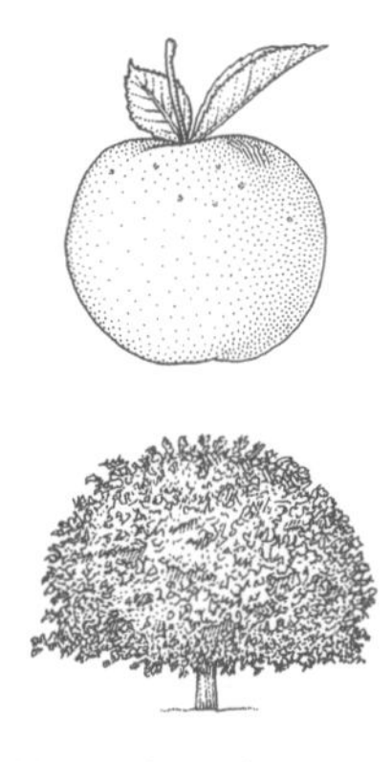
cultivated apple
Malus domestica

The Rosidae, of course, includes the Rosaceae, the rose family, with many of the greatest fruits such as apples, pears, plums, peaches, raspberries, and blackberries. But it also includes the Droseraceae (sundews and the Venus' fly-trap); Crassulaceae (stonecrops and houseleeks); Saxifragaceae (saxifrages, currants, and their like); Leguminosae (the stupendous family of peas, beans, groundnuts and lentils, which as food plants for human beings are second in importance only to the Poaceae, and also provide clovers and acacias, which are of enormous significance to grazing and browsing animals); Rhizophoraceae (mangroves and their like); Myrtaceae (cloves and the amazing genus of *Eucalyptus*, whose 600 or so species dominate Australian dryland); Puniaceae (pomegranate); Onagraceae (clarksias, fuchsias, and willow-herbs); Cornaceae (dogwoods); the amazing Gondwanan family of the Proteaceae (with *Protea*, *Banksia*, and *Grevillea*); Aquifoliaceae (hollies); the huge and varied family of the Euphorbiaceae (which includes the spurges, the staple crop cassava, and some remarkable cactus look-alikes in the genus *Euphorbia*); Vitaceae (grapes and the Virginia creeper); Hippocastanaceae (horse chestnuts); Aceraceae (maples); Burseraceae (frankincense and myrrh); the glorious Anarcardiaceae (cashew and mango); Meliaceae (mahoganies); Rutaceae (citrus fruits); Juglandaceae (walnuts, hickories, and pecan nuts); Linaceae (flax and linseed); Geraniaceae (cranesbills and pelargoniums); Balsaminaceae (balsams); Tropaeolaceae (nasturtiums); Araliaceae (which includes ivies, ginseng, and the fig-like garden bush *Fatsia*); and Umbelliferae (carrots, parsnips, angelica, fennel, celery, coriander, and a great many more besides).

spearmint
Mentha spicata

white dead-nettle
Lamium album

Lamiidae

The Lamiidae include the Labiatae, which is the greatest family of garden herbs: mints, thyme, marjoram, and sage are all labiates.

Asteridae

Finally, the Asteridae, as traditionally defined, includes a huge swatch of families including the Loganiaceae (buddleias and *Strychnos*, the source of strychnine); Gentianaceae (gentians); Apocynaceae (periwinkles and oleanders); Oleaceae (with olive, ash, lilac, and privet); the extraordinary Solanaceae (potato, tomato, capsicums, aubergines, garden favourites such as Chinese lantern, and a host of drug plants; Convolvulaceae (bindweeds, morning glory, and the sweet potato); Boraginaceae (borage, comfrey, and forget-me-not); Plantaginaceae (plantains); Scrophulariaceae (foxgloves and antirrhinums and also some notorious parasites of which the most significant is *Striga*, or witchweed, which lays waste a wide variety of crops in the tropics); Orobranchaceae (broomrapes); Bignoniaceae (*Catalpa* or 'Indian bean'); Campanulaceae (bellflowers); the extremely variable and probably polyphyletic Rubiaceae (with gardenias, coffee, and cinchona); Caprifoliaceae (with elders and honeysuckles); and the extraordinary Compositae or Asteraceae (the ubiquitous family of daisies, dandelions, and thistles, whose economically desirable members include sunflowers, artichokes, both globe and Jerusalem, and lettuce).

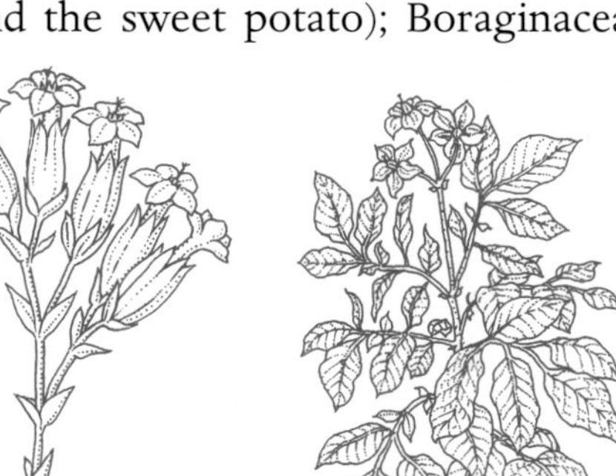

bladder gentian
Gentiana utriculosa

potato
Solanum tuberosum

It is hard at present to provide a satisfying classification of the angiosperms: one that is tidy, comprehensive, and is 'natural' in the post-Darwinian sense that it tracks phylogeny. Pleasingly, though, the biggest of all the angiosperm families and one of the most derived, the Compositae or Asteraceae, does now seem to be reasonably stable. To end this part of the book on a tidy and satisfying note, I look at this one family in greater detail. In a few years' time, with luck, it might be possible to treat *all* the families of flowering plants, or at least the main ones, with as much confidence.

25

DAISIES, ARTICHOKES, THISTLES, AND LETTUCE

FAMILY COMPOSITAE (OR ASTERACEAE)

About 10 per cent of all flowering plants belong to the Compositae, or Asteraceae,[1] the chief family of the great group of the Asteridae. The composites include daisies and sunflowers, marigolds and chrysanthemums, camomile and tansy, groundsel and ragwort, dandelions, dahlias, lettuce, thistles and globe thistles, and artichokes both Jerusalem and globe. Most are herbs and some, like many species in the genus *Senecio,* are succulent. Some are large, like the wonderful giant senecios from South Africa. Many are shrubs, and although the family is often said to be relatively short of aquatics and trees, there are quite a few in total. Some have been called 'treelets'. Many are annuals whereas others are persistent perennials, including many 'geophytes', which abandon all aerial growth when the weather is dry and tough it out below ground as tubers or rhizomes.

There are about 25 000 composite species in around 1500 genera—more than in any other angiosperm family. Even the marvellously various Orchidaceae has only 18 000 species, even though orchids have evolved largely in tropical forest, which fosters speciation. The Leguminosae (or Papilionaceae), the family of peas, beans, acacias, clovers, gorse, and so on, has around 17 000 species, while the ubiquitous grasses in the family Gramineae (or Poaceae) have 9000.

Ecologically, the herbaceous composites include many of the most accomplished pioneers—invaders of newly broken ground. Thus, like cockroaches and sparrows, they belong in that dubious elite of creatures that thrive on human industriousness. But although human beings have scattered the composites far and wide, they had already invaded every habitable continent—all except Antarctica—long

[1] Revised botanical convention demands that every plant family should be named after the first genus to be formally described within that family, and so the Compositae now enjoys the alternative name 'Asteraceae', named after *Aster* (which, as we will see, is not the genus of the florists' 'asters', but that of the Michaelmas daisy). Many botanists speak formally of the 'Asteraceae' but refer informally to 'the composites'.

before we came on the scene. The fossil record of early composites is sparse but best guesses suggest that they began in the Palaeocene, around 60 million years ago, when the dinosaurs had gone and the modern orders of mammals were coming into their own. Although the fossils are scarce we can infer that composites originated around South America or the Pacific, for they show the greatest genetic variation in South America and some tribes of composite—including the most primitive—are exclusively South American. They reach something of a pinnacle in South Africa, where they contribute gloriously to the colours of spring.

Composites have become so widespread largely because they disperse their seeds so efficiently, but they may have been helped on their way by the drift of continents. During the past 60 million years the former Gondwanan island of India sidled north across the Southern Ocean and into the south of Asia; landbridges came and went between North America and Eurasia, both via Siberia and Alaska and via Greenland and Iceland; and the Laurasian continent of North America and the Gondwanan continent of South America finally met in Panama.

Many composites are useful to human beings, although only a small proportion of the whole. Many are prized and valuable garden flowers—chrysanthemums, dahlias, zinnias, cosmos, marigolds—while many others, like the succulent senecios, are the cults of specialists. Many are good to eat, like artichokes and lettuce, and a few, notably the sunflower, are a serious source of nutrition. Several hundred species (or so it is estimated) provide infusions such as camomile tea, many of which are medicinal. Dandelion and chicory contain caffeine-like materials and so can provide ersatz coffee, which has baled out many a besieged society in wartime. Many senecios and others produce sesquiterpenes, which tend to be immensely bitter. The safflower and others provide dyes. Some provide insecticides—notably *Tanacetum cinerariifolium*, the chief natural source of pyrethrum. A few provide timber including *Brachylaena* of Africa—not high grade, but good for fences. Many produce latex and several, including a species of dandelion (*Taraxacum*), have provided rubber. The guyale shrub from the New World, *Parthenium argentatum*, is considered a very promising substitute for *Hevea*, the rubber tree (which is one of the Euphorbiaceae).

The speciosity and ubiquity of the composites, their ecological influence, and their prominence in human affairs, seems to depend primarily on two outstanding qualities. First, their intricate flowers are marvellously efficient, both in the fertilization of ova and in the dispersal of seed. Second, the composites are superb chemists, producing an unrivalled array of oils and other recondite materials both to support their own offspring and to repel herbivores and pathogens.

THE INTRICATE COMPOSITE 'FLOWER'

Typical flowers, as seen in their primitive state in, say, buttercups or water lilies, have an outer, supportive **calyx** composed of sepals, which are commonly green; an inner **corolla** composed of petals, which are often brightly coloured; and a central assemblage of sexual elements—the male stamens, each composed of filament and anther, which dispenses pollen, and the female styles, each composed of an ovary and a stigma, to receive pollen. Flowers that are pollinated by insects, birds, or bats also carry **nectaries** at the base of their petals, as lures. The whole flower structure is mounted on a **receptacle**.

But the 'flower' of a composite is in truth a whole pack of flowers acting as one—an entire, compact **inflorescence** fused into one unit. Each of the individual flowers within the inflorescence is called a **floret**, and the whole group together, the thing that most of us would call 'the flower', is known botanically as a **capitulum** (which means 'little head'). The green supporting structure that surrounds the capitulum of a daisy looks like a calyx, and does the same job, but it is made up of bracts, and is called an **involucre**.

The individual florets within a capitulum commonly lack a calyx, although each floret sometimes carries a ring of bristly hairs at the base that collectively form the **pappus**, which is generally believed to represent the calyx. But often the different florets in any one capitulum are dimorphic, meaning that they take two distinct forms: each form is a specialist, adapted to a different purpose, and together they carry out the functions of a more typical flower. Such consortia of specialist individuals are a common theme in nature: compare *Physalia*, the Portuguese man-o'-war, a colony of specialized medusae and polyps that acts like a single jellyfish.

In a typical capitulum, like a daisy, the florets are divided into two obvious groups. **Disc florets** form the centre, and **ray florets** form the outer ring, acting like the petals that encircle a buttercup. In the disc florets, each corolla is radially symmetrical, or 'actinomorphic': a simple tube with five equal lobes at the top. The ray florets are lop-sided or 'zygomorphic', with one side extended into a kind of strap that serves as a petal. In some, as in dandelion and chicory, one side of the disc floret is extended and so resembles a ray floret (even though it still retains the five teeth of a disc floret). Some composite capitulae, as in some senecios, have only disc florets. In other senecios, including groundsel and some succulent types, some individuals have ray florets in their capitula whereas others do not. In groundsel, the presence or absence of rays is determined by a single gene so that those with, and those without, commonly grow side by side.

There are further, less immediately obvious specializations. Notably, the typical composite floret is bisexual—like most flowers of any kind. But in some florets in some species the male elements are suppressed, so that the flower is functionally female; in others the style is present but non-functional, so that the floret is male. Ray florets are sometimes exclusively female and sometimes totally sterile.

The mechanism of pollen dispersal is remarkable, and unique to composites and a few other families like the Calyceraceae, which are thought to be closely related to them. Thus the filaments of the five stamens are attached to the base of the corolla tube, and the anthers are joined along their edges to form a tube. The anthers dehisce—split to release their pollen—inwards. Within each individual floret, the stamens develop and mature before the styles; as the styles develop, they must push their way like a piston through the circle that has already been formed by the anthers.

The styles grow until they protrude above the edge of the corolla, and as they do they push the pollen out with them. Typically, the style arms divide as they emerge to form what looks like a two-tined fork.

This method of dispersing pollen, known as the 'pump' mechanism, is typical of composites; but, in some, the styles have a ring of hairs attached, and push out the pollen like a sweep's brush. In either case, the styles are immature at the time they are pushing past the anthers, and cannot be fertilized by the pollen that they pick up as they pass—so at this point they are acting purely as agents of dispersal. By the time the styles are mature, the anthers beneath them will already have shed most of their pollen. But should the stigmas fail to capture pollen from another capitulum, then in some species they carry on growing until the tines of the fork double back on themselves, and make contact with any pollen that is still lingering within their own or a nearby floret. Thus, when cross-pollination fails, many composite flowers can pollinate themselves instead. We might suggest that whereas it is better in general to be cross-pollinated, self-pollination is better than none at all. Perhaps, though, a combination of cross-fertilization and self-pollination is the best of all strategies: the outcrossing ensures that genes are recombined and so provide variety, while self-fertilization ensures that successful individuals are propagated *without* so dramatically scrambling the partnerships of genes that obviously work well already.

The consortium of florets in a single capitulum may between them be active over many days or weeks—much longer than any one single flower; as we have seen, they typically encourage cross-pollination but can self-fertilize if outcrossing fails. Then again, because each floret has its own nectary, a visiting insect has to probe each one to feed, each time thrusting its proboscis past the stigma and anthers; while in a more conventional flower with one open, generous nectary, an insect may feast without great effort and pollination is likely to be less efficient. Thus the composite flower is a miracle of packaging, like triple-wrapped chocolate biscuits. Furthermore, the pump mechanism ensures that pollen is released over long periods—the most efficient way to distribute it—and in some, as in *Bellis*, the daisy, it is released *only* when an insect makes contact. There are secondary advantages, too, with all that packaging; for example, the nectaries are most unlikely to be flooded by rain.

Although a few composites seem to be wind-pollinated (especially some Australian types), animal pollination is the norm: primarily by insects, but also by birds and bats. Gordon Rowley, a great composite specialist formerly at the University of Reading, has speculated that when rays are present in fixed numbers and the flowers are yellow or mauve, and there is little or no scent, then bees or bumble bees are likely to be the pollinators. When rays are absent and the disc florets are long, and are brilliant red or mauve but again with little scent then, he suggests, butterflies are the likely pollinators—or, if the flowers are pendulous, perhaps birds: *Senecio amaniensis* may be bird-pollinated. But when rays are absent and the disc florets are small, white, and inconspicuous, but the flowers are scented, then beetles or flies are the likely pollinators (if the scent is musty) or moths (if the scent is sweet). He emphasizes that these ideas are speculation; but they are testable hypotheses.

Composites also spread their seeds extremely efficiently. The single seed remains within the ovary wall and so what most of us would call a seed is really a fruit, sometimes called a 'cypsela', although the dry types are often referred to as **achenes** and the fleshy types as **drupes**. Some composite 'seeds' are sticky or barbed, and are spread by animals. Others are dispersed by wind: some are winged while in many, like the dandelion, the pappus serves as a parachute. Some island types have lost their pappus, however—perhaps for the same kind of reason that many island birds have lost their wings. On islands it can be dangerous to take to the air, and be blown out to sea.

Several other angiosperm families have evolved compact inflorescences that at least roughly resemble those of composites. Some, like the Calyceraceae (a small South American family) and the Dipsacaceae (which includes teasels and the scabious) seem to be related to the Compositae, and may have evolved their compound flowers from the same common ancestor. But others with superficially composite-like flowers, like the Aizoaceae (the family of mesembryanthemums and the extraordinary 'stone-plant' *Lithops*) are not at all closely related to composites, and must have evolved such 'flowers' completely independently. This is yet another excellent example of convergence.

COMPOSITES AS CHEMISTS

All living things, perforce, are excellent organic chemists. All must produce proteins, sugars, fats, nucleic acids, and a host of other subsidiary materials including that loose category that we call 'vitamins'. Composites typically stock their seeds with nutritious oil, which human beings squeeze from sunflower and safflower; and all are in-clined to store carbohydrate in the form of inulin, which is a polymer of the simple sugar fructose—as opposed to the more common starch, which is a polymer of glucose. Inulin helps to give special flavour to Jerusalem artichokes, and the special flatulence. Inulin is unusual, but not unique to composites. For example, the Campanulaceae (bellflower family), which seem to be related to the composites, produce it too.

But most plants also produce a range of 'secondary metabolites' that are not involved in the second-by-second operations of the plant. These include tannins, alkaloids, terpenes, and essential oils (where 'essential' means 'of the essence', as opposed to 'necessary'). The total is vast, though all are variations on themes of fats, alcohols, phenols, amides, and so on. Their production may require a great deal of energy; and synthesis often involves intricate chains of enzymes, contained in special structures, so the mechanisms of production must also be genetically complex. If these secondary metabolites did not contribute commensurately to survival, then natural selection could not have favoured such expenditure of effort, and such genetic elaboration. Clearly they do serve to repel herbivores and parasites: without them, composites could not take on the soft-tissued, herbaceous, quick-growing habit that makes them such successful pioneers. And, of course, it is because the composites are such good chemists that we are able to make such use of them—as sources of dyes, pesticides, drugs, rubber, and so on, as well as food and timber.

For the following notes on classification, and the hypothetical phylogenetic tree, I am grateful to Spence Gunn of the Royal Botanic Gardens, Kew, and to Nicholas Hind, who is Kew's principal specialist in composites.

A GUIDE TO THE COMPOSITAE

The classification of flowering plants is in a state of flux as cladistic and molecular techniques each exert their influence, and it is not easy to reconcile latest ideas with traditional classifications. A constant flood of conferences are held for just this purpose. Thus in some traditional classifications the superorder **Asteridae** includes nine orders that between them include more than 40 families, with the Compositae being the sole member of the order Asterales. As shown in the tree, however, modern studies suggest that the Compositae are most closely related to the **Campanulaceae** (broadly defined), which are the bellflowers; the **Dipsacaceae**, which include teasels and scabious; the **Goodeniaceae**, which include the highly prized greenhouse shrubs, *Leschenaultia*, and *Scaevola*; and the **Calyceraceae**, a small family of scabious-like flowers from South America. In traditional classifications these apparent close relatives of the composites have generally been scattered among several orders. Clearly the Asteridae as a whole needs more sorting out.

Campanula rapunculus *Dipsacus fullorium* *Goodenia quadrilocularis* *Calycera horrida*

But the family **Compositae** itself forms a clearly defined and coherent group: as the great English botanist George Bentham (1800–84) commented in 1873, 'I cannot recall a single ambiguous species as to which there can be any hesitation in pronouncing whether it does or does not belong'. It is indeed monophyletic. But because the family is so extensive and so various, it has long been subdivided. In the early nineteenth century the French botanist Henri Cassini split the family into 19 tribes, and although these have subsequently been much regrouped, many of his original names still survive. Nicholas Hind now acknowledges just 14, as shown in the tree.

Nicholas Hind adopts cladistic criteria, as does everyone on whose ideas I have emphasized in this book; so the present tree (like all in the book) is a cladogram (without the footnotes!). But a more traditional subdivision of Compositae also persists alongside the cladistic one. This breaks Compositae into three subfamilies: Barnadesioideae, which is the most primitive group, and includes the Barnadesieae (and in some classifications also includes the Mutisieae); the Cichorioideae (which includes the Lactuceae, the Vernonieae, the Liabeae, and the Arctotideae); and the Asteroideae, which includes all the rest. In fact, if the tree shown here is correct, then, as you can see, only the subfamily Asteroideae is truly monophyletic, although the other two subfamilial names are clearly useful for reference purposes (or they would not have been coined in the first place). Cichorioideae would be monophyletic (the sister group of the Asteroideae) if it did not also contain the thistle tribe, Cardueae. Two related tribes, the Heliantheae and Eupatorieae, are regarded as the most 'derived' composites, although the Heliantheae (the family of sunflowers) was once thought to be among the most primitive of composites, and even to be the ancestor of all the rest.

Chuquiraga and relatives: tribe Barnadesieae

The Barnadesieae are South American. This tribe (or subfamily) includes *Barnadesia* and *Chuquiraga*—one of the few composites that forms a tree big enough to supply timber that is useful, albeit primarily for fences.

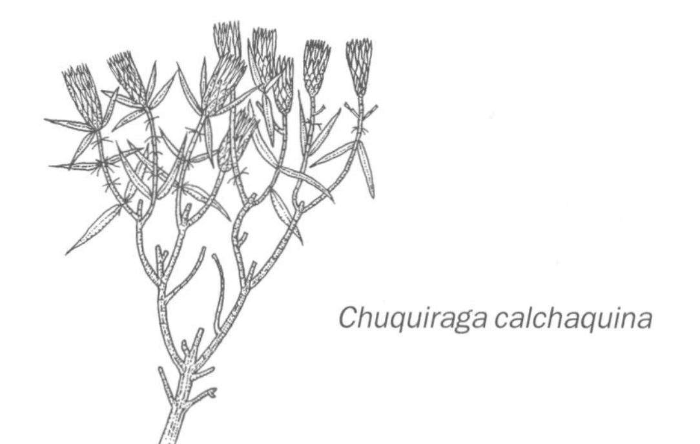

Chuquiraga calchaquina

Mutisia and relatives: tribe Mutisieae

The Mutisieae in some traditional classifications is included in the subfamily Barnadesioideae. Mutisieae includes about 1000 species in 90 genera—mostly South American, like *Stifftia* and the ornamental *Mutisia*. But it also includes African and Asian types such as *Gerbera*, another favourite ornamental.

Mutisia decurrens

related families
ASTERIDAE
COMPOSITAE
Asteroideae
25

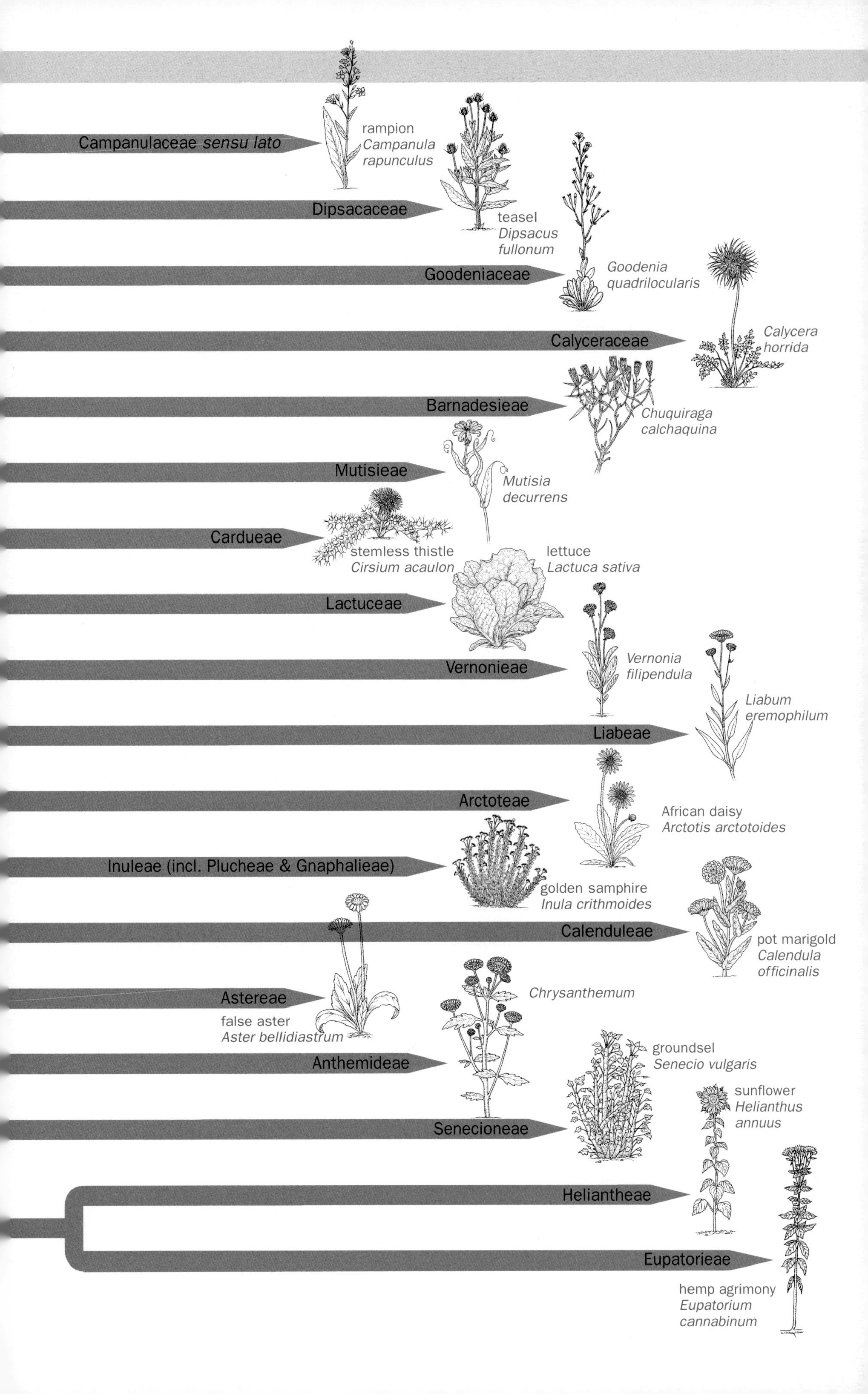

Campanulaceae *sensu lato*
rampion
Campanula rapunculus
Dipsacaceae
teasel
Dipsacus fullonum
Goodeniaceae
Goodenia quadrilocularis
Calyceraceae
Calycera horrida
Barnadesieae
Chuquiraga calchaquina
Mutisieae
Mutisia decurrens
Cardueae
stemless thistle
Cirsium acaulon
lettuce
Lactuca sativa
Lactuceae
Vernonieae
Vernonia filipendula
Liabeae
Liabum eremophilum
Arctoteae
African daisy
Arctotis arctotoides
Inuleae (incl. Plucheae & Gnaphalieae)
golden samphire
Inula crithmoides
Calenduleae
pot marigold
Calendula officinalis
Astereae
false aster
Aster bellidiastrum
Chrysanthemum
Anthemideae
groundsel
Senecio vulgaris
Senecioneae
sunflower
Helianthus annuus
Heliantheae
Eupatorieae
hemp agrimony
Eupatorium cannabinum

Carduus and relatives: tribe Cardueae

Much more familiar to north Europeans are the highly distinctive Cardueae, whose 2600 or so species in 80 genera include the thistles, globe thistles, globe artichoke, and the safflower. The thistles—British genera are *Carduus* and *Cirsium*—are simply wonderful, their prickly leaves superbly equipped for repelling herbivores, their opulently pappused fruits marvellous agents of dispersal; on favourable days in thistle country the air may be filled with their seeds. The globe thistles, *Echinops*, have taken the composite trick of flower condensation one step further. Their heads are inflorescences composed of many different capitula. Because the receptable of an ordinary capitulum, like a daisy, actually serves to support many different florets, it is a secondary receptable; the receptable of *Echinops* supports many capitula, so is a tertiary receptacle. Sometimes, too, several or many *Echinops* heads are grouped, thus forming a quaternary inflorescence. It seems that once the composites had hit on the trick of condensation (which presumably requires mutation of the genes that are high in the organizational hierarchy and direct the activities of other genes) they saw no reason to stop. Globe artichokes are of the genus *Cynara*, originally from the Mediterranean and Southwest Asia. The safflower, *Carthamus tinctorius*, yields an oil that is high in polyunsaturates and thus is one of the few composites of true nutritional (as opposed merely to gastronomic or medicinal) significance; its tender young shoots are also edible, and its orange-red flowers yield dyes, which are sometimes used to adulterate saffron (and hence the name).

stemless thistle, *Cirsium acaulon*

Lactuca and relatives: tribe Lactuceae

The Lactuceae contains about 2300 species in 70 genera, including the lettuce (genus *Lactuca*), endive (*Cichorium endivia*), and chicory (*Cichorium intybus*); the hawk's beards (*Crepis*) and the hieraciums; the dandelion *Taraxacum* (*T. bicorne* is a minor source of rubber) and the marvellous oyster-like edible root *Scorzonera*, plus the goat's-beard and another brilliant root, salsify, which are both in the genus *Tragopogon* (from the Greek *tragos*, 'goat', and *pogon*, 'beard'). Some Lactuceae are notorious weeds, including *Taraxacum* and the skeleton weed, *Chondrilla juncea*.

lettuce, *Lactuca sativa*

Stokesia and relatives: tribe Vernonieae

Vernonia filipendula

The Vernonieae has 1200 or so species in 50 genera, most of them tropical and including *Vernonia* and *Stokesia*, 'Stokes' aster', from the southeastern USA.

Liabum and relatives: tribe Liabeae

Liabum eremophilum

The Liabeae is a small group (by composite standards—just 120 or so species in 15 genera) from the New World, which includes *Liabum* from Central and South America. The Liabeae have not been well studied, because most of the species live in out-of-the-way places, but some are very beautiful.

Arctotis and relatives: tribe Arctotideae

African daisy
Arctotis arctotoides

The Arctotideae, with about 200 species in 15 genera, mostly South African, include those excellent ornamentals *Gazania* and *Arctotis*, one of which has acquired the soubriquet *Arctotis* 'Nicholas Hind'.

Helichrysum and relatives: tribe Inuleae

The first of the seven asteroidean tribes is the Inuleae (which Nicholas Hind takes to include the Plucheae and the Gnaphalieae). This is another enormous group with around 2100 species worldwide, in around 180 genera. These include the genus *Helichrysum*, which are treasured as everlasting flowers, and the curry plant, *Helichrysum italicum*, whose leaves do indeed impart the flavour of curry. *Helichrysum krausii*, however, is a noxious weed.

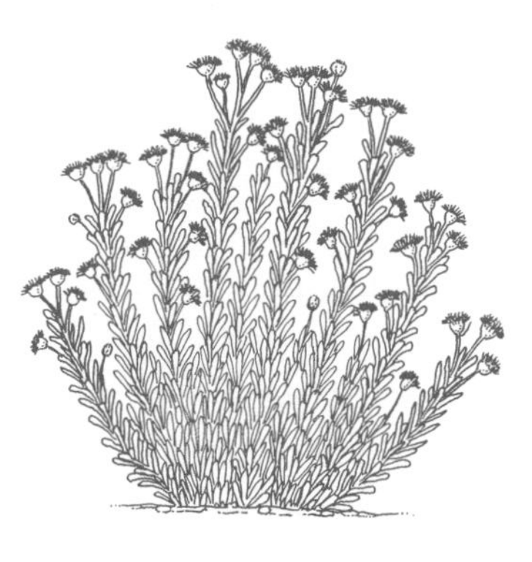
golden samphire, *Inula crithmoides*

Calendula and relatives: tribe Calenduleae

The Calenduleae contains only about 100 species in seven genera, but includes the pot marigolds, *Calendula*.

pot marigold
Calendula officinalis

Aster and relatives: tribe Astereae

The Astereae is the huge tribe of the asters and daisies—2500 species in about 120 genera. *Aster* includes the Michaelmas daisy; the garden daisy is *Bellis*; *Callistephus* is the genus that florists call 'asters'; *Olearia*, from Australia, is the 'daisy bush'; and *Solidago* is the golden rod.

false aster
Aster bellidiastrum

Chrysanthemum and relatives: tribe Anthemideae

The great tribe of the Anthemideae has about 1200 species in 75 genera: many are decorative, some are weeds, and not a few are noted chemists, featuring both in traditional medicine and in pest control. Probably at least a hundred composite flowers and leaves are made into teas, or 'tisanes', and many of them are from the Anthemideae. Genus *Chrysanthemum* includes *C. segetum*, the corn marigold. Many others that used to be included in *Chrysanthemum* have now been reallocated: thus the ox-eye daisy (once *C. leucanthemum*) is now *Leucanthemum vulgare*; the feverfew, or 'bachelor's buttons', was once known as *C. parthenium* but is now *Tanacetum parthenium*; and tansy, the erstwhile *C. vulgare*, is now *Tanacetum vulgare*. The London herbalist John Parkinson (1567–1650) in the seventeenth century declared that feverfew was 'very effectual for all paines in the head' and some now recommend it as a specific remedy for migraine; it does lower body temperature and its name may well derive from 'febrifuge'. Tansy, too, has been much favoured by apothecaries. The plants that florists call 'chrysanthemums', however, have mostly been transferred to the genus *Dendranthema*. *Artemisia* is the genus of mugworts

Chrysanthemum

and wormwoods—more apothecaries' herbs. *Tanacetum cinerariifolium* is the chief commercial source of natural pyrethrum, one of the world's most-important insecticides.

Senecio and relatives: tribe Senecioneae

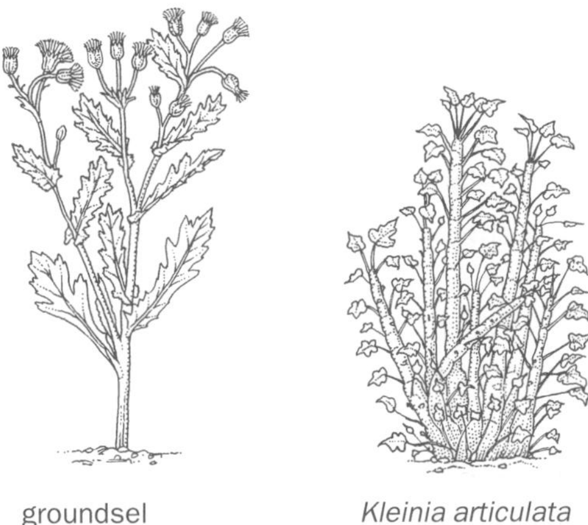

groundsel
Senecio vulgaris

Kleinia articulata

The Senecioneae is another vast tribe—3000 species in 85 genera—that touches human fortunes in many ways, good and evil. Most significant is the vast, many faceted genus *Senecio*, which Gordon Rowley describes as 'protean': only *Euphorbia* is comparably various. Some senecios are weeds, like *S. vulgaris* (groundsel) and *S. jacobaea* (ragwort). Poisonous senecios kill more livestock than all other toxic plants put together. But many senecios have adapted in various ways to drylands (that is, are 'xerophytes') with several groups independently evolving succulence (which is only one of several ways to be a xerophyte). Succulent senecios are much favoured by enthusiasts, together with the closely related *Kleinia* (indeed it is hard to tell where *Kleinia* ends and *Senecio* begins) and *Othonna*. Among the most wondrous of all composites are the tree senecios of South Africa and the 80 or so species of 'tree groundsels' of the genus *Espeletia*, from the high Andes of South America.

Helianthus and relatives: tribe Heliantheae

sunflower
Helianthus annuus

The Heliantheae includes about 4000 species in 250 genera; mostly but by no means exclusively hailing from the New World. They include *Ambrosia*, which despite its hints of divinity includes *Ambrosia artemisiifolia*, the pestilential ragweed, whose pollen, together with that of *A. trifida*, is a principal cause of hay fever in the USA, and more weeds in the form of *Xanthium spinosum* and *X. strumarium* (the cocklebur). *Bidens* is the genus of the delightful garden cosmos, but also of the weed *B. pilosa*, or black Jack. *Rudbeckia*, *Zinnia*, and *Dahlia*, originally from Central America, are more garden favourites. But the most economically significant of all composites, by virtue of its highly unsaturated oil, is *Helianthus annuus*, the sunflower; while the root tubers of the closely related *Helianthus tuberosus*, manifest as Jerusalem artichokes. Guyale, too, a significant substitute source of rubber, is one of the Heliantheae.

Eupatorium and relatives: tribe Eupatorieae

Finally, most of the 1800 or so species (in 120 genera) in the tribe Eupatorieae are from the New World, but are known in Britain for the genus *Eupatorium*, and especially *E. cannabinum*, the hemp agrimony, common in ditches and by the sides of streams.

hemp agrimony
Eupatorium cannabinum

It is easy to take the composites for granted. Too many are too familiar as weeds, and none has the majesty of a big tree or is quite so extraordinary as a cactus or a giant lily. Yet they are among the pinnacles of evolution: common because they are so successful; unspectacular because they are so efficient. Truly, they are worth seeking out.

PART III

EPILOGUE

SAVING WHAT IS LEFT

This book is in part a celebration of life present and of its even more glorious past. Must it be retrospective? Are the glories all in the past tense? What of the future? Books like this so often end in gloom, but is there an alternative? Modern conservation biologists typically suggest that at least half of the large (non-protist) terrestrial eukaryotes are likely to go extinct within the next few decades or centuries—yet a century is a twinkling in the 4500 million years of this planet's life.

How could it not be so, since most terrestrial species live in tropical forest, which is now disappearing so rapidly? The sea is not immune either, of course: many whales and dolphins and even the larger and more commercial fish, such as the blue-finned tuna, are clearly on the brink. Andrew Dobson, of Princeton University, has estimated that present extinction rates exceed the normal background loss of species loss by at least a hundredfold. In the past 550 million years, there have been at least five 'mass extinctions', in which up to 90 per cent of marine species disappeared; the last such wave, around 65 million years ago, saw off the dinosaurs (except, of course, for the birds). But those mass extinctions took thousands of years to unfold. The present wipe-out is happening within our own lifetimes.

Is this loss inevitable? Should we just wave our hands and say, 'That's life!'? Isn't extinction merely 'natural'; shouldn't we just admit that rhinos, say, have had their day? Should Miranda's goodly creatures simply give way to what she called 'the brave new world'? Can we really do anything significant to turn this tide? Is it right for us to intervene? By doing so, would we not be playing God, and isn't that an arrogant, and even a blasphemous thing to do?

Well no: the loss of species is not inevitable. We are bound to lose some, and if we try to cling to them all throughout the next few centuries—and to each fine subdivision within them, all the meticulously identified subspecies and varieties—then we will lose more than we need. But with commitment—serious investment, sharp technique, and, above all, a change of attitude—we could save a large proportion.

We, the human species, need to make concessions even though we cannot and should not allow the needs and rights of other creatures to prevail over our own. We should, though, expand present concepts of civilization. The task for humankind is not simply to create a world that is good for us, but to arrange this Earth to accommodate our own reasonable needs and those of our fellow creatures, or at least a good proportion of them, for as long as the future lasts. It will be harder to do this, at least in the short term, than it would be simply to allow our fellow species to go by the board. But the prize, if we save them, is incomparably greater. A world that merely contained human beings and our increasingly debased domesticates (plus flies and cockroaches, no doubt) would be a sad and lonely place, and probably precarious.

So what is the task, and how can we go about it—and what role, if any, can an interest in phylogeny play in shaping conservation policy?

THE PROBLEM OF HUMAN NUMBERS

The greatest and most inescapable ecological fact of the present Earth is human numbers. There is an ecological law that says that 'big, fierce animals are rare'—inevitably because they take up an enormous amount of room, and need so many other animals to feed on. We are big, by the standards of most animals (and animals are big by the standards of most other creatures). We are also the most efficient predators of all; not so strong as lions or as patient as crocodiles but far more cunning than either and able, as they are not, to kill at a distance, without personal risk; and able, too, through agriculture, not simply to predate but to shape the entire landscape to our needs and whims. Yet at the time of writing—around the start of the twenty-first century—there are 6000 million of us, swarming on every continent except Antarctica, and on most islands. Very few large animals, let alone predators, can be counted in millions, or even in hundreds of thousands; the crabeater seal of Antarctica is the only one that comes to mind. Only a few thousand wild tigers are left at most; four out of the five species of rhinoceros are reduced to a few hundreds each. Among mammals, only house mice and brown rats—creatures that cling to our own coat-tails—are remotely comparable to us in numbers.

Around 10 000 years ago, as the last Ice Age ended and people began farming on a scale that was large and intrusive enough to show up in the archaeological record, there were probably around 10 million human beings worldwide. We were worldwide by then—people seemed to have reached Australia from Southeast Asia at least 40 000 years ago, and North America (via Siberia) by around 15 000 years ago—and were already wiping out our rival mammals, such as the mammoths, ground sloths, and sabre-tooths of the Americas. But the world population was only that of, say, present-day Moscow. At the time of Christ, after 8000 years of arable agriculture, numbers reached an estimated 100–300 million. The thousand million (billion) mark was passed around 1800.

Since 10 000 years ago, the rise in population has been exponential: a word that does not simply mean 'fast', but does imply acceleration—the total numbers added each year grow more and more. At present, the world population is on course to double roughly every half century. Because individuals are living longer, and centenarians are now almost commonplace, human numbers worldwide could increase four times—double and then double again—within a person's lifetime. The issue of human population has become a bit of a cliché, and is even seen to be politically incorrect. Distaste, however, will not make it go away. The rising mass of humanity is one of the most extraordinary ecological phenomena in the turbulent history of this planet, and it is happening before our eyes. We are part of it.

This increase seems to be taking us inevitably towards disaster. If the world population is 6000 million now, then at the present rate of increase it should reach 12 000 million by the year 2050 (I will probably not be alive by then, but many readers surely will). By the end of the twenty-first century, the world population would reach 24 000 million. Many babies, now in their cots, will live to see this. By the middle of the twenty-second century world numbers would reach 48 000 million. By the end of that century they would be nudging towards 100 000 million. The twenty-second century is not so far away. Many of our children will see it. Some of our great grandchildren will see almost to the end of it. The twenty-second century is a family affair.

Many agronomists and ecologists suggest that we will be hard-pressed even to feed present numbers sustainably. The world is losing topsoil at a rate too depressing to record; and heroic efforts to irrigate desert are producing salination on an extraordinary scale as evident, for example, in South Australia—not in some nether place too remote to contemplate but on the doorstep of the staid colonial city of Adelaide. Present-day agriculture, say the Jeremiahs, is unsustainable; and no one knows enough (the calculations seem too difficult) to say that the Jeremiahs are wrong. Others, in more bullish vein, suggest, by contrast, that present-day agriculture is for the most part primitive and that if we applied even present-day technique a little more conscientiously we could feed 20 000 million or more, comfortably and forever.

I find these higher figures a little implausible. To be sure, with modern intensive horticulture yields can be stupendous; and if we exploited the wonderful phenomenon of photosynthesis to anything like its full potential we could feed a human being from a few square metres of ground. The reality, though, would be a degree of investment and organization that seems to exceed what is remotely likely. The Ukraine is already covered in wheat. Do the optimists envisage that it should be covered, end to end, in polythene? And how quickly? The decades flash past and each one is adding a thousand million people. By the second half of the twenty-second century we would be adding a thousand million people a year. How could we possibly keep up?

Even the optimists have to admit that high-tech and brilliant organization can only buy time. A population of 20 000 million might be sustainable. But I know no one who suggests that the world could, in reality, even begin to feed 100 000 million. Yet that is the number that could be reached within the lifetimes of people who might keep our photographs on their pianos; our children's children's children; our own people.

So what is going to happen? Some of the possibilities are so horrendous as to be unthinkable; nonetheless what we choose not to think about may come to pass. In some animals the population rises to enormous heights over a few years or decades and then crashes. Lemmings are the most famous example, but it is at least possible that in prehuman times populations of elephants in Africa may have risen over centuries, and then crashed. Somehow the mass death of little animals seems less horrible than the collapse of large ones—perhaps because scavengers clear the corpses quicker. In any case, a crash in human numbers from some vast height—one inadequate harvest, and then another, and then another—is not inconceivable, some time within the next two centuries. Indeed if numbers simply continue rising at present numbers, it is hard to see the alternative. Some populations of animals never reach constancy, or anything like it: they are always in some cycle of boom and crash. Perhaps human beings are doomed to follow such a path. This seems unlikely, however, because each rise seems bound to erode the world a little further, so that each ensuing collapse would plumb new depths.

Then there is the problem of population 'momentum'. Our descendants may foresee that disaster is liable to strike within the following, say, 30 years—as water tables drop, or topsoil disappears—and may perfectly well perceive that the way out of trouble would be to reduce numbers. But most of the people in a rising population are young, indeed are children, which means that they have not yet had children of their own. Even if some future world government introduced population control policies as rigorous as those of recent China, where couples are confined to one child each; and even if people stuck to the rules (an enormous 'if') the birth rate over the ensuing decades would still exceed the death rate, and total numbers would continue to rise. In short, rising populations cannot be stopped in their tracks; they are like large ships, on full steam ahead, with feeble reverse engines. We may see the trouble coming even decades ahead, and yet be unable to do a thing about it.

And what price our fellow creatures, if human numbers really do follow the simple demographic trend? They seem almost to be an afterthought. Present-day Indian farmers are remarkably tolerant of the elephants who take their year's crops in an hour; regarding them, with gentle Hindu patience, as fellow sufferers, forced into bad ways. But when India's population exceeds 2000 million, which it is on course to do in about 40 years time, patience even for them will seem too great a luxury.

Yet there is one significant and perhaps critical get-out clause. The absolute rate at which human numbers are rising continues to increase. Every year, as things are, we add more people than we added the year before. But the percentage rate of increase is falling. The percentage rate peaked in the 1960s at around 2.0 per cent a year; and now it is down to about 1.6 per cent. Around 2 per cent a year represents a doubling every 40 years or so. The new, lower rate represents a double every 50 years or so. If the percentage rate of increase continues to fall as it is as present, then by the year 2050 or thereabouts the percentage rate of increase would reach 0 per cent.

A STABLE POPULATION?

In other words, so United Nations demographers suggest, the world population could stabilize by the middle of the twenty-first century. By then, to be sure, it will still roughly have doubled; so it could be 10 000–12 000 million, and it will not be easy to feed that many people well on a sustainable basis. Nevertheless, when the human population is stabilized we can at last take breath. The task becomes definable. In principle, at least, the task should not go on getting bigger. Our children should not need to cater, as we do, for an ever-increasing demand.

There is also one highly significant bonus in this scenario. It is commonly supposed that human numbers can be reduced only by disaster: famine, war, pestilence. There is a certain crude logic behind this. Agriculture, industry, and medicine allowed human numbers to rise and their removal, presumably, would bring about a fall. There are many historical precedents: the potato famines of the 1840s dramatically reduced the population of Ireland; the Black Death of the 1340s is said to have cut Europe's population by about a third; World War II removed almost an entire generation of Russian men; and so on. Population reduction, in short, is associated with cruelty and mayhem.

But logic, and history, can be deceptive. Disasters have only a temporary impact on human numbers. People bounce back after epidemics, wars, and famines. Indeed, like most animals, human beings tend when they have the opportunity to breed their way out of trouble. In truth, the only measures that have any chance of stabilizing human numbers in the long term are all benign. Modern sociological studies reveal that people tend to have fewer children when things are going well: when they are affluent, secure, and they and their children are healthy. Before German reunification, the population of West Germany seemed poised to go down. Italy has the lowest rate of population growth in Europe—even though Italy is essentially Catholic and Catholic orthodoxy forbids contraception.

At first sight this may seem a little odd: people have fewer children at the times when they can most easily afford them; and more, when it seems they cannot. Yet this makes perfect biological sense. For one thing, population reduction over the long term—stabilization, and then a steady fall in numbers over decades or centuries—does not require dramatic restraint. The Chinese in recent years have been trying to bring their rise in population to a dead stop, and so rationed couples to one child each. But if couples average just two children each, total numbers will come down in time. After all, it takes two to make a baby, so two children per couple represents minimum replacement. But among those babies, some regrettably will die before adulthood even in the best regulated societies, some will be sterile, and some for whatever reason will choose not to have babies; and experience shows that if populations average fewer than about 2.3 children per couple, then total numbers are liable to fall after a time. So if people aim for two children per couple, and if three is positively rare, populations will indeed stabilize and diminish. In many modern, affluent societies two per couple seems enough.

Why, in fact, does anybody have more? Well, there are many reasons. Some people just like children; and we can envisage that pioneer families, desperate to get to grips with some new piece of 'virgin' territory (like the old American West), would have wanted as many as possible, partly for the company and partly to provide more pairs of hands. But often people have large numbers of children for nothing more than insurance, or for status. When infant mortality is high, you need a lot to be sure that some might live. When you have no pension, you need children for your old age. When women have no standing in their society except as mothers, they must reproduce. The factors that militate against these pressures are all benign: reduce infant mortality; provide security in old age; liberate and empower women. All these measures seem socially desirable even if they had no effect on overall human numbers. The fact that they are the prime policies for long-term, effective population control is a bonus indeed. Pensions seem dull. But they provide one of the principal means to secure the future of the planet because they reduce people's need to produce children for their old age. The world's bankers and financiers have even more responsibility than they know.

The same demographic projection that says that human numbers could and should level out by the year 2050 also suggests that the population will stay at around 10 000–12 000 million for several centuries (because human lifespan is increasing, this being one of the components of 'population momentum') but should then begin to decline. Modest projection of present-day trends suggests that the decline may begin in about 500 years time, when people will probably expect to live to at least 100, and people worldwide will consider it eccentric, and indeed antisocial, to have more than two children per couple. Human numbers might then fall to whatever level our descendants feel is sensible. There are no totalitarian overtones in this. Our descendants will (in this optimistic scenario!) be in full control of their own affairs, and will be having fewer children through their own personal choice. I once asked various ecologists what they thought a comfortable world population would be. Paul Ehrlich from California, who has warned of pending human disaster since the 1970s, felt that a world population of 2000 million would be sensible. The American conservation biologist Michael Soule felt that 300 million would be a good number. This, after all, was the possible number at the time of Christ—and at that time, as Soule points out, there was no shortage of great civilizations and of cultural variety: Greek, Roman, Egyptian, Jewish, Indian, Chinese and many others in Asia, North and South American, African, Australian. Ehrlich and Soule are not anti-humanity—far from it. They both point out that if human beings do reduce their numbers, and get their population within sustainable limits, then our species could last for an extremely long time; but if we do not, then we could crash. In the end, more human beings will be able to experience this Earth if we spread ourselves through time.

REWARD WITHOUT WEALTH

Finally, of course, we must acknowledge that numbers are only half of the issue. The other half is human consumption—with expectations rising, it seems, without conceptual ceiling. Paul Ehrlich has pointed out that a model family in Los Angeles, with Mom, Pop, and two beautiful children, may feel well that they take no more than their share of this planet and yet may consume far more than a fair-sized Bangladeshi village.

So we have to modify our aspirations. The world is not growing, and we cannot all hope to be indefinitely rich. It is unrealistic to hope that all of humanity can live as the middle classes do in California or, indeed, in the old West Germany. We need economies that recognize upper limits. We have to find ways of rewarding effort and merit, and conferring status, that do not simply involve addition of wealth. If our descendants can accept that it is reasonable to live at the material level of, say, a modern-day Greek villager—supplemented by the electronics that puts each of us in touch with the culture of the whole world—then their own chances of survival, and those of other creatures, will be enormously enhanced.

This, then, is the overall optimistic but nonetheless realistic demographic scenario. Human numbers seem bound to rise for the next 50 years, to reach a maximum of 10 000–12 000 million. They are liable to stay at that level for around 500 years; and should then fall again to whatever level our descendants think acceptable. Within 1000 years numbers should again have returned to present levels—around 5000 or 6000 million; and within a few centuries after that, total numbers could be as low as, say, 2000 million.

If the human species does follow such a demographic curve, then life for our descendants in 1000 years time could be good, and getting steadily better. The 2000 million or so lucky souls will live modestly (at least; more like Pythagoras than Bill Gates) but each would probably expect to enjoy good health for a century and a half or so. Clearly, though, the next 100 years are going to be very tense indeed. Either numbers will rise beyond countenance, with possible ecological collapse; or they will stabilize—but stabilization also brings problems in the short term (because a stable population is one with fewer children, which means that the average age of the citizens is steadily rising). Indeed the next 1000 years will be difficult because through all that time (assuming there is no ecological collapse) human numbers will reach 10 000 million or more, and remain at such levels for several centuries; and it will be a thousand years before they again fall to present levels. Conservation biologists accordingly have coined the expression 'demographic winter' to describe the next 500–1000 years, when times will be difficult for the human species—and even harder, because we will inevitably pass our troubles on, for our fellow creatures.

This scenario is hopeful nonetheless. A thousand years may seem a long time, but compared with the evolutionary times that have featured throughout this book—the times required to shape the many lineages of creatures—it is hardly measurable. Unfortunately, our minds are evolved, too, and natural selection has not equipped us to think easily much beyond the next few seasons, or at least beyond the lifetimes of our children. If we are serious about conservation we need to think in far greater intervals. A thousand years, even a million years, must come to seem a reasonable unit of political time.

More to the point, though, is that the thousand years of pending demographic winter is not an infinity of time. It will grind on, but it will end. We need not envisage that life will inexorably grow worse for our fellow species, for ever and ever, until they finally collapse. We can see the task is finite: to take our own species, and as many as possible of the other species on Earth, through the demographic winter. After that, life should be easier—and could be idyllic. So how, in reality, are we to get through the next millennium?

HOW TO SURVIVE THE DEMOGRAPHIC WINTER

I will speak in this section mainly of animals, for what applies to them also applies in principle to other creatures—and, in general, if we can save animals, we can save anything. Logically, there are two possible approaches to their salvation. We can seek to conserve the places where animals live—their habitats; or we can seek to save them species by species. Species-by-species conservation can mean, though it should not exclusively imply, captive breeding: breeding animals more or less intensively in dedicated reserves or in zoos.

These two approaches are entirely complementary; and it is a huge pity that many who think of themselves as conservationists see them in opposition. At least, I know of no conservationist who would not seek to save all possible scraps of worthwhile habitat—and one of the most serious pursuits in conservation is to identify those that are most worth preserving. No one doubts that habitat must be conserved as energetically as possible. But I do know many who oppose the whole notion of captive breeding, for all kinds of reasons.

On the face of things, habitat protection seems to offer so many advantages that any other approach, including captive breeding, simply seems fatuous. If you can save an entire habitat then you save not just one species, but many. This seems bound to be true even if you focus on just one or a few species within the habitat. After all, if you create a reserve for tigers then you must also save antelope and deer, or there is nothing for the tigers to feed on; and so you need a rich vegetation to support the prey animals, and this, in turn, will feed and protect a host of other creatures. If, by contrast, you merely breed tigers in a zoo—well, you save just tigers. Salvation in the wild seems much cheaper, too; it has been calculated, for example, that it costs 100 times more to keep an individual rhino in a zoo, than in the wild.

Wild habitat is, of course, an essential requirement—for if an animal cannot live in the wild, is its survival worthwhile? If the only tigers left in the world were in zoos, would tigers have any point at all? Finally, there's the arithmetic. If half the terrestrial eukaryotes are in danger of extinction, then this means that several million species are endangered. We might set up special protection schemes for black rhinos and tigers and a few more such 'charismatic megavertebrates' (as the bigger creatures are sometimes somewhat derisorily referred to) but we cannot establish a million such schemes. In short, all species-by-species schemes—or, at least, all captive-breeding schemes—are a drop in the ocean; hopeless gestures.

These arguments seem powerful, not to say inexorable; especially when we contrast the glorious herds of wild creatures that can still be seen on Africa's savannah, with the sometimes miserable pairs or even individuals that have all too often sweated out their dreary days in urban menageries. But although we can all fight for habitat protection, and although we may all hate the pointlessness and cruelty of bad zoos, it would be a huge mistake to write off the species-by-species approach in general, or to disdain the possibilities of captive breeding in particular.

Thus, although habitat protection is so desirable, it may not be an option; and when it is, the best that can be done may be insufficient. For a variety of reasons (which I discuss in *Last Animals at the Zoo*, and O. H. Frankel and Michael Soule treat at greater length in *Conservation and Evolution*) populations of sexually reproducing creatures such as animals need to be large if they are to survive in the long term. Small populations are liable to be wiped out sooner or later by disease or 'demographic stochasticity'—the skewing of sex ratios, for reasons purely of chance, so that an entire generation may finish up without breeding females or virile males. In the longer term, small populations lose genetic variation through genetic drift, and become inbred, which itself may be fatal; and in the longer term, they lose the genetic variation needed to evolve by natural selection if and when the environment changes further. For such reasons, rough-and-ready but nonetheless sensible calculations suggest that a 'safe', minimum population must generally contain around 500 individuals. This may not seem too many, but each individual tiger, say, needs between 10 and 100 square kilometres in the wild, depending on the quality of the terrain; which means that a viable population needs up to 50 000 square kilometres. The Indian government in particular has created excellent reserves for tigers, which is a noble thing to do in that crowded country. But there is no reserve in India or anywhere else big enough for 500 individual animals. In short, there is no population of wild tigers in the world that can truly be considered 'safe' in the long term. Any one wild population, and perhaps all of them, are liable to be wiped out by various forms of misfortune within the next century. Wild tigers, in short, are already living on borrowed time; and for the time being, a radical improvement in their lot seems inconceivable.

India is, of course, particularly crowded. There is much more room in, say, the United States, which has a much bigger area and only about a third of the population; and is far richer. Even so, there seems to be too little room for grizzlies in Yellowstone, which is the greatest of the national parks in the USA outside Alaska; and it is not yet obvious that the grey wolf will be re-established in Yellowstone as a permanent resident.

Of all the reserves that we may create for wild creatures, the national park is the greatest: the areas with the most protected status, portentously enshrined in law. But do laws last forever? Can we tell people in the twenty-second or twenty-third centuries that they should continue to protect the areas that we are now choosing to allocate? Of course not. And are the parks that exist now really the wild places that they were intended to be? Sometimes, but sometimes not. Most wild species live in tropical countries, and most tropical countries are poor—relative at least to those of the north. Most people in tropical countries live by small-scale agriculture. Thus many national parks worldwide contain more cattle than wild creatures, and many that appear on the maps as 'forest' contain no trees. (I do not want to cite cases, because I do not want to appear to criticize from this rich, European vantage point—especially as my own country, Britain, has one of the worst conservation records in the developed world. It is a fact, nonetheless, that many 'national parks' worldwide are a sham.)

The pending, or perhaps actual, greenhouse effect seems to provide the final blow. Some politicians and a few scientists deny its reality but most feel that the world's climate will change significantly over the next half-century—within our own lifetimes, or those of our children. The world in general will get hotter and wetter but some parts, caught in eddies, will be drier or perhaps colder. Britain could freeze if the Gulf Stream went into reverse.

Some time within the next thousand years, well within our demographic winter, the climate within many of the existing national parks will change beyond the tolerance of its inhabitants. Some—like wolves and the big cats—are extremely versatile but herbivores tend to rely on particular vegetation types, and plants in general are extremely sensitive. The world's climate has changed dramatically over 100 000-year intervals throughout the past million years as the ice ages have come and gone—sometimes interrupted by bursts of tropicality; as it has done so, the world's animals have taken themselves towards the equator, and then back towards the poles. There is cave near where I used to have a caravan in Yorkshire, in the north of England, that has yielded the fossil bones both of hippopotamuses, which are inveterately tropical, and of reindeer, which feed on the lichen that grows beneath snow. Both passed through Yorkshire as the weather changed, following the expanding and retreating ice at a greater or lesser distance.

The pending greenhouse effect could warm the world as dramatically as the ice ages cooled it, but the vast, north–south migrations that once enabled the world's wild animals to adjust are no longer feasible. Many conservation biologists have suggested that we need to construct vast north–south 'corridors' to enable creatures to shift, but this can rarely be done on the scale required; Nairobi and Paris and suchlike places are in the way. The animals that may have seemed secure in their national parks will die like rats in traps, unless we shift them wholesale into new locations and re-create their ecosystems. This is feasible, in principle; but it is not the laissez-faire conservation that some seem to favour and along the way it will certainly involve hands-on treatment that is liable to include spells of captive breeding.

What of cost? How can we justify keeping rhinos, say, in European, American, or Australian paddocks, when they could be supported for a fraction of the price in their native lands? But can they? At the start of the 1980s Zimbabwe had an estimated 1500 black rhinos. They were protected by platoons of brave and expensively armed wardens, deployed with maximal efficiency. But the poachers slipped through anyway, and by the mid-1990s most of the rhinos were gone. The 400 or so Indian rhinos in the Chitwan park of Nepal are protected by a somewhat greater number of soldiers. So how much does it really cost to keep a rhino safe and free in the wild?

The answer is unknown because the sums have never really been done. At least, it might be done in Chitwan; but the costings are hardly realistic because Nepalese soldiers are virtually unpaid. By contrast, breeding in captivity may be expensive, but it can at least be audited. Although some rhinos so far breed badly in captivity (whites are difficult, no one has succeeded with the Sumatran, and the Javan has not been attempted) the problems have apparently been solved at least with the blacks and perhaps with the Great Indian. Besides, because of the hazards of the wild, it has been calculated that individual populations of wild rhinos should contain at least 2500 individuals if they are to be viable long term because they must be able to withstand fluctuations. But such populations would be far larger than any that now exists in the wild, except perhaps of South African whites. Wild Sumatrans are down to a few hundreds, and the largest known population of Javans in Ujong Kulon has fewer than 50 individuals. Captive populations of rhinos, by contrast, should be safe from population fluctuations and could succeed long term with 150 individuals at any one time.

All in all, the alleged 100 to 1 contrast between the costings of captive and wild populations seem less convincing. Rhino protection may be expensive in captivity, but in theory, at least, it could work. In the short term, in the wild, there is little evidence that it can work at all, with the sole exception (so far) of South Africa.

As for total numbers: clearly, we cannot hope to create special breeding programmes for all the probable millions of endangered species. So we just have to admit that captive breeding is a feasible strategy in some cases, but not in others. But is this so ridiculous? The same argument applies, after all, to habitat protection. Desirable though it is, it cannot alone provide adequate protection for all creatures, including, as things stand, for tigers or probably for rhinos. We just have to use captive breeding where it is appropriate.

HORSES FOR COURSES

Captive breeding does seem particularly (though not exclusively) appropriate for large terrestrial vertebrates, such as the tiger and rhinos, which for the time being cannot feasibly be protected solely in the wild. In fact, although living things as a whole are hugely various, there are only about 23 000 living species of land vertebrate; around 4500 mammals, 5000 or so each of reptiles and amphibians, and about 8500 birds. Many are not especially endangered—the brown rat and the house mouse are obvious examples—and of those that are, many are best conserved in the wild; like the numbat of Western Australia, which breeds badly in captivity but is small, and can be conserved in reserves of modest proportions. In fact, only about 10 per cent of the extant land vertebrates probably needs support by captive breeding, including the big cats, primates, and the rhinos. Captive-breeding programmes for around 2000 species, therefore, could provide all the back-up necessary for all the terrestrial vertebrates (assuming that the best possible habitat protection was also undertaken). The terrestrial vertebrates are not the only important creatures on Earth, but it is perverse to deny their significance; and it is certainly feasible, even within the confines of present-day zoos if they truly put their minds to the task, to set up 2000 breeding programmes. In short, captive breeding is a necessary and feasible tactic that must be written in to our conservation strategies.

Finally, we may recall the point made earlier: that the immediate, 'short-term' task is to take humanity, and as many as possible of our fellow creatures, through the next thousand years. After that, things should get better. It may not be desirable, aesthetically, to keep tigers in captivity (although captivity need not imply a concrete box!) but such confinement does not represent the end of the line. The descendants of big cats that now while away their days in New York or Wiltshire may one day track the deer again in their native India. Of course, critics argue that it is impossible to return animals to the wild, but that is simply wrong. Some captive-bred creatures, including some parrots, have proved recalcitrant but most serious attempts to return wild creatures have succeeded sooner or later. The many species of thoroughly domestic animals that have escaped to become 'feral', and sometimes dominated their new environments, show in general how thin is the veneer even of committed domesticity. Successful—often too successful—feral animals worldwide include cats, dogs, horses, asses, camels, water buffalo, elephants, pigs, monkeys, pigeons—and several species of parrot. And, of course, the task for conservationists breeding wild animals in captivity is not to 'domesticate' them but to preserve as much of their wildness as possible, both their genes and their behaviour.

Overall, over the next 1000 years, while our own population settles into its next, huge, but nonetheless settled phase, we must treat the world as a whole as a mosaic. We need areas of wilderness, dedicated to wild creatures; areas of intensive arable agriculture and horticulture, so that we produce the maximum crops on the minimum area; but also places that are good both for wildlife and for human beings, where we can rub along together—extensive pastureland for livestock and cities that are as wildlife-friendly as possible. Cities are increasingly important to wild animals; the urban fox is one of the wildlife phenomena of Britain (along with the urban carrion crow and Canada goose) while Delhi has the greatest concentration of birds of prey of anywhere in the world.

The wilderness itself must be managed. This is not gratuitous officiousness, jobs for the boys and girls. The ecology and behaviour of animals in Africa or North America, say, is geared to the entire continent, or at least to a fair slice of it. Even the greatest of the national parks—Kruger, Yellowstone—cannot encapsulate the whole. The parks are not microcosms, they are fragments. Each is surrounded by hostile country, notably by agriculture. The populations of creatures within these fragments of wild country cannot be allowed to fluctuate as they did when the animals roamed the entire continent. Delicate trees must be sustained; weeds excluded. To some, 'wildlife management' is an oxymoron; but without it, mass extinction is inevitable. In general, wilderness should probably be managed in a 'stratified' way: some exploited for local industries; some kept 'wild', but organized essentially as a showpiece to provide income from tourists, and to keep the rich foreigners involved; and some kept sacred and sequestered (except for the professional managers and scientists) so that the wild creatures at least have some opportunity to be themselves. Australia's Great Barrier Reef has this kind of tiered management. Details must differ from place to place, but the overall logic seems inescapable.

This, then, is the general pattern of conservation strategy that we need to impose if we and our fellow creatures are to survive the next 1000 years in reasonable harmony; and this thousand years, we may hope, should be a prelude to better times.

Even with the best will, however, and with investment and expertise that far exceed what so far seem likely, we will not be able to save all of the creatures that are now endangered. If we try to save everything, or if we simply grab at whatever happens to take our eye, then we and the world will lose far more than we need to. We need a strategy, to decide on which creatures we should focus our energies. Phylogeny can help to shape our thoughts.

PHYLOGENY AND CONSERVATION

Conflict in conservation is inevitable. What is good for some species is not good for others. Insects, for example, flourish in long grass while many birds (the chough comes to mind) like short grass that is heavily grazed. Often it is good to create 'corridors' between otherwise isolated habitats—hedgerows between copses are a homely example—but rare and delicate species may sometimes do best in isolation. So we do have to manage wilderness and wildlife; but we have first to decide priorities. There is also the matter of sheer numbers—for we cannot save everything with space and funds that are inevitably limited. So what should our priorities be? What should be the criteria?

Different people have different views. To some, the most spectacular creatures are the obvious priorities: elephants, big cats, gorillas, parrots, orchids. To some extent this preference is merely anthropocentric: we target those creatures that please our senses. But such a view is not merely self-indulgent. The big showy animals also include most of the most intelligent, and intelligence is a rare biological quality that we surely ought to respect. The big animals, too, often set the tone of entire ecosystems. A forest that contains elephants is very different, ecologically, from one that does not. Creatures that have such an all-pervasive impact have been called 'keystone species'; and many ecologists have argued that such species, whether large or not, should be especially favoured because of their influence on all the rest.

Others, however, suggest that to focus on large animals is merely self-indulgent, and is therefore bad. Such critics commonly refer to the elephants and tigers as 'cuddly animals' while the expression, 'charismatic megavertebrate' is generally intended to be derisory. These critics often use ecological arguments that seem to me highly dubious, not to say plain wrong. Thus I have heard biologists in regrettably high places argue that the tone of ecosystems is set by the small creatures at the bottom of the food chain—the prokaryotes in the soil, then the insects, and so on—and that the bigger animals are merely hangers-on. Even if this were true it would be an odd argument: it is strange to suggest that we should strive particularly hard to save beetles, and forget about elephants, simply because beetles in the end are more influential. But the argument is not true. In an ecosystem, the arrows of influence flow all ways. The soil prokaryotes affect the flora, which influences the animals, right enough; but the creatures at the tops of the food chains, too, enormously influence the ones beneath. Elephants decide whether dung beetles live. Dung beetles do not determine the fate of elephants. Elephants cannot live without vegetation, but they are remarkably versatile feeders: within broad limits, they do not greatly mind what kinds of plants are on offer. The big carnivores are even more versatile. Leopards, for example, can live virtually anywhere on any creatures of suitable size—dogs, baboons, gazelles, deer; just line them up, and the leopard will take care of the rest.

Discussions on what creatures to save occupy entire conferences, are the subject of endless reports and academic theses, and are among the chief preoccupations of those zoos that take conservation seriously. Many of the arguments, as already hinted, are heavily laden with emotion, which makes them harder to resolve. Are the lovers of 'charismatic megavertebrates' merely sentimental and, if they are, so what? Conservation, after all, in the end must be driven by emotion.

Yet we can cut through the wrangling with arguments that are as nearly 'objective' as is possible in a field like this; and those arguments at bottom are phylogenetic.

Picture a phylogenetic tree showing all the branches of living creatures—preferably down to family level. It is impossible to provide such a tree in this book, both comprehensive and detailed. We would need a canvas the size of a football field. (Perhaps some well-endowed museum might provide a suitable wall!). The eukaryote tree (with the two prokaryote domains painted in) perhaps gives the best idea of the overall shape, but you really need to superimpose all the other trees to get a true feel for the whole thing—although many of these should ideally be shown in far greater detail than has been possible here. Thus I have shown only two families out of many thousands—the Hominidae and Compositae; and some entire important phyla (or their plant equivalent), such as the annelid worms and the ferns, have been treated only in passing. Nonetheless: imagine a vast and wonderful tree that shows everything, at least down to the level of the family.

The task, I suggest, is to conserve as much of that tree as possible. We need at least to ensure that representatives of all the thickest branches remain to us. Of course, some representative of the three great domains will always be with us, until Earth itself is reduced to a cinder. There will always be some bacteria, archaes, and eukaryotes. We need to ensure that all the kingdoms survive, too; so even if we do not like *Giardia*, that importunate protistan parasite, we ought to take good care of it. The World Health Organization in a couple of centuries time may find ways to wipe it out, just as it has wiped out the smallpox virus; but examples of *Giardia* should be saved in some micro-zoo for conservation purposes.

Indeed we need representatives from all the phyla. The newly discovered *Symbion*, unprepossessing though it may be ('a sack of guts', as Prince Hal said of Falstaff), must be a key target. So the process continues. We should certainly be able to save members of every class, and indeed of every order, even if we cannot in reality save every last family. We need at least to maintain an impression of the present tree. In a thousand years time that impression may be a shadow of present glory; but if life really does become easier after that, then that shadow should flesh itself out again. The taxa that have so far taken 3500 million years to unfold should resume their evolution. Present details will be lost—not every species can pull through—but if we conserve the rough phylogenetic outline, new details will take their place, just as has always happened in the past.

This view of conservation seems to provide the criteria we need. Is an elephant more important that a beetle? The point is not that the elephant is large and clever, and able to relate to us, whereas the beetle is small and on the whole unresponsive. It is that the order Proboscidea now contains only two distinct species: *Loxodonta africana*, the African elephant, and *Elephas maximus*, the Asian elephant. Lose those two and we will lose the entire, glorious lineage of proboscideans, which first appeared in a distinct form around 50 million years ago and is known, since that time, to have included at least 150 different species. Lose either of the present elephants and we have lost half the remaining Proboscidea. But the order of the beetles, the Coleoptera, contains at least 300 000 living species; and there are probably many millions more in the tropical forests (though many will disappear before we get to see them). Lose any one beetle and we lose the minutest fraction of the whole. On these grounds alone, there seems to be no contest. If we really have to choose between a species of elephant and a particular beetle, then the elephant wins hands down. By the same token, the tiger is more important than any one shrimp. The same principle clearly applies in all groups. Any one quillwort is surely more significant than any one member of the Compositae, even if the composite is prettier.

There are, of course, many other considerations. Cost and feasibility are among them. A keen individual might be able to set up a conservation programme for a beetle (although this is a lot harder than it may seem) whereas they would need to be very rich to set up a serious scheme for elephants. Yet it would be foolish and sad to despise a beetle-conservation plan simply because individual beetles seem on phylogenetic grounds to be less important. Policies must be geared to feasibility. We may decide, too, that some at least of the world's 300 or so species of parrots, many of which are now endangered, would need more resources than can realistically be allotted. We may nonetheless decide that some of the more extreme forms of parrot —like the almost flightless giant kakapo of New Zealand—ought to receive special attention, because it demonstrates the range of biological possibilities that parrots have explored. Similarly we might decide that, say, the extraordinary giant rhinoceros beetle is a particularly worthy coleopteran, precisely because it reveals the scope of beetledom.

We must acknowledge, too, that some of the groupings that biologists have called 'orders' are far more various than others. Thus there is clearly far more genetic variation within the order Coleoptera than there is within the living members of the class Aves. Nevertheless, the taxa that biologists have discerned and ranked over several centuries should not be taken lightly. They clearly reflect some of the reality of nature; and it seems to me far from foolish to take that aspect of reality as a guide. In general, therefore, I suggest that we should root a global conservation strategy in phylogeny. There are many other possible criteria. But none is better.

But why should we bother at all? Why not just let the creatures go, and enjoy the rest of Earth's party while it lasts? I cannot end this book without a few words on this.

WHY CONSERVE?

We live in a secular and fiscal age. Conservationists accordingly try to offer political and economic reasons why our fellow creatures should be preserved. This is fair enough, and indeed necessary. Conservation policies that do not take account of the needs of local people are undoubtedly doomed, and probably should be. Some conservationists seem to argue that human beings should simply be swept aside to make way for wild species. Such arguments are at least unattractive. The needs of wild creatures and those of human beings must be reconciled, and although this is difficult—it would be technically easier to sacrifice one or the other—that is the nature of the task. Somehow or other, then, conservationists do need to show that particular conservation policies will not wreck local economies, and can indeed enhance them. But although such arguments are necessary, they are not sufficient. Conservation needs something else, some factor *X*. That factor *X* is desire and attitude: a deep conviction that we should look after our fellow creatures, that it is our obligation, and we must be prepared to make sacrifices to achieve it. We must be prepared to make our own sacrifices, however; not propose simply that some other people, in some other country, should give up their livelihood (and in some cases their lives) simply to make us feel good, far from the action in the affluent north.

Why, though, are the secular and economic arguments 'necessary but insufficient'? At first glance they look sound enough. Kenya earns the lion's share of its foreign currency, and of its national income, through tourism; and although the tourists come in part for the beaches they mainly come for the wildlife. 'Ecotourism', largely based on wildlife, has become a huge force the world over. Many point out, too, that wild plants are the world's greatest source of medicinal drugs, and suggest that there must be many more such treasures out there, as yet undiscovered. In this age of biotechnology the countless millions of genes in the world's many millions of living creatures are surely among our greatest riches, to be compared with oil or precious metals.

But these arguments do not stand up as well as we might anticipate. Kenya's wildlife is wonderful for tourists and so is that, for example, of the Great Barrier Reef. But other environments, even richer in species, are much less favourable. Most wild species live in tropical forest and it is very hard indeed to establish viable ecotourism in forests. I know people who have tried. Indonesia has a fabulous array of native species—elephants, rhinos, the Sumatran tiger, orang-utans, and so on—but you cannot see them from a minibus, as you can watch the lions in the savannah of East Africa. Indeed you may spend a lifetime in a forest, surrounded by hundreds of thousands of species, and see nothing except leaves. I know a biologist who did a PhD on the Sumatran rhinoceros yet claims never to have seen one. The Javan rhinos of Ujong Kulon have been filmed convincingly only once. A botanist in Queensland told me that he once watched a sleeping tree kangaroo for 6 hours. He was a true naturalist. Most tourists would not pay much for such a privilege. It surely would not be impossible to lay on satisfying tourism in tropical forests, with treetop walks and semi-captive creatures, but it would be extremely expensive, especially if, as would surely be the case, the tourists wanted more than a hammock and a tarpaulin; something closer to, say, a four-star hotel.

Besides, if the economic argument is all that protects the wild places, what happens when something more profitable comes along? What price conservation when prospectors find oil or gold beneath the last reserve of some precious creature? Mangroves are wonderful havens of wildlife—but as I can attest, as my boat filled with spiders and the mosquitoes mustered their forces, they are extraordinarily hostile. Florida's Everglades and a few other such places are good for tourists, but if you really wanted to make money you would simply dig a few holes and raise prawns; or, better still, drain the whole lot and build a golf course and a casino. It is true, too, that wild genes should be a source of wealth, but not as true as some of its advocates like to claim. The useful plants that have been identified can be cultivated and although there may be many more, it may not be cost-effective to look for them.

So even the most 'romantic' of conservationists must seek economic reasons to protect the creatures they love; and they must discover and promote all the economic advantages they can. But economics cannot, in the end, carry the day. Except perhaps in Kenya, and the Great Barrier Reef, and a few such miraculous places, it is nearly always liable to be more economical to sweep the wildlife away, and do something else. There have to be other reasons for not doing this. So what are these 'other reasons'? They may loosely be called 'spiritual', though 'spiritual' is hard to define. They can certainly be called 'aesthetic', and 'moral'. What do these terms imply?

'Aesthetic', of course, has to do with what we find beautiful. Nature is beautiful; so if we like beauty, we ought to like nature. There is a useful economic twist to aesthetics, too. What we find beautiful we will pay for. A house with a good view is worth a lot more than one without. But, again, aesthetics on its own is not sufficient. For one thing—at least when 'aesthetics' is defined simply—it implies anthropocentricity, which is a fancy word for self-centredness. If aesthetics is all that drove us, then we would presumably save the pretty and striking species, like the tiger and gorilla, and allow the less flashy to go by the board; like all those endangered marsupials that few of us will ever see, and in any case tend to look much of a muchness. Besides, the aesthetic sense in most people is very uninformed. An arboretum, with scores of beautiful exotic trees, may at first glance seem more pleasing than a temperate wood, with oak and ash and very little else. But the wildlife—the fungi, wild herbs, invertebrates, birds, and mammals—prefer the natural wood. They, after all, are adapted to the native trees; not to the exotica. Michael Soule recalls that as a boy he used to listen to the wild birds in the shrubby valleys of the chaparral around San Diego. The valleys are still full of birdsong. But in the old days the song came from scores of species, while nowadays only the generalists are left; the mockingbirds and common finches. Most of us cannot tell the difference. The songs of common birds may be as beautiful as the rarest: in Britain, for example, no bird, not even the nightingale, out-sings the common blackbird. So if 'aesthetics', as commonly defined, is all that guides us, then most of us could happily preside over the extinction of most wild species, and not know that they had gone. Aesthetics is a poor arbiter unless our sensibility is trained; unless we feel the loss of particular birdsong, and regret its passing; or see that a wild and melancholy swamp is more beautiful than the manicured fields that replace it when it is drained.

Morality, however, implies more than aesthetics. At least in some interpretations (notably that of Immanuel Kant) morality implies that each of us as individuals is prepared to put the well-being of others before ourselves. This does not mean we have to be prepared to die for others—although we may choose to die for some. But it does imply that we ought to be prepared to make some sacrifice. There is no virtue (Kant said) in helping somebody else if it costs you nothing to do so, or offer help only in the expectation of reward; but there is virtue in helping if it requires effort, and if there is no such expectation. If we are serious about conservation, then we need to approach it in this spirit. We must be prepared to pay—through taxes, or some more direct payment: and we must be prepared, at least up to a point, to suffer. I met a woman in Greece (she wasn't a Greek) who wanted to drain the local pond because the frogs kept her awake at night. Well, frogs can keep you awake. But we should put up with it. Insects eat your garden plants. So what? That is no cause to saturate the ground in toxin. Insects are beautiful too; and they are fellow creatures (which in this context is more to the point). Unless we see them as such then they have no hope. Seventeenth-century Dutch flower painters commonly painted in the insects that were eating them, acknowledging that they were beautiful too. It is hard to know where to draw the line—it is grotesque, for example, to allow the bugs that carry Chagas' disease in South America to flourish in people's houses—but, in general, we should err on the side of generosity. We have to be sensible, but we have grown effete.

Still, though, the central question seems unanswered. Why should we make sacrifices? If golf courses really are more profitable than mangroves (and provide more fun for more people), then why save mangroves? If common and robust species can be just as pleasing to the eye as the rarer and more delicate ones, why bother with the latter?

Actually I do not believe there is a 'good' reason that will satisfy everybody—or even one that many people will find convincing. In the end you just have to feel that it is right to conserve our fellow creatures, and that it is a sin to wipe them out. We have to see ourselves, each of us individually and our species as a whole, as guardians. The argument that conservation is 'playing God' seems to me perverse; it implies, after all, that it is acceptable simply to do nothing and watch things die, or indeed to acquiesce in their destruction, but that it is not acceptable to lend a hand. Some religions do indeed take this passive line, but some do not. Christ was notably interventionist (and commonly got into trouble with the rabbis as a result). I think we need such interventionism. Simply to stand by and watch things die is a sin of omission.

But the notion that we should feel obliged to help our fellow creatures is, in the end, an emotional stance, an attitude. Such an attitude cannot simply be extrapolated from 'rational' argument. It cannot be a product of pure intellect. Religious people—or at least, conventionally religious people who believe in a God who is both creator and moral arbiter—may support such an attitude by an appeal to their God: that we are his representatives on Earth, and that it is our duty to protect his works. In truth, though, the Jewish–Christian–Islamic religions that most clearly embrace the notion of an omnipotent creator-God do not seem explicitly to demand such concern for our fellow species. The Ten Commandments of Moses do not tell us to take care of wild creatures. They might have done, but in fact they do not. God, according to

Genesis, gave us 'dominion' over the beasts of the fields, which people through the ages have interpreted in a wide variety of ways, and often to justify insouciance. Even those who follow conventional religion, therefore, cannot simply appeal to the word of God: and, of course, those who are not conventionally religious, would not want to do so.

So this emotional stance, this attitude, has no foundation at all except what we choose to believe in. But I do not feel that we need worry on this score. David Hume pointed out that, in the end, all ethical positions are rooted in emotion, and that moral philosophers merely find arguments to support whatever attitude they held in the first place. Emotion is what matters. The moral task for each of us is to explore our own feelings, refine our own emotional responses, and then follow our convictions.

For my part, I want simply to suggest that it is a privilege to be conscious in this Universe, to live on this particular planet, and to share it with so many goodly creatures. We can destroy them easily enough but with somewhat greater effort we could save them, just as we save ourselves. It has to be worth doing. I cannot demonstrate that it has to be done, and neither can anybody else. But it is hard to think of anything more worthwhile.

SOURCES AND FURTHER READING

Three main sources of information have contributed to this book. My own education, predilections, and day-to-day conversations have provided the general drift (although I have not relied on them for details). I have enjoyed many excellent discussions and pleasant correspondence with outstanding authorities worldwide, but particularly in Britain and the United States; and, of course, I have relied heavily on the written word—textbooks and learned papers, including reviews and original research.

Those who have helped me directly are acknowledged separately. Here I am listing my principal written sources. Like all reference lists—apart from those that aspire to be comprehensive—the following is somewhat arbitrary. I have recorded texts that have played a particular part in my own education, even if they are not particularly relevant now—such as my ancient 'BEPS' (*The Invertebrata*) of 1959, and my 1962 edition of J. Z. Young's *Life of Vertebrates*, which is a classic; and papers that I perceive to be seminal, both for the subject as a whole and for my own enlightenment—such as the 1993 *Nature* paper on zootypes by Jonathan Slack, Peter Holland, and C. F. Graham, or the paper on Hox genes and the body plans of insects and crustaceans by Michael Akam and Michalis Averof in *Nature* in 1995. I have also listed one or two texts published in 1999 that appeared too late for me to discuss in the text, but are clearly significant.

The lists for most of the chapters and sections begin with general texts and then continue with particular articles. References marked with an asterisk are especially recommended. They include occasional classics and, more significantly, include my key sources for the taxonomic summaries and the 'trees' in Part II.

PART I

CHAPTER 1 'So many goodly creatures' (pp. 3–15)

HALDANE, J. B. S., quoted by R. C. Fisher (1988). An inordinate fondness for beetles. *Biological Journal of the Linnean Society*, **35**, 131–319.

MAY, R. (1988). How many species are there on Earth? *Science*, **241**, 1441–9.

CHAPTER 2 Classification and the search for order (pp. 17–31)

Books

*DARWIN, C. (1859). *The origin of species by means of natural selection: or the preservation of favoured races in the struggle for life* (reprinted 1985, Penguin Books, Harmondsworth). Darwin beautifully presents the case for a 'natural' classification based on phylogeny in Chapter 13 (pp. 397–434 of the Penguin edition).

*MAYR, E. (1988). *Toward a new philosophy of biology.* Harvard University Press, Cambridge, MA.

STEVENS, P. F. (1994). *The development of biological systematics.* Columbia University Press, New York.

Articles and chapters

BERLIN, B., BREEDLOVE, D. E., and RAVEN, P. H. (1973). General principles of classification and nomenclature in folk biology. *American Anthropologist*, **75**, 214–42.

Krings, M., Stone, A., Schmitz, R. W., Krainitzi, H., Stoneking, M., and Pääbo, S. (1997). Neanderthal DNA sequences and the origin of modern humans. *Cell*, **90**, 19–30.

Martin, R. D. (1981). Phylogenetic reconstruction versus classification: the case for clear demarcation. *Biologist*, **28**, 127–64.

Patterson, C. and Rosen, D. E. (1977). Review of ichthyodectiform and other Mesozoic teleost fishes and the theory and practice of classifying fossils. *Bulletin of the American Museum of Natural History*, **158**, 154–64.

Raven, P. H., Berlin, B., and Breedlove, D. E. (1971). The origins of taxonomy. *Science*, **174**, 1210–13.

CHAPTER 3 The natural order: Darwin's dream and Hennig's solution (pp. 33–62)

Books

*Avise, J. (1994). *Molecular markers, natural history and evolution*, p. 35. Chapman and Hall, New York.

Corbet, G. B. and Hill, J. E. (1991). *A world list of mammalian species* (3rd edn). Natural History Museum Publications / Oxford University Press, London and Oxford.

Hennig, W. (1966). *Phylogenetic systematics*. University of Illinois Press, Urbana.

*Ridley, M. (1986). *Evolution and classification: the reformation of cladism*. Longman, Harlow.

Schwartz, J. H. (1984). *The red ape: orang-utans and human origins*. Elm Tree Books, London.

*Smith, A. B. (1994). *Systematics and the fossil record*. Blackwell Scientific, Oxford.

Sokal, R. R. and Sneath, P. H. A. (1963). *The principles of numerical taxonomy*. W. H. Freeman, San Francisco.

Articles and chapters

Funk, V. A. and Brooks, D. R. (1990). Phylogenetic systematics as a basis of comparative biology. *Smithsonian Contributions to Botany*, No. 73. Smithsonian Institution, Washington, DC.

*Platnik, N. I. (1979). Philosophy and the transformation of cladistics. *Systematic Zoology*, **28**, 537–46.

de Queiroz, K. and Gauthier, J. (1994). Towards a phylogenetic system of biological nomenclature. *Trends in Ecology and Evolution*, **9**, 27–30.

Stewart, C.-B. (1993). The powers and pitfalls of parsimony. *Nature*, **361**, 603–7.

Swofford, D. L. and Olsen, G. J. (1990). Phylogeny reconstruction. In *Molecular systematics* (ed. D. M. Hillis and C. Moritz), pp. 411–501. Sinauer Associates, Sunderland, MA.

Van Valen, L. M. and Maiorana, V. C. (1991). HeLa, a new microbial species. *Evolutionary Theory*, **10**, 71–4.

*Wyss, A. R. (1988). Evidence from flipper structure for a single origin of pinnipeds. *Nature*, **334**, 427–8.

CHAPTER 4 Data (pp. 63–79)

Books

Avise, J. (1994). *Molecular markers, natural history and evolution*. Chapman and Hall, London.

*Hillis, D. M., and Moritz, C. (1990). *Molecular systematics*. Sinauer Associates, Sunderland, MA.

*KEMP, T. S. (1999). *Fossils and evolution*. Oxford University Press, Oxford.

RAFF, R. A. and KAUFMAN, T. C. (1983). *Embryos, genes, and evolution*. Indiana University Press, Bloomington.

Articles and chapters

AKAM, M. (1989). Hox and HOM: homologous gene clusters in insects and vertebrates. *Cell*, **57**, 347–9.

DOBZHANSKY, TH. (1973). Nothing in biology makes sense except in the light of evolution. *American Biology Teacher*, **35**, 125–9.

FEDONKIN, M. and WAGGONER, B. (1997). The late Precambran fossil *Kimberella* is a mollusc-like bilaterian organism. *Nature*, **388**, 868–71.

GRAUR, D., DURET, L., and GOUY, M. (1996). Phylogenetic position of the order Lagomorpha (rabbits, hares, and allies). *Nature*, **379**, 333–5.

*SLACK, J., HOLLAND, P., and GRAHAM, C. F. (1993). The zootype and the phylotypic stage. *Nature*, **361**, 490–2.

WRAY, G. A. (1994). Developmental evolution: new paradigms and paradoxes. *Developmental Genetics*, **15**, 1–6.

CHAPTER 5 Clades, grades, and the naming of parts: a plea for Neolinnaean Impressionism (pp. 81–90)

LONG, J. (1995). *The rise of fishes*. Johns Hopkins University Press, Baltimore.

PATTERSON, C. and ROSEN, D. E. (1977). Review of ichthyodectoderm and other Mesozoic teleost fishes and the theory and practice of classifying fossils. *Bulletin of the American Museum of Natural History*, **159**, 85–172.

SIBLEY, C. and AHLQUIST, J. (1990). *Phylogeny and classification of birds: a study in molecular evolution*. Yale University Press, New Haven.

WILEY, E. O., SIEGEL-CAUSEY, D., BROOKS, D. A., and FUNK, V. A. (1991). *The complete cladist*, p. 2. University of Kansas Press, Kansas.

PART II

SECTION 1 From two kingdoms to three domains (pp. 95–106)

DOOLITTLE, W. F. (1999), Phylogenetic classification and the universal tree. *Science*, **284**, 2124–8.

WOESE, C., KANDLER, O., and WHEELIS, M. L. (1990). Towards a natural system of organisms: proposal for the domains Archaea, Bacteria, and Eucarya. *Proceedings of the National Academy of Sciences, USA*, **87**, 4576–9.

SECTION 2 The domains of the prokaryotes: Bacteria and Archaea (pp. 107–26)

DELONG, E. F., KE YING WU, PREZELIN, B. B., and JOVINE, R. V. M. (1994), High abundance of Archaea in Antarctic marine picoplankton. *Nature*, **371**, 695–7.

*Fox, G. E., Stackebrandt, E., Hespell, R. B., Gibson, J., Maniloff, J., Dyer, T. A., *et al.* (incl. Woese, C. R.) (1980). The phylogeny of prokaryotes. *Science*, **209**, 457–63.

Hershberger, K. L., Barns, S. M., Reysenbach, A.-L., Dawson Scott, C., and Pace, N. R. (1996). Wide diversity of Crenarchaeota. *Nature*, **384**, 420.

Huber, R., Burggraf, S., Mayer, T., Barns, S. M., Rossnagel, P., and Stetter, K. O. (1995). Isolation of a hyperthermophilic archaeum predicted by *in situ* RNA analysis. *Nature*, **376**, 57–8.

Kane, M. D. and Pierce, N. E. (1994). Diversity within diversity: molecular approaches to studying microbial interactions with insects. In *Molecular ecology and evolution: approaches and applications* (ed. B. Schierwater, B. Street, G. P. Wagner, and R. DeSalle), pp. 509–24. Birkhauser, Basel.

Mojzsis, S. J., Arrhenius, G., McKeegan, K. D., Harrison, T. M., Nutman, A. P., and Friend, C. R. L. (1996). Evidence for life on Earth before 3,800 million years ago. *Nature*, **384**, 55–9.

Nelson, K. E., Clayton, R. A., Gill, S. R., Gwinn, M. L., Dodson, R. J., Haft, D. H. *et al.* (1999). Evidence for lateral gene transfer between Archaea and Bacteria from genome sequence of *Thermotoga maritima*. *Nature*, **399**, 323–9.

Olsen, G. (1994). Archaea, Archaea, everywhere. *Nature*, **371**, 657–8.

Olsen, G. J., Lane, D. J., Giovannoni, S. J., and Pace, N. R. (1986). Microbial ecology and evolution: a ribosomal RNA approach. *Annual Review of Microbiology*, **40**, 337–65.

Olsen, G. J., Woese, C. R., and Overbeek, R. (1994). The winds of (evolutionary) change: breathing new life into microbiology. *Journal of Bacteriology*, **176**, 1–6.

Pace, N. R. (1991). Origin of life—facing up to the physical setting. *Cell*, **65**, 531–3.

Pace, N. R., Stahl, D. A., Lane, D. J., and Olsen, G. J. (1986). The analysis of natural microbial populations by ribosomal RNA sequences. *Advances in Microbial Ecology*, **9**, 1–56.

Rivera, M. C. and Lake, J. A. (1992). Evidence that eukaryotes and eocyte prokaryotes are immediate relatives. *Science*, **257**, 74–6.

Woese, C. R. (1987). Bacterial evolution. *Microbiological Reviews*, **51**, 221–71.

Woese, C. R. (1994). There must be a prokaryote somewhere: microbiology's search for itself. *Microbiological Reviews*, **58**, 1–9.

Woese, C. R. and Fox, G. E. (1977). Phylogenetic structure of the prokaryotic domain: the primary kingdoms. *Proceedings of the National Academy of Sciences, USA*, **74**, 5088–90.

Woese, C. R., Stackbrandt, E., Macke, T. J., and Fox, G. E. (1985). A phylogenetic definition of the major eubacterial taxa. *Systematics and Applied Microbiology*, **6**, 143–51.

SECTION 3 **The realm of the nucleus: domain Eucarya** (pp. 127–57)

Books

Borradaile, L. A., Potts, F. A., Eastham, L. E. S., and Saunders, J. T. (1959). *The Invertebrata: a manual for the use of students* (3rd edn, rev. G. A. Kerkut). Cambridge University Press, Cambridge.

*Dyer, B. D. and Obar, R. A. (1994). *Tracing the history of eukaryotic cells.* Columbia University Press, New York.

Margulis, L. and Schwartz, K. (1988). *Five kingdoms: an illustrated guide to the phyla of life on Earth* (2nd edn). W. H. Freeman, San Francisco.

Articles and chapters

FRESHWATER, D. W., FREDERICQ, S., BULTER, B. S., and HOMMERSAND, M. H. (1994). A gene phylogeny of the red algae (Rhodophyta) based on plastid *rbcL*. *Proceedings of the National Academy of Sciences, USA*, **91**, 7281–5.

LAKE, J. A. (1991). Tracing origins with molecular sequences: metazoan and eukaryotic beginnings. *Trends in Biochemical Sciences*, **16**, 46–50.

*SOGIN, M. (1994). The origin of eukaryotes and evolution into major kingdoms. In *Early life on Earth*, Nobel Symposium No. 84 (ed. S. Bengtson), pp. 181–92. Columbia University Press, New York.

TURNER, S., BURGER-WIERSMA, T., GIOVANNONI, S. J., MUR, L. R., and PACE, N. R. (1989). The relationship of a prochlorophyte *Prochlorothrix hollandica* to green chloroplasts. *Nature*, **337**, 380–2.

WAINWRIGHT, P. O., HINKLE, G., SOGIN, M. L., and STICKEL, S. K. (1993). Monophyletic origins of the Metazoa: an evolutionary link with Fungi. *Science*, **260**, 340–2.

SECTION 4 Mushrooms, moulds, and lichens; rusts, smut, and rot: kingdom Fungi (pp. 159–79)

*BERBEE, M. L. and TAYLOR, J. W. (1993). Dating the evolutionary radiations of the true fungi. *Canadian Journal of Botany*, **71**, 1114–27.

BOWMAN, B., TAYLOR, J. W., and BROWNLEE, A. G., LEE, J., SHI-DA LU, and WHITE, T. J. (1992). Molecular evolution of the Fungi: relationships of the Basidiomycetes, Ascomycetes, and Chytridiomycetes. *Molecular Biology and Evolution*, **9**, 285–96.

BOWMAN, B., TAYLOR, J. W., and WHITE, T. J. (1992). Molecular evolution of the Fungi: human pathogens. *Molecular Biology and Evolution*, **9**, 893–904.

BRUNS, T. D., FOGEL, R., WHITE, T. J., and PALMER, J. D. (1989). Accelerated evolution of a false-tuffle from a mushroom ancestor. *Nature*, **339**, 140–2.

BRUNS, T. D., VILGALYS, R., BARNS, S. M., GONZALEZ, D., HIBBETT, D. S., LANE, D. J. *et al.* (1992). Evolutionary relationships within the Fungi: analyses of nuclear small subunit rRNA sequences. *Molecular Phylogenetics and Evolution*, **1**, 231–41.

BRUNS, T. D., WHITE, T. J., and TAYLOR, J. W. (1991). Fungal molecular systematics. *Annual Review of Ecology and Systematics*, **22**, 525–64.

CURRAH, R. S. and STOCKEY, R. A. (1991). A fossil smut fungus from the authors of an Eocene angiosperm. *Nature*, **350**, 698–9.

DEPRIEST, P. T. (1995). Multiple origins of lichen symbioses in fungi suggested by SSU rDNA phylogeny. *Science*, **268**, 1492–5.

EDMAN, J. C., KOVACS, J. A., MASUR, H., SANTI, D. V., ELWOOD, H. J., and SOGIN, M. L. (1988). Ribosomal RNA sequences shows *Pneumocystis carinii* to be a member of the Fungi. *Nature*, **334**, 519–22.

GARGAS, A., DEPRIEST, P. T., GRUBE, M., and TEHLER, A. (1995). Multiple origins of lichen symbioses in fungi suggested by SSU rDNA phylogeny. *Science*, **268**, 1492–5.

*HAWKSWORTH, D. L., KIRK, P. M., SUTTON, B. C., and PEGLER, P. N. (1995). *Ainsworth & Bisby's Dictionary of the Fungi* (8th edn). CAB International, Oxford.

JOHNSON, C. N. (1966). Interactions between mammals and ectomycorrhizal fungi. *Trends in Ecology and Evolution*, **11**, 503–7.

SELOSSE, M.-A. and LE TACON, F. (1998). The land flora: a phototroph–fungus partnership? *Trends in Ecology and Evolution*, **13**, 15–20.

Simon, L., Bousquet, J., Levesque, R. C., and Lalonde, M. (1993). Origin and diversification of endomycorrhizal fungi and coincidence with vascular land plants. *Nature*, **363**, 67–9.

Swann, E. and Taylor, J. W. (1993). Higher taxa of Basidiomycetes: an 18S rRNA gene perspective. *Mycologia*, **85**, 923–6.

Swann, E. and Taylor, J. W. (1995). Phylogenetic diversity of yeast-producing basidiomycetes. *Mycological Research*, **99**, 1205–10.

SECTION 5 The animals: kingdom Animalia (pp. 181–210)

Books

Borradaile, L. A., Potts, F. A., Eastham, L. E. S., and Saunders, J. T. (1959). *The Invertebrata: a manual for the use of students* (3rd edn, rev. G. A. Kerkut). Cambridge University Press, Cambridge.

*Brusca, R. C. and Brusca, G. J. (1990). *Invertebrates*. Sinauer Associates, Sunderland, MA.

*Nielsen, C. (1995). *Animal evolution: interrelationships of the living phyla*. Oxford University Press, Oxford.

*Willmer, P. (1990). *Invertebrate relationships*. Cambridge University Press, Cambridge.

Articles and chapters

Aguinaldo, A. M. A., Turbeville, J. M., Linford, L. S., Rivera, M. C., Garey, J. R., Raff, R. A., and Lake, J. A. (1997). Evidence for a clade of nematodes, arthropods, and other moulting animals. *Nature*, **387**, 489–93.

Arendt, D. and Nubler-Jung, K. (1994). Inversion of dorsoventral axis? *Nature*, **371**, 26.

Carroll, S. B. (1995). Homeotic genes and the evolution of arthropods and chordates. *Nature*, **376**, 479–85.

*Conway Morris, S. (1993). The fossil record and the early evolution of the Metazoa. *Nature*, **361**, 219–25.

Conway Morris, S. (1995). A new phylum from the lobster's lips. *Nature*, **378**, 661–2.

Dilly, P. N. (1993). *Cephalodiscus graptolitoides* sp. nov.: a probable extant graptolite. *Journal of the Zoological Society of London*, **229**, 69–78.

Funch, P. and Kristensen, R. M. (1995). Cycliophora is a new phylum with affinities to Entoprocta and Ectoprocta. *Nature*, **378**, 711–14.

Halanych, K. M., Bacheller, J. D., Aguinaldo, A. M. A., Liva, S. M., Hillis, D. M., and Lake, J. A. (1995). Evidence from 18S ribosomal DNA that the lophophorates are protostome animals. *Science*, **267**, 1641–3.

Knoll, A. H. and Carroll, S. B. (1999). Early animal evolution: emerging views from comparative biology and geology. *Science*, **284**, 2129–36.

Lake, J. A. (1990). Origin of the Metazoa. *Proceedings of the National Academy of Sciences, USA*, **87**, 763–6.

Manton, S. M. and Anderson, D. T. (1979). Polyphyly and evolution of arthropods. In *The origin of major invertebrate groups* (ed. M. R. House), pp. 269–321. Academic Press, London.

Martindale, M. Q. and Kourakis, M. J. (1999). Hox clusters: size doesn't matter. *Nature*, **399**, 730–1.

Rigby, S. (1993). Graptolites come to life. *Nature*, **362**, 209–10.

Robertis, E. M. (1997). The ancestry of segmentation. *Nature*, **387**, 25–6.

ROBERTIS, E. M. and SASAI, Y. (1996). A common plan for dorsoventral patterning in Bilateria. *Nature*, **380**, 37–40.

ROSA, R. DE, GRENIER, J. K., ANDREEVAS, T., COOK, C. E., ADOUTTE, A., AKAM, M. *et al.* (1999). Hox genes in brachiopods and priapulids and protostome evolution. *Nature*, **399**, 772–6.

*SLACK, J., HOLLAND, P., and GRAHAM, C. F. (1993). The zootype and the phylotypic stage. *Nature*, **361**, 490–2.

WINNEPENNINCKX, B. M. H., BACKELJAU, T., and KRISTENSEN, R. M. (1998). Relations of the new phylum Cycliophora. *Nature*, **393**, 636–7.

SECTION 6 Anemones, corals, jellyfish, and sea pens: phylum Cnidaria (pp. 211–24)

Books

*BRUSCA, R. C. and BRUSCA, G. J. (1990). *Invertebrates*. Sinauer Associates, Sunderland, MA.

*NIELSEN, C. (1995). *Animal evolution: interrelationships of the living phyla*. Oxford University Press, Oxford.

Articles and chapters

BRIDGE, D., CUNNINGHAM, C. W., DESALLE, R., and BUSS, L. W. (1995). Class-level relationships in the phylum Cnidaria: molecular and morphological evidence. *Molecular Biology and Evolution*, **12**, 679–89.

CHEN, C. A., ODORICO, D. M., LOHUIS, M. T., VERON, J. E. N., and MILLER, D. J. (1995), Systematic relationships within the Anthozoa (Cnidaria: Anthozoa) using the 5'-end of the 28S rDNA. *Molecular Phylogenetics and Evolution*, **4**, 175–83.

FRANCE, S. C., ROSEL, P. E., AGENBROAD, J. E., MULLINEAUX, L. S., and KOCHER, T. D. (1996). DNA sequence variation of mitochondrial large-subunit rRNA provides support for a two-subclass organisation of the Anthozoa (Cnidaria). *Molecular Marine Biology and Biotechnology*, **5**, 15–28.

SCHMIDT, H. (1974). On evolution in the Anthozoa. *Proceedings of the Second Coral Reef Symposium. 1. Great Barrier Reef Committee, Brisbane, October 1974*, pp. 533–60.

SCHUCHERT, P. (1993). Phylogenetic analysis of the Cnidaria. *Zeitschrift für Zoologische Systematik und Evolutionsforschung*, **31**, 161–73.

SECTION 7 Clams and cockles, snails and slugs, octopuses and squids: phylum Mollusca (pp. 225–46)

Books

*BRUSCA, R. and BRUSCA, G. (1990). *Invertebrates.* Sinauer Associates, Sunderland, MA.

*TAYLOR, J. D. ed. (1966). *Origin and evolutionary radiation of the Mollusca.* Oxford Science Publications, Oxford.

*WILLMER, P. (1990). *Invertebrate relationships.* Cambridge University Press, Cambridge.

Articles and chapters

FEDONKIN, M. A. and WAGGONER, B. M. (1997). The late Precambrian fossil *Kimberella* is a mollusc-like bilaterian organism. *Nature*, **388**, 868–71.

*PONDER, W. F. and LINDBERG, D. R. (1996). Gastropod phylogeny—challenges for the '90s. In *Origin and evolutionary radiation of the Mollusca* (ed. J. D. Taylor), pp. 135–54. Oxford Science Publications, Oxford.

SECTION 8 Animals with jointed legs: phylum Arthropoda (pp. 247–67)

AKAM, M. and AVEROF, M. (1995). Hox genes and the diversification of insect and crustacean body plans. *Nature*, **376**, 420–3.

AKAM, M., AVEROF, M., CASTELLI-GAIR, J., DAWES, R., FALCIANI, F., and FERRIER, D. (1994). The evolving role of Hox genes in arthropods. *Development, 1994 Suppl.*, 209–15.

AVEROF, M. and AKAM, M. (1993). HOM/Hox genes of *Artemia*: implications for the origin of insect and crustacean body plans. *Current Biology*, **3**, 73–8.

AVEROF, M. and AKAM, M. (1994). Insect–crustacean relationships: insights from comparative developmental and molecular studies. *Philosophical Transactions of the Royal Society*, **B347**, 293–303.

BALLARD, H. W. O., OLSEN, G. J., FAITH, D. P., ODGERS, W. A., ROWELL, D. M., and ATKINSON, P. W. (1992). Evidence from 12S ribosomal RNA sequences that onychophorans are modified arthropods. *Science*, **258**, 1445–7.

FRIEDRICH, M. and TAUTZ, D. (1995). Ribosomal DNA phylogeny of the major extant arthropod classes and the evolution of myriapods. *Nature*, **376**, 165–7.

KUKALOVÁ-PECK, J. (1992). The 'Uniramia' do not exist: the ground plan of the Pterygota as revealed by Permian Diaphanopteroidea from Russia (Insecta: Paleodictyopteroidea). *Canadian Journal of Zoology*, **70**, 236–55.

*MANTON, S. (1974). Arthropod phylogeny—a modern synthesis. *Journal of the Zoological Society of London*, **171**, 111–30.

MARDEN, J. H. and KRAMER, M. G. (1995). Locomotor performance of insects with rudimentary wings. *Nature*, **377**, 332–4.

POPADIC, A. *et al.* (1996). Origin of the arthropod mandible. *Nature*, **380**, 395.

TELFORD, M. J. and THOMAS, R. H. (1995). Demise of the Atelocerata? *Nature*, **376**, 123–4.

*WILLMER, P. (1990). *Invertebrate relationships.* Cambridge University Press, Cambridge.

SECTION 9 Lobsters, crabs, shrimps, barnacles, and many more besides: subphylum Crustacea* (pp. 269–307)

*BRUSCA, R. and BRUSCA, G. (1990). *Invertebrates.* Sinauer Associates, Sunderland, MA.

HESSLER, R. R. (1992). Reflections on the phylogenetic position of the Cephalocarida. *Acta Zoologica* (Stockholm), **73**, 315–16.

SECTION 10 The insects: subphylum Insecta (pp. 287–307)

Books

*BRUSCA, R. and BRUSCA, G. (1990). *Invertebrates.* Sinauer Associates, Sunderland, MA.

*WILLMER, P. (1990). *Invertebrate relationships.* Cambridge University Press, Cambridge.

Articles and chapters

AVEROF, M. and COHEN, S. M. (1997). Evolutionary origin of insect wings from ancestral gills. *Nature*, **385**, 627–30.

*KRISTENSEN, N. P. (1991). Phylogeny of extant hexapods. In *The insects of Australia* (2nd edn), pp. 125–40. Melbourne University Press, Melbourne.

KUKALOVÁ-PECK, J. (1991). Fossil history and the evolution of hexapod structures. In *The insects of Australia* (2nd edn), pp. 141–79. Melbourne University Press, Melbourne.

SHEAR, W. A. (1992). End of the 'Uniramia' taxon. *Nature*, **359**, 477–8.

SECTION 11 Spiders, scorpions, mites, eurypterids, horseshoe crabs, and sea spiders: subphylum Chelicerata and subphylum Pycnogonida

(pp. 309–26)

BRIGGS, D. E. G. (1986). How did eurypterids swim? *Nature*, **320**, 400.

SELDEN, P. (1989). Orb-web weaving spiders in the early Cretaceous. *Nature*, **340**, 711–13.

*SELDEN, P. (1990). Fossil history of the arachnids. *Newsletter of the British Arachnology Society*, **58**, 4–6.

SELDEN, P. (1993). Fossil arachnids—recent advances and future prospects. *Memoirs of the Queensland Museum*, **33**, 389–400.

SELDEN, P. (1996). Fossil mesothele spiders. *Nature*, **379**, 498–9.

SELDEN, P. and JERAM, A. J. (1989). Palaeophysiology of terrestrialisation in the Chelicerata. *Transactions of the Royal Society of Edinburgh, Earth Sciences*, **80**, 303–10.

*SHEAR, W. (1994). Untangling the evolution of the web. *American Scientist*, **82**, 256–66.

SHEAR, W. A., PALMER, J. M., CODDINGTON, J. A., and BONAMO, P. M. (1989). A Devonian spinneret: early evidence of spiders and silk use. *Science*, **246**, 479–81.

SHULTZ, J. W. (1989). Morphology of locomotor appendages in Arachnida: evolutionary trends and phylogenetic implications. *Zoological Journal of the Linnean Society*, **97**, 1–56.

*SHULTZ, J. W. (1990). Evolutionary morphology and phylogeny of the Arachnida. *Cladistics*, **6**, 1–38.

*WEYGOLDT, P. and PAULUS, H. F. (1979). Untersuchungen zur Morphologie, Taxonomie, und Phylogenie der Chelicerata. *Zeitschrift für Zoologische Systematik und Evolutionsforschung*, **17**, 85–200.

SECTION 12 Starfish and brittlestars, sea urchins and sand dollars, sea lilies, sea daisies, and sea cucumbers: phylum Echinodermata

(pp. 327–37)

BAKER, A. N., ROWE, F. W. E., and CLARK, H. E. S. (1986), A new class of Echinodermata from New Zealand, *Nature*, **321**, 862–4.

*BRUSCA, R. and BRUSCA, G. (1990). *Invertebrates*. Sinauer Associates, Sunderland, MA.

WRAY, G. A. (1994). The evolution of cell lineage in echinoderms. *American Zoologist*, **34**, 353–63.

WRAY, G. A. (1995). Punctuated evolution of embryos. *Science*, **267**, 1115–16.

SECTION 13 Sea squirts, lancelets, and vertebrates: phylum Chordata

(pp. 339–53)

Books

ALEXANDER, R. M. (1975). *The chordates*. Cambridge University Press, Cambridge.

*BENTON, M. J. (1997). *Vertebrate palaeontology* (2nd edn). Chapman and Hall, London.

GOODRIDGE, E. S. (1958). *Studies on the structure and development of vertebrates*, Vols 1 and 2. Dover Publications, New York.

*NIELSEN, C. (1995). *Animal evolution: interrelationships of the living phyla*. Oxford University Press, Oxford.

*YOUNG, J. Z. (1962). *The life of vertebrates* (2nd edn). Oxford University Press, Oxford.

Articles and chapters

AHLBERG, P. E., CLACK, J. A., and LUKSEVICS, E. (1966). Rapid braincase evolution between *Panderichthys* and the earliest tetrapods. *Nature*, **381**, 61–4.

CHEN, J.-Y., DZIK, J. EDGECOMBE, G. D., RAMSKOLD, L., and ZHOU, G.-Q. (1995). A possible early Cambrian chordate. *Nature*, **377**, 720–2.

GABBOTT, S. E., ALDRIDGE, R. J., and THERON, J. N. (1995). A giant conodont with preserved muscle tissue from the Upper Oligocene of South Africa. *Nature*, **374**, 800–3.

JANVIER, P. (1995). Conodonts join the club. *Nature*, **374**, 761–2.

JEFFERIES, R. P. S. (1990). The solute *Dendrocystoides scoticus* from the Upper Ordovician of Scotland and the ancestry of chordates and echinoderms. *Palaeontology*, **33**, 631–79.

MIN ZHU, XIAOBO YU, and JANVIER, P. (1999). A more primitive fossil fish sheds light on the origin of bony fishes. *Nature*, **397**, 607–10.

POUGH, F. H., JANIS, C. M., and HEISER, J. B. (1999). Origin and radiation of tetrapods in the late Paleozoic. In *Vertebrate life*, Ch. 10. Prentice Hall, Upper Saddle River, NJ.

PURNELL, M. A. (1995). Microwear on conodont elements and macrophagy in the first vertebrates. *Nature*, **374**, 798–800.

SECTIONS 14 and 15 **Sharks, rays, and chimaeras: class Chondrichthyes** (pp. 355–67); **The ray-finned fish: class Actinoptergygii** (pp. 369–88)

BEMIS, W. E. (1995). Lecture outlines for ichthyology at the University of Massachusetts, Amherst (unpublished).

*BENTON, M. J. (1997). *Vertebrate palaeontology* (2nd edn). Chapman and Hall, London.

SECTION 16 **Lobefins and tetrapods: the Sarcoptergygii** (pp. 389–405)

Books

*BENTON, M. (1997). *Vertebrate palaeontology* (2nd edn). Chapman and Hall, London.

LONG, J. A. (1995). *The rise of fishes*. The Johns Hopkins University Press, Baltimore and London.

Articles and chapters

AHLBERG, P. E. and MILNER, A. R. (1994). The origin and early diversification of tetrapods. *Nature*, **368**, 507–13.

COATES, M. I. and CLACK, J. A. (1990). Polydactyly in the earliest known tetrapod limbs. *Nature*, **347**, 66–9.

FOREY, P. L. (1988). Golden jubilee for the coelacanth *Latimeria chalumnae*. *Nature*, **336**, 727–32.

Meyer, A. (1995). Molecular evidence on the origin of tetrapods and the relationships of the coelacanth. *Trends in Ecology and Evolution*, **10**, 111–16.

SHUBIN, N. H. and JENKINS, F. A. JR (1995). An early Jurassic jumping frog. *Nature*, **377**, 49–52.

TABIN, C. and LAUFER, E. (1993). Hox genes and serial homology. *Nature*, **361**, 692–3.

SECTION 17 **The reptiles: class Reptilia*** (pp. 407–31)

Books

BAKKER, R. (1986). *The dinosaur heresies*. Longman Scientific & Technical, Harlow.

*BENTON, M. (1997). *Vertebrate palaeontology* (2nd edn). Chapman and Hall, London.

CHARIG, A. (1979). *A new look at the dinosaurs*. Heinemann / Natural History Museum, London.

FRASER, N. C. and SUES, H.-D, ed. (1994). *In the shadow of the dinosaurs: Early Mesozoic tetrapods*. Cambridge University Press, Cambridge.

Articles and chapters

CALDWELL, M. W. and LEE, M. S. Y. (1997). A snake with legs from the marine Cretaceous of the Middle East. *Nature*, **386**, 705–9.

CORIA, R. A. and SALGADO, L. (1995). A new giant carnivorous dinosaur from the Cretaceous of Patagonia. *Nature*, **377**, 224–6.

FRASER, N. (1991). The true turtles' story. *Nature*, **349**, 278–9.

HEDGES, S. B. and POLING, L. L. (1999). A molecular phylogeny of reptiles. *Science*, **283**, 998–1001.

LEE, M. S. Y. (1996). Correlated progression and the origin of turtles. *Nature*, **379**, 812–15.

MOTANI, R. Y. H. and MCGOWAN, C. (1996). Eel-like swimming in the earliest ichthyosaurs. *Nature*, **382**, 347–8.

PATON, R. L., SMITHSON, T. R., and CLACK, J. A. (1999). An amniote-like skeleton from the Early Carboniferous of Scotland. *Nature*, **398**, 508–13.

REISZ, R. R. and LAURIN, M. (1991). *Owenetta* and the origin of turtles. *Nature*, **349**, 324–6.

RIEPPEL, O. (1999). Turtle origins. *Science*, **283**, 945–6.

RIEPPEL, O. and DEBRAGA, M. (1996). Turtles as diapsid reptiles. *Nature*, **384**, 453–5.

*SERENO, P. C. (1999). The evolution of dinosaurs. *Science*, **284**, 2137–47.

SERENO, P. C., FORSTER, C. A., ROGERS, R. R., and MONETTO, A. M. (1993). Primitive dinosaur skeleton from Argentina and the early evolution of Dinosauria. *Nature*, **361**, 64–6.

SWISHER, C. C., YUAN-QING WANG, XIAO-LIN WANG, XING XU, and YUAN WANG (1999). Cretaceous age for the feathered dinosaurs of Liaoning, China. *Nature*, **400**, 58–61.

VARRICCHIO, D. J., JACKSON, F., BORKOWSKI, J. J., and HORNER, J. R. (1997). Nest and egg clusters of the dinosaur *Troodon formosus* and the evolution of avian reproductive traits. *Nature*, **385**, 247–50.

XIAO-CHUN WU, SUES HANS-DIETER, and AILING SUN (1995). A plant-eating crocodyliform reptile from the Cretaceous of China. *Nature*, **376**, 678–80.

SECTION 18 The mammals: class Mammalia (pp. 433–60)

Books

*BENTON, M. (1997). *Vertebrate palaeontology* (2nd edn). Chapman and Hall, London.

CORBET, G. B. and HILL, J. E. (1991). *A world list of mammalian species* (3rd edn). Natural History Museum Publications / Oxford University Press, London and Oxford.

Articles and chapters

ARNASON, U. and GULLBERG, A. (1994). Relationship of baleen whales established by cytochrome *b* gene sequence comparison. *Nature*, **367**, 726–8.

D'ERCHIA, A. M., GISSI, C., PESOLE, G., SACCONE, C., and ARNASON, U. (1996). The guinea-pig is not a rodent. *Nature*, **381**, 597–600.

GINGERICH, P. D., WELLS, N. A., RUSSELL, D. E., SHAH, S. M. I. (1983). Origin of whales in epicontinental remnant seas: new evidence from the early Eocene of Pakistan. *Science*, **220**, 403–5.

GINGERICH, P. D., RAZA, S. M., ARIF, M., ANWAR, M., and XIAOYUAN ZHOU (1994). New whale from the Eocene of Pakistan and the origin of cetacean swimming. *Nature*, **368**, 844–7.

GRAUR, D. (1993). Molecular phylogeny and the higher classification of eutherian mammals. *Trends in Ecology and Evolution*, **8**, 141–7.

GRAUR, D., DURET, L., and GOUY, M. (1996). Phylogenetic position of the order Lagomorpha (rabbits, hares, and allies). *Nature*, **379**, 333–5.

*JANIS, C. M. and JAMUTH, J. (1990). Mammals. In *Evolutionary Trends* (ed. K. J. McNamara), Ch. 13, pp. 301–43. Belhaven Press, London.

JIN MENG and WYSS, A. R. (1995). Monotreme affinities and low-frequency hearing suggested by multituberculate ear. *Nature*, **377**, 141–4.

DE JONG, W. W. (1998). Molecules remodel the mammalian tree. *Trends in Ecology and Evolution*, **13**, 270–75.

KEMP, T. (1982). The reptiles that became mammals. *New Scientist*, 4 March, 581–4.

*MARTIN, R. D. (1993). Primate origins: plugging the gaps. *Nature* **363**, 223–34.

MAYR, E. (1986). Uncertainty in science: is the giant panda a bear or a raccoon? *Nature*, **323**, 769–71.

MILINKOVITCH, M. C. (1995). Molecular phylogeny of cetaceans prompts revision of morphological transformations. *Trends in Ecology and Evolution*, **10**, 328–4.

MILINKOVITCH, M. C., ORTI, G., and MEYER, A. (1993). Revised phylogeny of whales suggested by mitochondrial ribosomal DNA sequences. *Nature*, **361**, 346–8.

DE MUIZON, C. (1994). A new carnivorous marsupial from the Palaeocene of Bolivia and the problem of marsupial monophyly. *Nature*, **370**, 208–11.

DE MUIZON, C., CIFELLI, R. L., and PAZ, R. C. (1997). The origin of the dog-like borhyaenoid marsupials of South America. *Nature*, **389**, 486–9.

NORELL, M. A. and NOVACEK, M. J. (1992). The fossil record and evolution: comparing cladistic and paleontological evidence for vertebrate history. *Science*, **255**, 1690–3.

*NOVACEK, M. J. (1992). Mammalian phylogeny shaking the tree. *Nature*, **356**, 121–5.

*NOVACEK, M. J. (1993). Reflections on higher mammalian phylogenetics. *Journal of Mammalian Evolution*, **1**, 3–30.

NOVACEK, M. J. (1993). Genes tell a new whale tale. *Nature*, **361**, 298–9.

NOVACEK, M. J., MCKENNA, M. C., MALCOLM, C., NEFF, N. A., and CIFELLI, R. L. (1983). Evidence from earliest known erinaceomorph basicranium that insectivorans and primates are not closely related. *Nature*, **306**, 683–4.

POUGH, F. H., JANIS, C. M., and HEISER, J. B. (1999). The synapsida and the evolution of mammals. In *Vertebrate life*, Ch. 19. Prentice Hall, Upper Saddle River, NJ.

ROUGIER, G. W., WIBLE, J. R., and NOVACEK, M. J. (1996). Multituberculate phylogeny. *Nature*, **379**, 406.

*SIMPSON, G. G. (1945). The principles of classification and a classification of the mammals. *Bulletin of the American Museum of Natural History*, **85**, 1–350.

THOMAS, R. H. (1994). What is a guinea-pig? *Trends in Ecology and Evolution*, **9**, 159–60.

WYSS, A. R. (1987). The walrus auditory region and the monophyly of pinnipeds. *Nature*, **334**, 427–8.

WYSS, A. R., FLYNN, J. I., NORELL, M. A., SWISHER, C. C., CHARRIER, R., NOVACEK, M. J., and MCKENNA, M. C. (1993). South America's earliest rodent and recognition of a new interval of mammalian evolution. *Nature*, **365**, 434–7.

SECTION 19 Lemurs, lorises, tarsiers, monkeys, and apes: order Primates (pp. 461–88)

Books

*Benton, M. (1997). *Vertebrate palaeontology* (2nd edn). Chapman and Hall, London.

Corbet, G. B. and Hill, J. E. (1991). *A world list of mammalian species* (3rd edn). Natural History Museum Publications / Oxford University Press, London and Oxford.

Fleagle, J. G. (1999). *Primate adaptation and evolution* (2nd edn). Academic Press, San Diego.

Articles and chapters

*Delson, E. (1992). Evolution of Old World monkeys. In *The Cambridge encyclopedia of human evolution* (ed. S. Jones, R. D. Martin, and D. Pilbeam), pp.217–22. Cambridge University Press, Cambridge.

Ford, S. M. (1986). Systematics of the New World monkeys. In *Comparative primate biology*, Vol. 1: *Systematics, evolution, and anatomy* (ed. D. R. Swindler and J. Erwin), pp. 73–135. Alan R. Liss, New York.

*Martin, R. D. (1993). Primate origins: plugging the gaps. *Nature*, **363**, 223–34

Martin, R. D. (1994). Bonanza at Shanghuang. *Nature* **368**, 586–7.

*Rosenberger, A. L. (1992). Evolution of New World Monkeys. In *The Cambridge encyclopedia of human evolution* (ed. S. Jones, R. D. Martin, and D. Pilbeam), pp. 209–16. Cambridge University Press, Cambridge.

SECTION 20 Human beings and our immediate relatives: family Hominidae *s.s.* (pp. 489–513)

Books

*Darwin, C. (1871). *The descent of man and selection in relation to sex*, Vols 1 and 2 (2nd revised edn, 1874). John Murray, London.

Foley, R. (1987). *Another unique species*. Longman Scientific & Technical, Harlow.

Jones, S., Martin, R., and Pilbeam, D. ed. (1992). *The Cambridge encyclopedia of human evolution*. Cambridge University Press, Cambridge.

Kohn, M. (1999). *As we know it*. Granta Books, London.

*Lewin, R. (1987). *Bones of contention: controversies in the search for human origins*. Simon & Schuster, New York.

Ridley, M. (1996). *Origins of virtue*. Viking, London.

Stringer, C. and Gamble, C. (1993). *In search of the Neanderthals*. Thames and Hudson, London.

*Tudge, C. (1995). *The day before yesterday*. Cape / Pimlico, London (published 1966 in the USA as *The time before history*, Scribner / Touchstone, New York).

Tudge, C. (1998). *Neanderthals, bandits and farmers*. London, Weidenfeld & Nicolson.

Articles and chapters

Brunet, M., Beauvilain, A., Coppens, Y., Heintz, E., Moutaye, A. H. E., and Pilbeam, D. (1995). The first australopithecine 2,500 kilometres west of the Rift Valley (Chad). *Nature*, **378**, 273–5.

Culotta, E. (1999). A new human ancestor? *Science*, **284**, 572–3.

Day, M. H. and Wickens, E. H. (1980). Laetoli hominid footprints and bipedalism. *Nature*, **286**, 385–7.

GABUNIA, L. and VEKUA, A. (1995). A Plio-Pleistocene hominid from Dmanisi, East Georgia, Caucasus. *Nature*, **373**, 509–12.

KRINGS, M., STONE, A., SCHMITZ, R. W., KRAINITZI, H., STONEKING, M., and PÄÄBO, S. (1997). Neandertal DNA sequences and the origin of modern humans. *Cell*, **90**, 1–20.

LEAKEY, M. G., FEIBEL, C. S., MACDOUGALL, I., and WALKER, A. (1995). New four-million-year-old hominid species from Kanapoi and Allia Bay, Kenya. *Nature*, **376**, 565–71.

MORWOOD, M. J., AZIZ, F., O'SULLIVAN, P., NASRRUDDIN, HOBBS, D. R., and RAZA, A. (1999). Archaeological and palaeontological research in central Flores, east Indonesia: results of fieldwork 1997–98. *Antiquity*, **73**, 273–86.

RUFF, C. B., TRINKAUS, E., and HOLLIDAY, T. W. (1997). Body mass and encephalization in Pleistocene *Homo*. *Nature*, **387**, 173–6.

WARD, C., LEAKEY, M., and WALKER, A. (1999). The new hominid species *Australopithecus anamensis*. *Evolutionary Anthropology*, **7**, 197–205.

WHITE, T., SUWA, G., and ASFAW, B. (1994). *Australopithecus ramidus*, a new species of early hominid from Aramis, Ethiopia. *Nature*, **371**, 306–12; and corrigendum (1995), *Nature*, **375**, 88.

*WOOD, B. (1992). Origin and evolution of the genus *Homo*. *Nature*, **355**, 783–90.

WOOD, B. and COLLARD, M. (1999). The human genus. *Science*, **284**, 65–71.

SECTIONS 21 and 22 **The birds: class Aves** (pp. 515–28); **The modern birds: subclass Neornithes** (pp. 529–45)

Books

*AUSTIN, A. L. JR (1961). *Birds of the world*. Hamlyn, London.

FEDUCCIA, A. (1996). *The origin and evolution of birds*. Yale University Press, New Haven.

SIBLEY, C. and AHLQUIST, J. (1990). *Phylogeny and classification of birds: a study in molecular evolution*. Yale University Press, New Haven.

Articles and chapters

BOLES, W. E. (1995). The world's oldest songbird. *Nature*, **374**, 21–2.

*CHIAPPE, L. M. (1995). The first 85 million years of avian evolution. *Nature*, **378**, 349–53.

CHIAPPE, L. M. (1995). A diversity of early birds. *Natural History*, **104**, 52–5.

CHINSAMY, A., CHIAPPE, L. M., DODSON, P. (1994). Growth rings in Mesozoic birds. *Nature*, **368**, 196–7.

CHINSAMY, A., CHIAPPE, L. M., DODSON, P. (1995). Mesozoic avian bone microstructure: physiological implications. *Paleobiology*, **21**, 561–74.

CRACRAFT, J. (1986). The origin and early diversification of birds. *Paleobiology*, **12**, 383–99.

CRACRAFT, J. (1987). DNA hybridization and avian phylogenetics. *Evolutionary Biology*, **21**, 47–96.

CRACRAFT, J. (1988). Early evolution of birds. *Nature*, **331**, 389–90.

*CRACRAFT, J. (1988). The major clades of birds. In *The phylogeny and classification of the tetrapods*, Vol. 1: *Amphibians, reptiles, birds*, Systematics Association Special Volume, No. 35A (ed. M. J. Benton), pp. 339–61. Clarendon Press, Oxford.

JI QIANG, CURRIE, P. J., NORELL, M. A., and JI SHU-AN (1998). Two feathered dinosaurs from northeastern China. *Nature*, **393**, 753–61.

MARTIN, L. D. and ZHONGHE ZHOU (1997). *Archaeopteryx*-like skull in enantiornithine bird. *Nature*, **389**, 556.

MAYNARD SMITH, J. (1953). Birds as aeroplanes. *New Biology*, **14**, 62–81. Penguin Books, Harmondsworth.

MILNER, A. (1993). Ground rules for early birds. *Nature*, **362**, 589.

*NORELL, M., CHIAPPE, L., and CLARK, J. (1993). New limb on the avian family tree. *Natural History*, September, 37–42.

PADIAN, K. (1998). When is a bird not a bird? *Nature*, **393**, 729–30.

PADIAN, K. and CHIAPPE, L. M. (1998). The origin of birds and their flight, *Scientific American*, February, 28–37.

STAPEL, S. O., LEUNISSEN, J. A. M., VERSTEEG, M., WATTEL, J., and DE JONG, W. W. (1984). Ratites as oldest offshoot of avian stem—evidence from α-crystalline A sequences. *Nature*, **311**, 257–9.

WALKER, C. A. (1981). New subclass of birds from the Cretaceous of South America. *Nature*, **292**, 51–3.

SECTION 23 The plants: kingdom Plantae (pp. 547–75)

Books

INGROUILLE, M. (1992). *Diversity and evolution of land plants*. Chapman and Hall, London.

*KENRICK, P. and CRANE, P. (1997). *The origin and early diversification of land plants*. Smithsonian Institution Press, Washington DC.

*RAVEN, P., EVERT, R., and EICHHORN, S. (1992). *Biology of plants* (5th edn). Worth, New York.

Articles and chapters

CHALONER, W. (1989). A missing link for seeds? *Nature*, **340**, 185.

CHASE, M. W., SOLTIS, D. E., OLMSTEAD, R. G., MORGAN, D., LES, D. H., MISHLER, B. D. *et al.* (1993). Phylogenetics of seed plants: an analysis of nucleotide sequences from the plastid gene *rbcL*, *Annals of the Missouri Botanical Garden*, **80**, 528–80.

*DONOGHUE, M. J. (1994). Progress and prospects in reconstructing plant phylogeny. *Annals of the Missouri Botanical Garden*, **81**, 405–18.

GALTIER, J. and ROWE, N. P. (1989). A primitive seed-like structure and its implications for early gymnosperm evolution. *Nature* **340**, 225–7.

KATO, M. and INOUE, T. (1994). Origins of insect pollination. *Nature*, **368**, 195.

*KENRICK, P. and CRANE, P. (1997). The origin and early evolution of plants on land. *Nature*, **389**, 33–9.

MARTIN, W., GIERL, A., and SADLER, H. (1989). Molecular evidence for pre-Cretaceous angiosperm origins. *Nature*, **339**, 46–8.

MISHLER, B. (1994). Phylogenetic relationships of the 'green algae' and 'bryophytes'. *Annals of the Missouri Botanical Garden*, **81**, 451–83.

SECTION 24 The flowering plants: class Angiospermae (pp. 577–91)

Books

FRIIS, E. M., CHALONER, W. G., and CRANE, P. R. ed. (1987). *The origins of angiosperms and their biological consequences*. Cambridge University Press, Cambridge.

*HEYWOOD, V. H. ed. (1978). *Flowering plants of the world*. Oxford University Press, Oxford.

Articles and chapters

*CRANE, P., FRIIS, E. M., and PEDERSON, K. R. (1995). The origin and early diversification of the angiosperms. *Nature*, **374**, 27–33.

DOYLE, J. A., DONOGHUE, M. J., and ZIMMER, E. A. (1994). Integration of morphological and ribosomal RNA data on the origin of angiosperms. *Annals of the Missouri Botanical Garden*, **81**, 419–50.

EDWARDS, D., DUCKETT, J. G., and RICHARDSON, J. B. (1995). Hepatic characters in the earliest land plants. *Nature*, **374**, 635–6.

HERENDEEN, P. and CRANE, P. (1995). The fossil history of the monocotyledons. In *Monocotyledons: systematics and evolution* (ed. P. J. Rudall, P. J. Cribb, D. F. Cutler, and C. J. Humphries), pp. 1–21. Royal Botanic Gardens, Kew.

LIDGARD, S. and CRANE, P. (1988). Quantitative analyses of the early angiosperm radiation. *Nature*, **331**, 344–6.

MANHART, J. R. and PALMER, J. D. (1990). The gain of two chloroplast tRNA introns marks the green algal ancestors of land plants. *Nature*, **345**, 268–70.

SECTION 25 Daisies, artichokes, thistles, and lettuce: family Compositae (pp. 593–606)

BREMER, K. (1996). Major clades and grades in the Asteraceae. In *Compositae: systematics*, Proceedings of the International Compositae Conference, Kew, 1994, Vol. 1 (ed. D. J. N. Hind and H. J. Beentye), pp. 1–7. Royal Botanic Gardens, Kew.

PART III

EPILOGUE Saving what is left (pp. 609–627)

FRANKEL, O. H. and SOULE, M. (1981). *Conservation and evolution*. Cambridge University Press, Cambridge.

TUDGE, C. (1991). *Last animals at the zoo*. Hutchinson Radius, London.

GEOLOGICAL TIMESCALE

era	period	epoch	duration in millions of years	millions of years ago
Cenozoic	Quaternary	Holocene		0.01
		Pleistocene	1.8	1.8
	Neogene	Pliocene	3.5	5.3
		Miocene	18.5	23.8
	Palaeogene	Oligocene	9.9	33.7
		Eocene	21.1	54.8
		Palaeocene	10.2	65
Mesozoic	Cretaceous		77	142
	Jurassic		63.7	205.7
	Triassic		42.5	248.2
Palaeozoic	Permian		41.8	290
	Carboniferous Pennsylvanian		33	323
	Mississippian		31	354
	Devonian		63	417
	Silurian		26	443
	Ordovician		52	495
	Cambrian		50	545
Precambrian			*c.*4055	*c.*4600

GENERAL INDEX

Bold indicates the defining entry for a given term

D

E

F

INDEX OF ORGANISMS

B

C

D

E

F

G

H

N

O

P

Q

R

S

T

U

V

Please remember that this is a library book, and that it belongs only temporarily to each person who uses it. Be considerate. Do not write in this, or any, library book.

DATE DUE

LORD PRINTED IN U.S.A.